"Jeff Halverson is a gifted writer and brings meteorology to life in this comprehensive and engaging textbook. The highly accessible scientific explanations and marvelous illustrations make it an invaluable resource that will foster much greater understanding of severe storms."

Jason Samenow, Chief Meteorologist,
Washington Post *Capital Weather Gang*

"This is a textbook that doesn't read like a textbook. The way Dr. Halverson has written this book is almost conversational, as if I'm listening to him really tell the story of the atmospheric processes. There are so many great examples for students to really learn about the science. The details of events with examples can really put weather into perspective for students to understand these concepts in an easy way."

Aubrey Urbanowicz, Chief Meteorologist,
WHSV, Harrisonburg, Virginia

An Introduction to Severe Storms and Hazardous Weather

This book presents a deep and encompassing survey of severe weather in all its forms. *An Introduction to Severe Storms and Hazardous Weather* is an exciting new textbook that allows students to learn the principles of atmospheric science through the drama, exhilaration, and even tragedy of severe weather.

Balancing breadth and depth, Jeffrey B. Halverson adeptly combines a short, accessible introduction to the basic principles of meteorology with detailed coverage on large- and small-scale weather hazards. He draws on specific up-to-date case studies from North America to illustrate the cause of meteorological events including hurricanes, heavy snow and ice, floods, and tornadoes. Unlike existing books on the market, Halverson delves deep into the societal impacts of these events, drawing on examples from agriculture, utility infrastructure, and commercial aviation. Each chapter also features high-quality, customized color artwork by Thomas D. Rabenhorst that helps to enhance and embed learning.

Thorough in its scope, and written with an impeccable focus on the science, this book will be an essential resource for introductory undergraduate courses in severe weather, natural hazards, and extreme meteorology. It is also an excellent supplemental textbook for courses on meteorology and atmospheric science.

Jeffrey B. Halverson received his PhD in Environmental Science at the University of Virginia in 1994, then assumed a post-doc under Dr. Joanne Simpson (the first woman in the United States to receive a PhD in Meteorology) at NASA's Goddard Space Flight Center. He is currently Professor at the University of Maryland, Baltimore County (UMBC), where he teaches courses on physical geography, water science, natural hazards, meteorology, severe storms, climate change, and Earth's natural history. He and his team of graduate students investigate severe storms, particularly hurricanes. In 2000 he helped pioneer a new type of technology for measuring air temperature at high altitudes in the eyes of hurricanes. Halverson has authored or coauthored more than 60 scientific publications on severe storms and has appeared in science documentaries aired by NOVA, National Geographic, The Weather Channel, and The Discovery Channel. He is a columnist, feature writer, and assistant editor for *Weatherwise Magazine*. He is a writer and Severe Weather Expert for the *Washington Post*'s Capital Weather Gang. In 2015 he adapted a new teaching technology – a large, digitally projected sphere of the Earth, called Magic Planet – for teaching his courses.

An Introduction to Severe Storms and Hazardous Weather

Jeffrey B. Halverson
Illustrations by Thomas D. Rabenhorst

Designed cover image: © Getty Images

First published 2024
by Routledge
4 Park Square, Milton Park, Abingdon, Oxon OX14 4RN

and by Routledge
605 Third Avenue, New York, NY 10158

Routledge is an imprint of the Taylor & Francis Group, an informa business

British Library Cataloguing-in-Publication Data
A catalogue record for this book is available from the British Library

ISBN: 978-1-032-38423-8 (hbk)
ISBN: 978-1-032-38424-5 (pbk)
ISBN: 978-1-003-34498-8 (ebk)

DOI: 10.4324/9781003344988

Typeset in Times New Roman
by Apex CoVantage, LLC

Foremost, for my friend, colleague and coauthor Tom, who with great perseverance completed the vast majority of artwork in this text. We did this together, for many years, and it was a joy to work with you. You are greatly missed.

And for my family. Many in my immediate family are gone and will never be able to hold this weighty book in their hands. For those able to turn the pages – my loving wife Kathleen, son Matthew and mother Margaret – the next page, wherever you choose, is for you.

Contents

Acknowledgments

I am most grateful for Mark Rabenhorst, Tom's skilled son, who completed the artwork contained herein with great conviction and dedication. You ably rushed in to fill the void.

I am extremely grateful to my editor at Routledge, Annabelle Harris, who helped me expertly turn a rough but nearly complete manuscript into a polished gem . . . and who shares my vision that this book will make a difference.

I am very grateful to the long succession of brilliant mentors in my academic career, who constantly encouraged me and provided me with so many opportunities to grow: Professor Steve Colucci, Professor Michael Garstang, and Dr. Joanne Simpson.

Many thanks to the legions of academic reviewers who provided a steady stream of rich commentary, helping to improve this textbook.

PART I

Introductory Principles

CHAPTER 1
Introduction to Severe Storms and Societal Impacts

Learning Objectives

1. Describe the typical categories of billion-dollar weather disasters that strike the United States each year.
2. Describe the geographical distribution of common weather hazards and severe storms that affect the United States.
3. Discuss key trends in weather disasters across the United States over the past several decades.
4. Identify trends in weather and storm-related fatalities in the United States over the past half century.

Introduction to Severe and Hazardous Weather

Severe storms are headline makers. How many times have you read or watched news coverage of a devastating hurricane along the Gulf Coast or a destructive multiday tornado outbreak in the Mid-South? Even in winter, these storms persist, with blizzards and extreme wind chill raking the Dakotas or lake effect snowfall accumulating by the feet in the Great Lakes region. During the warmer months, flash floods inundate many communities – and once again, the media feature images or videos of harrowing rooftop rescues and cars being swept away in muddy torrents. Perhaps even you have been impacted by a severe storm, experiencing some degree of suffering or loss.

Severe storms and other weather hazards such as extreme heat waves and droughts are examples of the natural hazards that affect people and societies across the globe. Other types of natural hazards are earthquakes, volcanoes, landslides, tsunamis, and wildfires. Natural hazards tend to occur intermittently and episodically, many without warning. At times, the death toll and destruction are staggering. The exponential growth of the human population and the expansion of infrastructure and property have turned many of these severe-weather events into multibillion-dollar catastrophes, with cumulative fatalities in the tens to hundreds of thousands across the globe.

Figure 1.1 shows presidential disaster declarations across the United States from 1964 to 2013, stemming from all types of natural hazards. Note that severe storms and floods occur almost everywhere and account for the great majority of all geophysical natural hazards in the United States. In the southeastern United States and along the East Coast, hurricanes contribute significantly to the toll of disaster declarations.

This book will help you understand the causes and effects of severe storms and weather hazards. We focus on two general categories of severe storm. The first category includes cyclones (large systems of low pressure, rotating wind, and precipitation), which affect widespread regions with high wind and heavy precipitation, sometimes lasting days. Examples of cyclones include wintertime Nor'easters along the East Coast and hurricanes that can strike anywhere between Texas and Maine. The second general category includes severe local storms. These are much smaller, shorter-lived events that focus their destructive energies on individual counties or cities. They last only a few minutes to a few hours. Examples of these events include severe thunderstorms, tornadoes, and flash floods. Quite often, severe local storms are embedded in a much larger and longer-lived cyclone (Figure 1.2).

We discuss these two general categories of severe storms in separate sections of this textbook (Part II covers cyclonic storms; Part III handles all types of severe local storms), and along the way, we also discuss other high-impact weather events, including heat waves and arctic, cold air outbreaks. Our geographical focus is the United States and North America, but most of the principles discussed apply readily around the globe. Part I of the book introduces important general concepts about meteorology and the technology that we use to observe and investigate storms. Each chapter expands on this core knowledge as the need arises.

DOI: 10.4324/9781003344988-2

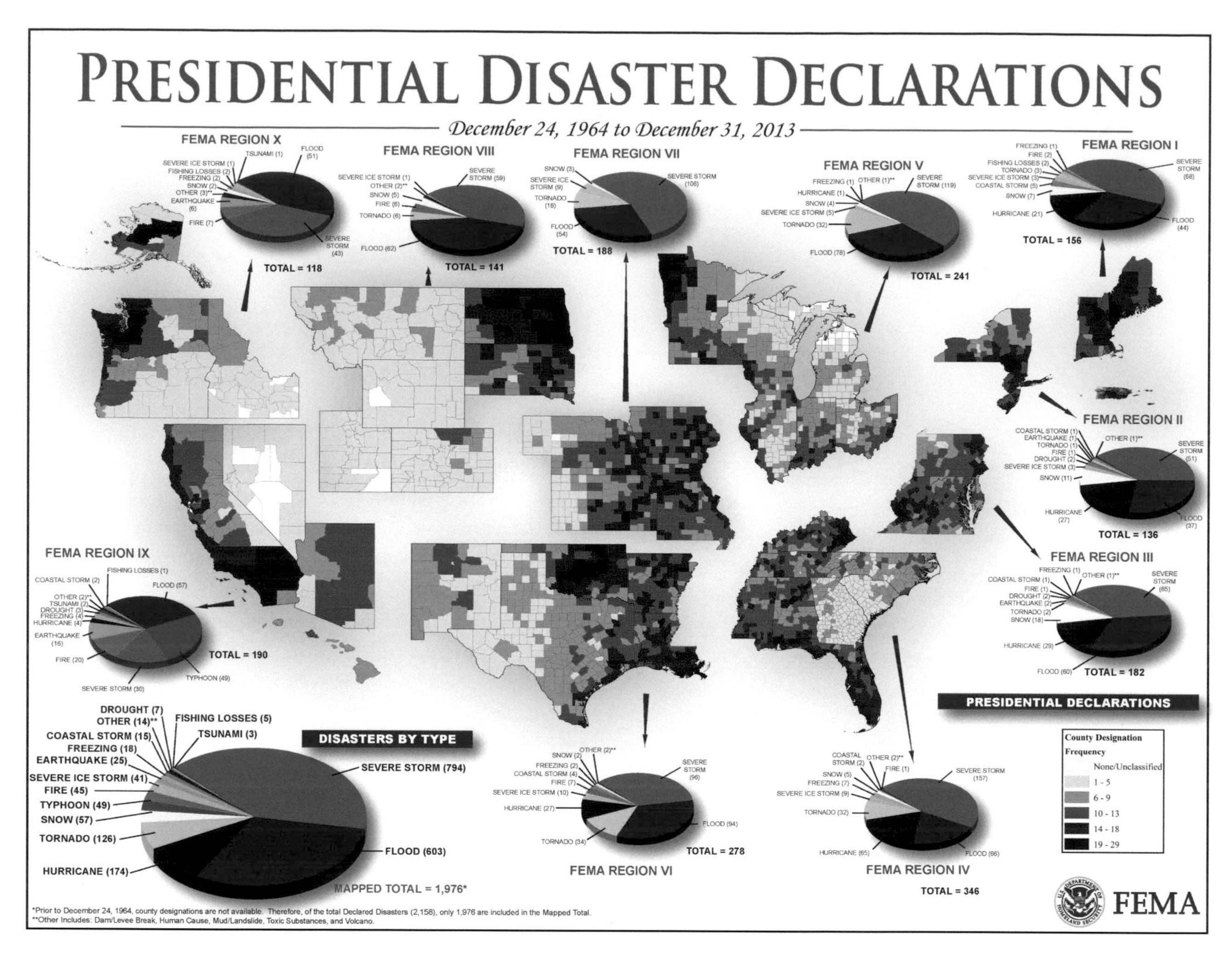

Figure 1.1 Presidential disaster declarations in the United States, 1964–2013. These maps show the great variety of natural disasters that took place across the United States for nearly half a century, broken down by region. Many of these events caused billions of dollars in damage. In all regions, most geophysical hazards are meteorological events. (National Oceanographic and Atmospheric Administration [FEMA].)

Billion-Dollar Weather Disasters in Recent Years

Thanks to the exponential growth of property (both insured and uninsured) across the United States, some weather hazards, such as tornado outbreaks and hurricanes, have become multibillion-dollar disasters. Figure 1.3 presents national maps showing the types and locations of billion-dollar weather and climate disasters for 2019–2021. Note the large year-to-year variation of these disasters – including both the total number of disasters in the United States and the geographical distribution of various disaster types. For instance, in 2019 there were a number of devastating floods across the Central Plains and only two hurricane and tropical cyclone landfalls. In 2020, however, while there were no severe and widespread floods, six hurricanes and tropical cyclones made landfall in the United States. While some geographical regions are more prone to certain types of weather hazards than others, there is no guarantee that the same type or severity of hazard will occur every year.

During this three-year time frame, typical weather hazards included severe local storms such as tornadoes, hailstorms, flash floods, and derechos (severe windstorms generated by a line of fast-moving thunderstorms). Floods generally fall into two categories. Aerial floods may be widespread and long-lasting, in contrast with flash floods, which are more localized and short-lived. Tropical cyclones include both hurricanes and their weaker cousins, tropical storms.

Winter weather hazards typically involve a combination of heavy snow and/or ice with brutally cold arctic air (termed a

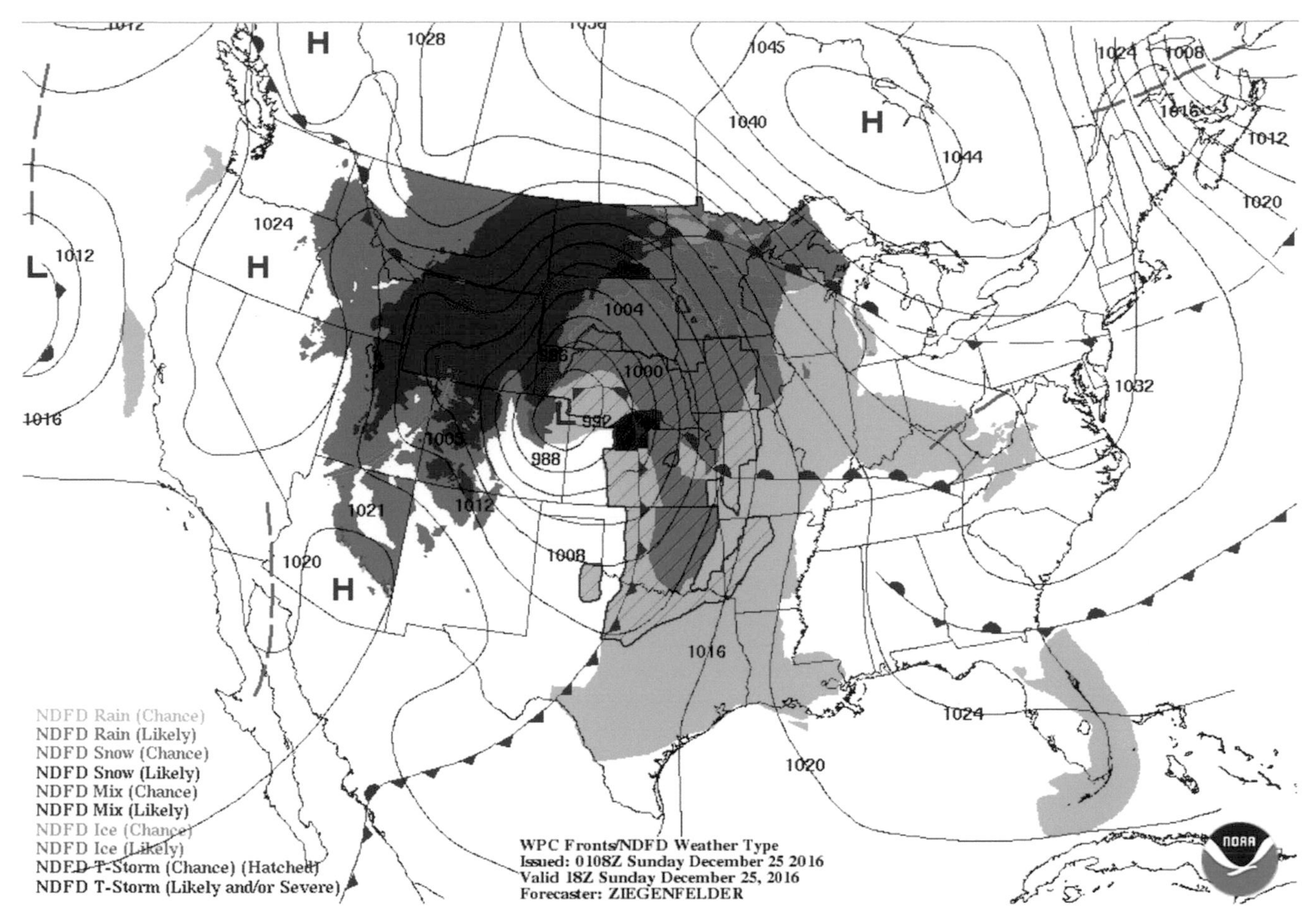

Figure 1.2 Multiple weather hazards stemming from a single cyclone in early winter. The center of the parent low-pressure system is shown by the red "L." Hazards include a small zone of severe thunderstorms (red shade), heavy rain and flash flooding (green shades), accumulating ice (sleet and freezing rain, orange shade), and heavy snow and blizzard conditions (dark blue). (National Oceanographic and Atmospheric Administration [NOAA].)

cold snap); this combination frequently creates widespread, life-threatening conditions. High winds contribute to wind chill and also promote blizzards in which windblown snow creates whiteout conditions. Droughts are a type of climate disaster, particular when the drought is long-lived and severe. Note also the geographic distribution of wildfires in Figure 1.3. Although wildfires are not a meteorological phenomenon per se, droughts often set the stage for wildfires, and wildfires are often triggered by lightning strikes and then fanned by severe local windstorms.

Figure 1.4 shows the annual total damage caused by weather-related disasters from 1980 to 2021, broken down by month. While there are a great many curves on this diagram, we've highlighted those years in which the annual damage has exceeded $100 billion. These five years have all happened since 2004. Across all years, the annual damage totals vary by a factor of three to four. Note the huge upturn in disaster costs in late summer through early autumn; this upturn is due to one or more major hurricane landfalls in the United States. Hurricanes are the single, costliest type of natural disaster (including all geophysical events) in the United States. In 2005, Hurricane Katrina struck Louisiana and Mississippi with a price tag of $180 billion (2022 adjusted dollars). In 2017, three major hurricanes made landfall in the United States: Hurricane Harvey along the Texas and Gulf coasts ($143 billion); Hurricane Maria over the U.S. commonwealth of Puerto Rico ($104 billion); and Hurricane Irma over Florida ($58 billion).

Between 1980 and 2016, there were over 200 U.S. weather and climate disasters that caused more than $1 billion in damages, amounting to a 36-year total loss of $1.2 trillion. Half of this trillion-dollar total comes from tropical cyclones making landfall along the U.S. coastline. Droughts, heat waves, wildfires, tornado outbreaks, floods, and earthquakes have each accounted for losses in the tens of billions between 1980 and 2016, but the

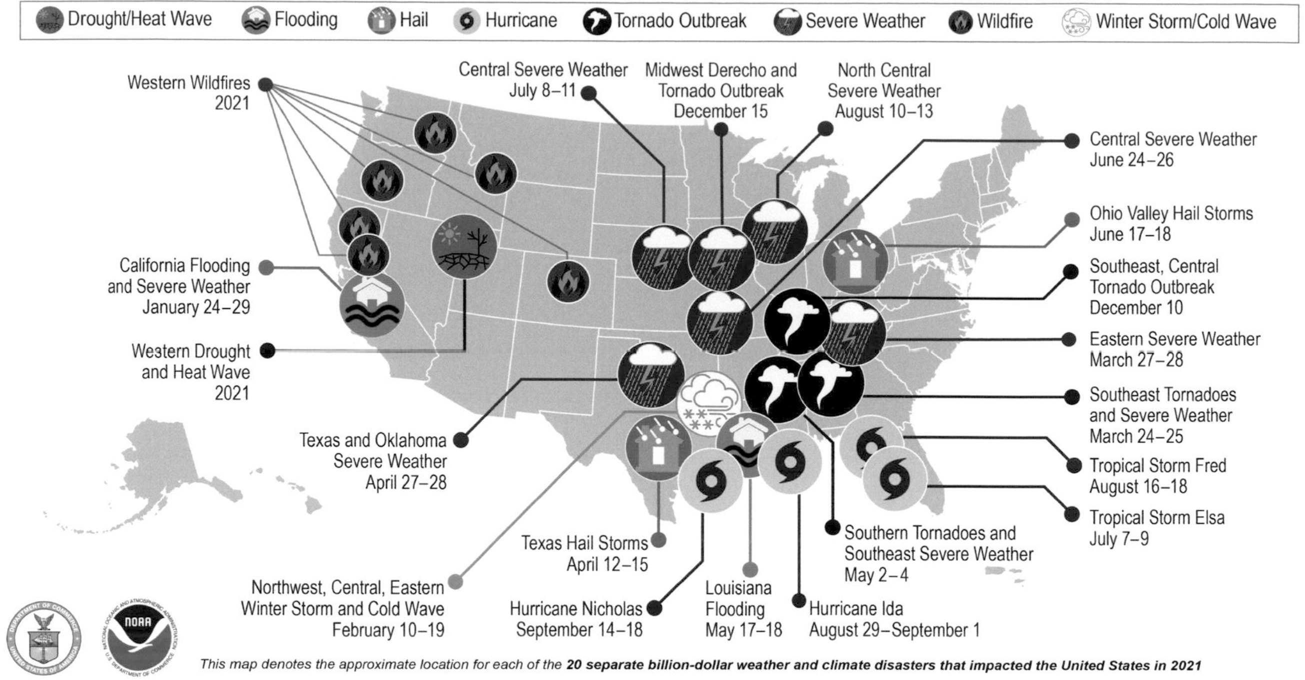

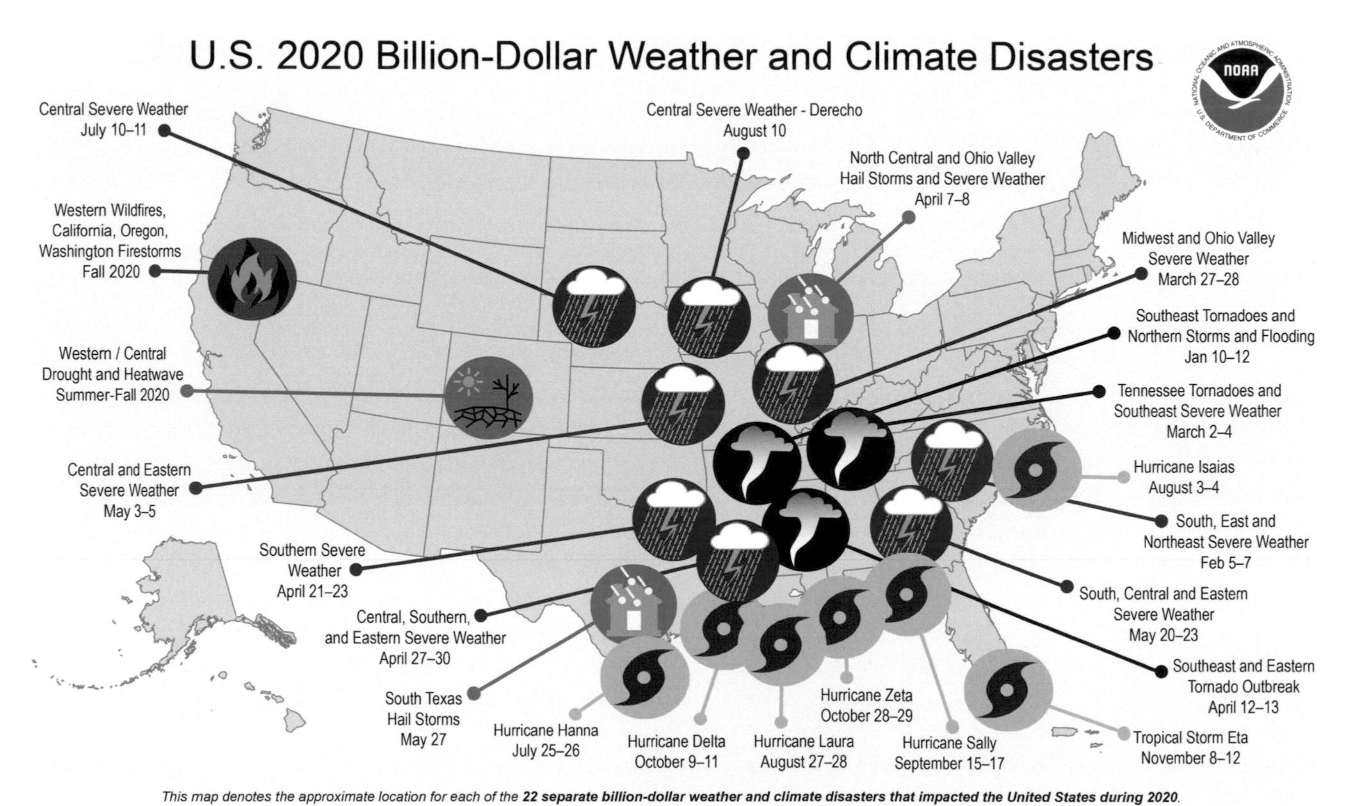

Figure 1.3 Billion-dollar weather and climate disasters, 2019–2021. The tally includes droughts because they frequently contribute to heat waves and make regions prone to wildfires. On these maps, the term "severe weather" can refer to a wide variety of these storm types, including tornadoes, hailstorms, flash floods, and derechos. (National Oceanographic and Atmospheric Administration [NOAA].)

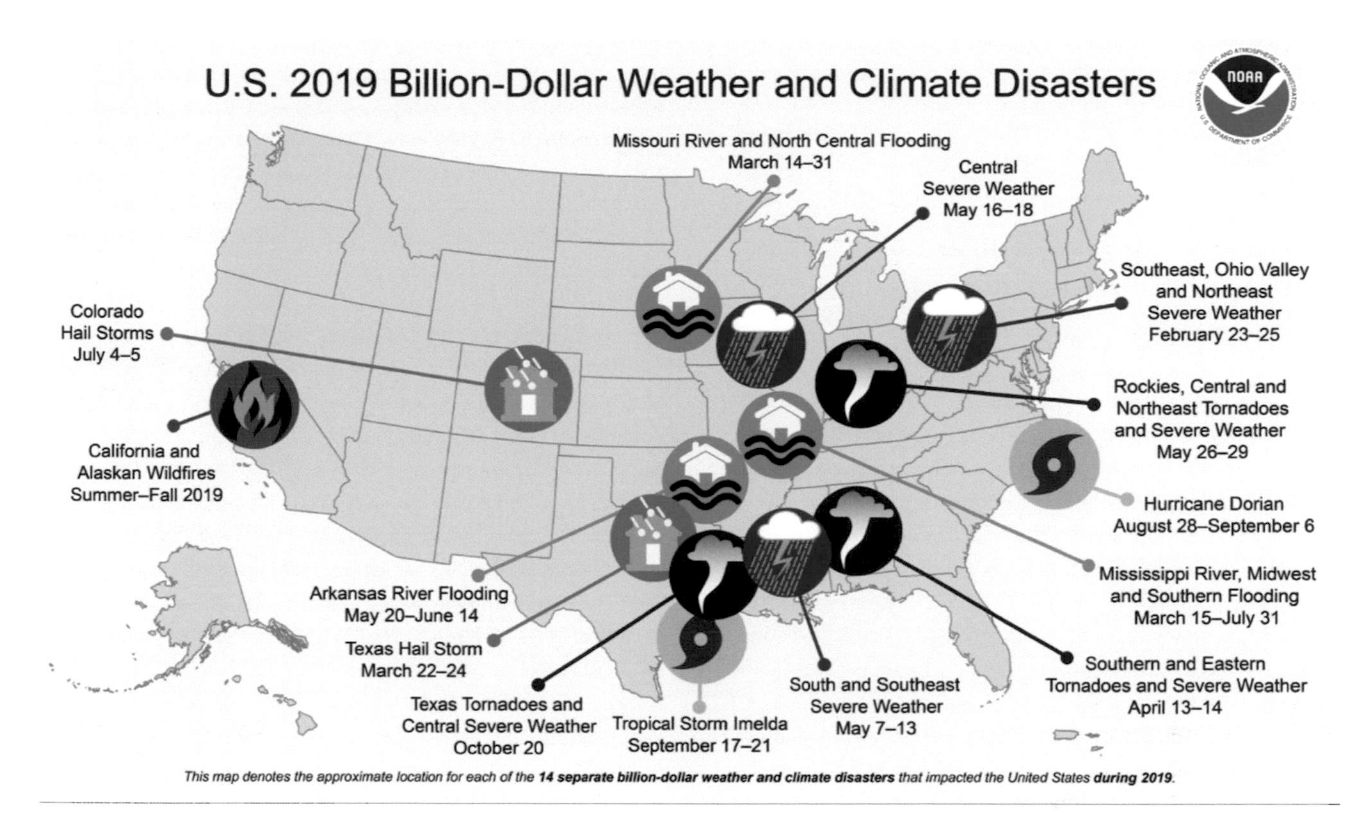

Figure 1.3 (Continued)

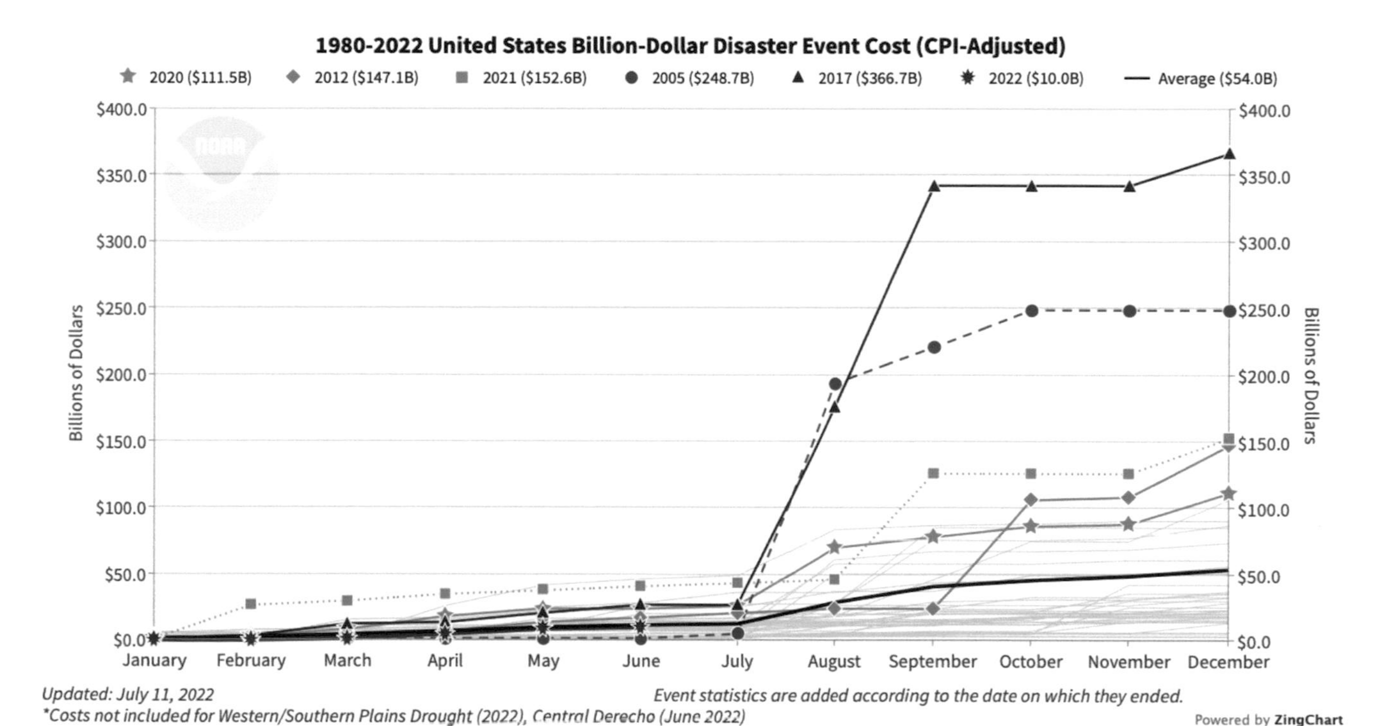

Figure 1.4 Total annual damage from weather-related disasters, 1980–2021. The figure shows dollar totals (accounting for inflation) for five sample years from the period in which annual losses exceeded $100 billion in the United States. (National Oceanographic and Atmospheric Administration [NOAA].)

most frequently occurring type of billion-dollar disaster is the hurricane. Accordingly, several chapters in this book are devoted to these large, cyclonic storms.

Geographical Distribution of Storms and Weather Hazards Across the United States

Some U.S. regions face a regular onslaught from multiple types of weather hazards; other regions experience just a few types of hazards. In terms of large cyclonic storms, there is a seasonality tied to geographical location. In the cool months (November–April) cyclonic storms called extratropical cyclones (a migrating system of rotating wind and precipitation in the mid-latitudes, usually accompanied by one or more weather fronts) create many types of weather hazards across the United States. On the northern flanks of these cyclones, hazards include widespread heavy snow, ice, strong wind, and extreme cold. On their warmer, southern portions, severe local storms including tornado outbreaks develop across the southern and central Plains, Mid-South and Ohio Valley – particularly during early and middle spring. The coastal areas of the United States can be hit by especially vicious extratropical cyclones during winter, called Nor'easters, from the Outer Banks of North Carolina to Maine. Meanwhile, along the Pacific Northwest, powerful oceanic cyclones called Big Blows deliver phenomenal amounts of precipitation and very strong wind to the Cascades and Sierra Nevada.

Tropical cyclones generate immense societal cost and disruption during the summer and early fall months from Texas to New England. Two states in particular, Florida and North Carolina, have experienced a large number of extreme hurricanes and tropical storms. In the case of Florida, the state's unique shape contributes greatly to the frequency of hurricane hits, because much of it extends southward into the tropics. In North Carolina, the state's eastern region (Outer Banks) projects eastward into the western Atlantic. A fair number of tropical cyclones cross the eastern portion of the state as they recurve out of the tropics, arcing northward into the middle latitudes.

Severe local storms, including tornadoes and the damaging hail and straight-line wind of thunderstorms, impact a broad region between the Rockies and Appalachians, with a maximum centered over the southern and Central Plains states. Another maximum extends from the Mid-South northward into the Ohio Valley. In fact, there are multiple "tornado alleys" or concentrated zones of tornado activity within the United States, and you may be surprised to learn that the greatest annual number of violent and fatal tornadoes occurs in Louisiana, Mississippi, Alabama, Georgia, Arkansas, and Tennessee (a region collectively known as "Dixie Alley"). The United States leads the world in terms of annual tornado counts. Severe local storms are most common in the warmer months of the year because sun-heated surface air creates an unstable situation, with air rising rapidly in the cores of thunderstorm clouds.

The most widespread type of severe weather hazard in the United States is flooding – including both aerial and flash floods. Nearly every state from coast to coast has dealt with a billion-dollar flood event, including desert states such as Arizona, New Mexico, and Utah (here, a summertime monsoon flow frequently triggers heavy rainfall). The Gulf Coast states are among the most flood-prone – with extreme rain totals from individual storm events totaling 30–40 inches – and many of these floods stem from landfalling tropical cyclones. The U.S. record rainstorm of 62 inches occurred in 2017 from the remnants of Hurricane Harvey in the Houston region. Other vulnerable locations include the Appalachians and mountains of the Pacific Northwest; here, extreme rains become focused where mountains lift moist, oceanic air – a process termed orographic rainfall. Flash floods are the deadliest of all severe storm-related natural hazards.

Long-Term Trends in Weather Disasters

The steady increase in U.S. population, built infrastructure, and insured property over past decades is one explanation for why the number of billion-dollar weather and climate disasters (as well as the total dollar loss of individual events) has been steadily increasing. Figure 1.5 shows this trend over the period 1980–2020. While there is considerable variation in total annual losses from year to year, upward trends in annual costs and the frequency of several types of weather disasters are unmistakable.

Another cause for increased numbers of certain weather disasters is climate change, at least in part. However, as we shall see in coming chapters, the cause-and-effect relationship between increased storminess and global warming is not clear-cut. Extreme weather events are short-lived phenomena that generally impact small regions (compared to the entire globe), whereas the Earth's warming trend is truly global in scope and has been steadily unfolding over many decades. There is increasing evidence, however, that droughts, heat waves, and floods are becoming more frequent and severe as a direct consequence of global warming. Much more controversial is

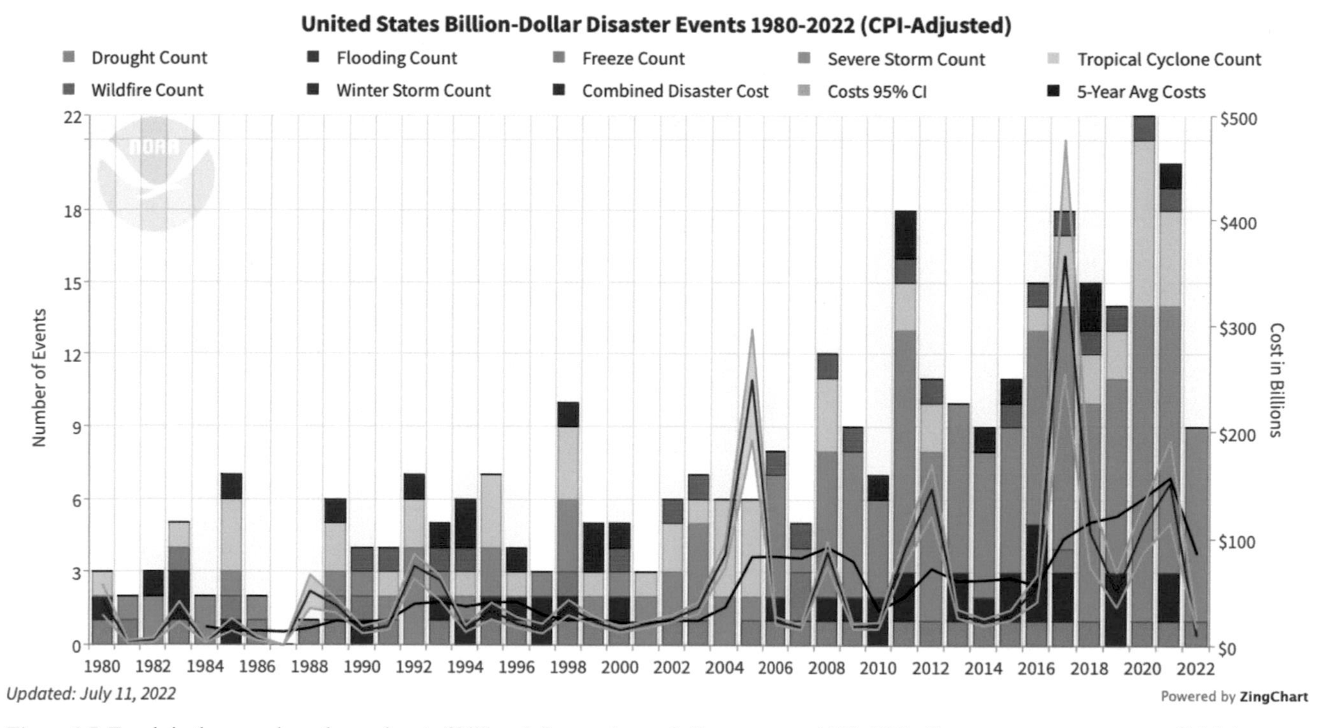

Figure 1.5 Trends in the annual number and cost of billion-dollar weather and climate events, 1980–2020. Shown are numerous types of high impact events (colored bars) and changes in long-term costs (solid lines). (National Oceanographic and Atmospheric Administration [NOAA].)

the link between hurricanes and a warming planet. In addition, the most recent research fails to show any long-term increase in strong and violent tornado frequency with global warming trends.

Increasing amounts of property and increased property value are indeed major drivers of escalating property losses due to severe weather. This is particularly true for several high-growth regions of the United States, including the Gulf Coast and the Northeastern megalopolis, where regional populations have been experiencing exponential growth rates.

Figure 1.6 presents a powerful conceptual idea, which relates our societal vulnerability to severe weather – in a specific location – to two key factors: the incidence of the natural hazard and the risks of human behaviors, that is, our choices to locate and concentrate in certain spots. Over decades, more people and property have been placed in harm's way. With suburban sprawl increasing, the "targets" for severe storms such as tornadoes have been expanding in size. Natural hazards pose threats; people and property placed in their path turn those hazards into costly, fatal disasters and outright catastrophes. The burden of severe weather on society has risen substantially in the past several decades, and extrapolations of trends suggest the situation will only worsen. Throughout this textbook,

Paradigm of Extreme Weather Vulnerability

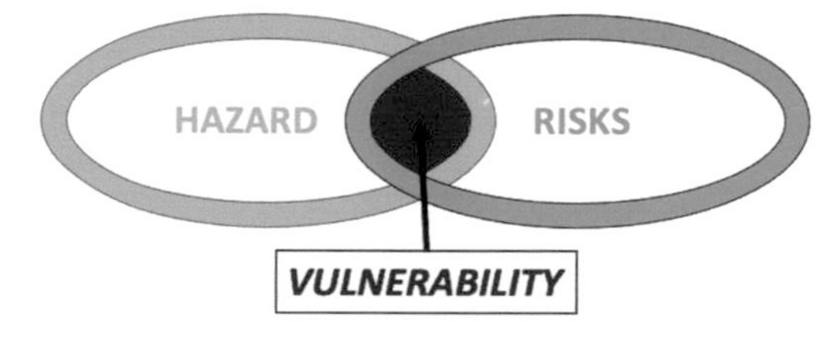

Figure 1.6 Paradigm of extreme weather vulnerability. This conceptual model applies to any natural hazard, including severe and hazardous weather events. Societal vulnerability comes from a combination of the natural hazard and risky behavior that puts humans in harm's way.

we encourage you to use this paradigm as a framework for understanding our vulnerability across the great spectrum of severe and hazardous weather.

U.S. Severe Weather Fatalities

Many types of weather extremes take a terrible toll on human lives. In Figure 1.7, which summarizes U.S. fatalities by severe weather type, we are most interested in the 30-year average

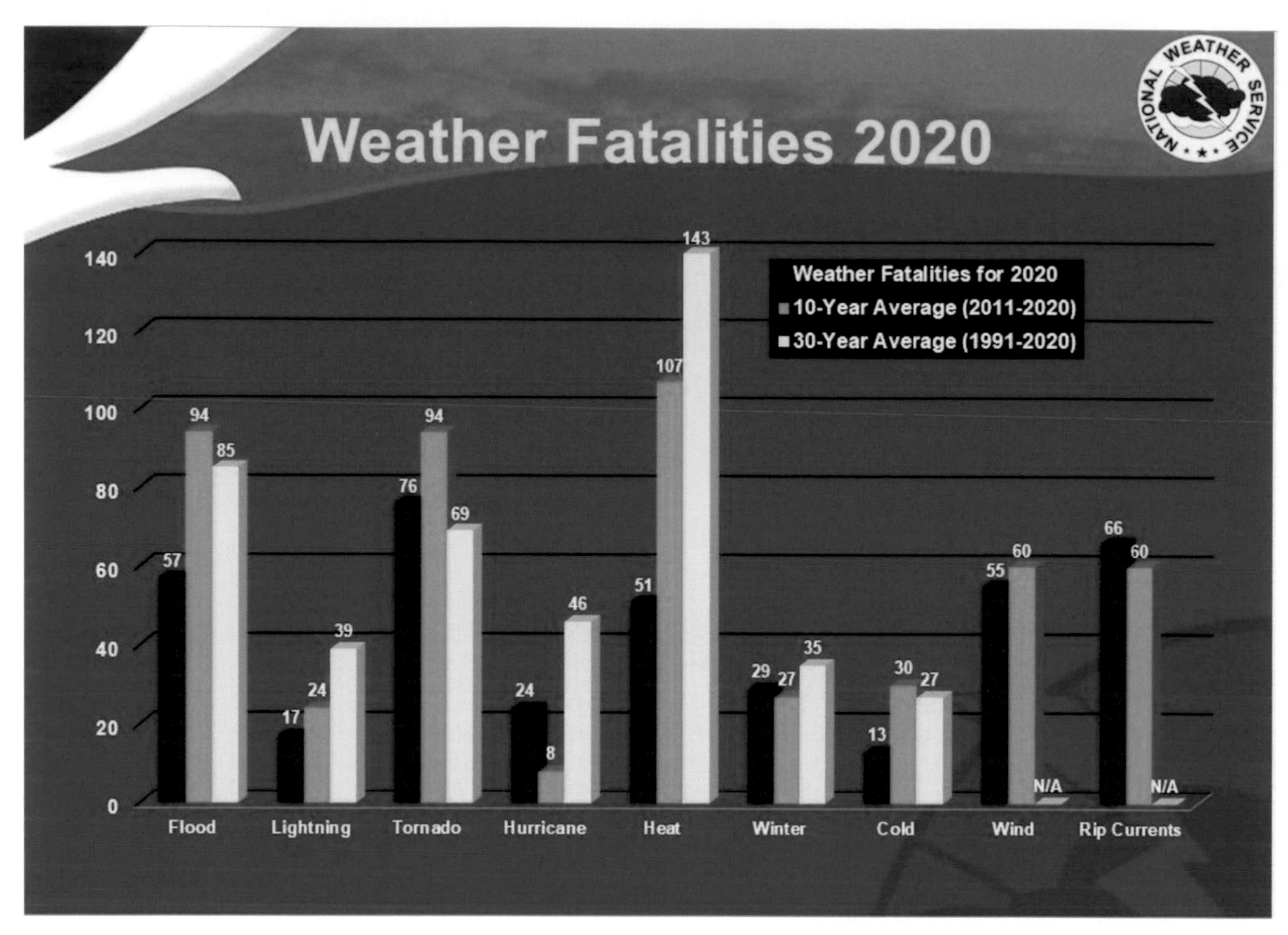

Figure 1.7 U.S. weather fatalities by severe weather category (2020, 5- and 30-year averages). (National Oceanographic and Atmospheric Administration [NOAA].)

annual fatality rates (yellow bars). Overall, heat waves account for the greatest share of weather-/climate-related fatalities. In various parts of the globe, the most extreme heat spells have killed many thousands per episode in the past few decades. Considering *just* severe storms, flash floods account for the greatest annual loss of life in the United States. (This fact may surprise you; after all, rain is most frequently beneficial!) Among severe storms, tornadoes rank second in terms of overall fatality rate, but the number of tornado deaths per year varies considerably, depending on (1) the number of tornadoes in any given year and (2) whether strong or violent tornadoes strike major population centers. For instance, although the average number of tornado fatalities is close to 70 per year, 2011 saw nearly 560 U.S. tornado fatalities! Lightning ranks third, close to that of tropical cyclones. In terms of annual fatalities, hurricanes are perhaps the single most variable type of severe storm. In some years, there have been no U.S. fatalities, yet a single blockbuster storm making landfall can extinguish thousands of human lives, as Hurricane Katrina did in 2005 with over 1800 U.S. fatalities.

Figure 1.8 illustrates trends in U.S. severe weather fatalities over the past 65+ years (Ritchie and Roser, 2014). Here there is both good news and bad news. Many of the severe weather hazards – lightning (orange), tornadoes (green), and floods (blue) – have been trending decidedly downward over decades. These trends are the result of more effective preparedness, detection, and warning programs; improved scientific understanding; and increased public education. Most importantly, the average annual fatality rate has declined in spite of aggressive U.S. population growth.

However, there are a few troubling aspects. At the very end of the graph, there is an enormous spike in tornado fatalities for 2011 (spiked green curve): close to 560 deaths, compared to the long-term average value of nearly 70. This sobering

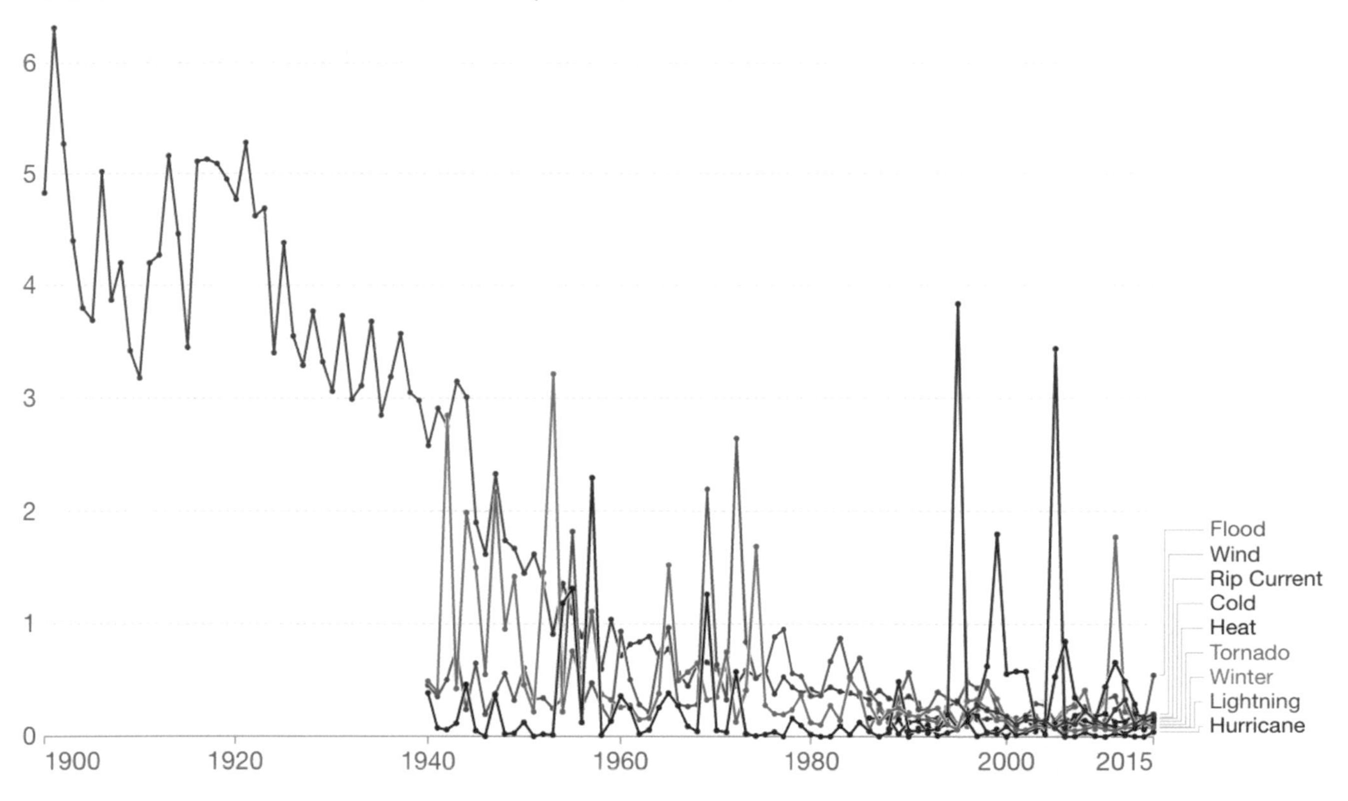

Figure 1.8 Long-term trends in U.S. weather hazard fatalities. The figure is organized according to type of weather hazard, expressed as fatalities per million persons. (National Weather Service [NWS].)

statistic reflects both a record tornado year, with numerous super outbreaks, and several violent tornadoes that struck densely populated towns and cities. In fact, the long-term tornado fatality rate may have reached a plateau, starting around 1990. It suggests that even with the best warnings, the strongest tornadoes are inherently unsurvivable, no matter the type of refuge, including well-fortified structures such as wood-frame homes and steel-framed public buildings such as schools.

Equally concerning is the large number of fatalities due to heat waves in 1995, 2005, and 2012 (magenta curve spikes). Large cities such as Chicago suffer disproportionate shares of heat-related fatalities. Finally, throughout the past several decades, the United States has seen highly intermittent death tolls due to hurricanes. In many years, the loss of life is small. But the nearly 1,800 fatalities in 2005 (dark brown spike) derive from a single storm: Hurricane Katrina, which severely flooded New Orleans and obliterated large portions of the Mississippi coastline. No U.S. hurricane death toll has been this large since 1900!

How This Book Is Organized

This text is organized into three main sections. Part I introduces a variety of basic science concepts, terms, and principles pertaining to general meteorology. Meteorology is an applied science concerned with the processes and phenomena of the atmosphere, especially as they relate to weather forecasting. The chapters in Part I cover the forces governing pressure and winds, general weather systems such as jet streams and cyclonic vortices, atmospheric stability, and the formation of various types of precipitation. Additionally, Part I explains the technologies used to observe weather systems, including Doppler radar and satellites. Overall, Part I provides a foundation for the more detailed expositions on severe storms in Parts II and III.

Part II examines large-scale, high-impact weather systems, including extratropical and tropical cyclones. Extratropical cyclones and their associated weather fronts cross the United

States daily and affect most regions dozens of times a year. When these storms reach severe levels, they unleash widespread weather hazards during most months of the year. An entire chapter is devoted to winter storm hazards stemming from these storms, including blizzards, crippling ice storms, and arctic air outbreaks. You will also learn about the properties of the volatile, warm-air regions of extratropical cyclones – which generate tornado outbreaks, especially in the spring months. Part II also treats in detail hurricanes – the costliest and deadliest weather events of all.

Part III analyzes severe local storms in great detail, starting with introductory material on thunderstorms – including lightning, hail, downdraft formation, squall lines, and the factors that push thunderstorms to severe levels. Next we examine supercell (rotating) thunderstorms and tornadoes from both a scientific standpoint and their societal impacts, with significant coverage of tornado outbreaks, tornado detection and warning, and tornado intensity rating scales. An entire chapter is devoted to downbursts and derechos; studies reveal that the social disruption, damage, and fatality rates created by derechos are comparable to tornadoes and landfalling hurricanes. Part III ends with a discussion of flash floods and their associated hazards, including mud and debris flows. We pick up the topic of heat waves in an Appendix.

This unique textbook – written at a scientifically rigorous but introductory level – will help you better understand the weather headlines you read and the newscasts you watch. We sincerely hope that you will gain a valuable new understanding from both this text and the course in which you are enrolled.

Summary

LO1 Describe the typical categories of billion-dollar weather disasters that strike the United States each year.

1. Many types of storm and weather hazards strike the United States every year, ranging from large cyclones to short-lived, severe local storms.
2. The damage caused by heavy storms varies considerably from year to year, depending on the types of severe weather and the geographical distribution of population density.
3. Given drastic population growth, single storm events in the United States, such as a powerful hurricane, can create nearly $100 billion in damages and thousands of human fatalities.

LO2 Describe the geographical distribution of common weather hazards and severe storms that affect the United States.

1. Billion-dollar winter storms occur most commonly in the Mid-Atlantic and Northeast, while tropical cyclones do the most damage along the Gulf Coast and East Coast. Of all the states, Florida and North Carolina have weathered the greatest hurricane damage totals.
2. Billion-dollar tornadoes and damage tend to localize over the southern and central Plains, but a pronounced secondary maximum extends across the Mid-South, the so-called Dixie Alley.
3. Billion-dollar flood events, including both river floods and flash floods, are the most widespread severe storm type. Nearly every state in the United States contends with this hazard.

LO3 Discuss key trends in weather disasters across the United States over the past several decades.

1. Severe local storms have produced a dramatic increase in total dollar damages (in excess of several billion dollars) over the past several decades.
2. The number of hurricanes, the costliest type of severe storm faced by the United States, varies from year to year but can account for the majority of any single year's total extreme weather damage.
3. Long-term trends toward increased damage from severe local storms (for example, tornadoes and hurricanes) can have multiple causes, including climate change in some categories of weather hazard. Importantly, the higher levels of damage strongly stem from dramatic increases in property and public infrastructure tied to U.S. population growth.
4. The framework of societal vulnerability to natural hazards underscores contributions from both the frequency and intensity of the specific natural hazard and from human factors such as actions that place large concentrations of people and property in harm's way.

LO4 Identify trends in weather- and storm-related fatalities in the United States over the past half century.

1. In the United States, heat waves account for the greatest number of annual deaths resulting from extreme weather, followed by flash floods, tornadoes, lightning, and hurricanes.
2. Several weather and storm hazards show decades-long reductions in annual fatalities, including lightning and tornado deaths. However, several recent years have been marked by exceedingly large death tolls associated with heat waves, tornadoes, and hurricanes.

Reference

Ritchie, H. and M. Roser, 2014. *Natural Disasters*. Published online at OurWorldInData.org. https://ourworldindata.org/natural-disasters

CHAPTER 2

Meteorological Primer, Part I: Pressure and Wind Relationships

Learning Objectives

1. Define atmospheric pressure and explain how pressure changes with altitude.
2. Understand how the pressure gradient force creates wind, and determine wind strength by examining the spacing of isobars on a surface weather map.
3. Explain how the geostrophic wind, which blows parallel to isobars, is created by a balance between the pressure gradient force and the Coriolis effect.
4. Describe how horizontal variations in air temperature create a pressure gradient and geostrophic wind.
5. List the principal features of general atmospheric circulation, and discuss how global air circulation is driven by the movement of excess heat energy from the equator toward the poles.
6. Discuss how the types of wind forces differ among extratropical cyclones, hurricanes, and tornadoes.
7. Describe the origins and geographical distribution of various regional-scale wind circulations, including the sea breeze, mountain–valley circulation, nocturnal low-level jet, North American Monsoon, and severe mountain winds (Santa Ana and Chinook).

Introduction

Much of the destruction caused by severe storms arises from strong winds blowing around atmospheric vortices. For example, tornado winds are highly localized and may approach 217 kts (250 MPH). The circulation of air around hurricanes is about 1000 times larger than that in a tornado, and extreme winds in the hurricane's core may exceed 130 kts (150 MPH), with higher gusts. And the everyday, extratropical variety of cyclone, which brings changeable weather to many parts of North America, may contain winds approaching 52 kts (60 MPH). Tornadoes inflict damage to engineered structures that is highly localized and often near-total, while much larger extratropical cyclones and hurricanes down tree limbs and create power outages across widespread areas. In both cases, intense winds arise from atmospheric forces that combine in predictable ways.

In this chapter, we introduce these forces and the types of weather systems that produce strong winds. As you begin reading, keep the following key point in mind: Wind (the movement of air) arises from differences in atmospheric pressure across some unit of distance. Over a pressure gradient – that is, a change in surface pressure over a specified distance – air will accelerate from a region of high pressure toward an area of low pressure. A wind is born!

To set the stage for our study of wind, let's begin with some important definitions. Cells of low pressure, around which winds spiral counterclockwise, are the common extratropical cyclones that move from west to east across North America. Some 50–70 of these storms, which can produce destructive winds across large areas, may impact any given U.S. region in the course of a year. Hurricanes are a type of cyclone that form in the tropics. Embedded within these cyclones, particularly those in the mid-latitudes, are intense thunderstorms that spawn tornadoes. Tornadoes are very small-scale cyclones that create narrow corridors (typically a few kilometers long by a few hundred meters wide) of extreme wind destruction.

Now let's examine the basics of atmospheric pressure and winds.

Sea-Level Atmospheric Pressure Is the Weight of the Overlying Air Column

Atmospheric pressure is the weight of the overlying atmosphere per unit area. The concept of atmospheric pressure is quite simple, as Figure 2.1 shows. Air is composed of a mixture of

DOI: 10.4324/9781003344988-3

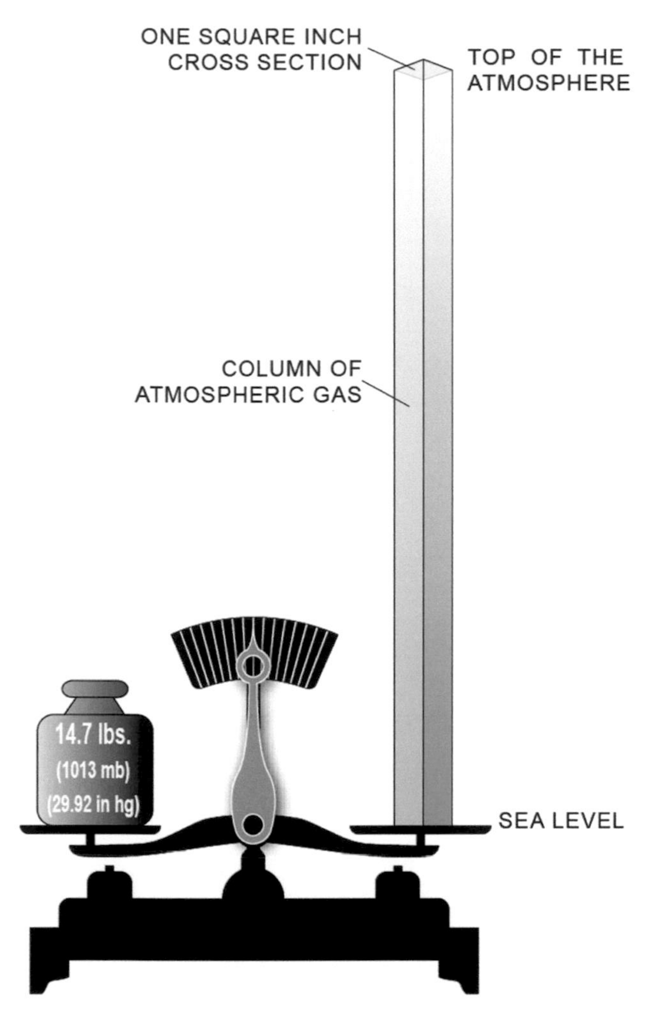

Figure 2.1 Atmospheric pressure is the accumulated weight of the atmosphere. We seldom think of atmospheric (air) pressure unless we are changing altitude rapidly (taking off or landing in an airplane, for example). However, air does have weight. The average weight of 1 inch^2 column of air measured from the top of the atmosphere to sea level weighs 14.7 pounds (1013 hPa or 1013 mb or 29.92 inches hg).

gases, mostly nitrogen (78%) and oxygen (21%). Although invisible, air has mass. If you designed a scale to measure the weight of a 1×1 inch vertical column of the Earth's atmosphere, extending from surface to space, you will find that it weighs 14.7 pounds per inch^2 (psi) at sea level. The metric unit of atmospheric pressure is the pascal (Pa), and 14.7 psi equals 1013 hPa (hectopascals). American meteorologists commonly use a unit called the millibar (mb), equating 1 mb with 1 hPa, or 1 mb = 1hPa. So we can link all of our equivalent sea level pressure units together in the following manner:

$$14.7 \text{ psi} = 1013 \text{ hPa} = 1013 \text{ mb} = 29.92 \text{ inches hg}$$

The gases that compose the atmosphere are compressible; that is, they can be squeezed together. Molecules become squeezed together close to the Earth's surface, due to the weight of the overlying air column. If we were to take a balloon ride upward through the atmosphere, we would discover that pressure decreases with altitude. But it does not decrease in a steady manner. Instead, the pressure decreases exponentially with height. (An exponential change is one in which the change is proportional to the amount of a substance.) Near the Earth's surface, where the pressure is greatest, the decrease with an incremental change in altitude is rapid. Halfway up the troposphere – say, at a level of about 5486 m (18,000 feet) – atmospheric pressure is about 500 mb, or about half of what it is at sea level. Here, the decrease in air pressure with altitude is much smaller. Figure 2.2 illustrates the exponential drop in pressure with altitude.

A measurement closely related to air pressure is air density. Density is the ratio of mass to volume. Because air molecules are squeezed close together at sea level, air density is largest at sea level. Density drops off with height in a very similar manner to pressure – that is, it follows an exponential decrease.

Pressure Gradient Is the "Motor" That Generates Wind

Substances tend to move from regions of high concentration to regions of low concentration. Think about what happens when you add a drop of dye into a tank of still water or when a smokestack releases pollutants into still air. Those heavy concentrations are soon dispersed. A similar process is at work with variations in atmospheric pressure at the Earth's surface or at any level in the atmosphere for that matter. Air will flow freely from a region of high pressure toward a region of low pressure. Nature likes to even out, or equilibrate, these regions of uneven pressure.

An important force arises between areas of unequal pressure, called the pressure gradient force. This force accelerates the movement of air from high pressure to low pressure. Air that is moving is simply wind. Wind, in turn, exerts a stress or force of its own on structures, such as the leaves of a tree or a flag on a flagpole.

Figure 2.3 illustrates the pressure gradient force. Imagine two hollow boxes separated by a partition. The partition has an aperture opened and closed by a sliding door. On the left side, we allow air pressure to build to a higher value than on the right side. The difference in pressure, or ΔP, is 6 mb. Now we quickly slide open the aperture. Air will immediately flow from the high-pressure side to the low-pressure side. Now let's increase the size

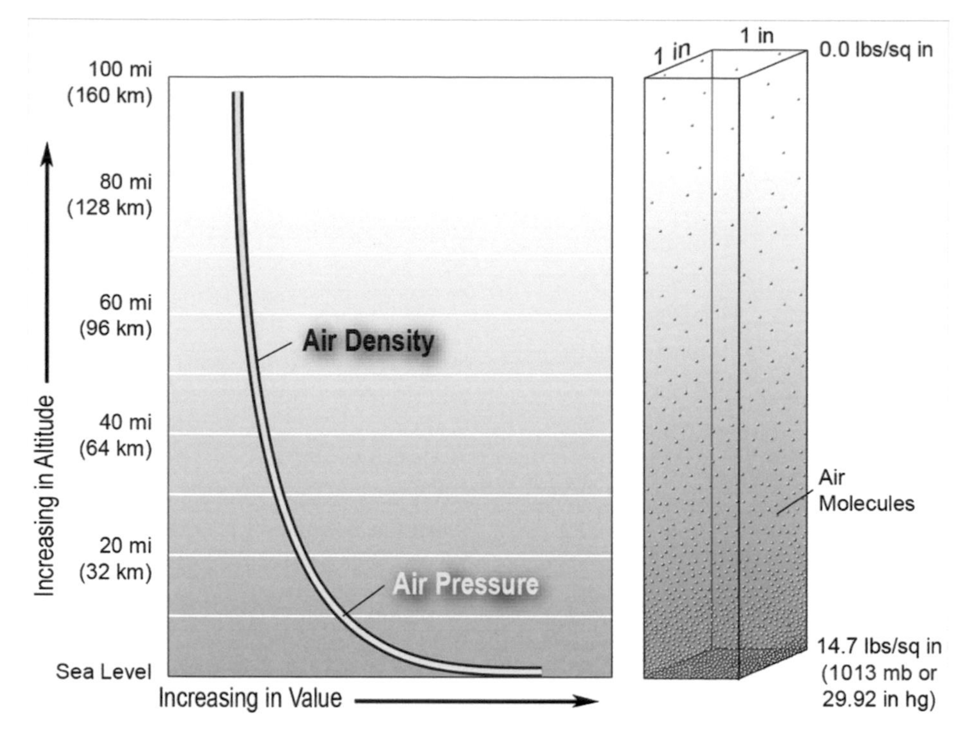

Figure 2.2 The atmosphere is a compressible gas. It is densest at the Earth's surface and thins with increasing altitude. This graph illustrates how air density and air pressure change exponentially in a one-to-one correlation with change in altitude.

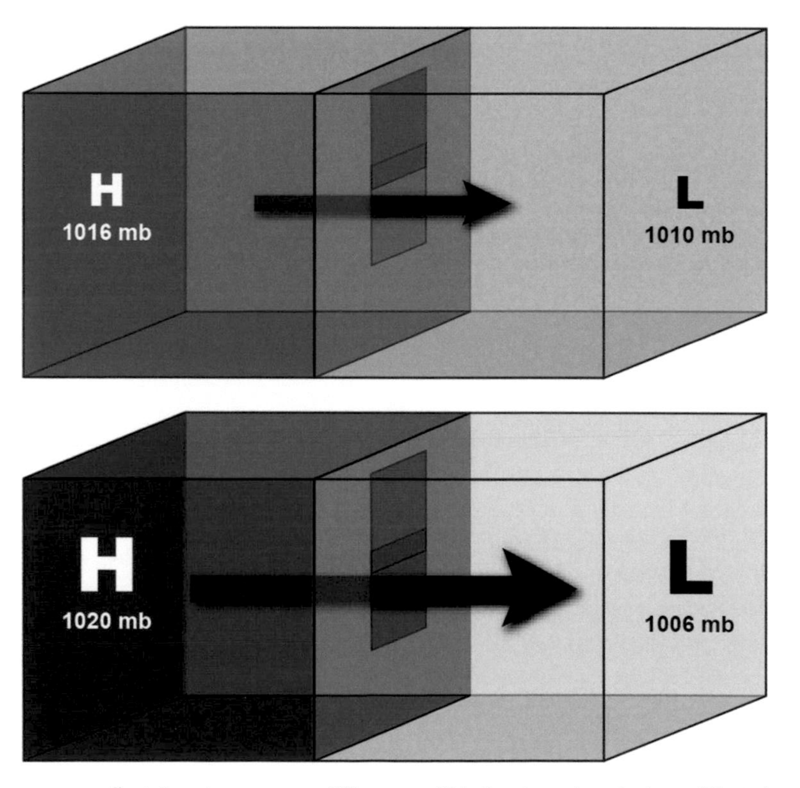

Figure 2.3 Illustration of the pressure gradient. In nature, pressure differences within the atmosphere try to equilibrate by moving air from areas of high pressure to areas of low pressure. We know these air movements as wind. As illustrated here, the greater the pressure difference over distance (pressure gradient), the stronger the wind.

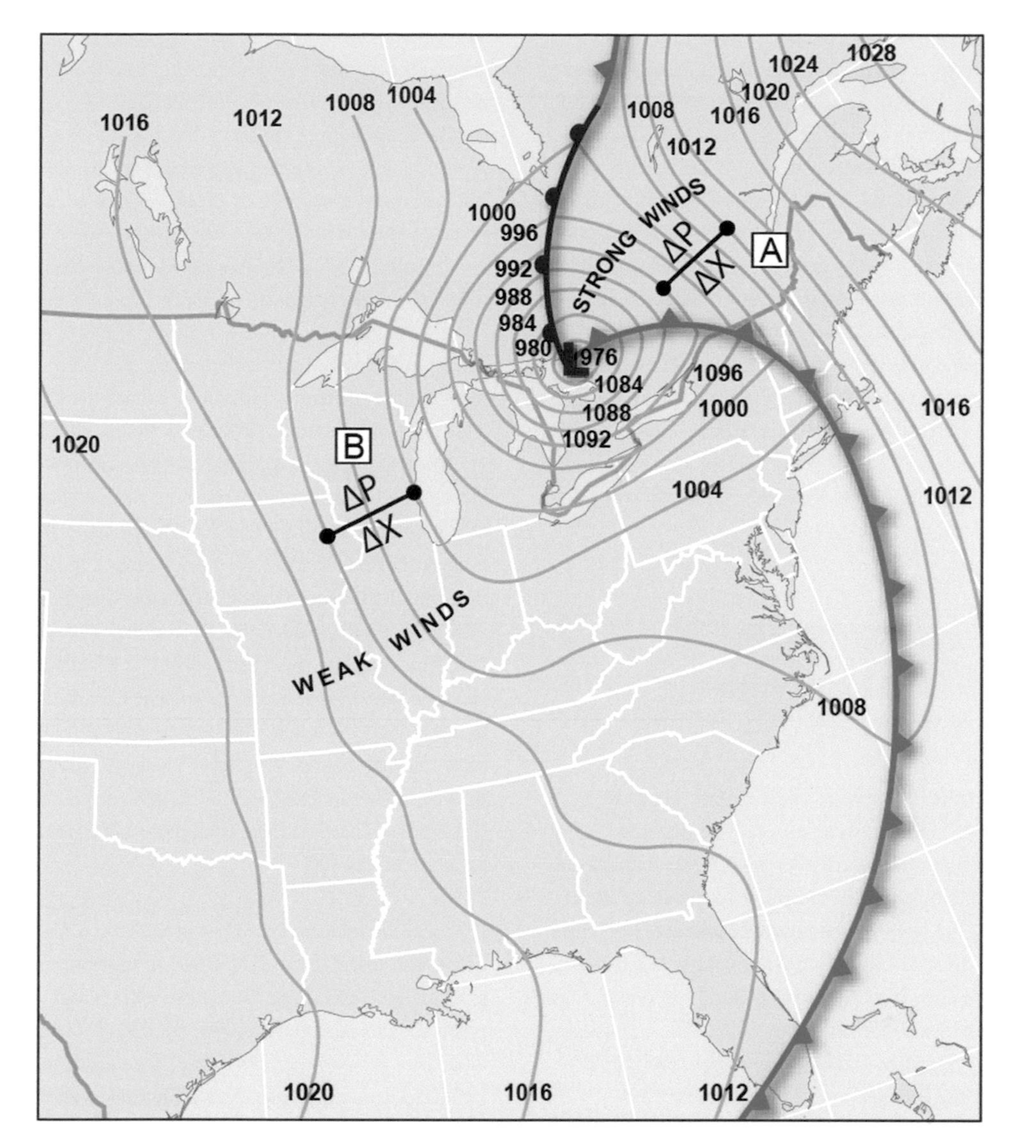

Figure 2.4 Map displaying isobars distributed around a low-pressure system. The areas with higher pressure gradients are indicated by closely spaced isobars. Where isobars have greater separation, the pressure gradient will be less. The relative strength of wind depends on the pressure gradient. The higher the gradient, the stronger the wind. By definition, the pressure gradient is equal to $\Delta P/\Delta X$, where ΔP = difference in pressure, and ΔX = increment of distance.

or magnitude of the pressure difference, such that the $\Delta P = 14$ mb. When we slide open the aperture, air will flow more vigorously from high pressure to low pressure. We expect a stronger wind will flow through the aperture in the second example.

From this simple illustration, let's jump to a surface weather map, as shown in Figure 2.4. We know that the magnitude of the pressure gradient force (and strength of the wind) is proportional to the size of the pressure difference, ΔP. A larger difference in pressure will generate stronger winds. But the pressure gradient is related to *both* the difference in air pressure *and* the distance over which this difference is measured. By definition, the pressure gradient is equal to $\Delta P/\Delta X$, where ΔP = difference in pressure and ΔX = increment of distance. We can assess the strength of the pressure gradient by examining the isobars – lines of constant pressure – drawn on a weather map. In Figure 2.4, we see a region of high pressure, 1020 mb, separated from a region of low pressure, 976 mb. The area of low pressure is in fact an intense extratropical cyclone over the Great Lakes during the fall. This particular storm had such an intense pressure gradient that it produced widespread wind damage within a few hundred miles of the storm's center.

Let's examine the pressure gradient at a couple of locations. For a given difference in pressure, ΔP, a large increment of distance,

ΔX, gives rise to a small overall pressure gradient. If the two regions are separated by a small increment of distance, the gradient will be large. To avoid doing this type of mental math, we can simply count the number of isobars between adjacent regions of high and low pressure. By convention, isobars on weather charts are spaced every 4 mb. If the isobars are close together, such that a large number of isobars lie between the region of high pressure and low pressure, the pressure gradient is large, and the wind will blow strongly, as in Region A in Figure 2.4. If, on the other hand, the isobars are spaced far apart over the same distance, both the pressure gradient and the wind are much weaker (Region B). In Figure 2.4, very strong northeasterly winds (winds that blow *from* the northeast) are flowing into the storm north of the Great Lakes. However, to the west of the storm and farther from its center, much slower winds blow from the west (westerly winds).

But There's a Twist: Earth Spins, So We Must Introduce the Coriolis Effect

You should now be able to examine any standard chart of surface pressure and (1) identify regions of high and low pressure and (2) estimate the relative strength of the wind at any location on the map. However, there's more to the story about wind. Until now, we've neglected another important effect, one that determines the direction in which winds blow. Winds in the northern hemisphere circulate clockwise around regions of high pressure, and counterclockwise around regions of low pressure. This important effect is called the Coriolis effect, and it arises from the rotation of the Earth.

The Coriolis effect causes moving air to deflect to the right of its path in the northern hemisphere (and to the left in the southern hemisphere). The Coriolis effect arises only when the pressure gradient accelerates the air; it has no influence on air that is perfectly still. This effect explains why air does not flow straight outward from the center of high-pressure cells, like spokes on a wheel, but rather circulates clockwise around the high-pressure cell. Similarly, air circulates counterclockwise around a low cell rather than flowing straight inward from all directions.

The winds through the deep atmosphere arise from a balance between the pressure gradient force and the Coriolis effect. This state of balance gives rise to the geostrophic wind. Figure 2.5 illustrates the geostrophic wind, with arrows that show the size and direction of the pressure gradient force, the Coriolis effect, and the resultant wind; such arrows expressing both magnitude and direction are called vectors. In Figure 2.5, you will note that the pressure gradient force is directed from the region of high pressure toward that of low pressure, along the pressure gradient. Because the geostrophic wind is a state of balance, the Coriolis effect must equal the pressure gradient force, operating in an opposite manner. The geostrophic wind, then, does not blow directly from high to low pressure but is instead oriented *parallel* to the isobars, with high pressure located to the right (facing the direction from which the wind blows) and low pressure on the left.

It's useful to make a few more points about the Coriolis effect. The effect increases in proportion to the magnitude of the pressure gradient force. When the wind blows strong (that is, when the pressure gradient force is large), the Coriolis effect is also large. Also, the magnitude of the Coriolis effect varies with latitude across the globe. This effect arises because of the Earth's

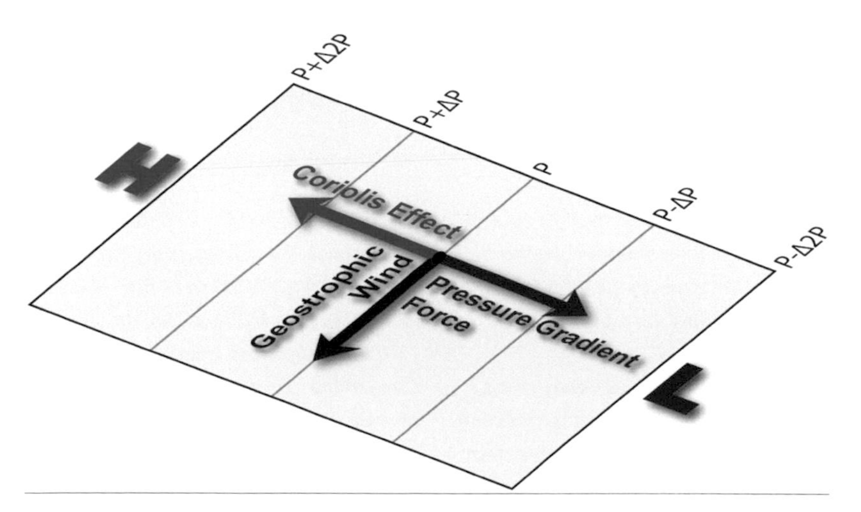

Figure 2.5 Geostrophic (Earth-turning) wind. Geostrophic winds are the result of the interaction between the Coriolis effect and the pressure gradient force. Air wants to move from high to low pressure. However, due to the Earth's rotation, the Coriolis effect causes the wind to be deflected to the right in the northern hemisphere (pictured here) and to the left in the southern hemisphere.

spin about its axis. Earth's spin is maximum at the poles, and zero along the equator. The Coriolis effect therefore vanishes at 0° latitude – in the tropics. The tropics are the one place on Earth where wind does indeed blow in nearly a straight path from high pressure toward low pressure. Finally, the Coriolis effect comes into play only when the pressure gradient operates over fairly large distances and over a long time, typically in excess of several hundred kilometers and several hours. Thus many small-scale, short-lived wind circulations and vortices – such as tornadoes – are not directly influenced by the Coriolis effect.

Horizontal Variations in Air Temperature Give Rise to a Pressure Gradient

Now we arrive at a very important concept in introductory meteorology, one that will enable you to understand how horizontal variations in air temperature give rise to a horizontal gradient in air pressure. Examine Figure 2.6, which portrays three vertically tall columns of atmosphere, extending from the surface to the 500 mb level (5486 m [18,000 feet]). All three air columns contain the same mass of air and therefore exert the same pressure on the surface (Panel A). Now let's uniformly cool the air inside the left-hand column by several degrees and warm the air inside the right-hand column. Like most substances, air contracts when it is cooled and expands when it is heated. So the cold column shortens, and the warm column lengthens, relative to the middle column with no temperature change (Panel B). Here is another way to think about these changes: In the warm column, air molecules spread apart. The air in the column becomes less dense. The column must get taller (deeper), in order to achieve the same weight (surface pressure). In the cold air column, air molecules move closer together. The air in the column is more dense. The column height decreases to achieve the same surface pressure.

Now let's tie all this into the concept of the pressure gradient. Panel (b) shows something interesting that happens to the 500 mb pressure level. We can think of this level as a two-dimensional surface. In the middle column, the height of the 500 mb pressure level lies at 5486 m (18,000 feet). But above the warm column, the 500 mb pressure level is pushed upward to 6046 m (20,000 feet). Over the cold column, the 500 mb pressure level has dropped to 4877 m (16,000 feet). The 500 mb pressure surface takes on a slope, tilting downward toward the cold air column.

In Panel (c), we've left the sloping 500 mb pressure surface in place, but we've taken away the air columns. We've added another surface, a surface of constant altitude, that cuts across

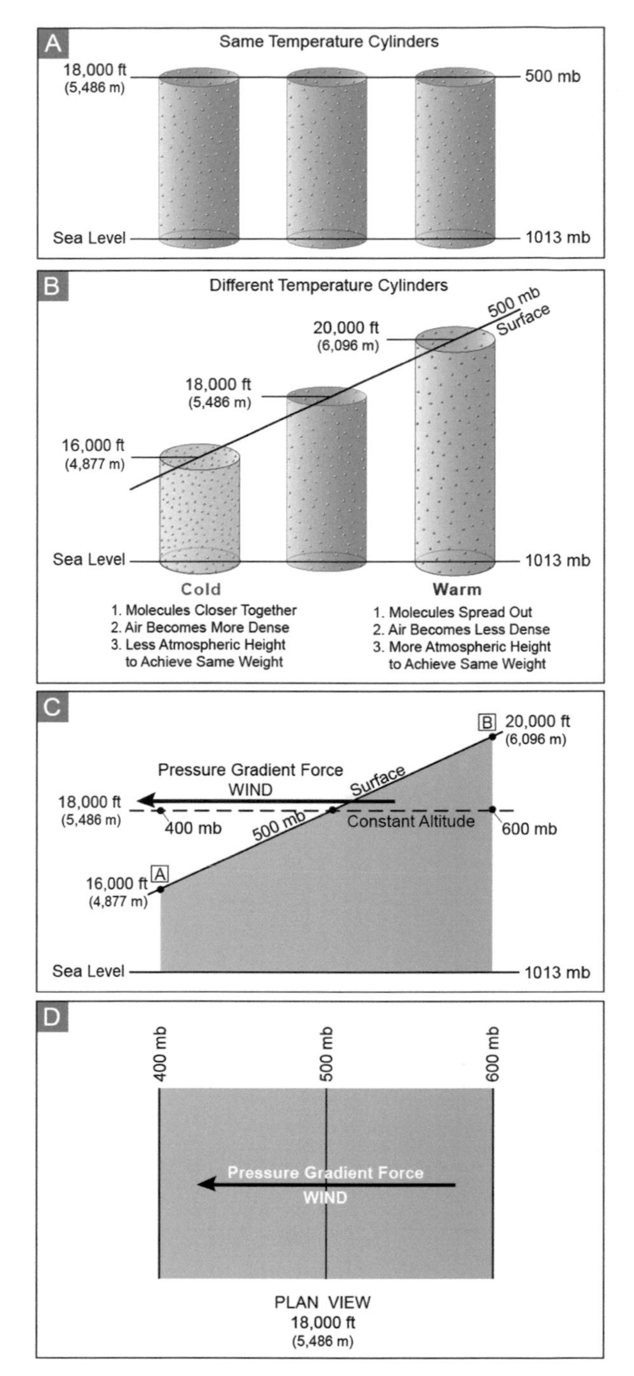

Figure 2.6 How air mass temperature variations give rise to a pressure gradient. (a) Three identical columns of air – same volume, temperature, and pressure. (b) What happens when one column of air (left) is cooled: One (center) remains unchanged, and the third (right) is warmed. The cool air contracts as the molecules move closer together. The warm air expands as the molecules move farther apart. The diagonal line represents the point where the air is at a constant pressure. Of course, air in the atmosphere is not restricted to columns but rather is continuous. (c) As temperatures vary from place to place, the level of equal air pressure changes altitude in concert with air temperature. (d) A plan view of the plane represented by the dashed line in (c). Notice that, with constant altitude, air pressure changes in response to air temperature.

all air columns at exactly 5486 m (18,000 feet). Notice that the surface of constant pressure and surface of constant altitude do not coincide; rather, the constant-altitude surface cuts through the plane of constant pressure. Along the constant-altitude surface, the air must accelerate horizontally because a pressure gradient lies along it. At Point A on the constant-altitude plane, the pressure is lower than 500 mb (it is 400 mb, to be exact). At Point B, the pressure is higher than 500 mb (it is 600 mb, to be exact).

Exactly at the plane's midsection, the pressure is 500 mb. So along the horizontal plane, there exists a pressure gradient force, and the air accelerates from right to left, from a high (600 mb) to a low (400 mb) pressure. The plan view at the bottom of Panel (d) shows the pressure gradient, and the pressure gradient force, directed from high pressure toward low pressure. The Coriolis effect must balance this force and deflect the flowing air to its right. The wind on the constant-altitude surface will turn into the page and is in fact a geostrophic wind. And this geostrophic wind has developed simply because adjacent air columns have heated and cooled! In other words, you can now see that a horizontal gradient in temperature generates a horizontal pressure gradient.

You are now armed with some very powerful concepts in meteorology, and you are ready to apply this knowledge to learn how the Earth's general atmospheric circulation arises.

Global Atmospheric Circulation: A Giant Atmospheric Heat Engine

The Earth is in a state of radiative imbalance, with a surplus of solar radiation in the tropics and minimal solar radiation at the poles. You might ask why the tropics do not steadily get hotter and hotter, and the poles colder and colder. The answer lies in the fact that Earth's atmosphere (and oceans) continuously transport heat energy from the equator toward the poles. Nature attempts to constantly equilibrate this imbalanced state, to even out the excesses and deficits of temperature. The simplest picture we can portray of the atmospheric general circulation is this: Hot air over tropical oceans and land masses rises into the upper troposphere, then moves poleward at high altitude. There is a return flow of cool air from the polar regions advancing toward the tropics. When a fluid such as air moves excess heat vertically upward, the process is called *convection*. At the hemispheric scale, one can imagine the need for a giant, overturning air cell acting as a heat pump for the entire planet (Figure 2.7).

Figure 2.8 illustrates the atmospheric heat engine in more detail. In each hemisphere, there is a very broad temperature contrast between the tropics and the pole. The heated air column over the equator is deeper than the cold air column over the pole

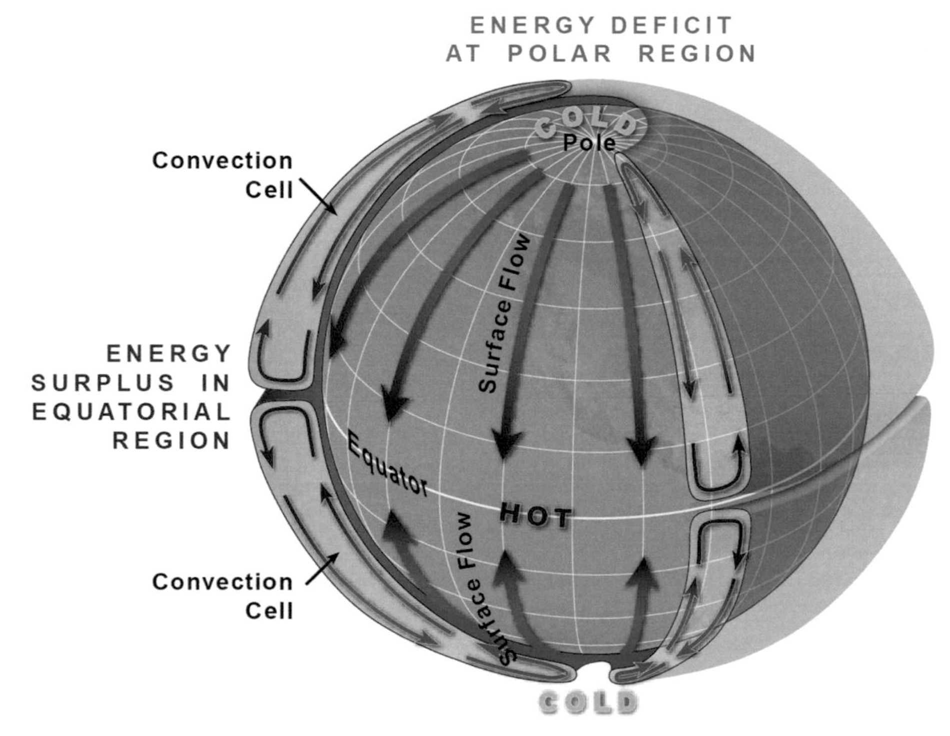

Figure 2.7 Radiation imbalance across the Earth. The Earth is constantly attempting to equilibrate imbalances. Heat from the tropics is moved poleward and cold air at the poles moves toward the tropics. Without this natural adjustment, the zone of habitation would be greatly restricted because the tropics would heat dramatically, and the polar regions would move severe cold significantly toward the tropics.

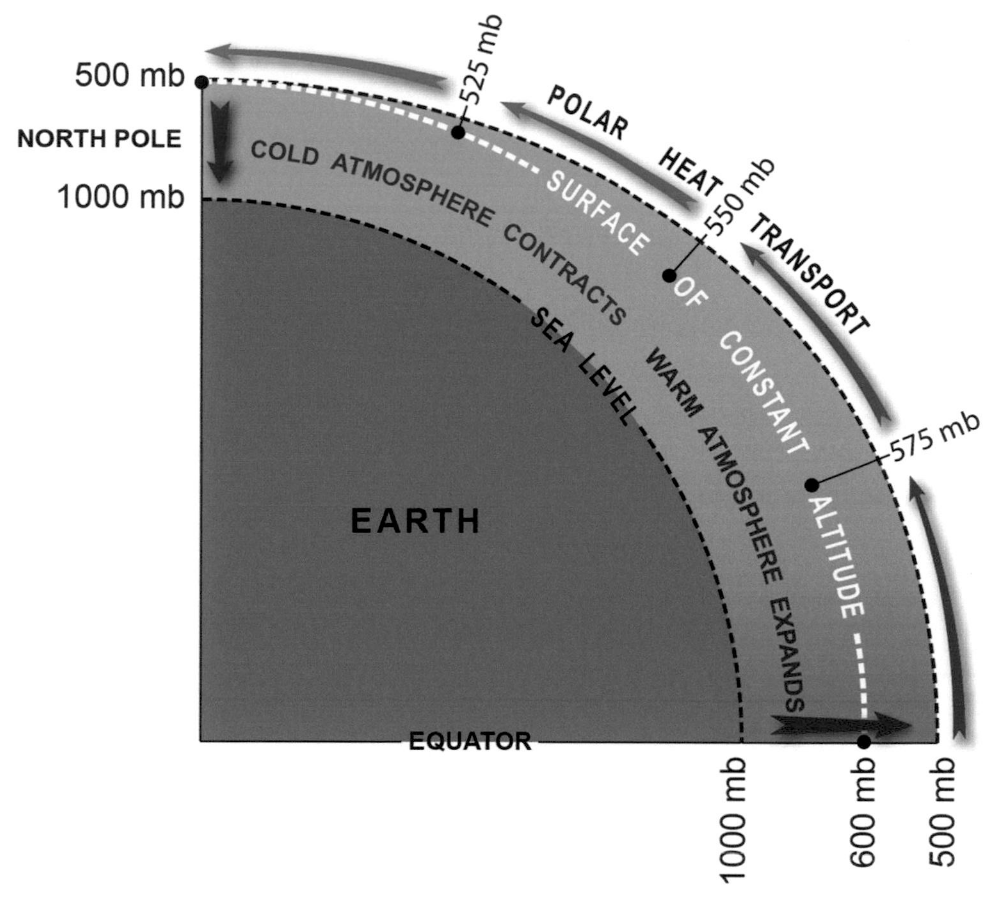

Figure 2.8 Cause of air movement from the tropics toward the poles. Notice that, at a constant altitude, the air pressure decreases going poleward, even though air pressure remains constant at the surface. This occurs because the weight of the atmosphere on average is uniform around the world. However, contracting air at the poles and expanded air at the equator create different pressures at a constant altitude.

(as discussed in Section 2.4). As in Figure 2.6, we see how the 500 mb pressure surface slopes (arcs) downward from the equator toward the pole. If we choose a surface of constant altitude – say, at 4877 m (16,000 feet) – the pressure on this surface over the equator (Point A) is higher than 500 mb, and that over the pole (Point B) is exactly 500 mb. Along this surface, there is a horizontal pressure gradient force directed from high pressure over the equator, toward the north pole, at high altitudes. The air over the tropics accelerates toward the pole. The wind moves excess heat from the equator, and this general atmospheric circulation attempts to even out temperature across all latitudes.

Now this is a highly simplified model of how the Earth balances its uneven heat distribution. The reality is more complicated. First, as Figure 2.9 shows, the general circulation actually consists of three separate overturning cells, or "heat pumps," in each hemisphere. The largest one over the tropics is called the Hadley Cell; the one over the middle latitudes is the Ferrel Cell, and the high-latitude circulation is called the Polar Cell. All three work like individual cogs or gears in the global atmospheric heat engine. But let's not forget that Earth is also rotating, so we must consider the Coriolis effect. Coriolis deflects the poleward flow of air at high levels toward the right, creating a westerly current of air over mid-latitudes. Conversely, the low-level return flow of polar air flowing back toward the equator is also deflected to the right, causing easterly winds to stream from the east across the tropics.

Figure 2.10 portrays the observed general circulation with all its nuances. This is a busy figure, so we'll break it down into several geographic regions, starting with the tropics (20°N to 20°S). Throughout this discussion, keep in mind the two guiding principles: (1) Equatorial heat must be moved poleward, through a series of stepwise circulations, and (2) the Coriolis effect deflects air streaming north or south into

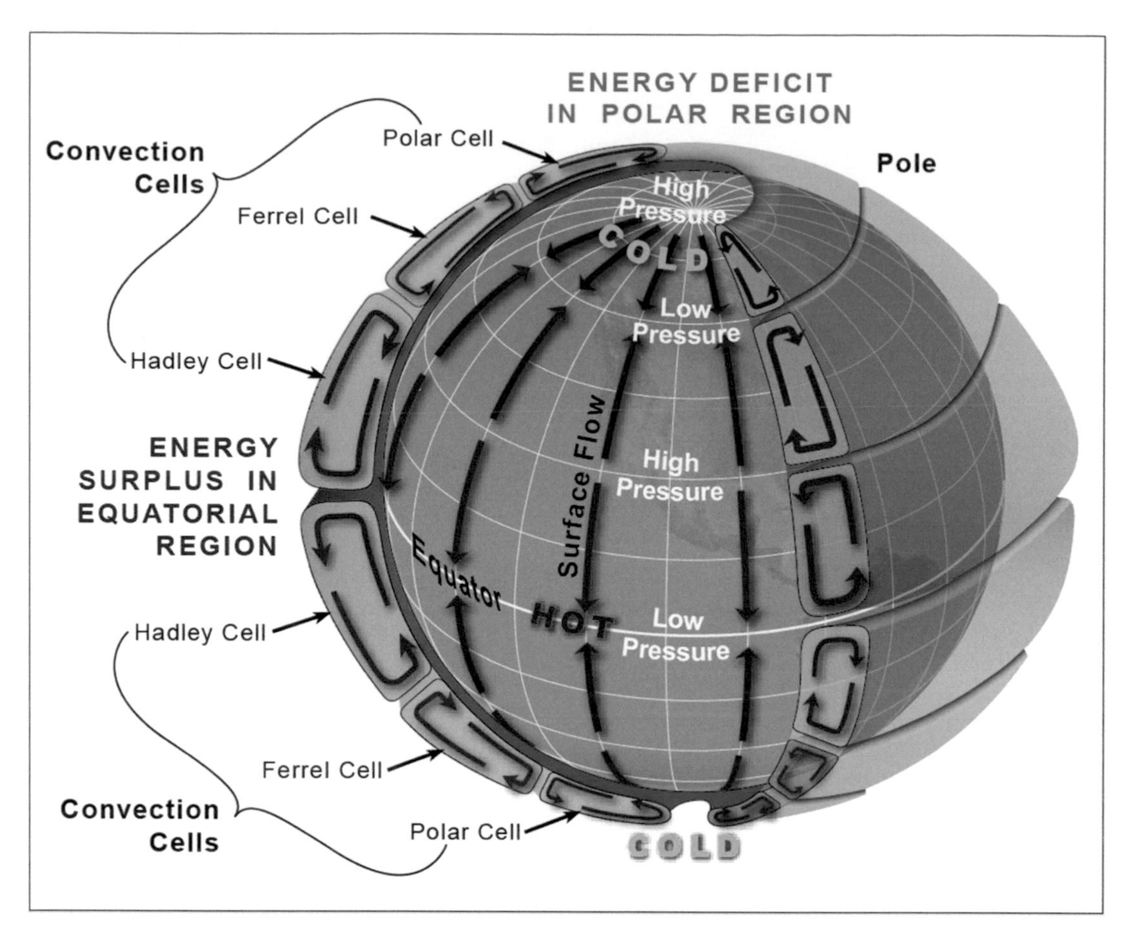

Figure 2.9 The Earth's atmospheric circulation system is composed of three convection cells in each hemisphere. Alternating high- and low-pressure regions mark the boundary of the cells. Beginning at the equator, there is a region of low pressure. Then, moving poleward, there are a high-pressure region, a low-pressure region, and finally high-pressure regions at the poles.

westerly and easterly flows, respectively. In the following discussion, we describe features in the northern hemisphere, but the same reasoning applies to understanding circulation patterns that develop across the southern hemisphere (where the Coriolis effect deflects air to the left).

Circulation Features in the Tropics

The Hadley Cell (one on either side of the equator) is often referred to as the "firebox of the tropics." Within this firebox, warm air over the low latitudes ascends in giant rain clouds. The air cools as it rises, and, upon encountering the stratosphere (at about 16 km [10 mile] altitude), it flows laterally toward the pole. The air at high altitude radiates some of its heat energy to space. This increases its density, causing it to sink back toward the surface around 30° N. Along the equator, there is a persistent band of deep rain clouds. On either side of the equator, trade winds at the surface converge from both hemispheres. When air converges, it is forced to ascend and cool. Moisture in the ascending air condenses into deep rain clouds all around the equator; this area spanning the equator is termed the intertropical convergence zone (ITCZ). As air rises from the surface, pockets of low pressure are created within the ITCZ. Some of these pockets develop into large cloud clusters containing intense tropical thunderstorms. Because the humidity content of warm, tropical air is very high (from the evaporation of warm seawater), these tropical clouds frequently generate torrential downpours and localized flash flooding. Occasionally, during the summer and fall months, cloud clusters near the equator may intensify into highly destructive tropical cyclones (a general term that includes hurricanes) that move toward the west and north, sometimes threatening the United States.

Circulation Features in the Subtropics

The subtropical latitudes are centered at about 30° N and 30° S. High-altitude air on the northern edge of the Hadley Cell sinks over these latitudes. As the air piles up at the surface, high pressure develops. The mass of air is forced to diverge, or spread apart, across the surface in all directions. These large, elongated cells of high pressure are called subtropical anticyclones (STACs), and they are permanent features at this latitude. An anticyclone is a high-pressure cell in which air spirals clockwise outward from the center of the high. The Coriolis effect deflects air flowing southward out of the STAC toward the west, creating a belt of northeasterly trade winds. Air streaming northward from the STAC is deflected toward the east, giving rise to a belt of mid-latitude westerlies. Some of the air at high altitude flowing northward out of the Hadley Cell is deflected to the east by the Coriolis effect. This narrow belt of fast westerly winds in the upper troposphere is called the subtropical jet stream, with winds of 65–87 kts (75–100 MPH) in its core.

Circulation Features in the Mid-Latitudes

The mid-latitude belt extends from roughly 40–60° N and 40–60° S. Westerly flow prevails at all levels in this belt. Most of the continental United States lies within this belt. Along its southern border, warm and humid air masses arrive from the south as air streams northward out of the STAC. Along the northern part of the mid-latitudes, polar easterlies move cold and dry air southward from the north pole. Thus the mid-latitudes are a location in which warm and cold air masses repeatedly clash. The convergence of disparate air masses creates a deep, narrow zone of strong temperature contrasts (a strong temperature gradient) called the polar front. Air along the polar front is in a constant state of being mixed. Large vortices, spanning 1000 km (620 miles) or more, constantly form and decay along the polar front. These vortices include mid-latitude cyclones and anticyclones. These cyclones and anticyclones migrate along the polar front, swept toward the east by the mid-latitude westerly air currents. They have a lifetime of a few days, and as they cross any given location, they bring changeable weather conditions. In fact, these traveling disturbances are responsible for much of the daily weather variation and storminess across the United States.

Extratropical cyclones, also called mid-latitude cyclones, are very different from hurricanes, which are their tropical latitude, cyclonic counterparts. These cyclones are called extratropical because they form outside the tropical belt. Within an extratropical cyclone, air spirals inward toward the center of low pressure in a counterclockwise manner. As the air converges, it is forced to ascend. As air ascends, it cools and the water vapor condenses into clouds and precipitation. Thus extratropical cyclones are associated with disturbed weather, including clouds and precipitation. Extratropical cyclones may generate various forms of severe weather, such as violent thunderstorms (during the warm months), heavy snow and ice (during the cold months), and widespread strong winds (if the pressure gradient is especially large). The most severe weather is often concentrated along one or more fronts associated with the cyclone. In Figure 2.10, note how each extratropical cyclone tends to distort the polar front into a wavelike shape, causing the polar front to become fragmented as it encircles the northern hemisphere.

Extratropical anticyclones, in contrast, are characterized by divergence at the surface caused as air sinks from aloft. When air descends, it warms, and any cloud or precipitation evaporates. The skies clear and fair weather prevails. At any given location in the United States, the repeated passage of cyclones and anticyclones brings days of fair weather and foul over the course of the year.

There is one other crucial general circulation feature of the mid-latitudes: the polar jet stream. Like the subtropical jet stream, the polar jet is a narrow, high latitude "river" of air that blows from west to east. It develops as a consequence of the intense temperature gradient along the polar front combined with the Coriolis effect. The polar jet (and polar front) are strongest during winter, when winds in the core of the polar jet can approach 174 kts (200 MPH). The polar jet develops a series of meanders (or waves) that oscillate north-south and move through the jet stream from west to east. Figure 2.11 illustrates a top-down view of the northern hemisphere, showing the typical configuration of these waves, which can number from three to six. Note that extratropical cyclones and anticyclones are associated with waves in the polar jet. In fact, the low- and high-pressure cores in these vortices are created and maintained by changes in the direction and speed of airflow in the jet stream's waves.

The polar front and polar jet migrate north during the northern hemisphere summer and south during its winter. These wind patterns shift along with the rest of the Earth's general atmospheric circulation, which in turn follows the Sun's wander across the equator through the seasons. During the winter, the core of the polar jet lies across the southern and central United States. Vigorous cyclones and anticyclones develop in the strong polar jet's meanders and sweep from west to east across the country. During the summer, the polar jet weakens and moves north, across the northern Plains, Great Lakes, and southern Canada. Cyclones and anticyclones develop less frequently and

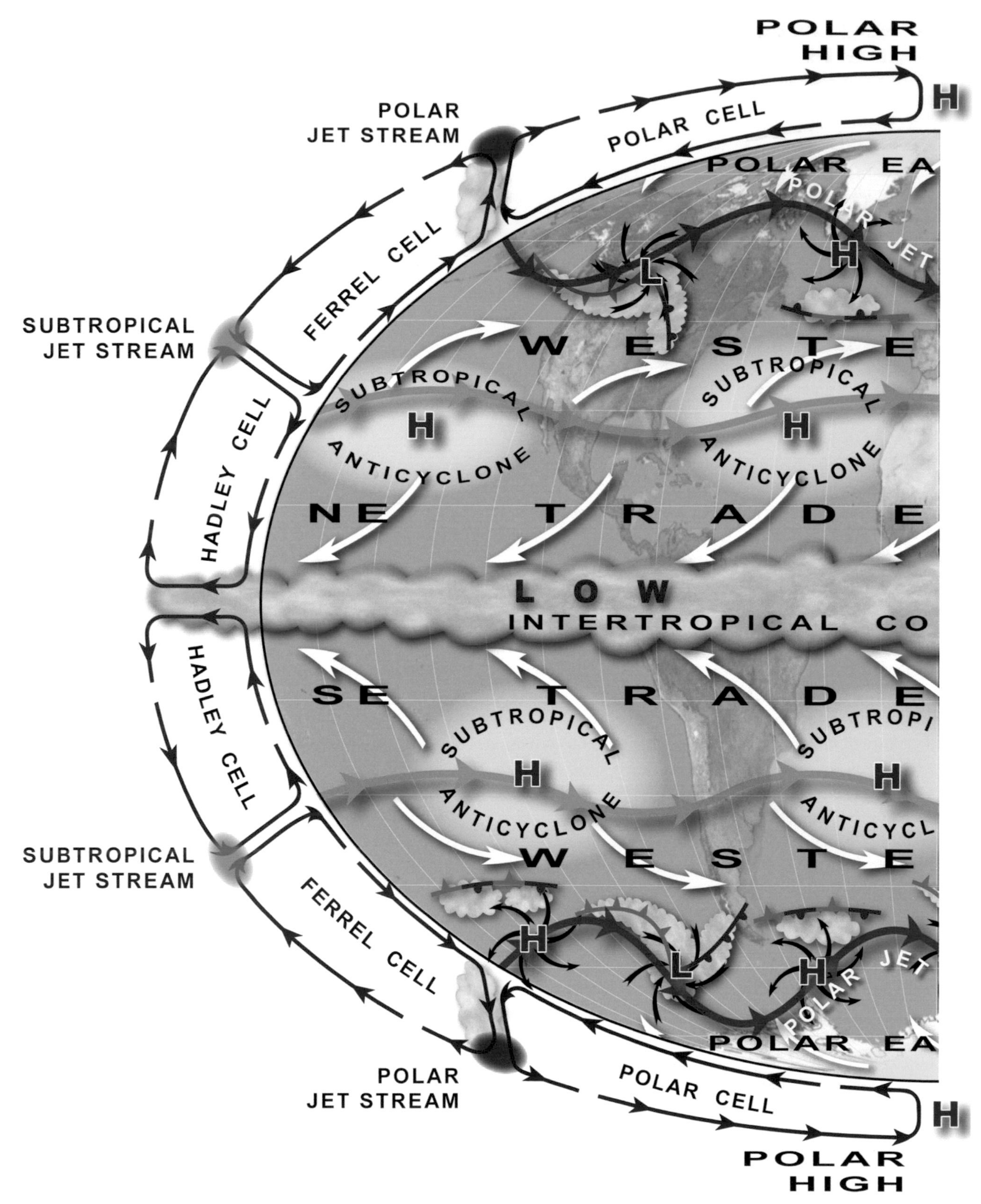

Figure 2.10 The complex Earth's atmospheric system. The effect of one component often dictates the response of another. Shown are three cells of circulation in each hemisphere. They are a response to and a cause of alternating regions of low and high pressure. The movement of air from regions of high pressure to regions of low pressure is responsible for winds that are, in turn, deflected by the Coriolis effect. All of these elements are dynamic, not stationary, thereby producing an almost endless combination of effects that influence the weather.

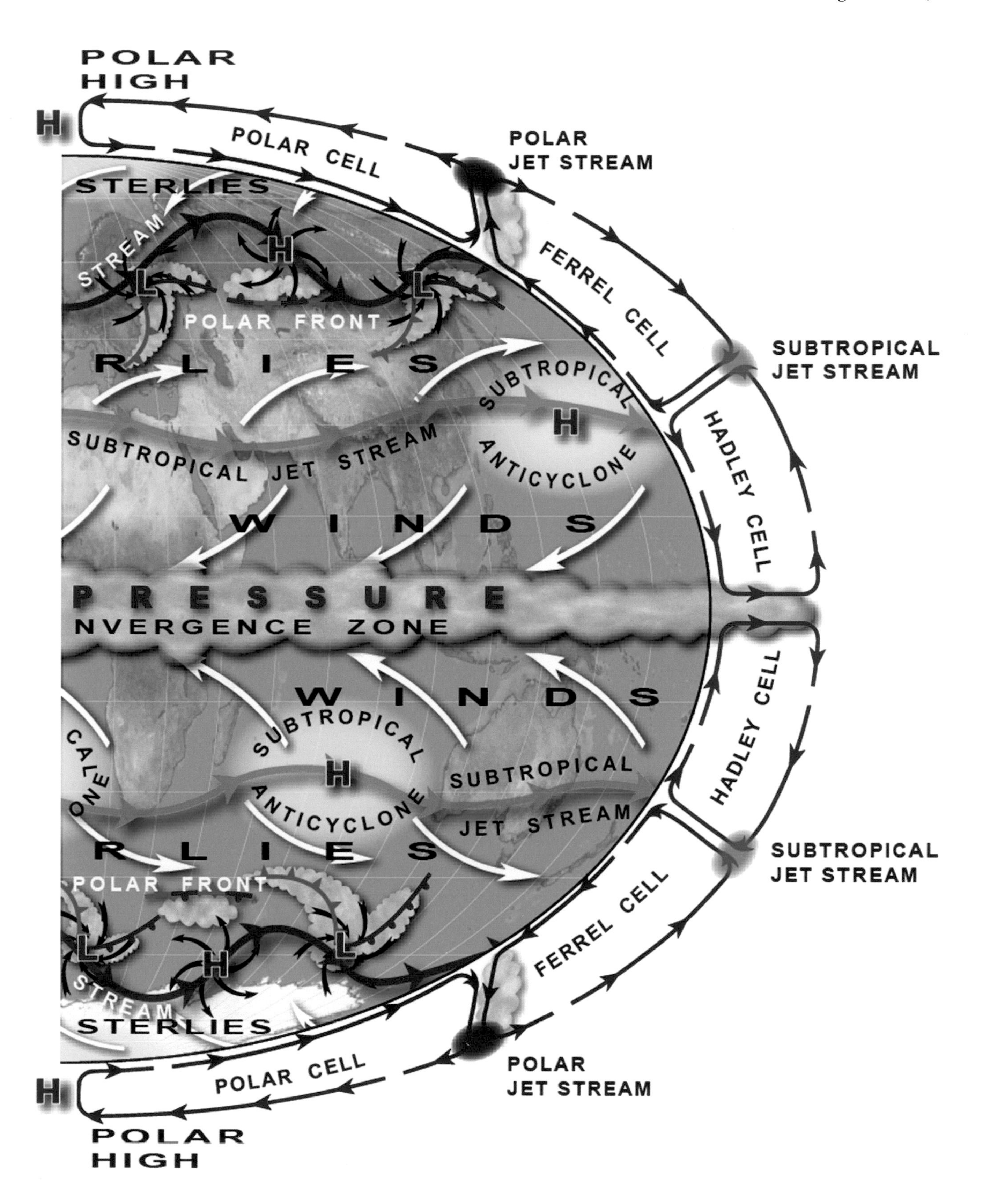
POLAR HIGH
H
POLAR CELL
POLAR JET STREAM
STERLIES
STREAM
H
L
L
POLAR FRONT
FERREL CELL
SUBTROPICAL JET STREAM
R L I E S
SUBTROPICAL
H
SUBTROPICAL JET STREAM
ANTICYCLONE
HADLEY CELL
W I N D S
P R E S S U R E
NVERGENCE ZONE
W I N D S
HADLEY CELL
SUBTROPICAL
CAL
ONE
H
ANTICYCLONE
SUBTROPICAL
JET STREAM
SUBTROPICAL JET STREAM
R L I E S
POLAR FRONT
FERREL CELL
L
H
L
STREAM
STERLIES
POLAR CELL
POLAR JET STREAM
H
POLAR HIGH

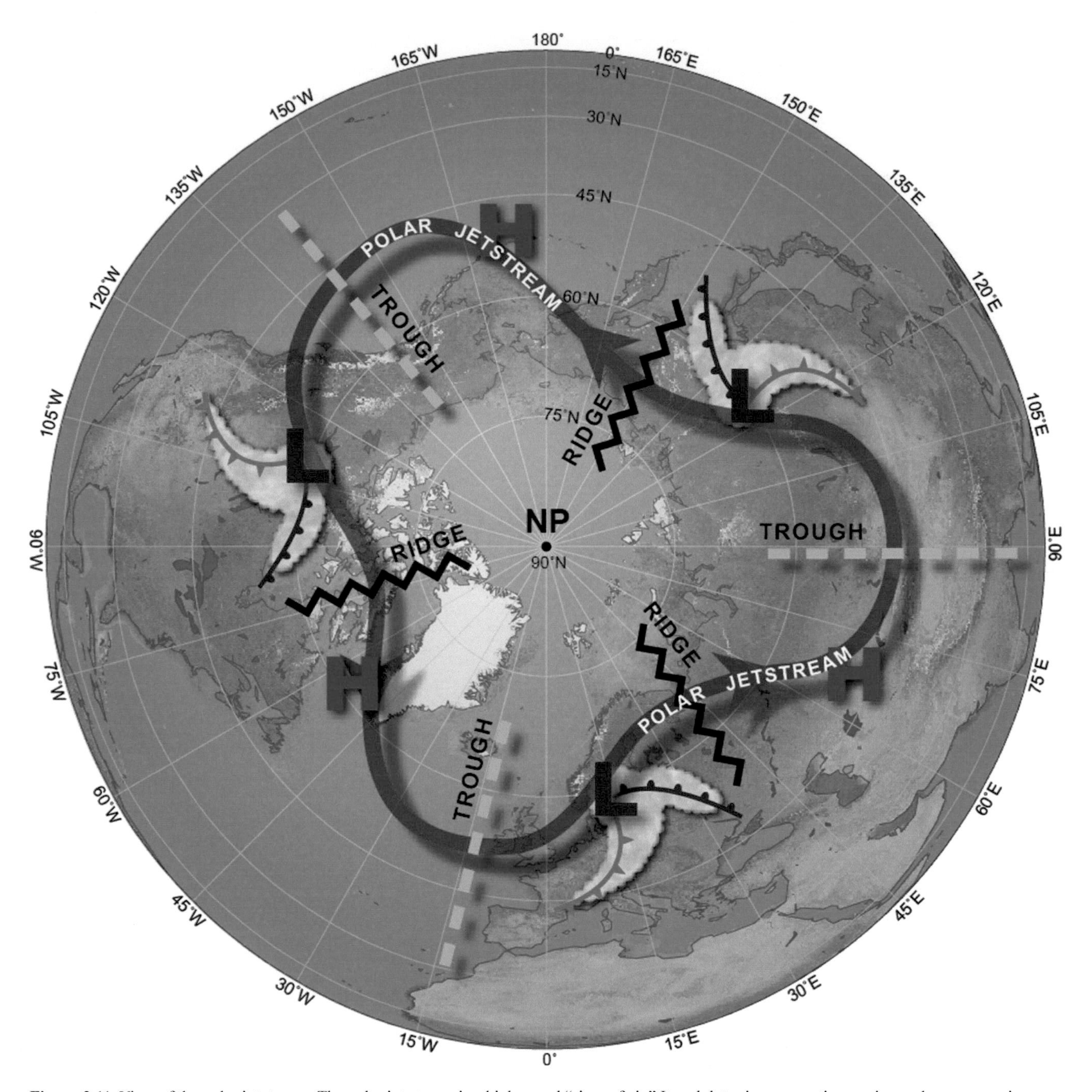

Figure 2.11 View of the polar jet stream. The polar jet stream is a high-speed "river of air." It undulates in a serpentine motion and on average is centered about 50° latitude. The polar jet stream influences much of the weather that occurs in the mid-latitudes as well as the tracks of storm systems.

are not as strong. Additionally, the subtropical anticyclone in the north Atlantic strengthens and moves north. This feature, often called the Bermuda High, ushers in a hot, humid, tropical airflow across much of the eastern two-thirds of the United States. Heat waves are common. Of all the severe weather hazards affecting the United States, heat waves lead to the largest average annual loss of life.

Circulation Features in Polar Latitudes

Finally, we arrive at latitudes north of 60°. The planet's coldest air masses straddle the northern hemisphere's cap, chilled by outgoing long-wave radiation and sustained contact with snow and ice. Air sinks over the polar region and is very dry (because cold air contains very little water vapor). As a permanent high-pressure cell builds over the polar regions, air at its center diverges outward, then becomes deflected into a large clockwise spiral. This flow pattern is known as the polar easterlies. During the northern hemisphere winter, extremely frigid air masses break free from the polar latitudes and surge south across the United States as part of an extratropical anticyclone. This process ushers in arctic air outbreaks. Extremely low temperatures often lead to widespread loss of crops and livestock, damage to critical infrastructure such as water mains, and the loss of numerous human lives.

Giant Wind-Producing Systems, Called Extratropical Cyclones and Anticyclones, Are Generated by Waves in the Jet Stream

Overall, the atmospheric heat engine accomplishes roughly two-thirds to three-quarters of the required heat export from the tropics; the remainder is handled by surface ocean currents. The amount of energy involved in this atmospheric transfer is impressive indeed, approximately 5 petawatts of power. A petawatt is a quadrillion (10^{15}) watts (the humble light bulb burns at a steady 60 watts), and this is the energy being processed every second! So in the grandest sense, the role of large atmospheric vortices such as extratropical cyclones is to transport heat. The severe weather that they sometimes generate is a by-product of this heat-transfer process.

Before we continue, let's take a moment to review some key phenomena. A large low-pressure cell at the surface is called a cyclone and features a counterclockwise, inward spiral of air. A large high-pressure cell at the surface is called an anticyclone and features clockwise, outward spiral of air.

Forces Behind the Wind

You now understand that wind is created because various forces act upon the air, first setting the air in motion (pressure gradient force) and then deflecting the direction in which the wind blows (Coriolis effect). This balanced state – the geostrophic wind – strictly applies to winds blowing in a straight path at high altitudes. For winds that flow in curved paths (such as air circulating around a vortex) and for winds that blow across the Earth's surface, there are a few additional forces to consider.

Figure 2.12 presents a colorful depiction of the various types of wind and the forces acting in balance. Panel (a) is the pure geostrophic wind. Panel (b) shows a different balance of forces when air circulates around a vortex, such as an extratropical cyclone. In addition to the pressure gradient (red) and Coriolis effect (blue), we must include a force that keeps the air spinning in a curved path – the centripetal force (yellow). To understand this force, imagine twirling a rock attached to the end of a string. To keep the rock moving in a circle, your arm must exert part of the force as an inward-directed pull. If you eliminate the inward pull by letting go of the string, the rock flies off in a straight direction. Around a low-pressure vortex, both the centripetal force and pressure gradient force are directed inward; both must be balanced by the Coriolis effect. The resulting circular wind is called the gradient wind. The gradient wind blows parallel to curved isobars around a vortex: counterclockwise around a large, low-pressure vortex (cyclone) and clockwise around a large, high-pressure vortex (anticyclone).

Panel (c) illustrates the most complicated type of wind – that of a large vortex at the surface. Our force balance retains all of the elements as described for the gradient wind. But now we must add a fourth force: friction (green). Friction slows the air in immediate contact with the surface. Friction acts in the opposite direction of the wind (purple). However, as the wind slows, the Coriolis effect becomes smaller (recall from Section 2.3 that the magnitude of the Coriolis effect is proportional to wind speed). The smaller Coriolis effect allows the combination of pressure gradient and centripetal forces – both of which are directed inward – to exert a stronger influence. The stronger inward pull causes the wind direction to cross isobars, at an angle, from high pressure toward low pressure. In the case of the extratropical cyclone shown in Panel (c), the surface wind blows counterclockwise and inward, an inflowing spiral around the entire vortex (Figure 2.13). Note that the Coriolis effect (blue arrow) always remains perpendicular to the observed wind (purple).

Extratropical Cyclones and Anticyclones Are Created Beneath Waves in the Polar Jet Stream

Now that we have learned about the forces balancing different types of wind, let's examine another key concept. We can study the location and intensity of cyclones and anticyclones on a surface weather map. But these weather systems are inherently three-dimensional. Their vortices extend upward through the depth of the troposphere (16 km [10 miles]). In fact, their circulations are connected at high levels to meanders in the polar jet stream. Cyclones and anticyclones develop and are maintained by changes in the speed and direction of wind as it flows through the jet stream. The meanders in the jet stream move from west to east, so they move the cyclones and anticyclones along with them. Let's explore the vital connection

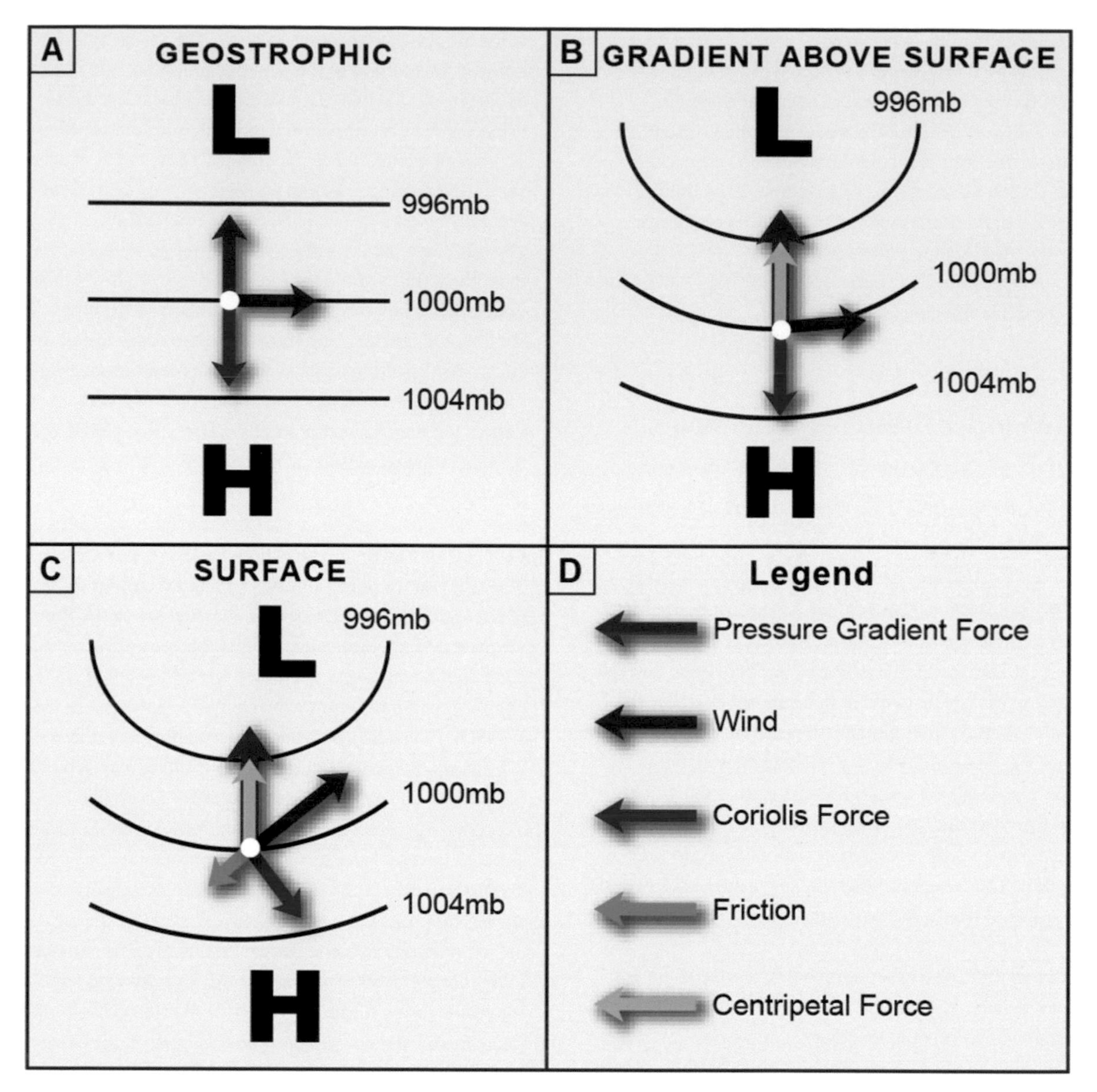

Figure 2.12 Various forces that create wind. (a) The forces that influence geostrophic wind; (b) the forces at work in regions of curved airflow; and (c) the impact of friction on a surface wind.

between pressure at the surface, as well as the twists and turns of air streaming along at 10,670 m (35,000 feet).

Figure 2.14 illustrates a set of meanders in the westerly jet stream. Where the jet stream dips to the south and the airflow curves counterclockwise, the region is called a trough. Within the trough, the deep cold air mass dips down from Canada and the Arctic. When the jet arcs northward, such that the curved airflow is clockwise, the region is called a ridge. A deep warm air mass from the tropics intrudes northward inside a ridge.

Now let's think about the type of wind balance that must operate inside these meanders. In the straight section of airflow connecting the trough and ridge, the wind is purely geostrophic. In the base of the trough, where the flow curves, we must have a gradient wind. Likewise, in the crest of the ridge, the curved flow is also gradient. But there's an additional detail: The gradient wind inside the trough is *weaker* than the geostrophic wind, and the gradient wind inside the ridge blows *stronger* than the geostrophic wind. You can see this difference by noting the length of the white arrows in each region. The wind in the trough is called subgeostrophic; that in the ridge, supergeostrophic.

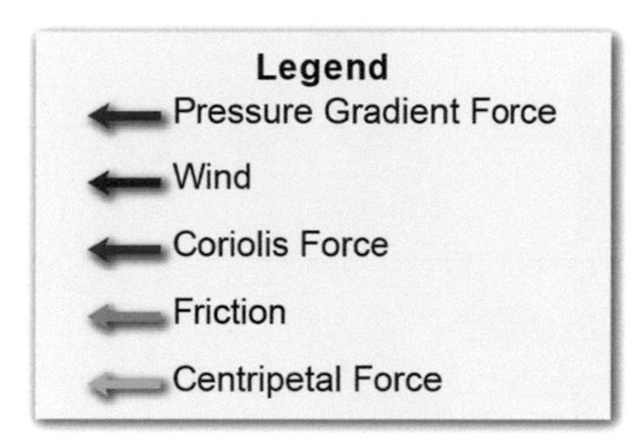

Figure 2.13 Top diagram: Various forces that work in concert to create surface winds around cyclones (purple arrows). Pressure gradient force is generated by differences in pressure, with air moving from higher to lower pressure. Centripetal force maintains flow in a curved path. Coriolis effect is a product of the Earth's rotation on its axis. Friction force opposes the forward movement of the wind. The combined effect of these forces causes the winds to move inward and counterclockwise (northern hemisphere, as illustrated in the lower diagram) or clockwise (southern hemisphere).

In Section 2.5, we saw how deep, vertical air columns with different temperatures experience changes in pressure. The pressure near the surface can also be changed by physically adding air to or removing air from inside the air column. If air is allowed to exit the top of a tall air column, the surface pressure will drop (the weight of the overlying air decreases). If air is added to the top of the column, the surface air pressure must rise.

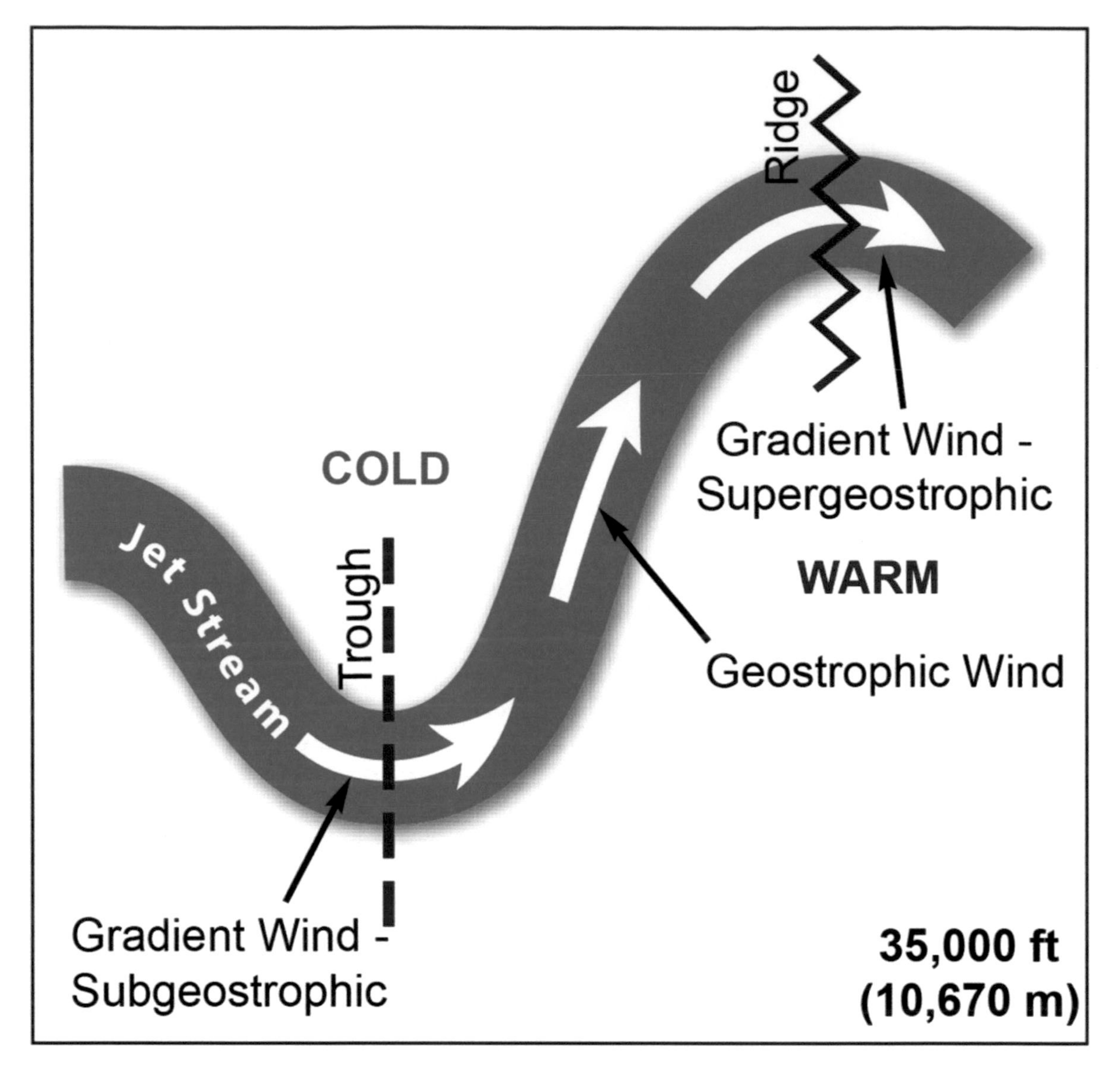

Figure 2.14 Ridges and troughs in the jet stream. In the northern hemisphere, troughs are places along the jet stream where the wind turns in a counterclockwise direction. Ridges are located where the wind turns in a clockwise direction.

Figure 2.15 shows a trough and ridge in the polar jet. Imagine that these meanders represent the lanes of a superhighway. High-speed packets of airflow along the jet like cars on the road. Just before entering the curve of the trough, there is a construction zone, and cars abruptly slow. The slow traffic moving through the trough is subgeostrophic. Coming out of the bend, traffic speeds up to geostrophic speeds. The traffic (air packets) diverge, or spread apart, in the straightaway. Traffic continues to accelerate through the curve into the ridge, reaching supergeostrophic speeds. But coming around the bend in the ridge, we see brake lights once again – perhaps due to an accident. Traffic abruptly slows going into the next straightaway. Cars (air packets) squeeze together; we say that the flow "converges" in this region.

So, along the meandering jet stream, there is a region where the airflow diverges (speeds up) between the trough and the ridge and converges (slows down) between the ridge and the trough. These are two-dimensional changes in airflow, but recall that the atmosphere is three-dimensional. Where the airflow diverges aloft, air in a vertical column beneath will rise upward, ascending to fill the void. The rising air will create a large region of low pressure at the surface, which in turn develops into an extratropical cyclone. Where the airflow aloft converges, air in a vertical column will be forced to sink downward, removing the buildup of air. The sinking air generates a large region of surface high pressure, developing a surface anticyclone. In this way, air inside the jet stream is connected to patterns of surface pressure – the extratropical

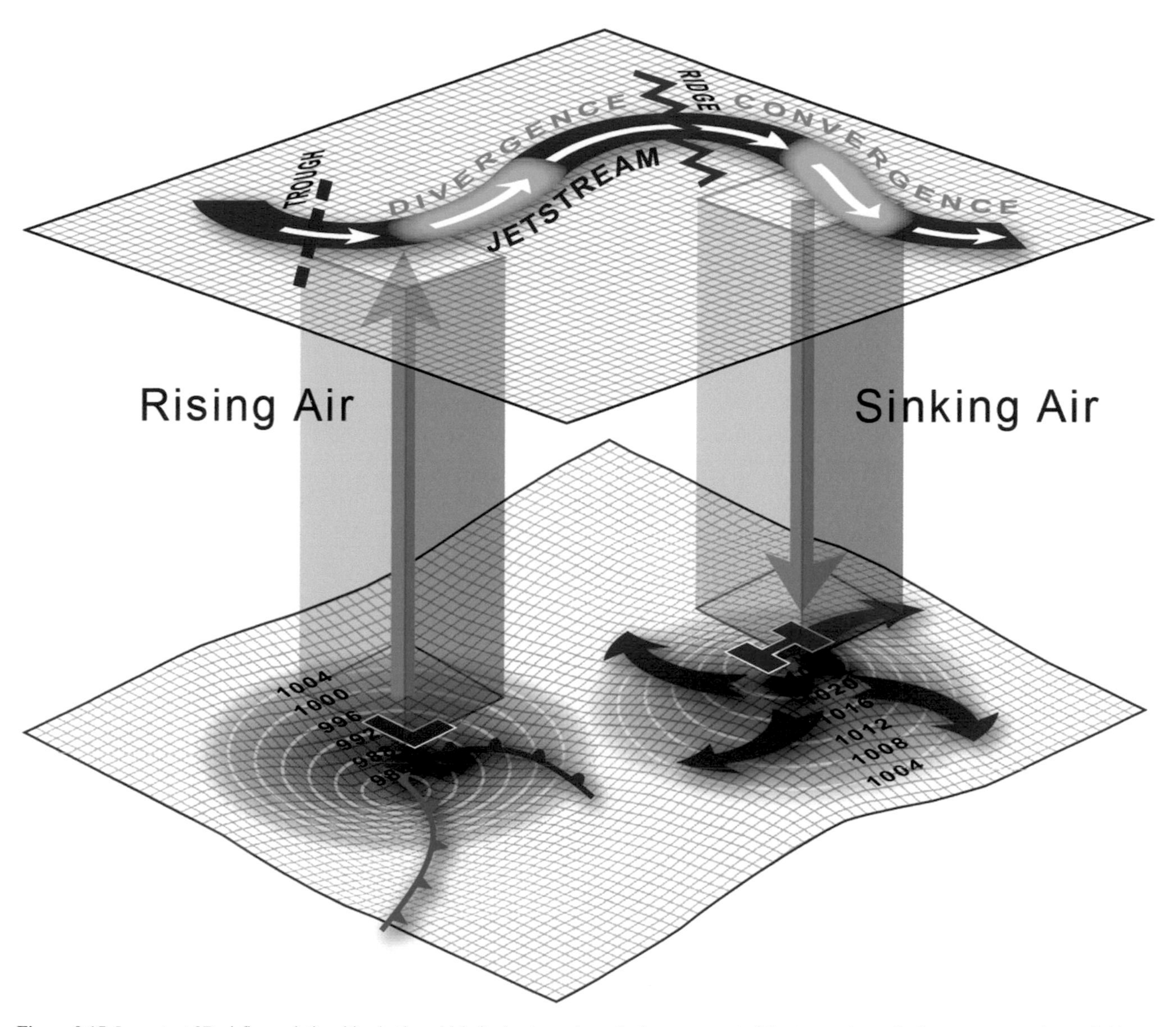

Figure 2.15 Important 3D airflow relationships in the mid-latitude atmosphere. In the upper part of the troposphere, the jet stream, troughs, and ridges are linked with events taking place at the surface. Wind coming out of a trough will speed up, creating lower pressure that draws air upward from the surface, creating a region of low pressure. Where convergence occurs at the ridge, air accumulates and pressure increases. Some of the air is forced downward, contributing to a region of high pressure at the surface.

cyclones and anticyclones. Knowing what is happening at the jet stream level is critical to predicting how daily weather will evolve in any part of North America.

Changeable Daily Weather in Mid-Latitudes Is Tied to the Formation and Movement of Cyclones and Anticyclones

Extratropical cyclones and anticyclones are associated with significant changes in everyday weather – including the arrival and departure of many types of severe storms and other weather hazards. With the help of Figure 2.16, let's briefly examine the characteristic weather changes caused when a cyclone-anticyclone pair passes through a location (labeled "A").

The first thing you will notice is two large, contrasting air masses: (1) a very cold and dry polar air mass, originating from northern Canada (blue), and (2) a warm, humid air mass sitting over the subtropics (red). At 30° N, a subtropical anticyclone (red H, with central pressure of 1013 mb) is embedded in this air mass; its clockwise circulation sweeps warm and humid air northward, on its western side. Farther north, cold polar air circulates southward in the clockwise outflow of an extratropical anticyclone (blue H, located at 50° N, with central pressure of 1013 mb). But while the subtropical anticyclone remains stationary at 30° N, the extratropical anticyclone is on the move, headed east, because it's embedded within the upper-level jet stream (not shown).

The polar front lies along the border of these two contrasting air masses, spanning a latitude of 35–50°N. However, an extratropical cyclone (area of low pressure, shown by the red "L," with minimum central pressure of 988 mb) develops along the polar front, at 40° N. Above and just to the west of the cyclone lies a trough in the jet stream, and the region of low pressure is maintained by rising air "filling the void" aloft. The cyclone tracks toward the east, in tandem with the extratropical anticyclone. The anticyclone to the north feeds its cold air southward into the cyclone, and the subtropical cyclone circulates warm air northward. The cyclone's counterclockwise inflow helps to draw these contrasting air masses together. The cyclone's circulation distorts the polar front into a wavelike configuration. A cold front develops where the cold and dry air advances toward the east around the low; this dense, chilly air undercuts and lifts warm air to the south of the low. A warm front forms along the polar front, close to the center of the low, where a warm, southerly flow of air runs up and over cooler air to the north and east. Along both of these fronts, moist air is forced to ascend. The air cools as it rises, the moisture condenses, and extensive clouds and precipitation develop. Showery rain clouds and thunderstorms also erupt in the warm, tropical air mass, between the subtropical anticyclone and extratropical cyclone.

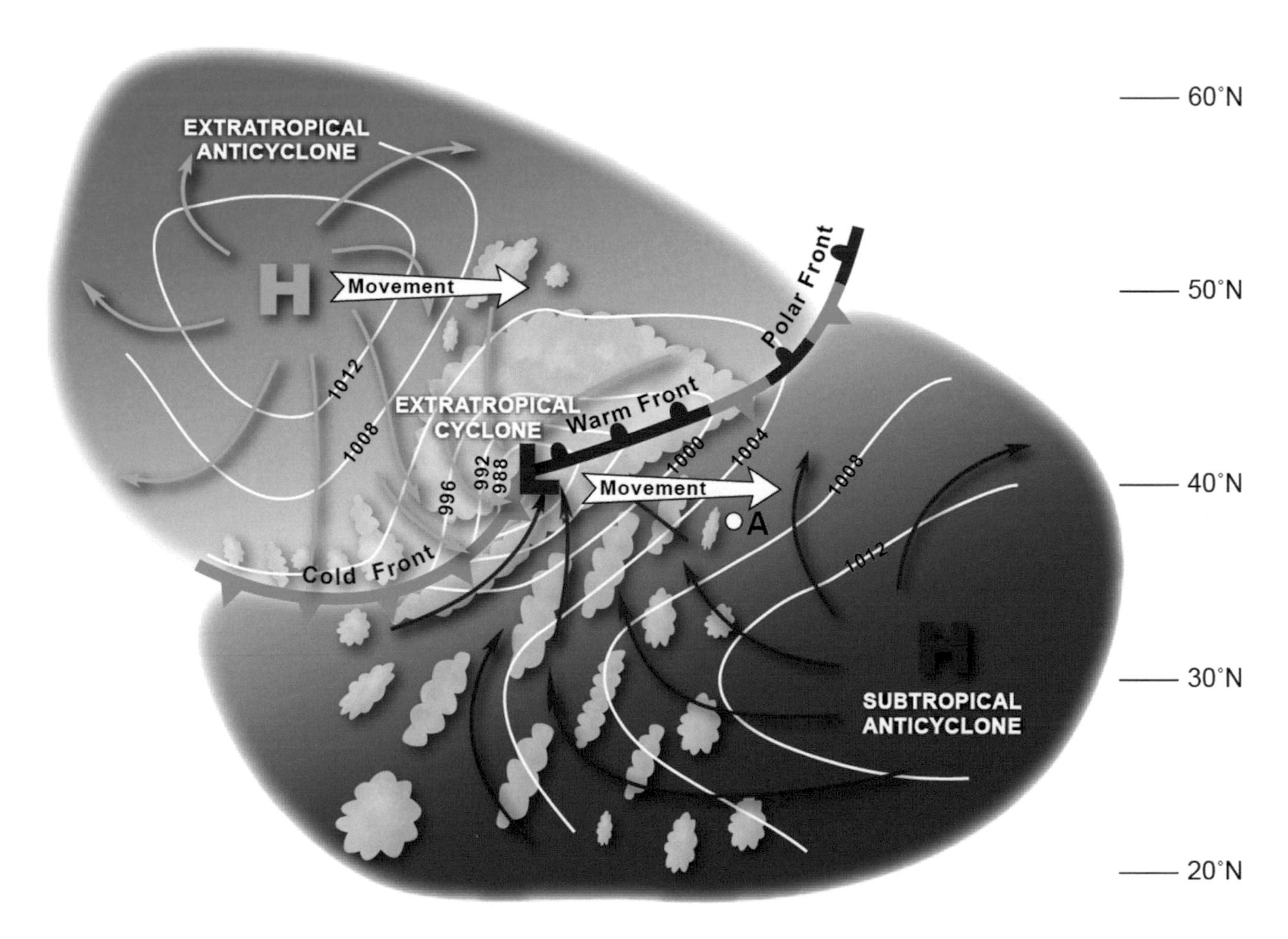

Figure 2.16 Structure of cyclones and anticyclones, which form an interacting wind system. Extratropical anticyclones feed cold dry air toward warm moist air driven by subtropical anticyclones. When these air masses clash, the warm moist air is shoved upward as the cold dry air advances. A region of low pressure forms and feeds into this region in a counterclockwise direction (northern hemisphere). The unstable warm air rises, cools, and condenses to form clouds, often accompanied by precipitation.

The extratropical cyclone features a region of extensive cloud cover, and heavy precipitation falls along its fronts and over the center of low pressure. In the cool season (November through March), air mass temperatures < 0° C (32° F) often support the development of heavy snow and ice. During the spring and early summer, extratropical cyclones are notorious for producing bands of severe thunderstorms. These thunderstorm cells, in turn, produce damaging wind gusts, flash floods, large hail, and tornadoes. If the pressure gradient between the extratropical cyclone and anticyclone is intense, damaging winds develop to the north and west of the cyclone.

Figure 2.16 shows how an extratropical cyclone and one or more anticyclones "team up" to generate various forms of severe and hazardous weather. The most intense weather generally develops where warm and cold air masses are forced together, along one or more fronts. You will note that the extratropical anticyclone contains few if any clouds within its core. As the air descends into it from jet stream level, the air warms and dries. Any clouds and precipitation tend to evaporate into the warm and dry air. Extratropical anticyclones, therefore, are generally associated with fair and slowly changing weather.

However, there are exceptions. During the middle of winter, these anticyclones can circulate extremely frigid air southward, along with gusty winds. The resulting arctic air outbreak and associated subzero wind chills pose a serious weather hazard in their own right. And, during summer, the hot, humid air mass circulated northward by persistent subtropical anticyclones leads to extended heat waves. Anticyclones also create a number of other localized weather hazards, including damaging Santa Ana and Chinook winds across the mountains of western North America, dense fog, localized severe thunderstorms and flash floods, and localized lake effect snow across the Great Lakes.

Table 2.1 presents a concise summary of the characteristics of extratropical cyclones and anticyclones. In the mid-latitudes, they are key players in severe weather, and much of this textbook will explore the widespread and localized weather hazards contained within them.

Table 2.1 General Properties of Extratropical Cyclones and Anticyclones

Attribute	Low Pressure	High Pressure
Common names	Cyclone Depression Extratropical cyclone Mid-latitude cyclone Wave cyclone	Anticyclone
Surface wind	Counterclockwise, inward spiraling	Clockwise, outward spiraling
Develops downwind of...	Jet stream trough	Jet stream ridge
Vertical wind	Upward	Downward
Weather	Widespread clouds, precipitation	Clear to partly cloudy; fair
Movement	Toward E-NE	Toward E-SE
Dimensions	1000+ km (620 miles)	1500–2000 km (930–1240 miles)
Pressure gradient	Compact, intense	Broad, weak
Sfc temperature	Narrow zones of strong temperature contrast	Widespread, homogeneous temperature
Sfc humidity	Concentrated moist and dry regions	Uniform humidity levels
Weather hazards	Severe thunderstorms, heavy rain, high wind, sleet, heavy snow	Temperature extremes (record low or high), dense fog, severe thunderstorms, lake effect snow, Chinook and Santa Ana winds

Notes: E-NE = east-northeast; E-SE = east-southeast; km = kilometers; Sfc = surface.

Comparison of Winds in Atmospheric Vortices of All Sizes

Thus far, we've studied several types of winds, based on different forces acting in balance: geostrophic, gradient, and friction. In Figure 2.17, Panel (a) summarizes how these various force balances come into play within the mid-latitude circulation system (jet stream trough and ridge, extratropical cyclone and anticyclone). The particular type of wind depends on (1) the presence of straight vs curved airflow and (2) whether the wind occurs close to the surface.

In comparison, Figure 2.17, Panel (b), shows the various types of winds contained within a mature hurricane. While we haven't yet discussed tropical cyclones in detail, the hurricane is a type of cyclone but smaller in diameter than the typical extratropical cyclone. A hurricane contains a broad region of gradient winds circulating around low pressure and the frictional wind at the surface. In the upper levels, the airflow actually reverses direction, spiraling outward from the storm's top, in a gradient wind around high pressure. The inner core of the hurricane contains a special type of wind, called the cyclostrophic wind, which represents a balance between the inward-directed pressure gradient force and outward-directed centrifugal force. At the very small scale of the hurricane's inner core (~20–30 km [12–18 miles]), the Coriolis effect is negligibly small and is not included in the force balance.

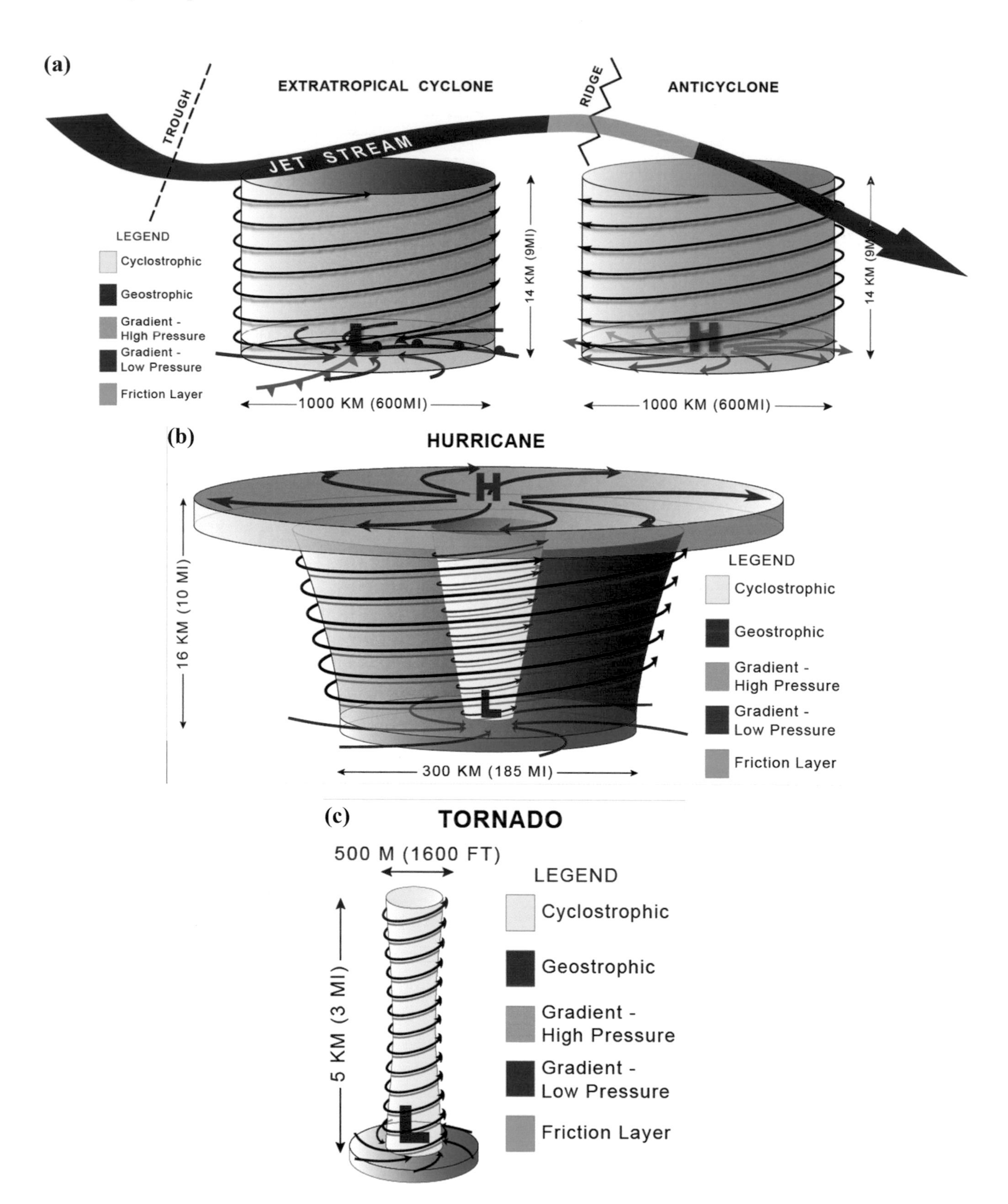

Figure 2.17 (a) A three-dimensional comparison of an extratropical cyclone and an anticyclone. They are color-coded to reflect the various wind and pressure conditions associated with each. (b) The various winds and pressures found within a hurricane. (c) Due to the small geographic scale of a tornado, no geostrophic winds are associated with this kind of storm. Furthermore, the scale prevents the development of gradient high or low pressures. However, the central core of the tornado has cyclostrophic flow.

A tornado is an intense cyclonic vortex, much smaller in scale than a hurricane – a tall, narrow tube of violently spinning air, typically only 500 m (1,650 feet) across. The most intense tornadoes are generated within severe thunderstorm cells, which, in turn, are found within the broader circulation of an extratropical cyclone. As you might surmise, tornadoes are so small that they, too, do not directly "feel" the effect of the Earth's spin, so there is no Coriolis effect. Tornadoes, like the very inner core of hurricanes, are in a state of cyclostrophic balance (except at the surface, where friction must be included).

Many Small-Scale Wind Circulation Systems Are Associated With the Generation of Severe Storms and High-Impact Weather

Figure 2.18 shows the geographic distribution of several types of wind circulations, tied to terrain, across the United States. One caution, however: These circulations do not occur simultaneously all the time. The two most widely distributed wind patterns include the coastal sea breeze, stretching from the Gulf Coast to New England (green shading), and the mountain–valley circulations (MVCs, pink shaded regions), which are generated along elevated terrain – including the Appalachians in the east and numerous mountain ranges of the western United States. Both of these wind patterns are fairly localized and are driven by solar heating during the late morning and afternoon hours. They are also quite shallow, confined to the lowest 2–3 km (1.2–1.8 miles) of the troposphere.

Understanding the Sea Breeze Is . . . a Breeze!

The sea breeze, shown in Figure 2.19 (top), develops during the daytime, particularly in the warm season, because the land mass adjacent to the ocean warms to a greater degree than the water. Water has a very high heat capacity, meaning it can absorb a tremendous amount of solar energy through a deep layer of water but experience very little temperature change. Nor does it cool quickly as heat is radiated. In contrast, land surfaces – including soil, sand, and vegetation – absorb solar energy in the uppermost 1–2 mm and don't conduct heat through a deep layer. The "skin" of the land surface thus experiences a large, rapid rise in temperature. As the land quickly warms, so does the overlying air, and the air density decreases. In response, low pressure develops in the lowest 100 m (330 feet) of atmosphere. Air flows onshore, from the cooler ocean surface toward land, to fill this void. This generates a breeze along the coast. The inrush

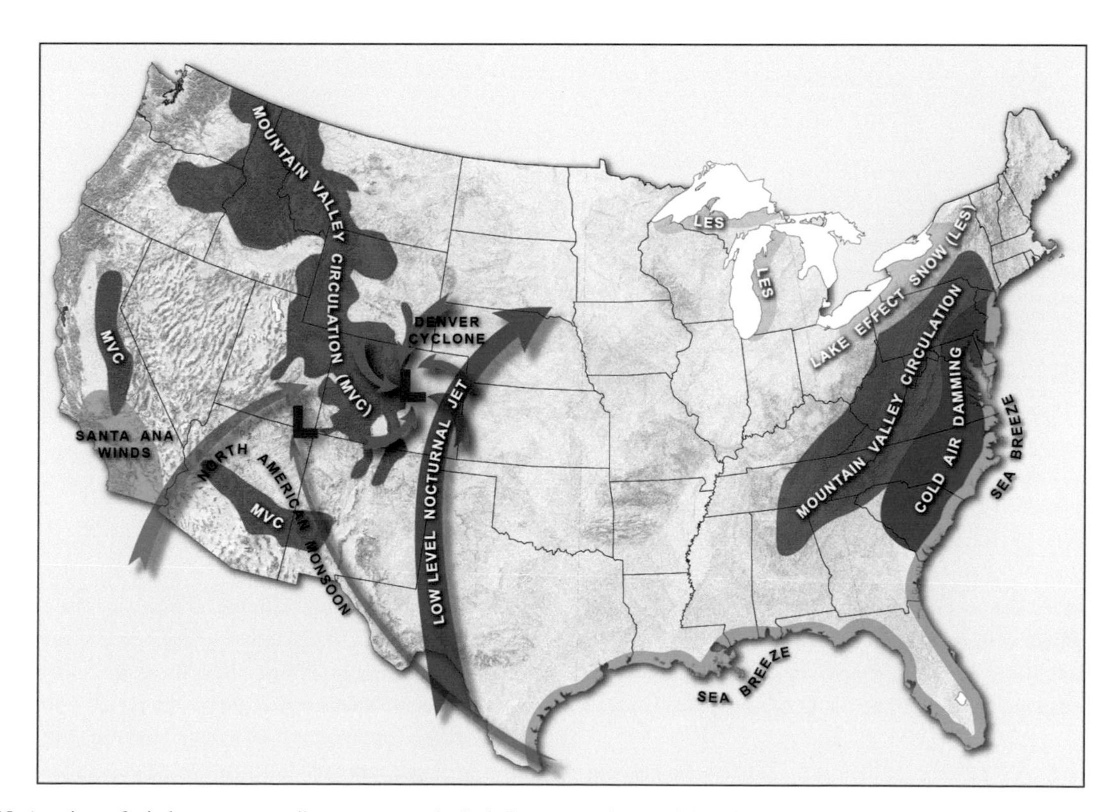

Figure 2.18 A variety of wind patterns contribute to meteorological phenomena that can bring about hazardous weather. While the distribution shown on the map is general, these conditions are not restricted to these areas alone. Rather, they are widely distributed throughout the United States.

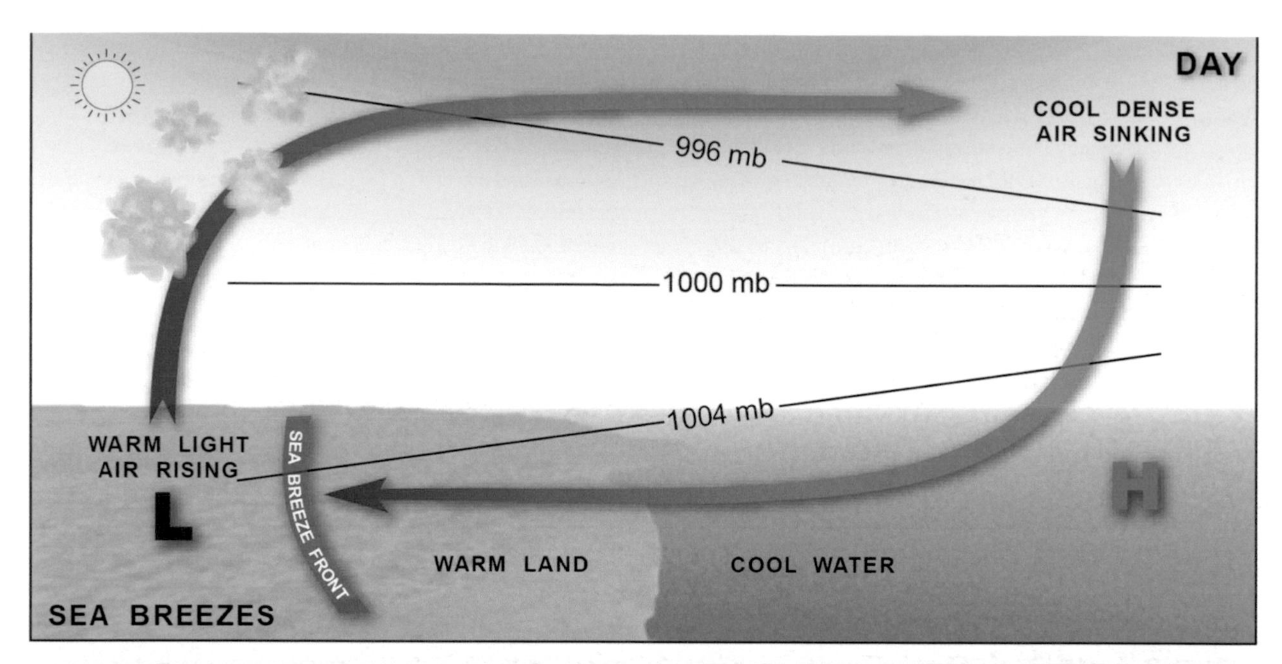

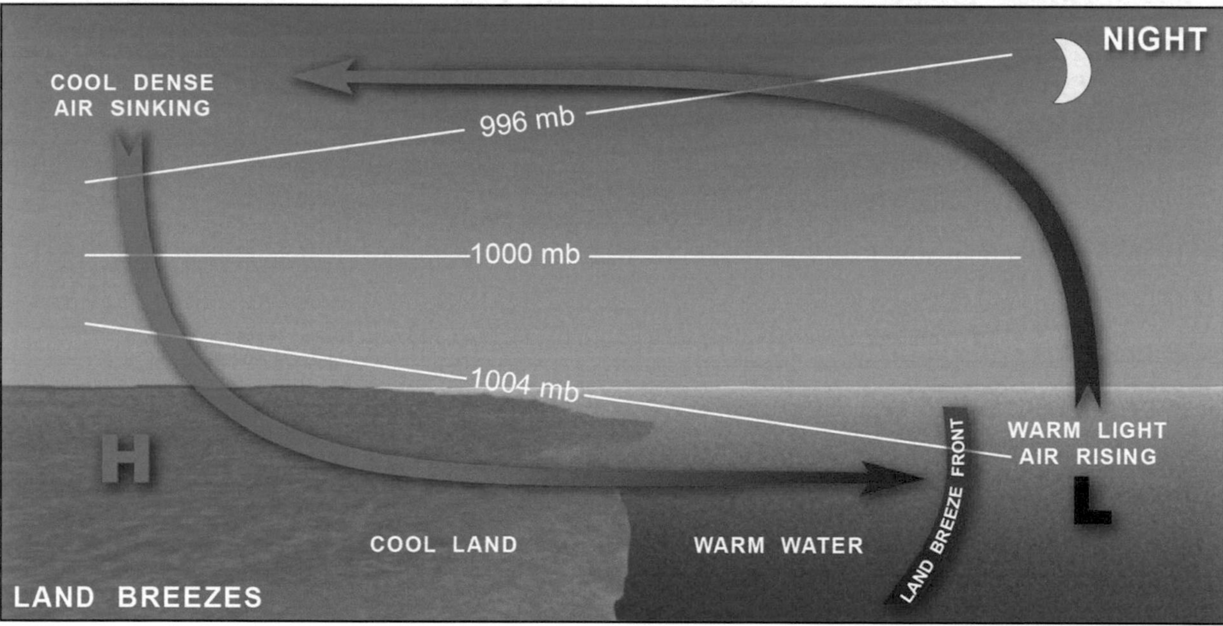

Figure 2.19 Formation of the sea breeze. Day: Sea breezes occur during the day when water temperatures are cooler than land temperatures. High pressure develops over the water and a low pressure forms over the land. The air moving from high to low pressure creates the sea breeze. Night: Relative temperature differences reverse over the land and water, causing the wind to blow from the land out over the water.

of cool air behaves like a cold front, lifting the air ahead of it. The rising air flows back toward the ocean, where it subsides – creating a closed, fairly shallow loop of overturning air along the coast. Clouds tend to form in the rising air, on the land side.

Figure 2.20 shows a weather satellite view of clouds developing along the inland side of the sea breeze. On hot summer days, lines of thunderstorms frequently erupt along the edge of the sea breeze, just a few kilometers inland of the ocean. Occasionally, these thunderstorms may become severe. At night, the direction of airflow along the coast reverses: The land surface cools faster than the ocean, which remains warm. The resulting land breeze blows from land out over the water. Lines of clouds and thunderstorms often develop in the early morning hours, offshore, as humid ocean air is lifted along the land breeze front.

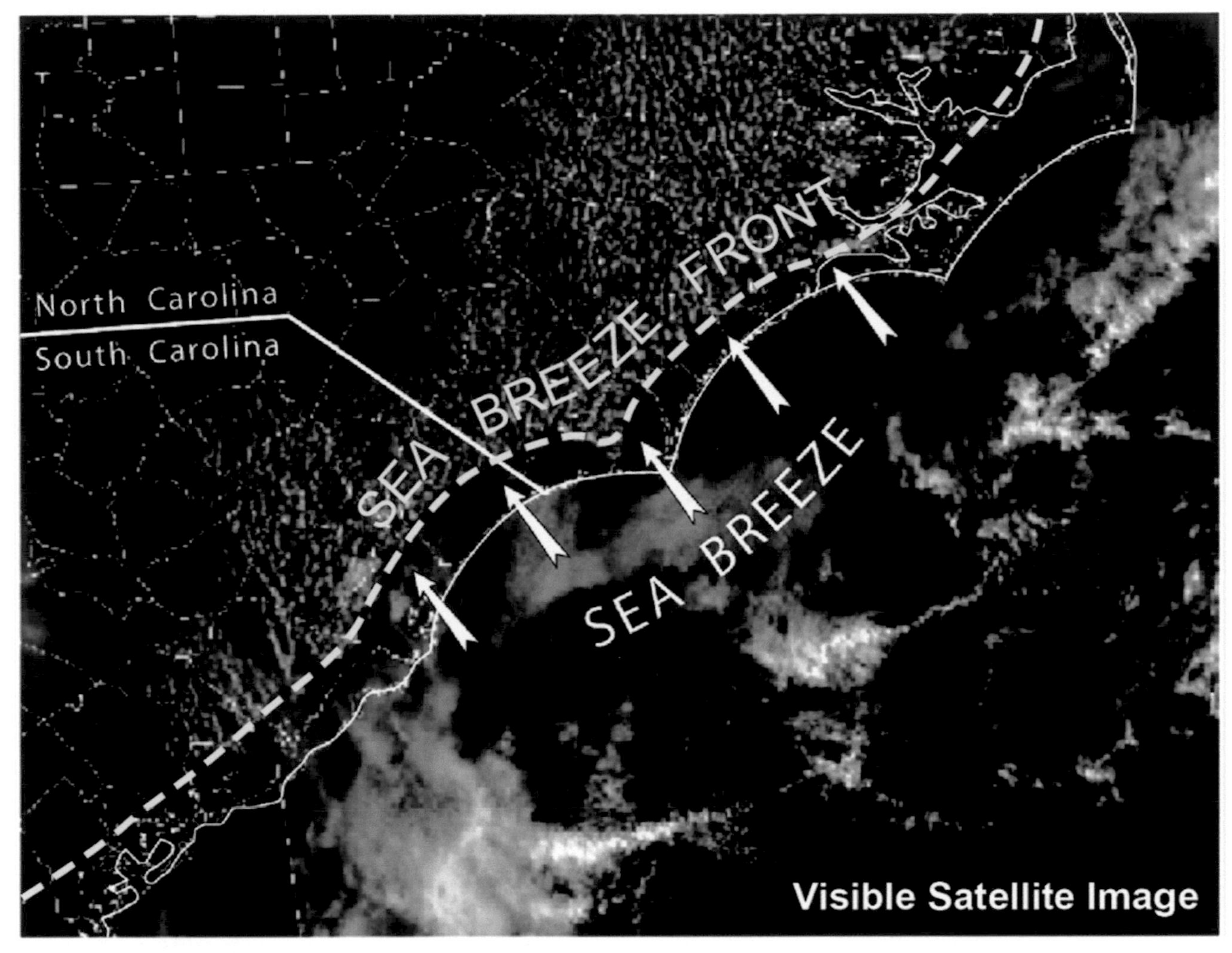

Figure 2.20 How thunderstorms are connected to the sea breeze. As sea breezes blow onto the land, they carry moist air inland. As the moist air encounters the warmer temperatures over the land, the air becomes lighter. Subsequently, the air rises and cools, bringing about the formation of clouds and often local thunderstorms. (Annotated from NOAA image.)

Air Goes Up the Mountain and Then Comes Back Down

Mountain–valley circulations are quite widespread across higher terrain but are often localized to the taller peaks. Valley breezes blow upslope during the day, and air flows downslope at night. During the day, the land surface on a hill slope warms more quickly than the valley because the slope receives the Sun's rays more directly, particularly in the morning hours. The warm hill slope heats the overlying air. A small region of lowered density and decreased pressure develops over the hill slope. Air rises up from the valley floor to fill the void. As air is drawn out of the valley bottom, eventually it converges and sinks back down over the valley center, but at a higher altitude. The overturning of air due to terrain generates the shallow-circulation cells shown in the top panel of Figure 2.21. But as in the case of the sea breeze, warm and humid air that is forced to ascend forms deep clouds along the ridge tops. On hot summer days, afternoon thunderstorms frequently develop along the ridges of the Appalachians and the Rockies. At times, these storms reach severe levels. At night, the circulation pattern reverses: Hill slopes cool quickly by radiating heat to space. This chills the overlying air, which becomes dense and flows down into the valley bottom.

Figure 2.22 shows the combined effects of the sea breeze and the valley breeze along the mountain slopes of the big island of Hawaii. Warm, humid air over the land is forced upward along the inward-moving sea breeze front, augmented by a valley wind flowing up the mountain slope. Deep clouds in these circulations can produce heavy rain showers; they are a frequent, almost daily, occurrence just inland and along the hill slopes of Hawaii.

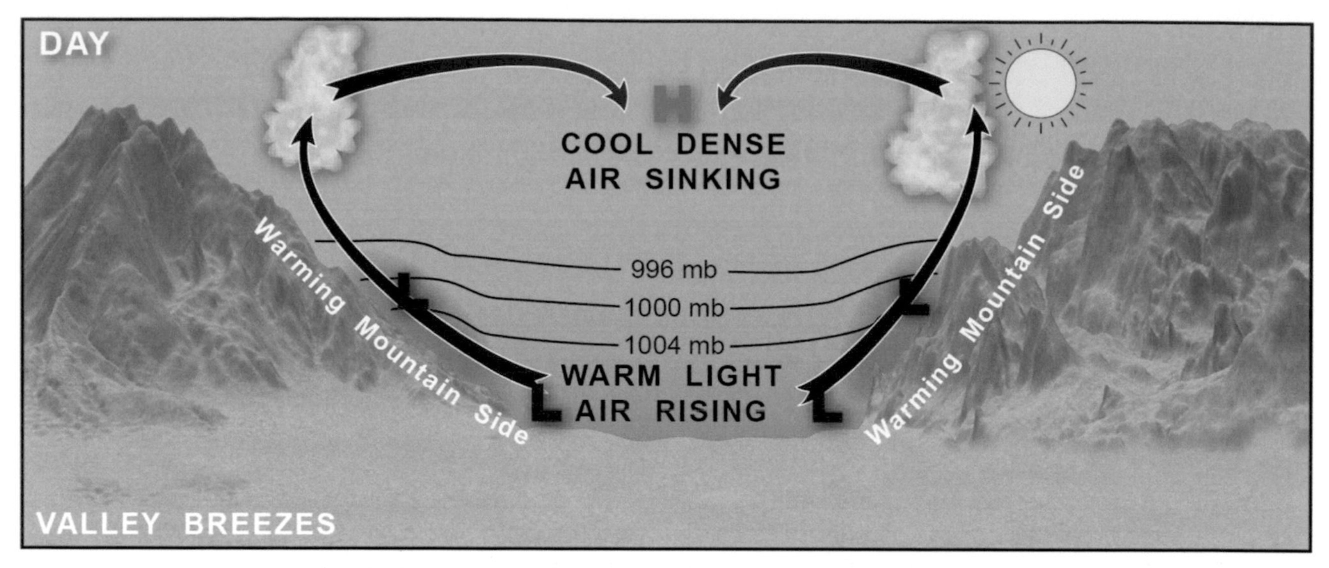

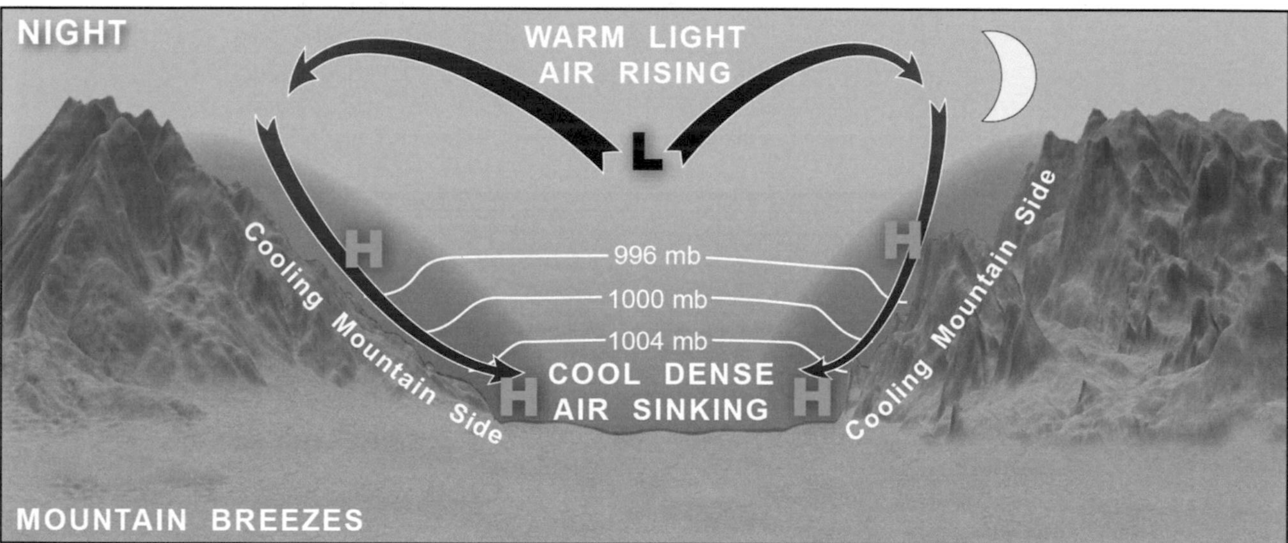

Figure 2.21 Local mountain–valley wind circulation. Similar mechanics that form sea and land breezes are also at work in the valley and mountain breezes of the mountain–valley circulation. Day: The valley surface heats up, transferring the warmth to the air. The warm air then rises up the valley wall, creating a valley breeze. Night: The air in direct contact with rapidly cooling mountain slopes becomes dense and, under the pull of gravity, flows down the hill slopes into the valley, creating a mountain breeze.

Nocturnal Low-Level Jet: Pumping Gulf Moisture Over the Great Plains

In Figure 2.18, a narrow swath of wind is shown sweeping inland off the Gulf of Mexico (purple arrow), moving from south to north over the central Plains. This corridor of moist air is a frequent nighttime feature over the Plains, east of the Rockies. In the warm season, it helps fuel enormous complexes of nocturnal thunderstorms that deliver much of the annual rainfall to the Great Plains. At times the rainfall is so heavy that these cloud systems generate extensive flooding. The nocturnal low-level jet develops as consequence of temperature differences along the Rocky Mountains and Great Plains (Figure 2.23). At night, air over mountain slopes cools and drains eastward onto the Plains. Further east, the air above the surface remains relatively warm. A large horizontal gradient in temperature above the surface becomes established. The temperature gradient creates a horizontal pressure gradient (as described in Section 2.4). The pressure gradient, in turn, causes air to accelerate toward the Rockies at low levels, but the air becomes deflected to the right by the Coriolis effect. During the nighttime hours, a narrow, low-altitude ribbon of fast air streams northward from the Gulf of Mexico, across the Plains. Within this ribbon, large nighttime complexes of thunderstorms erupt. One such complex

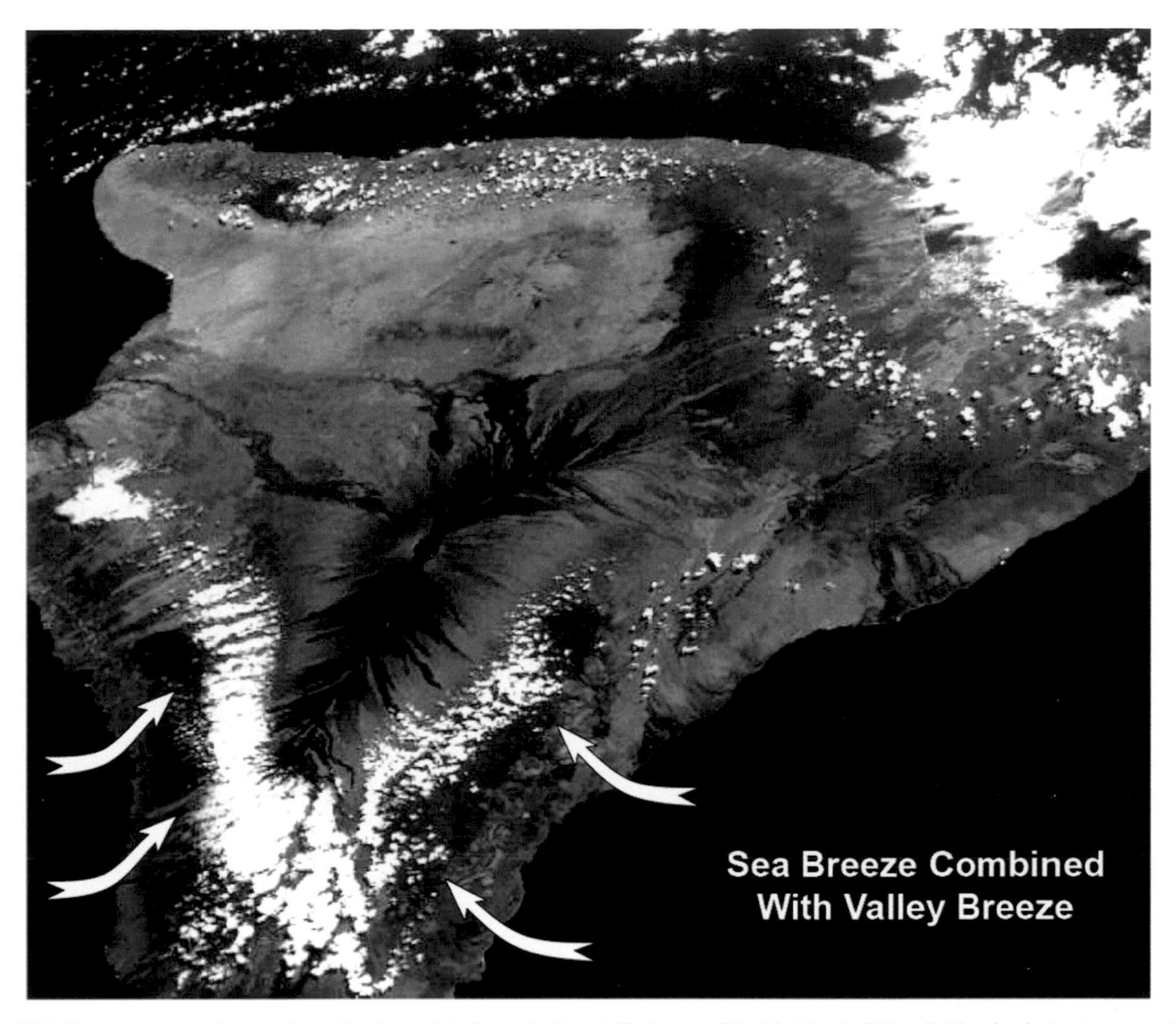

Figure 2.22 The sea breeze and mountain–valley breeze join forces in this satellite image of the Big Island of Hawaii. The clouds (next to arrows) reveal the combined effect of sea breezes and valley breezes as moist air is carried inland and then up the mountain, where the air cools and condenses. (Annotated from NASA image.)

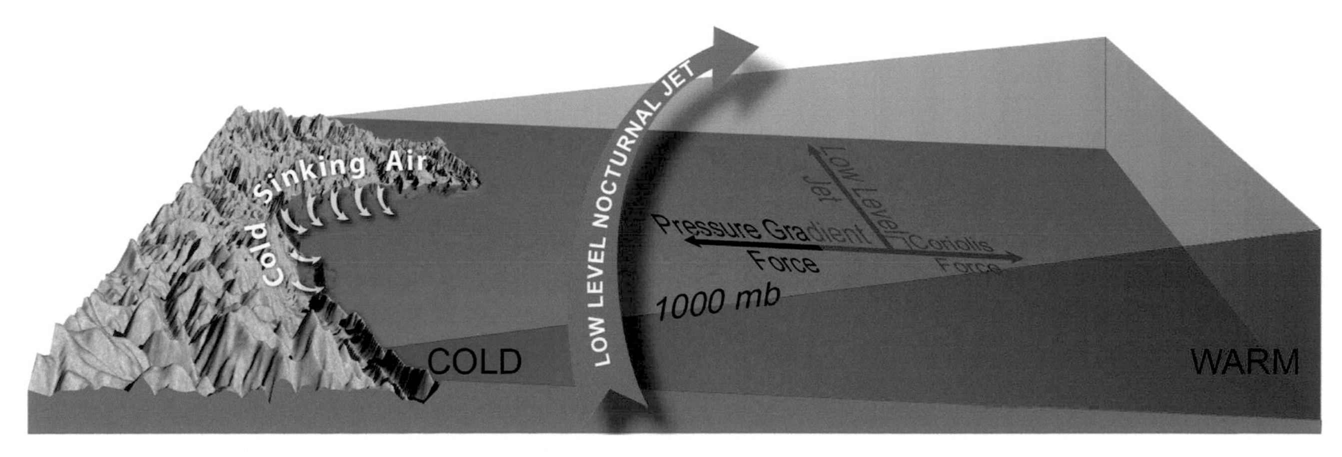

Figure 2.23 Formation of the nocturnal low-level jet. This wind system forms when cold mountain air flows eastward out of the mountains at night. This creates a lens of cold air that warmer air existing east on the plains rides upon. The result is a horizontal pressure gradient that accelerates a stream of air toward the Rockies. The Coriolis effect deflects the wind to the east.

Figure 2.24 Example of how nighttime thunderstorms over the Plains feed off Gulf moisture. The nocturnal low-level jet is a low-altitude ribbon of fast air that streams northward from the Gulf of Mexico, across the Plains. Within this ribbon, large nighttime complexes of thunderstorms erupt. One such complex is shown in the image. (NASA.)

is shown in Figure 2.24, which depicts a satellite image of an enormous nocturnal thunderstorm clustered over the heart of the Plains. The colors indicate the height of the clouds, with red hues showing extremely deep thunderstorms.

Potpourri of Other Regional Circulations

Figure 2.18 shows a few additional types of wind circulations, each of which creates storminess and other types of weather hazards. The purple shaded region east of the Appalachians identifies the zone of cold air damming. This phenomenon occurs when cold, dense air from eastern Canada and New England moves south but becomes trapped up against the high terrain, just above the surface. Moist air flowing off the Atlantic (often pulled westward by passing extratropical cyclones) flows up and over the cold "wedge," generating widespread layers of thick cloud, heavy snow, and ice storms.

To the northwest of the Appalachians, the light blue shaded regions denote areas that are frequently impacted by lake effect snows. These heavy snows, which tend to be highly localized along the lee shores of the Great Lakes, occur when cold winds blow from the west-northwest, under the influence of a strong anticyclone over the northern Plains. As the cold air moves over the relatively warmer water surface, the air warms and moistens. This makes the air mass unstable. The air ascends into clouds that produce heavy snow showers.

Farther west, along the Front Range of the Rockies across Wyoming and Colorado, cold air damming also occasionally develops during the winter months. The uplift of warm, humid Gulf air over the dome of cold air can produce extremely heavy snows. During the warm months, a different type of mountain circulation feature, called the Denver Cyclone, spins up. The cyclone is a small, shallow vortex created by southeasterly winds interacting with the complex terrain in the Denver region. Convergence of unstable air into the vortex during the summer months triggers severe thunderstorms, and the cyclone is a notorious breeding ground for weak tornadoes.

Moving to the southwestern United States, there are two regional wind patterns of interest. The first is a large, persistent circulation that develops across Arizona, Nevada, Utah, Colorado, and New Mexico during the summer months. This is called the North American Monsoon. Because the landmass over the desert of the southwest becomes intensely heated, a large low-pressure cell forms in the Four Corners region (junction of Colorado, Utah, Arizona, and New Mexico). The low pulls in plumes of moisture from the Gulf of California and Gulf of Mexico. As moist, heated air is forced to rise above the low-pressure area, widespread clusters of afternoon and evening thunderstorms develop on a near-daily basis. These storms frequently attain severe levels, producing damaging winds, haboobs (desert dust storms), and torrential rains that lead to flash flooding.

Severe Downslope Winds in Mountainous Terrain

Three specific types of windstorms develop in connection with high terrain, often described in the media during western U.S. wildfires. These windstorms have no connection to cloud systems, such as thunderstorms or cyclones. The Santa Ana is a type of downslope wind that develops when a strong high pressure cell (anticyclone) becomes established over Nevada and surrounding states. A strong pressure gradient accelerates airflow from the east, where it crosses the Mojave desert and then abruptly descends the Coast Ranges west of the L.A. Basin. The air becomes heated and dried crossing the desert, then warms further while descending the mountains (due to adiabatic compression, a principle discussed in Chapter 3). The air may warm 10° C (18° F), and the relative humidity (discussed in Chapter 3) may plummet to less than 10% as the Santa Ana wind fans over the Los Angeles region. Wind may gust to 43–61 kts (50–70 MPH), and even higher, and is strongest as it funnels through narrow mountain passes (such as the Soledad and Cajon Passes) and along the ridgelines (Figure 2.25).

Figure 2.25 Schematic showing the formation of the Santa Ana wind. High pressure over Nevada directs airflow westward. The air becomes funneled and channeled through mountain passes, moving downslope onto the coastal plain adjacent to the Pacific Ocean. The air compresses, warms, dries, and accelerates on its journey toward the L.A. Basin.

The strong, hot and dry wind breeds the perfect fire weather, or Red Flag, conditions. Many destructive and deadly wild fires have spread uncontrollably in and around Los Angeles, under the Santa Ana's influence, most commonly during the fall months (Figure 2.26). The winds can create widespread destruction, including damage to thousands of trees, roof damage, and extensive power outages. A particular notorious Santa Ana event during December 2011 generated sustained winds of

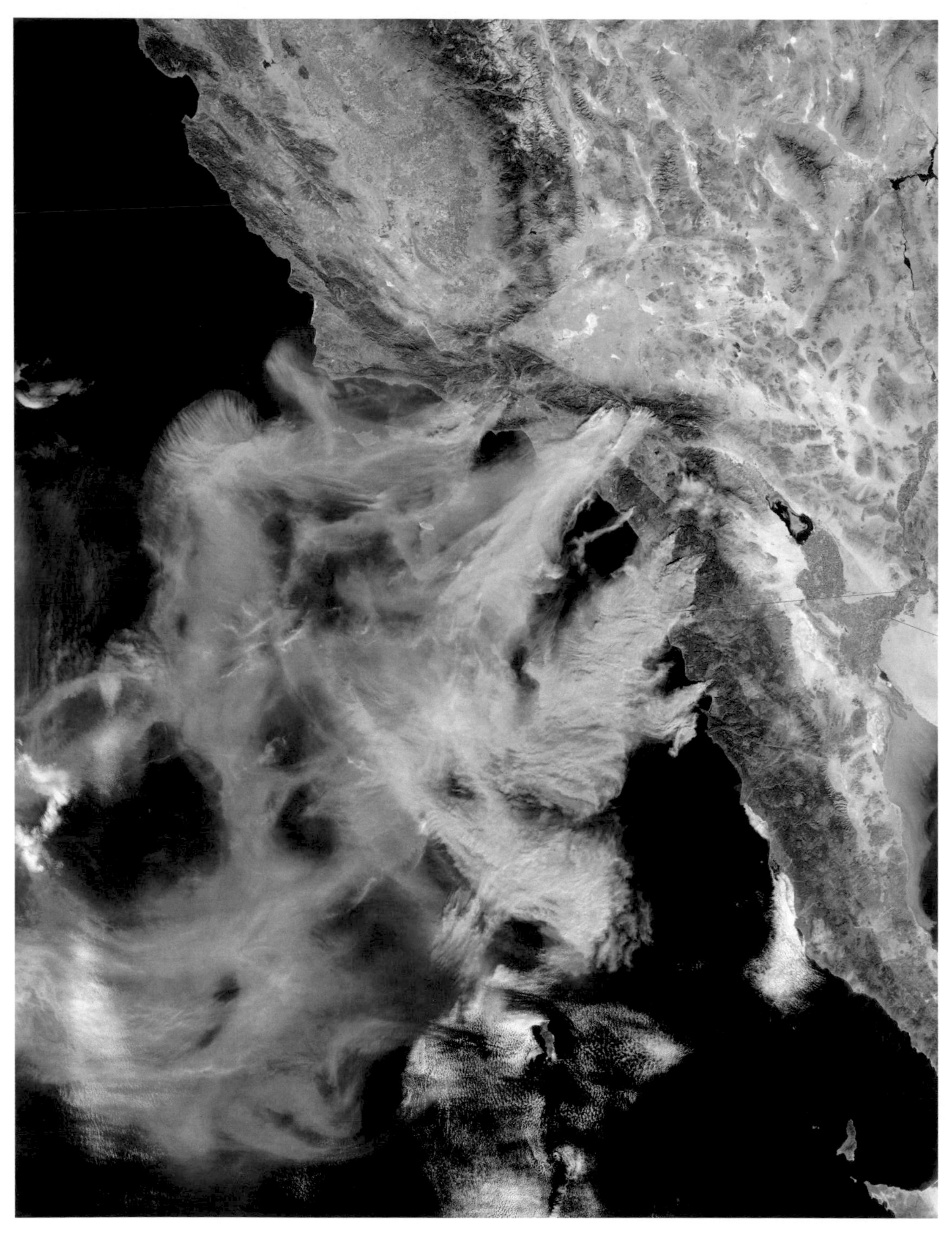

Figure 2.26 Smoke from wildfires blowing toward the southwest over southern California. Santa Ana winds blowing from the northeast spread multiple smoke plumes over the California coast and adjoining Pacific Ocean. (NASA satellite image.)

84 kts (97 MPH), and gusts to 145 kts (167 MPH), near Pasadena and Altalena, California. Flights at Los Angles International were grounded due to debris blown onto taxiways and runways. Fugitive dust lofted by the wind reduced visibility and posed health risks to asthmatics.

The Diablo wind features the same meteorological setup as the Santa Ana, except that this downslope wind is localized to the ridges in and around the San Francisco Bay Area. The wind was so named by the local media around the time of the 1991 Oakland wildfires because the wind blew from the direction of the Diablo Valley and Diablo Mountain.

A third important type of downslope wind develops along the eastward-facing slopes of the U.S. and Canadian Rockies and over the adjoining Prairies. The Chinook wind is created when high pressure builds over the intermountain region of the western United States, and a strong pressure gradient accelerates air toward the east and down the imposing rampart of the Front Range. Similar to the Santa Ana, the wind accelerates along ridgetops and through narrow mountain passes; gusts can exceed 87 kts (100 MPH) and create considerable damage. A Chinook wind during June of 1972 created a gust to 119 kts (137) MPH in Boulder. The warming that attends the downslope movement of air can be quite abrupt and dramatic (the term "Chinook" means "snow eater"). In January of 1972, Chinook winds over Montana caused the air to warm from −48° C (−54° F) to +9° C (+49° F) in 24 hours! While Colorado, Wyoming, and Montana all experience Chinook winds, the region of greatest annual Chinook activity is located in the southern corner of Alberta, Canada.

Turbulent airflow generated by Chinook winds poses an occasional hazard in the airspace lee of the Rockies. During Chinook wind events, wavelike oscillations can develop in the airflow streaming off the Front Range. Under conditions of a stable atmosphere (discussed in Chapter 3), these waves can amplify and travel more than a hundred miles downrange. Within about 33–48 km (20–30 miles) of the mountains, large horizontal vortices can develop, called rotors (Figure 2.27). Arrayed like giant "rollers" above the ground, these spinning tubes of air have been likened to "horizontal tornadoes" and pose a grave danger to aircraft for the severe turbulence they contain. Rotors are often invisible; at times, they may be identified by parallel rows of wavelike, billowing clouds. A rotor circulation may have contributed to the demise of a United Airlines flight over Colorado Springs in March 1991.

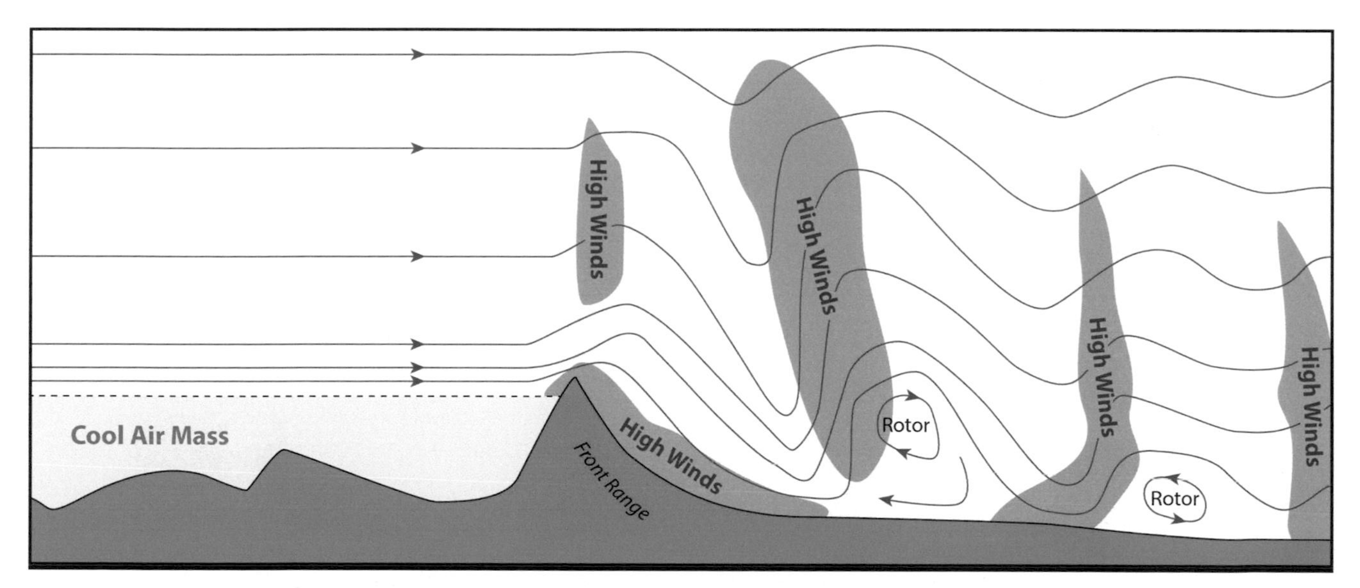

Figure 2.27 Formation of rotor circulations in the Chinook wind. As strong winds from the west blow across the tops of the Colorado Front Range, the flow often becomes wavelike on the lee side (eastward-facing slopes). Numerous rotor circulations develop close to the ground in this wavelike flow. (Adapted from image provided by Tim Vasquez.)

Summary

LO1 Define atmospheric pressure and explain how pressure changes with altitude.

1 Atmospheric pressure is the weight of the overlying atmosphere, per unit area. At sea level, the atmospheric pressure is 1013 mb (14.7 pounds per inch^2, or 29.92 inches of mercury).
2 Pressure (and air density) decrease exponentially with increasing altitude.

LO2 Understand how the pressure gradient force creates wind, and determine wind strength by examining the spacing of isobars on a surface weather map.

1 The pressure gradient force, which is related to the pressure difference between two locations at Earth's surface, sets the air into horizontal motion, creating wind.
2 On a surface weather map, the larger the pressure gradient (isobars more closely spaced together between a region of high pressure and a region of lower pressure), the faster the wind.

LO3 Explain how the geostrophic wind, which blows parallel to isobars, is created by a balance between the pressure gradient force and the Coriolis effect.

1 On the rotating Earth, the Coriolis effect balances the pressure gradient force and gives rise to the geostrophic wind. Because the geostrophic wind is a state of balance, it does not blow directly from high to low pressure. Rather, it is oriented parallel to the isobars.
2 The Coriolis effect in the northern hemisphere deflects air (set in motion by the pressure gradient) to the right of its path. The deflection effect is proportional to wind speed, is maximum at the poles, and vanishes at the equator.

LO4 Describe how horizontal variations in air temperature create a pressure gradient and geostrophic wind.

1 Horizontal variations in air mass temperature create a horizontal pressure gradient force, accelerating the wind from a warm column of air toward a cold column of air. The Coriolis effect balances the pressure gradient force. The wind on the constant-altitude surface is thus a geostrophic wind.

LO5 List the principal features of general atmospheric circulation, and discuss how global air circulation is driven by the movement of excess heat energy from the equator toward the poles.

1 The general atmospheric circulation, in the simplest sense, operates as a giant convective (overturning) air cell to move excess heat energy from the tropics toward the poles.
2 The observed general atmospheric circulation in the tropics includes the Hadley Cell, intertropical convergence zone (ITCZ), and sometimes tropical cyclones. Features of atmospheric circulation in the subtropics include the subtropical jet stream, subtropical anticyclones, and easterly trade winds. In the polar regions, surface high pressure generates polar easterlies.
3 In the mid-latitudes, the polar front denotes the zone of contrast between tropical and polar air masses. Above this front, a polar jet stream containing intense westerly winds undulates around the mid-latitudes. Extratropical cyclones and anticyclones occur in the mid-latitudes.
4 Waves in the polar jet stream (troughs and ridges) contain regions of curved airflow. The curved flow induces rising air downstream of a trough and sinking air downstream of a ridge. Areas of surface low pressure and high pressure develop beneath these regions, respectively.

LO6 Discuss how the types of wind forces differ among extratropical cyclones, hurricanes, and tornadoes.

1 A large low-pressure cell at the surface is called a cyclone and features a counterclockwise, inward spiral of air. A large high-pressure cell at the surface is called an anticyclone and features a clockwise, outward spiral of air.
2 Around cyclones and anticyclones, the wind is in gradient balance, which is a three-way force balance involving the pressure gradient force, Coriolis effect, and centripetal force. At the surface, a fourth force, friction, is also included.
3 Extratropical cyclones and anticyclones form along the polar front, beneath troughs and ridges in the polar jet stream, and move from west to east across the mid-latitudes. These vortices are responsible for the day-to-day weather variations

(alternating periods of storminess and fair weather) and frequently contain smaller, embedded regions of severe and hazardous weather.

4 Extratropical cyclones feature ascending air in their core. As the air ascends and cools, water vapor condenses into various forms of cloud and precipitation. Extratropical anticyclones contain air that sinks down their center, warming and drying as it does so, leading to clear skies and fair weather.

5 Outbreaks of severe local storms – including summertime thunderstorms containing damaging wind, torrential rain, large hail, and tornadoes – frequently concentrate along the fronts (warm front and cold front) contained in extratropical cyclones. During the winter months, extensive regions of heavy snowfall, ice, and strong winds develop and move across the United States within extratropical cyclones;

6 Extratropical anticyclones are most often associated with extremes in air mass temperature, creating hazardous weather during both the winter (arctic air outbreak) and summer (heat waves).

7 Extratropical anticyclones are also sometimes associated with weather hazards. These include localized regions of severe thunderstorms and flash flooding, heavy lake effect snows, and damaging windstorms down steep mountain slopes (Santa Ana and Chinook winds).

8 Hurricanes and tornadoes are the most intense forms of low-pressure cyclones. The most intense winds in these vortices are called cyclostrophic winds, and they are created by a balance of pressure gradient and centrifugal forces.

LO7 Describe the origins and geographical distribution of various regional-scale wind circulations, including the sea breeze, mountain–valley circulation, nocturnal low-level jet, North American Monsoon, and mountain winds (Santa Ana and Chinook).

1 Across the continental United States, regional wind circulation patterns occur in geographically preferred regions. Many of these circulations, including the sea breeze, North American Monsoon, and mountain–valley circulation, are quite shallow and are driven when the surface is heated by the afternoon sun. These circulations can generate thunderstorms, some of which reach severe levels.

2 Severe downslope winds include the Santa Ana of southern California's western mountain slopes, the Diablo along west-fast slopes around San Francisco, and the Chinook winds along the Front Range of the Rockies. All are capable of damage to trees, utilities and structures . . . and both the Santa Ana and Diablo are implicated in wildfire spread.

Key Scientific Principles Covered in This Chapter

Note to the student: Many of these "first principles" are needed for later chapters, as a reminder of an important physical process or definition, to aid in your understanding of more advanced material.

2.1 Atmospheric pressure is the weight of the overlying atmospheric gasses per unit surface area.

2.2 Pressure decreases rapidly with altitude in the lowermost atmosphere and less rapidly at higher levels (this variation describes an exponential decrease with height).

2.3 Differences in atmospheric pressure across a horizontal distance give rise to the pressure gradient force, which accelerates air from high-pressure toward low-pressure regions – creating wind.

2.4 On a surface weather chart, isobars are drawn as contour lines of constant pressure, typically spaced every 4 mb. The more closely spaced the isobars, the stronger the wind.

2.5 The Coriolis effect arises from the spinning Earth; it deflects air that begins to move; the deflection is to the right in the Northern Hemisphere.

2.6 The geostrophic wind is a balanced state between the pressure gradient force and Coriolis effect; this wind moves at constant speed, in a straight path, above the surface.

2.7 Horizontal variations in atmospheric temperature, over long distances, give rise to a horizontal pressure gradient force (low pressure is associated with warm air; high pressure with cold, dense air).

2.8 A system of global atmospheric winds, called the atmospheric general circulation, arises from a strong imbalance in radiational heating between the equator and poles. There are several components to the general circulation: in the tropics, mid-latitudes, and polar regions.

2.9 The tropical Hadley Cells circulate warm, equatorial air toward the higher latitudes; these cells give rise to tropical easterly winds (trade winds), the ITCZ, subtropical anticyclones, and the subtropical jet stream.
2.10 The intertropical convergence zone (ITCZ) is where easterly trade winds from each hemisphere converge near the equator, causing air to rise. This forms a nearly continuous belt of showers and thunderstorms, often organized into large clusters of storms.
2.11 Subtropical anticyclones (high pressure cells) are situated near 30° N and S; these cells are formed by air descending on the periphery of the Hadley Cells. As the air descends, it warms and dries, creating large cloud-free zones.
2.12 The subtropical jet stream forms in in the upper atmosphere, in each hemisphere; it forms when poleward-moving air, exiting the Hadley Cell, becomes deflected toward the east by the Coriolis effect.
2.13 Mid-latitude circulation features of the atmospheric general circulation include the polar front, polar jet stream, extratropical cyclones, and anticyclones.
2.14 The polar front is a deep, wavelike discontinuity between tropical and polar air masses, lying across the mid-latitudes. Extratropical cyclones frequently develop from waves along this boundary.
2.15 The polar jet stream develops along the polar front at high altitudes. It is maintained by the strong north–south temperature gradient across the polar front. The polar jet constantly develops large meanders (Rossby waves) that are inherently unstable, tied to the formation of extratropical cyclones and anticyclones near the surface.
2.16 The upper-atmosphere polar vortex and band of deep, polar easterlies are found above the Arctic Circle in each hemisphere.
2.17 Extratropical cyclones and anticyclones are large, transient vortices that migrate from west to east across the mid-latitudes. Air converges in a counterclockwise spiral around these cyclones, which feature low pressure in their core ("L" on a weather map). Air diverges in a clockwise spiral around anticyclones, with high pressure in the core ("H" on a weather map). These systems bring changeable weather (extremes in temperature, precipitation, fair weather) to most mid-latitude locations.
2.18 Around extratropical cyclones and anticyclones, a type of circular wind called the gradient wind develops as a consequence of the pressure gradient, Coriolis effect, and centripetal (inward-directed) force.
2.19 Near the surface, the friction force slows the wind and creates an imbalanced flow in the gradient wind, causing air to spiral inward toward the center of an extratropical cyclone and away from the center of an extratropical anticyclone.
2.20 The principle of mass conservation requires air that spirals inward toward the center of an extratropical cyclone to rise and air that spirals outward in an extratropical cyclone to draw down air from above.
2.21 The meanders in the jet stream, called Rossby waves, take the form of troughs (equatorward-curved segments) and ridges (poleward-curved segments). Cold, polar air moves south in a trough, while warm, tropical air moves north in a ridge.
2.22 Air flowing through a trough in the jet stream is subgeostrophic (slower than geostrophic). As the air moves away from the trough, it accelerates (diverges); this effect draws up air from the surface (via the principle of mass conservation), leading to the development of a large region of low pressure, and subsequent formation of an extratropical cyclone.
2.23 Air rising in the core of an extratropical cyclone cools and becomes saturated, leading to widespread cloud cover and precipitation (i.e., a storm) within the cyclone.
2.24 Air flowing through a ridge in the jet stream is supergeostrophic (faster than geostrophic). As the air moves away from the ridge, it deaccelerates (converges); this effect causes air to descend toward the surface (via the principle of mass conservation), leading to the development of a large region of high pressure and the subsequent formation of an extratropical anticyclone.
2.25 Air sinking in the core of an extratropical anticyclone warms and becomes unsaturated, leading to widespread fair weather and generally clear skies.
2.26 In an extratropical cyclone, fronts develop in the lower atmosphere where contrasting air masses (i.e., tropical and polar) are drawn together. The warm, buoyant air is forced to ascend vigorously along fronts, developing bands of deep cloud, heavy precipitation, and sometimes severe weather.
2.27 A tropical cyclone (hurricane in the Atlantic, typhoon in the Pacific) is an intense cyclonic vortex in the tropics, with a compact, inner core of extremely low pressure. Very strong winds around the low are in a state of cyclostrophic balance between the pressure gradient force and outward-directed centrifugal force.
2.28 A tornado is the most intense type of cyclonic vortex, typically less than half a mile across, and it is generated within a severe, rotating thunderstorm called a supercell. Its winds, which may top 300 MPH, are in a state of cyclostrophic balance.
2.29 Regional and local wind systems are driven by uneven heating of the atmosphere. These winds include the sea breeze, mountain/valley breeze, nocturnal low-level jet, and North American Monsoon. Under certain conditions, thunderstorms develop along a narrow zone of rising air in each case.

2.30 The sea breeze develops between heated land and cooler, adjacent ocean water during the daytime. It is a shallow circulation cell with rising air over the land and sinking air over the ocean.

2.31 The mountain/valley breeze is created by the strong heating of sloped, elevated terrain during daytime. Air rises from the valley floor, up the mountain, creating a steady wind. At night, the elevated land cools more strongly than the surrounding air, causing a cold, dense airflow to settle into the valley floor.

2.32 The nocturnal low-level jet is a wind current of the Great Plains. It is a narrow, elongated ribbon of air that blows moist, unstable air from the Gulf of Mexico northward over the Plains – leading to large complexes of heavily raining thunderstorms and frequent severe weather.

2.33 The North American summer monsoon is a seasonal (summertime) wind pattern that develops over the U.S. southwestern desert and adjoining portions of northern Mexico. The large expanse of exceptionally hot air develops a zone of low pressure, which accelerates airflow from the Gulf of California northward. Late afternoon and evening thunderstorms blossom in this unstable air mass across Arizona, New Mexico, Utah, Nevada, and Southern California from June through September.

CHAPTER 3

Meteorological Primer, Part II: Moisture and Precipitation in Storms

Learning Objectives

1. Distinguish among humidity, dew point temperature, and relative humidity, and describe how these values typically vary with time of day, by season, and across geographic regions.
2. Discuss two ways in which air cools adiabatically when lifted and becomes saturated.
3. Explain how supersaturation, cloud condensation nuclei, and collision-coalescence lead to the growth of cloud droplets.
4. Define supercooled water and explain how it relates to the formation of small ice crystals in subfreezing clouds; distinguish between vapor deposition and riming.
5. Explain why an unstable air mass leads to rapid growth of the deep convective clouds associated with heavy rain showers and hail.
6. Describe how regions of precipitation are developed and arranged in a typical extratropical cyclone (cool-season weather pattern) and during the summer months.

Introduction

In Chapter 2, we examined the forces that give rise to moving air (wind) on the Earth, including local breezes, cyclones, anticyclones, and jet streams. Except in desert settings, most air masses on the Earth contain some degree of moisture in the form of water vapor. Additionally, layers and billows of clouds – some of which become quite extensive – frequently develop in association with cyclonic storms. Many kinds of precipitation, including rain, snow, and hail, develop from persistent zones of cloudiness. The connection between extreme precipitation and storm hazards takes many forms: Rapid accumulation of rain causes flash floods; heavy snow and ice can paralyze entire states; and large hailstones heavily damage crops and property.

In this chapter, we focus on the processes and mechanisms that create precipitation. First we look at a few ways that water vapor is measured in the atmosphere. We then examine the processes that form clouds. From clouds we discuss the formation of raindrops and frozen precipitation such as snow. We discuss the vital concept of atmospheric stability and what this means for the growth of deep, vertical clouds called convective clouds, which include thunderstorms. Finally, we distinguish between processes leading to horizontal layers of precipitating cloud, called stratiform precipitation, from the more intense and localized convective showers and thunderstorms.

Common Measures of Atmospheric Humidity: Dew Point Temperature and Relative Humidity

Most of us are familiar with the concept of atmospheric humidity. Think about the oppressive, clinging stickiness of the air on a hot, humid summer day . . . or the skin-chapping dryness of a cold winter wind. The simple idea of changeable moisture content in the air has surprisingly large effects on our daily lives.

Vertical and Horizontal Variations of Atmospheric Moisture

There are many scientific measures of water vapor, all of which can be loosely categorized as measures of the air's humidity (water vapor content). Water vapor is a trace gas (an atmospheric constituent present in minute quantities), derived from the evaporation of surface water from oceans, lakes, inland seas, and other bodies of water. Water vapor tends to be most concentrated close to the Earth's surface, decreasing with height through the troposphere. Water vapor concentrations may range from 10 to 20 parts per thousand (ppt) near the surface and over large water bodies and can comprise up to 3% of the air's total mass

DOI: 10.4324/9781003344988-4

over warm tropical seas and rain forests. Many miles above the surface, the concentration of water vapor drops to parts per million – values so small they are difficult to measure. Water vapor content can also be expressed as a gas concentration – for example, 16 g of water per kilogram of dry air.

As Figure 3.1 shows, the amount of water vapor in the atmosphere varies greatly across North America. This image is a color-enhanced satellite view of mid-tropospheric water vapor, in the 3–8 km (2–5 mile) altitude range. Regions of air that are saturated contain the maximum amount of water vapor for a given temperature and are light blue. Very dry (low humidity) areas are orange or red.

The gradient (rate of change with distance) of water vapor across a region can be quite dramatic, This zone denotes a weather front, across which air mass properties (temperature, humidity, and winds) vary markedly over a short distance. In fact, water vapor is a remarkable tracer of motion in the atmosphere. For instance, note the large cyclone over New England. Its giant, counterclockwise swirl is readily apparent by wisps of humid and dry air converging into the low-pressure center. This area of low pressure, called an extratropical cyclone (Chapter 2), draws in a narrow corridor of very humid air from the south. The tropical cyclone near Cuba is also very apparent in this imagery.

Dew Point Temperature Relates Directly to the Air's Moisture Content

As air cools, dew, fog and clouds can form – as commonly observed before sunrise on an autumn night. When the air cools, water vapor molecules slow down (they lose kinetic energy) and are more likely to clump – or condense – into liquid droplets. This condensation occurs either on (1) grassy surfaces, forming dew, or (2) around microscopic dirt particles in the air, forming fog or cloud. The dew point is the temperature to which the air must cool for the first traces of condensed water to appear.

The drier the air, the more the air must cool to reach the point of saturation. Saturation is a requirement for condensation and the formation of liquid water droplets. When air is saturated, the rate of condensation exceeds evaporation, and liquid droplets begin growing. Dry air with a dew point temperature of, say, 35° F (2° C) requires more cooling to cause saturation than air with a dew point temperature of 70° F (21° C).

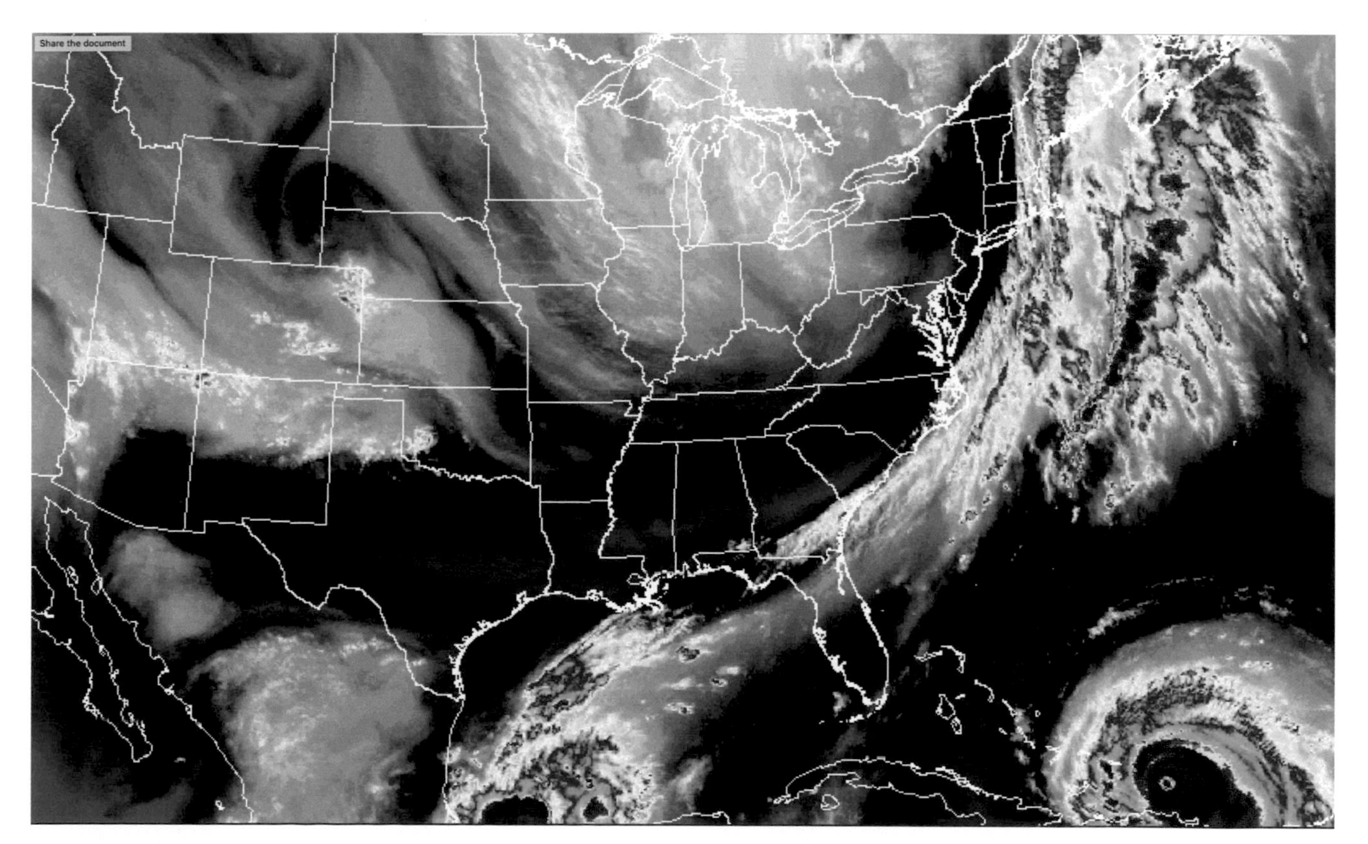

Figure 3.1 Atmospheric water vapor across North America. False-color water vapor imagery, obtained from a weather satellite, depicts the large variation of humidity (atmospheric moisture) in the upper atmosphere across the United States on a typical day. (NOAA.)

For example, morning fog rarely forms during the early summer months. Let's say the dew point temperature at sunset is 65° F (18° C). During the short nights of midsummer, the air simply cannot cool sufficiently to reach the dew point temperature. But on a midautumn night, starting again at 65° F (18° C), additional cooling hours afforded by the longer night allows the air to reach its dew point temperature – and a fogbank or dew-covered grass develops.

Another way of expressing the relationship between dew point temperature and humidity is to state that the dew point temperature is *proportional to* (varies directly with) the air's moisture content.

Dew point temperature varies considerably across the United States, from season to season, and with geographic location on any given day. Figure 3.2 provides maps showing characteristic dew point temperature values for mid-summer and midwinter across the continental United States.

These maps portray the average distribution of dew point temperature over many weeks. Daily and even hourly values depart from these averages, due to migrating weather systems such as extratropical cyclones, fronts, and high-pressure cells. Looking first at summer (Figure 3.2[a]), we note very high dew point values, approaching 70–75° F (21–24° C), along the Gulf Coast region and southeastern United States. These high values are due to intense evaporation off the very warm Gulf of Mexico and Caribbean Sea.

Dew points generally decrease poleward and westward across the continental United States for two reasons: (1) increasing distance from the oceanic moisture source and (2) progressively cooler temperatures at higher latitudes. Dew point temperature tends to decrease as air temperature decreases, for the simple reason that there is less heat energy to supply vapor molecules with the kinetic energy they need to remain in the gaseous state.

Note also a significant reduction in summertime dew point values over the intermountain west. The high elevations and sparse vegetation contribute to low moisture content of the air, even in summer. Recall that water vapor decreases as altitude increases. In the desert southwest, where fewer trees grow and grasslands are sparse, there is less transpiration of water vapor into the air. Transpiration is caused by plant roots that draw soil moisture through vascular stems. The moisture then diffuses through small pores in the leaves.

Average dew point temperature during wintertime (Figure 3.2[b]) reflects a decrease in moisture values everywhere. We expect lower moisture because air temperatures are much lower countrywide during the cold season. The highest wintertime values are found along the Gulf Coast, where the Gulf of Mexico has cooled significantly since summer but

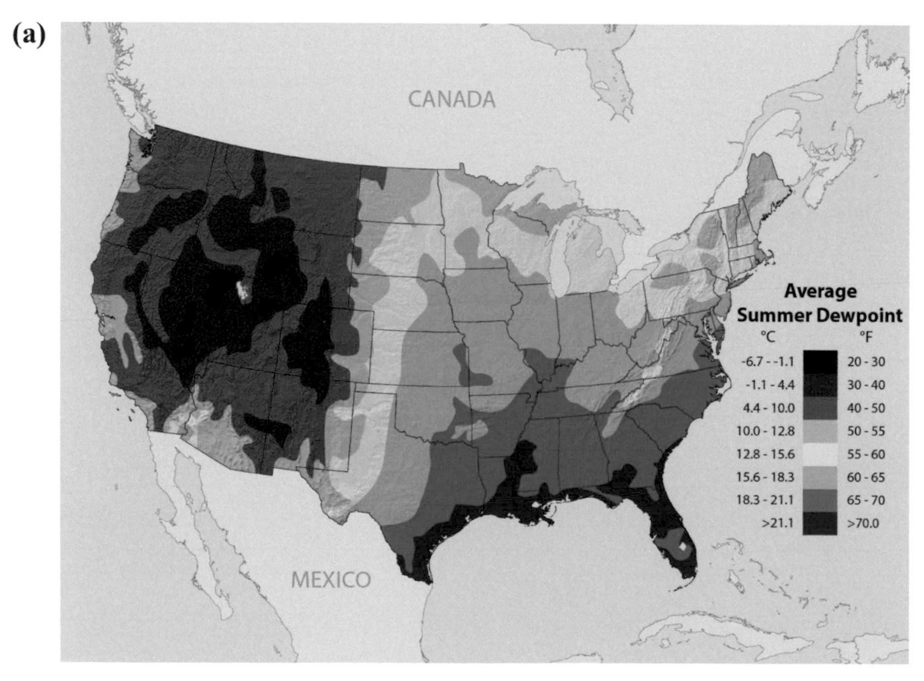

Figure 3.2 Seasonal variation in dew point temperatures. The spatial variation of surface dew point temperatures across the United States during (a) summertime (July) and (b) wintertime (January).

(b)

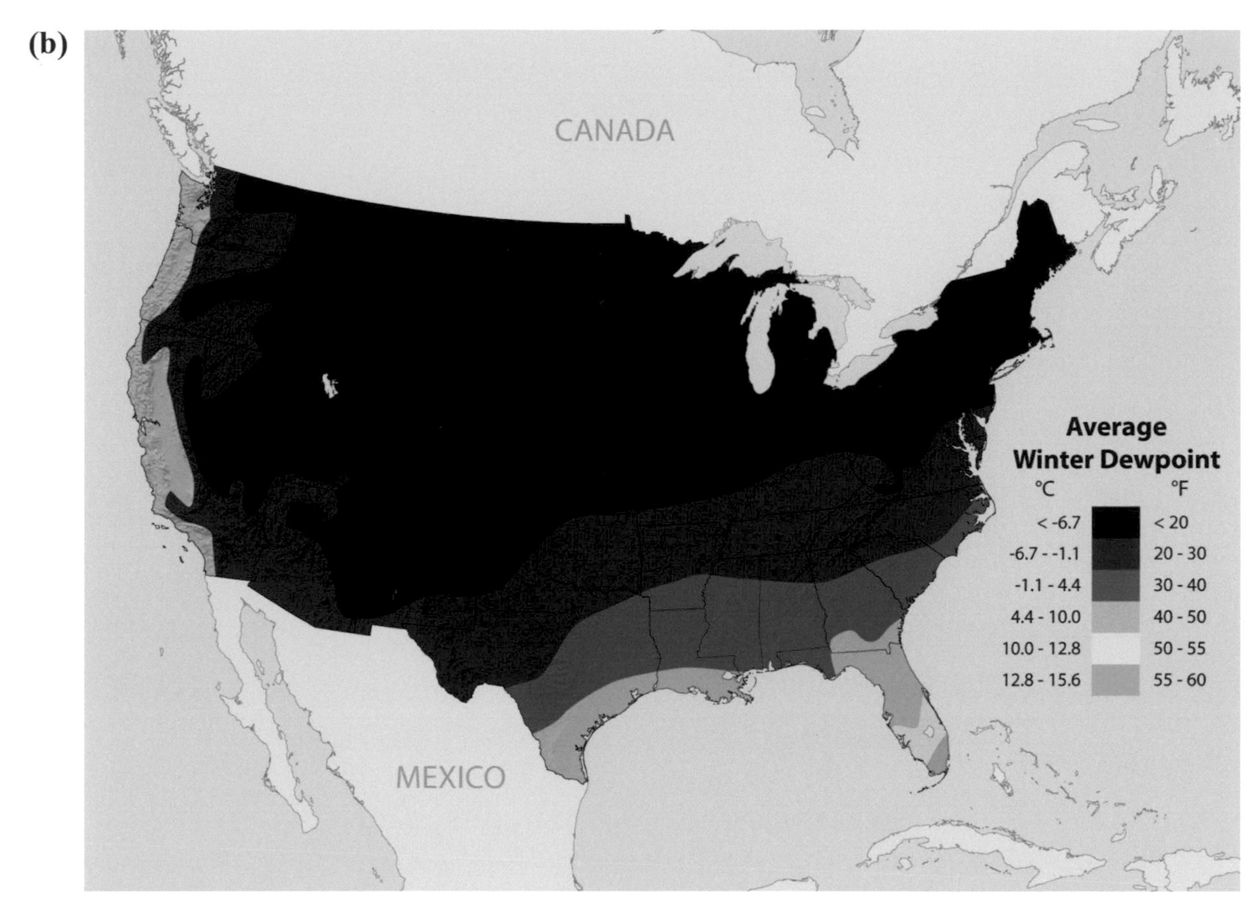

Figure 3.2 (Continued)

nonetheless remains a source of evaporating water. Extremely low dew point values are found over the high-elevation, chilly air of the western United States.

Dew point temperature also exhibits significant day-to-day variability, caused by migrating extratropical cyclones and anticyclones in the mid-latitude westerlies (Chapter 2). Figure 3.3 illustrates the striking contrast in dew point temperature across a typical springtime extratropical cyclone. Within these large storms, a converging, clockwise spiral of low-level wind brings dissimilar air masses into close proximity. Humid air streams north from the Gulf of Mexico, while very dry air circulates south out of Canada. There is a 40–50° F (4–10° C) variation in dew point temperature across the storm system, leading to very changeable weather conditions as the storm system moves through.

Dew point temperature directly affects how we feel in the outdoors. During the hot days of summer, high temperature is exacerbated by the presence of high-humidity (high-dew point) air. As the dew point temperature climbs above 70° F (21° C), the air feels quite heavy and oppressive. This perception stems from the body's inability to efficiently cool on very hot, humid days. Efficient cooling is reduced because evaporation from skin is curtailed when the vapor content of the surrounding air is already high – there is less room in the air for additional vapor. Evaporation is a cooling process (heat is extracted from the skin to vaporize liquid water), so when evaporation rates are slowed, our bodies cannot shed their metabolic heat, and we become in danger of overheating.

Relative Humidity: An Imperfect Measure of Water Vapor

Another measure of the air's water vapor is relative humidity. The definition hinges on the word *relative*: Relative humidity expresses the ratio of the air's actual humidity to the maximum amount of humidity that completely saturates the air at the given temperature. In other words, it's a *percentage* showing how close the air comes to achieving saturation. It

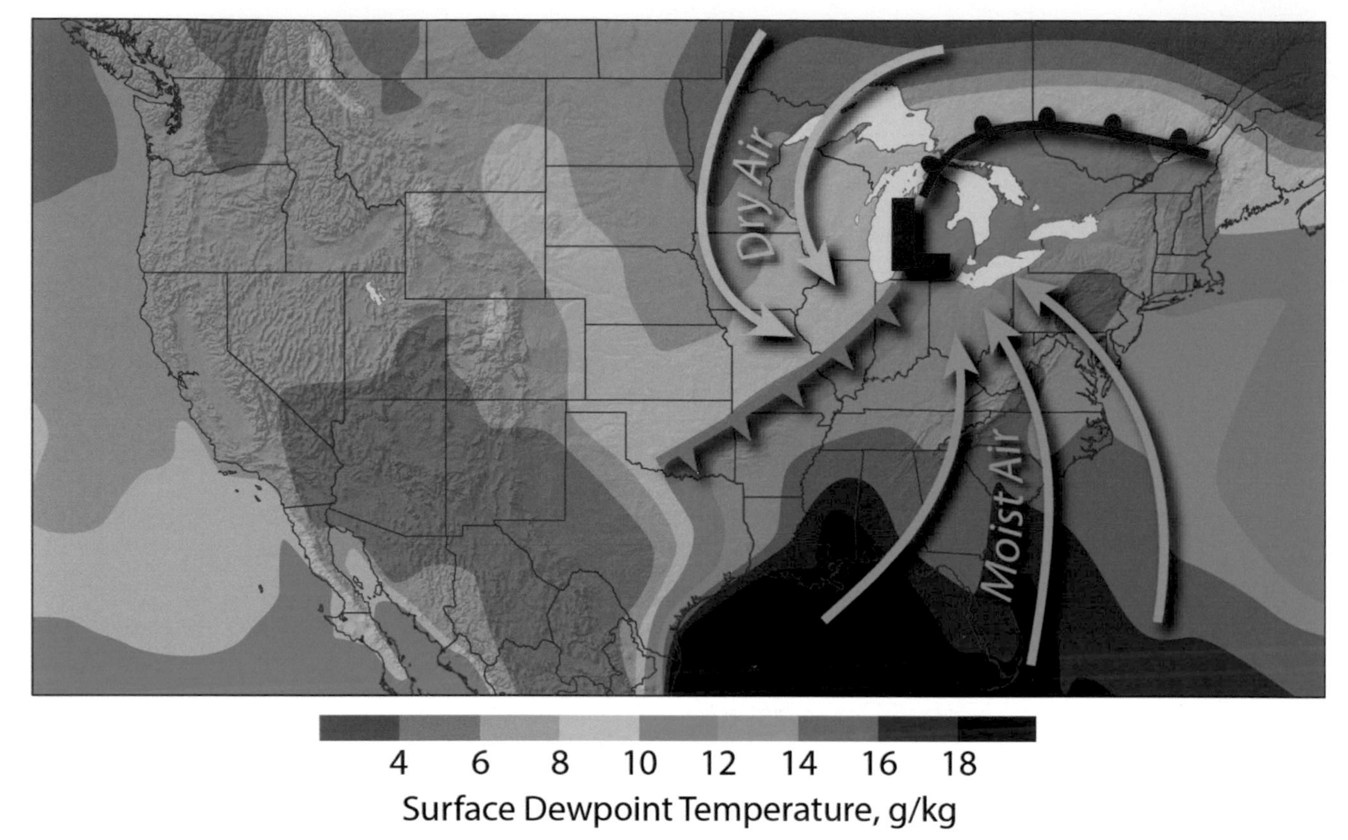

Figure 3.3 Surface dew point temperature variations across a typical extratropical cyclone—the moist and dry sides of a typical low-pressure system over the United States. Warm, humid air is circulated north by southerly winds on the east side of the system, while cold and dry air of Canadian origin circulates toward the south on the west side.

says nothing about the actual, physical amount of water vapor present!

The problem is that this percentage is a function both of the ambient air's moisture content and its temperature. The higher the air temperature, the higher the saturation value of humidity. So if the temperature of a moist blob of air changes, so will its relative humidity, even though we have not changed the air's actual moisture content. For water vapor content to remain constant when temperature is raised, relative humidity must *decrease*. When the air cools, relative humidity must *increase* for water vapor to remain constant.

As Figure 3.4 shows, relative humidity commonly undergoes cyclic changes through the course of the day. The true moisture content of the air – expressed by the dew point – remains constant (green curve). In the afternoon, as temperature (red curve) rises, relative humidity (blue curve) must decrease. During the early morning hours, normally the coolest time of day, relative humidity must increase. Yet the physical amount of water vapor in the air has not changed at all.

Many meteorologists consider relative humidity to be a "second tier" measure of air's vapor content because relative humidity is not an unambiguous measure of humidity. Temperature changes throughout the day profoundly affect the value of relative humidity. If the ambient humidity varies as well (often the case), there is no way of telling whether moisture, or temperature, or both, caused the observed change in relative humidity.

How Air Becomes Saturated

Air achieves supersaturation (relative humidity exceeding 100%) in two ways. When air becomes supersaturated, condensation of vapor exceeds the rate of evaporation on surfaces, enabling small droplets to begin growing. Supersaturation can therefore occur in one of two ways: (1) Air temperature remains constant, but humidity increases, or (2) humidity remains constant, but the air cools until it reaches its dew point temperature. The first process sometimes occurs over very wet land surfaces or over bodies of water, where the source of moisture is essentially

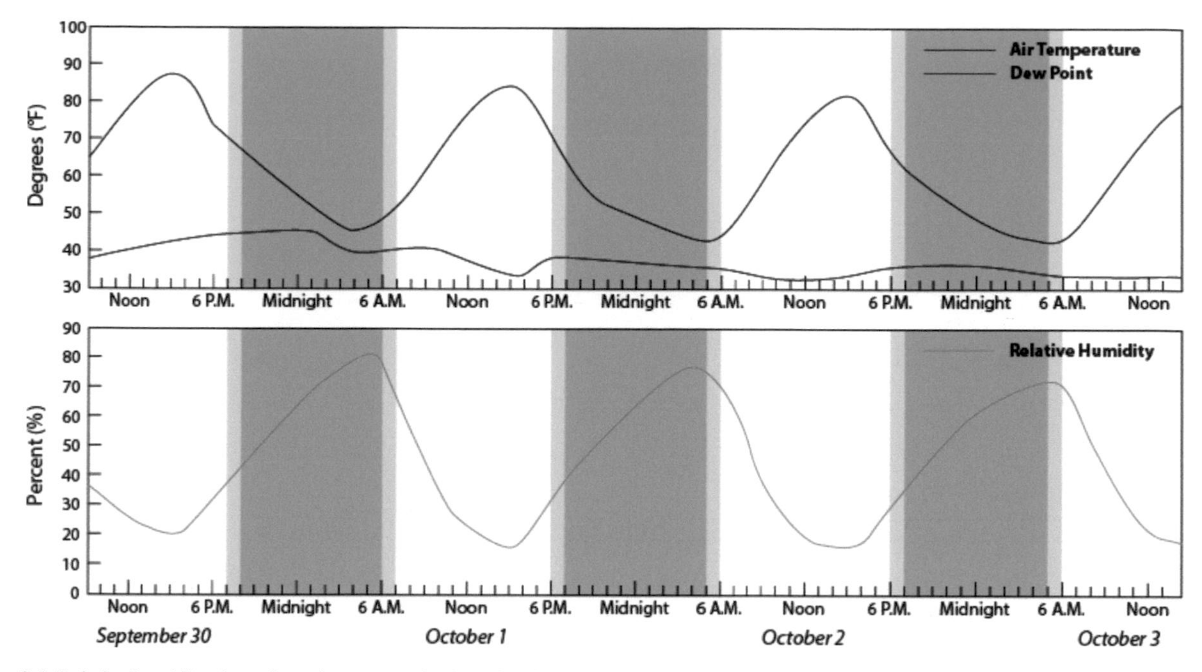

Figure 3.4 Relative humidity throughout the course of a day. The diurnal (daily) variation of relative humidity, compared to constant dew point temperature, illustrates how relative humidity is inversely related to air temperature.

unlimited, and other processes facilitate vigorous evaporation into the air. But the much more common process involves cooling air.

You may implicitly understand that rising air cools, but you may not understand exactly why. Air cools as it rises not because rising blobs of air encounter cooler air aloft; rather, the blobs *expand* as they rise, due to lower atmospheric pressure. Visualize how a helium-filled balloon must get larger as it ascends, expanding into the thinner air. For a rising air bubble to expand, it must work on its surroundings, pushing the surrounding air outward. The work depletes some of the bubble's internal energy, and the air temperature inside the bubble falls (temperature is a measure of the average kinetic energy of the gas molecules, which must be slowing down as they lose internal energy). You can experience this type of cooling by letting the air out of a tire; the air stream will feel very cool against your hand because the air expands in volume as it escapes.

Adiabatic cooling is the expansion and cooling of air, under lower pressure aloft. The term derives from the laws of thermodynamics, which state that the bubble of rising air remains a closed system and has no heat exchange with its surroundings; all temperature changes must come from changes in the bubble's own internal energy. It turns out that rising, unsaturated blobs of air cool adiabatically at a constant rate, which is −10° C/km (−18° F/3300 feet) of ascent.

Let's say that a rising blob of air initially has a surface temperature of 90° F (32° C) and a dew point temperature of 63° F (17° C). This is a fairly moist but unsaturated air bubble. Now the air is allowed to ascend freely. It immediately encounters reduced air pressure aloft and begins expanding, cooling adiabatically. At a rate of −10° C/km, the air chills to 72° F (22° C) after the first kilometer of ascent. After another half kilometer, it has chilled a further 5° C (41° F), but then its internal temperature equals the dew point temperature, 63° F (17° C). At this point, the air bubble has cooled to the point of saturation. With the slightest additional bit of cooling, condensation begins exceeding evaporation, and water vapor molecules cling to form the first liquid cloud droplets. Voilà, a cloud appears! This process is illustrated in Figure 3.5.

Air is commonly lifted and undergoes adiabatic cooling by four processes, as shown in Figure 3.6: (1) free ascent of buoyant air blobs called thermals; (2) forced ascent of air up the side of a hill or mountain, or orographic lifting; (3) lifting of air along frontal boundaries, such as when a cold, dense air mass wedges beneath a warm, less dense air mass (frontal uplift); and (4) convergence of air into a low-pressure system (extratropical cyclone) near the ground. Convergence in this context refers to airstreams that swirl inward toward a more compact spiral. The air in the center begins flowing upward, away from the ground.

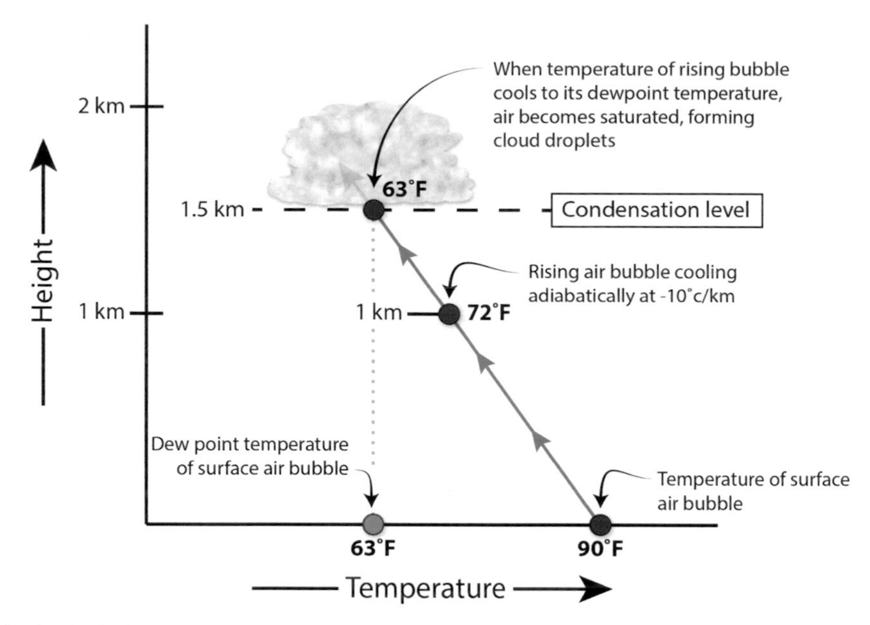

Figure 3.5 Formation of a cloud. The figure shows the key processes leading to formation of a cloud, following ascent of initially unsaturated air that cools adiabatically to its dew point temperature.

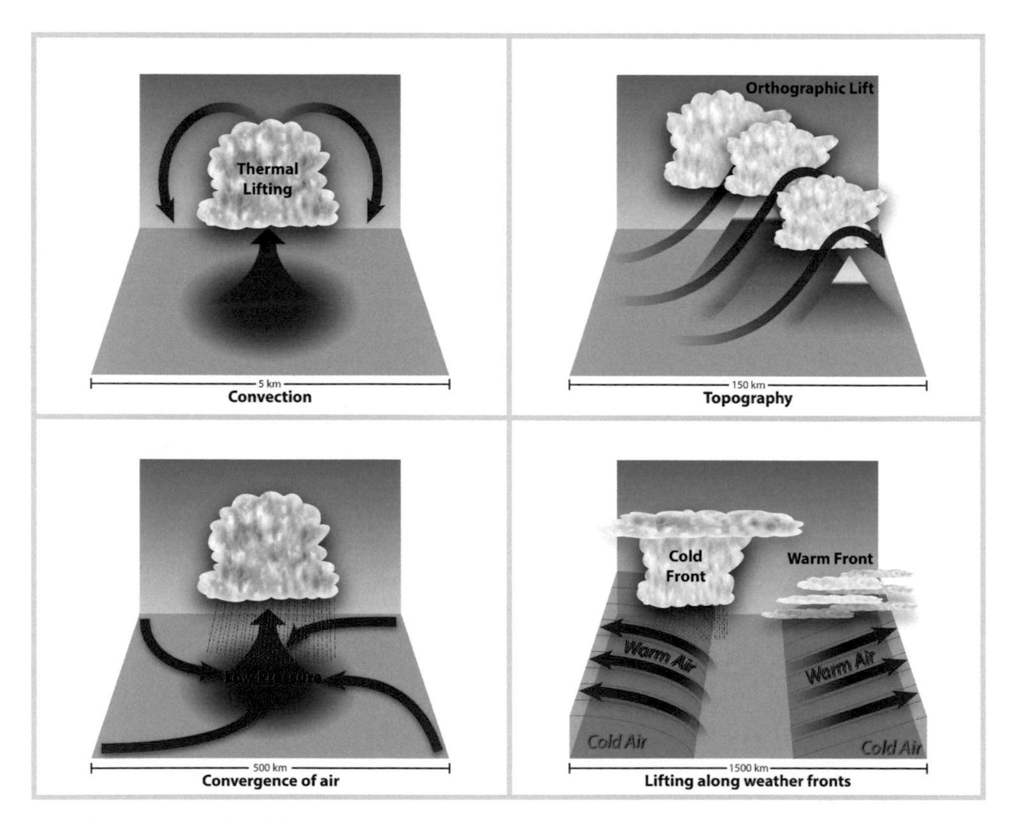

Figure 3.6 Adiabatic cooling processes. Air is lifted in the atmosphere in the four ways illustrated here, producing adiabatic cooling that leads to saturated air and cloud formation.

Thermals frequently develop when surface air is much warmer than the cooler air aloft. Small regions of ground heat to higher temperatures than others, enabling some bubbles of air to develop slightly warmer temperatures. Through the principle of buoyancy, these blobs begin ascending, like bubbles rising in a lava lamp. As a result of adiabatic cooling and condensation, puffy cumulus clouds develop atop each thermal. The upward transfer of warm, moist air is called convection. If the thermals become especially strong and rise to great heights, deep convective clouds, called cumulonimbus, develop. The cumulonimbus is also known as a thunderstorm cloud (Figure 3.7), and it is the harbinger of severe local storms.

Orographic lifting or ascent commonly occurs along the Appalachians and mountains of the western U.S. coast. In both locations, moist air flowing off the ocean often leads to saturated air and extensive cloud formation. These clouds often generate very heavy precipitation (both rain and snow), which may create flash floods and blizzards.

Frontal uplift takes two forms. The first occurs when cold air wedges beneath warmer, more humid air, which occurs along a cold front. The second form occurs when warm, humid air glides gently atop a colder air mass hugging the surface, establishing a warm front. As Figure 3.8 shows, both cold fronts and warm fronts commonly develop in association with extratropical cyclones across the United States. The converging, clockwise spiral of wind brings dissimilar air masses into abrupt opposition. Cold, dry air circulates south out of Canada, while warm, humid air is drawn northward out of the Gulf of Mexico or western Atlantic. Especially large gradients of temperature and dew point become established across the cold front and warm front (Figure 3.9). Fronts are the location of very active kinds of weather, given that humid air is forced to rise along these narrow zones.

The uplift of air occurs along a narrow zone ahead of an advancing cold front, and it is often quite vigorous. This air develops into a narrow band of convective showers and sometimes thunderstorms. North of the storm's warm front, clouds and precipitation become more widespread because the uplift of air is more gradual and spread out over a large, gently sloping region. Figure 3.10 depicts a weather radar view of a springtime extratropical cyclone; the figure shows the locations of the cold front and warm front rain bands.

The convergence of inward-streaming air masses helps to focus the uplift of air, along with cloud formation and precipitation, near the center of an extratropical cyclone. This broad shield of heavy weather is termed the comma head when viewed from the perspective of a weather satellite (Figure 3.11).

Figure 3.7 Cumulonimbus cloud. The lower portion of the cloud is composed of liquid water, the uppermost portion of ice particles. The mid-regions feature a mixture of supercooled water and ice particles. (Photo by astronauts aboard the International Space Station. NASA.)

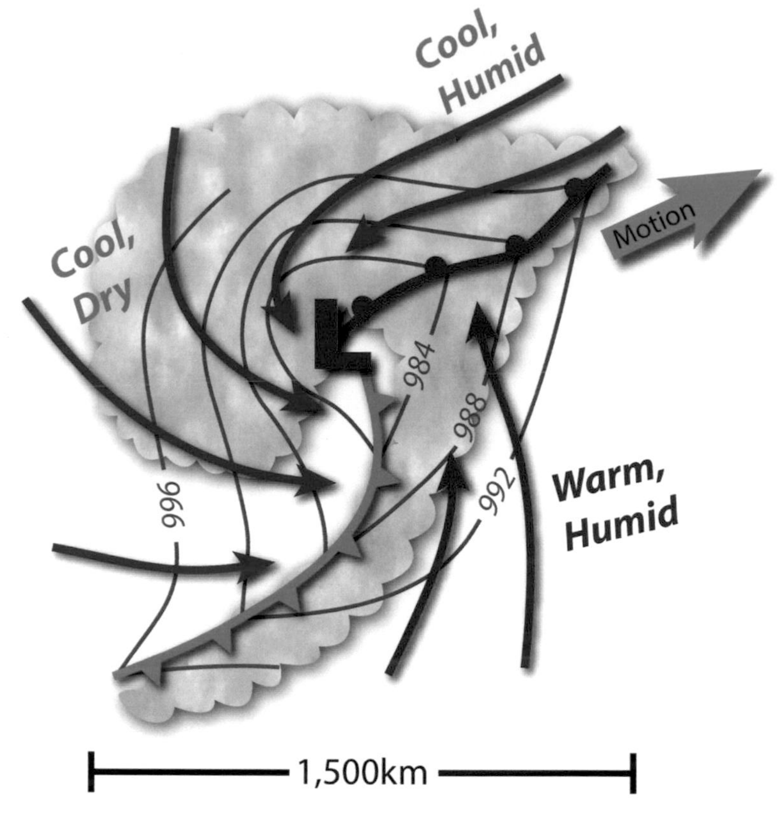

Figure 3.8 Schematic weather map of the general structure of an extratropical cyclone. The general structure of an extratropical cyclone, with types of air masses and frontal structures, is labeled. Arrows indicate airflow, and solid thin lines denote isobars (lines of constant pressure).

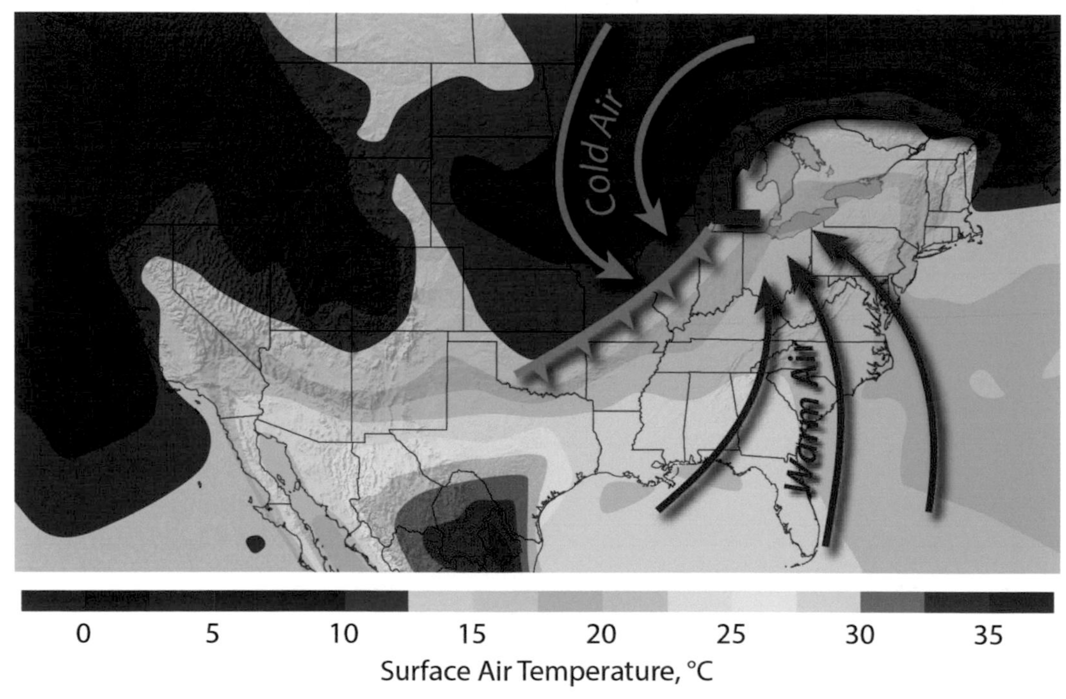

Figure 3.9 Surface temperature variations across a typical extratropical cyclone. Compare this figure with the dew point temperature variations in Figure 3.3. The contrasting temperature and humidity values derive from the different air mass source regions.

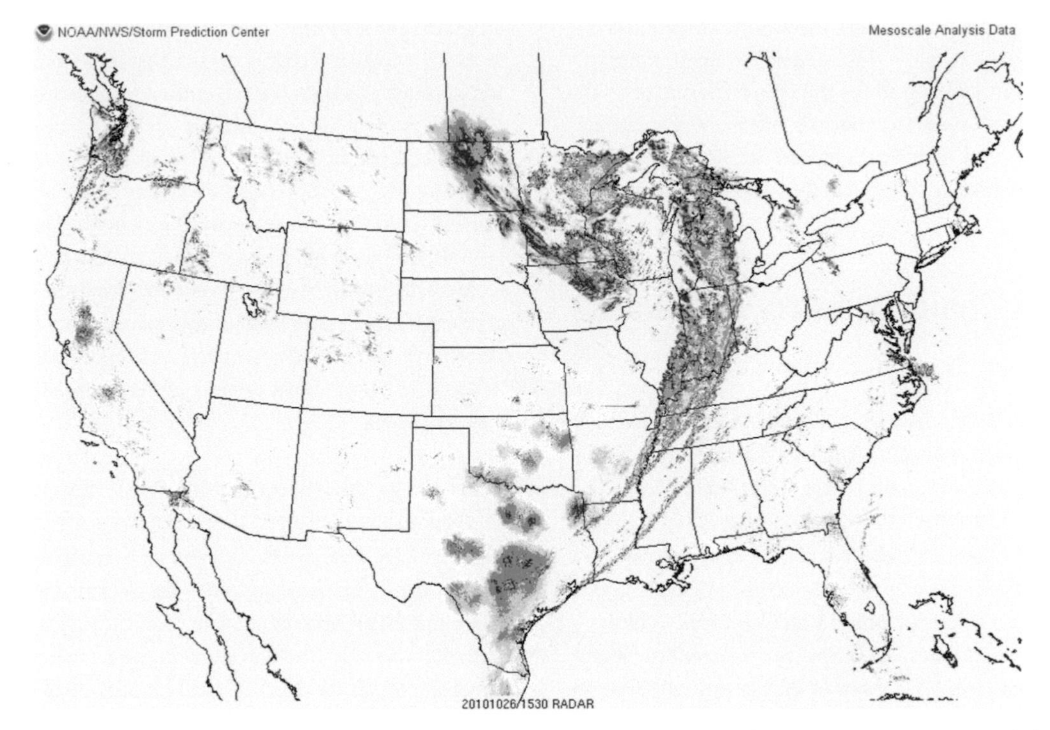

Figure 3.10 Precipitation in a mature extratropical cyclone. This weather radar shows a broad view of the precipitation structure (rain location and intensity) in a mature extratropical cyclone. Blue and green denote light rain; yellow shows moderate rain; and red indicates heavy rain. (NWS.)

Figure 3.11 Comma head. This satellite view of a mature extratropical cyclone reveals the comma shape of the cloud top structure. (NOAA.)

Finally, convergence is mainly why tropical cyclones of low-latitude origin form concentric rings and spiral bands of heavy convective showers. Lacking fronts (there is no source of cold air in the tropics – the air mass is uniformly warm and humid), the intense convergence of humid air supplies the vigorous uplift and cooling of air that develops a tropical cyclone's torrential rains.

Growth of Cloud Droplets and Rain in Clouds

Clouds are formed from cloud droplets, which are the tiniest forms of liquid water in the atmosphere, with diameters measured in microns (millionths of a meter). A microscope is needed to see cloud droplets. For cloud droplets to form, moist air must first cool enough to become slightly supersaturated. In addition, tiny particles called aerosols must be present; these act as nuclei (centers) around which water vapor molecules condense into spherical droplets. These aerosols, called cloud condensation nuclei (CCN), consist of dust grains, which might be derived from windswept soil particles or ashes from volcanic eruptions and wildfires. Additionally, sea salt particles lofted by breaking waves and chemical emissions from oceanic plankton can nucleate cloud droplets. Anthropogenic aerosols (pollutants) deriving from industrial emissions, biomass burning, and agricultural practices can also serve as cloud droplet nuclei.

The difference in volume between a cloud droplet and a raindrop is a millionfold or more. Raindrop diameters range from 1 to 6 mm. The growth of cloud droplets to raindrop size, through condensation of water vapor, takes many hours, in contrast to convective rain showers, which are often generated in as little as 15–20 minutes! What process can vastly speed up this rate of droplet growth?

Warm clouds (those above freezing temperature) undergo a kind of chain reaction, whereby the rate of droplet growth grows exponentially. The process is based on a distribution of cloud droplet sizes – that is, many small droplets mixed in with several larger ones. This size variation occurs naturally, as a result of different sizes of CCN. The larger droplets move through the cloud slightly faster than the small droplets. As a result, the larger droplets overtake and stick to the small droplets (Figure 3.12).

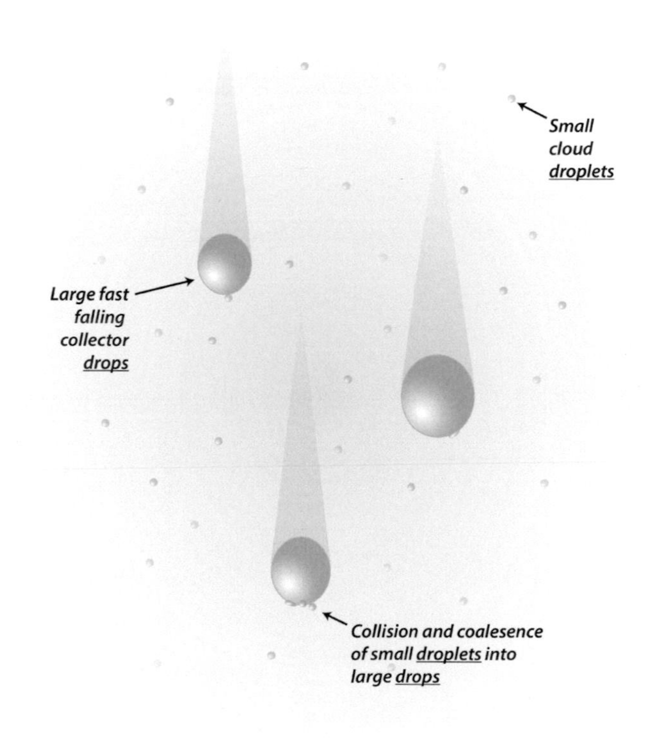

Figure 3.12 Collision-coalescence. The process of collision-coalescence is based on large falling cloud droplets overtaking and sticking to many smaller cloud droplets. The larger droplet grows at an exponential rate, quickly reaching raindrop size.

As these big droplets, called collector drops, rapidly overtake small droplets in their path, their mass (volume) rapidly increases to raindrop size. This process is the essence of the chain-reaction-type growth.

The chain reaction continues even after falling raindrops have formed. When the diameter of falling raindrops grows larger than 2–3 mm, their (relatively) large size makes them unstable as they are buffeted by air turbulence. Big raindrops vibrate and deform into shapes that resemble inverted hamburger buns, and their wobbling becomes so intense that they disintegrate into many smaller raindrops. The cascade of smaller drops thus released grows through continued collisions with smaller drops and cloud droplets.

This whole process, summarized in Figure 3.12, is called collision-coalescence (coalescence means "sticking together"), and the process is inherently unstable, leading to rapid generation of moderate to heavy rain showers in just a few tens of minutes. This process occurs as long as there is a steady supply of rising, humid air within the cloud – in other words, a continued supply of moisture that quickly cools to supersaturation. During the summer months, when the air mass is very humid and unstable (discussed later in this chapter), cumulonimbus clouds develop and generate torrential rain, partly through the collision-coalescence process (there are additional mechanisms at play, which we discuss next). These deep convective clouds are notorious for manufacturing heavy rain at the rate of several inches per hour, leading to flash floods – the number one cause of U.S. storm fatalities.

Growth of Ice Crystals and Snow in Clouds

The processes just described account for precipitation in warm, low-altitude clouds. What about deeper clouds, extending many miles (kilometers) above the Earth's surface, where temperatures are well below freezing? The formation of precipitation at subfreezing temperatures has a few surprising aspects. First, liquid cloud water will not spontaneously freeze to form ice crystals as the temperature falls below 32° F (0° C). Water may exist in a supercooled state, remaining liquid all the way down to −40° F (−40° C). Freezing of water in the atmosphere requires the presence of microscopic freezing nuclei – aerosol particles that mimic the crystalline, lattice-like configuration of frozen water. Many of these aerosols are not "activated" until temperatures fall considerably below freezing.

Clouds between the temperature of 0° C (32° F) and −20° C (−4° F) contain a mixture of ice crystals and supercooled liquid water drops. In this mixed phase condition, the tiny ice crystals tend to grow at the expense of the liquid water droplets. Vapor deposition is the process by which water vapor molecules evaporate from the surface of cloud droplets and deposit directly onto either ice nuclei or any preexisting ice crystals. Through this process, small snowflakes build up mass until they become heavy enough to fall out of the cloud. Collisions among small snowflakes enables them to stick together, forming even larger flakes in a process is called aggregation. During the cold winter months, aggregation is responsible for the large flakes of a heavy snowfall. At other times of the year, the flakes completely melt as they descend into air above freezing near the ground, striking the ground as raindrops.

When supercooled water abounds in the presence of ice crystals, the supercooled drops may immediately freeze upon contacting an ice crystal. This process, called riming, rapidly builds growing shells of frozen water, to the point where small, spherical graupel particles consisting of opaque, mushy ice develop. Riming is more common during the summer months, high up in cumulonimbus clouds, where powerful updrafts acting on very humid air deliver large quantities of supercooled water into the mixed-phase region of the cloud (0° to −20° C [32° to −4° F]).

Graupel particles serve as the building blocks of larger hailstones. The formation of hail occurs by continuous accumulation of supercooled water onto a growing, frozen sphere that remains suspended in the cloud updraft. The stronger the updraft, the longer the stones remain suspended and the larger they can grow, sometimes to the size of softballs. It may seem paradoxical that the largest pieces of precipitating ice that fall from clouds occur during the warmest season of the year! But the intense updrafts necessary to levitate large stones occur only during summer, when the air is very unstable and strongly buoyant (discussed later in this chapter).

Figure 3.13 summarizes the processes of cold-cloud precipitation formation, including vapor deposition, aggregation, and riming. It's important to recognize that a significant fraction of rain in the mid-latitudes reaches the surface via these cold-cloud processes, with subsequent melting of frozen precipitation into liquid drops close to the surface. This is even the case for summer thunderstorms. High up in the frozen reaches of cumulonimbus clouds, a veritable blizzard of heavy snow rages even when surface temperatures exceed 90° F (32° C)!

The mixture of supercooled water drops, ice crystals, and graupel in warm season convective clouds leads to cloud electrification and lightning. More detailed discussion of lightning appears in a later chapter, but cloud and precipitation particles become

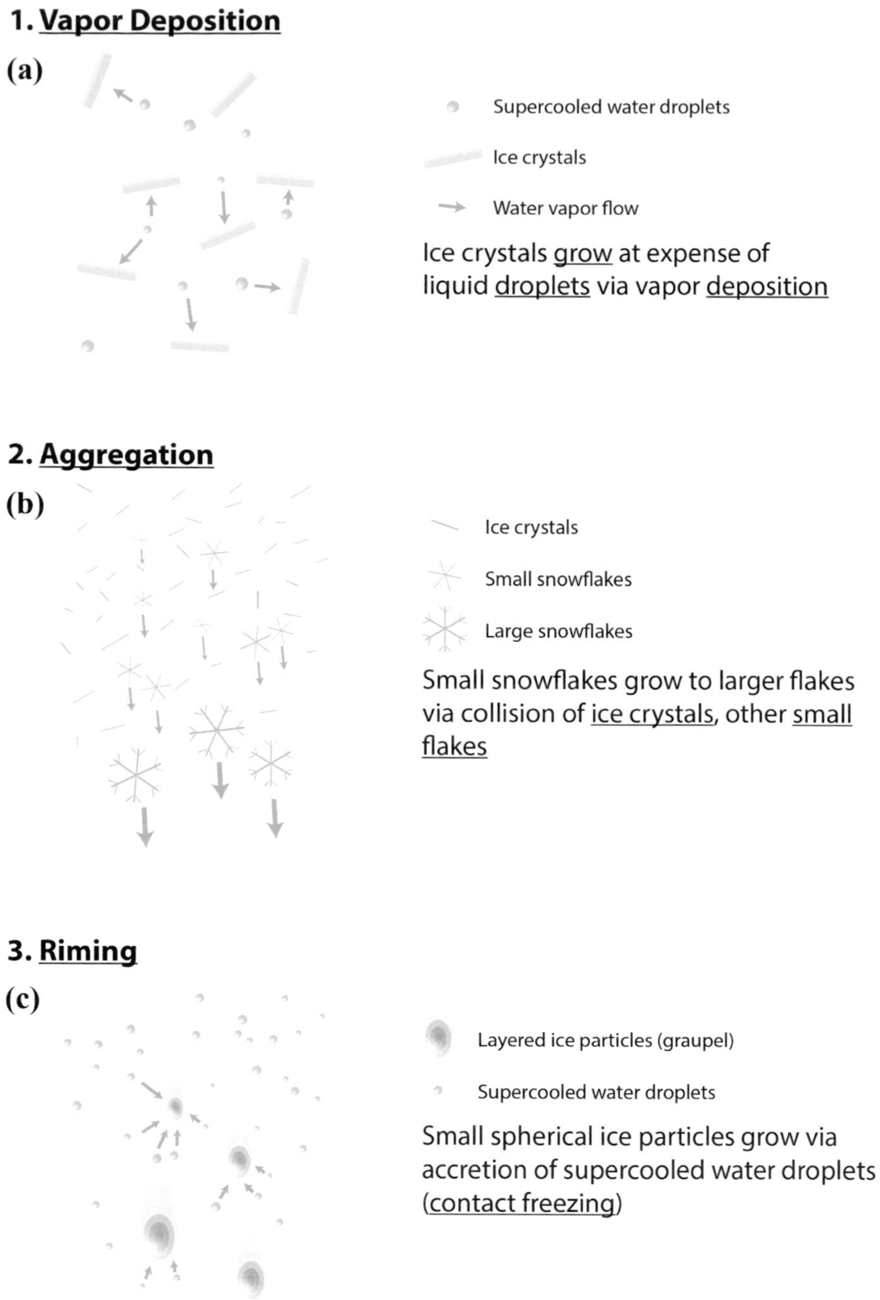

1. Vapor Deposition

(a)

Supercooled water droplets

Ice crystals

Water vapor flow

Ice crystals grow at expense of liquid droplets via vapor deposition

2. Aggregation

(b)

Ice crystals

Small snowflakes

Large snowflakes

Small snowflakes grow to larger flakes via collision of ice crystals, other small flakes

3. Riming

(c)

Layered ice particles (graupel)

Supercooled water droplets

Small spherical ice particles grow via accretion of supercooled water droplets (contact freezing)

Figure 3.13 Cold-cloud precipitation formation. Frozen precipitation can grow in clouds through a number of processes: (a) vapor deposition, (b) aggregation, and (c) riming.

electrically polarized depending on their phase (liquid vs ice) and size. Cloud updrafts and downdrafts redistribute or sort these particles by size, leading to opposing charge centers within different regions of the cloud. The process is akin to charging a battery. When the voltage (difference in charge) between negative and positive regions becomes large enough, an electrical discharge (in this case, lightning) restores electrical equilibrium within the cloud (or between cloud and ground).

Atmospheric Stability and the Formation of Thunderstorms and Layer Clouds

Everyone knows that a hot summer day frequently breeds afternoon and evening thunderstorms. Inherently, we understand that the intense summer sun strongly heats the surface, causing thermals of moist air to rise and condense into thunderclouds. Convection is the process by which excess heat energy is moved upward through the atmosphere, into the cold upper layers. Convection occurs spontaneously, as long as the atmosphere is unstable – but what exactly does "unstable" mean?

An unstable atmosphere is one in which rising air cools at a slower rate than the surrounding atmosphere. The large-scale layer of air extending many miles (kilometers) above the surface normally experiences a decrease in temperature with altitude; this lapse rate, on average, is around −6.5° C/km (−12° F/3,300 feet). But recall that rising bubbles of warm, humid air cool as they expand. This is adiabatic cooling, which occurs at a constant rate of −10° C/km (−18° F/3,300 feet). As long as the rate of cooling in rising air is *less* than that of the surrounding air, the blob of air remains warmer than its surroundings. The blob remains buoyant and continues to ascend. This condition is the hallmark of an unstable atmosphere. If the atmosphere is unstable through a great depth, up to around 15 km (50,000 feet), a deep thundercloud will develop.

Figure 3.14 shows how we can assess the degree of instability by comparing two rates of temperature change – that of the average environment and that of a bubble of air rising through the larger environment – by comparing the slopes (mathematically defined as the "rise over the run") of the two lines representing these processes.

The slope of the line representing the environmental lapse rate can be altered in two ways. Either the surface layer can warm up, or the upper atmosphere can cool. In both cases, the slope of the line gets *steeper*, but the adiabatically cooling (rising) thermal always follows the same path. The lower atmosphere commonly heats up over the course of a summer afternoon, and the upper atmosphere will cool if winds aloft move colder air over the region. Under these conditions, the atmosphere will become more unstable, as Figure 3.15 shows.

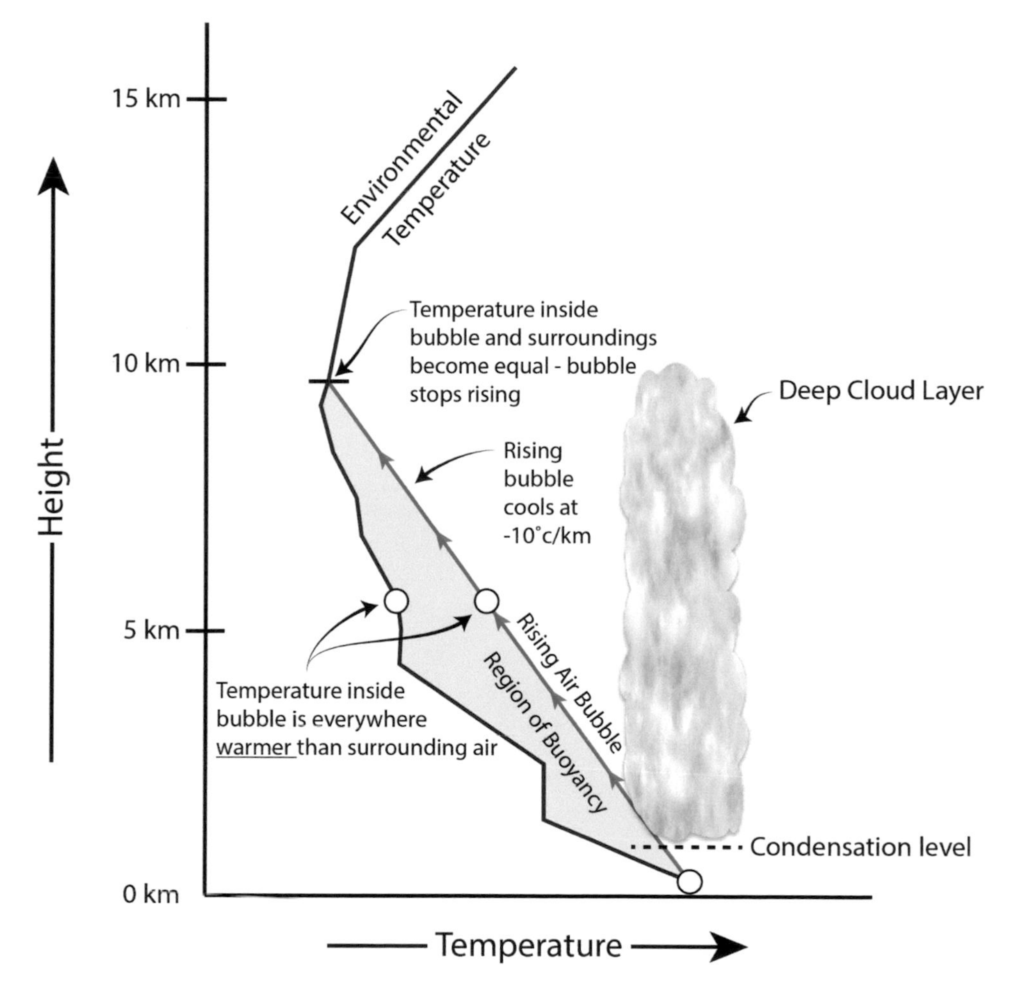

Figure 3.14 Unstable atmosphere. These illustrations compare environmental temperature decreasing at a characteristic lapse rate and temperature in a rising bubble of air cooling adiabatically. As long as the temperature inside the blob remains warmer than its environment, the bubble is buoyant and continues to rise. Under these conditions, the atmosphere is unstable.

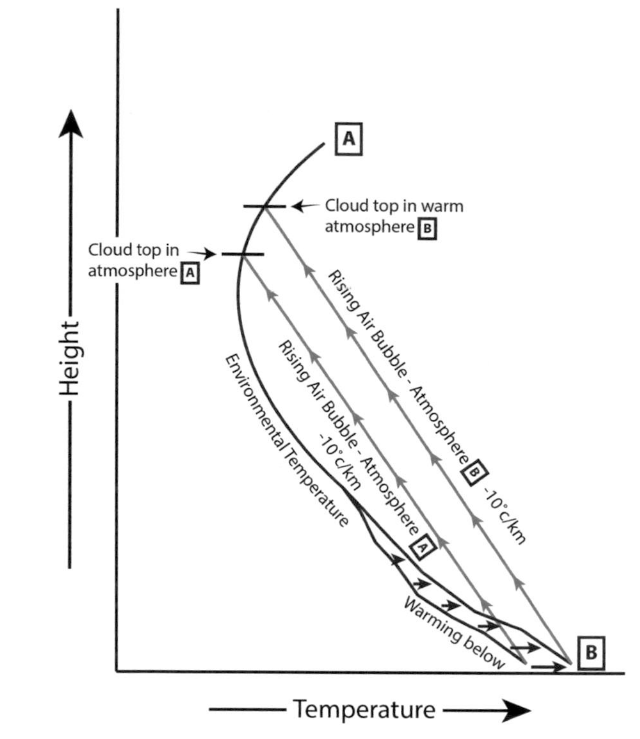

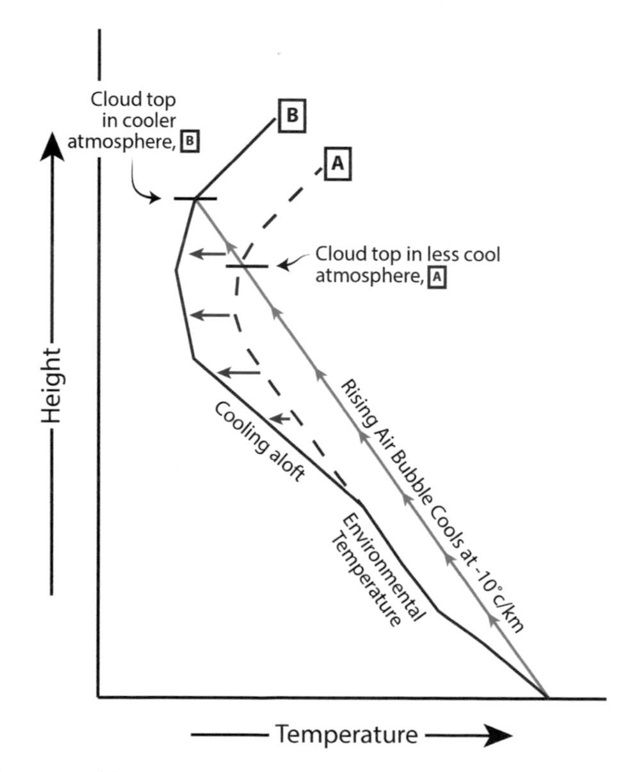

Figure 3.15 Common processes that lead to an unstable atmosphere. (a) Strong warming of the surface. (b) Cooling of the upper atmosphere.

What happens if instead the surface cools, which normally occurs overnight? The slope of the environmental lapse rate becomes less steep. Now, rising bubbles arrive at upper levels *cooler* than their surroundings. The bubbles are colder and denser and thus sink back to the surface. This is a stable atmosphere. In a stable atmosphere, air cannot rise very far on its own, if at all; there is zero buoyancy. A stable air mass cannot develop deep clouds. This is not the same as saying no clouds; cool but nearly saturated air can be *forced* to ascend the short distance needed to saturate the air and form shallow layer clouds (by orographic lifting, frontal lifting, or convergence). This is a common situation during the cool season, particularly during the winter months. Figure 3.16 illustrates the temperature profiles for a stable atmosphere and the formation of horizontal layer clouds.

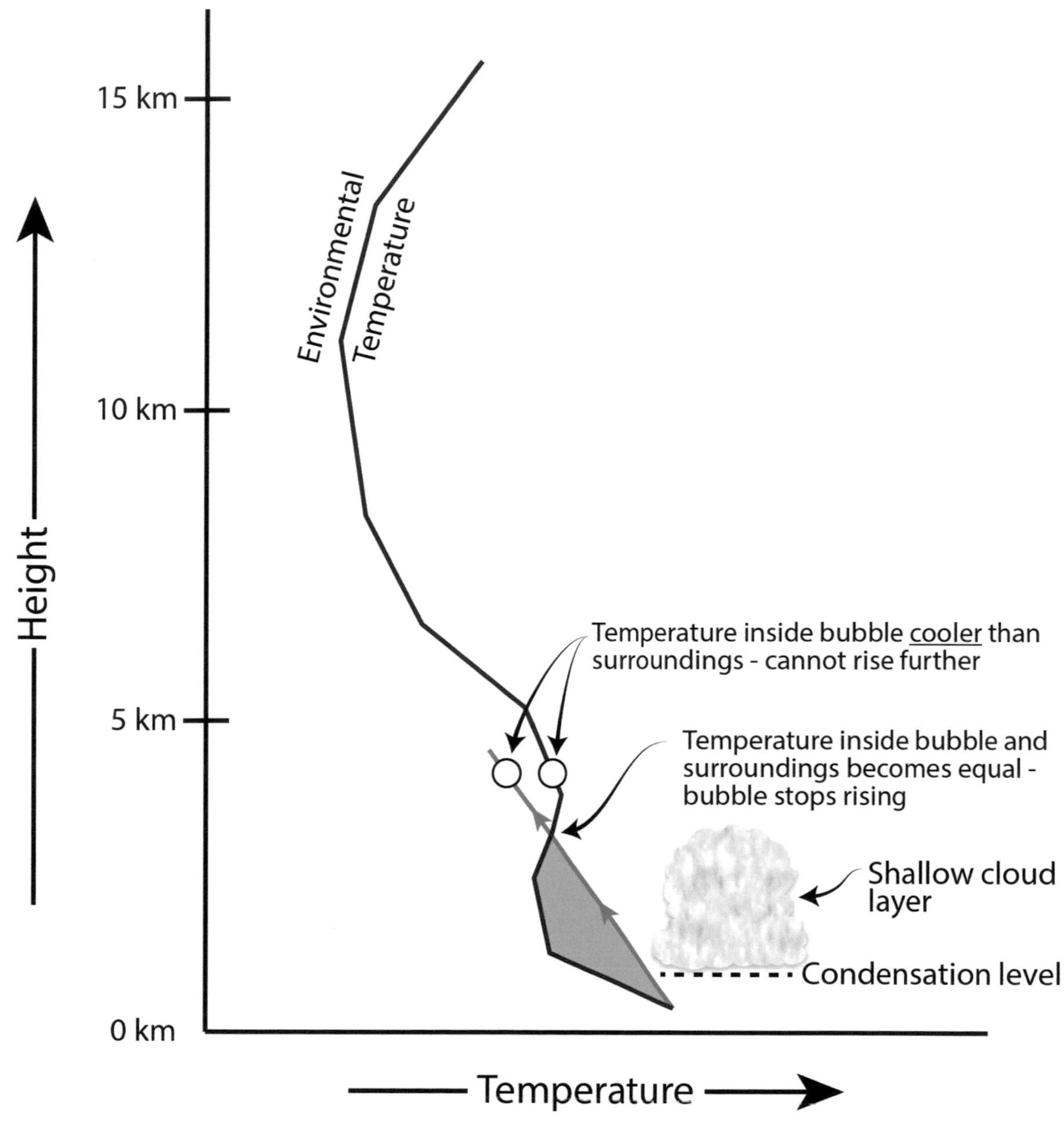

Figure 3.16 Stable atmosphere. Environmental temperature decreasing at a characteristic lapse rate, and temperature in a rising bubble of air cooling adiabatically at various altitudes. When the temperature inside the blob becomes cooler than its environment, the bubble sinks back toward the surface, inhibiting deep cloud growth. Under these conditions, the atmosphere is stable.

For the sake of simplicity, the diagrams in Figures 3.14–3.16 omit part of the more complete story, which we cannot ignore. When water vapor condenses to form rain, heat energy is released during the water's phase change. Rapidly vibrating water vapor molecules possess kinetic energy, a form of heat. When the air containing the vapor cools, vapor molecules lose heat and slow down, enabling them to coalesce into liquid drops. The heat that escapes into the surrounding air is termed latent heat. The heat is "hidden" or latent, not becoming manifest until the water changes phase. As Figure 3.17 shows, there are six forms of latent heat because water can undergo a total of six phase changes.

Both the condensation of vapor into liquid and the freezing of liquid into ice are associated with the release of latent heat. This heat warms the rising bubbles of air, such that the adiabatic rate of cooling (due to expansion of air) must be *reduced*. For a bubble of rising air, the adiabatic lapse rate reduces to approximately −6° C/km (−11° F/3,300 feet). In other words, some of the adiabatic cooling is offset by the steady release of latent heat.

So it's a simple matter to reassess the definition of an unstable atmosphere, accounting for this more complete understanding of the processes involved. Figure 3.18 illustrates a deep, unstable atmosphere allowing for the release of latent heat inside the growing convective cloud. Figure 3.19 diagrams the process for a stable air mass and layered clouds.

To summarize, the summer months frequently develop an unstable atmosphere, one that enables the formation of deep cumulonimbus clouds and severe local storms. The winter months, in contrast, feature a stable atmosphere permitting only shallow clouds. Rain or snow falls from widespread layer clouds called nimbostratus. Compared to thunderstorms, the vertical currents in nimbostratus are very gentle.

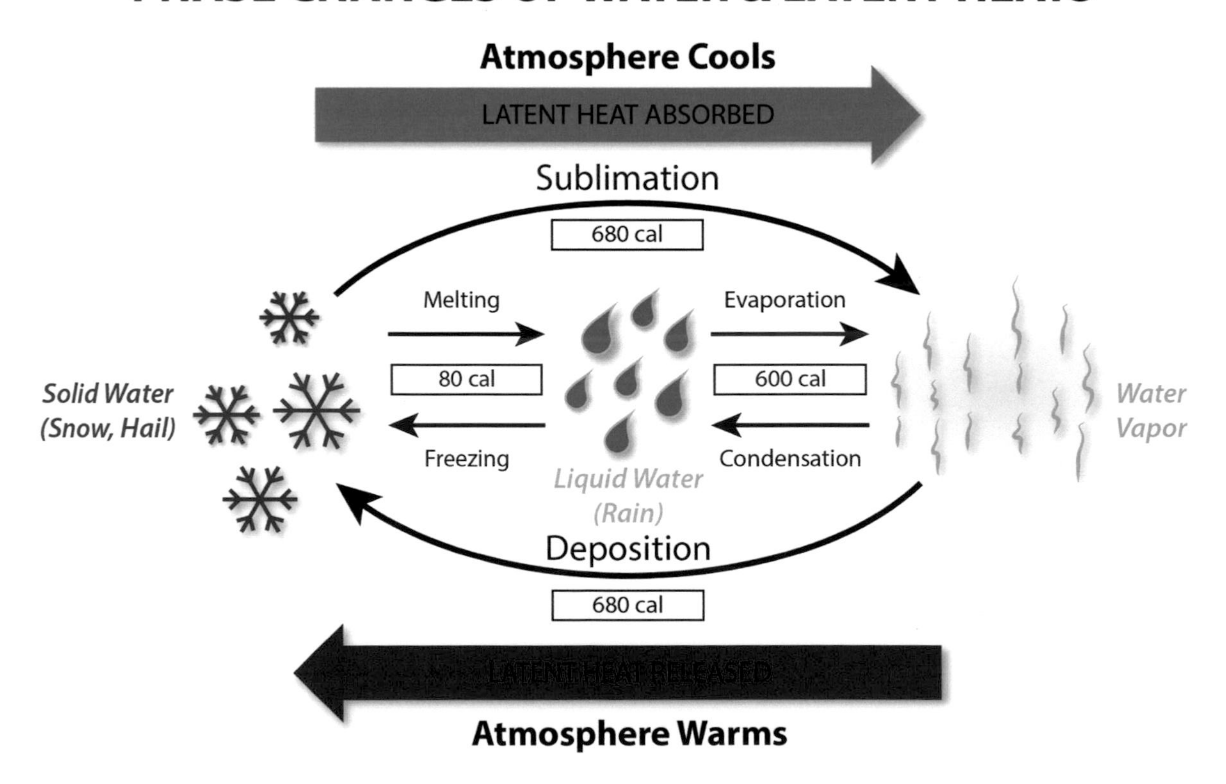

Figure 3.17 Six forms of latent heat—the latent heat associated with the six-phase changes of water. Heat is absorbed (the air cools), when ice melts and liquid water evaporates. Heat is released (the air warms), when vapor condenses and liquid water freezes.

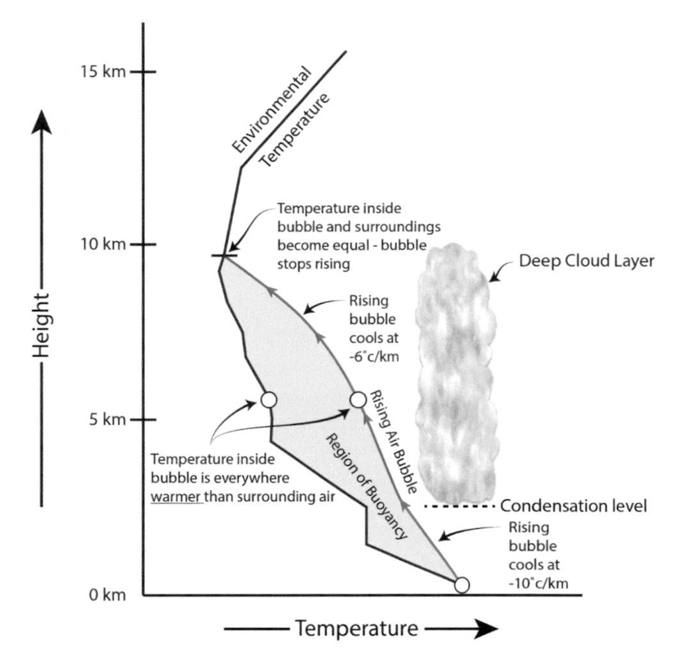

Figure 3.18 Latent heat and deep cloud formation in an unstable atmosphere. The processes leading to deep cloud formation in an unstable atmosphere, taking into account the release of latent heat during condensation.

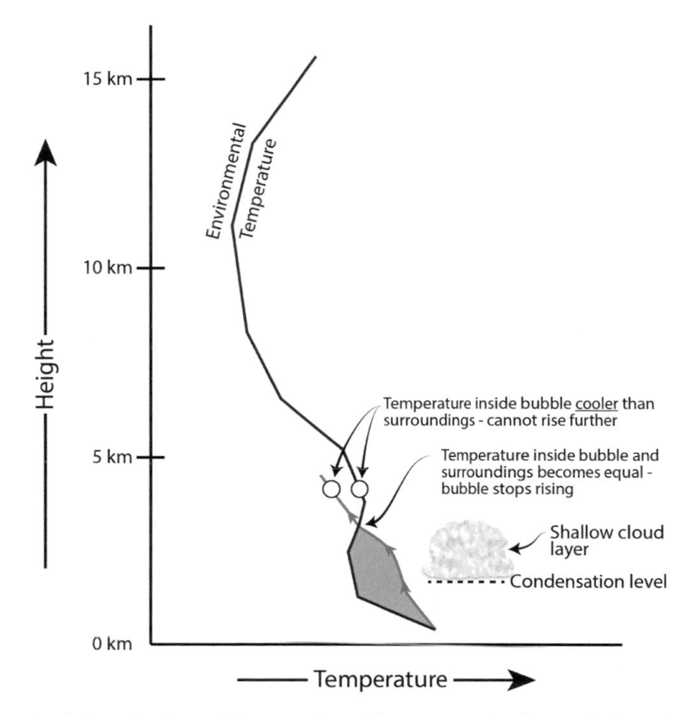

Figure 3.19 Latent heat and shallow cloud formation in a stable atmosphere. The processes leading to shallow cloud formation in a stable atmosphere, taking into account the release of latent heat during condensation.

Common Precipitation Types in Summer and Winter Settings

In this section, we examine common types of precipitation during the warm and cool seasons across the United States and their association with typical weather systems. We discuss general precipitation patterns, which can develop into the severe local storms discussed throughout this text.

Precipitation Associated With Extratropical Cyclones

Figure 3.20 presents a conceptual model of an extratropical cyclone during the November–December time frame in the United States. Recall that these large regions of low pressure are associated with the widespread uplift of air and weather fronts. Clouds and precipitation develop along the fronts and near the center of the extratropical cyclone, where air rises most vigorously. This air is cooled to its dew point temperature, leading to saturation, cloud development, and the formation of both liquid and solid precipitation.

The triangular-shaped region on the south and east side of the extratropical cyclone is the warm sector. Within the warm sector, low-level airflow streams from the south, causing temperatures and humidity to rise. The warm, humid air mass comes from tropical latitudes. Because the air warms strongly at low levels, the atmosphere becomes unstable. The warm sector is thus conducive to convective showers and thunderstorms. The showers and storms may organize into one or more bands in which rainfall is enhanced. A particularly prominent and intense band develops along the cold front. The cold frontal band may develop along the front, immediately behind the front, or up to several hundred km (miles) in advance of the front. Heavy rain sometimes develops along a narrow corridor of strong, low-level southerly flow, called a warm conveyor belt. Very humid air in this corridor can generate prolonged heavy rain and flash flooding.

Figure 3.21 shows examples of these rain patterns using weather radar images. If surface temperatures fall below freezing (for example, in January–February), heavy snow showers develop along the cold front, leading to snow squalls. During the spring, the warm sector becomes a place where violent weather is born. Severe thunderstorms may develop, accompanied by large hail, high winds, and tornado outbreaks.

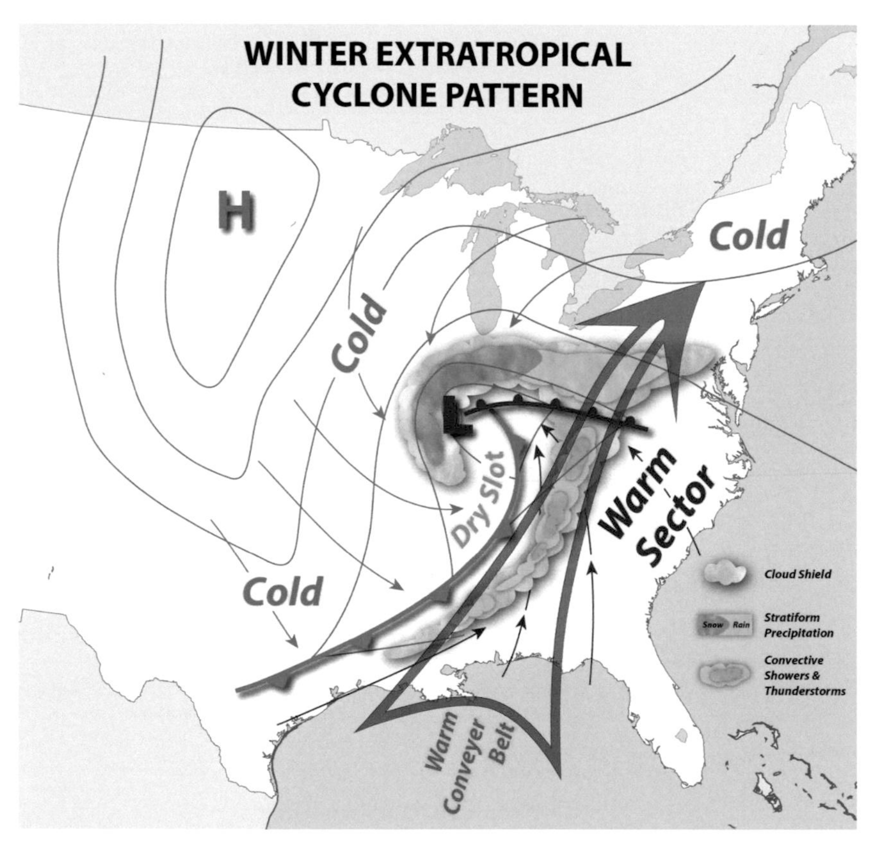

Figure 3.20 Key features of a mature extratropical cyclone during the late fall or early spring. Thin arrows represent general airflow. Thick arrows show the warm conveyor belt and dry slot. The cold front and warm front are also indicated, along with general regions of precipitation.

Another significant region of precipitation lies along and north of the warm front. Along this front, air from the warm sector glides above a layer of colder, denser air north of the front. The cold and humid air mass approaches the center of the extratropical cyclone as a low-level, easterly air current. Because cool surface air lies beneath warm air aloft, the atmosphere is stable and resists the strong, buoyant air motions associated with convective storms. However, the forced ascent of air up the incline or "wedge" of cold air does lift the air until it cools to saturation. Horizontally extensive layers of nimbostratus cloud develop along and north of the warm front. Widespread gentle to moderate precipitation, called stratiform precipitation, falls from these clouds. This type of rain is unlikely to lead to flooding. During winter, with subfreezing air beneath the cloud deck, various forms of frozen precipitation develop, including snow, sleet, and freezing rain. Extensive layers of nonprecipitating cloud often extend far to the north and east of the stratiform precipitation shield, leading to overcast sky conditions many hours before the precipitation's arrival. Figure 3.22 shows a radar depiction of stratiform rain contained within a larger extratropical cyclone.

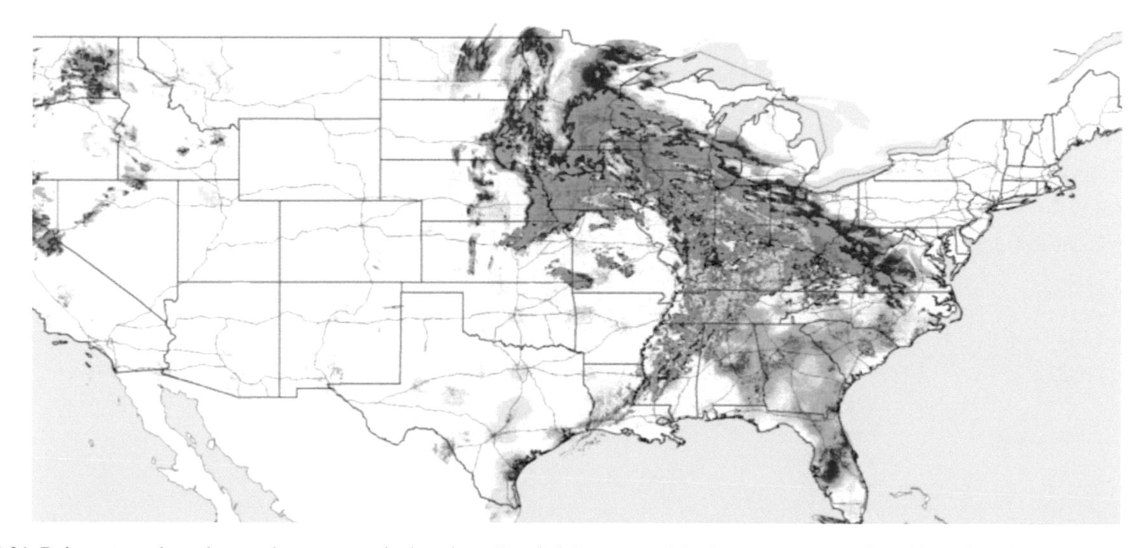

Figure 3.21 Rain patterns in and around an extratropical cyclone. Banded, heavy precipitation structures are found in various locations in and around an extratropical cyclone, as this weather radar image illustrates. (NWS.)

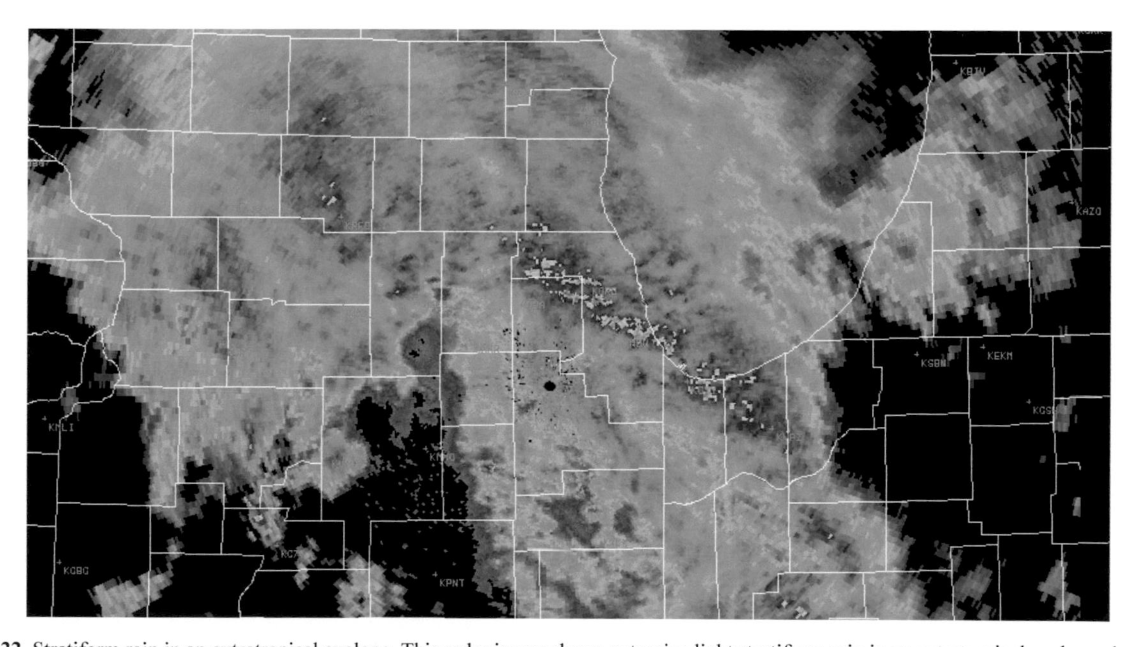

Figure 3.22 Stratiform rain in an extratropical cyclone. This radar image shows extensive light stratiform rain in an extratropical cyclone along with some banding due to heavier (convective) clouds (yellow tones along southern Lake Michigan). (NWS.)

Wrap-around clouds and precipitation usually extend to the north and west on the cold air side of the extratropical cyclone. Here, cold, dry, low-level winds stream down from the north. This air near the surface is often too dry to support precipitation, but residual moisture entering the storm from aloft and from the east is sufficient to saturate the cold air. In subfreezing conditions, one or more bands of heavy snow develop about 240–320 km (150–200) miles to the north and west of the extratropical cyclone center.

At times, banded precipitation also develops within the large mass of stratiform precipitation. These narrow bands arise if small pockets of mildly unstable air develop, and they can lead to brief periods of heavier precipitation. The embedded bands are most often arranged parallel to the main warm front.

As Figure 3.20 shows, a cloud-free zone called the dry slot often develops behind the cold front and comma head cloud structure. The dry slot is caused by very dry air sinking and warming adiabatically from high altitudes. The warm, dry air strongly suppresses any tendency for uplift, cooling, and saturation. At the surface, immediately after the cold front passes, rain tapers, clouds clear, and the winds kick in from the west-northwest. The air also abruptly cools and dries as Canadian air begins to arrive from the northwest.

Precipitation Associated With Summertime Convective Storms

During the summer months in the United States, the polar jet stream weakens and retreats to the north, over southern Canada. Extratropical cyclones are fewer in number and not as intense as during the cool season, and they track along the northern tier of the United States, impacting the Upper Midwest, Great Lakes, and New England. Much of eastern United States comes under the influence of a broad dome of high pressure called the Bermuda High, which also extends across the entire western and central Atlantic at low latitudes (Figure 3.23).

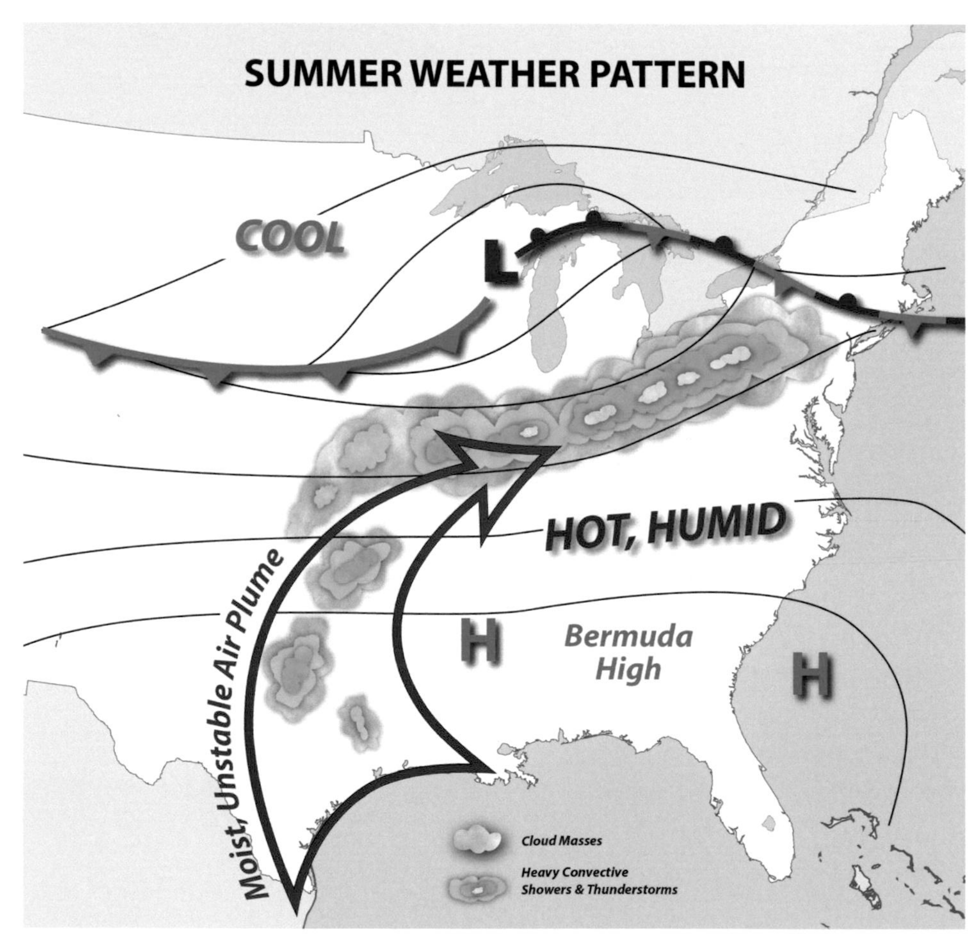

Figure 3.23 Principal summertime U.S. surface weather systems. This schematic illustrates principal surface weather systems on a typical June or July day across the United States, including the large Bermuda High and weak extratropical cyclones moving west to east across southern Canada.

Within the core of this high-pressure cell, air sinks for many miles, leading to a dry, warm, and stable air mass across its center. But on the western edge of the Bermuda High, low-level air spirals outward toward the north and west. The lower levels of this southerly current are warmed and humidified by the Caribbean Sea and Gulf of Mexico, while the hot land surface further heats the air. A strongly unstable air mass develops over the entire central and eastern United States. Localized, heavy convective showers and thunderstorms become the prevailing weather maker, leading to a hit-or-miss variety of rain (Figure 3.24).

Prospects for stratiform rain are greater along the northern tier of the United States, associated with the passage of multiple, weak extratropical cyclones. Infrequently, the cold front from one of these extratropical cyclones will extend southward into the Central Plains and Mid-Atlantic, triggering a narrow band of strong or even severe thunderstorms.

Summertime thunderstorms can still organize into intense clusters and lines by other processes. The largest of these systems produces copious rain and flooding across the Midwest and Ohio Valley, especially at night. This type of storm, called a mesoscale convective complex, is illustrated in Figure 3.25. Another type of especially violent convective storm system called a derecho (Figure 3.26) commonly develops east of the Rockies, bringing widespread wind damage across regions up to 1,610 km (1000 miles long).

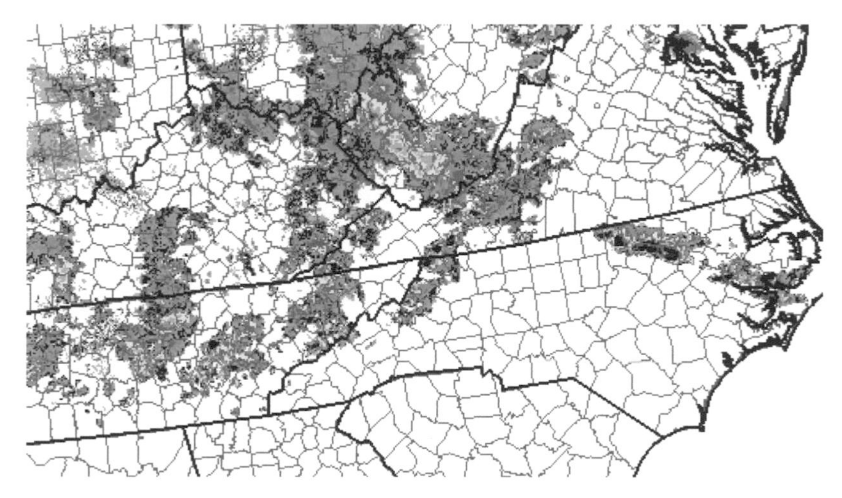

Figure 3.24 Convective shower and thunderstorm activity across the United States on a typical summer day. Contrast this figure with Figure 3.22, which shows typical precipitation patterns associated with a cool season weather system. (NWS.)

Figure 3.25 Mesoscale convective complex (MCC). This satellite image shows a typical Midwestern mesoscale convective complex (MCC). This enormous aggregate of thunderstorms frequently produces widespread flooding and other forms of severe weather during the summer. (NOAA.)

In the western United States, summertime convective storms are more limited, given the lower prevailing dew point values. But the high, mountainous terrain triggers afternoon thunderstorm growth, as heated surface air rises along mountain slopes. This orographic convection (or mountain-induced convection) leads to spotty rain. Occasionally severe thunderstorms erupt along Colorado's Front Range and the U.S. Southwest, associated with a monsoon rain pattern that develops from June through August. Figure 3.27 provides a radar snapshot of orographic summer rains.

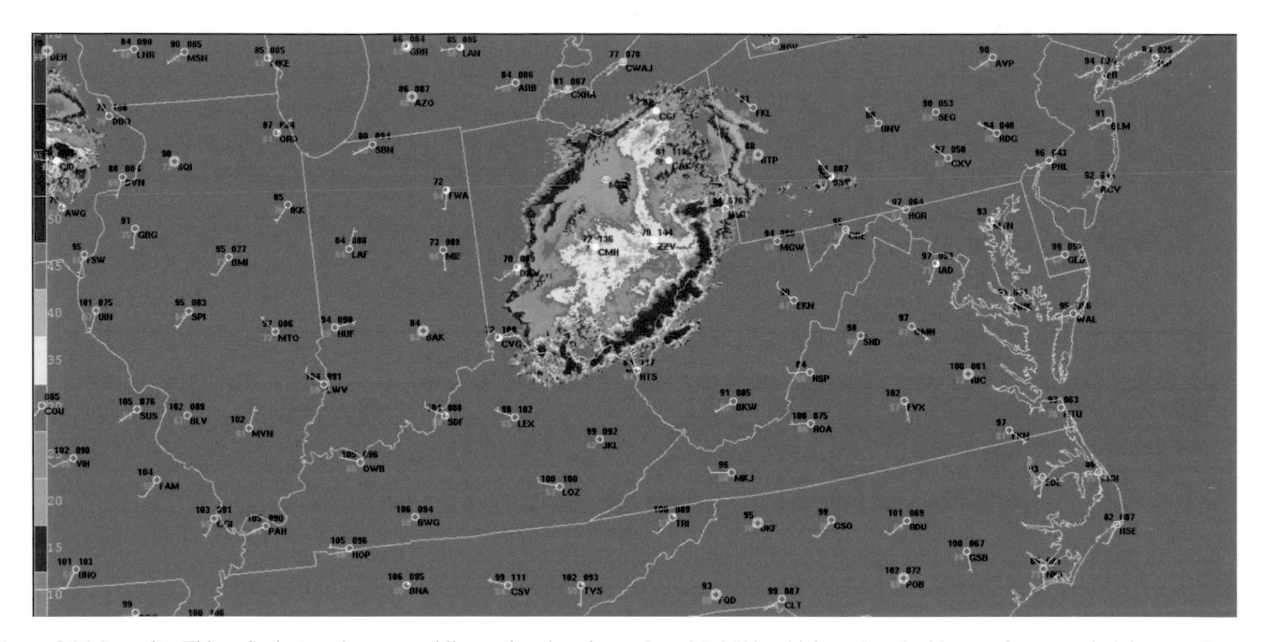

Figure 3.26 Derecho. This radar image shows a rapidly moving derecho on June 30, 2012, which produced widespread severe wind damage along a 1000 mile long corridor, knocking out power to nearly 5 million customers. (NWS.)

Figure 3.27 Orographic summer rains. This satellite image shows summer thunderstorms over Palomar Mountain, California, and vicinity. Prevailing winds force warm and humid air to rise along mountain slopes. (NWS.)

Summary

LO1 Distinguish among humidity, dew point temperature, and relative humidity, and describe how these values typically vary with time of day, by season, and across geographic regions.

1. Humidity refers to any measure of atmospheric water vapor content. Dew point temperature describes the degree of cooling needed to fully saturate the air. Relative humidity is a ratio that reflects the percentage saturation of air, for a given temperature and water vapor content.
2. Dew point temperatures are uniformly higher across the United States during summer than they are during winter. Within each season, the dew point tends to be lower (drier air) over the continental interior, away from oceans, and at high altitude.
3. Relative humidity varies strongly with air temperature, being lowest during the afternoon and highest during early morning, as long as water vapor content is held constant.

LO2 Discuss two ways in which air becomes saturated, and describe adiabatic cooling of air.

1. Air can be saturated by increasing the evaporation rate at constant temperature or by cooling the air at constant dew point temperature.
2. Air most commonly cools to saturation by rising and undergoing an adiabatic expansion.
3. During an adiabatic expansion, the internal energy and hence temperature of the air decreases, cooling at a rate of 14° F/0.6 mile (−10° C/km) of ascent.
4. Air rises by four main mechanisms: (1) flowing up a mountain slope, (2) lifting along a front, (3) converging and rising into a low-pressure center, and (4) rising freely due to its internal buoyancy.

LO3 Explain how supersaturation, cloud condensation nuclei, and collision-coalescence lead to growth of cloud droplets.

1. Condensation of cloud water involves the aggregation of water vapor molecules around a microscopic condensation nucleus suspended in air, often a grain of dust or an ash particle.
2. Condensation also requires that the air be cooled to the point of slight supersaturation.
3. Raindrops are over a million times more massive than cloud droplets and require an efficient growth process occurring in as little as 15–20 minutes.
4. The process of collision-coalescence enables an exponentially increasing rate of cloud droplet growth to the size of raindrops.

LO4 Define supercooled water, and explain how it relates to the formation of small ice crystals in subfreezing clouds; distinguish between vapor deposition and riming.

1. Liquid cloud droplets do not spontaneously freeze. Rather, they remain as supercooled liquid droplets and require a microscopic ice nucleus upon which to freeze into ice crystals.
2. Small ice crystals grow by direct deposition of vapor, at the expense of liquid water droplets in a cloud. Ice crystals, in turn, aggregate into larger crystal clumps called snowflakes.
3. Small supercooled water drops in a cloud may spontaneously freeze into large ice particles upon colliding with an ice crystal, a process called riming.

LO5 Explain why an unstable air mass leads to rapid growth of the deep convective clouds associated with heavy rain showers and hail.

1. An unstable air mass, typical of summertime, occurs when a rising bubble of air cools at a slower rate than the surrounding air, such that the bubble remains buoyant.
2. An unstable air mass permits growth of convective clouds to deep altitudes, forming thunderstorm (cumulonimbus) clouds.
3. The large amounts of humid air entering into a thunderstorm cloud permit large amounts of condensed water to form, leading to heavy but localized rain showers.
4. A stable air mass is one that suppresses deeply rising air. In a stable atmosphere, cloud formation requires air to be forcefully lifted upward a short distance, resulting in thin but widespread stratiform clouds.

LO6 Describe how regions of precipitation are developed and arranged in a typical extratropical cyclone (cool season weather pattern) and during the summer months.

1. During the winter months, extratropical cyclones create widespread, forced uplift of relatively stable but moist air, leading to extensive layer clouds and regions of precipitation (including snow and ice when the surface air is below freezing).
2. The warm sector of an extratropical cyclone may feature unstable air, with bands of convective showers and thunderstorms, particularly during the spring months – leading to outbreaks of severe weather (e.g., tornadoes).
3. The summertime weather pattern over the eastern United States is mainly dominated by the large Bermuda High, featuring unstable air with hit-or-miss, heavy convective rains. The mountainous West also features an unstable air mass with thunderstorms controlled by warm air rising up steep mountain slopes.

Key Scientific Principles Covered in This Chapter

Note to the student: Many of these "first principles" are important in later chapters, as a reminder of an important physical process or definition, to aid in your understanding of more advanced material.

3.1 Atmospheric humidity is any measure of the amount, concentration, ratio, or percentage of water vapor in the air.

3.2 Evaporation is a change in phase of water from liquid to vapor, requiring the input of heat (latent heat of evaporation) from a free water surface, or the air. Evaporation varies as a function of temperature, in an exponential manner; that is, warmer temperatures can evaporate vastly larger amounts of vapor than cooler air.

3.3 Transpiration is the diffusion of water vapor into the air from small openings in leaf surfaces from grass, trees, corn stalks, etc.

3.4 Air becomes saturated when the maximum amount of water evaporates into the air, from a free water surface; any further increase in vapor mass will lead to condensation.

3.5 Air can become saturated in two ways: (1) cool the air, while holding the amount of water vapor constant, and (2) evaporate additional vapor into the air, while holding air temperature constant.

3.6 The dew point temperature is that temperature at which air must be cooled in order to reach saturation. Dew point temperature is proportional to the amount of vapor mass in the air.

3.7 The dew point temperature commonly decreases with increasing latitude and altitude across the United States and reaches higher values during the summer rather than during winter. It is largest along the Gulf Coast states during summer and lowest at high elevations in the Desert Southwest during winter.

3.8 The relative humidity is a percentage expressing how close the air is to saturation. It is the ratio of the amount of vapor in the air, to the maximum amount of vapor if the air were saturated. Relative humidity depends on *both* vapor content and air temperature.

3.9 Throughout a 24-hour period, relative humidity varies strongly with the change in air temperature (assuming constant values of vapor content); it reaches its highest value in the early morning, when the air is coolest, and its lowest value during the late afternoon high temperatures.

3.10 The most common way that air cools to saturation involves the ascent, expansion, and adiabatic cooling of a humid air parcel. Adiabatic cooling comes about as an expanding parcel loses internal energy. The parcel cools at a rate of 10° C/km of ascent (as long as it remains unsaturated).

3.11 Four ways that moist air can be lifted to saturation include isolated rising bubbles or thermals, orographic lift (forced ascent up a mountain slope), frontal uplift, and convergence of air into a region of low pressure.

3.12 Air that has been lifted becomes slightly supersaturated before the first condensed cloud droplets form, on the order of 100.1–100.5%.

3.13 Condensation describes the phase change of water vapor, in supersaturated air, to liquid cloud droplets. The surrounding air is warmed by the release of the latent heat of condensation.

3.14 The formation of microscopic cloud (or fog) droplets requires a slight amount of supersaturation and the presence of cloud condensation nuclei, which serve as tiny "staring points" upon which a small sphere of liquid water can condense.

3.15 Growth of microscopic cloud droplets to macroscopic, precipitating size occurs rapidly by a chain reaction (unstable) type of process within the cloud, termed collision-coalescence (warm cloud process).

3.16 Cloud droplets and raindrops commonly exist in a supercooled state, remaining liquid down to temperatures of −40° C (−40 °F).

3.17 Between temperatures of 0° C (32 °F) and −40° C (−40 °F), thick clouds contain a mixture of liquid (supercooled) and frozen forms of precipitation; the layer is termed the mixed-phase region, and it plays a critical role in the electrification of thunderclouds and the formation of hail inside thunderstorms.

3.18 In the very cold, upper regions of a summertime thundercloud, or in the shallow, layered clouds of winter, microscopic ice crystals in subfreezing air develop when vapor deposits directly onto microscopic ice condensation nuclei. This process is termed vapor deposition.

3.19 Snowflakes in a subfreezing cloud commonly develop when ice crystals aggregate or clump together, while being jostled by turbulent air motions, settling through the cloud or being lifted by rising air.

3.20 When a mixed phase region of the cloud contains abundant supercooled water, small, conical ice particles called graupel develop from a process called riming (accretion and instant freezing of liquid water onto a tiny ice grain). Continued growth of graupel to hailstones occurs when a strong cloud updraft levitates ice particles within the mixed phase region for long periods of time.

3.21 The cold cloud processes in subfreezing clouds, described by vapor deposition, aggregation, and riming, lead to a significant production of rain in cloud systems during both summer and winter; the rain forms as descending ice particles (crystals, snowflakes, graupel, hail) that melt upon encountering a warm air layer above the surface.

3.22 Air mass stability describes the tendency for a deep air mass to spontaneously "overturn" – that is, the tendency for parcels of warm air to rise away from the surface, and keep rising, through a deep layer.

3.23 An unstable atmosphere is one in which parcels of warm air can rise to high levels, promoting the formation of convective showers and thunderstorms.

3.24 Convection describes vertical currents of air, in which a buoyant thermal or updraft rises through a deep layer, and a downdraft of cooler air sinks back toward the surface. A thunderstorm or cumulonimbus cloud is the deepest, most vigorous form of convection in an unstable atmosphere.

3.25 Convective rain describes heavy showers, or cloudbursts, generated by deep convective clouds and thunderstorms. This type of rain tends to be spotty (isolated) and intense, lasting only for brief periods (tens of minutes), and is characteristic of summertime weather systems across the United States.

3.26 A stable atmosphere is one in which surface-based air parcels lack buoyancy to rise spontaneously; moist parcels may be forced upward, for short distances, reach saturation, and form extensive layers of shallow cloud.

3.27 Stratiform precipitation describes rain or snow falling from stable, horizontally oriented cloud layers (nimbostratus) that cover widespread areas. The precipitation tends to be moderate or light and often persists for many hours.

3.28 Atmospheric stability is assessed from weather balloon data. The vertical change of air temperature, called the atmospheric lapse rate, is compared to the adiabatic lapse rate of a rising air parcel, at many levels.

3.29 To more completely describe whether a rising air parcel is stable or unstable, the adiabatic lapse rate must be adjusted once the parcel achieves saturation. Above the saturation level, the adiabatic lapse rate changes to −6° C/km (−10.8 °F/3,300 m on average) because the cooling due to expansion is partly compensated by the release of latent heat of condensation.

3.30 The atmosphere commonly destabilizes (becomes more unstable through a deep air layer) from two processes: (1) strong surface heating from the Sun (mid-late afternoon) or arrival of a warm air mass near the surface; and/or (2) strong cooling or arrival of a cold air mass in the upper air layers.

3.31 During the cool season (October–April), extratropical cyclones are the dominant precipitation-producing weather systems over much of the United States. These large, traveling storms in a stable atmosphere create regions of stratiform precipitation, often heaviest along fronts.

3.32 Cloud and precipitation features of a typical extratropical cyclone, as seen from satellite or weather radar, include the comma head, cold frontal rain band, dry slot, and warm frontal rain shield. Heavy rains and/or thunderstorms occasionally develop in the more unstable warm sector of the storm system, ahead of the cold front, and are fed by a deep river of moist air termed the warm conveyor belt.

3.33 During the warm season (May–September), summertime precipitation-generating weather systems are more convective in nature, developing in an unstable atmosphere. The Bermuda High, a subtropical anticyclone located across the Atlantic Ocean, pumps high-dewpoint air across the eastern two-thirds of the United States.

3.34 Convective, flash-flood generating complexes of summertime include mesoscale convective complexes (MCCs), derechos, squall lines, and clusters of orographic thunderstorms; many of these systems also generate other forms of severe weather, including tornadoes, damaging wind gusts, and large hail.

CHAPTER 4
Meteorological Observations and Forecasting

Learning Objectives

1. Explain the various networks that make surface observations of weather phenomena, and list the basic types of information portrayed on the synoptic weather chart.
2. Describe the instruments that are used to create upper-air analysis charts and the types of information that these charts portray.
3. Distinguish between geostationary satellites and polar orbiting satellites, and discuss the types of information about cloud systems revealed by visible, infrared, and microwave wavelengths.
4. Discuss how weather radar identifies the location and intensity of precipitation and lightning.
5. Explain the basic principles behind weather forecast models, including the different types of models and how ensemble forecasts can improve model certainty.
6. Describe the basic functions of NOAA's Centers for Environmental Prediction and the network of Weather Service Forecast Offices, explaining the difference between a watch and a warning.

Introduction

No doubt you have seen weather charts on the evening news, in the newspaper, or online. The most common type of weather chart is the surface analysis chart or synoptic weather map, which is a snapshot of major pressure systems, fronts, and precipitation across the United States or North America. Have you ever wondered where the data used to create these charts comes from? In this chapter, we explore the common instrument systems used to analyze weather patterns, and not only those limited to the surface.

Across the continental United States, the National Weather Service (NWS) and other agencies maintain a network of approximately 900 surface weather stations. Many of these are located at major airports, and most are completely automated. In regions where severe thunderstorms are common, denser networks of surface stations provide detailed, minute-by-minute updates of observations at the resolution of individual counties. Also, a network of approximately 100 upper-air observing stations across the United States tracks GPS weather balloons (called radiosondes) twice a day and more often when severe weather threatens. The balloon data, combined with other datasets provided by satellites and wind profilers, form the basis of upper-air analysis charts used to study wind systems and temperature patterns leading to outbreaks of severe weather.

These datasets provide a heavy flow of data streaming into analysis centers and forecast offices across the country. Powerful computers quickly analyze the data and input them into a sophisticated array of numerical weather prediction (NWP) models. These models are now the mainstay of modern weather forecasting, and their output is freely available on the Internet. In this chapter, we discuss how these models generate weather forecasts, the major types of models, and some of their limitations. We conclude by looking at the forecast process used by the National Oceanographic and Atmospheric Administration (NOAA), with an overview of specialty forecast centers such as the Storm Prediction Center and National Hurricane Center.

Surface Observations and Analysis Charts

Each surface weather station, operated solely by the National Weather Service or jointly with the Federal Aviation Administration (FAA), is called an ASOS (automated surface observation system). As Figure 4.1 shows, each station is actually a small site containing a cluster of instruments designed to measure several atmospheric variables: cloud amount and height, visibility, precipitation type and accumulation, pressure, temperature and dew point temperature, and wind (both speed and direction and sometimes peak wind gust).

DOI: 10.4324/9781003344988-5

Figure 4.1 ASOS (automated surface observation system) station. ASOS sites, many of which are located at major airports around the United States, are the source of hourly synoptic weather charts. (NOAA.)

Observations are collected every few minutes, and the data are relayed to central locations for quality control and archiving. On a schedule varying from hourly to every 3 hours, all observations at a particular time are plotted and analyzed to produce synoptic weather maps at the surface. The synoptic scale covers the entire United States and spans several thousand kilometers. Figure 4.2 shows all major weather systems across the United States at 8 a.m. on March 14, 2016.

Digitally generated maps like those in Figure 4.2 can be combined with other datasets to create composite charts, which relate pressure systems and fronts to regions of precipitation. Figure 4.3 is an example of this type of blended analysis, which provides a much more complete picture of active weather areas, including the areal coverage and intensity of precipitation.

Mesoscale means "middle scale" – tens of kilometers to a few hundred kilometers. Smaller networks of weather stations closer together, called mesonetworks, provide a more detailed analysis over smaller regions. Figure 4.4 shows an example from the State University of New York (SUNY) Albany Mesonet.

Some mesonets, such as the Oklahoma Mesonet, were first installed in parts of the country that frequently experience severe local storms, such as supercell (rotating) thunderstorms and tornadoes. These mesonets help forecasters locate intense gradients in temperature, moisture, and wind that give rise to strong or severe thunderstorms. These gradients occur on scales of individual counties and require a dense observing network to identify large changes over short distances. With this high-resolution information, weather fronts can be located with pinpoint accuracy, as well as small circulations (disturbances in pressure and wind) that can initiate new thunderstorm development. To make better sense of the data and quickly identify patterns such as large gradients, a computer-generated analysis creates colored contours of equal numerical value for a particular variable of interest. In Figure 4.4, we see shaded color contours of surface temperature at a particular time. In Figure 4.5, we have shown the relative humidity distribution around the state for the same time. Cooler, more humid areas indicate small regions where pockets of heavy rain from recent thunderstorms have occurred. These cool zones can trigger small wind circulations which, in turn, may trigger additional thunderstorms.

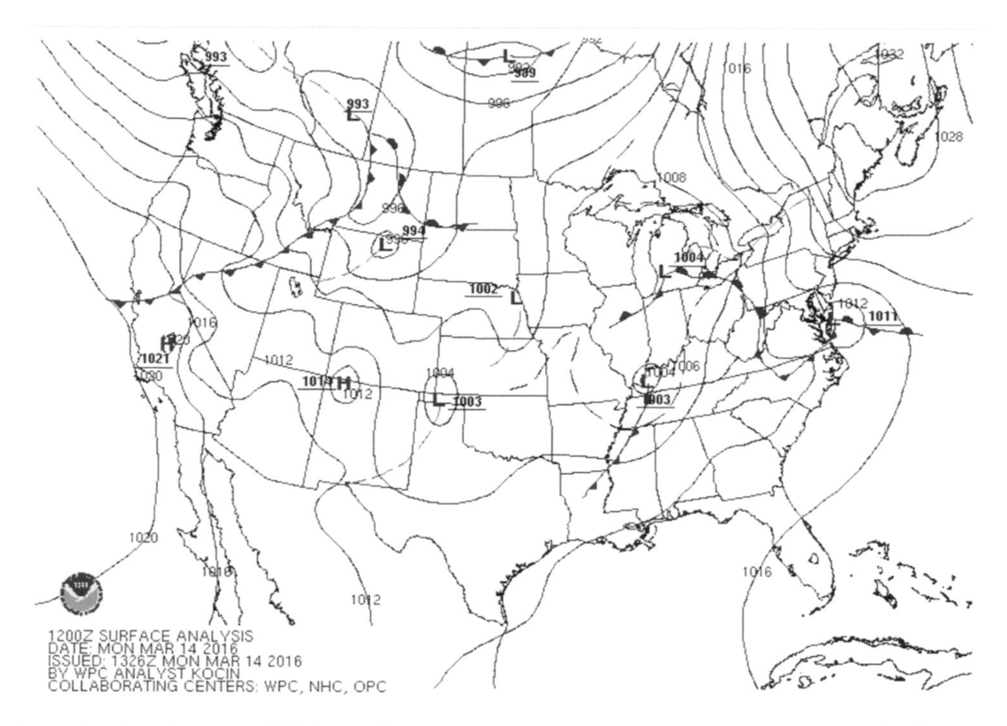

Figure 4.2 Synoptic weather chart, March 14, 2016, 8 a.m. These charts show the locations of weather fronts (red and blue lines), lines of constant pressure (or isobars; thin brown lines), cyclones (red "L"), and anticyclones (blue "H"), along with corresponding central pressure values (shown in millibars). All the map's weather features are constructed from ASOS observations of temperature, pressure, humidity, and winds. (NOAA.)

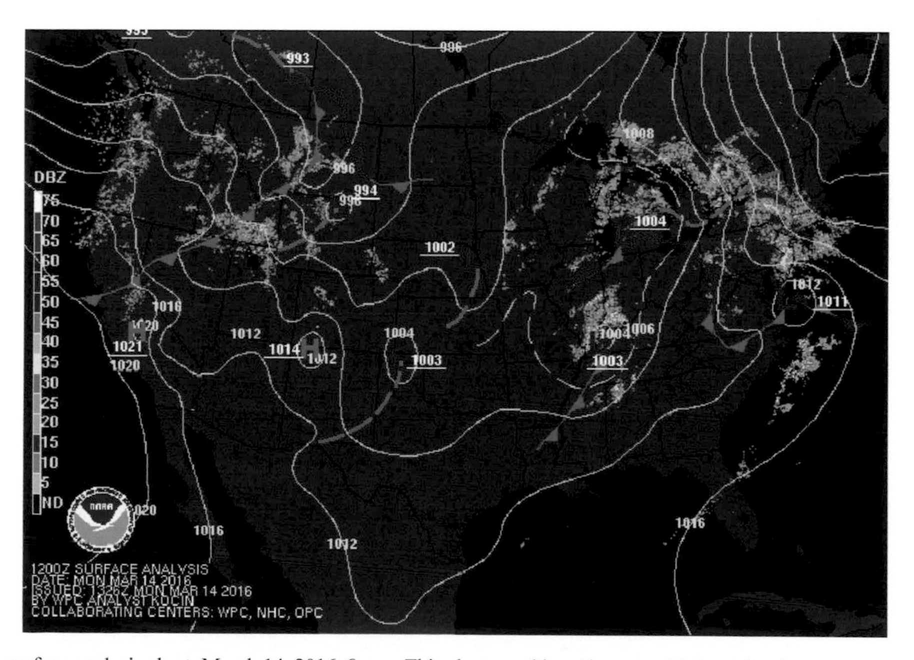

Figure 4.3 Composite surface analysis chart, March 14, 2016, 8 a.m. This chart combines the essential frontal and pressure systems with radar information depicting precipitation coverage and intensity (with yellow-orange shading representing the heaviest precipitation). (NOAA.)

Figure 4.4 Output from the dense network of surface weather stations comprising the SUNY Mesonet. Each station reports a standard set of meteorological observations; shown here is the statewide distribution of surface temperature at 3 p.m. EDT on August 18, 2022. (SUNY.)

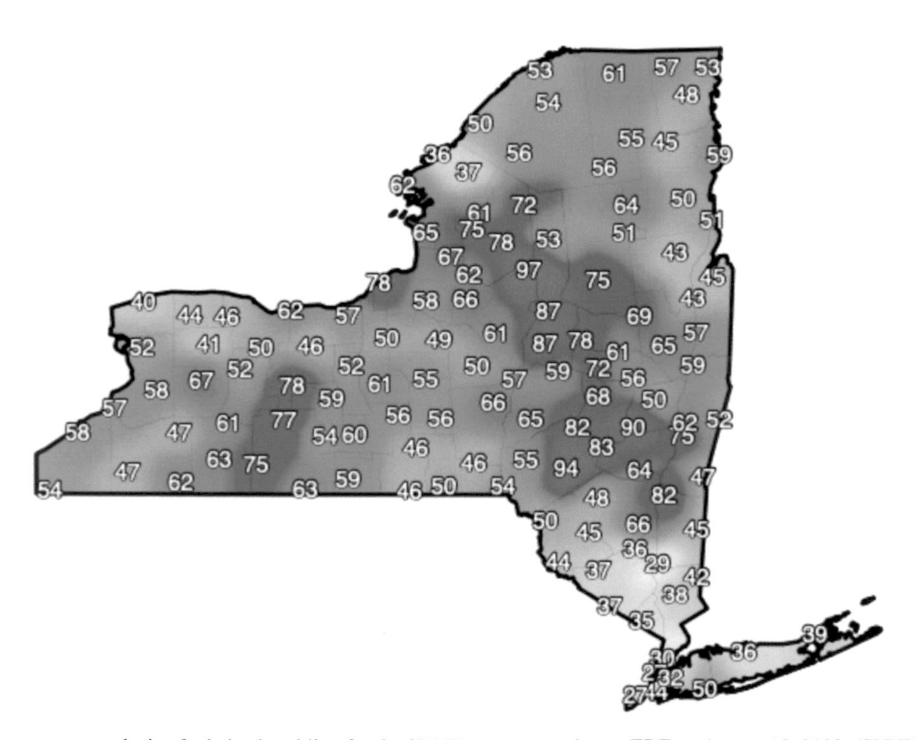

Figure 4.5 Color contour analysis of relative humidity, for the SUNY mesonet, at 3 p.m. EDT on August 18, 2022. (SUNY.)

Many large storm systems, including extratropical cyclones and tropical cyclones, frequently impact the U.S. coastal regions. Large coastal storms called Nor'easters frequently rake the Eastern Seaboard during winter, and hurricanes make landfall during summer and fall. As Figure 4.6 shows, networks of ocean buoys are stationed at more than 150 offshore sites. These sites automatically measure and report many of the same atmospheric variables as land-based ASOS

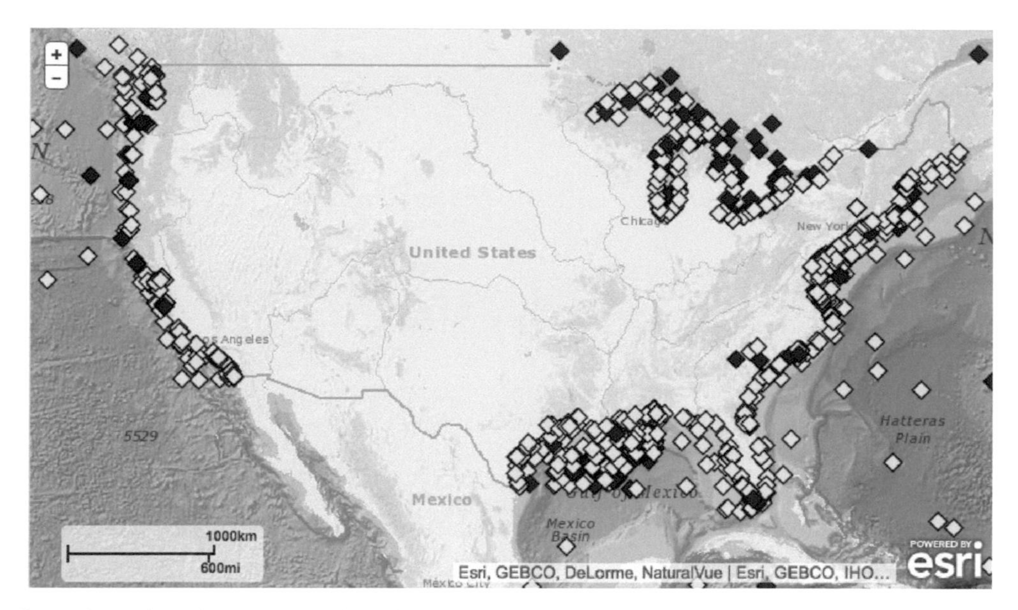

Figure 4.6 Map of coastal water buoy locations across the United States, March 2016. Colors represent the reporting times of each site. Data from these buoys flow automatically into the National Data Buoy Center (NDBC), an arm of NOAA. Coverage in the northern Gulf of Mexico is particularly extensive; this region also harbors a great concentration of deepwater oil drilling platforms. (NOAA.)

sites, but they also report data on water temperature, tides, and wave heights. Ocean temperature is useful when predicting the intensity of large oceanic storms, which derive significant energy from heat stored in the water column. Wave height measurements provide critical data on storm surge and on open-ocean waves generated by a cyclone's strong winds. In Figure 4.6, note also the dense coverage of buoys around the shores of the Great Lakes. These provide wave height information for large, Midwestern cyclones during autumn, called November Witches, that have historically sent thousands of ships to their watery graves.

Upper-Air Observations and Charts

When you study a surface weather map, it's easy to conclude that all significant weather occurs at the surface. In fact, many weather systems – especially cyclonic storms and thunderstorm complexes – extend to great heights, up to 15–16 km (49,000–52,000 feet). The mid-latitude westerlies, including the polar jet stream, tend to be strongest at high altitude. Weather fronts often extend upward for many kilometers. Dry and moist air masses stream along at multiple levels. For almost 100 years, the radiosonde (Figure 4.7) has provided accurate upper-air atmospheric observations. The modern radiosonde is an ultra-lightweight, battery-powered sensor package lofted by a helium-filled balloon. It contains a GPS antenna, and as the package drifts in the wind (as it simultaneously rises), it measures wind speed and direction with great accuracy. A small electronic thermometer and hygrister (which measures humidity), along with a sensitive barometer, complete the instrument suite. Data regarding temperature, moisture, and winds are relayed to the ground launch station every second via a radio communication channel.

Radiosondes are launched twice a day at approximately 100 sites across North America, (Figure 4.8). Standard release times are 7 a.m. and 7 p.m., and it takes about 90 minutes for radiosondes to terminate at about 30 km (95,040 feet). Note that the radiosonde network is relatively sparse compared to the 900+ ASOS observing stations across the United States. Radiosonde observing systems are relatively costly and labor-intensive to operate. Like ASOS, most radiosonde stations are located at major airports. It's always desirable to increase the radiosonde's coverage and to launch them more frequently because significant variations (gradients) in atmospheric wind and temperature systems often go undetected. Fortunately, several other technologies help fill the gaps between radiosonde locations. These include satellite remote sensors, weather instruments flown on commercial aircraft, and Doppler radar units called wind profilers.

The wind profiler (Figure 4.9) is a vertically pointing Doppler radar (the Doppler principle is described later in this chapter). It consists of a flat antenna array that directs narrow beams of pulsed microwave energy vertically. The beam angles are varied

Figure 4.7 Radiosonde attached to a helium balloon, ascending through the troposphere. The instrument package transmits data on temperature, humidity, and winds to the ground as the radiosonde ascends to great heights. (NOAA.)

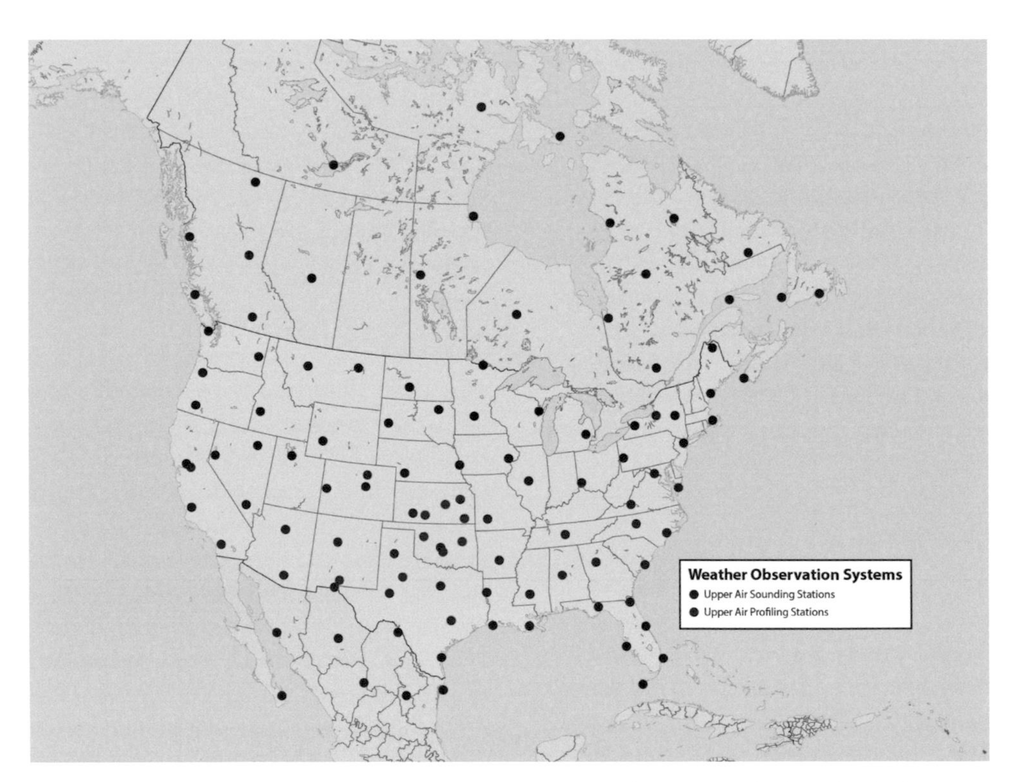

Figure 4.8 Map of all North American upper-air observations sites, from which radiosondes are released twice a day. Stations are separated by an average distance of approximately 200 km (124 miles).

in such a way that is possible to geometrically identify and measure the horizontal winds at multiple levels. In this manner, a vertical wind profile is obtained, showing how wind speed and direction vary with altitude. Figure 4.10 shows an example of these wind data.

An extensive wind profiler network was operated by NOAA until 2014, covering much of the U.S. Midwest (Figure 4.11). The profiler stations enhanced the monitoring of hourly wind variations associated with the development of severe local storms, including supercells and tornadoes. The network was decommissioned in

Figure 4.9 Wind profiler antenna array. The antennas transmit energy pulses vertically. Turbulent motions in the atmosphere scatter these energy pulses. The energy returned to the wind profiler provides information on wind speed and direction at various altitudes. (NOAA.)

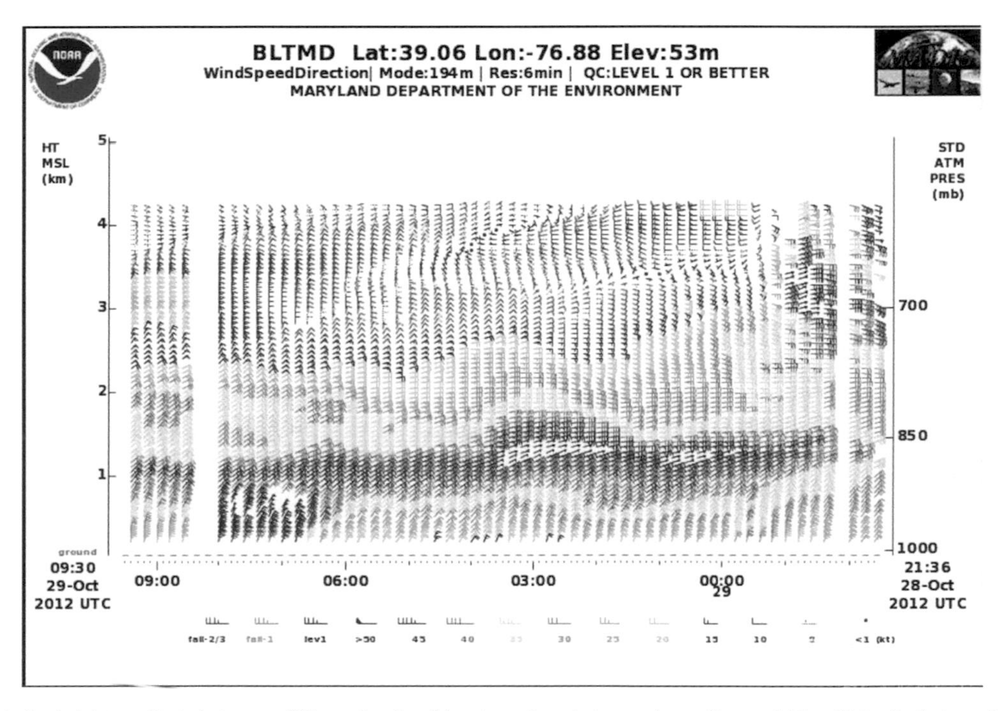

Figure 4.10 Vertical wind data collected at many different levels of the atmosphere between the surface and 4 km (2.5 miles). A vertical wind profile for Baltimore, Maryland between 6 p.m. and 4 a.m. on October 28–29, 2012. This wind profile is an example of a height–time graph, with time increasing from right to left across the diagram and altitude increasing along the vertical axis. Individual wind barbs are shown at each analysis level. Colors depict relatively fast flow from east-northeast, exceeding 60 MPH close to the surface, during an intense Nor'easter storm. (NOAA.)

2014, in lieu of using a combination of wind observations from Doppler radars, commercial aircraft, and satellites.

A sample upper-level analysis chart appears in Figure 4.12. This synoptic map shows the wind speed and direction at the 9 km (30,000 feet) level at 7 a.m. on March 14, 2016, across North America. The map, one of several made at different altitudes and synthesized from radiosonde and wind profiler observations, reveals the main core of the westerly jet stream. There are two jet streams at this time: a northern branch, with core winds

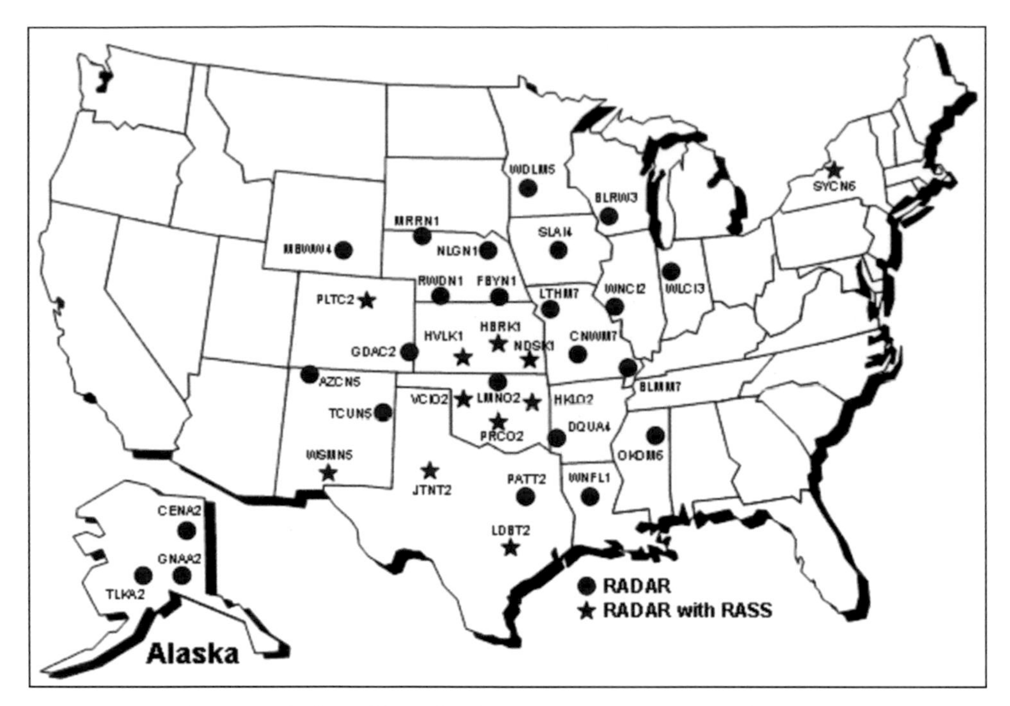

Figure 4.11 Map of the U.S. wind profiler network operated by NOAA. Profiler coverage was fairly dense in the Midwest, in order to identify important wind circulations associated with the severe weather outbreaks common in the Plains states. (NOAA.)

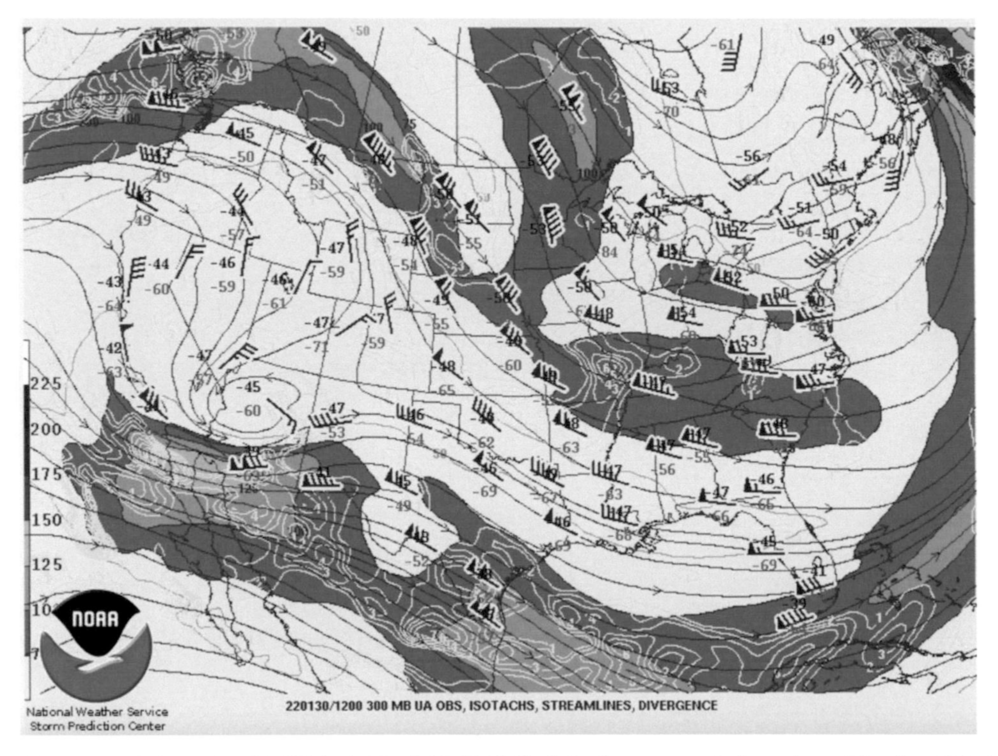

Figure 4.12 Map of jet stream winds at 9 km (30,000 feet), March 13, 2016. Shaded color contours depict regions of fast flow (light blue). The confluence of two jet stream branches is shown over the southern Plains. Individual wind barbs show that most of the airflow at this level is from the west. (NOAA.)

exceeding 120 kts (140 MPH) across the western United States, and a more southerly branch situated across the Gulf of Mexico and Florida, with winds exceeding 100 kts (115 MPH). During the spring, strong jet stream patterns like this one combine with an unstable air mass, creating outbreaks of severe thunderstorms and tornadoes across the Midwest and Mid-South regions.

Satellites Used in Meteorology

Some of the observing tools previously described, such as the wind profiler, are called remote sensing instruments because they collect information about weather phenomena from a distance, by detecting one or more wavelengths of energy. Passive remote sensors intercept electromagnetic energy emitted from the Earth and the atmosphere. Active remote sensors ping parts of the atmosphere with pulses of energy, a fraction of which is scattered back to the sensor.

Weather satellites are mostly passive remote sensors. They have been routinely placed in orbit since the early 1970s. Today dozens of weather satellites form a "sensor web" around Earth. Most are launched and operated by NOAA and NASA.

There are two types of satellites, geostationary satellites and polar orbiting satellites. A geostationary satellite is placed in a very high orbit, at 35,786 km (22,236 miles), and spins at the same rate as Earth. Geostationary satellites are relatively far away from Earth, and they view the same geographical region the entire day. Snapshots taken every 15 minutes or so are routinely animated to create movies showing the evolution of cloud and storm systems. Polar orbiting satellites are much closer to Earth (1,000 km [620 miles]) and circle the planet every 90–100 minutes. Unlike geostationary satellites, they do not image the entire Earth but rather a narrow swath that encircles the globe. Earth spins beneath the satellite's orbital track, such that a new geographical strip gets imaged with each pass. Typically the same region will be imaged twice a day. Because of their lower orbit, polar orbiting satellites capture much higher-resolution images (that is, they have better clarity and granularity than geostationary satellites). Polar orbiting satellites cannot monitor a particular storm system hour by hour, but they do create image details superior to that of the images captured by geostationary satellites.

Many different energy bands comprise the electromagnetic spectrum. This spectrum describes various wavelengths or "bands" of energy, as a function of wavelength and frequency. Passive remote sensors commonly employ the visible light channel, detecting the Sun's visible light reflected from cloud tops and Earth's surface (Figure 4.13). The images provide information on cloud coverage, growth rate, and structure, particularly convective clouds (thunderstorms) and large cyclonic storms (extratropical cyclones and hurricanes). Some satellites, including the GOES-14 (Geostationary Operational Environmental Satellite) can focus on small geographical sectors using RapidScan mode, collecting images every minute. Animations at 1-minute intervals provide astounding detail in regions that experience severe thunderstorm outbreaks, such as the explosive growth of individual supercells. RapidScan is also useful for monitoring changes in the eyewall region of approaching hurricanes or the central core of wintertime coastal cyclones.

Another type of passive remote sensor detects infrared radiation emanating from cloud tops, land, and water. Infrared radiation has a longer wavelength than visible light, lying just beyond the color red. One advantage of infrared is that it provides 24-hour coverage (visible images can only be collected from Earth's sunlit side), thereby allowing continuous monitoring of weather systems. Another advantage is that infrared sensing can determine the altitude of different cloud layers. Because temperature decreases with height in the troposphere, cloud tops emitting small amounts of infrared energy are found high up, whereas warm clouds have a lower altitude. Thus ground fog can be readily distinguished from mid-level stratus clouds and high-level cirrus clouds. Infrared sensing is particularly useful for monitoring trends in thunderstorm growth; the highest, most intense thunderstorm cells can have temperatures as low as −80° C (−112° F)! For a sample of infrared imaging, see Figure 4.14.

Beyond the infrared lies another band, called the microwave. Satellite sensors began exploiting this channel to study meteorological phenomena in the 1990s. Microwaves can penetrate thick cloud layers, allowing the surface features relevant to meteorology (including the extreme winds generated by hurricanes and other cyclonic storms) to be imaged. Microwave imaging relies on active remote sensing, whereby the orbiting satellite pings Earth's surface with pulsed microwaves many times a second. The rougher the ocean surface, the stronger the wind at the surface. Sophisticated mathematical algorithms calculate both wind speed and direction at discrete points along the satellite swath. A stunning example of this capability is shown in Figure 4.15. Here, an active remote sensor called (flown on the International Space Station) revealed a swath of hurricane force (> 65 kts [74 MPH]) winds in the core of a large ocean cyclone in the North Atlantic. Without this satellite passing overhead, it would be very difficult to ascertain the peak, surface-level winds in such a remote location – information that is vital to shipping and fishing interests.

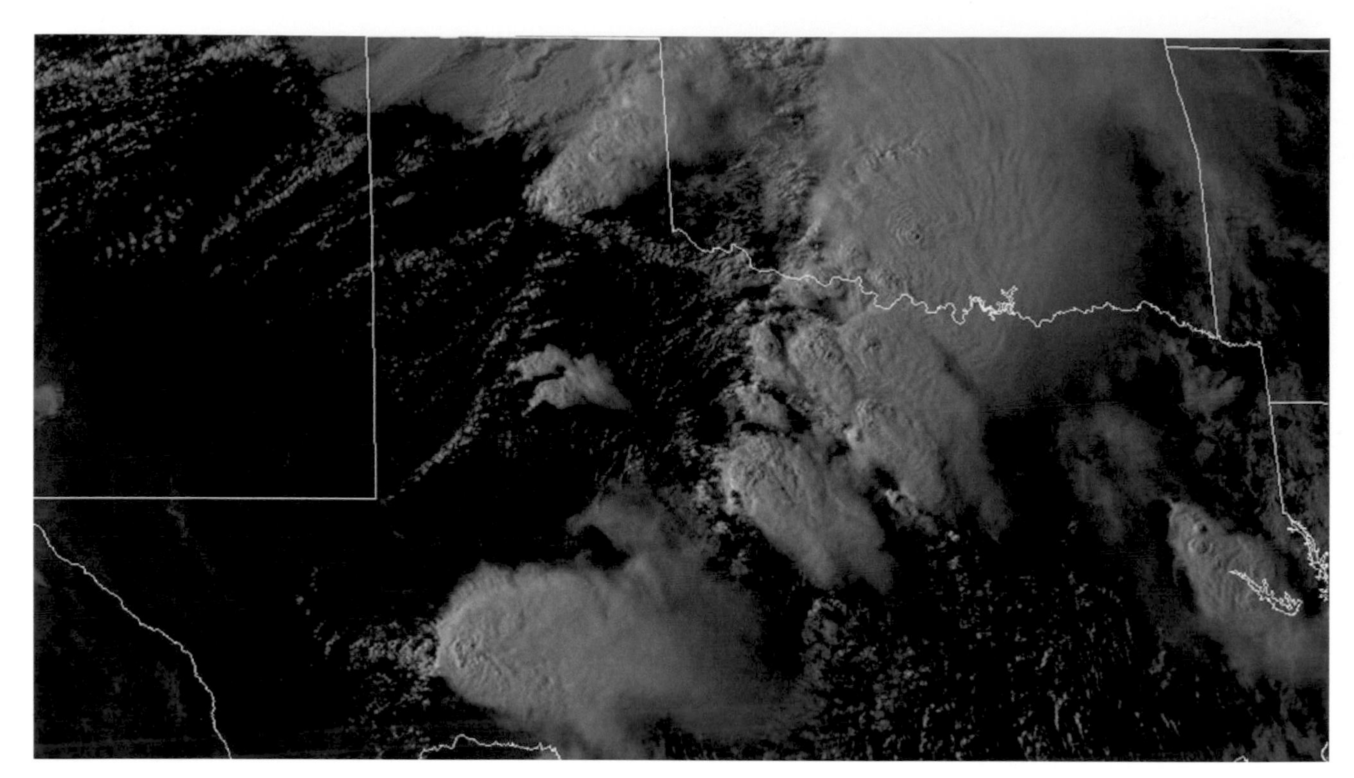

Figure 4.13 Still image from a GOES-14 RapidScan movie of a large severe thunderstorm over western Texas on May 19, 2015. The RapidScan is constructed from 1-minute snapshots recorded in the visible wavelength. (NOAA.)

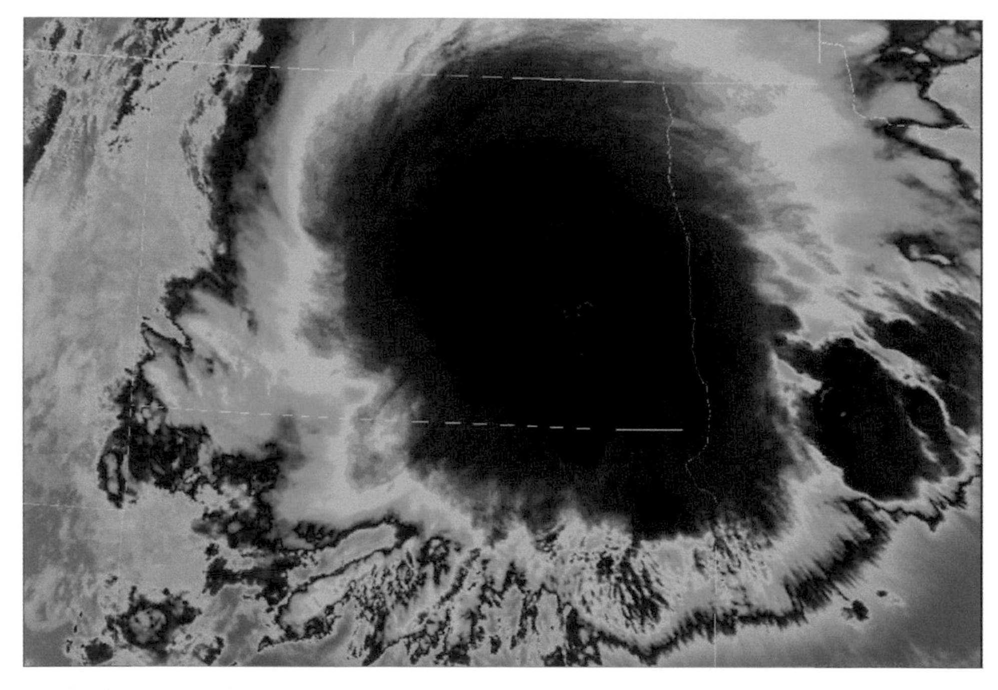

Figure 4.14 Enormous thunderstorm complex over the northern Great Plains, as imaged in the infrared satellite band, July 6, 2012. This type of storm system, called a mesoscale convective complex (MCC), frequently causes warm-season flash flooding during the nighttime hours. False color is used to highlight the coldest cloud top regions (shown in black). Note that this massive complex of thunderstorms covers much of North Dakota. (NOAA.)

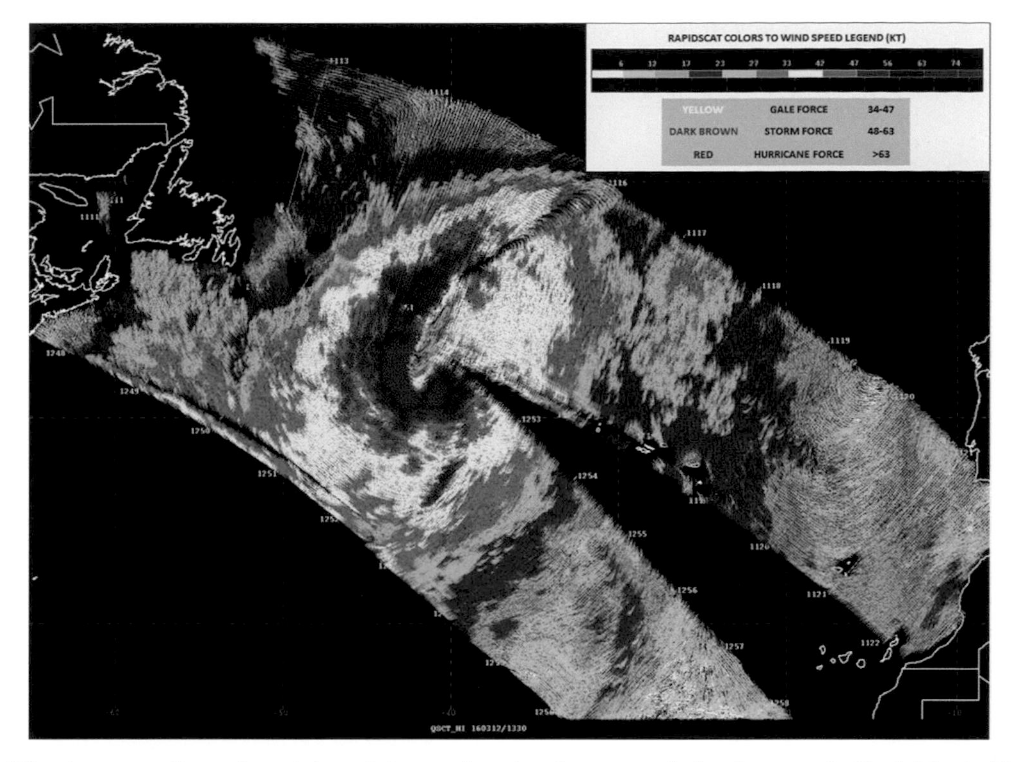

Figure 4.15 RapidScat image revealing surface wind speeds in a massive wintertime extratropical cyclone over the North Atlantic, March 12, 2016. The small red region indicates sustained hurricane force wind. (NASA.)

Figure 4.16 3D image of towering thunderclouds inside developing Hurricane Rita, imaged by the TRMM radar, September 2005. White-gray shades show the outline of the storm's clouds. Like a CAT scan, colored regions reveal the vertical rain structure, including two intense chimney clouds producing extremely heavy rain (red) near the storm's center. At this time, Rita was entering the Gulf of Mexico and beginning to rapidly intensify to Category 5 status. (NASA.)

Active microwave sensors on satellites are also used to study precipitation regions within storm clouds, particularly over tropical oceans, where conventional radar observations are impossible to acquire. NASA's Tropical Rainfall Measurement Mission (TRMM) satellite was the first satellite to carry a weather radar, pinging the atmosphere with pulsed microwave. For over a decade, this satellite provided remarkable 3D images of the rain regions inside developing hurricanes, akin to taking a CAT scan of the storm (Figure 4.16). Intense vertical towers of extremely heavy rain, called chimney clouds, were commonly found to precede periods of rapid storm intensification. TRMM's radar could also accurately locate the eye of a developing hurricane, which was otherwise obscured by thick layers of upper-level cloud.

Weather Radar and Lightning Detection

In 1991 the National Weather Service began installing a comprehensive network of Doppler radars covering most of the contiguous United States (Figure 4.17). This network was designated the Next Generation Radar (NEXRAD). Since then, additional radars have been added at major airports, enabling more sensitive detection of wind shear over runways and along aircraft approach/departure corridors. Numerous television stations, particularly in the Midwest, now have their own Doppler radars, to enhance local detection of tornadic circulations and to "scoop" their competitors. How exactly does radar detect weather systems and storms?

How Radar Detects Precipitation

Doppler radar is an active remote sensor. Figure 4.18 illustrates the iconic Doppler radar installation. Externally, a spherical radome (a housing made of material transparent to microwaves) protects the radar assembly. It is elevated on a tall tower to prevent the radar beam from being blocked by nearby trees and structures. Inside the radome is a dish-shaped antenna. A transmitter in the middle of the antenna emits microsecond-duration pulses of microwave energy, and the antenna dish focuses and transmits these pulses into a narrow beam. The antenna swivels 360° and also tilts up and down. In this manner, the antenna systematically scans a volume of atmosphere surrounding the radome every 5–8 minutes. If targets in the form of precipitation particles are present within the scan volume, some of the microwave energy is scattered back to the radar. This energy is collected and amplified by the antenna. Distance to the targets is computed based on the speed of light and the elapsed time between transmitted and received pulses. Intensity of the precipitation is determined by the density of precipitation particles, as well as their size and composition (water vs ice).

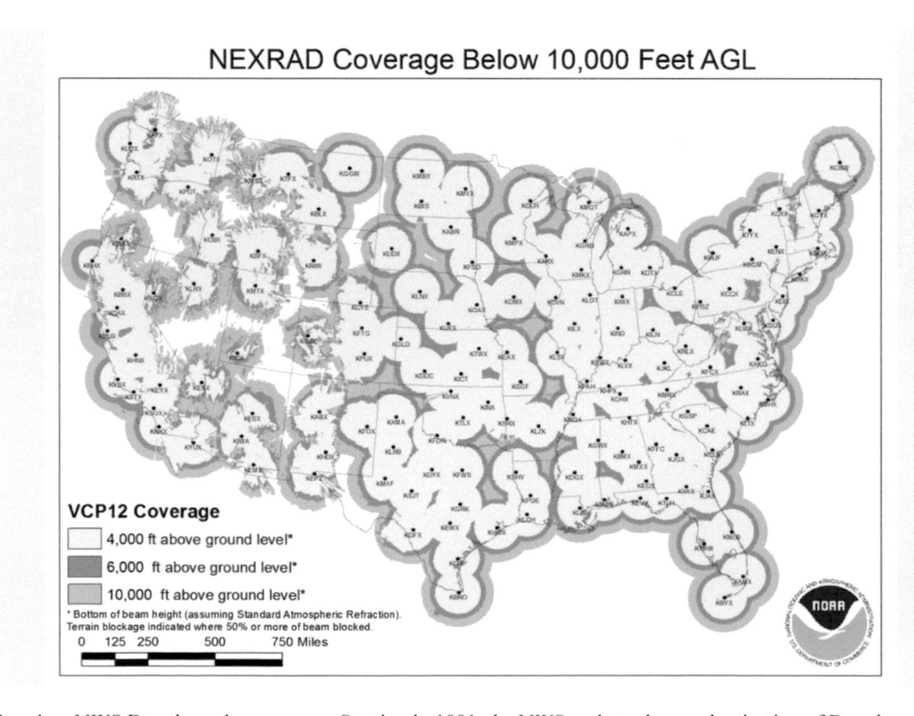

Figure 4.17 Map of modern NWS Doppler radar coverage. Starting in 1991, the NWS undertook a modernization of Doppler radar coverage. This program is called NEXRAD (Next Generation Radar). (NOAA.)

Figure 4.18 Typical NEXRAD installation at NWS sites. Inside the dome is the radar transmitter and antenna dish, which swivels on a motorized pedestal. A separate motor tilts the dish assembly in small increments, above the horizontal, after each 360° sweep. (NWS.)

Figure 4.19 Radar scan from the Columbia, South Carolina, NEXRAD site showing an approaching squall line of heavy showers and thunderstorms. Rain intensity is shown by the various colors, with red shades indicating the heaviest rain cores. (NWS.)

Empirical (statistical) relationships have been developed that relate the power of returned microwave energy to the intensity of precipitation at the surface. The results are digitized and mapped, and a color scale is used to denote various precipitation intensity levels. In terms of rain, blue and green shades connote very light rain, such as drizzle; yellows indicate moderately heavy rain; and oranges and reds imply a torrential downpour, up to several inches per hour. Figure 4.19 depicts a typical radar image, revealing a squall line of heavy showers and thunderstorms. The radar site is located at Columbia, South Carolina, and is shown in the center of the image by a small red dot. The squall line appears as a nearly continuous orange-red band, approaching rapidly from the northwest. Behind these thunderstorms, a more extensive region of light to moderate rainfall extends for approximately 80 km (50 miles).

Radar provides a host of information about precipitation falling around the radar site, including rain and snow rates, the presence of hail (which shows up as magenta-colored regions within the center of intense thunderstorm cells), the location and movement of precipitation regions, and whether the precipitation is intensifying or diminishing, expanding or shrinking. Most scans show precipitation at a single elevation angle surrounding the radar, closest to the surface. Radar scans also allow meteorologists to view different elevation angles, slice through higher levels of the storm system, and display complete vertical slices taken at specific compass angles. Information on storm structure aloft helps distinguish ordinary thunderstorms from severe cells, particularly those that generate large hail and strong tornadoes.

Radar mosaics are images encompassing large regions (or even the entire United States), rendered by digitally stitching together the smaller areas scanned by each radar. Figure 4.20 shows an example. Here, an extratropical cyclone over the southeastern United States was generating extensive heavy rain on December 18, 2009. Green, yellow, and red connote various intensities of rain, with reds indicating the heaviest rain rates.

Doppler Techniques Reveal Winds Within Storms

The Doppler principle describes the frequency shift of electromagnetic energy emitted or scattered by a moving target. Raindrops move horizontally through a cloud because they are embedded in cloud winds, even as they fall. Microwave pulses that reflect off these moving targets will experience a frequency shift (like the sound of a train horn, which shifts to a higher pitch or frequency as the train approaches). Thus Doppler radar can detect air motions moving toward or away from the radar, along each beam. As the radar beam sweeps out a circle, a picture of horizontal wind surrounding the radar emerges. It's important to understand, however, that only air motions moving directly *toward* or *away from* the radar are measured. For this reason, the detected winds are termed radial winds. Air motion that is perpendicular to the beam cannot be ascertained. Identification of the complete two-dimensional wind field (that is, winds of all compass directions) requires two Doppler radars placed in close proximity.

Still, radial winds obtained at different scan altitudes contain a tremendous amount of information. In a severe thunderstorm, radial wind can be used to infer the presence of rotating vortices, called mesocylones, which are the precursor to most strong and violent tornadoes. In other situations, a powerful downward and outward blast of downdraft air, called a downburst, can be identified. Downbursts create significant amounts of structural damage, and they caused numerous commercial aviation accidents prior to the 1990s, before sensitive Doppler radars, called Terminal Doppler Weather Radar (TDWR), were installed at major airports.

Figure 4.21 shows the power of Doppler radar. The image shows a supercell thunderstorm over Oklahoma. The precipitation scan is shown on the left. The supercell is the large red mass north of Norman, which was producing a violent tornado at the time. Note that the storm cell has a peculiar hook-shaped

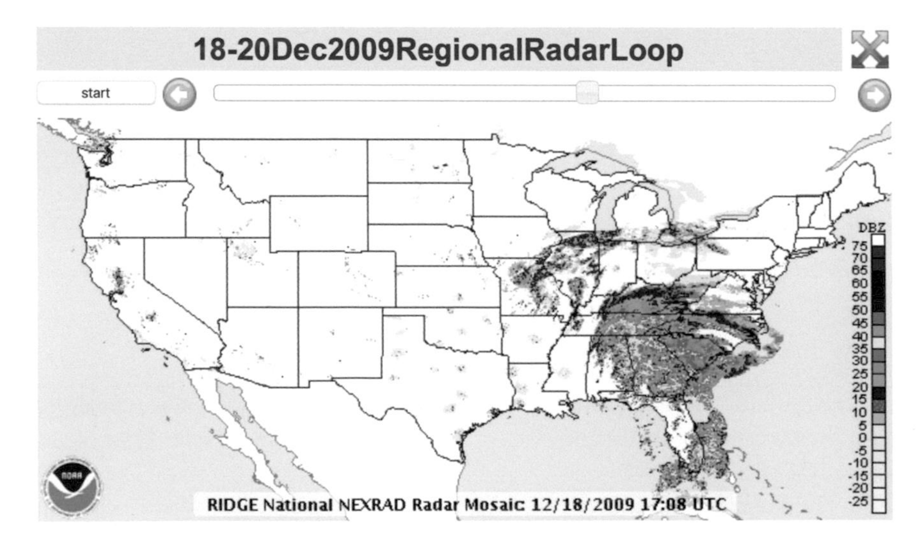

Figure 4.20 Radar mosaic of the Southeast, showing precipitation generated by a large cyclonic storm, December 18, 2009. Rain intensity is revealed by the color scale on the right, proceeding from light intensity (blue and green colors) to heavy (orange and red). (NOAA.)

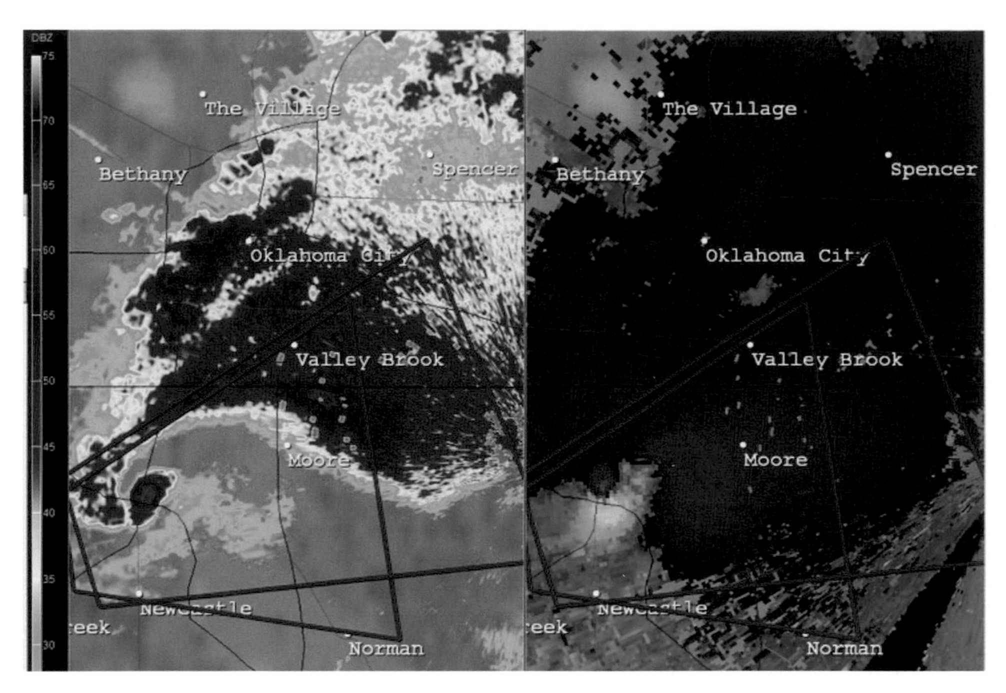

Figure 4.21 Exceptionally large and violent tornado as detected by Doppler radar near Norman, Oklahoma. The left panel shows a conventional precipitation map, with the most intense precipitation (heavy rain and hail) denoted by the cyan (light blue) colors. In the prominent hook-shaped appendage, magenta colors likely indicate large pieces of debris lofted by the tornado. The right panel is the corresponding display of Doppler-derived radial wind. The small and intense velocity couplet implies the location of the tornadic circulation. (NWS.)

appendage, implying that precipitation was being circulated counterclockwise by storm winds. The magenta "ball" located at the apex of the hook, representing large values of returned (scattered) microwave energy, was in fact created by large pieces of debris lofted by the tornado! The panel on the right is the Doppler view of the same supercell. The colors here indicate the direction and speed of the radial wind. Red is airflow moving away from the radar, while green and blue indicate airflow streaming toward the radar. (The radar's location is off the image to the left) Note the "couplet" of blue next to red, shown by the arrows, connoting fast outbound flow adjacent to fast inbound flow. The two opposing, intense flows suggest that there must be a small region of counterclockwise spin. This region was, in fact, the location of the violent tornado, corresponding exactly to the debris ball shown in the left-hand panel. The red trapezoidal boxes indicate regions placed under a tornado warning; this warning was issued in large part as a result of these Doppler observations.

Polarimetric Principles Better Distinguish Among Rain, Hail, and Snow

In 2013 the National Weather Service began upgrading its Doppler radars with a new capability, one that uses polarized microwave pulses. A polarimetric radar is one capable of transmitting and receiving energy in two planes, one horizontal and one vertical (you may have learned that polarized sunglasses filter out the horizontal component of sunlight). As Figure 4.22 shows, the schematic radar beam is composed of vertical and horizontal components of microwave energy. Not all precipitation particles are spherical. Large raindrops become flattened as they fall, and snowflakes have an even larger horizontal aspect. In contrast, hailstones are largely spherical. Snowflakes and large raindrops thus scatter or return more horizontally oriented than vertically oriented microwave energy, while hailstones scatter both components equally. By measuring the ratio or difference between the horizontal to vertical components, computer algorithms can determine the most likely type of precipitation contained within the radar beam.

Figure 4.23 is a side-by-side comparison of a severe thunderstorm cell imaged using conventional radar (left panel) and polarized energy (right panel). The left panel implies a very intense precipitation rate, as suggested by the red and magenta colors. In severe thunderstorm cells, there is some ambiguity whether the high precipitation intensity is due to extremely torrential rain or to a mixture of heavy rain and large hailstones. The ambiguity arises because both heavy rain and hail return large amounts of energy to the radar (think of the radar volume as filled either with an extraordinarily large number of large

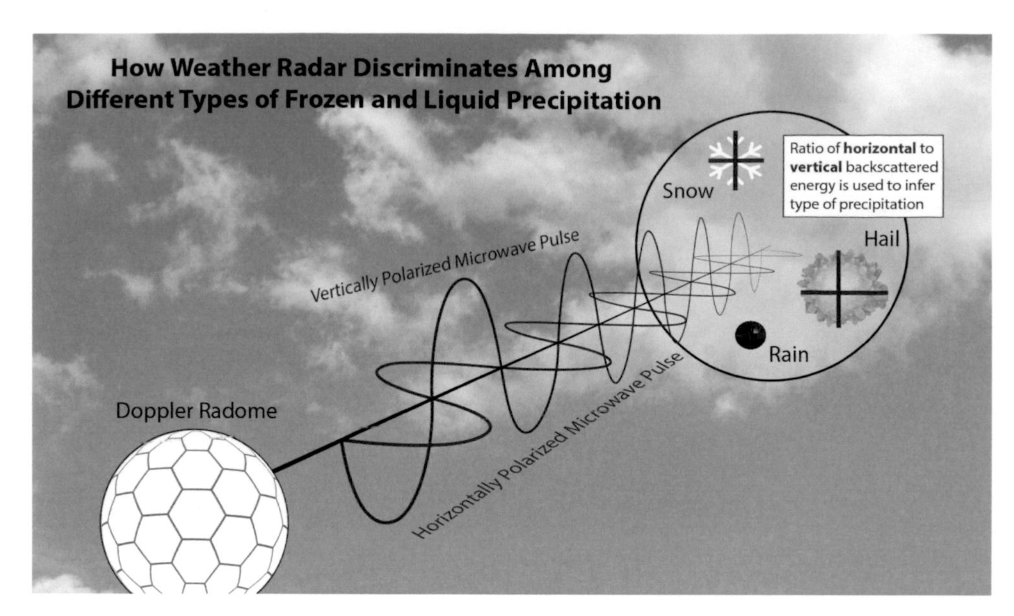

Figure 4.22 Basic principle of polarimetric radar. Polarimetric radar transmits and receives microwave energy in both the horizontal and vertical planes. Meteorological targets have different aspect ratios (ratio of horizontal to vertical dimensions) and thus scatter the horizontal and vertical parts of the radar beam differently, enabling meteorologists to better distinguish among rain, hail, and snow. (NWS.)

Figure 4.23 Example of how polarimetric radar is used to identify probable regions of large hail within thunderstorm cells. The thunderstorm cell in the left panel is imaged using conventional precipitation mapping; in the classification map in the right panel, polarimetric radar has assigned a hail designation (red region) to the most intense precipitation core. (NWS.)

raindrops or with several large, water-coated hailstones, which essentially appear as enormous drops to the radar's computer). Using polarimetric principles, it becomes possible to classify the *type* of precipitation falling in different regions of the same storm cloud. In the right panel, the computer has shaded part of the storm cell a red color, which is the specific designation for hail. The dark green shades surrounding the hail denote heavy rain, the lighter green shades moderate rain. Blue shades imply wet snowflakes, which commonly develop at higher altitudes inside thunderstorm cells, even on hot summer days. A yellow trapezoid denotes the area placed under a severe thunderstorm warning, issued on account of the large hail.

Combining Radar with Citizen Science to Better Measure Rain

The closely spaced national network of Doppler radars provides (Figure 4.17) fairly dense coverage over most of the United States, but there are gaps, particularly across the West, where high terrain blocks radar beams. Radars encounter problems: They have difficulty measuring precipitation at great distances from the radar; some of the radar energy is absorbed or "attenuated" by rain and snow; some of the precipitation detected high up in the cloud may evaporate before it reaches the surface; and Earth's curve causes the relatively straight radar beam to overshoot cloud tops at large distances. Because of these difficulties, the statistical relationships used to link returned microwave energy to observed surface rain rate suffer from a degree of imprecision. The radar can under- or overestimate the actual rain (as measured by rain gauges) by a factor of two or more. While providing excellent areal coverage, radar does not have the inherent accuracy of a closely spaced network of rain gauges.

Meteorologists have long been aware of this shortcoming, and over the past decade, a cooperative network of "citizen science" observers has come to the rescue. A program called the Community Collaborative Rain, Hail, and Snow Network (CoCoRAHs), run by the Colorado State University, has built a volunteer network of nearly 20,000 citizen rain observers across the United States and its territories. Using relatively simple but accurate rain gauges, observers report daily rain accumulation (and liquid equivalent by melting snowfall captured in the gauge). A website updates national and regional rain accumulation maps every 10 minutes. NWS meteorologists, along with other government, university, and private agencies, use the CoCoRAHs data to help refine radar estimates of rain. The CoCoRAHs program is a wonderful way to involve students, teachers, and aficionados in the science of meteorology, while at the same time providing a valuable service to operational meteorologists, particularly in regions of sparse radar coverage and during extreme rain events leading to river flooding.

The Promise of Lightning Detection Networks

Thunderstorms emit significant amounts of energy in the formation of electromagnetic fields and lightning discharges. They repeatedly charge and discharge in the manner of giant capacitors. Only in the past couple of decades have meteorologists begun to understand the mechanisms involved in lightning creation, and we now realize that lightning detection provides much useful information about the severity of storm cells. For instance, it's been discovered that an abrupt increase or "jump" in the frequency of internal cloud discharges often heralds imminent generation of severe weather, including tornadoes, large hail, and downbursts. Intracloud lightning activity may provide 20 or more minutes of valuable lead time before the onset of severe weather. Commercial agencies such as Earth Networks (based in Germantown, Maryland) now issue Dangerous Thunderstorm Alerts to clients, based on changing levels of lightning activity.

Meteorological instrument and data provider Vaisala, Inc. operates national and global networks of lightning detectors, using detection of very high frequency (VHF) signals emitted by lightning discharges. Data from their National Lightning Detection Network (NLDN) are often combined with radar images to locate the most intense or energetic cores within lines and clusters of thunderstorms. Because the NLDN updates every few minutes, meteorologists can determine whether a storm system is intensifying or weakening, based on trends in lightning activity. Especially dense regions of lightning activity enable forecasters to identify particularly severe thunderstorms.

Experimental, high-resolution lightning detection networks are being deployed around the country to ascertain their usefulness for issuing lightning warnings. Figure 4.24 shows a map of lightning detected by a special lightning mapping array (LMA) over Alabama. The array uses precision GPS techniques to map out individual lightning channels within storm clouds, and it accurately locates all electrical discharges, including cloud-to-ground and intracloud. Lightning density maps such as the one shown here dramatically portray the location of active thunderstorm cores down to extremely small scales, similar to the fine level of detail provided by Doppler radar.

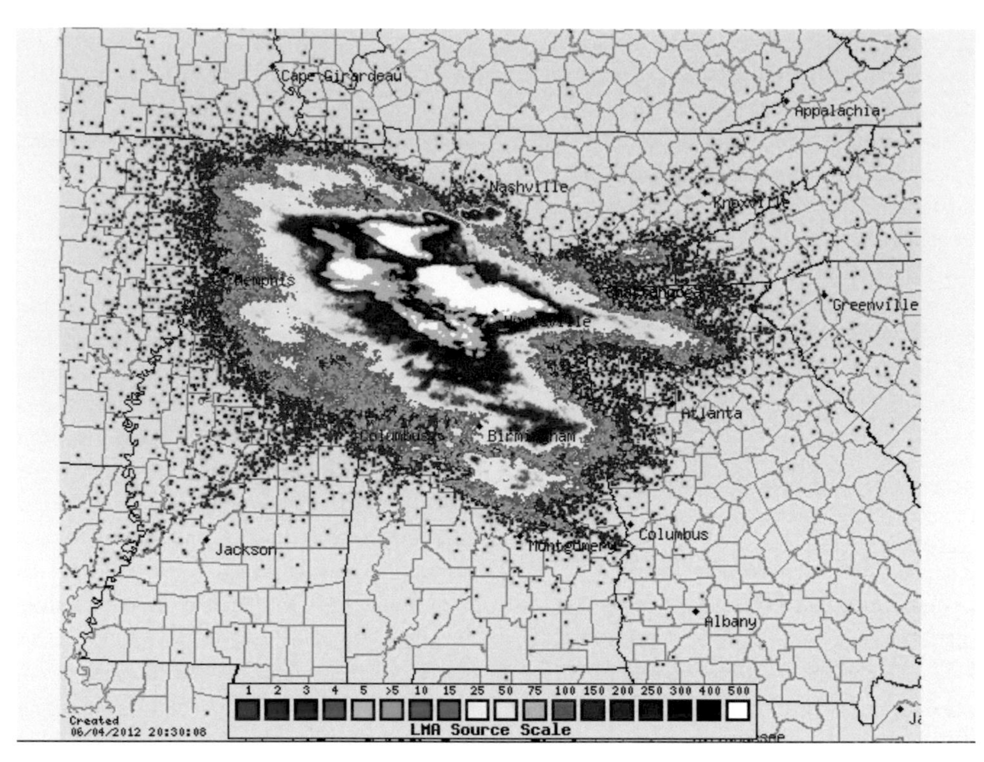

Figure 4.24 Intense thunderstorm activity over northern Alabama detected by an experimental lightning mapping array (LMA). Each dot specifies the exact location of a detected electrical discharge, either within the cloud or between cloud and ground. The color scale shows the density of discharges per square kilometer. (NASA/NOAA.)

Numerical Weather Prediction

How are weather forecasts are generated? Much of the forecast process is built on mathematical representations or models of the evolving atmosphere called numerical weather prediction (NWP) or simply as weather forecast models. Greater emphasis is now being placed not just on the results of these predictions but also on the uncertainty they contain.

What Is Numerical Weather Prediction (NWP)?

In forecast models, the most fundamental variables of the atmosphere – temperature, moisture, pressure, and wind – are represented by a set of mathematical equations that predict their future state after a short time interval. These equations are not exact and must be solved approximately, at a number of discrete, 3D points within the atmosphere. The set of points forms a type of spherical grid that envelops Earth. A typical model may use a 16 km (10 miles) horizontal spacing between points and 64 vertical levels from the surface to the top of the atmosphere. The model thus presents a formidable number of computational points. Simultaneously at each point, the computer must solve a set of extremely complex equations for time steps of 10–15 minutes. Only the most powerful supercomputers are capable of running these forecasts out to 10 or so days, with a new forecast made every 6 hours, every day of the year. There is a trade-off between the model's spatial resolution, or grid spacing, and the computational efficiency or "timeliness"; finer grid intervals entail more points at which to solve equations, which takes more time, much longer than the typical 6-hour forecast cycle.

The data flowing into the model forecast every 6 hours come from many sources, including surface and upper-air observations, satellites, commercial aircraft, and sometimes weather radar. These data comprise the model's initial conditions. Quality checks ensure that the initial conditions contain minimal errors and accurately represent the "starting state" of the atmosphere. One problem is that the observations scattered around the globe do not conform to the model's rigidly spaced mathematical grid, so an analysis is required to reconcile the variables. This analysis introduces a degree of error because a best guess must be made wherever a model grid point lies between multiple observation sites. The error can be significant with regard to upper-air observations, which have a much coarser spacing than surface weather stations.

Furthermore, much of the globe is covered with sparsely inhabited surface, including ocean, desert, and dense rain forest. Few if any surface and upper-air observations exist for large portions of Earth. Satellite data and aircraft observations help to provide data for these remote regions, and the results of other model simulations are used as a "first guess" starting point. Commercial aircraft provide about 450,000 automated weather observations a day, at various altitudes, helping to fill many of the gaps. Nevertheless, significant uncertainty in the initial conditions cannot be avoided. The first guess introduces error that tends to grow with time, as the model simulations are run out over a 10-day period. Thus, a 10-day prediction of weather patterns contains much more uncertainty than a 48-hour forecast.

Types of Weather Prediction Models

The United States and other countries have spent decades developing sophisticated forecast models, with billions of dollars invested in both scientific development and supercomputing hardware. NOAA is the agency responsible for U.S. model development, which takes place at the Weather Prediction Center (WPC) in College Park, Maryland (on the campus of the University of Maryland). These so-called American Models include a 10-day (medium-range) prediction model called the Global Forecast System (GFS); a shorter-range model, the North American model (NAM), which runs weather predictions out to 2.5 days; and a very short-range model, the High Resolution Rapid Refresh (HRRR) model, which only goes out to 18 hours, but is run every hour of the day. The GFS uses a grid point spacing of 13 km (8 miles), the NAM runs at 4 km (2.5 miles), and the HRRR uses 3 km (2 mile) grid point spacing. The HRRR is particularly well suited for forecasts of severe thunderstorms, which tend to be much smaller-scale and shorter-lived phenomena than large cyclonic storms.

The NWS and private forecasters use the entire U.S. suite of models and also a few models developed outside the United States. These include the European Center for Medium Range Weather Forecasting (ECMWF) model, which is comparable to the GFS in terms of its forecast duration (10 days) and grid resolution. Often simply termed the European Model, it is quite sophisticated and has outperformed the GFS in certain situations, such as the historic landfall of Superstorm Sandy (2012) in New Jersey. Another medium-range model frequently compared with the American and European models is run by Environment Canada.

The models generate many types of output, including fields of temperature, accumulated precipitation, snow depth, wind gusts, pressure, cloud cover, and even simulated weather radar. Figures 4.25 and 4.26 show an example of a high resolution set of predictions for shower and thunderstorm coverage on August 19, 2022. The figure compares the HRW (High Resolution Window) and NAM model predictions for 6 p.m. EDT. The two models run for the same future time show significant differences in the distribution and intensity of storm cells in various regions of the United States. These differences are due to varying grid resolutions, the manner in which physical processes are mathematically represented, and the type of data that is used to initialize each model. Forecasters are often faced with model simulations that differ, and this is where experience in understanding the various model shortcomings and strengths becomes paramount.

Conveying Uncertainty in the Forecast

Forecasts run by different models often differ significantly in terms of their placement of low- and high-pressure systems, the intensity and track of those systems, and the type and magnitude of weather impacts generated. Nothing is more frustrating to forecasters than model runs that portray large differences, or "diverge," when it comes to a potential winter storm impacting the Northeastern megalopolis, which can impact up to 50 million people. One model might suggest cold rain, while another predicts a foot of snow in the same location. Additionally, storm tracks and intensities can shift markedly from one 6-hour run to the next. As forecasters gain experience in their particular regions of responsibility, they learn the models' characteristic strengths and weaknesses, called biases. Basically, predictions of certain weather patterns are considered more trustworthy than others.

To better constrain the model uncertainty, it's now customary to examine suites of model simulations that start off with slightly different initial conditions. Doing so effectively captures some of the error that can lead to different predictions of evolving weather patterns. The suite of slightly varying model runs is called an ensemble forecast. The American (GFS) model generates an ensemble forecast every six hours, based on 20 model members (individually varying runs). The European model runs an ensemble forecast consisting of 51 members. It's important to note that ensemble runs are made at roughly *half* the grid resolution or spacing compared to the full model; recall the trade-off between computational efficiency and grid spacing. The full-resolution version that is run every forecast cycle is called the deterministic run.

Two different types of ensemble forecast products are shown in Figure 4.27. The first illustrates the spread in forecast position and intensity of a wintertime coastal cyclone called a Nor'easter. This storm on February 24, 2016, was threatening to create

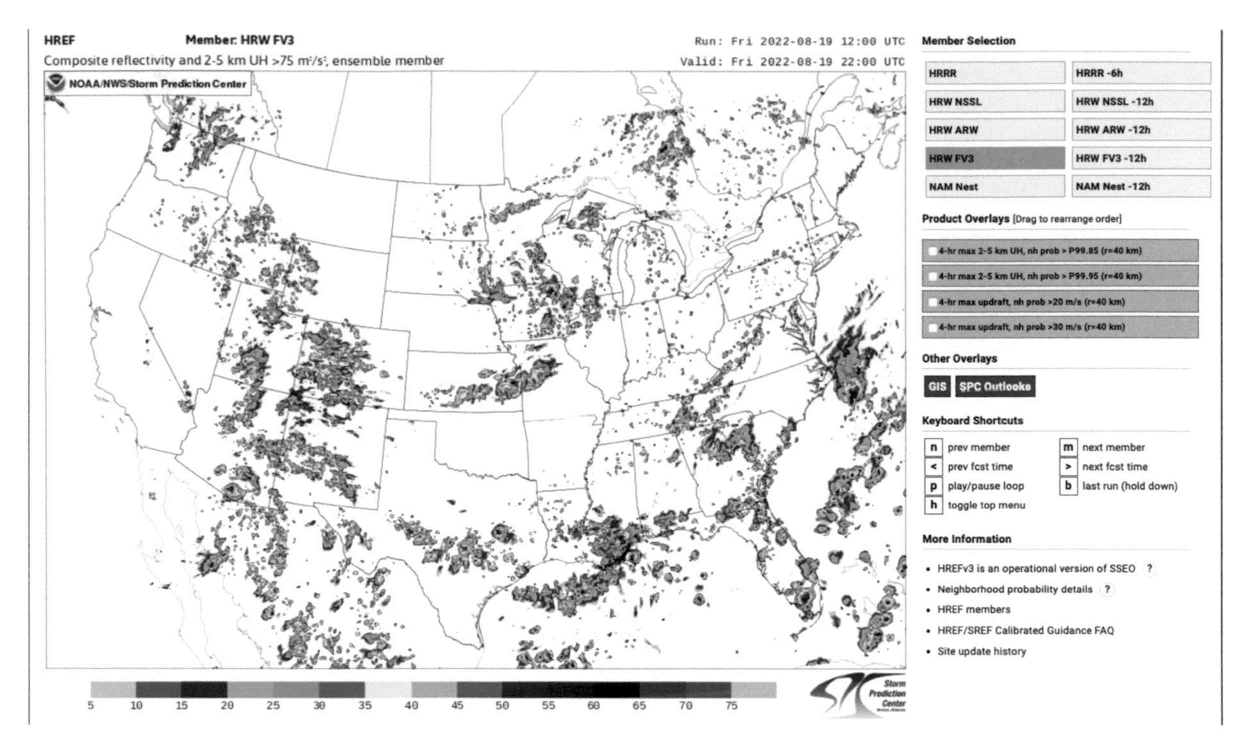

Figure 4.25 HRW model simulation of shower and thunderstorm activity on a summer day across the United States (valid at 6 p.m. EDT, August 19, 2022). The output field illustrates simulated radar reflectivity; the rain intensity scale is shown across the bottom of the image, with intensity values increasing from left to right along the color bar. (NOAA.)

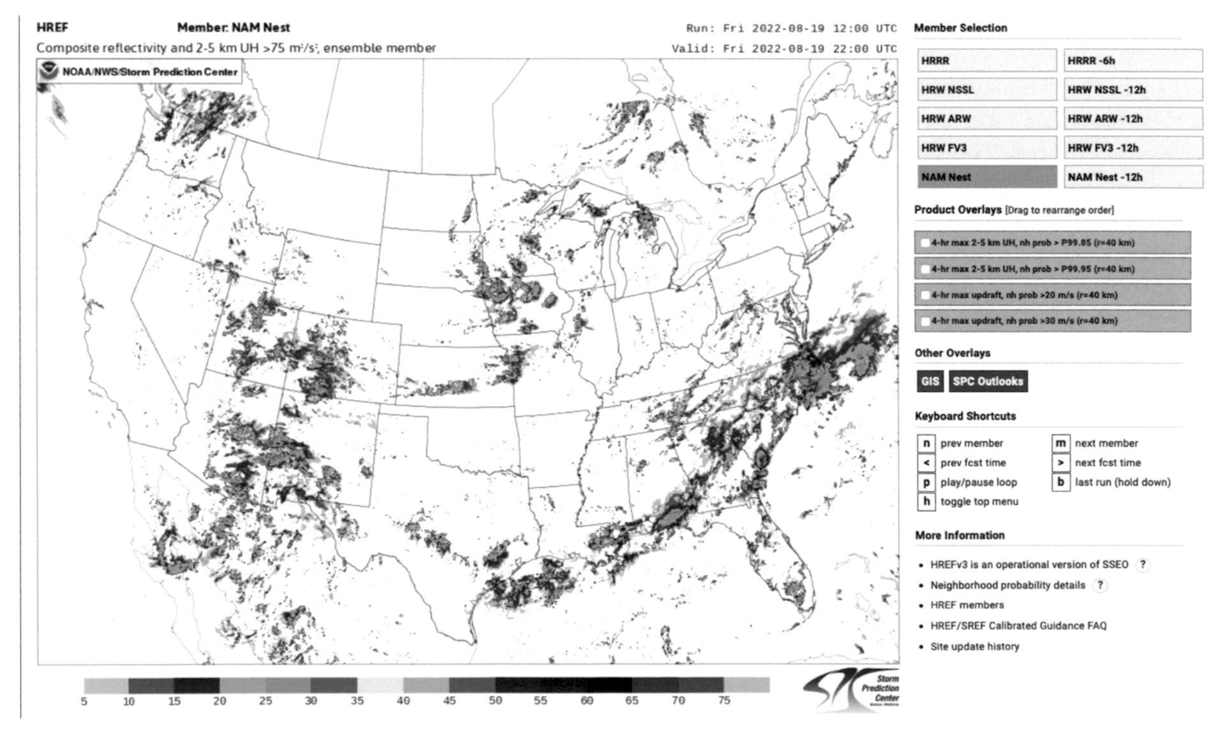

Figure 4.26 NAM model simulation of shower and thunderstorm activity on a summer day across the United States (valid at 6 p.m. EDT, August 19, 2022). The output field illustrates simulated radar reflectivity; the rain intensity scale is shown across the bottom of the image, with intensity values increasing from left to right along the color bar. (NOAA.)

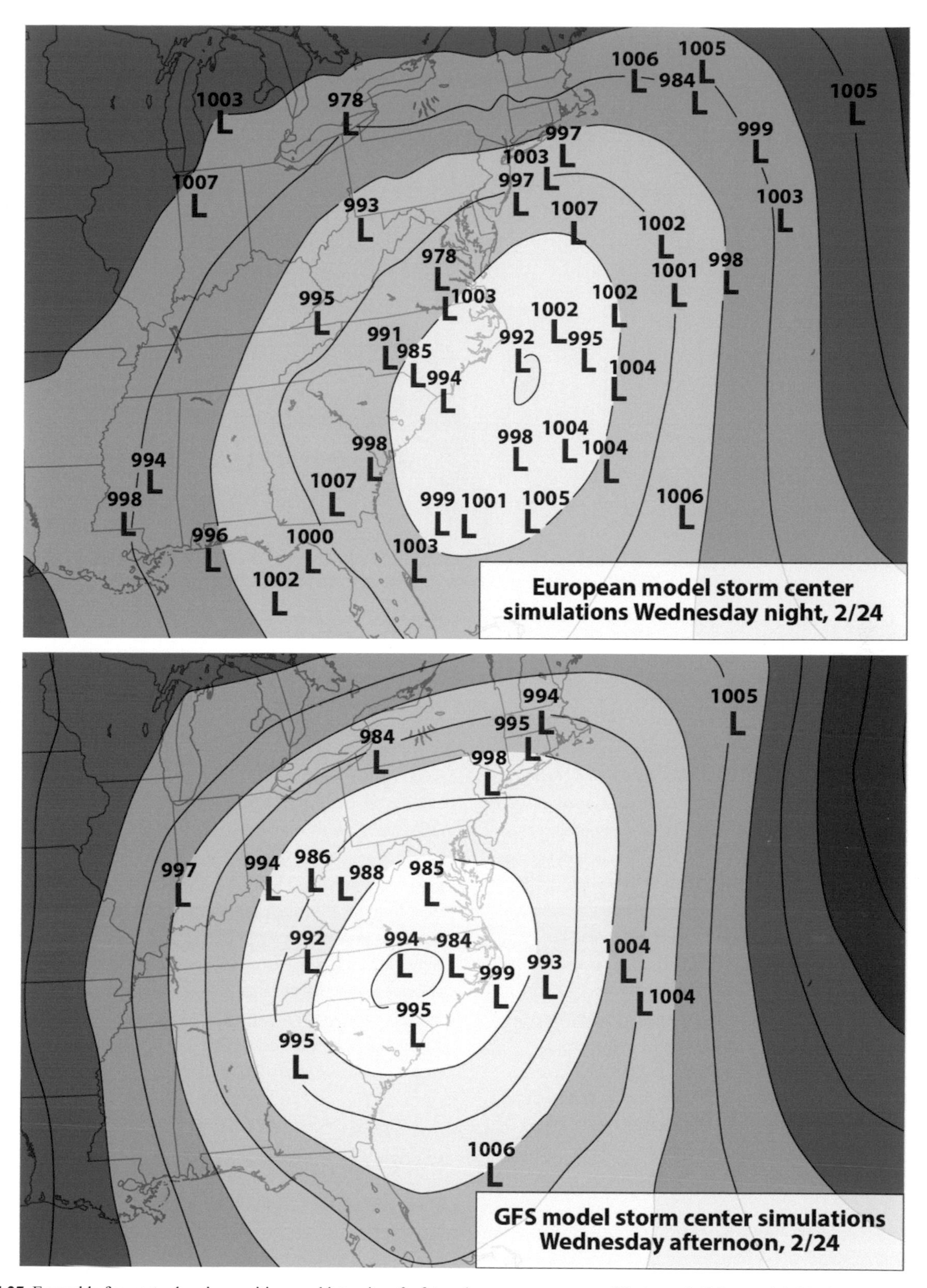

Figure 4.27 Ensemble forecasts showing positions and intensity of a future low-pressure system (Nor'caster), February 24, 2016. Intensity is expressed as minimum central pressure in millibars (mb). The yellow color ring indicates the highest probability of the storm's track – that is, where the most ensemble members cluster. Ensemble predictions for the European model are shown in the top panel, ensemble predictions for the GFS (American model) in the bottom panel.

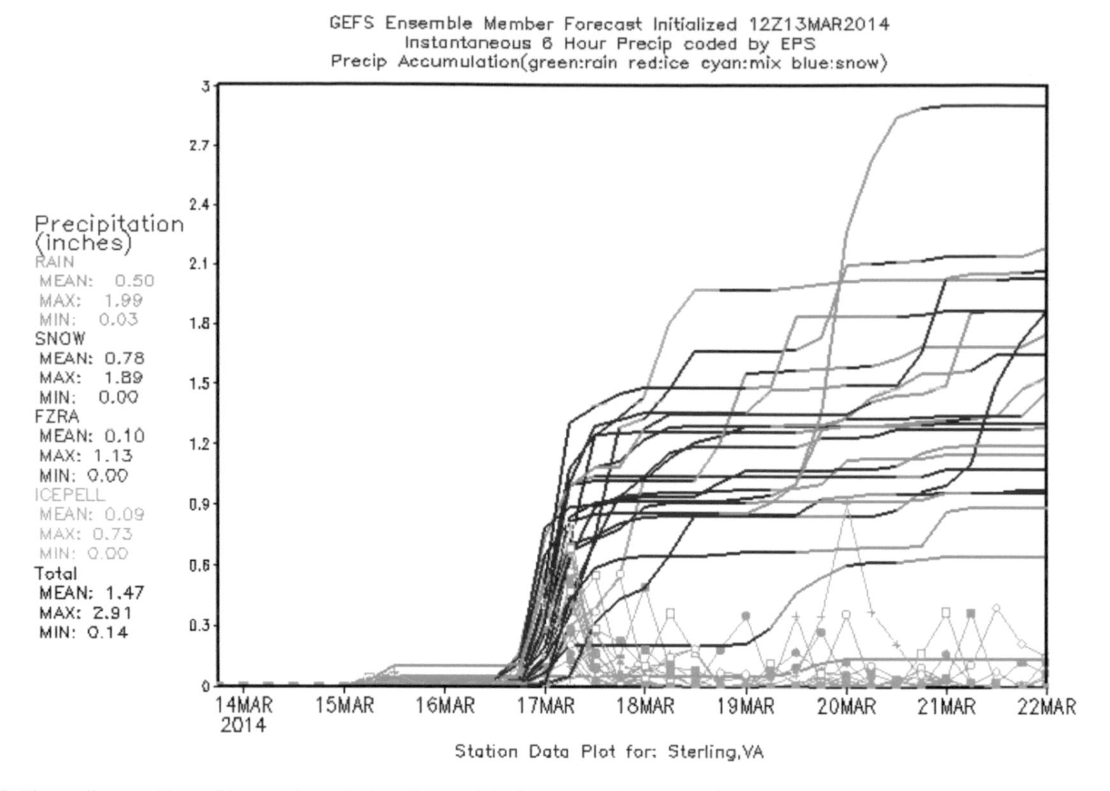

Figure 4.28 Plume diagram. Ensemble model predictions for precipitation type and accumulation for an East Coast storm over Washington, D.C., March 17 through March 22, 2014. All members of the American (GFS) model ensemble are shown here. (NOAA.)

a major snowstorm over parts of the Mid-Atlantic and New England. As we will explain in Chapter 7, the likelihood of heavy snow very much depends on the track and intensity of the low. The top panel depicts the position of the low for each European model member. There is significant variation, or *spread*, in the low's center. The American ensemble runs are shown in the bottom panel. Not only is there less geographic spread between various ensemble members, but the highest concentration of members is shifted considerably further west, over North Carolina. The American model thus predicted an inland track for the Nor'easter, while the European model suggested an offshore track. Whenever the ensemble members are clustered tightly together, we say that the uncertainty for the forecast cycle is low; widely divergent or spread out members imply higher uncertainty. In the latter situation, the model is having trouble handling the particular weather system's evolution.

Figure 4.28 shows another way to express ensemble model predictions. The precipitation forecast for each GFS ensemble member is plotted on a diagram, showing the accumulation of precipitation over time. For winter storm situations, the *type* of precipitation is also color coded: green for rain, red for ice, and blue for snow. The mass of lines for these types of plots visually resembles a set of plumes, hence the name plume diagram. In this example, the GFS ensemble members are plotted for a Nor'easter predicted to impact the Washington, D.C. region from March 17 through March 22, 2014. Many of the ensemble members predicted a significant amount of total precipitation, between 2.3–5.2 cm (0.9 and 2.1 inches). Most ensemble members also suggested that the precipitation would start as snow (blue lines) but then shift to rain or a rain-snow mix on March 19. Again, the tighter the clustering among plumes, the greater the confidence that can be placed in that particular model run.

Forecast Offices and Specialty Prediction Centers

The NWS, operating as an arm of NOAA, provides forecasts for specific U.S. regions. The map in Figure 4.29 illustrates the area of responsibility for each Weather Service forecast

Figure 4.29 Regions of forecast responsibility for all National Weather Service forecast offices in the contiguous United States. (NOAA.)

office (WSFO). Many offices are located on or near major airports. Some large states such as California and Texas have 10–12 forecast offices. Some offices handle different portions of multiple states. WSFOs transmit all routine forecasts, updated several times a day. Warnings are occasionally issued for severe weather, such as flash floods, tornadoes, severe thunderstorms, high winds, excessive heat or cold, winter storms, ice storms, and blizzards. Advisories are also issued for similar weather threats when the impacts are expected to be less – for example, a winter weather advisory, wind advisory, or heat advisory.

NOAA also operates numerous National Centers for Environmental Prediction (NCEP), staffed by experts in a particular forecasting specialty. As Figure 4.30 shows, the NCEP offices provide national coverage for specific types of weather threats and coordinate closely with individual WSFOs. The NCEP nexus is located in College Park, Maryland, at the Weather Prediction Center (WPC). Here, predictions regarding heavy snow and heavy rain are made for the entire country. The WPC also houses the Climate Prediction Center (CPC), providing national guidance on short-term climate fluctuations such as El Niño and long-term global effects such as global warming. The Environmental Modeling Center (EMC), also at WPC, develops and improves numerical weather prediction models. The Ocean Prediction Center (OPC) at WPC issues warnings and forecasts for stormy conditions across the Atlantic and Pacific Oceans.

Perhaps better known are the Tropical Prediction Center (TPC, also known as the National Hurricane Center) in Miami, Florida, and the Storm Prediction Center (SPC) in Norman, Oklahoma. TPC provides expert guidance on all tropical cyclone threats affecting the U.S. mainland, issuing tropical storm watches and hurricane watches. SPC handles severe local storms, including severe thunderstorm and tornado watches. Watches are issued when conditions over a broad geographic region favor development of severe weather in the next 3–6 hours. Watches means conditions are ripe, but severe weather is not a certainty. Warnings are issued for small regions/communities, last approximately 30 minutes, and are not issued until the weather phenomena are detected or observed.

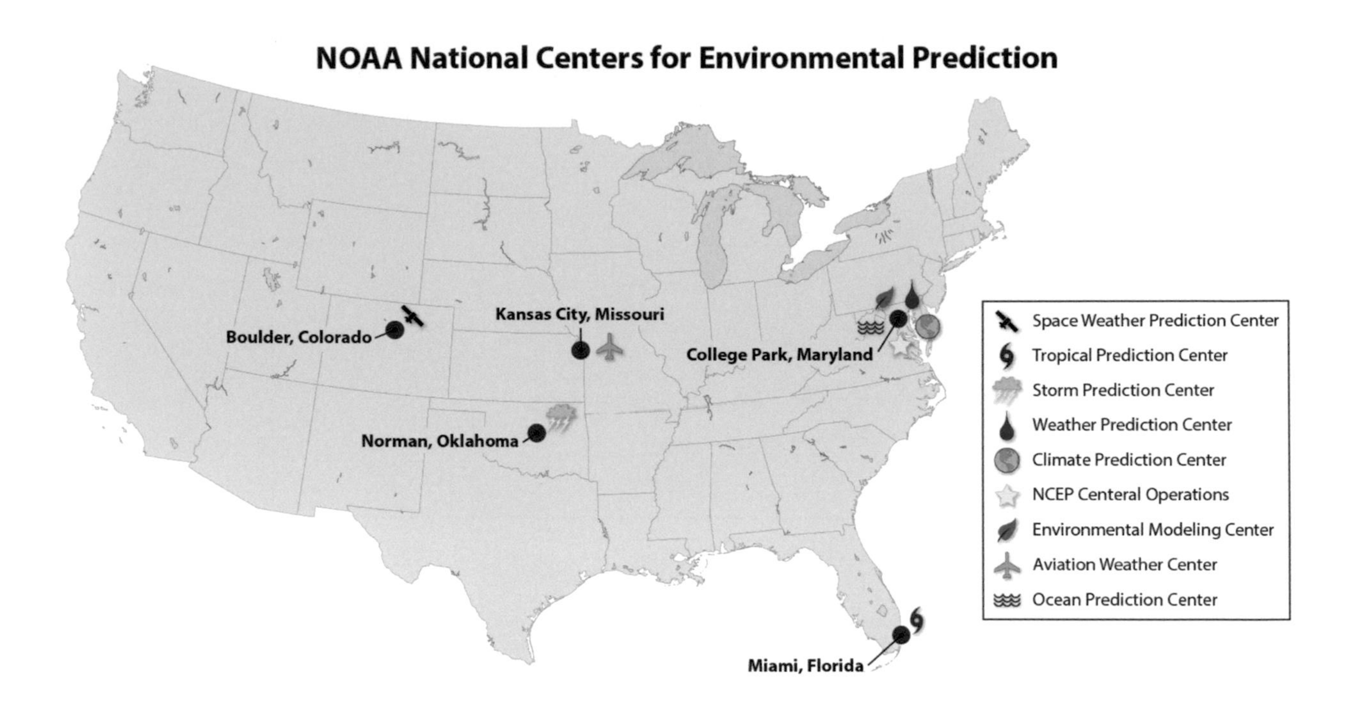

Figure 4.30 Locations and designations of the various National Centers for Environmental Prediction (NCEP) operated by NOAA.

Summary

LO1 Explain the various networks that make surface observations of weather phenomena, and list the basic types of information portrayed on the synoptic weather chart.

1. In meteorology, the synoptic scale covers the entire United States and spans several thousand kilometers. The National Weather Service (NWS) employs a national network of automated surface weather sites; the data collected at each site in the Automated Surface Observation System (ASOS) are synthesized into synoptic surface weather charts every 1–3 hours.
2. Mesoscale refers to "intermediate scale" phenomena, such as clusters or lines of thunderstorms and weather fronts, with distance scales on the order of tens to hundreds of kilometers. High-resolution networks of surface weather sensors, called mesonetworks, are found in some parts of the country, particularly where severe thunderstorms frequently occur.

LO2 Describe the instruments that are used to create upper-air analysis charts and the types of information that these charts portray.

1. Twice-daily radiosonde releases at approximately 100 sites provide data on upper atmospheric conditions around North America, including temperature, moisture, and winds.
2. Wind profilers and satellite data are used to fill in the gaps between radiosonde stations spaced approximately 200 km (124 miles) apart.
3. Upper-air analysis charts are constructed at various levels in the troposphere. The chart at approximately 9 km (30,000 feet) reveals the location and intensity of the westerly jet stream, which is closely tied to the formation of surface weather systems, including severe thunderstorms and tornadoes.

LO3 Distinguish between geostationary satellites and polar orbiting satellites, and discuss the types of information about cloud systems revealed by visible, infrared, and microwave wavelengths.

1. Geostationary satellites orbit in such a manner as to provide a fixed view of a specific region of Earth's surface, collecting images every 15 minutes.
2. Polar orbiting satellites image narrow swaths or strips of Earth's surface and atmosphere as they complete a 90–100 minute orbit, revisiting each location on Earth twice a day.
3. The visible light channel on satellites is used to study the distribution of clouds and the structure of cloud tops, and in select regions, it is used to monitor rapid changes in thunderstorm organization (with images collected every minute).
4. The infrared satellite channel divulges information about cloud altitude and is used to monitor the intensity trend of large thunderstorms as well as more localized thunderstorm systems.
5. The microwave channel can "see" through thick cloud layers, providing information on wind speed and direction above the ocean surface, as well as the 3D precipitation structure inside storm clouds.

LO4 Discuss how weather radar identifies the location and intensity of precipitation and lightning.

1. The modern network of NWS weather radars uses microwave pulses to determine the location, intensity, and movement of precipitation.
2. The Doppler principle enables detection of horizontal winds inside storm systems, including small rotating regions called mesocyclones, which are the precursor to most strong and violent tornadoes.
3. Polarimetric radar is based on the principle that different types of precipitation targets (large raindrops, hail, and snowflakes) scatter radar energy in the horizontal and vertical planes differently, enabling polarimetric radar to distinguish among the various precipitation types in storm clouds.
4. Because weather radars suffer from a number of shortcomings, particularly at great distances from precipitation targets, a dense network of rain gauges is used to refine radar estimates of rain and snow accumulation.
5. Lightning detection networks reveal the most energetic cells contained with clusters and lines of thunderstorms, pinpointing the most likely regions of severe weather.

LO5 Explain the basic principles behind weather forecast models, including the different types of models and how ensemble forecasts can improve model certainty.

1. Weather models are created by representing the atmosphere and Earth's surface as a spherical grid of 3D points. Predictive equations for temperature, moisture, pressure, and wind are solved at each grid point for future time increments.
2. Weather models require a carefully constructed set of initial conditions, based on real observations at the surface and all atmospheric layers. Small errors in the initial conditions can mushroom into large errors in predicting the evolution of weather systems over a 10-day period.
3. The American suite of weather forecast models includes the Global Forecast System (GFS), which comprises the medium range (out to 10 days); the North American model, which is short range (with predictions out to 2.5 days); and the High Resolution Rapid Refresh model, which provides ultra-short range (18-hour) predictions of weather.
4. The European Center for Medium Range Forecasting (ECMWF) model is a direct competitor with the American medium-range model; the two models are often compared during the prediction of large, high-impact storm systems such as hurricanes and Nor'easters (wintertime coastal cyclones).
5. Ensemble forecasts (collections of slightly varying model runs) give information about the degree of error contained in member predictions, providing forecasters with the level of confidence in a particular forecast run.

LO6 Describe the basic functions of NOAA's Centers for Environmental Prediction and the network of Weather Service Forecast Offices, explaining the difference between a watch and a warning.

1. The NWS operates a national network of weather service forecast offices (WSFOs) that handle daily, routine forecasts as well as issue warnings for high-impact weather phenomena.
2. Several National Centers for Environmental Predication (NCEP) have forecasters specially trained in severe storms, hurricanes, and heavy precipitation. These offices provide guidance on the national level and issue weather watches.
3. Weather watches means conditions are favorable for the occurrence of a particular type of weather threat; they are issued for large geographical regions and time frames of 3–6 hours. Warnings are issued for small regions/communities, last approximately 30 minutes, and are not issued until the weather phenomena are detected or observed.

Key Scientific Principles Covered in This Chapter

Note to the student: Many of these "first principles" will be used in later chapters, as a reminder of an important physical process or definition, to aid in your understanding of more advanced material.

4.1 The synoptic scale refers to the large-scale atmospheric state, encompassing all of North America and adjoining oceans. A synoptic weather map is a snapshot of meteorological observations at the surface or upper levels of the atmosphere. Key weather systems include cyclones, anticyclones, and jet streams.

4.2 The term mesoscale refers to middle atmospheric scales, including the narrow region along fronts, clusters and lines of thunderstorms, and regional wind systems such as the sea breeze and Santa Ana.

4.3 The radiosonde is small, digital package lofted by a helium-filled balloon, providing measurements of temperature, humidity, pressure and winds from the surface to about 100,000 feet.

4.4 Vertical wind profilers use narrow microwave beams, pointed skyward, to ascertain variations in wind speed and direction with altitude.

4.5 Geostationary satellites image the full disc of the Earth, from a variety of wavelengths, with new images collected every few minutes. Polar orbiting satellites image narrow swaths of Earth atmosphere and surface, much less frequently (once or twice a day) but at higher resolution than geostationary satellites.

4.6 In satellite imagery, visible radiation is used generate picture-like images of cloud tops. Infrared radiation is used to determine cloud top height and the temperature of land surface and bodies of water. Microwave radiation can penetrate clouds, discerning areas of precipitation and winds over the open ocean.

4.7 Weather radar uses microwave energy pulses to determine distance to precipitating areas, precipitation intensity, and the vertical distribution of precipitation. Drizzle, rain, snow, and hail can all be imaged by radar. The animation of sequential radar images reveals the movement and intensity trends of precipitating cloud systems.

4.8 Doppler radar detects frequency shifts in precipitation particles moving toward or away from the radar beam. This information can be used to reconstruct air motions in weather systems, including severe thunderstorms, making possible the detection of small, rotating motions (mesocyclones and tornadoes) and small regions of shearing winds (downbursts and microbursts).

4.9 Polarimetric radar uses radar beams that alternate between horizontal and vertical polarization in order to better discriminate between rain/snow, rain/hail, and more accurate detection of heavy rain regions.

4.10 National lightning detection networks and smaller lightning mapping arrays enable instantaneous, precision mapping of cloud-to-ground lightning strokes, and in some cases, total lightning (both cloud-to-ground and intracloud discharges). Lightning detection pinpoints the most intense thunderstorm cells, can be used to monitor thunderstorm intensity trends, and provides lead time before the onset of severe weather.

4.11 Numerical weather prediction (NWP) is the mathematical simulation of the atmosphere's future state. There are many types of models, including global, regional and fine-scale (i.e., down to the scale of an individual thunderstorm cell).

4.12 Numerical weather prediction now embraces ensemble forecasting, in which a suite of model runs is made in parallel, each simulation starting with slightly different starting conditions. This takes into account the typical errors associated with the observations used to initiate the model. The spread or difference in ensemble members is used to gauge the reliability of the model run.

4.13 The National Oceanographic and Atmospheric Administration (NOAA) oversees operations of the National Weather Service (NWS). The NWS, in turn, operates numerous weather service forecast offices around the country. These offices handle daily forecasts and issue special advisories and warnings when severe conditions warrant. NOAA also operates a number of national centers including numerical weather prediction, prediction of flooding and heavy snowfall, severe thunderstorms/tornadoes forecasting, aviation weather, and hurricane forecasting.

PART II

Weather Hazards Generated by Large-Scale Atmospheric Vortices

CHAPTER 5

Structure, Energetics, and Climatology of Extratropical Cyclones vs Hurricanes

Learning Objectives

1. Discuss the principal features of a typical extratropical cyclone, in terms of its connection to the polar jet stream, the interaction of air masses with different origins, pressure and wind distributions, cyclone life cycles, and fronts.
2. Describe the typical origin of Atlantic tropical cyclones, including their stages of development (tropical depression, tropical storm, hurricane), and factors that determine the track of the storm.
3. Explain how extratropical cyclones and tropical cyclones lie along a continuum or spectrum of large-scale cyclonic vortices, which also includes subtropical cyclones, arctic cyclones, and maritime cyclones (coastal lows and Nor'easters).
4. Discuss the difference between a cold-core extratropical cyclone and warm-core tropical cyclone in terms of principal energy sources. What is the source of potential energy for each type of vortex, and how is this converted to the kinetic energy of wind?
5. How are tropical cyclones categorized? Where in the Atlantic are they most prevalent and at what time of year? What processes influence how many tropical cyclones make landfall in the United States? How common are extratropical cyclones compared to tropical cyclones? What are their characteristic tracks across North America? How do these cyclones vary seasonally?
6. Compare and contrast severe weather hazards for each type of vortex. What type of hazards are found in the cold and warm sides of an extratropical cyclone? How do winds and rainfall vary across a typical hurricane?

Introduction

The generic term cyclone refers to any large-scale (i.e., impacting several states simultaneously, or a large region of the United States) low-pressure vortex. A cyclone is a vortex in which winds spiral inward toward the center in a counterclockwise manner. Cyclones are responsible for a large percentage of all U.S. severe weather. They strike year-round, assume many forms, and are called a variety of names. Hurricanes, an intense form of tropical cyclone, are endemic to the tropics, but frequently move into the mid-latitudes during the late summer and fall.

There is a spectrum of cyclone types. On one end are large, extratropical cyclones of the mid-latitudes (Figure 5.1) with the wavelike frontal structures, introduced in Chapter 2. On the other end lie the compact but extremely powerful tropical cyclones of late and summer and fall (Figure 5.2), including named tropical storms (winds > 34 kts [39 MPH]) and hurricanes (winds > 65 kts [74 MPH]). Between these two extremes lie various hybrid varieties of cyclones, including coastal lows, intense marine cyclones, subtropical cyclones, and arctic hurricanes. Each type of vortex can transition from one type to another. For instance, during its post-tropical phase, a hurricane might transform into an extratropical cyclone and continue to produce severe weather as a strong frontal system hundreds of miles inland. The surface synoptic map in Figure 5.3, obtained on September 20, 2005, shows that more than one type of cyclone may be active in North America at any given time. While Hurricane Rita brought heavy rains and high winds to southern Florida, a strong extratropical cyclone unleashed gusty winds and thunderstorms across New England.

DOI: 10.4324/9781003344988-7

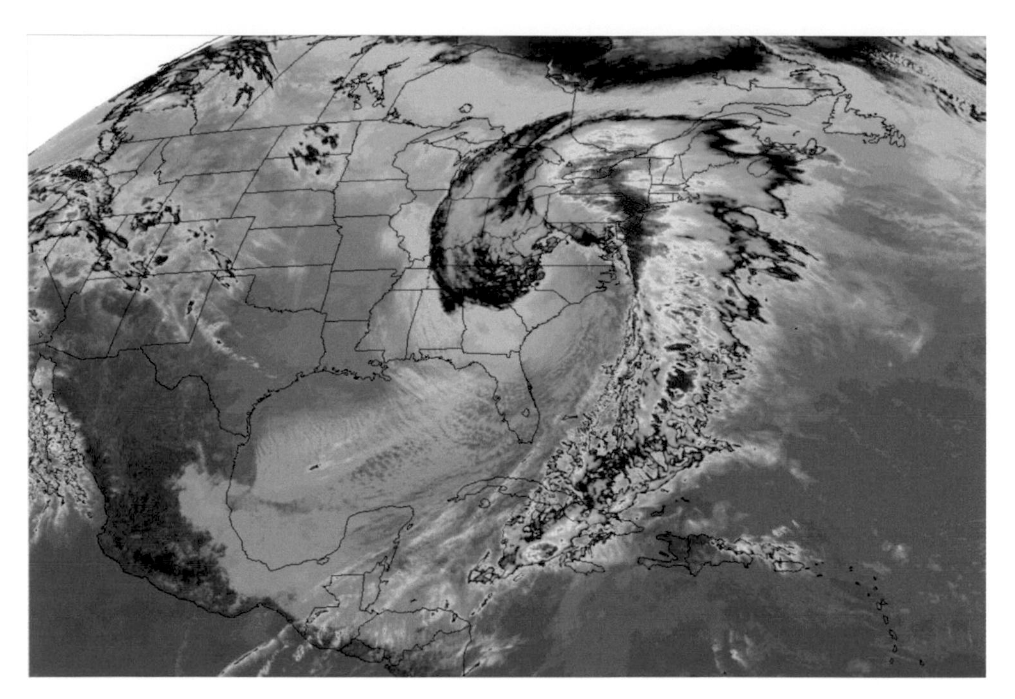

Figure 5.1 “Storm of the Century”: The March 1993 Nor’easter. The March 1993 Nor’easter is an example of a powerful extratropical cyclone that pummeled the U.S. Eastern Seaboard with heavy snow and high winds (false-color infrared satellite image). (NOAA.)

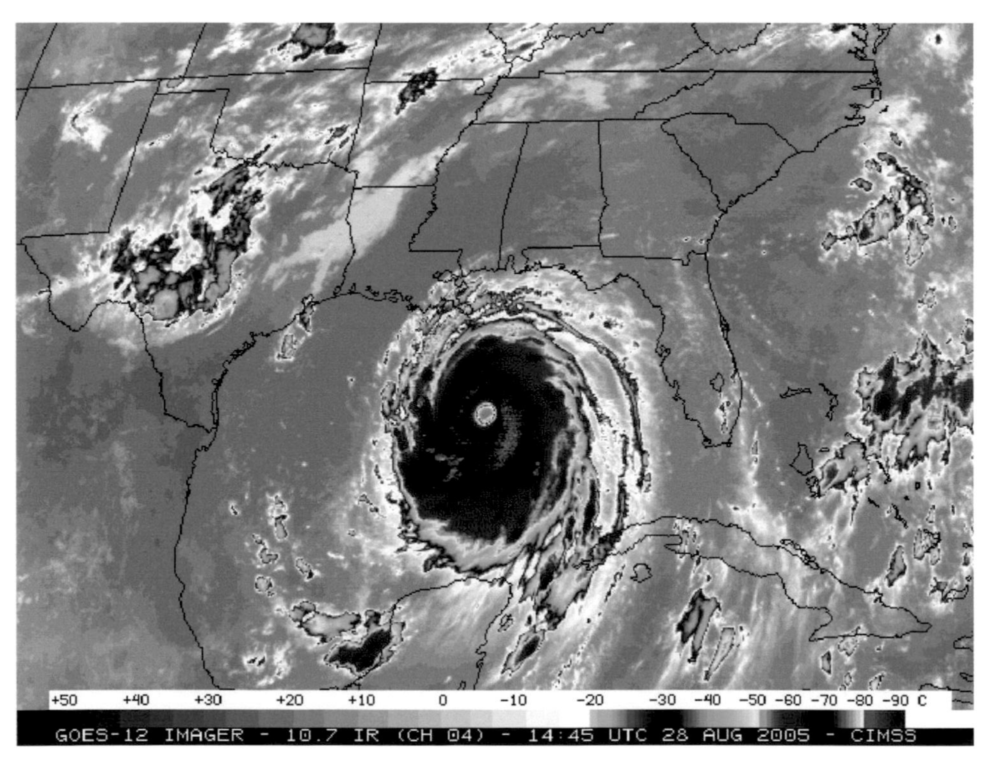

Figure 5.2 Hurricane Katrina as viewed from a false-color infrared satellite image. This storm dealt a devastating blow to the Gulf Coast region during September 2005, creating torrential rains and widespread wind damage and completely drowning the city of New Orleans. (NOAA.)

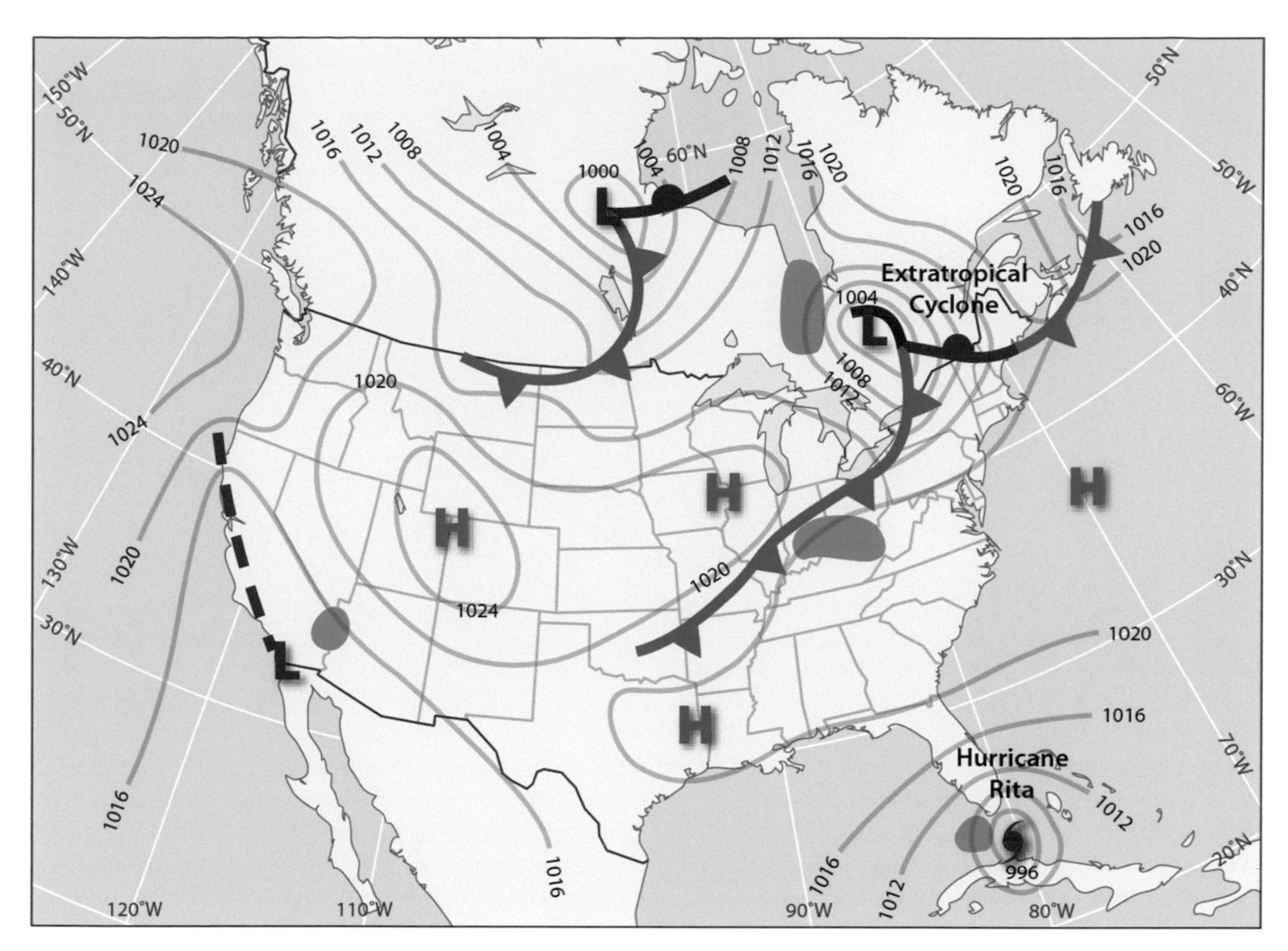

Figure 5.3 This surface weather chart illustrates both an extratropical and a tropical cyclone striking the United States at the same time. The hurricane is Rita; the extratropical cyclone bears no name (it is common practice not to name cyclones of extratropical origin).

In this chapter, we focus on extratropical cyclones and tropical cyclones. We broadly compare these kinds of storms in terms of physical structure, development and movement, energy sources, and characteristic weather hazards. In subsequent chapters we discuss, in much greater detail, each of these cyclone types and the severe weather they produce.

The Extratropical Cyclone: The Mid-Latitudes' Big Weather Maker

Extratropical cyclones are the most common synoptic-scale storm of the mid-latitudes, affecting every state in the continental United States. These large, intense vortices develop beneath patterns of curved airflow in the jet stream (discussed in Chapter 2) and feature fronts, including a warm front and cold front, across which air temperature, moisture, and winds vary markedly. Many areas of the country experience the effects of 50 or more of these systems in a given year, but they are most frequent and intense during the cool season (October–April). These disturbances, which also go by the names low, depression, and wave cyclone, are responsible for producing the majority of changeable weather conditions from day to day and much of the severe weather across the United States. Severe weather may include deadly tornadoes, widespread flooding, damaging winds, heavy snow and ice, fast-moving lines of severe thunderstorms, and coastal storm surges. Figure 5.4 shows the three-dimensional structure of a classic extratropical cyclone. The following sections examine its key characteristics.

Extratropical Cyclones Have Strong Air Mass Contrasts and Are Embedded in the Jet Stream

Extratropical cyclones are distinguished from other types of large, cyclonic vortex by (1) the presence of a warm front and

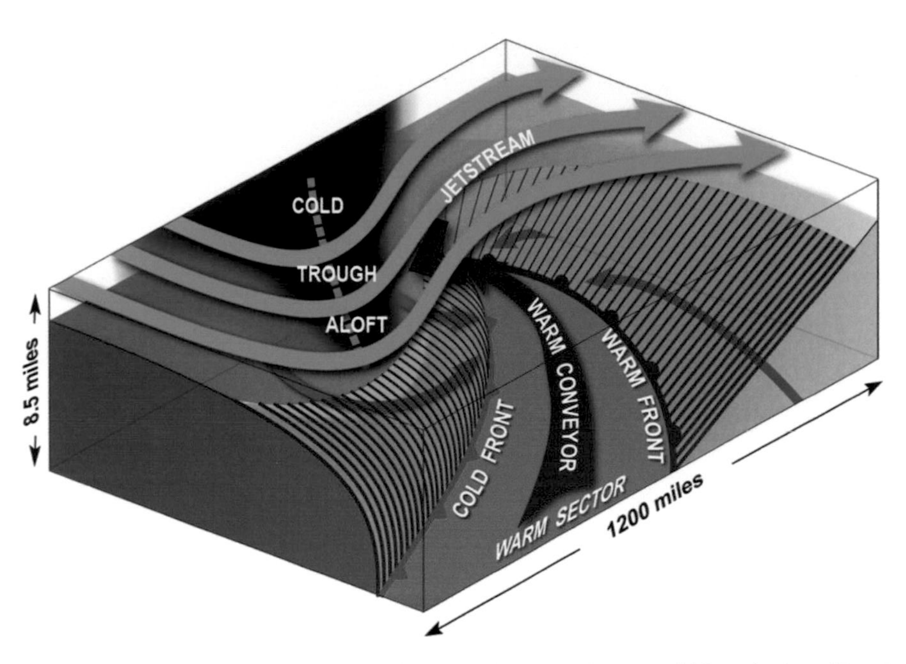

Figure 5.4 Three-dimensional structure of a typical extratropical cyclone. Note the relatively steep cold front dominated by advancing cold air and a more gentle warm front created by warm air overriding cooler dense air.

cold front and (2) the storm's intimate connection with the jet stream. The jet stream features regularly occurring ripples, or troughs, with cyclonic airflow (counterclockwise-turning in the northern hemisphere) and very cold air. Downstream of the trough, air ascends through the troposphere, creating a region of low pressure at the surface. When conditions are favorable, this cyclone intensifies, drawing in air masses from different locations. Cold, dry air arrives from the north, and warm, humid air arrives from the south. Where these air masses meet, frontal zones develop. The warm front occurs where warm, humid air overruns colder air to the north and east; the cold front develops along the leading edge of the cold air mass as it advances southward and eastward. Between these fronts lies the warm sector, a broad wedge of warm, humid air.

Within the warm sector, a narrow corridor of fast-moving airflow from the south, called the warm conveyor, imports subtropical and even tropical moisture into the system. Where this moisture is lifted along the warm front and cold front, bands of cloud and precipitation develop. Extratropical cyclones are typically quite large; their primary weather impacts extend across a region of 1000–1500 km (620–930 miles) or more in diameter. A core of very cold air resides in their upper levels, and winds within the vortex increase with height. These storms exemplify a type of large-scale weather disturbance characterized by large temperature and moisture gradients. These temperature contrasts are a source of potential energy that becomes converted into the kinetic energy of air. (Potential energy refers to a reservoir of energy that accumulates but has not yet realized motion in the atmosphere. Kinetic energy in the atmosphere refers to air in motion – e.g., the wind.) Because they are embedded in fast jet flow aloft, extratropical cyclones move eastward in connection with the trough, at about one-half the speed of air moving through the trough.

Extratropical Cyclones Undergo a Characteristic Life Cycle

Extratropical cyclones undergo a life cycle of stages including genesis, maturity, and decay. The life cycle unfolds over several days. Figure 5.5 shows the typical life cycle of an extratropical cyclone and its attendant upper-level trough as it moves across North America. These cyclones quite frequently are "born" just east of the Rockies. As the cyclone moves eastward and matures, note how the jet stream trough amplifies and changes its orientation in relation to the surface cyclone. On Day 1, formation day, the warm front and cold front become defined, and a shield of cloud and precipitation develops. It is important to note that the surface cyclone develops ahead of the upper-level trough, such that the vortex tilts westward with height. By Day 3, spiral inflow into the cyclone increases, and this tightens the temperature contrast. As the cyclone matures, the frontal precipitation becomes more intense. The mature stage (Day 3) is also called a wave cyclone.

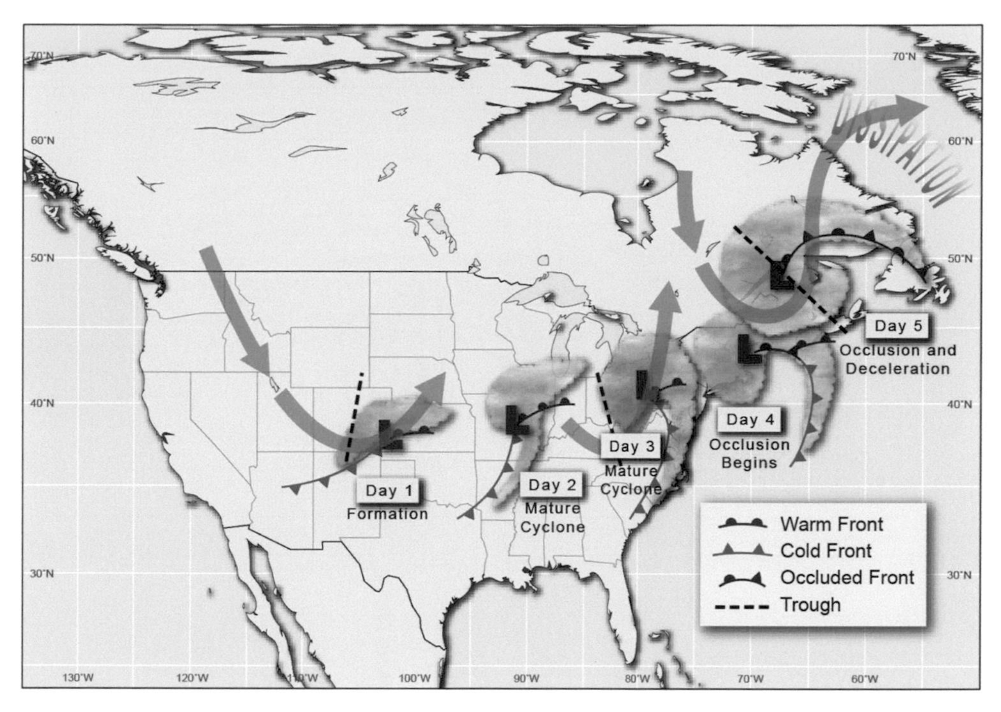

Figure 5.5 Typical life cycle of a wave cyclone and the attendant upper-level trough as it moves across the United States. Notice how the warm and cold fronts move together, much like a pair of scissors closing, as the system tracks across the country.

By Day 4, the process of occlusion begins as cold air wraps completely around the low. The cold front merges with the warm front, creating the occluded front in the northeastern part of the system. During occlusion, the warm sector diminishes in areal extent and becomes elevated. Additionally, the jet stream trough is now highly amplified. The center of surface low pressure migrates beneath the trough axis. In this configuration, the trough and surface cyclone move more slowly, and the entire system may become stationary. By Day 5, the surface system has completely occluded. Cold air completely envelops the low, and with the loss in temperature contrast, the storm weakens and its winds begin to spin down. But a large cloud shield remains and both heavy precipitation and strong surface winds may continue for several days.

Tropical Cyclones Include Both Tropical Storms and Hurricanes and Occasionally Impact the United States

Unlike extratropical cyclones, tropical cyclones develop in very low latitudes where temperature contrasts are insignificant. Compared to extratropical cyclones, tropical cyclones are quite rare, with only 80–90 forming worldwide each year. Within a uniformly warm and very humid tropical air mass, large and persistent clusters of thunderstorms erupt. Given proper environmental conditions – including a region of broad cyclonic rotation at the surface and warm ocean temperature – these thunderstorms may slowly coalesce into an intense vortex.

Tropical Cyclones Feature a Central Warm Core and Spiral Rain Bands

Figure 5.6 shows the three-dimensional structure of a mature tropical cyclone. Tropical cyclones are very intense storms containing sustained strong winds, heavy rain squalls, and damaging ocean waves. Their peak frequency in the Atlantic is mid-September. The main cloud system, which is more compact than an extratropical cyclone, features spiral cloud bands and a clear, central eye that remains vertically erect. The eye is the very core of the hurricane, featuring warm, calm, and clear air. The eye is surrounded by a ring-shaped eyewall of intense thunderstorms. The strongest winds in the hurricane vortex form a ring-like structure that coincides with the eyewall. Spiral rain bands arc inward toward the eyewall and are composed of numerous thunderstorm cells. Within these cells, air vigorously ascends; on the edges of the cells and within the eye, the air subsides (sinks).

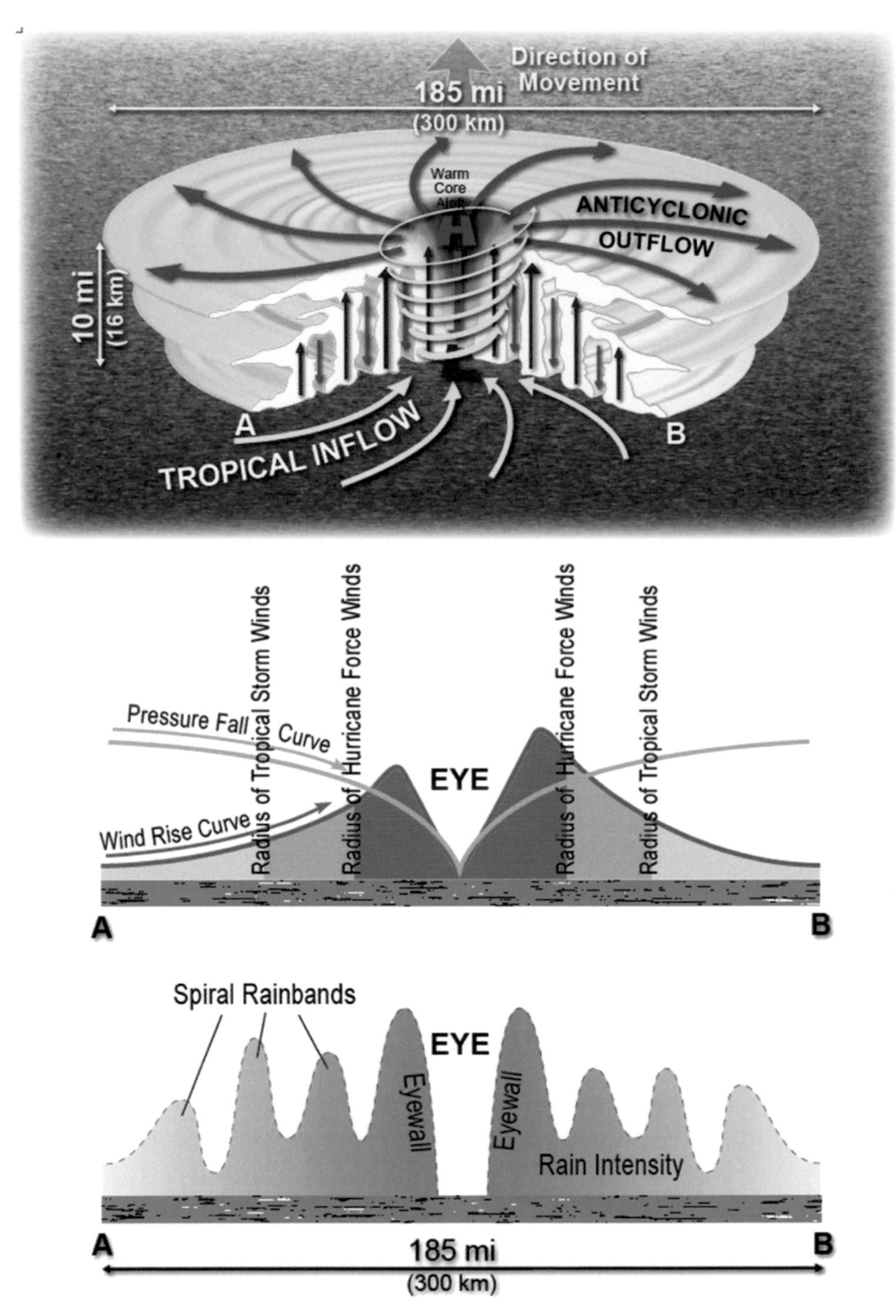

Figure 5.6 Three-dimensional structure of a mature hurricane. Wind speed increases, moving into the storm as pressure falls. Notice that rain intensity does not conform to the more uniform wind and pressure curves. Rather, it is more variable and linked to cloud structure and associated rain bands.

Through most of the storm's depth, the airflow describes a strongly cyclonic (counterclockwise) spiral, with air converging into the central low. The uppermost 1–2 km (0.6–1.2 miles) of the storm, however, feature high pressure. This high creates anticyclonic (clockwise) outflow away from the storm's eye. Unlike the extratropical cyclone, the hurricane's inner vortex is warm (particularly in the upper levels) relative to the surrounding atmosphere, and vortex winds decrease in intensity with altitude. The storm receives its energy not from temperature contrasts or the jet stream but from warm air and water vapor overlaying the ocean surface. Generally, the warmer the ocean surface, the stronger the storm may potentially become.

Tropical Cyclones Undergo Numerous Stages of Development as They Cross the Atlantic Ocean

Figure 5.7 shows the typical development and evolution of an Atlantic tropical cyclone. Tropical cyclones are typically longer-lived than their mid-latitude cousins and become much more intense. They are more compact, with intense winds typically restricted to a diameter of 150–300 km (93–186 miles). All Atlantic tropical cyclones begin as low-pressure regions embedded in the easterly trade winds. On Day 1 a type of elongated low-pressure system termed an African easterly wave emerges off the coast of West Africa. Within this disturbance, clusters of tropical thunderstorms develop. Given a source of preexisting spin in the lower and middle troposphere, these thunderstorms begin to coalesce into spiral rain bands by Day 3. At this stage, a circular region of low surface pressure and broad cyclonic wind circulation define a tropical depression. If the thunderstorms become more numerous and intense, the depression may strengthen further. Once sustained surface winds exceed 34 kts (39 MPH), the cyclone becomes a tropical storm and receives an official name. The tropical storm becomes a hurricane if winds intensify beyond 64 kts (74 MPH) and a closed eye develops. The sequence outlined thus far unfolds gradually, over the course of several days, but portions of this evolution may occur rapidly. Abrupt periods of rapid intensification may occur, with transitions taking place in a matter of hours!

Once a tropical cyclone has achieved minimal hurricane status, the storm may further intensify, or it might flounder and weaken. The outcome depends on the nature of the larger-scale atmospheric and oceanic environment through which the storm passes. Hurricanes are rated according to an intensity scale, ranging from Category 1 (65 kts [75 MPH]) through Category 5 (135+ kts [155+ MPH]).

The tropical cyclone is broadly embedded within and steered by the easterly trade winds along the southern flank of a giant high-pressure cell called the subtropical anticyclone. As the storm enters the western Atlantic, it begins the process of recurvature around the western limb of the anticyclone. Recurvature is the process by which a tropical cyclone changes direction, heading northeast or east, as it moves into higher latitudes. Recurvature may take place over the open ocean, along the eastern U.S. coastline, or inland. Some westward-moving tropical cyclones also pass straight into the Gulf of Mexico, their recurvature delayed until

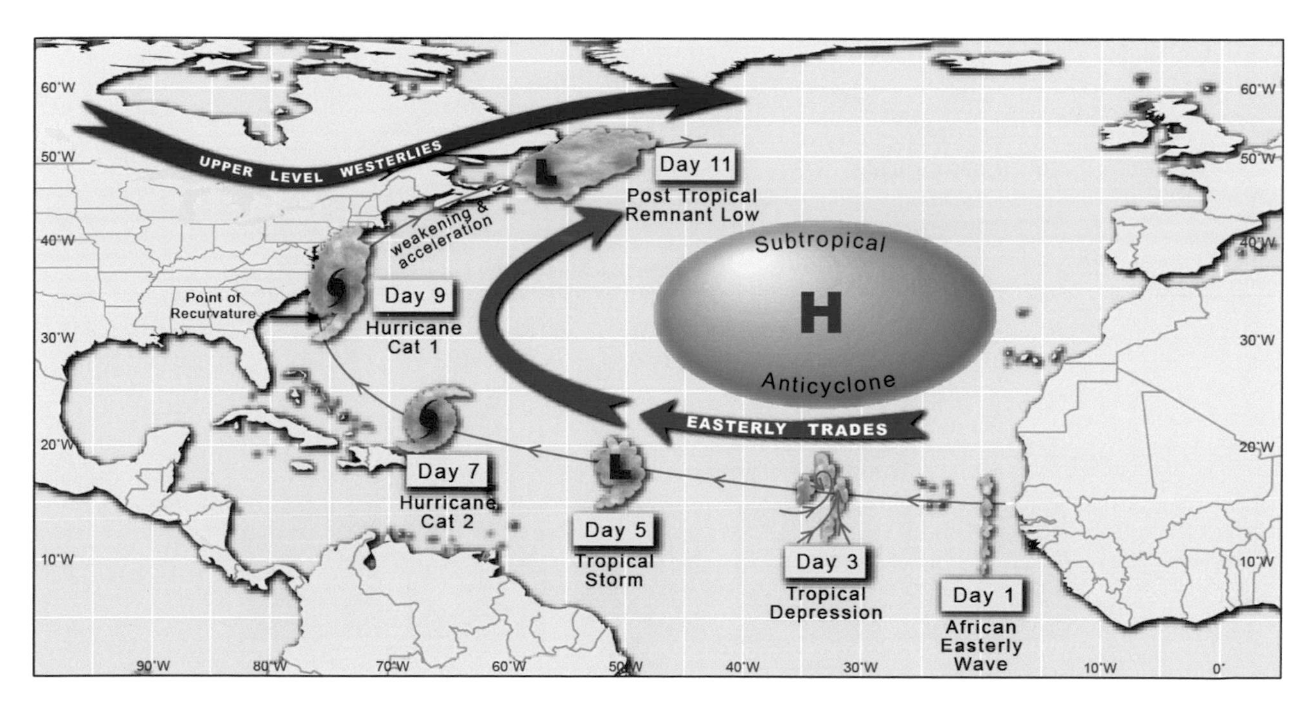

Figure 5.7 Typical life cycle of a tropical cyclone. Hurricanes are generally longer-lived than extratropical cyclones. Most begin with an east-to-west movement and then recurve and develop a west-to-east motion in high latitudes. The storm can undergo numerous transitions along the way leading to marked changes in wind speed and weather impacts.

after making landfall. Regardless of when the storm turns, its entry into higher latitudes marks a major transformation. Tropical cyclones derive all of their energy from the warm ocean, in the form of sensible heat (warm air) and water vapor. If over water, and the system encounters cooler sea surface temperatures, the storm receives less energy; if over land, the system is removed from its oceanic source of water vapor "fuel." Additionally, drier air resident over the United States may enter the storm and evaporate thunderstorm clouds. For all these reasons, the tropical cyclone rapidly weakens.

As the dissipating storm moves northward, it encounters the westerly jet stream. These strong upper-level winds deform or shear away the storm's cloud and precipitation shield. Thus, while weakening, the storm becomes asymmetric and accelerates toward the east-northeast. This highly transformed hurricane in mid-latitudes has entered its post-tropical phase. However, tropical remnants are still capable of producing heavy rains and high winds, perhaps several days and hundreds of miles inland. It is even possible for post-tropical remnants to rejuvenate and create a variety of severe weather catastrophes across the eastern United States and Canada.

Beyond Purely Extratropical and Tropical Cyclones, There Is a Broad Spectrum of Cyclone Types That Can Transform From One to Another

In the previous sections, we compared the two principal types of cyclonic storms that regularly impact North America: tropical cyclones and extratropical cyclones. However, these storms are both endpoints along a continuum or spectrum of large-scale atmospheric vortices. Between these two extremes lie "shades of gray" – that is, hybrid vortices that contain elements of both tropical and extratropical systems, many of which produce substantial weather impacts. Figure 5.8 presents a conceptual framework of cyclone morphology. The arrow indicates the temperature structure of the cyclone core. On the left side, the classic extratropical cyclone is representative of the deep cold-core, frontal-type wave cyclones so common in the mid-latitudes. "Cold core" refers to the prevailing temperature structure within the vortex. At the other extreme lies the deep warm-core hurricane. Between the extremes are intense marine cyclones in the mid-latitudes, arctic cyclones, and subtropical cyclones.

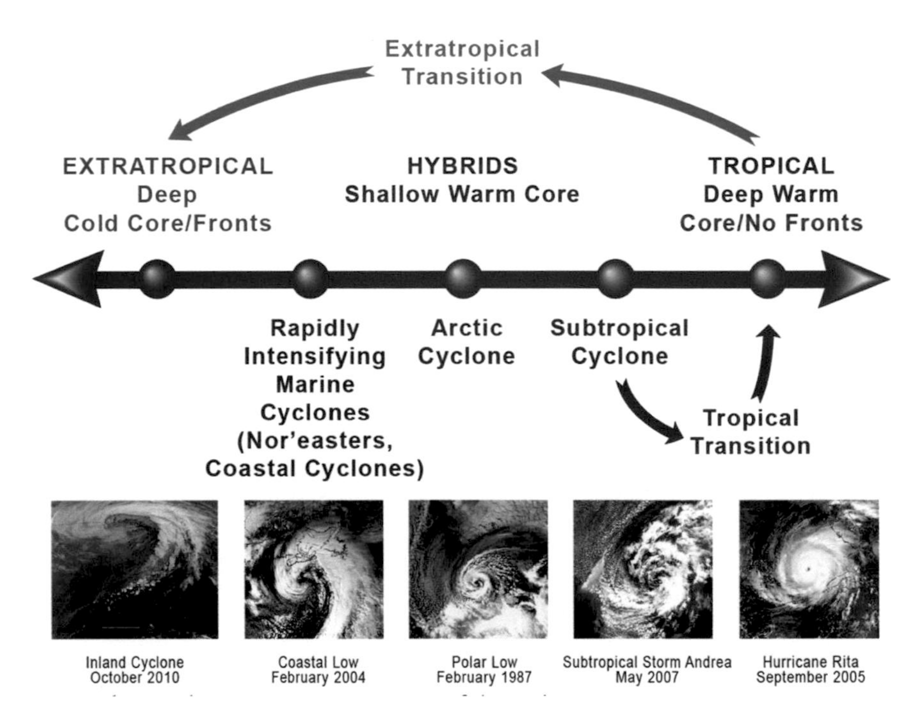

Figure 5.8 Spectrum of cyclone types. This conceptual framework illustrates a wide range of storm types found from tropical latitudes to high latitudes. Some of these storms transition from one kind to another as they progress through their life cycles.

Subtropical Cyclones Are Similar to Hurricanes but Larger and Weaker

Subtropical cyclones resemble hurricanes in many ways but are not as compact or as intense; nor do they develop in the "deep tropics" but rather in latitudes closer to 30° N and S. These storms feature a broad wind circulation that is often weaker than hurricane strength. However, they retain a central core of intense thunderstorms and a shallow warm core in the lower-middle troposphere. There are no fronts extending from the center of circulation. They often develop from extratropical cyclones that move offshore during the early summer. Like tropical storms and hurricanes, subtropical storms are given names (once winds exceed 39 MPH). Subtropical cyclones can form over waters cooler than the typical threshold for hurricane development, which is 26° C (79° F), typical of late summer through early fall. Thus subtropical cyclones can form in the late spring or early winter, outside the classic hurricane season (June 1–November 30). And if a subtropical cyclone moves over warmer waters or there is sufficient instability in the atmosphere, it may undergo tropical transition, whereby its warm core deepens and intensifies, developing into a classic hurricane.

Arctic Cyclones Are Very Compact, High-Latitude Storms With Hurricane-Force Winds

Arctic cyclones (also called arctic hurricanes) are another type of "hurricane-like" vortex of intense wind, including a central eye, but they develop at high-latitudes during winter. They tend to be very compact (560–725 km [348–450 miles] in diameter), and they feature spiral cloud bands, a central core of thunderstorms, and a clear eye. Surface winds are typically in the 30–39 kts (35–45 MPH) range but can exceed 56–61 kts (64–70 MPH) on occasion. Like true tropical cyclones, they contain a warm core – but unlike hurricanes, they are well connected to the polar jet stream aloft. They derive some energy from the jet stream and also extract limited amounts of heat from the ocean. Arctic cyclones represent a hazard to oceanic shipping and fishing, producing high seas, heavy snow, whiteout conditions, and even lightning. Sea spray and intense cold rain can rapidly ice up the ship's superstructure.

Maritime Cyclones Are Intense Extratropical Storms Over the Ocean

A final category of intense extratropical cyclones are the maritime cyclones, which form exclusively over open ocean and which include two subtypes: coastal lows and Nor'easters. These are primarily wintertime cyclones that derive some of their energy from relatively warm ocean water.

Coastal lows are extratropical cyclones over the North Atlantic that initiate along the U.S. Gulf Coast and Eastern Seaboard of North America. Coastal lows also impinge upon the North American West Coast as fully developed storms, tracking inland off the Pacific Ocean. Nor'easters are a subtype of coastal low that develop in the vicinity of Cape Hatteras, North Carolina, or the Delmarva Peninsula. They get their name from the intense winds that blow onshore from the northeast. A given winter season may feature up to 5–10 Nor'easters, some of which produce damaging storm surge, high winds, and heavy precipitation. These effects lead to beach erosion and extensive property damage, and their cumulative seasonal impact rivals that of a strong hurricane. Further inland, Nor'easters produce nearly all heavy snowstorms along the northeastern megalopolis (the densely populated urban corridor stretching from Washington, D.C., to Boston). Nor'easters contain fronts and are connected to the jet stream, yet they derive a substantial amount of energy from warm Gulf Stream waters off the East Coast. Some Nor'easters even have a hurricane-like eye, and sustained surface winds can exceed 44–52 kts (51–60 MPH). These storms generate many weather hazards, and these effects are notoriously hard to predict.

Extratropical Transition Is the Process by Which a Weakening Tropical Cyclone Morphs Into an Extratropical Cyclone

Figure 5.9 illustrates a type of metamorphosis called extratropical transition (ET), which occurs in the western Atlantic when tropical cyclones recurve northward and weaken. The transition occurs as post-tropical remnants combine with preexisting, mid-latitude weather systems, including frontal boundaries, troughs in the jet stream, and even other extratropical cyclones. During ET, the tropical cyclone's deep warm core evolves into a more asymmetric cold core and develops a frontal wave structure. On rare occasions, ET can lead to rejuvenation of the post-tropical cyclone over land. During rejuvenation, the vortex either maintains the pressure it had as a tropical cyclone, or the pressure decreases. Historically, ET is responsible for numerous inland weather disasters, including outbreaks of tornadoes and widespread flooding (particularly over the Appalachians).

Figure 5.9 illustrates the ET of Hurricane Floyd (1999) over the Eastern Seaboard. Floyd first became a large and powerful Category 4 hurricane over the western Atlantic. Floyd started as a shallow warm-core vortex in the tropics on September 8 but then intensified to Category 4 and expanded in size by September 13. After the 13th, it began to recurve, making landfall over the North Carolina Outer Banks as a Category 2 hurricane (Figure 5.10). It then tracked northeastward along the

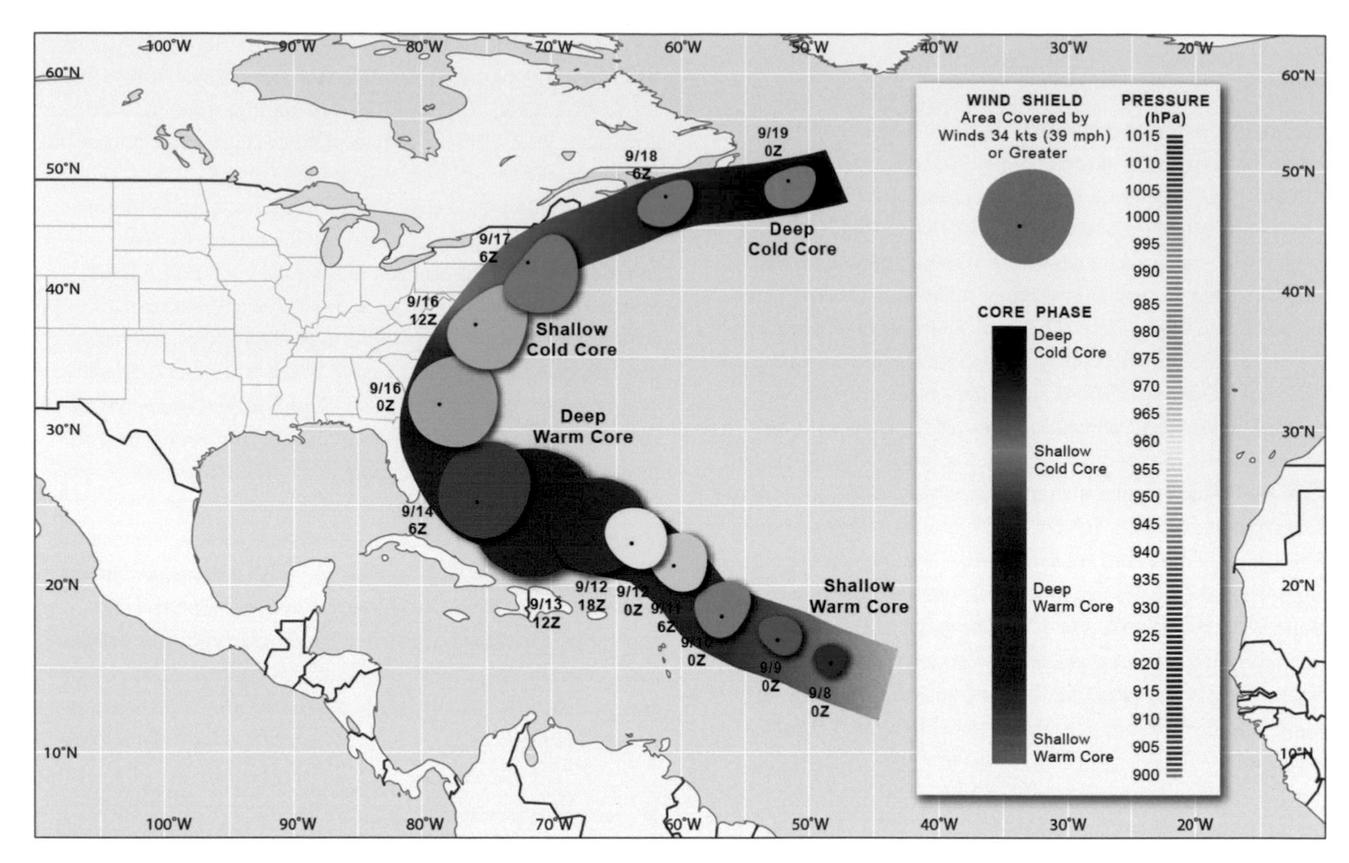

Figure 5.9 Extratropical transition of Hurricane Floyd. As Hurricane Floyd morphed from a warm-core to a cold-core storm, its symmetry also changed from a more uniform hurricane shape to the increased asymmetry associated with extratropical cyclones. Circular regions show size of wind footprint; circle color corresponds to central pressure scale in legend.

Figure 5.10 Satellite view of Hurricane Floyd making landfall along the East Coast of the United States. (NOAA.)

U.S. East Coast, undergoing ET. During this transformation, the storm evolved from deep warm core to deep cold core. It developed fronts, and its wind distribution became highly asymmetric. Extremely heavy rain spread over North Carolina, Virginia, and the Delmarva Peninsula (Figure 5.11), with the heaviest rainfall west of track. Rainfall exceeded 38 cm (15 inches) in many locations, resulting in devastating river flooding, in some cases exceeding a 500-year flood.

Extratropical and Tropical Cyclones Derive Their Energy From Different Sources

The atmosphere produces many types of large-scale vortices, and the processes that energize these storms are also varied. The processes operating in the mid-latitudes differ markedly from those operating in the tropics. In extratropical and tropical cyclones, winds blow because kinetic energy is generated from a source of potential energy. In extratropical cyclones, that potential energy accumulates when large-scale warm and cold

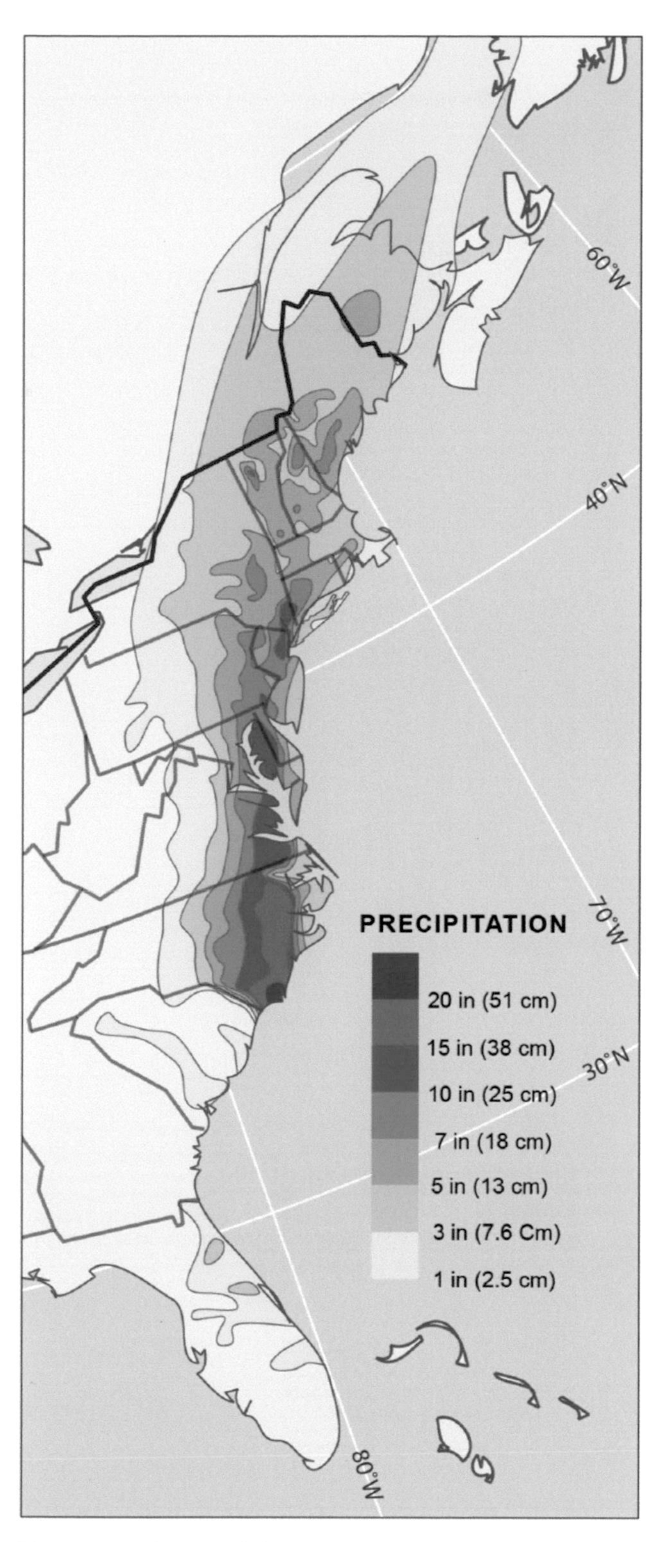

Figure 5.11 Rain accumulation map for Hurricane Floyd during extratropical transition. Hurricane Floyd deposited large quantities of rain along the East Coast of the United States, generating widespread river flooding and fatalities across North Carolina.

air masses are brought into juxtaposition, creating a frontal zone (a narrow region of strong temperature contrasts). In the environment that hosts tropical cyclones, there are no dissimilar air masses; instead, a vast reservoir of potential energy exists in the form of water vapor. Water vapor is created when ocean water evaporates – a process that extracts ocean heat (the energy level of liquid water must be "boosted" in order for molecules to break free of the water surface, as vapor). This stored heat is termed "latent heat." When vapor rising in the tropical thunderclouds of a hurricane condenses and subsequently freezes, latent heat is released into the air, warming it. The warm, buoyant air accelerates upward. Winds spiral inward to replace the air that is rising upward. In this manner, the fierce upward- and inward-flowing winds of a tropical cyclone are energized.

These sources of potential (and kinetic) energy are not mutually exclusive. For instance, when a hurricane moves over land and undergoes extratropical transition, the fraction of potential energy from latent heat diminishes, while the fraction derived from temperature contrasts increases. Thus for a time during ET, the storm may derive energy from two sources. In wintertime Nor'easters, much of the potential energy derives from strong temperature contrasts, but a substantial fraction arises from both sensible heat (air warmed from contact with the sea surface) and latent heat (in the form of water vapor) drawn off the warm ocean water.

Extratropical Cyclones Derive Their Energy Principally From Differences in Air Mass Temperature

Figures 5.12 and 5.13 contrast the processes by which extratropical and tropical cyclones derive their energy. The first diagram (Figure 5.12) shows the situation for extratropical cyclones, where deep warm and cold air masses are brought together along one or more frontal zones. The frontal zone is a narrow region of strongly contrasting temperature. Cold, dense air tends to sink as low as possible, while the more buoyant warm air rises upward. The process whereby cold air sinks and warm air simultaneously rises creates kinetic energy in the atmosphere. The frontal zone thus represents a reservoir of potential energy that is unleashed by these buoyant forces. In a mature extratropical cyclone, the conversion into kinetic energy drives the storm's vertical circulations and swirling wind and can be quite intense. By the time the cyclone occludes, cold air has completely wrapped around the low. With the loss of contrasts in air mass temperature, the storm weakens. But the weakening process is not necessarily abrupt; the inertia of the

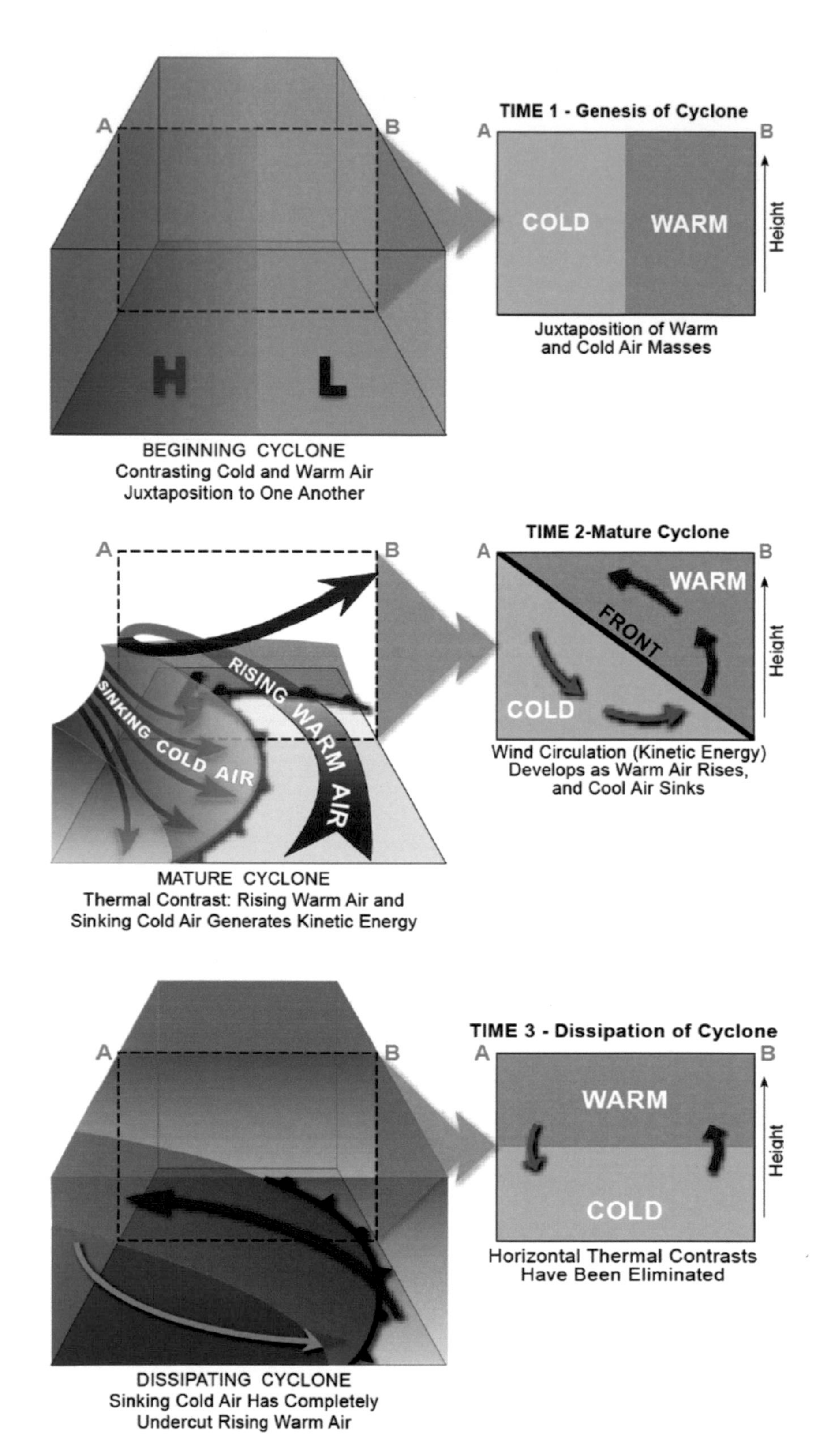

Figure 5.12 How extratropical storms derive their energy from contrasting air masses. As cold air interfaces with warm air (Time 1), the dense cold air sinks beneath the lighter warm air, forcing the warm air upward (Time 2). This action produces kinetic energy that drives the storm. Eventually, the cold air displaces the warm air at the surface, and the dynamic process slows or stops, resulting in the dissipation of the storm (Time 3).

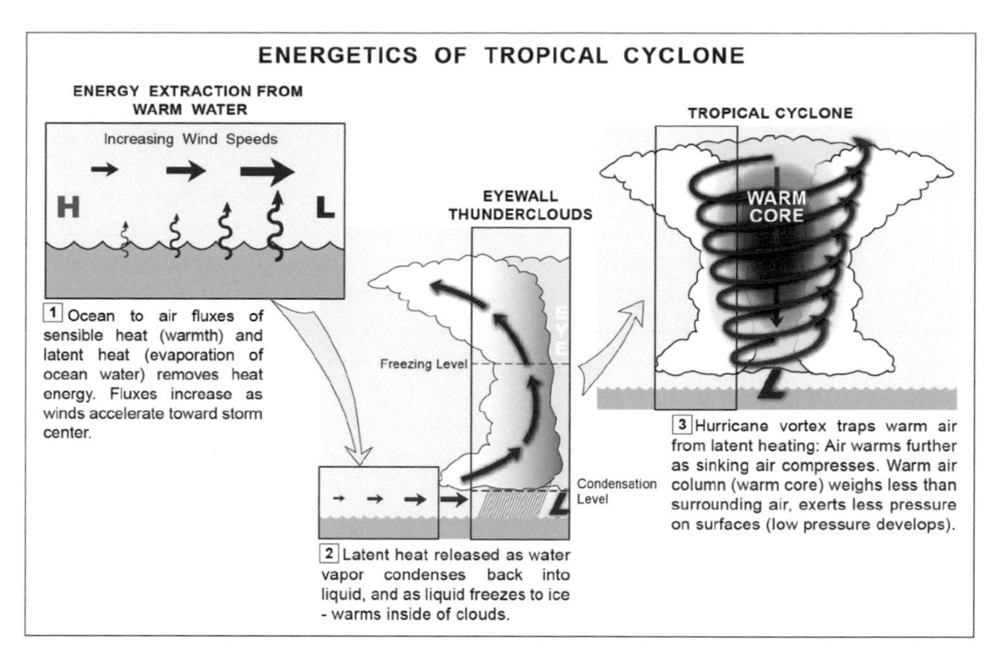

Figure 5.13 How tropical cyclones extract energy from warm water. As water is evaporated and carried aloft, cooling occurs, causing condensation and the release of latent heat, thus providing energy to the storm.

storm's winds and moisture held aloft may cause it to produce severe or hazardous weather for days after weakening begins.

Tropical Cyclones Are Powered by Latent Heat Released During Condensation

Figure 5.13 illustrates how a mature hurricane acquires energy from warm ocean water. There is an upward energy transfer from sea surface to overlying atmosphere. The movement of energy is called a flux, which is defined as a unit of energy transferred (a Joule) per unit area and unit of time. Fluxes are the means by which a hurricane "extracts" energy from the ocean.

There are two types of energy fluxes, one involving upward transfer of sensible heat (warm air rising) and one involving latent heat. The latent heat flux is closely related to the evaporation of seawater. Evaporation extracts heat from the ocean when liquid water is converted to water vapor. The water vapor is said to contain latent heat, which can be released inside clouds when the vapor condenses to liquid. Fluxes of sensible and latent heat are driven by differences in temperature and humidity, respectively, between the ocean surface and the atmosphere. Wind blowing above the ocean surface dramatically facilitates the upward transfer of heat and moisture into the air. Turbulence in the wind "stirs" the air layer above the ocean, enhancing the rate at which sensible heat and water vapor are extracted from the ocean. The latent heat flux tends to be the larger of the two forms of heat that are transferred upward.

Buoyant, warm air containing high water vapor content ascends in the hurricane eyewall. After ascending only a couple thousand feet, the air has cooled into condensation. During condensation, latent heat is released within the eyewall's convective updrafts. Updrafts are narrow, ascending currents within deep clouds. Latent heat warms the interior of the cloud towers, buoying updrafts to great heights (imagine turning up the heat inside a rising hot air balloon). At about 5490 m (18,000 feet) above the surface, liquid cloud droplets freeze, releasing additional latent heat. The addition of latent heat within eyewall updrafts ensures that the air remains warmer than its environment and thus continues to rise.

At the tropopause level (16.8 km [55,000 feet or higher]), air in the updrafts is forced to spread laterally. Much of the air flows outward, away from the storm, but a fraction converges from all directions over the storm's center. Convergence aloft drives a sinking air current down the center of the eye. As the air subsides, it compresses and warms (as discussed in Chapter 3). The result is pronounced warming in the storm's core. The rotating air column in the hurricane vortex prevents warm air from escaping. The warm, central air column – being less dense than its surroundings – exerts less pressure on the ocean surface. The reduction in sea level pressure accelerates inflowing air. The Coriolis effect deflects these winds inward, creating the strong, spiral inflow above the ocean surface. The high winds in turn extract more energy from the ocean surface. The process is somewhat self-sustaining. Through it, oceanic warmth and humidity are ultimately transformed into the kinetic energy of vortex winds.

Tropical Cyclones Are More Intense and Longer-Lived Than Extratropical Cyclones

Given a large enough region of very warm water, hurricanes can become extremely intense, with sustained surface winds in excess of 139 kts (160 MPH), and remain intense for days. In contrast, extratropical cyclones are self-limiting. While they can become very intense during the mature stage (wintertime Nor'easters occasionally develop sustained winds of 44–52 kts or 50–60 MPH), occlusion is an inevitable process that occurs within a day or two of the storm's reaching maturity.

Extratropical cyclones destroy the very temperature contrast that energizes them. During occlusion, cold air wraps around the entire storm. The source of potential energy is lost when the contrast in air mass temperature is eliminated. For this reason, extratropical cyclones can never become as powerful or as long-lived as the strongest hurricanes.

Digging Deeper: Vertical Structures of Tropical and Extratropical Cyclones

To provide further insight into the differences between tropical and extratropical cyclones, Figure 5.14 shows the vertical temperature structure of these storms. The upper panel illustrates what happens to horizontal surfaces of constant pressure in the troposphere when layers are either heated or cooled. During this discussion, bear in mind that hurricanes are vertically erect, warm-core vortices, while extratropical cyclones feature a central cold core that tilts with height.

Strong and concentrated warming is found in the middle troposphere of a hurricane (upper-left panel of Figure 5.14) and arises from subsidence-induced warming in the storm's eye. Warming causes air layers in the troposphere to expand vertically. When layers expand, the process distorts horizontal pressure surfaces, deflecting them downward beneath the region of concentrated heating. This reduces the pressure in the lower atmosphere, giving rise to low pressure at the surface. Where the pressure surfaces are most steeply tilted, the gradient wind around the vortex blows the strongest (the air must "flow downhill" faster toward the storm center). Within a hurricane core, the largest tilt occurs in the lower troposphere, and the tilt becomes less pronounced with height. Warm-core hurricane vortices, then, contain their strongest winds just above the ocean surface, and the vortex weakens with height.

In a mid-latitude cyclone (upper-right panel), a core of very cold air is found in the jet stream trough at high levels. This region of concentrated cooling contracts the air layers vertically, tilting horizontal pressure surfaces downward toward the trough from high up. A region of strong low pressure develops at high altitudes. Rising air induced downwind of the trough axis creates a region of low pressure at the surface. The cold cyclone vortex, then, must tilt westward with height. The horizontal pressure surfaces are tilted most steeply in upper levels. The strongest gradient winds are thus found aloft, and they weaken inside the vortex progressing toward the surface.

The middle panel of Figure 5.14 applies these concepts to real-world storms. Here we see the vertical temperature and pressure distributions inside a strong hurricane (left panel) and winter extratropical cyclone (right panel). The variation of winds with height is also shown (using white barbs and flags – described in the Appendix). The bottom set of panels shows the top-down or "plan view" for each storm. On these plan views, the orientation of the vertical slice through each storm is shown by the line segment AB.

The hurricane is a very compact system, featuring a concentrated warm core in the middle troposphere, and it is vertically erect. The pressure gradient at the surface is very large. Hurricane-force winds weaken with height. The extratropical cyclone is a much larger vortex and features a core of very cold air in the middle and upper troposphere. The structure is also highly asymmetric and characterized by strong horizontal temperature contrasts. Although the jet stream and its trough are not illustrated in this diagram, note that a wedge of frigid high-level air is pulled into the storm's circulation. While both storms have identical surface pressure (960 mb; recall that average sea level pressure is 1013 mb), the pressure gradient is weaker in the extratropical cyclone, leading to a more extensive but weaker zone of surface winds.

Why are hurricanes generally smaller vortices than extratropical cyclones? The upper-atmospheric disturbances in the jet stream that create extratropical cyclones (troughs in the jet stream) are larger than the precursor disturbances that form hurricanes, spanning several hundred kilometers. Hurricanes are thought to form from a number of small vortices (perhaps a few tens of kilometers in diameter) that aggregate and morph into the main storm vortex.

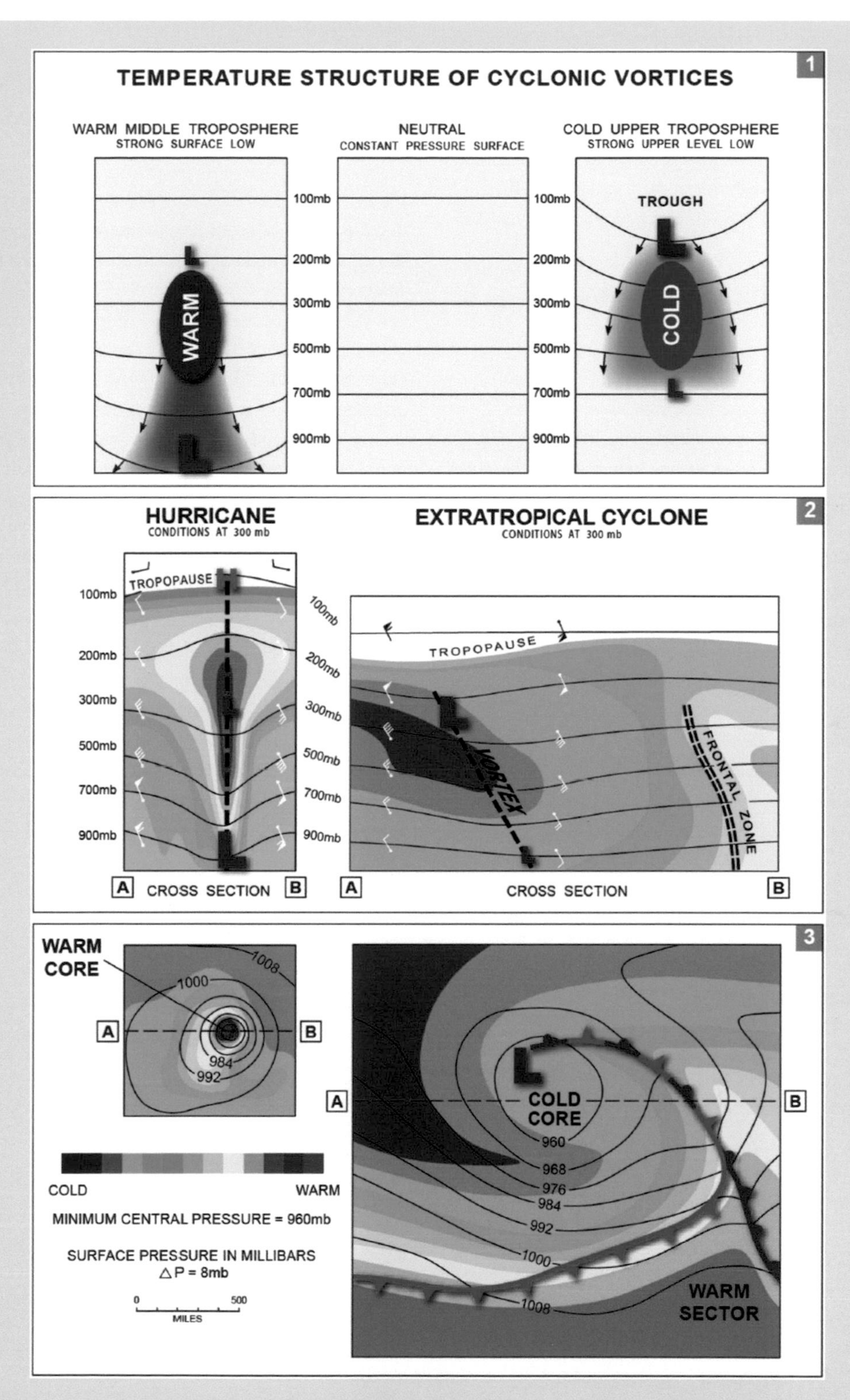

Figure 5.14 Different thermal structures of tropical and extratropical cyclones. Strong and concentrated warming is found in the middle troposphere of a hurricane (upper-left panel). In a mid-latitude cyclone (upper-right panel), a core of very cold air is found in the jet stream trough at high levels. In the middle panel, the vertical temperature and pressure distributions are illustrated inside a strong hurricane (left) and a wintertime extratropical cyclone (right). Variation of winds with height is also shown using barbs and flags. The bottom panel depicts the hurricane as a very compact system, featuring a concentrated warm core in the middle troposphere that is vertically erect. The extratropical cyclone shown in the lower panel has a much larger vortex and features a core of very cold air in the middle and upper troposphere.

Climatology of Tropical and Extratropical Cyclones

In this section we explore the climatology of tropical and extratropical cyclones; that is, we describe their geographical and seasonal variations. We start with the tropical cyclones, which can vary from year to year both in number and where they strike.

North Atlantic Tropical Cyclone Activity Peaks in Mid-September

The official hurricane season in the North Atlantic (including the Gulf of Mexico) runs from June 1 through November 30. Occasionally hurricanes can develop outside this time frame. But the likelihood of tropical cyclones occurring either very early or very late is small, due to the constraints of ocean temperature. Most tropical cyclones develop during August through October, peaking around September 10. Why does the peak occur three months after the summer solstice (the solstice is the time of year when the Sun is highest in the sky and the hours of daylight are greatest)? The reason is simple: Tropical cyclones, including hurricanes, are fueled by warm ocean water requiring large fluxes of sensible and latent heat. The fluxes increase with ocean temperature. Because of water's unusually large heat capacity (heat capacity is a measure of how much heat leads to a 1° C rise in temperature of a substance), many weeks under the summer sun are needed before the water reaches a temperature of at least 26° C (79° F). Above this threshold, the evaporation of seawater is vigorous enough to sustain the large energy requirements of hurricanes.

Hurricanes Are Categorized by the Saffir–Simpson Hurricane Intensity Scale

Figure 5.15 shows the frequency distribution of North Atlantic tropical cyclones according to intensity category. Tropical storms and hurricanes are rated according to a five-scale category called the Saffir–Simpson Scale, based on wind speed and minimum central pressure. Storms with maximum sustained winds in excess of 97 kts (112 MPH) (Categories 3–5) are termed major hurricanes. Figure 5.16 shows the incidence of storms by intensity, with all tropical cyclones (including tropical storms) shown in blue and the subset of cyclones making landfall shown in red. The distribution for both categories is consistent with the Law of Rare Events – that is, the weakest categories of storms are the most frequent.

There are only a few Category 4 or 5 hurricanes, accounting for 12% of the total. During the 20th century, only three Category 5 hurricanes struck the United States. But when the violent storms occur, they cause a disproportionate share of lives lost and property damaged. Based on the long-term average, the North Atlantic basin as whole experiences about 10 named tropical cyclones (tropical storms and hurricanes) and approximately 4 strong hurricanes each year. Historically, the number of storms per season has ranged from as few as 1 to as many as 30.

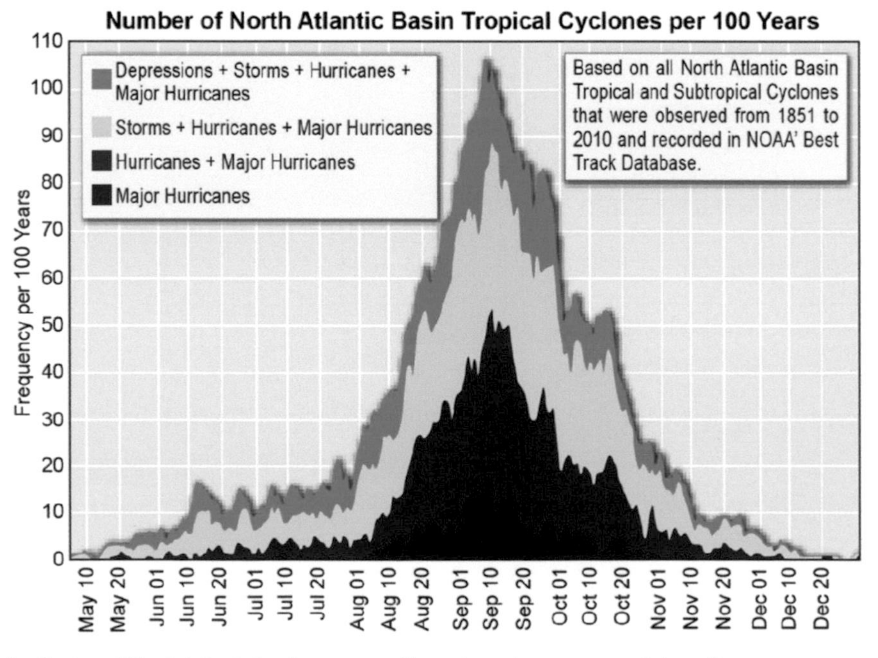

Figure 5.15 Frequency distribution of North Atlantic hurricanes according to intensity category and time of year.

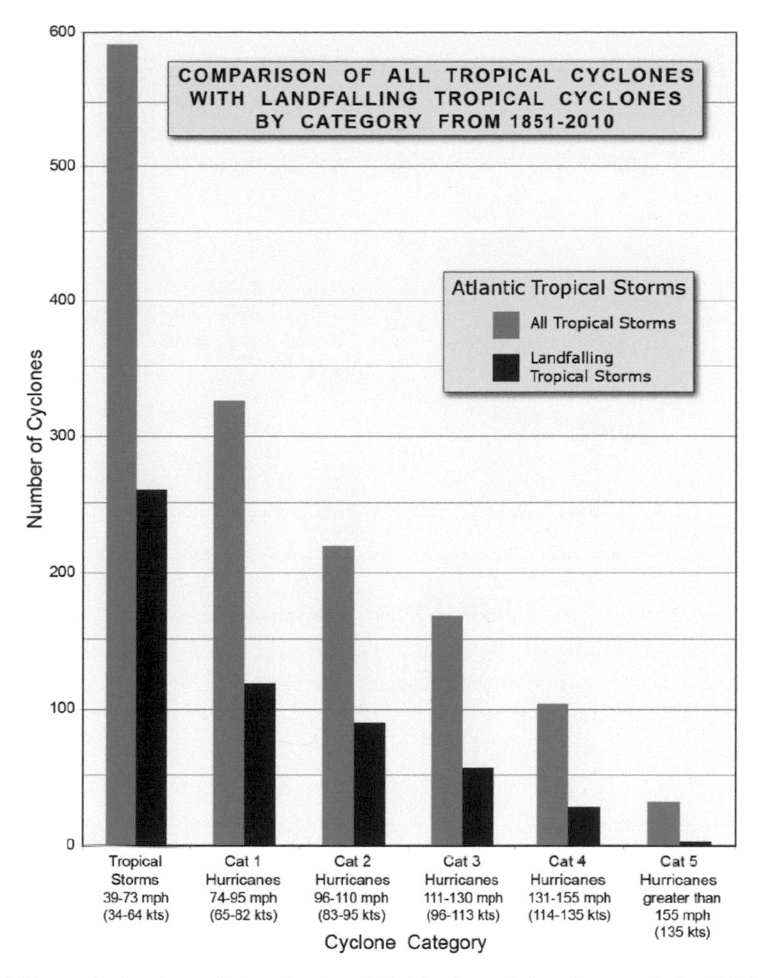

Figure 5.16 Frequency of landfalling tropical cyclones. Only a fraction of all Atlantic tropical cyclones make landfall in the continental United States. The long-term average for landfalling storms is 39%, or about 3.5 storms per year.

Number of Atlantic Tropical Cyclones in a Given Season Is a Weak Predictor of How Many Make Landfall

Figure 5.16 reveals that only a fraction of all Atlantic tropical cyclones undergo landfall in the continental United States. The long-term average for landfalling storms is 39%, or about 3.5 storms per year. It's natural to wonder whether years with more Atlantic hurricanes experience a greater number of tropical cyclone landfalls. The answer, as shown by Figures 5.17–5.19, is not straightforward. Figure 5.17 shows the total annual number of tropical cyclones versus the total annual number of landfalling tropical cyclones for the period 1851–2010. The positive slope in this scatterplot suggests a correlation, but not a very strong one ($R^2 = 0.3$). We must also allow for the fact that not all storms will take aim at the United States. Storms are steered by a number of larger atmospheric wind currents. Tropical cyclone movement is controlled by the strength, configuration, size, and location of the Atlantic subtropical anticyclone. These characteristics can vary markedly from year to year and even within a single season. Furthermore, the westerly jet stream of the mid-latitudes has a variable course and configuration during the hurricane season, strongly influencing when and where recurvature occurs.

Figures 5.18 and 5.19 illustrate two different situations, based on hurricane tracks for the 2004 and 2010 seasons. Both years featured high storm counts (15 and 21 tropical cyclones, respectively). In 2004, 8 storms made landfall (53%), while in 2010 – the more active season – only 3 storms (14%) made landfall. The reason for this paradox lies in steering currents. In 2004, a piece of the large subtropical anticyclone called the Bermuda High became established close to the U.S. East Coast, propelling storms into the Gulf Coast. Storms did not recurve until inland over the southeastern United States.

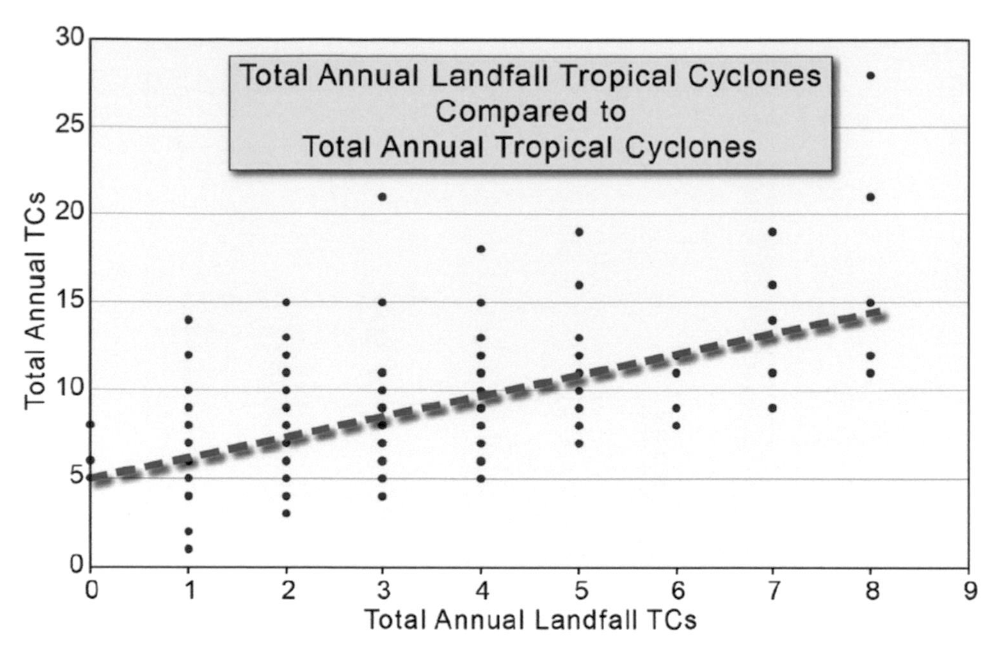

Figure 5.17 Statistical assessment of total vs landfalling tropical cyclones. The positive slope in this scatterplot suggests a correlation between the annual number of tropical cyclones and the annual number making landfall but not a very strong correlation ($R^2 = 0.3$).

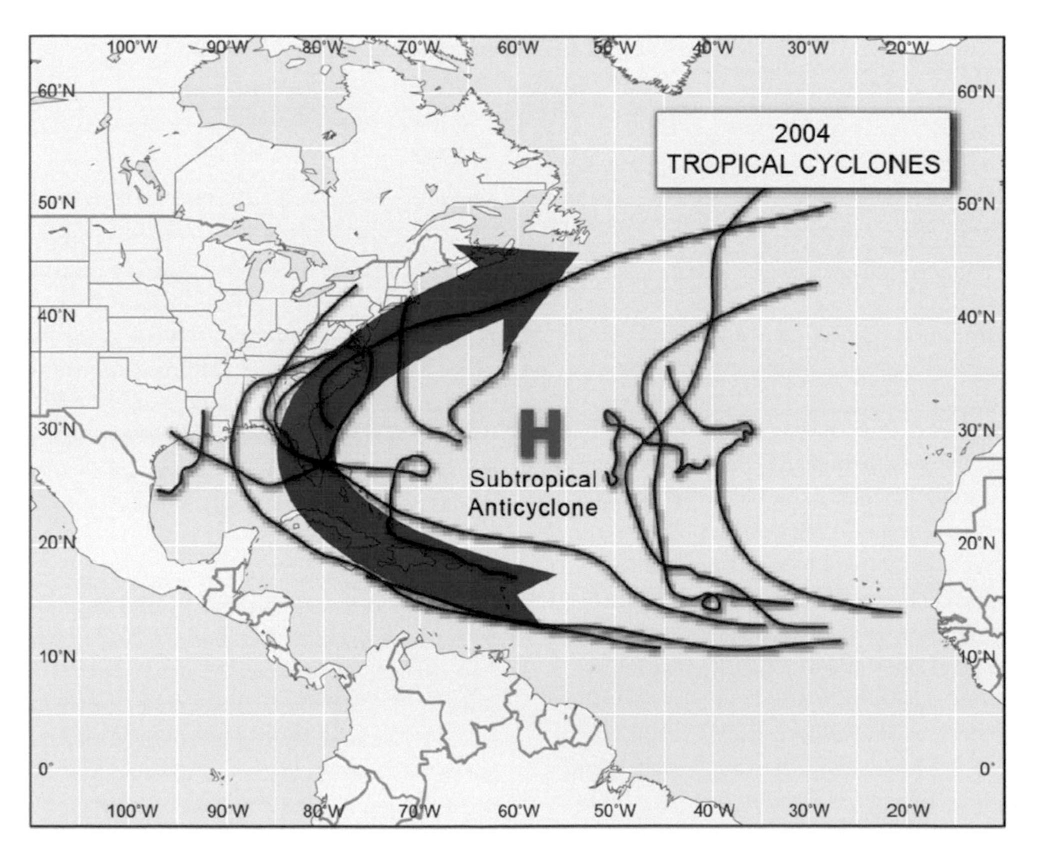

Figure 5.18 How the position of the Bermuda High influences Atlantic hurricane track. In 2004, a segment of this high (H) became established close to the U.S. East Coast, propelling storms into the Gulf Coast.

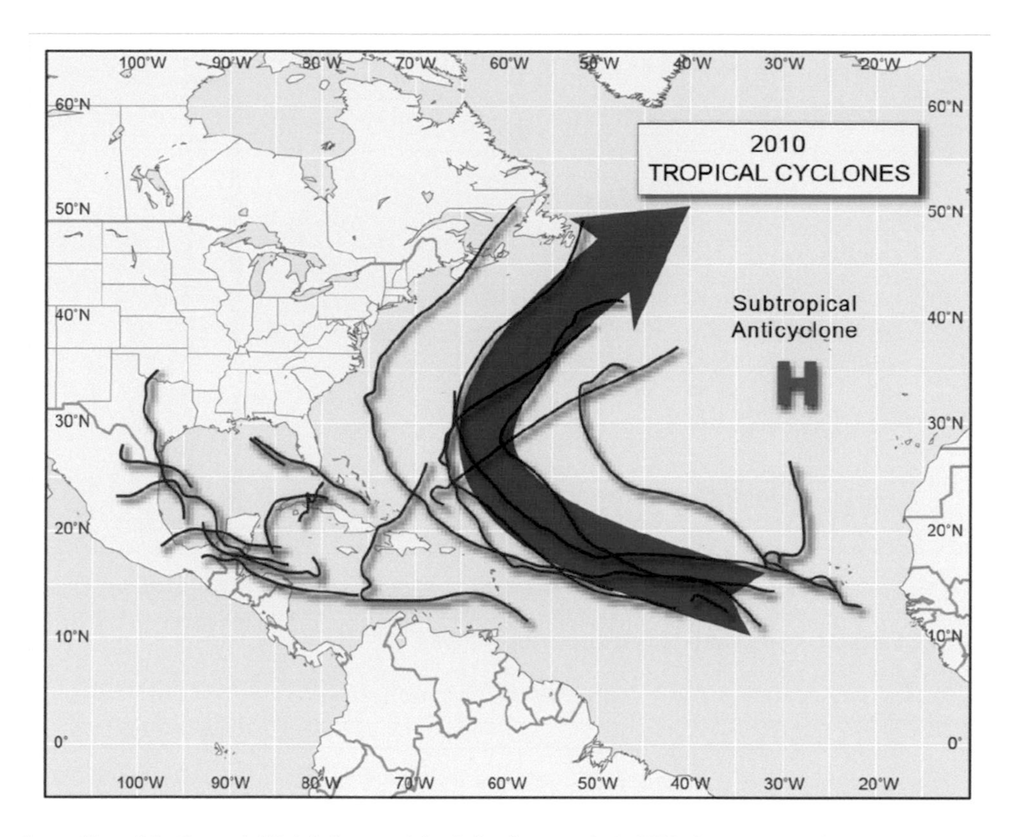

Figure 5.19 How the position of the Bermuda High influences Atlantic hurricane track. In 2010, the Bermuda High (H) was located farther east, with many of the storms recurving over open water.

In 2010, the Bermuda High was located farther east, and many of the storms recurved over open water. A subset of storms embedded in the easterly trade winds moved into the Gulf of Mexico but remained south of the U.S. Gulf Coast.

Where Tropical Cyclones Roam: Geographical Hot Spots in the North Atlantic Basin

Figure 5.20 summarizes the geographical distribution of tropical cyclones across the North Atlantic and the Gulf of Mexico. For each 2.5×2.5° latitude-longitude grid cell over the ocean, we have counted the number of times a hurricane (Categories 1–5; tropical storms are not assessed) crossed each grid box, with greater weight given to the more intense storms. Frequent storms and/or intense storms yield high levels of hurricane activity. We have also included characteristic tracks (white arrows) of hurricanes.

Several "hot spots" for hurricane activity are apparent. A broad maximum is found in the western tropical-subtropical region, north of the Bahamas. Many of these storms are born from tropical easterly waves emerging off the coast of Africa. These incubate across a long tract of warm Atlantic water and may get an extra boost from the warm Gulf Stream. A second maxima is noted over the north central Gulf of Mexico, along the Louisiana coastline. Many of these hurricanes are "home grown" over very warm Gulf waters, then quickly make landfall, leaving authorities very little time to warn and evacuate people along the coast. The apparent minimum in activity along Cuba and other islands of the Caribbean archipelago should not be interpreted as "no storms track through this region." Tropical cyclones frequently impinge on these islands, but the mountainous terrain and loss of oceanic heat weakens many tropical cyclones as they transit the region.

Extratropical Cyclones Are More Common Than Tropical Cyclones and Are Concentrated in the Middle Latitudes

Figures 5.21–5.25 display some general statistical properties of extratropical cyclones, related to their time, space variability, and characteristic storm track. Several scientific studies have examined cyclone frequency across North America, and Figure 5.21 illustrates the conclusions of one of these studies for 1958–1977. Compared to tropical cyclones, many more extratropical cyclones develop each year, ranging from about 35

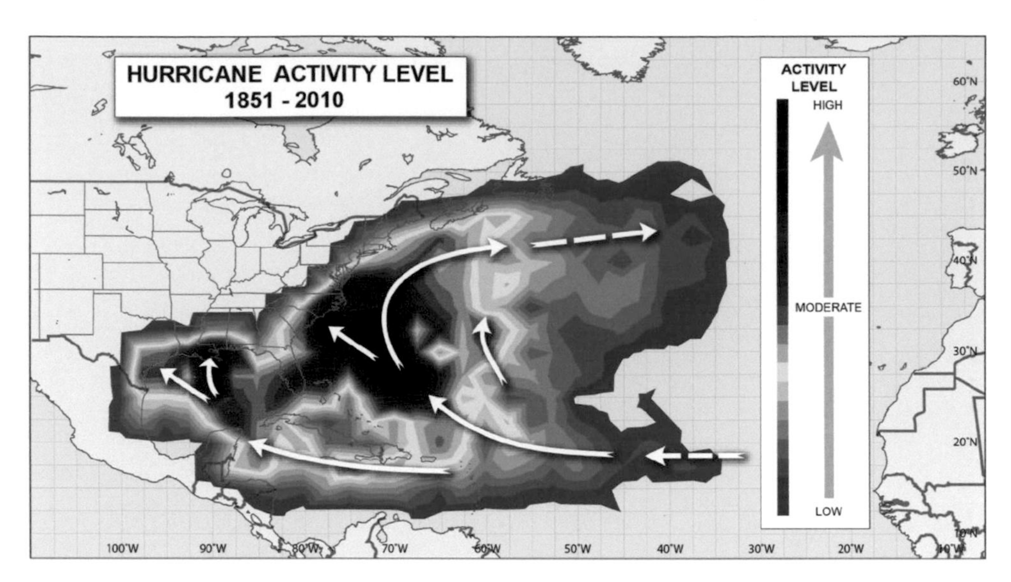

Figure 5.20 Summary of the spatial incidence of tropical cyclones across the North Atlantic and the Gulf of Mexico based on 2.5×2.5° grids. Several "hot spots" for hurricane activity are apparent. A broad maximum is found in the western tropical-subtropical region, north of the Bahamas. A second maxima is noted over the north central Gulf of Mexico, along the Louisiana coastline.

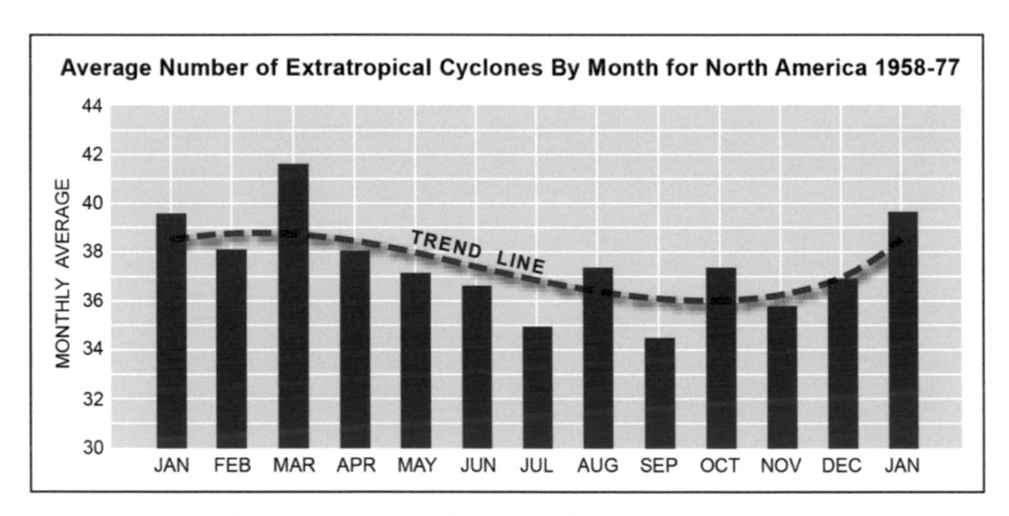

Figure 5.21 Frequency of extratropical cyclones across North America. From 1958 to 1977, on average, there were 35–42 extratropical cyclones per month in North America, depending on the time of year. (Adapted from Whittaker and Horn, 1982.)

to 42 storms per month on average. Extratropical cyclones occur year-round, and many locations within the continental United States may experience 50 or more in a given year. On any given day, there may be two or three extratropical cyclones active across North America. Storm frequency exhibits a seasonal cycle that is related to the overall intensity of the jet stream. The jet is strongest during the winter months, when hemispheric temperature contrasts are greatest, and jet stream troughs are large and vigorous. The strong ascent induced in these troughs, combined with the large temperature contrast, leads to very intense winter storms.

Figure 5.22 (top graph) illustrates the frequency distribution of surface central pressure. The majority of extratropical cyclones fall in the 980–1000 mb range. Storms with pressures lower than 970 mb are rare. The lowermost limit is about 950 mb. For comparison, recall from Figure 5.12 that a hurricane's minimum central pressure can drop below 900 mb in rare Category 5 storms. As is the case with tropical cyclones, the frequency distribution for extratropical cyclone intensity obeys the Law of Rare Events. From Figure 5.22 (lower graph), we also note that extratropical cyclones typically last approximately 2–4 days, while the average lifetime of a named tropical cyclone is 6 days.

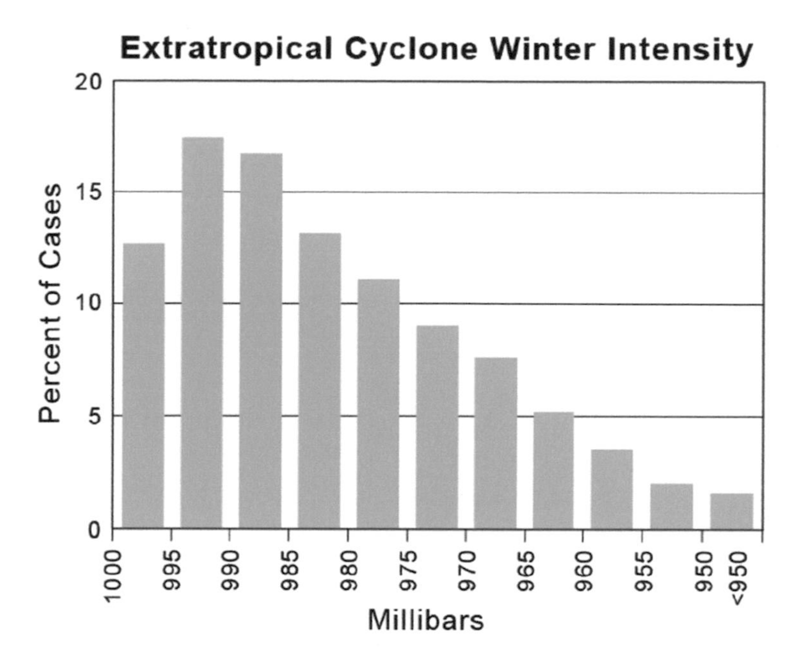

Figure 5.22 Characteristics of extratropical cyclone intensity and duration. The top graph shows the frequency distribution of winter cyclone intensity, based on vortex central pressure. The lower graph displays cyclone lifetime. The typical lifespan of an extratropical cyclone is 4–6 days, but it can extend to 10 days in the most extreme cases. (Adapted from Gulev et al., 2001.)

Figure 5.23 illustrates the geographical distribution of developing extratropical cyclones across the northern hemisphere. Here we plot storm frequency by latitude and month. The belt of extratropical cyclone formation is constrained to about 30–60° N latitude. This distribution is clearly related to the prevailing zone of strong latitudinal temperature contrasts, as well as the position of the westerly jet stream. However, the zone of storm activity shifts with the seasons. During winter months, activity is most intense between 35° and 40° N, a time when the jet stream is displaced southward (following the seasonal progression of the Sun, which moves into the southern hemisphere). March is the month of greatest activity at these latitudes, with up to 180 extratropical cyclones developing across the northern hemisphere early in the month. During summer, the jet stream shifts northward and weakens. The band of storminess moves into higher latitudes (45–55° N), and

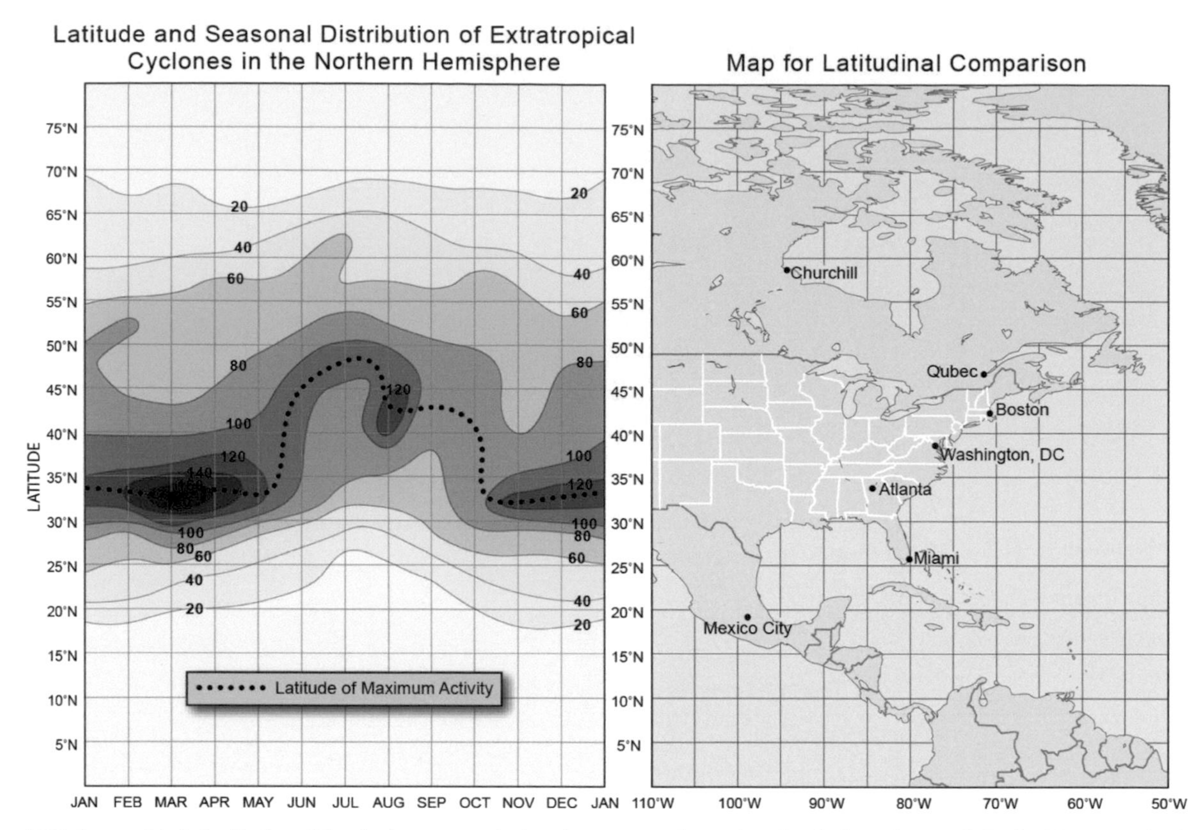

Figure 5.23 Geographical distribution of developing extratropical cyclones across the northern hemisphere. Average latitudinal frequency is plotted against time (month). Notice that the belt of extratropical cyclone formation is constrained to about 30–60°N latitude. (Adapted from Whittaker and Horn, 1982.)

the incidence of storm development decreases substantially. We know from experience that some of the most severe U.S. weather hazards associated with extratropical cyclones (specifically, blizzards and tornado outbreaks) occur during the late winter and early spring months, when extratropical cyclones are most intense and numerous.

Extratropical Cyclone Tracks Across North America

Figures 5.24 and 5.25 show characteristic extratropical cyclone tracks over North America for January and July (1950–1977). In general, storms move from southwest to northeast. During January, a set of tracks originate over the leeside Rockies. Storms bred over the Texas Panhandle (Panhandle Lows) and Montana (Alberta Clippers) converge over the Great Lakes and then traverse New England. The greatest number of winter storms over the continental United States is found over the Great Lakes region. Even larger numbers of storms occur over the Gulf Stream just off New England. These storms initiate as coastal lows or Nor'easters off coastal North Carolina then track northeastward. A third region of frequent storminess impacts the Pacific Northwest, arising from storms that develop in the Gulf of Alaska.

During June, fewer storms erupt across the entire domain, and all tracks shift northward. The main belt of storm formation shifts north of the Great Lakes, tracking from the lee slopes of the Canadian Rockies eastward to Hudson's Bay. During this time of the year, the jet stream is located across Canada and is much weaker than during the winter; as a result, extratropical cyclones are also weaker during summer.

Extratropical and Tropical Cyclones Produce a Characteristic Pattern and Type of Severe Weather Impacts

Table 5.1 summarizes many of the important attributes of tropical and extratropical storms.

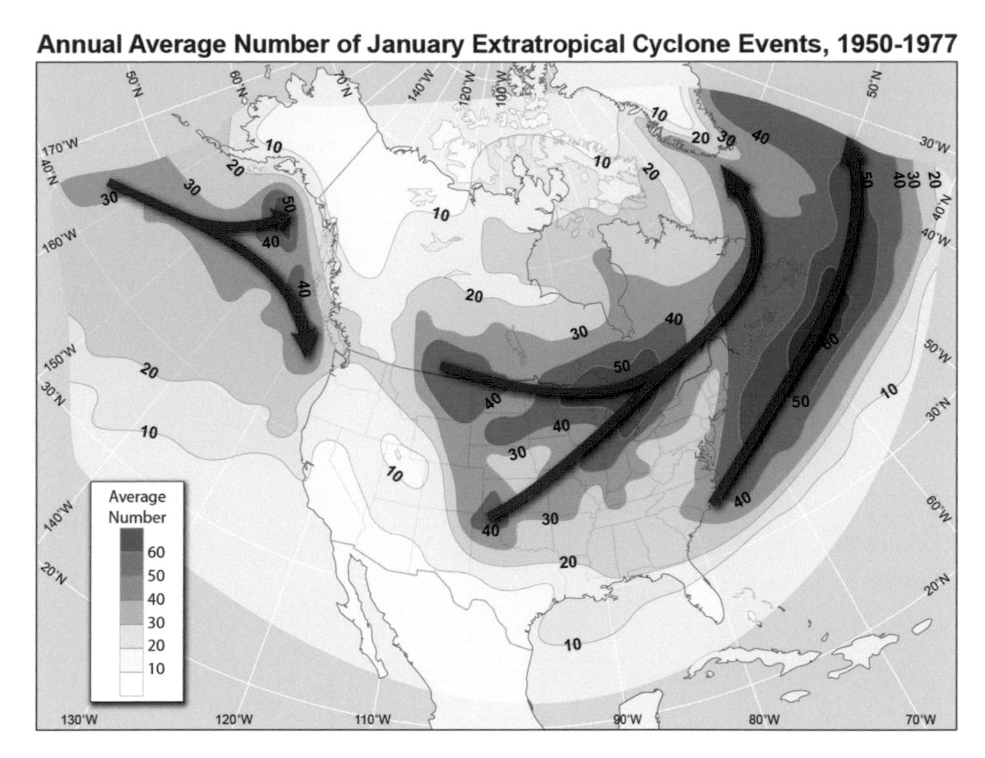

Figure 5.24 Characteristic wintertime tracks of extratropical cyclones. During January, a set of tracks originates over the leeside Rockies. The greatest incidence of winter storms over the continental United States is found over the Great Lakes region. Note that the greatest number of storms in the western and northern hemispheres occur over the Gulf Stream just off New England. (Adapted from Zishka and Smith, 1980.)

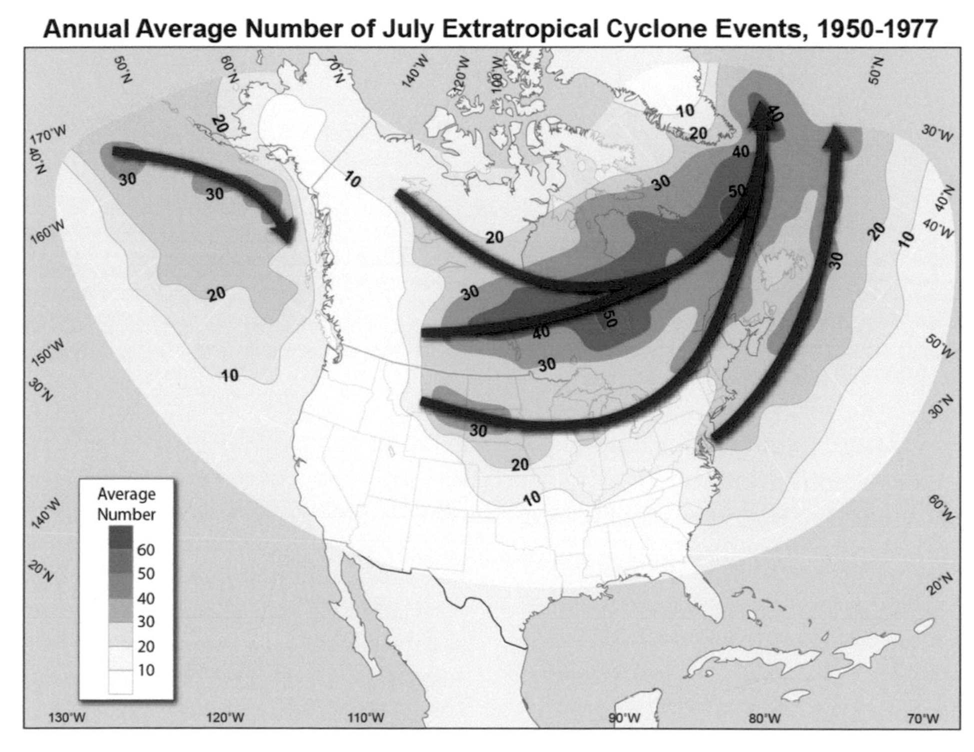

Figure 5.25 Summer tracks of extratropical cyclones. During July, there is a similar geographical coverage of storms, but few extratropical cyclones occur during this time of year. Additionally, the tracks have shifted northward with greater numbers of cyclones tracking over land areas and fewer over the northwestern Atlantic. (Adapted from Zishka and Smith, 1980.)

Table 5.1 General Properties of Extratropical Cyclones vs Tropical Cyclones

Property	Extratropical Cyclones	Tropical Cyclones
Energy source	Thermal gradient	Ocean heat and moisture flux
Dimensions	> 1609 km (1,000 miles)	< 805 km (500 miles)
Pressure gradient/wind maximum	Broad	Compact
Wind distribution	Asymmetric (cyclone is strongest NW of center)	Asymmetric (cyclone is strongest to the right of the track)
Storm movement	Toward east (26–35 kts; 30–40 MPH)	Toward west (8.7–13.0 kts; 10–15 mph)
Precipitation distribution	Asymmetric (left of center)	Symmetric
Weather contrasts	Significant (warm- and cold-side hazards)	Homogenous (high winds, heavy rain)
Thermal structure	Cold core	Warm core
Vortex winds	Increase with height	Are strongest just above the ocean surface

Notes: NW = northwest; MPH = miles per hour.

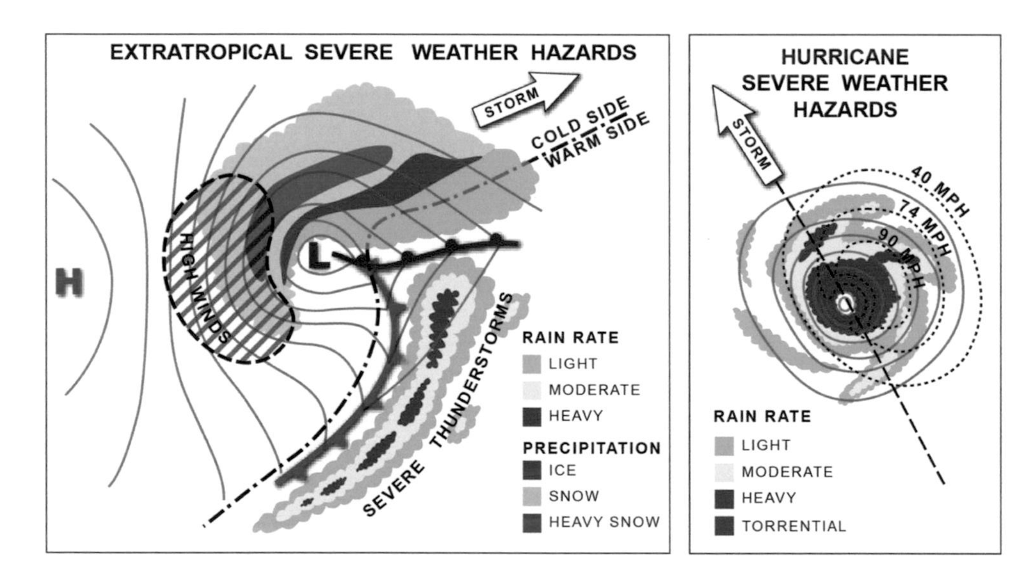

Figure 5.26 Most common types of weather hazards associated with tropical and extratropical cyclones. The right panel illustrates a typical Category 2 hurricane, with core winds of up to 110 MPH. The left panel depicts an extratropical cyclone, characteristic of a strong winter storm. This kind of storm features a warm side and cold side, each with its own set of weather hazards.

Tropical Cyclones, Most Notably Hurricanes, Contain a Compact Region of Intense Wind and Heavy Rain, but Winds Are Stronger on One Side

The right panel of Figure 5.26 illustrates a typical Category 2 hurricane, with core surface winds of up to 96 kts (110 MPH). This compact storm is moving toward the northwest. The two primary weather hazards associated with tropical cyclones are (1) strong, sustained winds and (2) intermittent but torrential rain showers. (Not illustrated are the principal oceanic hazards, including large wind-generated waves, ocean swell [storm waves emanating a great distance from the storm], and storm surge [the elevated dome of seawater traveling beneath the eyewall]. These are wind-driven phenomena that tend to be most damaging where the winds are also strongest.) Rainfall is most intense in the inner core of the tropical cyclone, within the eyewall and surrounding spiral rain bands. Heavy showers also occur in the spiral rain bands. As multiple bands cross a given location, rain totals rapidly add up. It is not uncommon for coastal regions to receive 15–30 cm (6–12 inches) of rain accumulation in a 12-hour period, with higher totals in mountainous terrain.

Hurricane-force sustained winds (those exceeding 65 kts [74 MPH]) decrease in intensity with increasing distance from the eye. Wind gusts typically exceed the sustained wind by 30% or more, especially within spiral rain bands. In a moving tropical cyclone, the overall "footprint" of high wind is asymmetric with respect to the storm track. This lack of symmetry occurs because the storm's forward speed must be adjusted to account for its swirling wind. To the right of track, we add the storm's movement and the wind because they are in the same direction. To the left of track, we subtract the storm's movement from the swirling wind because they are in opposite directions.

Extratropical Cyclones Have Characteristic Warm-Side and Cold-Side Weather Hazards Extending Over Large Areas

The extratropical cyclone in Figure 5.26, characteristic of a strong winter storm, features a warm side and cold side, each with its own set of weather hazards. The overall system is moving toward the northeast. Not all extratropical cyclones necessarily develop severe weather; the nature and intensity of weather hazards depend on many factors, including the temperature contrast, storm intensity, and jet stream strength and configuration. In general, though, the footprint (areal coverage) of severe weather tends to be fairly extensive. Along the storm's warm side (that is, within the warm sector and ahead of the cold front), lines of strong thunderstorms may contain locally heavy rainfall, hail, damaging straight-line winds, and tornadoes.

Figure 5.27 Heavy snowstorms frequently cripple transportation and commerce throughout parts of North America. A resident northwest of Baltimore plows out his driveway in the February 2010 snowstorm. (Carol Rabenhorst.)

Figure 5.28 Localized, severe wind damage from thunderstorms and tornadoes embedded within the warm sector of an extratropical cyclone. These are destroyed homes struck by a tornado in Joplin, Missouri, on May 22, 2011. (Carol Rabenhorst.)

At the same time, the broad cold side of the storm (to the left of track) generates ice and heavy snow. The heaviest snow usually falls from one or more narrow bands 150–300 km (92–186 miles) northwest of the storm center. During the coldest months, when temperatures in the warm sector are below freezing, heavy snow squalls with embedded thundersnow develop along the cold front. Snow accumulation rates of 3–8 cm/h (1–3 inches/hour) commonly develop in the strongest precipitation bands.

A strong anticyclone is often located to the west of the cyclone and moves eastward in tandem with the low. A very intense pressure gradient can develop between the two systems, creating a zone of strong winds on the back side of the low. Where these winds coincide with the band of heavy snowfall, blizzard conditions develop; elsewhere, gusty winds of up to 52–61 kts (60–70 MPH) can lead to structural damage and widespread power outages.

Amazing Storms: "Hurricane Huron" Over the Great Lakes

During the middle of September 1996, a very unusual storm developed over the Great Lakes. What began as an ordinary extratropical cyclone briefly evolved into a hurricane-like vortex over Lake Huron. Figure 5.29 shows a visible satellite image of this hybrid system, nicknamed "Hurricane Huron," on September 14. How unusual it was to watch a classic frontal storm rapidly transform into a tightly coiled spiral, complete with a 32 km (20 mile) diameter eyewall, and a clear, central eye! Heavy showers locally produced more than 10 cm (4 inches) of rain. Sustained winds near the center reached 31 kts (36 MPH), and wind gusts briefly exceeded tropical storm strength (40 kts [46 MPH]). What were the unique meteorological conditions that lead to the development of this hybrid cyclone?

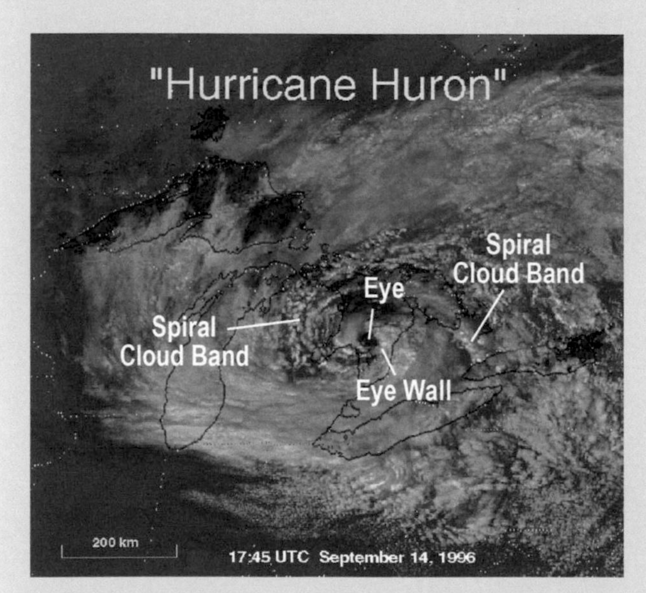

Figure 5.29 A Hurricane over a lake! This image of Hurricane Huron (captured on September 14, 1996, at a time close to its peak intensity of 993 mb) displays a well-developed eye and a storm diameter of over 480 km (300 miles). (NOAA)

From Figure 5.8, we recall that the transformation of an initially cold-core vortex into a warm-core system occurs via the process of tropical transition. Almost always, this is a process that unfolds over the subtropical or tropical oceans. But on September 11, 1996, a moderately intense wave cyclone migrated over the Great Lakes region. The system was embedded in a jet stream trough, and it featured a warm front and a cold front.

Showers and thunderstorms developed in the warm sector and moderate, steady rain fell north of the warm front over Ontario. During the next 1–2 days the wave cyclone began to occlude. During occlusion, the storm's temperature contrast weakened, and the system became stationary over the Great Lakes. Unexpectedly, the storm's central pressure began to drop, from 1004 mb on September 13 to 993 mb on September 14. What could explain this unusual behavior?

The answer, according to researchers at Pennsylvania State University, was large amounts of warmth and moisture contained in the lake's surface waters. The Great Lakes are warmest during the early fall, peaking at about 20–21° C (68–70° F) during September. As the weakening cyclone moved over the lake, its cool and comparatively dry air mass (swept off Canada) moved over the warm and humid lake surface. The temperature difference and strong winds created large fluxes of sensible and latent heat. Like a hurricane, the dying storm extracted a new source of energy. An analogous type of heat extraction powers localized and intense lake effect snow squalls during late fall and early winter.

As the air in the lowest layers warmed, the atmosphere became unstable, and deep thunderstorm clouds erupted in the storm's center. Condensation within the clouds released latent heat, and this additional warmth helped establish a shallow warm core in the center of circulation. The warm core deepened over time, and the storm grew progressively more intense from the bottom up. At the upper levels, the vortex winds actually weakened. So, during this unusual, inland example of tropical transition, an occluded cyclone with a deep cold core morphed into a deep warm-core system over a 2–3 day period. Because the storm remained stationary over Lake Huron for more than 2 days (a consequence of the original cyclone's occlusion), the storm extracted large amounts of heat from the lake water surface. In fact, when the storm migrated over the lake's eastern shore on September 15, it weakened.

For the original scientific study on Hurricane Huron, see Miner et al. (1999).

Digging Deeper: How West Pacific Typhoons Can Influence North American Weather

At first, the notion that a single tropical cyclone (typhoon) over the western Pacific can alter weather patterns across North America – thousands of miles away – may seem unrealistic. After all, typhoons are disturbances of the tropics, and North America is a middle latitude region, where weather patterns are dominated by perturbations of the polar jet stream. Oftentimes, though, a decaying typhoon over the western Pacific will recurve north, move into higher latitudes, than head eastward. Once embedded in the polar jet stream, some recurving typhoons dissipate, while others transition into extratropical low-pressure systems. While the jet stream transforms the storm, the remnant typhoon, in turn, can alter the structure of the jet stream.

Massive amounts of warm air exiting the upper levels of a typhoon can build a region of high pressure aloft, downstream (to the east) of the storm. This creates a ridge in the jet stream – a wavelike undulation in which the jet stream

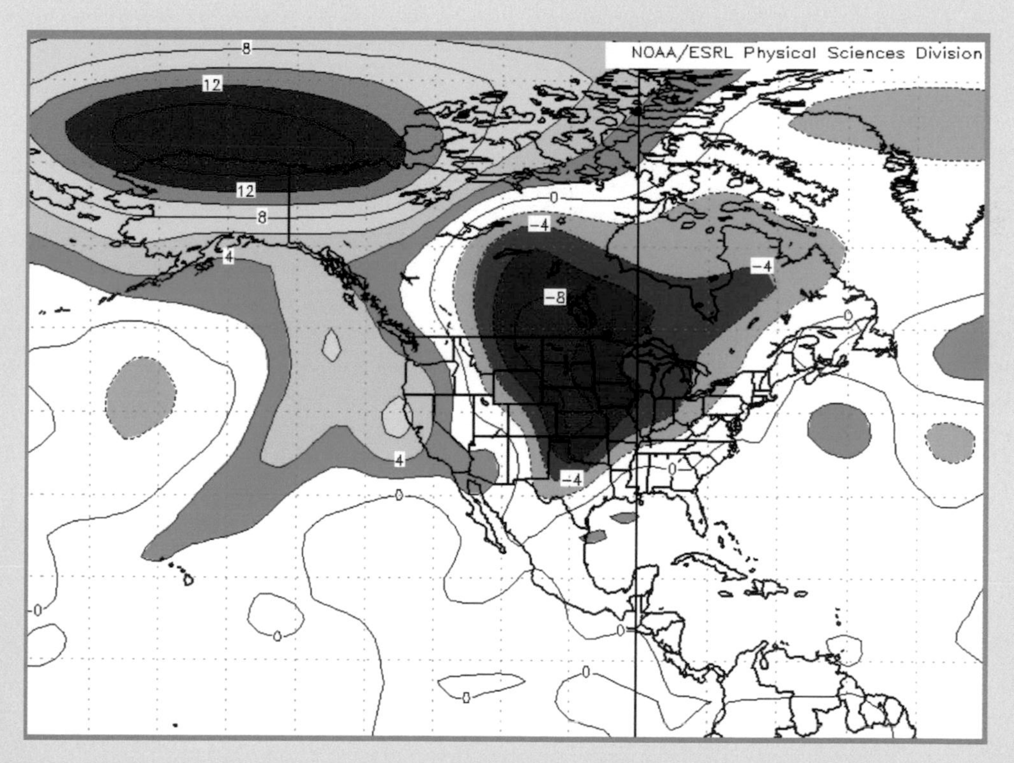

Figure 5.30 Extreme cold spreads across central United States in late September 1995. Blue and purple shaded regions show anomalously cold temperatures (°C) at an altitude of 1 km (3300 feet) in the atmosphere. (NOAA.)

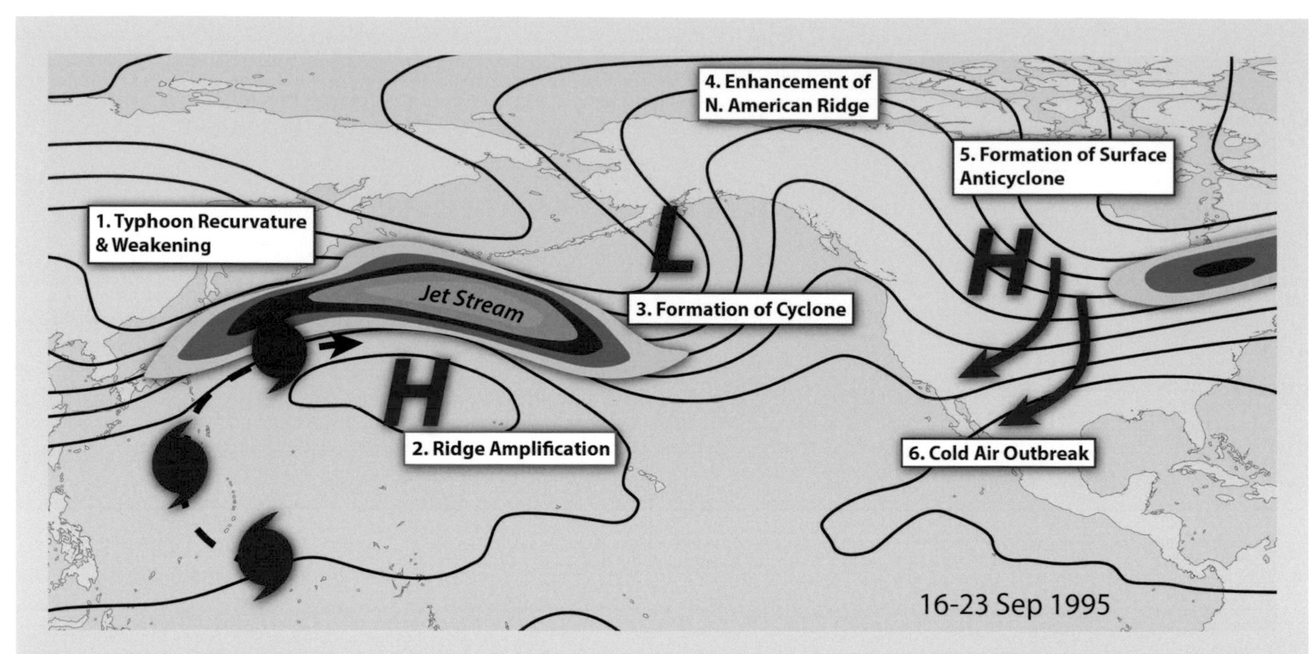

Figure 5.31 Schematic showing how a western Pacific typhoon can alter weather patterns across the central United States. The connection between the typhoon and this arctic air outbreak over the Plains is the jet stream at high altitudes. The typhoon perturbs the jet stream, forming large waves that move eastward, across the Pacific. One of these waves generates the surface high pressure that circulates arctic air southward. (Adapted from Archambault et al., 2007.)

curves anticyclonically (clockwise). In fact, the typhoon can excite a series of large waves (Rossby waves) in the jet stream, both ridges and troughs, that migrate eastward toward North America. Sometimes the wavelike pattern persists for several weeks and ultimately leads to some sort of extreme weather pattern, such as an arctic air outbreak, across the United States. The likelihood of this West Pacific–North American connection is greatest in late fall through spring.

To be clear, the process here does not involve western Pacific typhoon remnants directly making landfall along the North American West Coast; rather, the storm – as it moves into the jet stream – causes downstream changes in the very structure of the jet stream. These structural changes, in turn, induce a weather event, typically over the western half of the United States.

To illustrate, we turn to an early-season, arctic air outbreak over the central United States (Great Plains) in late September 1995. From September 20–22, anomalously chilly air swept southward over the Plains, from Canada, breaking more than 120 records for daily low temperature and leading to a costly, early end to the growing season (Figure 5.30). Research by Dr. Heather Archambault at SUNY Albany tied this cold air outbreak to the decay of two West Pacific typhoons, Oscar and Polly, during mid-September. The typhoons began interacting with and altering the jet stream nearly 100° of longitude to the west of the Plains!

As Figure 5.31 shows, many stages are involved. First, one or more typhoons recurve northward and become embedded in the jet stream. Upper-level winds amplify an existing ridge in the jet stream. The undulation induces further wavelike undulations farther downstream – in this case, the amplification of a large trough-ridge structure over Alaska and the Bering Sea. Over the Canadian Plains and northern U.S. Plains, a strong surface anticyclone develops (blue "H"), which in turn pumps extremely cold air southward across the United States.

Through this kind of "domino effect" in the jet stream, an extreme cold air outbreak can ultimately be traced to the actions of a tropical cyclone, half a hemisphere away. Sometimes it's difficult to predict these so-called Archambault Events because weather observations, such as radiosondes, are quite sparse across the Pacific Ocean. But it is fascinating that occasional weather events in the United States do indeed have such far-flung connections.

Summary

LO1 Discuss the principal features of a typical extratropical cyclone, in terms of its connection to the polar jet stream, the interaction of air masses with different origins, pressure and wind distribution, cyclone life cycle, and fronts.

1. Extratropical cyclones are the most common type of large-scale storm in the mid-latitudes, producing much of North America's severe and hazardous weather, especially during the cold season.
2. Extratropical cyclones feature strong temperature contrasts (cold front and warm front) and develop beneath jet stream troughs.
3. Extratropical cyclones undergo a 2–4 day life cycle as they track eastward. The life cycle ends with the process of occlusion, when the cold front overtakes and combines with the warm front, at which time the storm may slow down or stall.
4. Extratropical cyclones contain a core of cold air that tilts westward with increasing height and winds that increase in speed with height. These storms impact large areas of North America compared to tropical cyclones.

LO2 Describe the typical origin of Atlantic tropical cyclones, including their stages of development (tropical depression, tropical storm, hurricane), and factors that determine the track of the storm.

1. Tropical cyclones, which number 80–90 around the globe each year, include hurricanes (winds > 64 kts [74 MPH]) and tropical storms (winds < 64 kts [74 MPH] but greater than 34 kts [39 MPH]).
2. Tropical cyclones have a central core of warm air (the eye), develop in low latitudes, and feature a circular eyewall of intense thunderstorms and spiral rain bands. Compared to extratropical cyclones, these storms are fairly compact and generally move from east to west. Winds are strongest just above the surface, decreasing in intensity with height.
3. Many tropical cyclones that impact eastern North America develop within surface pressure waves that move off Africa. When a closed circulation develops, the disturbance is classified as a tropical depression. A tropical depression may further intensify into a tropical storm and then into a hurricane.
4. As tropical cyclones approach North America, many recurve northward then eastward, with or without making landfall.
5. The post-tropical phase of hurricanes occurs when the storm leaves tropical latitudes and becomes influenced by mid-latitude weather systems. The post-tropical remnant may undergo extratropical transition as it combines with preexisting fronts or the jet stream. Storms undergoing extratropical transition are still capable of producing a variety of hazardous weather, even hundreds of miles inland.

LO3 Explain how extratropical cyclones and tropical cyclones lie along a continuum or spectrum of large-scale cyclonic vortices, which also includes subtropical cyclones, arctic cyclones, and maritime cyclones (coastal lows and Nor'easters).

1. Hybrid vortices, which include subtropical cyclones, coastal lows, and Nor'easters, have cloud structures and circulations that contain elements of purely tropical and extratropical systems.

LO4 Discuss the difference between a cold-core extratropical cyclone, and warm-core tropical cyclone, in terms of principal energy sources. What is the source of potential energy for each type of vortex, and how is this converted to kinetic energy of wind?

1. Both tropical and extratropical cyclones derive their energy by converting various forms of potential energy into kinetic energy. In extratropical cyclones, the primary energy source is strong air mass (temperature) contrasts. In tropical cyclones, the energy is derived from fluxes (transfers) of heat and moisture from the warm upper layer of the ocean, including release of latent heat during condensation.
2. Tropical cyclones tend to last longer (6 or more days) vs extratropical cyclones and become more intense because extratropical cyclones quickly eliminate the strong temperature contrast that sustains them.

LO5 How are tropical cyclones categorized? Where in the Atlantic are they most prevalent, and at what time of year? What processes influence how many tropical cyclones make landfall in the United States? How common are extratropical cyclones compared to tropical cyclones? What are their characteristic tracks across North America? How do these cyclones vary seasonally?

1. Tropical cyclones in the North Atlantic reach seasonal peak activity in mid-September, after many weeks of intense solar heating of the ocean.
2. Hurricane intensity is ranked according to five categories (Saffir–Simpson Intensity Scale). Strong hurricanes (Category 3 and higher) are those in which sustained winds exceed 97 kts (111 MPH).
3. Hurricane intensity follows Poisson's Law of Rare Events: The weakest category storms are the most prevalent, the strongest are comparatively rare.
4. Some Atlantic hurricane seasons are much more active than others. An average of 9 named tropical cyclones and 6 hurricanes develop on an annual basis. The number of hurricanes making landfall over the eastern United States varies considerably from year to year. The number of landfalling storms depends not only on how many storms develop but also on the direction in which storms are steered.
5. Atlantic Basin hurricanes are most frequent in the western Atlantic and Caribbean and also in the Gulf of Mexico. Gulf storms either originate from storms farther east or develop entirely within the Gulf.
6. Approximately 35–42 extratropical cyclones impact North America each month, with greater numbers developing during winter. In the northern hemisphere, extratropical cyclones are most prevalent between 30 and 60° N latitude, in close connection with the westerly jet stream.
7. The frequency distribution of extratropical cyclones also obeys Poisson's Law of Rare Events (weak storms are the most common), with the most intense storms occurring during winter.
8. Extratropical cyclones frequently form east of the North American Rockies and along the Eastern Seaboard. Storm tracks are oriented west to east across the United States with many passing through the Great Lakes region. Storm tracks shift northward into Canada during the summer and south toward the Gulf Coast during winter.

LO6 Compare and contrast severe weather hazards for each type of vortex. What type of hazards are found in the cold and warm sides of an extratropical cyclone? How do winds and rainfall vary across a typical hurricane?

1. In tropical cyclones, the heaviest rains fall in the eyewall, the center, and in the outer spiral bands. Winds increase rapidly as one approaches the storm center, are calm in the eye, and are strongest to the right of track in a moving storm.
2. During winter and spring, extratropical cyclones produce heavy snow and ice on their northwest side and intense thunderstorms in the southeastern warm sector. Especially strong winds develop on the western side because of the large pressure gradient located between storm center and high pressure located to the north and west.

References

Archambault, H., L.F. Bosart and D. Keyser, 2007. *Recurving Typhoons as Precursors to an Early Season Arctic Outbreak Over the Continental U.S.* Oral presentation given at Northeast Regional Operational Workshop IX November 7, 2007.

Gulev, S.K., O. Zolina and S. Grigoriev, 2001. Extratropical cyclone variability in the Northern Hemisphere winter from NCEP/NCAR reanalysis data. *Climate Dynamics*, 17:795–809.

Miner, T., P.J. Sousounis, J. Wallman and G. Martin, 1999. Hurricane Huron. *Bulletin of the American Meteorological Society*, 81(2):223–237.

Whittaker, L.M. and L.H. Horn, 1982. *Atlas of Northern Hemisphere Extratropical Cyclone Activity, 1958–1977*. Madison, WI: University of Wisconsin Press.

Zishka, K.M. and P.J. Smith, 1980. The climatology of cyclones and anticyclones over North America and surrounding ocean environs for January and July, 1950–77. *Monthly Weather Review*, 108:387–401.

CHAPTER 6

Genesis, Evolution, and Intensification of Extratropical Cyclones and Hurricanes

Learning Objectives

1. Explain how hurricanes develop, intensify, and weaken.
2. Discuss several measures of hurricane strength and energy, including the Saffir–Simpson Hurricane Wind Scale and accumulated cyclone energy (ACE).
3. Summarize the controversy regarding the effects of global warming on changes in hurricane frequency and intensity.
4. Discuss the progress made in forecasting hurricane track and intensity.
5. Explain how extratropical cyclones develop.
6. Explain the origins and impacts of strong extratropical cyclones in four susceptible U.S. regions: the Pacific Northwest, Great Lakes, East Coast, and North Atlantic.

Introduction

In Chapter 5, we discussed the structure and behavior of large cyclonic storms, such as hurricanes and extratropical cyclones. We described these storms' energy sources and evolution (life cycle), and we saw how their numbers and spatial distribution vary throughout the year. Because cyclones affect so many lives and sometimes lead to catastrophic damage, this chapter explores cyclones in more detail. It answers the following questions:

- What controls these storms' intensity and size?
- Why do some hurricanes undergo rapid and unpredictable intensity changes, presenting forecasters with nightmarish scenarios (Figure 6.1)?
- How do extratropical cyclones achieve remarkable levels of fury in certain U.S. geographical locations, including the Pacific Northwest, Great Lakes, and East Coast?
- How might the intensity of hurricanes and extratropical cyclones change in the future?

Taken together, Chapters 5 and 6 provide a solid scientific foundation for understanding the specific types of severe weather that we examine in later chapters: heavy snow, ice, and cold air hazards stemming from extratropical cyclones; coastal hazards when hurricanes make landfall; delayed hazards when post-tropical remnants travel far inland; severe local thunderstorms and tornadoes; and flash floods.

Hurricane Genesis and Intensity

Tropical thunderstorm cells are extremely common; they erupt by the thousands each day around the globe. In contrast, tropical cyclones are extremely rare, averaging approximately 80–90 around the global annually. Of this total, the North Atlantic averages 10–11 named storms annually, while the greatest numbers form in the tropical western Pacific.

Tropical Cyclogenesis

Figure 6.2 shows the global distribution of tropical cyclones, from 1851 to 2006. The great concentration of storms in the Northwest Pacific readily stands out. Also note the complete absence of storms along the equator and at high latitudes. These characteristics underscore key factors required for tropical cyclogenesis, or formation of tropical cyclones:

1. sufficiently warm sea-surface temperatures (SST),
2. Coriolis deflection due to the spin of the Earth (see Chapter 3), and
3. weak wind shear.

Poleward of 30–35° latitude, the ocean surface is too cool to support the formation of tropical cyclones. Recall that tropical cyclones derive their primary heat source through evaporation and the transfer of heat from the ocean surface into the air.

DOI: 10.4324/9781003344988-8

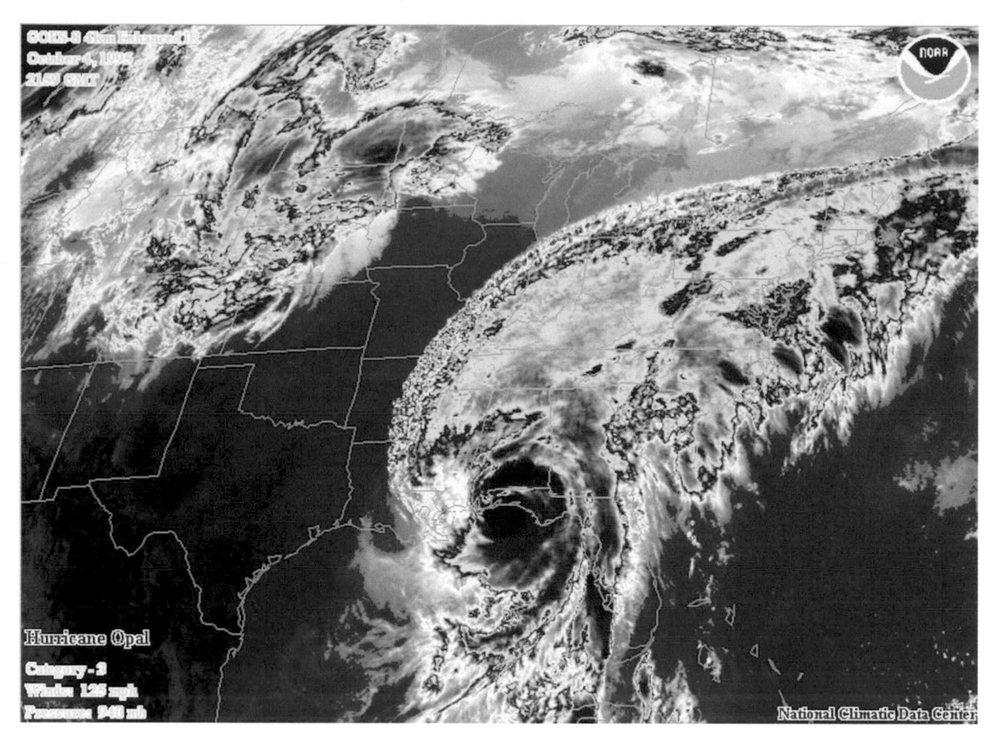

Figure 6.1 Hurricane Opal, October 1995. The unexpected, rapid deepening of Hurricane Opal in the final hours before landfall was a forecaster's worst nightmare. Later analysis revealed that the storm had moved over a small pocket of exceptionally warm ocean heat, causing the storm to escalate from Category 1 to Category 4 in approximately 18 hours. Opal inflicted $5.1 billion in damage and caused 63 fatalities throughout the Gulf of Mexico region. (NOAA.)

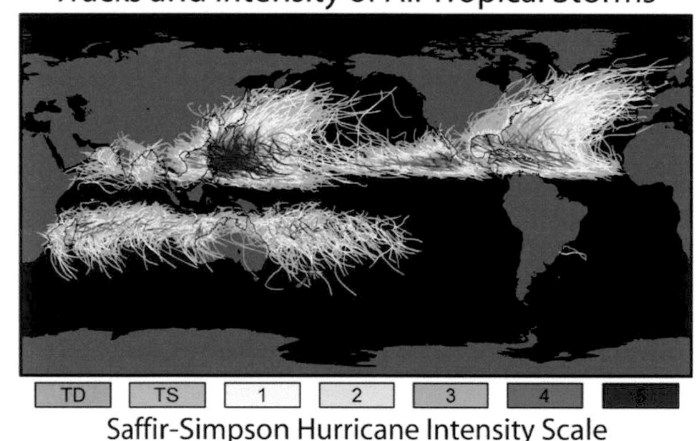

Figure 6.2 Global distribution of tropical cyclones, 1851–2006. Red and orange colors indicate stronger winds, green and blue weaker winds. (NOAA.)

And too close to the equator, within 5–10° of latitude on either side, the Coriolis deflection is insufficient to make the air rotate around low-pressure centers. Thus there is a relatively limited band of latitude over which tropical cyclones can develop.

Tropical cyclones struggle to form when SST values drop below 26° C (79° F). Additionally, wind shear (the change in wind speed and/or direction with altitude) must be minimal. The hurricane vortex is quite fragile in its early stages. It must remain cohesive and vertically oriented, such that it sequesters heat in the storm's inner core. (This warm air derives from condensation inside towering thunderstorms and from air that warms while sinking through the vortex center.) The warm column of air, which has a lower density than the surrounding atmosphere, weighs less, and it thus exerts reduced pressure at the sea surface. As a result, the storm's central pressure falls. Strong winds at high altitude or winds moving in the opposite direction cause the vortex to tilt sideways or become otherwise disrupted. A vertically erect, warm air column cannot be maintained (Figure 6.3), and, as a consequence, pressure at the bottom of the air column is not efficiently reduced. The storm thus fails to deepen and intensify.

Sufficient Coriolis deflection, critically warm ocean water, and weak wind shear often coexist over broad regions of the tropics. Why, then, do hurricanes not form all the time? These three conditions are necessary but insufficient. A sequence of smaller-scale processes, unfolding in just the right way, is needed to initiate the storm's intense region of concentrated spin. For decades, these processes remained a mystery. Now, thanks to an armada of weather satellites, aircraft missions

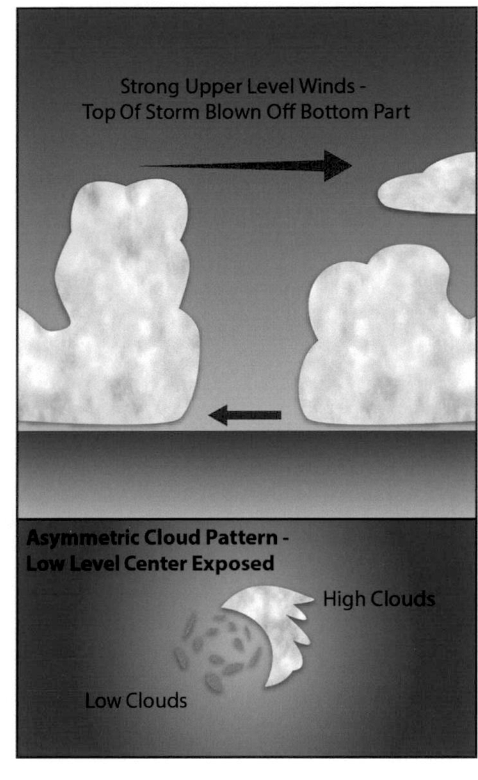

Figure 6.3 Effect of vertical wind shear on a developing hurricane vortex. The vortex tilts over and may even open up at high altitude, allowing warm air to escape and disperse over a broad region. As a result, the storm fails to intensify.

into developing hurricanes, and advanced computer models, we have developed a conceptual picture of how the initial spin develops.

Precursor Weather Disturbance: Top-Down vs Bottom-Up Development

Tropical cyclones start with a precursor weather disturbance, such as a low-pressure trough or wave in the tropical easterly current, where clusters of intense thunderstorms repeatedly develop over a multiday period. According to the top-down hypothesis, small regions of spin tens of kilometers in diameter develop within some of these cloud clusters and begin interacting. The vortices form high up, in middle levels of the cloud layers. The small vortices merge, coalescing into a deeper, stronger vortex that begins to develop toward the sea surface. In another hypothesis called the bottom-up hypothesis, even smaller vortices (just a few kilometers wide) develop within towering thunderstorm cells, close to the ocean surface. These vortical cloud towers merge in such a manner that a large, intense, mid-level vortex is built from the bottom up. Figure 6.4 summarizes these two hypotheses. In either case, the main cyclone vortex is constructed from a family of small, interacting vortices over many hours, through the interaction of numerous thunderstorms and tropical cloud clusters.

Once a persistent, vertically coherent vortex is established over a fairly large region, the circulation intensifies through a process of positive feedback. The whirling air near the ocean surface causes the air pressure to drop (think of the air as being pushed away from the vortex center, as the clothes are pushed toward the sides of a washing machine). The inflow of wind gains strength, accelerated by the low pressure and curved into a broad spiral by the Coriolis deflection. As the ocean surface winds blow stronger, more heat is extracted from the ocean. The winds effectively "fan" the ocean, drawing out the heat, and facilitate increased rates of evaporation. The air above the ocean surface rapidly warms and moistens. When this air is drawn into thunderclouds that ring the vortex, condensation in cloud updrafts releases tremendous amounts of latent heat (warming of the surrounding air due to the phase change of water vapor to liquid cloud droplets). The storm's interior warms further because the vortex traps the heated air. A cascade of air sinks back down through the center of the thunderstorm ring from great heights, warming further through adiabatic compression (descending air encounters increasing air pressure, squeezing together the molecules and increasing their internal heat). The ring of thunderstorms becomes the hurricane's eyewall, and within hours, the storm acquires a warm, inner core, which is the defining characteristic of a tropical cyclone. The warm, low-density air column causes sea level pressure to drop further, which tightens and intensifies the inward spiral of air, leading to further heat extraction and evaporation from the ocean, in a continuous positive feedback loop. Figure 6.5 summarizes these processes.

Tropical cyclogenesis in the Atlantic is complicated by an additional characteristic found nowhere else in the world – namely, giant plumes of African dust. Figure 6.6 shows an example of these dust plumes, viewed by satellite. The

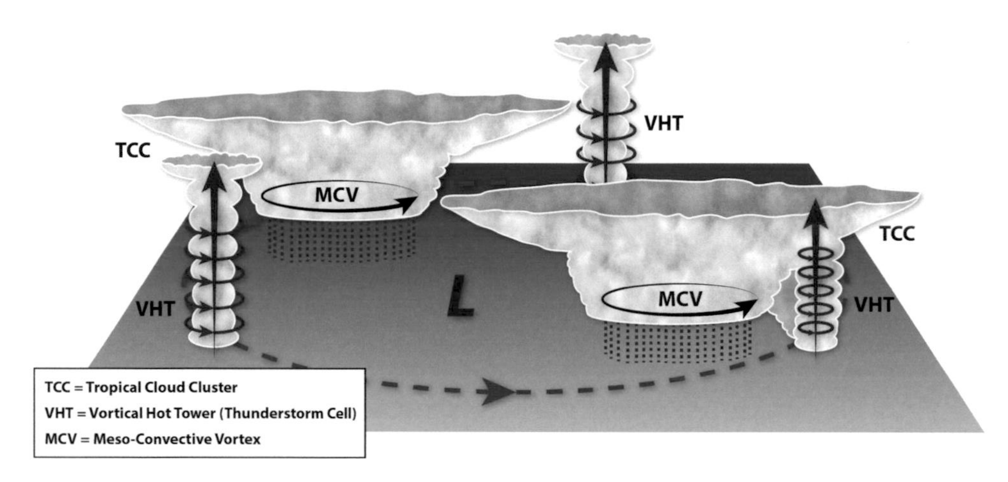

Figure 6.4 Top-down and bottom-up formation hypotheses of tropical cyclones. Scientists have hypothesized two processes that lead to the formation of a tropical cyclone's intense inner vortex. Both processes require the interaction of smaller vortices spawned in deep tropical clouds, including thunderstorm cells and larger cloud clusters. (Adapted from Houze et al., 2009.)

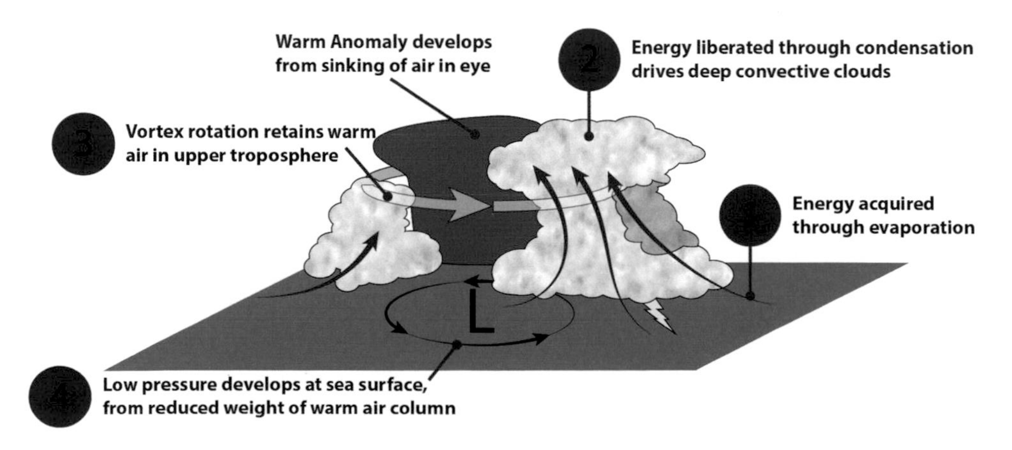

Figure 6.5 How energy is extracted from the ocean in a developing hurricane. The process is cyclic, forming a positive feedback loop that causes the storm to rapidly organize and intensify. (Adapted from Brueske, 2001.)

dust plumes are embedded in an elevated, dry, hot air mass characterized by strong wind shear, and they drift westward on the tropical trade winds. These clouds can remain intact for thousands of miles downwind (to the west of) the Saharan Desert. This windy air mass, called the Saharan Air Layer (SAL), may be hostile to eyewall thunderstorms, causing them to tilt off-vertical and evaporate. The thunderstorms power the hurricane heat engine, and if they are weakened and disrupted, the developing hurricane may fail to thrive. However, in some situations, the SAL can actually intensify storms over the eastern Atlantic. In fact, dry air masses over the tropical Atlantic may not even be associated with the SAL, developing instead within large high-pressure cells where extremely dry air sinks from high altitude.

Hurricane Intensification and Weakening

Most tropical cyclones develop at a somewhat leisurely pace, taking several days to spin up their sustained wind beyond 64 kts (74 MPH). Infrequently, tropical cyclones can deepen very rapidly, almost explosively. Rapid intensification is often defined as a 30 kt (35 MPH) or greater increase in sustained wind over a 24-hour period. Rapid intensification is often underway before meteorologists catch on to the process, making prediction difficult. In the case of Hurricane Opal (1995), the storm crossed a small mass of exceptionally warm ocean water in the Gulf of Mexico (Figure 6.7), but meteorologists discovered this fact only when conducting a careful, post-storm scientific study. The additional heat contained in that warm ocean pool, which extended to fairly deep levels, was akin to "fuel-injecting" Opal's heat engine with additional water vapor and warm air. Recent studies are finding an increase over the past decade in the frequency of rapidly intensifying hurricanes and in the rates of intensification in the North Atlantic, a behavior that may be linked to the increased heat content of the tropical ocean.

Exceptionally warm, humid ocean patches and currents make the overlying atmosphere highly unstable, particularly in the Gulf Stream off the southeastern U.S. coast and the Loop Current meandering through the Gulf of Mexico. Thunderstorms in such an air mass become exceptionally vigorous and long-lived. Persistent, intense clusters of thunderstorms called convective bursts – igniting within the inner core of developing tropical cyclones – often signify a period of rapid intensification. These regions, which are most often found over open ocean and far from coastal weather radar, are best detected with satellites. Figure 6.8 shows an example of these towering thunderclouds during the developing of Hurricane Rita (2005) over the Gulf of Mexico.

Figure 6.6 Saharan Air Layer (SAL). The SAL is shown blowing off the west coast of Africa, as viewed by satellite. (NASA.)

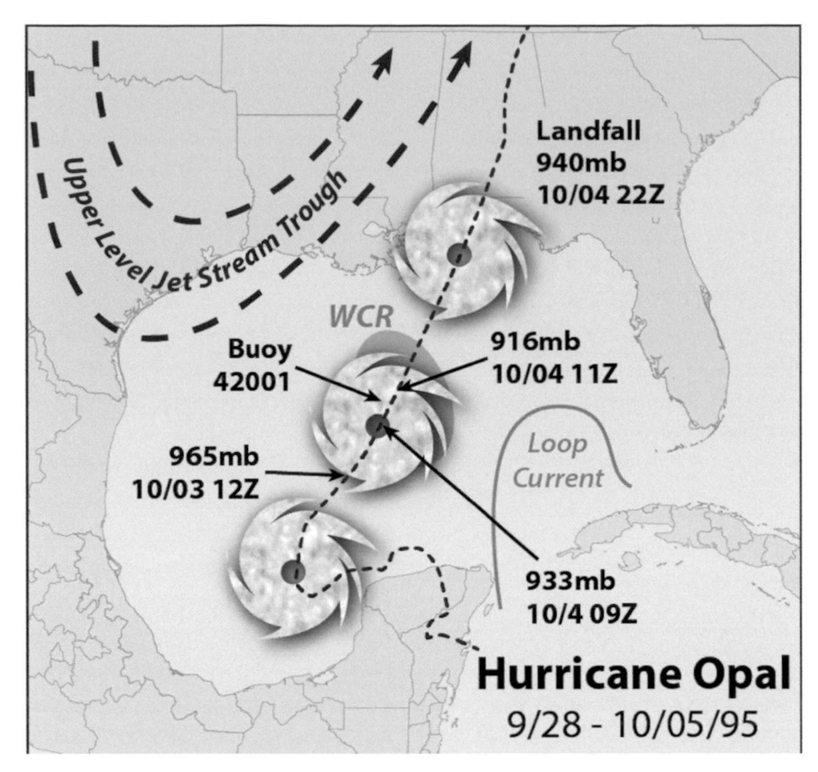

Figure 6.7 Interaction of Hurricane Opal with a warm eddy in the Gulf of Mexico. As the storm passed over this warm patch, additional heat and moisture were drawn into the storm, suddenly intensifying the hurricane as it approached the Gulf Coast. (Adapted from Bosart et al., 2000.)

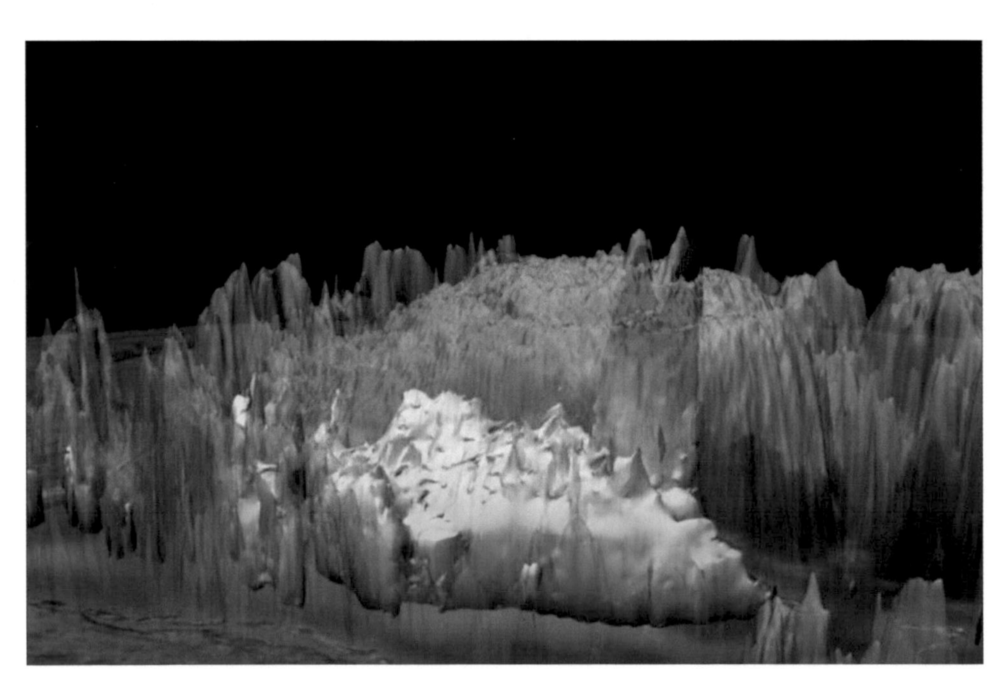

Figure 6.8 Hurricane Rita's twin thunderclouds, September 2005. Massive thunderclouds erupt in a convective burst in the early stage of Hurricane Rita's rapid deepening in September 2005. These twin thunderclouds rose to altitudes in excess of 18 km (60,000 feet). The image was obtained using NASA's Tropical Rainfall Measurement Mission (TRMM) satellite. (NASA.)

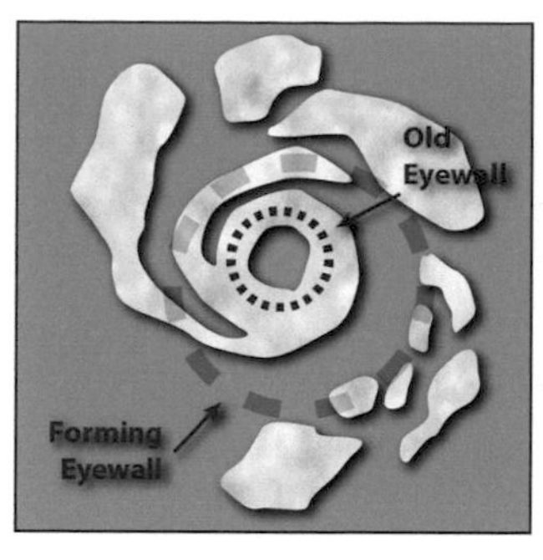

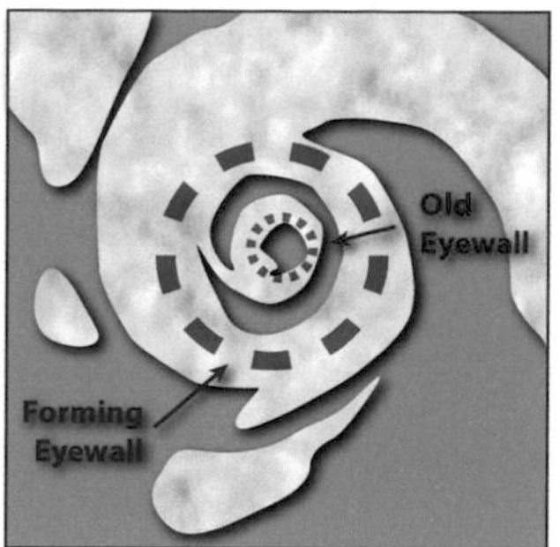

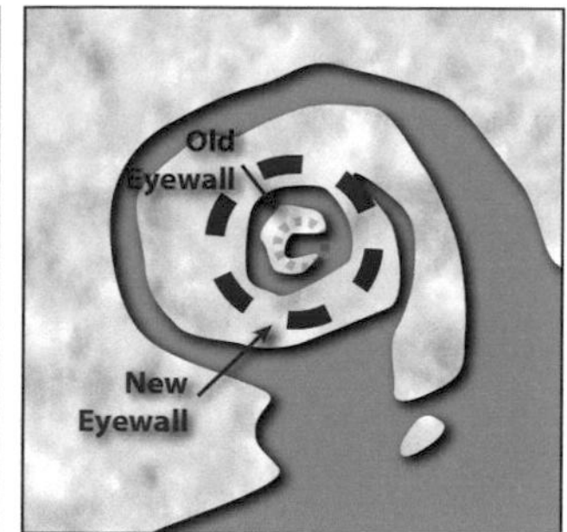

Time: 0 Hours
Wind Speed: 157 mph

Time: +28 Hours

Time: +34 Hours
Wind Speed: 116 mph

Figure 6.9 Eyewall replacement in an intense hurricane. The process leads to abrupt expansion of the eyewall and a decrease in the sustained winds.

Just as it is difficult to predict rapid intensification, it is equally difficult to predict periods of sudden weakening, particularly in stronger hurricanes. General, leisurely weakening happens when a storm moves over cooler ocean water or over an island land mass. But strong hurricanes seem to possess a sort of "safety valve" that kicks in when the sustained winds approach high values. During eyewall replacement, a process that usually spans 6–12 hours, a hurricane possesses a dual (concentric) set of eyewalls –that is, one ring of thunderstorms nestled inside the other (Figure 6.9).

The weakening is caused by the conservation of angular momentum. As Figure 6.9 shows, the hurricane initially contains a core of very fast, sustained winds. Suddenly, a new ring of thunderstorms begins to develop outside the main eyewall, with a larger radius. Once complete, this outer ring begins to contract inward. In doing so, it robs the inner eyewall of inflowing heat and water vapor. The inner eyewall begins to dissipate. For several hours during this transition, the effective radius of maximum wind shifts outward, to the location of the new eyewall. Like a spinning ice skater spreading her arms to slow down her spin, the winds must slow down at the larger radius. As the outer eyewall forms and replaces the inner one, the hurricane winds weaken, usually by 10–20 kts (12–23 MPH). Within a few hours, the new, outer eyewall contracts inward, causing some re-intensification (again, based on the conservation of angular momentum – think of the ice skater drawing her arms inward and spinning faster). Hurricane Andrew (1992) was one of the most intense hurricanes to ever strike the United States. An eyewall replacement cycle came too early, occurring far offshore; the storm was in the process of re-intensifying when it finally struck, subjecting South Florida to the most horrific wind damage ever inflicted during a U.S. hurricane strike.

Measures of Hurricane Strength, Size, and Energy

Since the early 1970s, the Saffir–Simpson hurricane intensity scale has been used to assess Atlantic and East Pacific hurricane strength, as well as rank storms historically (back to 1851). The Saffir–Simpson scale describes five intensity categories, based on maximum sustained wind (MSW, defined as the 1-minute averaged wind at the surface). Category 1 storms begin at 65 kts (75 MPH), and Category 5 storms start at 137 kts (155 MPH). Category 5 storms have no upper wind speed limit; there has never been a "Category 6" storm. Category 3 storms (96 kts [110 MPH]) and above are considered major hurricanes because of the potential for significant life loss and damage.

The Saffir–Simpson scale has been revised several times since its inception. Earlier versions incorporated values for minimum central pressure (MCP, the lowest observed sea level pressure in the center of the vortex) and height of the storm surge (the elevated mound of seawater blown onshore as the eye makes landfall). However, storm surge height depends on many factors besides maximum sustained wind. MCP was designed into the scale because wind speed measurements were not routinely available, but often barometric pressure was recorded and served as a proxy for storm intensity. Figure 6.10 shows that MCP and MSW are related in a relatively straightforward manner. With increased aircraft reconnaissance of Atlantic hurricanes, wind measurements became commonplace after 1990. Because the MSW is more directly related to inland storm damage and is thus an ideal measure of storm intensity, MCP was dropped from the Saffir–Simpson scale. In 2012, the scale was further updated to reflect minor changes to the Category 3–4 and 4–5

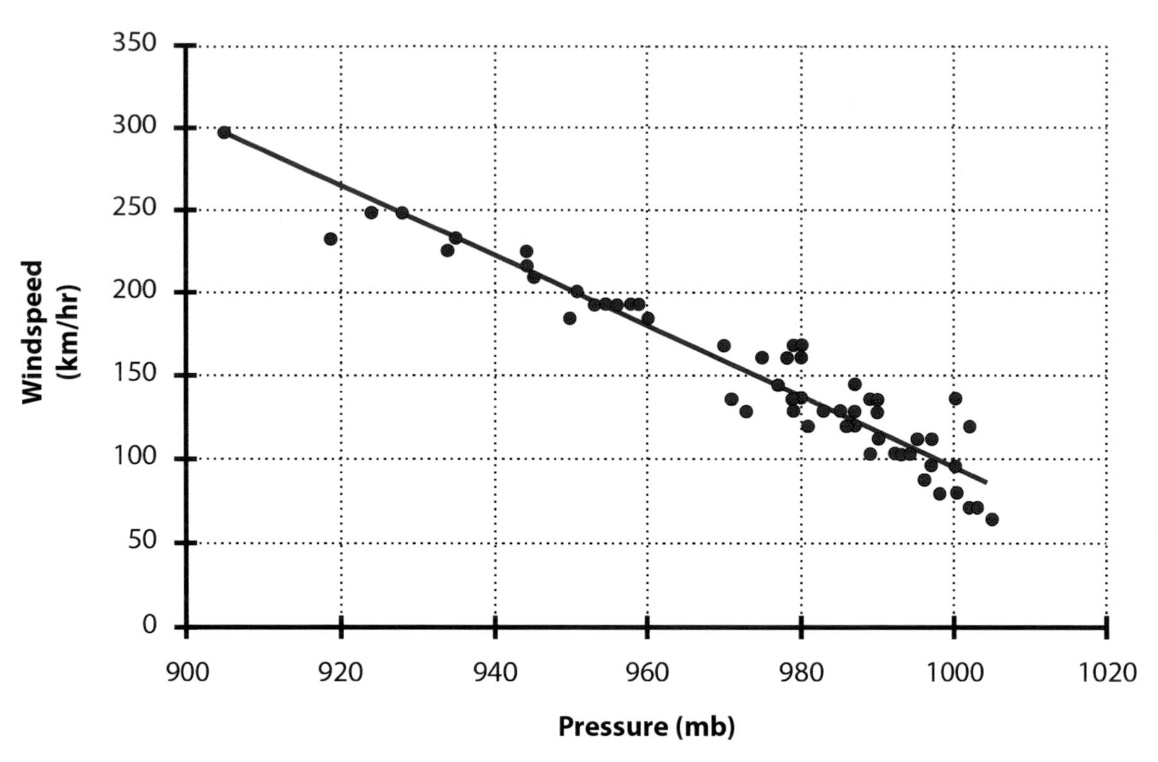

Figure 6.10 Maximum sustained (MSW) wind and minimum central pressure (MCP). The relationship between a hurricane's maximum sustained wind (MSW) and minimum central pressure (MCP) plots as a highly correlated, nearly straight line. However, there is some variation in MCP values for a given MSW.

Table 6.1 Saffir–Simpson Hurricane Wind Scale

Category	Wind Speeds
Five	≥ 70 m/s, ≥ 137 kts ≥ 157 MPH, ≥ 252 km/h
Four	58–70 m/s, 113–136 kts 130–156 MPH, 209–251 km/h
Three	50–58 m/s, 96–112 kts 111–29 MPH, 178–208 km/h
Two	43–49 m/s, 83–95 kts 96–110 MPH, 154–177 km/h
One	33–42 m/s, 64–82 kts 74–95 MPH, 119–153 km/h
Related Classifications	
Tropical storm	18–32 m/s, 34–63 kts 39–73 MPH, 63–118 km/h
Tropical depression	≤ 17 m/s, ≤ 33 kts ≤ 38 MPH, ≤ 62 km/h

Notes: m/s = miles per second; MPH = miles per hour; km/h = kilometers per hour.

boundaries. Table 6.1 shows the modern version of the intensity scale, now properly called the Saffir–Simpson Hurricane Wind Scale.

More recently, hurricane scientists and natural hazards managers recognize that hurricane size (diameter) is just as critical a parameter as MSW. Two historical examples underscore this idea.

The first example comes from a comparison of two extremely damaging hurricanes: Hurricane Ike, which made landfall in Texas in 2008 as a Category 2 storm, and Hurricane Charley, which struck western Florida in 2004 as a Category 4. Interestingly, Charley generated a 2.1 m (7 feet) high storm surge, while much weaker Ike generated a surge nearly three times higher (6.1 m [20 feet]). The major determinant of surge height was the significantly larger wind circulation of Ike, compared to tiny Charley. The second example is Hurricane Sandy, which pummeled the East Coast in late October 2012. Sandy's destructive wind field expanded to a greater than 1600 km (1000 mile) diameter. While the MSW at landfall (along the New Jersey coast) was only Category 1, 24 states east of the Mississippi river experienced strong (and often damaging) winds. As the storm moved inland, these winds even raised 3 m (10 foot) waves in the Great Lakes!

A large-diameter storm also tends to prolong the effects of a hurricane's worst elements – namely, heavy rain, high wind, and surge. The reason is simple: For a given forward speed, it takes longer for a bigger storm to transit a given location, thus drawing out the effects. More time spent beneath heavily raining cloud bands equates to higher rain totals. Structures are more likely to fail the longer they are buffeted by high wind. And a larger wind "footprint" readily translates into more damage to structures, vegetation, utilities, and property in the storm's path. A comparative example is the difference between Hurricanes Ivan (2004) and Dennis (2005), both striking near Pensacola, Florida, at Category 3 intensity. But Ivan was considerably larger, leading to more than five times more dollar damage than Dennis.

Very few scientific studies have examined the size distribution of Atlantic hurricanes. Many factors can be examined: the diameter of the storm's eye, the radius of hurricane-force wind, the radius of tropical storm-force wind, and radius of outer closed isobar (ROCI), which is a measure of how far outward the storm's pressure extends). Figure 6.11 is a histogram (frequency distribution) of storm sizes from 1988 to 2014 based on ROCI, recorded when each storm was at peak intensity. The average storm diameter was 587 km (365 miles), with the smallest being Tropical Storm Albert (2012) at 162 km (101 miles), and the largest being 1988's Hurricane Gilbert at 1624 km (1009 miles). There was no clear relationship between a storm's diameter and its maximum sustained wind.

The bottom line is that nature creates tropical cyclones with a fairly broad spectrum of sizes; Figure 6.12 compares some historically

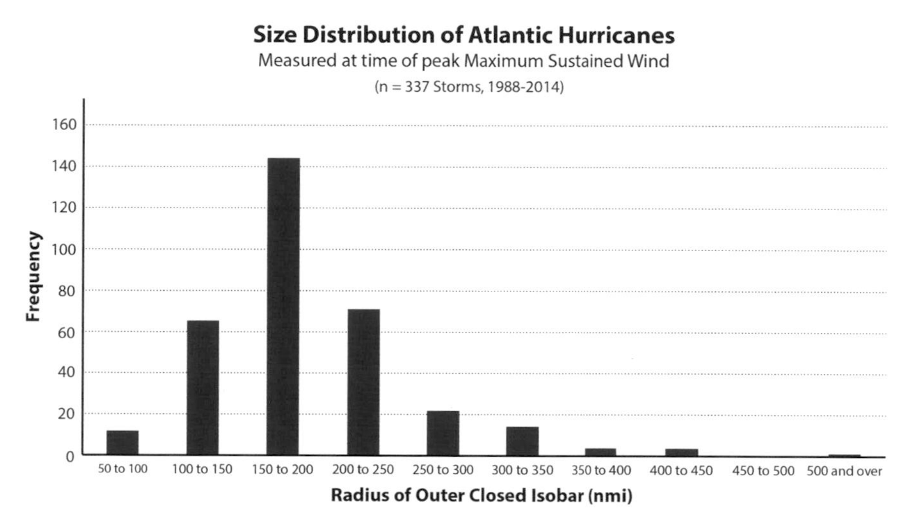

Figure 6.11 Size distribution of several North Atlantic hurricanes for the period 1988–2014 showing ROCI (radius of outer closed isobar) data at peak storm intensity.

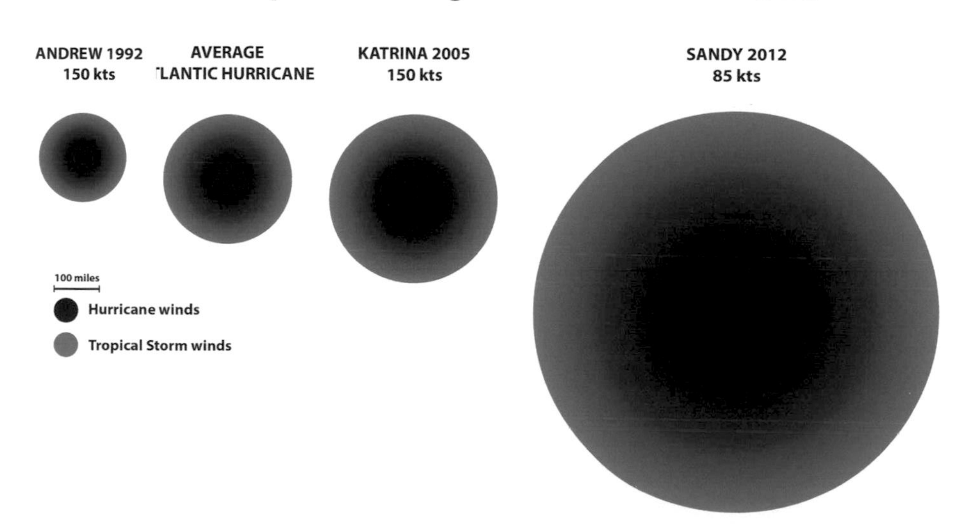

Figure 6.12 Size comparison of U.S. Hurricanes, 1988–2014. Storm diameters have little correlation to wind intensity.

significant U.S. hurricanes. Additionally, most hurricanes change diameter during their life cycle, with many storms becoming larger during the weakening and decay stages. Still, even with this wide variation, tropical cyclones in general are considerably smaller than extratropical cyclones of the middle latitudes.

Some hurricane scientists are finding novel ways of combining tropical cyclone size and maximum sustained wind into a single impact measurement. These new scales include the Hurricane Severity Index (HSI) and Cyclone Damage Potential (CDP) index. The CDP computes an index value based on the following three parameters: (1) storm size, which varies directly with damage potential; (2) MSW, in which wind damage is proportional the cube of MSW; and (3) speed of storm movement, which is inversely related to storm duration and thus total damage potential.

Hurricane activity levels in the Atlantic can be assessed according to the kinetic energy (energy of moving air, or wind, proportional to the square of MSW) contained in each storm. This measurement, termed accumulated cyclone energy (ACE), is based on an integrated measure of the storm's wind intensity and duration, as computed every 6 hours and then summed over the storm's lifetime. ACE does not include any measure of storm size. Single ACE values can be computed for individual tropical cyclones, measuring the total kinetic energy expended over each storm's lifetime. Long-duration storms can accumulate a larger ACE value than more intense storms of short duration.

ACE for all storms in a given hurricane season, over a single ocean basin, can also be summed. The result is a season-total ACE; the resulting large number must be divided by 10,000 to make it more manageable. For each season, ACE value takes into account the number, strength, and duration of all tropical cyclones in a particular ocean basin. Season-total ACE also shows each basin's contribution to the global total of all hurricane activity and helps scientists distinguish among normal, below-normal, above-normal, and hyperactive storm seasons. An example of a hyperactive season was the record-breaking 2005 Atlantic hurricane season, with 28 named storms – so many that the National Hurricane Center ran out of storm names! When seasonal ACE values are plotted for many years, scientists can examine the data for cycles and long-term trends in hurricane activity. Figure 6.13 shows seasonal ACE values plotted for all Atlantic hurricane seasons from 1851 to 2014.

The figure shows that seasonal ACE values jump considerably from year to year. But when the data are "smoothed" using numerical averaging techniques, several coherent cycles of low vs high activity emerge, on timescales lasting several decades. These cycles imply that some sort of strong climate variation, on the order of 40–50 years, controls hurricane activity levels, creating "dry" vs "hyperactive" eras. It turns out that this is a natural cycle, not one related to global warming, a topic we take up in the next section.

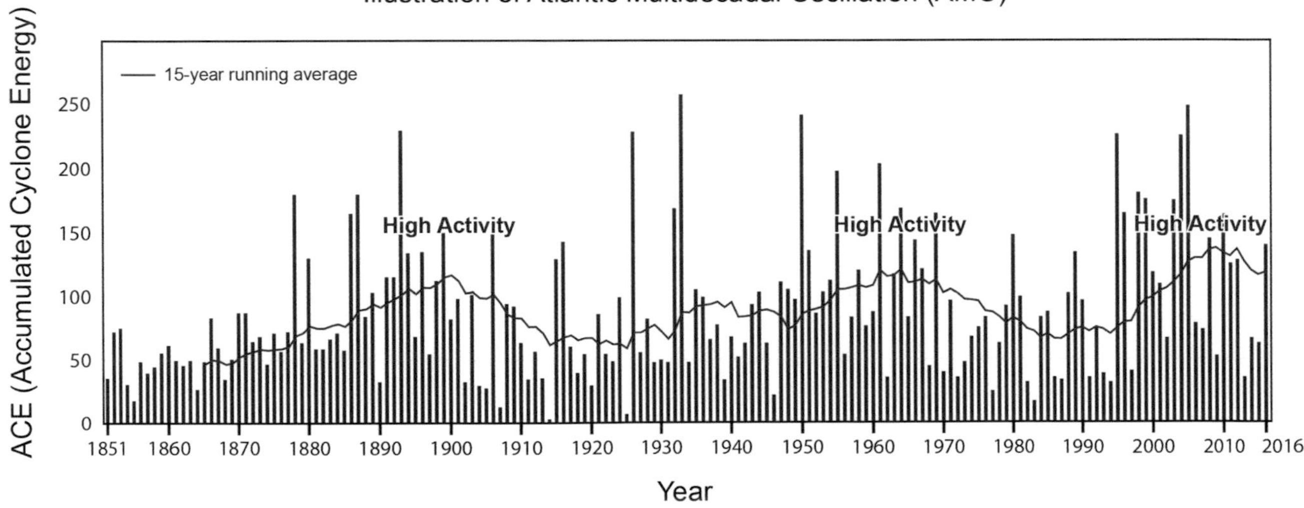

Figure 6.13 Seasonal U.S. accumulated cyclone energy (ACE) for the Atlantic, 1851–2014. Note the pronounced, multidecade active eras between decades-long lulls in total hurricane activity. (Adapted from Williams, 2004.)

Hurricanes and Climate Change

A great many studies have examined how hurricanes might "adjust" to a warmer world. Since the Industrial Revolution, the planetary average surface temperature has risen 1° C (1.8° F), and ocean surface temperatures have increased by 0.5° C (1° F). The degree of warming has not been equal across the globe; polar regions have experienced twice the amount of warming compared to the global average value.

According to basic thermodynamic principles, an increase in ocean temperature translates into greater evaporation. Indeed, the relationship between evaporation and temperature is very sensitive, such that a small, incremental rise in temperature leads to a disproportionately large increase in water vapor. Water vapor is the "fuel" that powers the hurricane's heat engine. These facts are central to the argument that hurricane intensity should increase, theoretically, as the planet warms.

However, the response of tropical cyclones to global warming is a multifaceted, complex problem, and ocean temperature is likely just one of several controlling factors. We must also examine the vertical change of atmospheric temperature, which influences air mass stability (i.e., how easily tropical thunderstorms can form), and changes in wind shear. Additionally, large shifts in global wind patterns will influence where storms form and track.

In Fall 2005, two pioneering scientific papers showed that hurricane activity has indeed increased, based on observations since 1970. Research by Peter Webster and colleagues indicated a 30-year trend toward more frequent and intense hurricanes, namely Category 4 and 5 storms, in multiple ocean basins around the globe (Figure 6.14). A second research study by Kerry Emanuel examined the power dissipation index (PDI) of hurricanes in the North Atlantic and North Pacific oceans. The power dissipation – a measure of the kinetic energy that is expended by the storm – is computed from the MSW (cubed), integrated over the total lifetime of the storm. Emanuel's results, summarized in Figure 6.15, suggest a marked rise in the PDI, which is highly correlated with the rise in ocean surface temperature. Longer storm lifetimes and greater storm intensity have both contributed to greater overall power dissipation.

Since 2005, many additional studies have been published. Some are based on observational data; others are rooted in numerical simulations using climate models. The emerging field of paleotempestology uses historical data from the natural environment to reveal hurricane activity levels over past millennia. However, some scientists caution that any

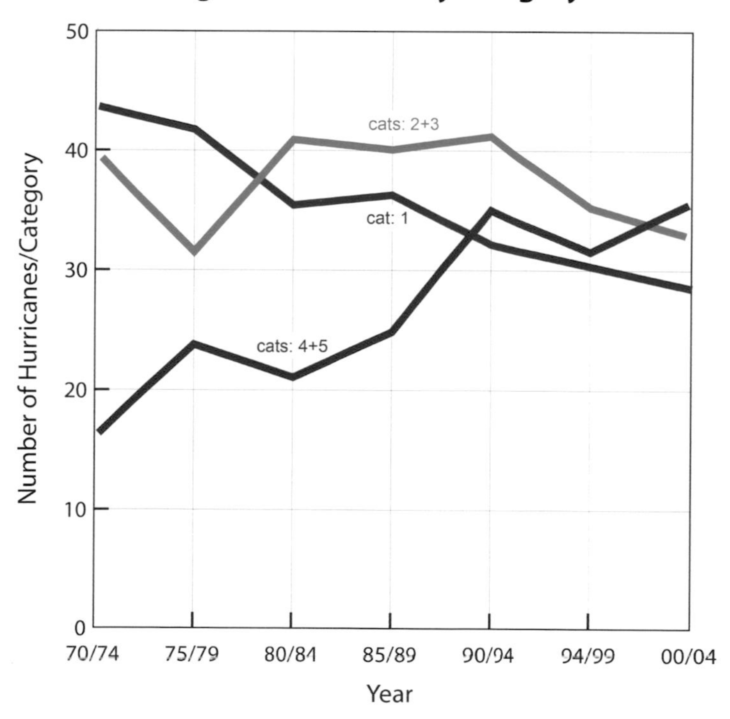

Figure 6.14 Large increase in the number of Category 4 and 5 tropical cyclones in ocean basins around the globe since the 1970s. This graph summarizes the pioneering work of Peter Webster and colleagues. (Adapted from Webster et al., 2005.)

examination of historical hurricane trends must account for the way observing technologies have evolved. For instance, we now live in an era with global satellite coverage, but during the 1970s satellites were few in number, and their imaging capabilities were relatively crude. Estimates of hurricane wind intensity based on satellite imaging techniques must therefore be carefully calibrated against other measures common at the time, including those from reconnaissance aircraft and buoys. If these techniques are not carefully cross-validated, biases can enter into the interpretation of the data, perhaps leading to inflated data with regard to both hurricane intensity and frequency.

Indeed, whether hurricane activity is increasing as a result of global warming remains the topic of intense scientific debate. Adding fire to the debate is long-standing evidence that hurricanes in the North Atlantic are strongly influenced by natural, decades-long climate cycles called the Atlantic Multidecadal Oscillation (AMO; see Figure 6.13). The AMO is driven by changes in Atlantic Ocean temperature and feedbacks involving large atmospheric wind patterns. During the AMO, the North Atlantic oscillates between persistent states of abnormally warm and cold ocean surface temperatures. The ocean oscillation is tied to cyclic variations of cloud cover, African dust, monsoon rainfall patterns, and wind shear intensity across the tropical and subtropical Atlantic (Figure 6.16).

Multidecadal periods of increased hurricane activity, most recently 1995–2012, correlate strongly with above-average-temperature ocean water and weak tropical wind shear. Both of these conditions favor the development of stronger and more frequent hurricanes. During intervening cold phases of the AMO, the Atlantic ACE drops. These active vs inactive hurricane phases reflect cyclic patterns due to naturally occurring processes and a complex interplay among the ocean, atmosphere, and land. Such strongly coherent cycles make detection of hurricane activity trends (as might be caused by steady global warming) difficult to discern. Some hurricane scientists agree that while such a trend may be present, it likely amounts only to a small percentage increase in storm intensity, on the order of 1–3%.

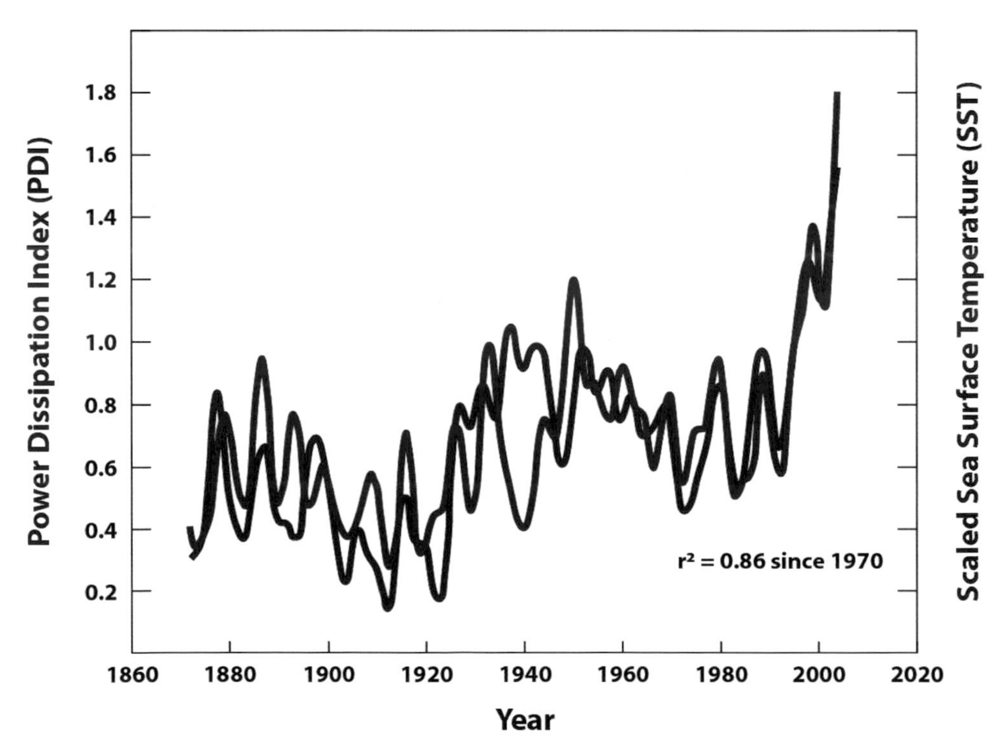

Figure 6.15 Large increase in power dissipated by tropical cyclones since the 1970s. This graph summarizes research by Kerry Emanuel. Power dissipated closely correlates with the rise in ocean surface temperature since the 1970s. (Adapted from Emanuel, 2005.)

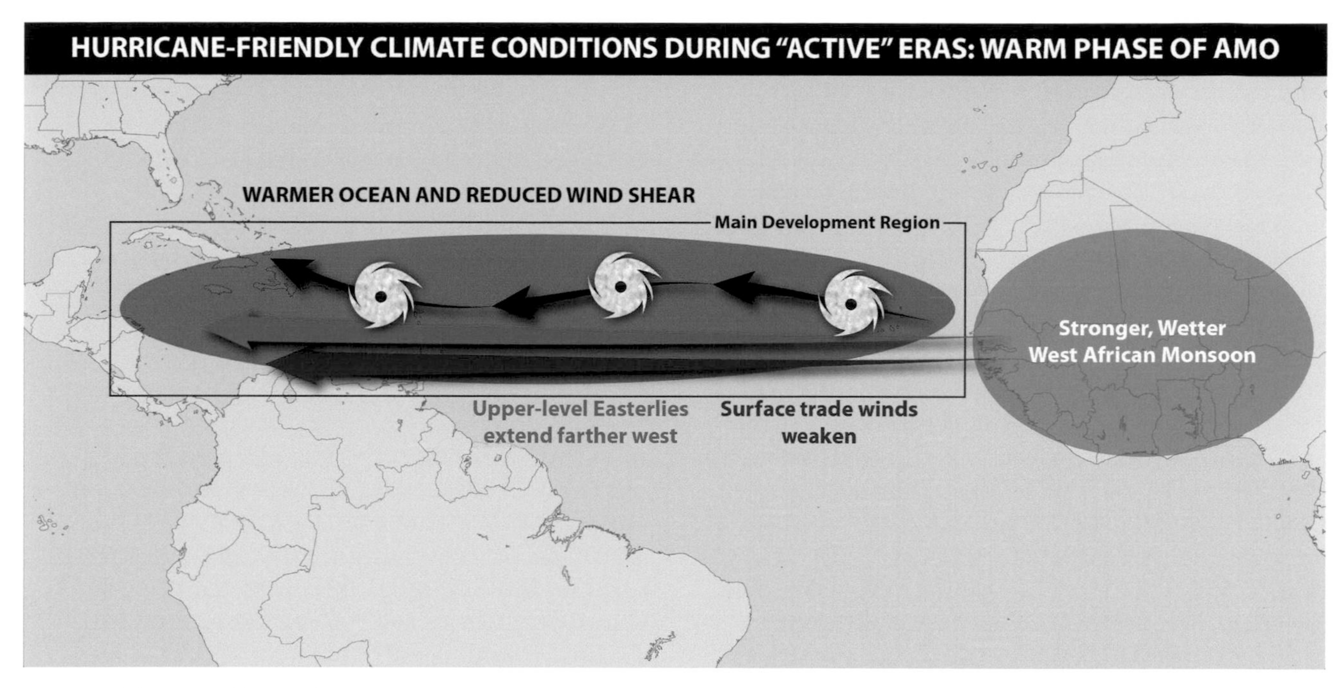

Figure 6.16 Hurricane-friendly climate conditions during the active phase of the AMO. Shown are an elongated zone of above-average ocean temperature coinciding with weak winds in the troposphere. (Bell, 2014.)

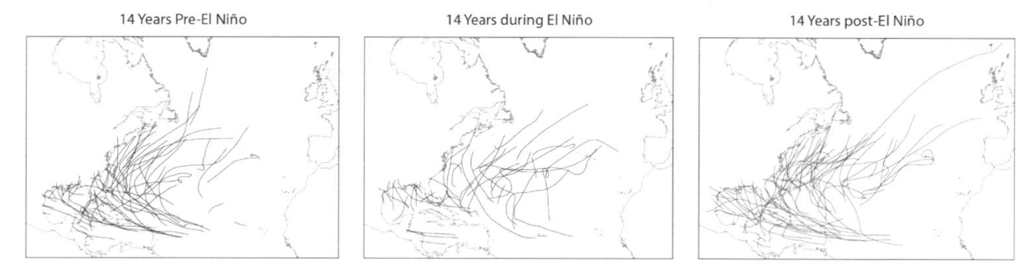

Figure 6.17 El Niño Southern Oscillation (ENSO). The hurricane tracks during El Niño (warm phase) years, compared with those in non-El Niño years, across the Atlantic Ocean. (Adapted from Gray, 1984.)

Another cycle is the El Niño Southern Oscillation (ENSO), which describes an interplay between the ocean and wind patterns across the tropical Pacific. The ENSO climate cycle leads to alternating warm and cold ocean states. The timescale of ENSO is much shorter than that of the AMO (years instead of decades), and ENSO operates in an entirely different manner than the AMO. But like the AMO, ENSO involves a strong feedback between tropical ocean temperature and atmospheric circulation patterns. However, strong ENSO phases lead to global disruption of wind circulation patterns around the globe. As a consequence, the wind shear frequently increases across the tropical Atlantic, inhibiting hurricane formation and development in the Atlantic. Statistically, El Niño years are associated with fewer Atlantic hurricanes, including fewer U.S. hurricane landfalls, as Figure 6.17 shows.

Challenges of Making Accurate Hurricane Forecasts

Several technologies are used to forecast and monitor hurricanes. Three key technologies are represented by the three legs on a sitting stool: (1) weather satellites, (2) aircraft reconnaissance, and (3) numerical prediction models. If a single leg fails, accurate hurricane prediction will suffer.

U.S. hurricane forecasting falls under the purview of NOAA's National Hurricane Center (NHC) located near Miami, Florida. Here, analysts and forecasters predict storm intensity and movement for the Atlantic and East Pacific ocean basins. NOAA's Hurricane Research Division (HRD) conducts research into hurricane structure and dynamics by flying instrumented aircraft into and around storms. The NOAA HRD occasionally partners with other agencies such as NASA to conduct more elaborate scientific missions incorporating high-altitude research aircraft and new types of satellite remote sensors.

The forecast process begins by identifying precursor tropical disturbances showing signs of increased organization, which includes persistent regions of tropical thunderstorms and rotation. Most disturbances, which include tropical depressions, are identified by satellite. Satellites provide images of cloud tops, the temperature (and hence altitude) of the clouds, the temperature of the ocean surface, the wind shear intensity, ocean surface winds, atmospheric humidity levels, and the location and intensity of tropical rainfall. Measurement of ocean surface winds is particularly important, often revealing vortical wind patterns hidden beneath thicker cloud canopies, as shown in Figure 6.18. Over the United States and adjacent waters, routine satellite observations come from a combination of geostationary and polar-orbiting satellites built and operated by NOAA, NASA, and the Defense Meteorological Satellite Program (DMSP).

Once an approaching disturbance in the Atlantic comes within range of air bases across the Southeastern United States and Caribbean, reconnaissance aircraft are dispatched to investigate the storm. The USAF provides a "workhorse" fleet of four-engine turboprop aircraft, called J-130s (Figure 6.19). The J-130s are part of the 53rd Weather Reconnaissance squadron based out of Keesler Air Force Base, Mississippi. The crews fly the J-130s at low levels through the developing storm, pinpointing the center's exact location, measuring minimum central pressure and maximum sustained winds. Flight plans are designed to survey all four quadrants of the storm circulation, including numerous passes through the eye region. Dropsondes, which are probes that measure air temperature, moisture, pressure, and winds, are ejected from the aircraft (Figure 6.20). Up to 30 or more dropsondes may be dropped during a single mission. The goal

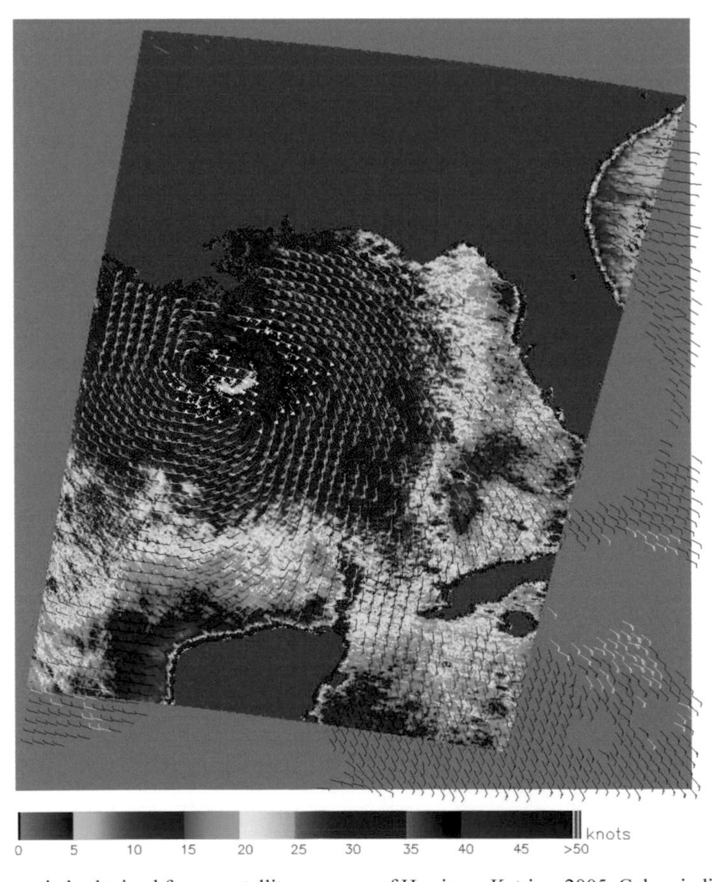

Figure 6.18 Image of ocean surface winds obtained from a satellite overpass of Hurricane Katrina, 2005. Colors indicate wind speeds, and small arrows denote wind direction. (NASA.)

Figure 6.19 USAF J-130 hurricane reconnaissance aircraft flying within the stadium-like eye of a mature hurricane. (USAF.)

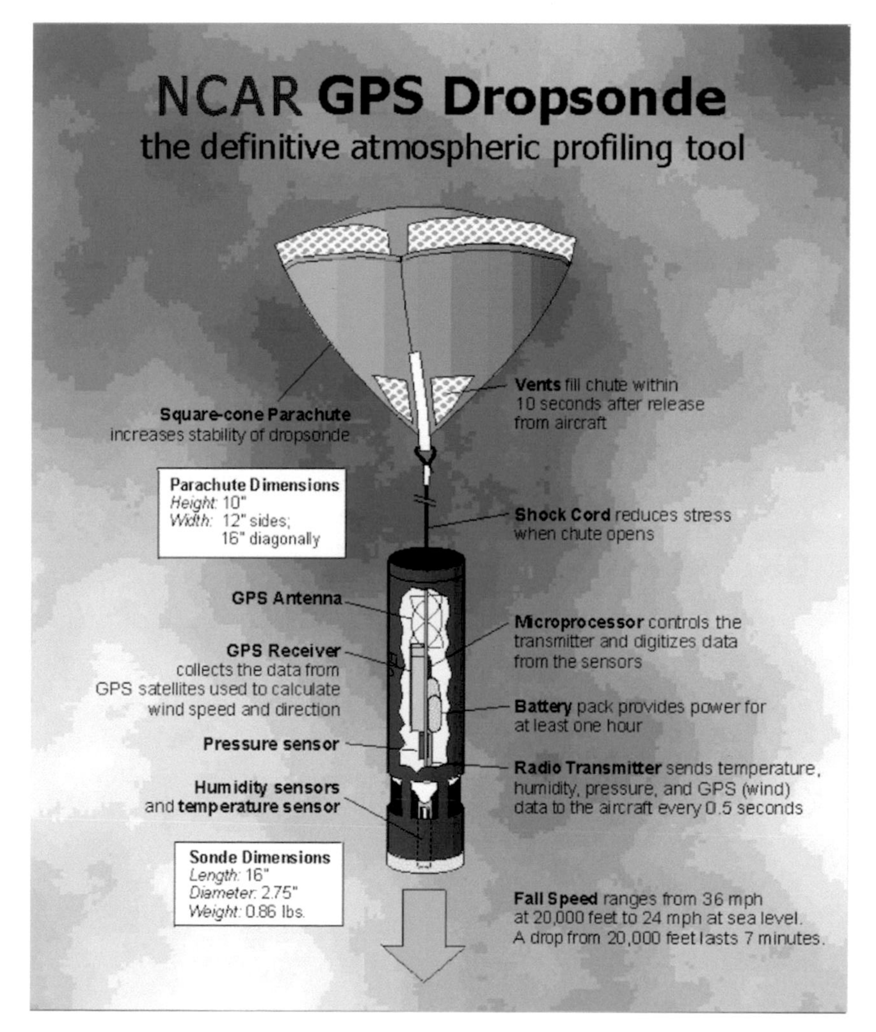

Figure 6.20 Schematic showing the components of a dropsonde instrument package, dropped from reconnaissance aircraft on parachutes. One of the authors of this textbook (Halverson) spent his early career serving as dropsonde scientist on NASA hurricane reconnaissance aircraft. (National Center for Atmospheric Research.)

of each aircraft mission is to obtain an accurate position "fix" of the storm's center and MSW. These data not only gauge intensity trends but also provide accurate initial conditions needed for the forecast models.

Data obtained during the reconnaissance flights, which may be run back to back, are transmitted in real time to the NHC. Those data as well as data from many other atmospheric observing networks are used in models that predict the future intensity, size, and track of the storm. There are quite a few of these models, which have been developed by myriad agencies, including NOAA, the Defense Department, and universities around the United States. Each numerical prediction model is coded around a different set of equations, operating assumptions, computational framework, and data assimilation method. Each model has its own unique strengths and weaknesses.

Superensemble Forecasting and the Cone of Uncertainty

Several predictive systems use ensemble forecasts, which quantify the uncertainty in observations of the atmosphere's initial state. Slight variations in initial atmospheric conditions are prescribed in order to bracket the inherent error in observations. Each slightly different run of the model is called an ensemble member; the suite of all members constitute an ensemble forecast. A recent innovation is superensemble forecasting, which further optimizes track prediction by constantly "learning" from the past performance of the ensemble members that comprise it. NHC staff examine the output from as many as 25–30 different forecast models, ensembles, and superensembles. Figure 6.21 shows an example of the track forecast for Hurricane Sandy, a vicious 2012 storm. This graph

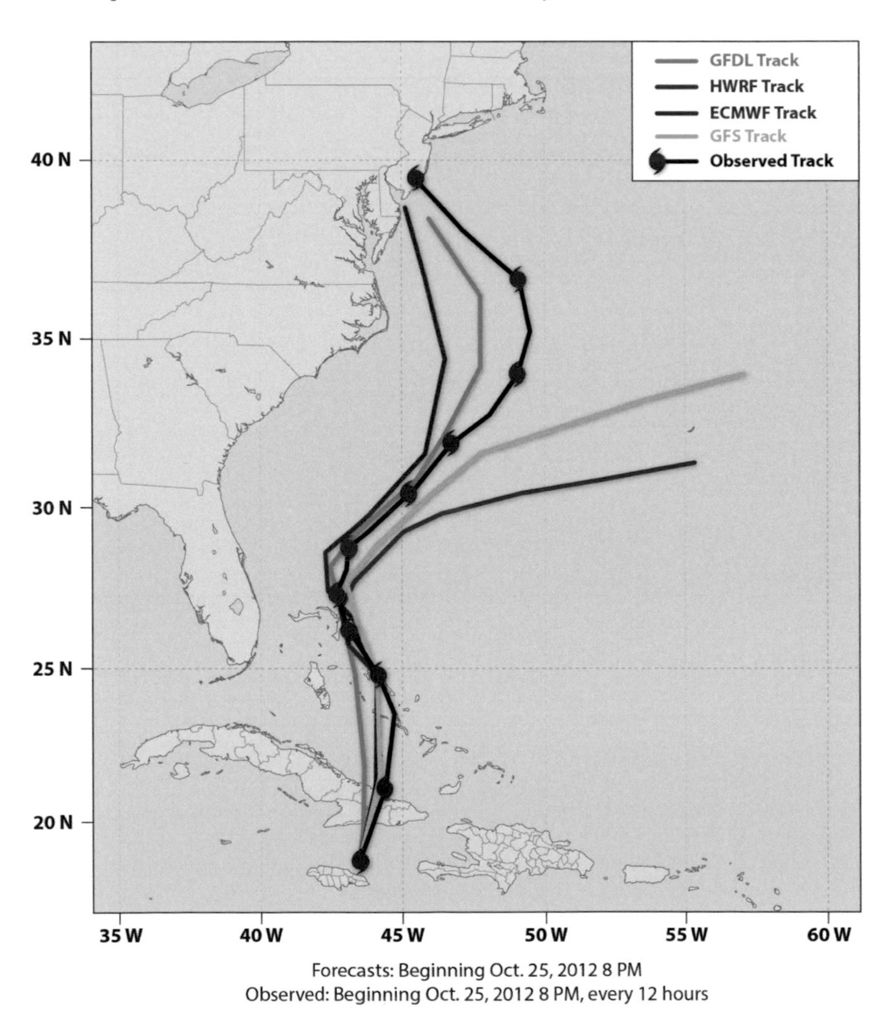

Figure 6.21 Track predictions for Hurricane Sandy, October 2012. The models pictured here include those from the European (ECMWF) and American (GFS) forecast models. The observed storm track is shown in black. (NOAA.)

displays just a few of the models compared at a single forecast time. Sandy's forecast proved particularly troublesome because the hurricane was transforming into a hybrid-type storm as it interacted with numerous mid-latitude weather systems. Two of the four models were predicting a highly unusual "left hook" into the Mid-Atlantic. The observed track of Sandy is shown by the solid black line with tiny hurricane symbols.

For named storms, including tropical storms, NHC issues a suite of forecast products and advisories every 6 hours. Probably the most widely accessed product is the cone of uncertainty. The example in Figure 6.22 shows Hurricane Joaquin's (2015) track along the U.S. East Coast, with 4 consecutive forecasts of storm position, each 12 hours apart. Each cone represents the spread of probable tracks. Joaquin's forecast was problematic in

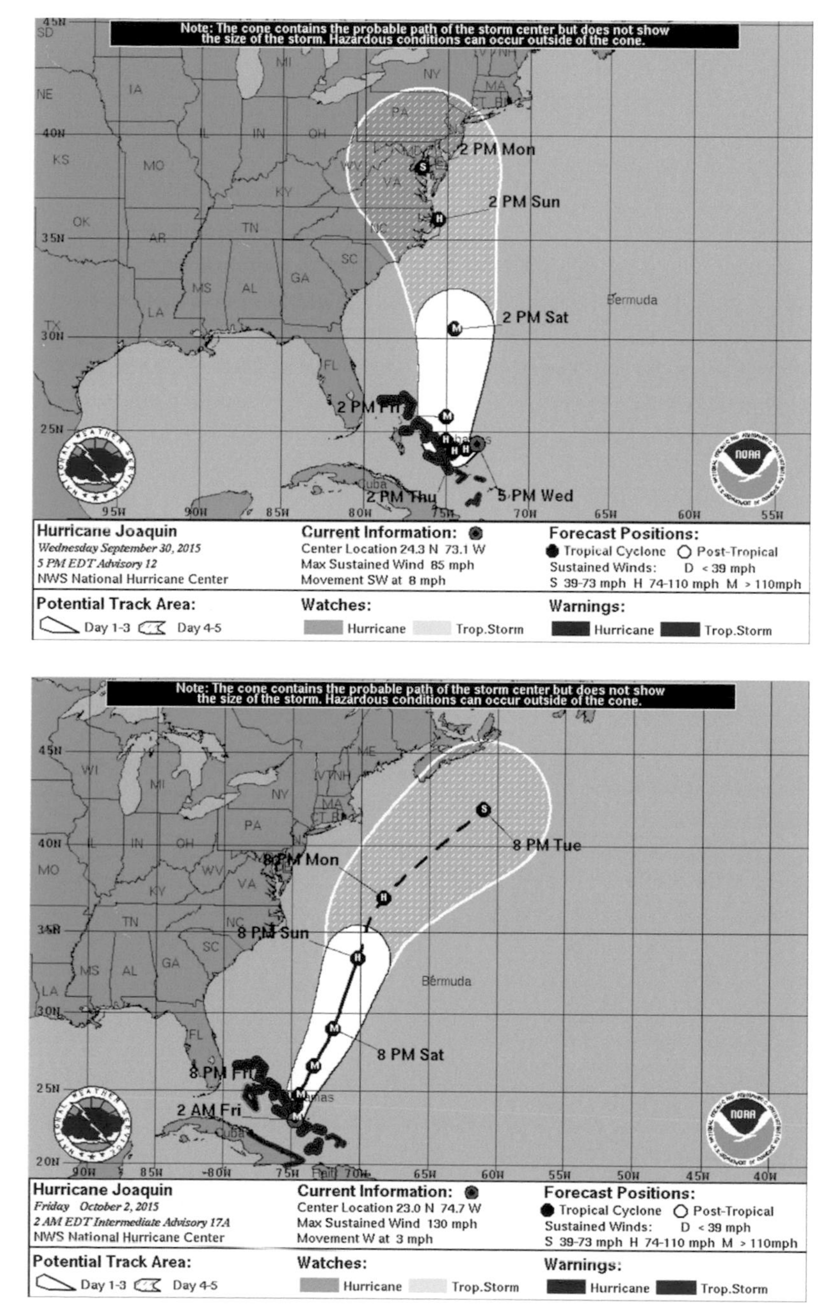

Figure 6.22 Cone of uncertainty for Hurricane Joaquin, 2015. The Figure provides a consecutive forecast of Hurricane Joaquin's predicted track along the Eastern Seaboard, showing a high degree of variability from one forecast cycle to the next. Each track prediction is bounded by an expanding cone of uncertainty. (NOAA.)

that the suite of model predictions showed great inconsistency from run to run. As a result, the error cones swung back and forth, a phenomenon dubbed "windshield-wipering." The inconsistency challenged preparation efforts all along the East Coast, keeping emergency planners on their toes for many days.

Let's discuss the cone of uncertainty further. The public tends to focus on the most probable track forecast, shown by the black solid (1–3 day) or dotted (4–5 day) centerline in Figure 6.22. However, the cone's width delineates the average track error culled from the past five years of forecasting efforts. The uncertainty envelope remains a fixed size for all storms, all season long. The NHC states that the actual track will lie somewhere within this cone 60–70% of the time. And it is very common for the cone's width to expand with time because forecast error typically increases the longer a storm is numerically simulated. Also, it's important to recognize that a tropical cyclone is not a point; rather, the effects of wind, rain, and surge can extend hundreds of miles from the center, well beyond the bounds of the cone of uncertainty. Nor does the cone describe the range of all possible track outcomes, even those by trusted and well performing models, as Figure 6.23 shows.

Improvements in Track Forecasting

How much have track forecasts improved in recent years? Figure 6.24(a) plots the official NHC track error from 1990 to 2020. The Figure shows various forecast time periods, ranging from 1 to 5 days (the NHC did not begin Day 4 and Day 5 track forecasts until 2001). The trend lines, which all slope downward toward smaller forecast errors, are certainly encouraging. They reflect steady improvement in forecasting, thanks to the powerful combination of satellites, aircraft reconnaissance, and numerical weather prediction. The typical 48-hour track error has decreased by half since the late 1990s. The most dramatic improvement is in the medium-range or Day 4–5 track forecasts, though these more advanced predictions tend to show the greatest fluctuation in annual skill.

The dramatic reduction in error directly translates into smaller uncertainty cones issued with new hurricane forecasts. It also has the effect of reducing "overwarning" and evacuations in coastal areas on either side of the cone. Such overwarned regions, which do not experience hurricane conditions, can be costly. The oft quoted statistic is that it costs approximately $1 million to evacuate each mile of coastline. Overwarning also erodes public confidence in the forecasts.

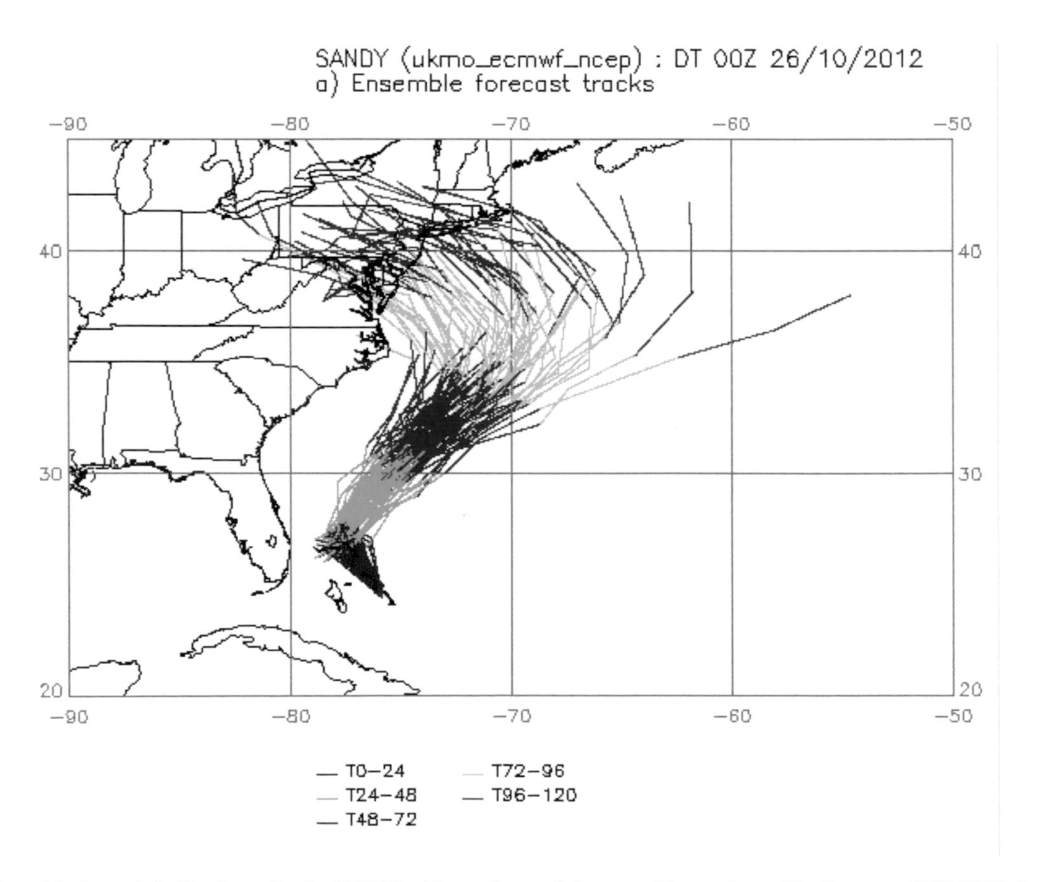

Figure 6.23 Ensemble forecast for Hurricane Sandy, 2012. The Figure shows all the ensemble members of the European (ECMWF), American (GFS), and UKMET models. The divergence of ensemble tracks later in the forecast suggests significant uncertainty in the track forecast. (NOAA.)

Unfortunately, the great progress in hurricane track prediction is not mirrored by a similar improvement in forecasts of hurricane intensity. Intensity forecasts, especially forecasts of particularly rapid changes in intensity, remain problematic. As the flat line in Figure 6.24(b) shows, there has been little improvement in 24-hour predictions of intensity. There has been more improvement in forecasting hurricane intensity at 48 hours and beyond. There are many explanations, including a basic lack of understanding about what causes hurricanes to develop and also the way that numerical forecast models simulate individual storms. These models must be designed to run at higher resolution, down to the scale of individual thunderstorm clouds. These requirements are very costly from the standpoint of requiring big advances in computing power and data storage. To become more effective at prediction, the models must begin to incorporate more detailed observations of the atmosphere's initial state. They must also improve the way they simulate physical processes, such as the energy transfer between the upper ocean layer and overlying atmosphere.

(a)

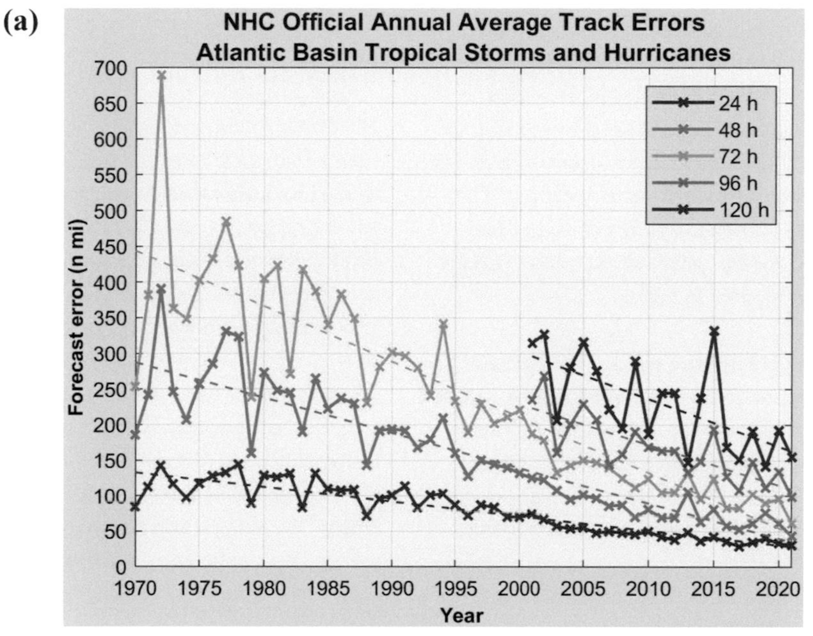

(b)

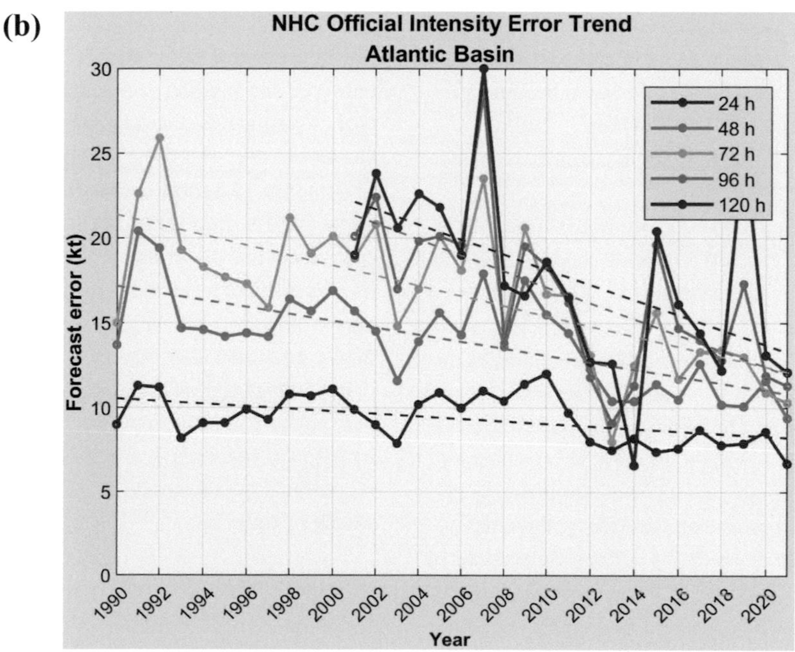

Figure 6.24 Improvements in hurricane forecasting. (a) Significant improvement in hurricane track prediction over a 15-year period and (b) less impressive improvements in predicting hurricane intensity level. (NOAA.)

Threats Posed by Hurricanes and the Need for Accurate Warnings

Starting a couple of decades ago, destructive tropical cyclones became commonplace in the United States. Hurricanes now pose a monumental threat to U.S. lives, property, and economy. In fact, hurricanes are the costliest natural disaster facing the United States. The top five costliest U.S. hurricanes on record are Katrina (2005, $161 billion); Harvey (2017, $125 billion); Maria (2017, $90 billion); Sandy (2012, $71 billion); and Irma (2017, $50 billion). It's especially noteworthy that three of these top five occurred as recently as 2017. In fact, the total insured losses from all U.S. hurricanes since 1986 approaches three-quarters of one trillion dollars! Hurricanes' increased destructiveness is driven mainly by population growth along the U.S. coastal regions. Such an exponential expansion of lives and property is likely to continue unabated in many locations. The overall U.S. vulnerability to hurricanes, which considers both the frequency of the hazard and the behavior that places people in harm's way, is at its highest point in history.

We must continue to improve public warnings and preparedness measures. A lack of basic preparation, including failures to evacuate and a botched government response in the aftermath of Hurricane Katrina, provided painful lessons that society must work harder to cope with these inevitable disasters. The National Weather Service is now engaging teams of social scientists to help the NWS better tailor warnings to various types of decision makers. Even a perfect forecast for a storm is a failure if members of the public do not correctly perceive the risk and therefore do not heed the warnings. During Hurricane Sandy, the warnings disseminated by the NHC and National Weather Service did not deliver a unified message concerning the nature of the threat because the storm became a hybrid system – part hurricane and part Nor'easter – before completely morphing into a Nor'easter prior to landfall.

All hurricane warnings were dropped before Sandy's landfall in New Jersey and New York. The NWS instead issued a confusing blend of gale, coastal flooding, and flood (rainfall) warnings. Collectively, these warnings did not convey the message that an enormous storm, once called a hurricane and now continuing to behave as a hurricane, was about to strike. The message received instead was that a Nor'easter – one of many that regularly hit the Mid-Atlantic in a given year – was on the way. This confusion, along with the fact that warnings for Hurricane Irene during the previous year were more ominous than justified by the actual storm event, caused many to discredit the serious threat posed by Sandy. Subsequently, the NHC and NWS reached an agreement to continue issuing hurricane warnings for hurricanes that have transitioned to extratropical systems. This new policy was first implemented in 2016.

Development and Intensification of Extratropical Cyclones

Hurricanes and extratropical cyclones are generated by fundamentally different processes. Extratropical storms develop in response to large low-pressure disturbances in the jet stream, called troughs. Extratropical storms become larger than most hurricanes, simply because the parent disturbance in the jet stream is so much larger than the precursor cloud clusters in the tropical trade winds.

Genesis of Extratropical Cyclones

The development and growth of an extratropical cyclone results from the interaction of a jet stream disturbance (trough) with a low-level frontal boundary. A broad region of air rises downwind (to the east of) a jet stream trough. This lowers surface pressure which, along with the Coriolis deflection, induces a counterclockwise spiral of air at the surface. These principles are reviewed in Chapter 2.

Along the front, two dissimilar (hot and cold) air masses are pulled toward each other, creating strong gradients in temperature (i.e., along the cold front and warm front). The close alignment of warm and cold air masses creates potential energy; the rising warm air and cold sinking air convert potential energy into kinetic energy, and in this manner the low-pressure vortex deepens and intensifies (see Chapter 5 for a review of cyclone energetics).

The development of the surface low and the jet stream trough reinforce one another; each amplifies the other, over time, in a type of positive feedback called self-development. This process can unfold over 1–2 days; infrequently it can occur more rapidly (in less than 12 hours), particularly during the winter season along the U.S. East Coast. As the storm's circulation draws cold air southward on its west side and warm air northward on its east side, the jet stream's meanders become distorted; these curves amplify (grow larger), forming a giant S-shape. The tighter jet stream curvature further increases the strength of air rising downstream of the trough, which lowers surface pressure, making the vortex stronger. Such self-amplification cannot occur indefinitely, however. Eventually, the cold air is drawn all the way around the south side of the storm, cutting it off from the supply of warm air.

When the storm's temperature gradient is weakened or even eliminated in this manner, the storm can no longer generate kinetic energy, and friction begins to rob the storm of its existing kinetic energy. As a result, the storm slowly spins down. During the process of occlusion (discussed in Chapter 5), the cold front catches up to, and merges with, the warm front. Thus, unlike hurricanes, extratropical cyclones sow the seeds of their own destruction and seldom last beyond 5–6 days. Hurricanes, in contrast, can persist longer – and reach higher intensity – as long as they have access to a deep layer of warm ocean water and avoid regions of destructive wind shear.

Other factors that limit an extratropical cyclone's wind speeds are the storm's large size (compared to tropical cyclones) and geographical setting. Extratropical cyclones are inherently larger vortices at birth compared to tropical cyclones; the broad region of uplift beneath and downstream of the jet stream trough and distortion of the underlying front leads to a large wind circulation, often exceeding 1600 km (1000 miles). In tropical cyclones, as previously shown, the contraction and merging of individual cloud vortices lead to an inherently smaller storm vortex. While the minimum pressure in an extratropical storm may be comparable to that of a hurricane, the winds will not blow as strongly because the pressure gradient (change in pressure with distance) is weaker in an extratropical system, leading to a slower wind circulation. And because these storms form at higher latitudes, the Coriolis effect is more pronounced than in the tropics. Because the Coriolis effect opposes the pressure gradient, the resulting wind must blow slower around an extratropical cyclone (vs a tropical cyclone with comparable pressure gradient).

Extratropical Cyclones and Global Warming

Can we assume that extratropical cyclones will become more intense in an era of global warming? Around the globe, extratropical cyclones are the bread-and-butter weathermakers in the middle latitudes. They generate rapidly changeable temperatures, heavy precipitation (floods, heavy snow, ice storms), strong wind, and outbreaks of severe local storms (tornadoes and severe thunderstorms). If these storms become more frequent and/or stronger in the coming century, those living in the middle latitudes are subject to more frequent weather extremes.

A few studies have noted a statistical uptick over the past few decades in the frequency and magnitude of extreme weather events in middle and high latitudes, including heat waves, arctic air outbreaks, devastating floods, and extreme cyclonic storms. Invariably, most of these events are tied to highly amplified jet stream patterns – that is, a jet with large curves (both troughs and ridges). Such an amplified state tends to "lock in" and persist, causing days of stagnating weather conditions, such as storminess or long-lived, intense heat and cold spells. These blocking patterns set up when the jet flow is highly contorted, causing the troughs and ridges to slow in their eastward progress or even to stop moving altogether.

Recent studies have examined arctic amplification, whereby global warming occurs twice as fast at the poles compared to elsewhere around the globe. Arctic amplification may be causing the northern hemisphere jet stream to change its character. As the poles warm more rapidly, the temperature gradient between polar regions and the equator weakens. Recall from Chapter 2 that the polar jet stream is powered by the hemispheric gradient in temperature. When this gradient weakens, it is hypothesized that the polar jet stream slows down. A weaker jet stream tends to meander more strongly, developing more amplified and slower moving loops. This causes more prolonged periods of temperature extremes, stronger extratropical storms, and overall more extreme weather events.

This hypothesis is controversial, however. More research is needed to prove that the jet stream is in fact slowing and becoming wavier. Additionally, is the "increase" in storminess and other weather extremes simply a matter of perception? Weather patterns in the middle and high latitudes are inherently very changeable, and a large degree of what we consider extreme weather is in fact the norm. Assertions that the past two decades have seen more extreme weather must withstand numerous tests of statistical confidence.

Extratropical Cyclone Mechanics: Fronts and Conveyor Belts

Let's review what we know about extratropical cyclones. The basic extratropical cyclone features a central region of low pressure with a cold-core vortex tilting toward the jet stream trough with altitude. The converging, inward spiral of air (counterclockwise in the northern hemisphere) draws in dissimilar types of air masses. Fronts, or strong zones of temperature and moisture contrast, develop at the boundaries of these air masses, including cold fronts and warm fronts. Surface air is lifted along the frontal zones, leading to concentrated regions of cloud cover and precipitation. For instance, bands of severe thunderstorms, called squall lines, frequently develop along and ahead of the cold front, in the storm's warm sector, particularly during the warmer months of the year. Extensive layers of moderate precipitation (including snow and ice during winter) frequently form along the warm front. High-impact weather, including extreme forms of precipitation and severe local storms (thunderstorms and tornadoes), tend to concentrate along fronts.

A "CAT scan" of a typical extratropical cyclone would reveal narrow ribbons or channels of airflow that enter the storm at different levels (Figure 6.25). Observing the motion and structure of cloud layers in satellite movies reveals these channels, which are called conveyor belts. A broad ribbon of warm and humid air enters the storm's warm sector from the south. This warm conveyor belt (WCB) gradually rises as it approaches the low-pressure center, but it remains in the warm sector during its ascent. The WCB rises above the warm front, then makes an abrupt turn toward the east-northeast, ascending to jet stream level (9 km [30,000] feet and higher). The WCB is the principal feed of humid air, and it is a major supplier of the storm's heavy precipitation within the warm sector and warm frontal zones.

One or more thunderstorm squall lines can erupt within the WCB, ahead of the cold front. Sometimes a secondary surge of cold, dry air will develop aloft and race out ahead of the surface cold front. This upper level cold front displaces the deep uplift of warm air hundreds of miles ahead of the surface cold front, which explains why squall lines are sometimes observed far ahead of the surface front (sometimes hundreds of kilometers ahead).

To the north of the warm front lies a second channel of air, termed the cold conveyor belt (CCB). This is a ribbon-like inflow of cooler, humid air (often drawn from the northern Atlantic in East Coast storms) from the east. The CCB, drawn from a more stable air mass, rises sluggishly but provides an important source of moisture as it moves along low-level cloud layers along the warm front. Where the warm front joins the center of low pressure, the CCB has risen several

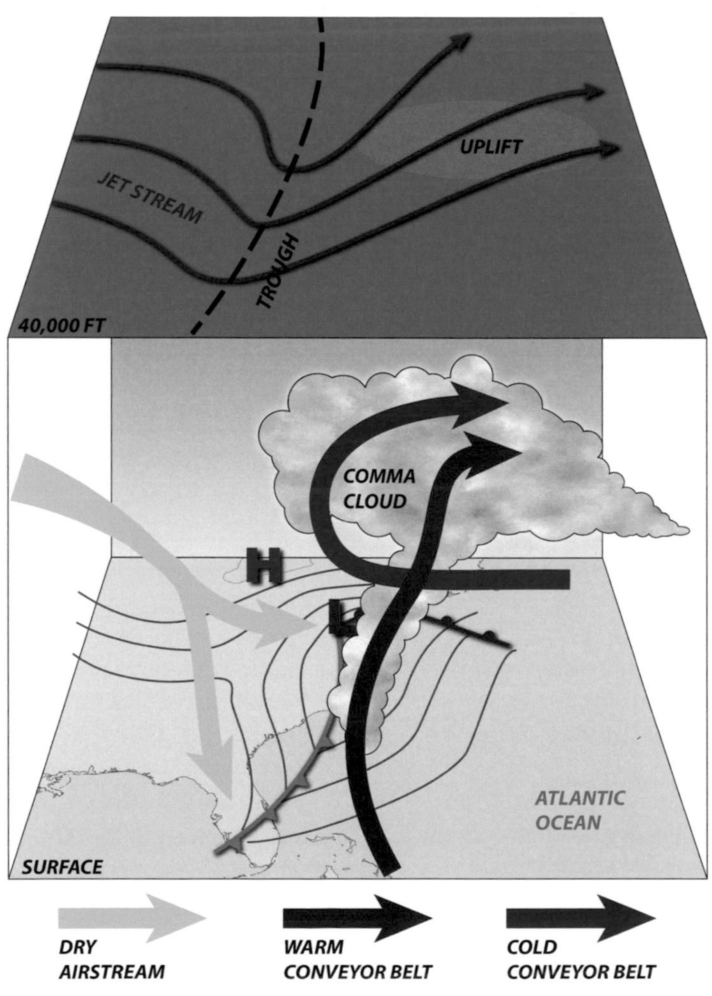

Figure 6.25 Conveyor belts in an extratropical cyclone. This 3D model of an extratropical cyclone shows three principal airflow channels, or conveyor belts. Vertical slices through the storm's warm sector and warm front are also depicted. (Adapted from Kocin and Ucellini, 1990.)

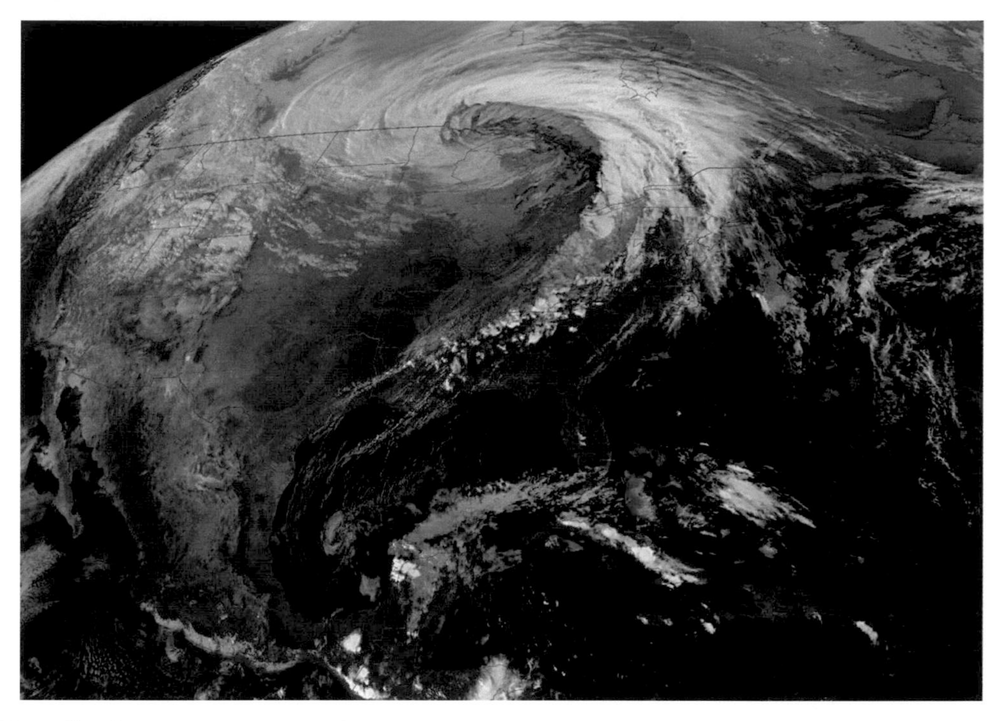

Figure 6.26 Visible satellite view of an intense extratropical cyclone. This satellite image shows the warm sector, WCB, comma head, dry slot, and a large squall line. (NASA.)

thousand feet and wraps around the back side of the vortex. Residing within a very deep, cold air layer, moisture in the CCB condenses into extensive cloud decks (stratiform cloud). These layers wrap around the back of the storm as a broad, comma-shaped cloud. During winter, the CCB provides a key feed of moisture for heavy snow bands that develop within the comma cloud.

It's not uncommon for heavier precipitation within widespread frontal clouds to become organized into narrow, parallel bands. Multiple processes lead to band-like organization. These include wavelike oscillations of rising and sinking air (imagine ripples on a pond) and a type of instability that causes the air to move along slanted (sloped) trajectories.

A third flow channel, the descending dry current (DDC), develops behind the cold front. Here, fast-flowing air in the jet stream trough descends, spiraling north into the center of low pressure. Descending ribbons create a sheetlike current that warms and dries while descending, thus evaporating any precipitation and cloud. On satellite images, this process is readily seen in a pronounced dry slot that wraps behind the cold front, spiraling into the comma head. In some storms, the DDC acts as a conduit for transporting the momentum of fast winds and even stratospheric ozone toward the surface. It has been implicated in enhancing violent wind gusts across the Great Lakes in the fall.

Figure 6.26 is a satellite view identifying many of the key features we have just discussed. Note the squall line of thunderstorm cells, which appears as a bright, triangular-shaped cloud structure within the WCB and warm sector and which arcs along the cold front. Its southernmost point is the burst point where new thunderstorm cells continuously erupt within the warm sector. Then the strong southerly winds of the WCB blow the upper cloud layers into a dense cloud stream that rapidly expands northward and eastward.

Intense Extratropical Cyclones by U.S. Region

We presented the basic structure, evolution and energetics of extratropical cyclones in Chapter 5. In this section, we explore some additional details related to the generation of high-impact weather. Then we examine severe weather associated with four extreme types of extratropical cyclone: intense storms of the Pacific Northwest, the Great Lakes, East Coast, and North Atlantic.

Powerful Cyclones of the U.S. Pacific Northwest

Exceptionally large and fierce extratropical cyclones, termed Pacific Northwest (PNW) storms, are frequently generated over the Bering Sea (south of Alaska) during the winter. During these months, the jet stream is most intense, with large, vigorous troughs that advance off the Pacific over western Canada and the Pacific Northwest (including Washington, Oregon, and northern California). The ocean remains warm relative to land and serves as a heat source for these storms; high winds generated around low pressure efficiently extract heat from the upper ocean and evaporate seawater, adding energy to the storm. These cyclones typically arrive along the coast either occluded or in the process of occluding, meaning they are at peak intensity. The storm's low-level winds have spent days accelerating over the relatively low friction ocean surface. This long fetch (distance over which winds blow steadily from the same direction) enables ocean waves to grow to mammoth heights, building into long trains of ocean swell. These waves pummel the coast, causing significant erosion of beaches and headlands, flooding, and property destruction.

PNW storms generate exceptionally high winds and extreme amounts of precipitation. The mountainous coastline of the western United States and Canada (that is, the Cascade Mountain Range) amplifies the uplift of moisture-laden air and channels airflow into extremely powerful winds. Many meteorologists argue that PNW storms are the most powerful type of extratropical cyclone to strike the United States. Their cumulative effect makes the Pacific Northwest the wettest location in the contiguous United States; the region receives 150–300 cm (60–120 inches) of precipitation every year. The precipitation falls in the form of frequent, heavy rains at low elevations and as deep, wet snow along mountain ridges.

There is no better example of a PNW storm than the Columbus Day Storm of October 12, 1962. Nicknamed "the Big Blow," its winds approached that of a Category 3 hurricane. However, the Big Blow projected those winds over a significantly larger area than a typical hurricane. The storm often serves as a benchmark against which all other 20th-century extratropical cyclones are compared. With a minimum central pressure of 955 mb, the storm claimed the lives of 46 and blew down an estimated 26 to 35 million cubic meters of timber – an amount greater than the entire region's annual timber harvest and ten times more timber felled than in any other U.S. storm. The Columbus Day Storm destroyed 53,000 homes and knocked out power to one million customers. Total economic losses were a staggering $1.8 billion to $2.2 billion.

The surface wind map in Figure 6.27 shows this storm striking the coastline on October 12. The storm actually stemmed from the remnants of Typhoon Freda near the Philippines, having undergone an extratropical transition while crossing the Pacific (see Chapter 5). During October 8–10, the storm tracked eastward over cooler ocean and became influenced by the strong jet stream, morphing into an extratropical system. As the transformed vortex approached the U.S. mainland from the southwest, it rapidly deepened. The storm moved northward under a vigorous jet stream at a fast clip, over 33 kts (40 MPH), while remaining just offshore. The rapid movement enhanced winds on the right (coastal side) of the track because southerly winds circulating around the storm got boosted by the storm's movement from the south.

As the storm made landfall, the strongest winds occurred behind the cold front, blowing from the south-southwest. Extremely powerful winds were concentrated along a large segment of coastline (Figure 6.27). Very strong winds also occurred farther inland at the high elevations. Levels of structural damage imply that the winds may have been even higher than those indicated on this map; many anemometers (instruments for measuring wind speed) were destroyed before they could register the highest wind speeds. For instance, Cape Blanco may have experienced sustained winds of 96 kts (110 MPH) with peak gusts to 157 kts (181 MPH).

Another element sometimes contained in PNW storms is a feature termed an atmospheric river, which can lead to extended periods of heavy precipitation (flooding rains and heavy snow) over West Coast states. Atmospheric rivers are narrow ribbons of high humidity sourced in the tropics that can become incorporated into the warm sector of a PNW storm. Atmospheric rivers are discussed in the context of flash floods in Chapter 14.

Great Lakes Gales: November Witches

Another U.S. region experiencing very intense cyclonic storms is the Great Lakes and Upper Midwest. This region is particularly vulnerable during fall, as a result of a combination of the strengthening jet stream and warm lake water. The lakes' warmth adds energy (in the form of heat and moisture) to low-level air streaming into the storm. These storms are sometimes called November Witches, and they are infamous for creating severe gales across the Great Lakes. The history of the Lakes is riddled with shipwrecks caused by these storms, such as the loss of the iron-ore hauling vessel *Edmund Fitzgerald* on November 10, 1975.

Minimum central pressures in a November Witch can reach the 955–960 mb level, equivalent to the low pressure in a

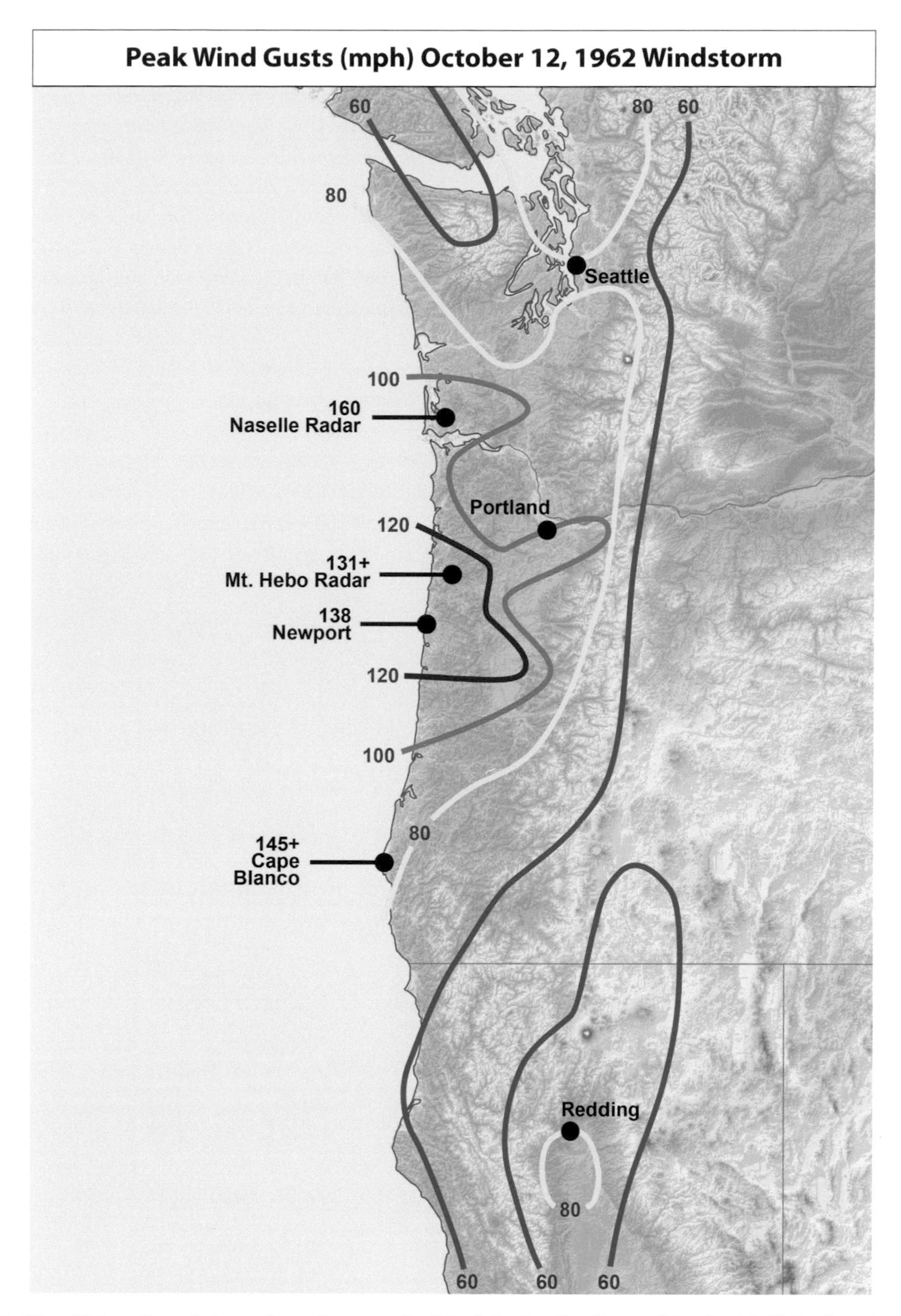

Figure 6.27 The Big Blow. This surface wind map shows the approach of the Columbus Day Storm of October 12, 1962, along the coast of the Pacific Northwest. (Adapted from Mass, 2008.)

Category 3 hurricane. Wind gusts of 61–70 kts (70–81 MPH) in parts of these systems are not uncommon, along with torrential rain, waves of 6–7 m (20–23 feet), and sometimes blizzard conditions (a combination of heavy snow and strong wind inducing whiteout conditions). A few notable examples of November Witches are the Armistice Day Blizzard of 1940, Great Appalachian Storm of 1950, and the "Chiclone" (Chicago Cyclone) of 2010.

Many meteorologists, when asked to rank top U.S. cyclonic storms of the 20th century, place the Great Appalachian Storm of 1950 at the very top. Figure 6.28 shows the surface weather map for this event. The storm's circulation was truly enormous. At the time shown, the storm was in a highly occluded state; cold air from the Upper Plains had circulated completely around the storm. In fact, the cold over the Great Lakes arrived from the south and east, a most unusual situation. The storm's developmental history was also unconventional. It first formed along the coast of North Carolina on November 24. Tracking initially northward, it underwent a period of rapid deepening over Washington, D.C., but it then took a highly anomalous turn westward, toward the Great Lakes. The unusual trajectory was caused by strong high pressure aloft over eastern Canada and a massive low-pressure system over the upper Midwest. Clockwise flow around the high and counterclockwise flow around the low steered this storm toward the Great Lakes.

On November 26, the intense surface low, spinning over the Great Lakes, moved beneath the upper low center. As a result, the entire system – now a single, deep, rotating air column – stalled over Ohio. Very warm air from the Atlantic Ocean fed into the storm from the east and south; at the same time, deep cold air wrapped around the storm from the west. While over the Great Lakes, a piece of this low-level warm-air parcel became sequestered (trapped) near the storm's center, such that the lower half of the storm took on a structure resembling that of a hurricane (recall that hurricanes are warm-core systems). In the final days of November, the storm rapidly occluded over the Great Lakes and drifted slowly toward the northeast, finally weakening over Canada.

The storm's impacts were both diverse and extreme across the entire Ohio Valley, Mid-Atlantic, and Northeast. The storm had a pronounced warm side and a pronounced cold side. As Figure 6.28 shows, an intense dome of high pressure to the storm's

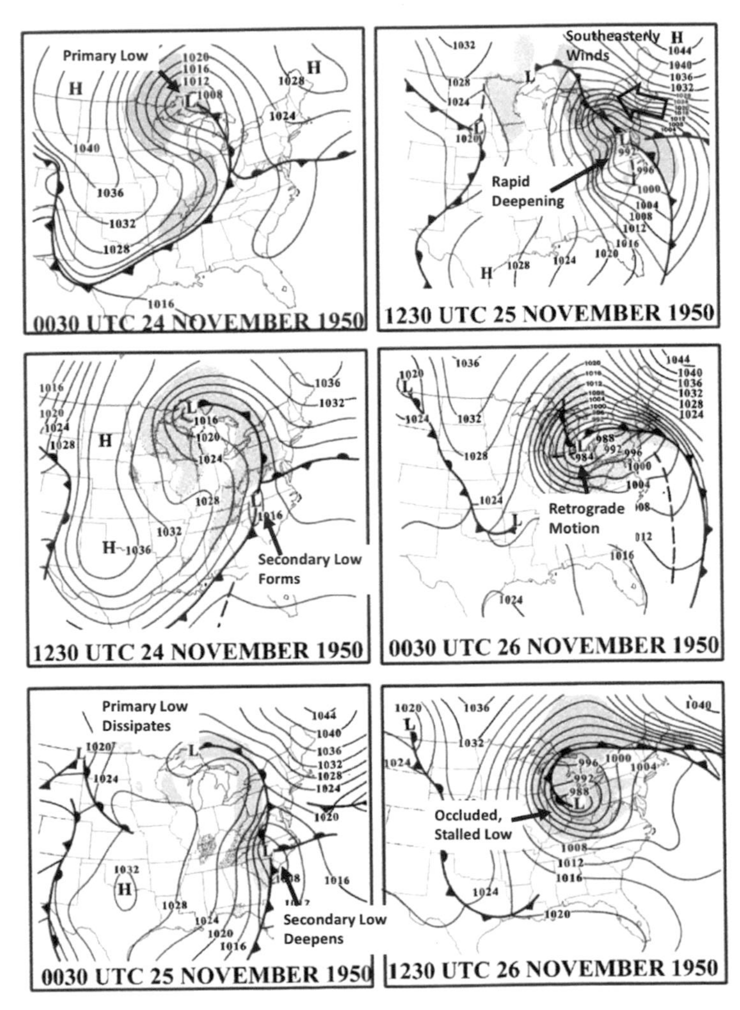

Figure 6.28 Great Appalachian Storm of 1950. This surface weather map depicts the evolution of isobars, fronts, principal winds, and motion over several days. (Adapted from Kocin and Ucellini, 2004.)

northeast tightened the pressure gradient, creating strong sustained winds of 43–52 kts (50–60 MPH) and stronger wind gusts of 70–78 kts (81–90 MPH) along the New England and Mid-Atlantic coasts. Massive ocean waves slammed into coastal regions, leading to severe inundation and structural damage. Record low temperatures across the entire Tennessee Valley and Southeast ensued from an unseasonably cold air mass that wrapped around the south side of the storm. The cold air surge generated a wide swath of heavy, early season snow, ranging from 50–88 cm (20–35 inches), across the higher elevations of West Virginia, western Pennsylvania, and eastern Ohio. The storm caused 353 fatalities, and total economic losses approached $600–700 million.

Nor'easters: Wintertime Devastation in the Mid-Atlantic and New England

Nor'easters are intense, wintertime extratropical cyclones that track northward along the U.S. East Coast. They are named because of the strong, sustained winds that blow from the northeast (northeasterly winds), off the North Atlantic, raking coastal communities and driving large ocean waves into the shoreline. Figure 6.29 provides a satellite view showing snapshots of a Nor'easter using visible, infrared, and moisture imagery. The large, comma head cloud swirl is unmistakable, blanketing all of the Mid-Atlantic and Northeast. The water vapor image in Figure 6.29(b) shows dual swirls of saturated air (green colors) and dry air (orange hues) mixing like spirals of stirred coffee and milk. This particular storm occurred in mid-April, 2007; it produced numerous tornadoes over the Carolinas, sustained 61–70 kt (70–81 MPH) winds along the New England coast, offshore waves to 10 m (33 feet), 0.6 m (2 feet) of snow over New York, and over 23 cm (9 inches) of rain in New Jersey.

Strong Nor'easters develop only during the cool months (October–April) and along the East Coast because they are powered by strong temperature contrasts. Their peak season is January through March, when arctic and Canadian air masses plunge southeast across the United States. When they arrive along the East Coast, a very pronounced temperature contrast develops with the mild ocean water just offshore (which has arrived courtesy of the western Atlantic's Gulf Stream current). When a jet stream trough interacts with this gradient, a coastal cyclone initiates (a favored spot is along North Carolina's Outer Banks) and rapidly deepens. In fact, Nor'easters sometimes undergo bombogenesis – that is, their central pressure drops more than 24 mb in 24 hours – and at times, the rate of pressure fall is two to three times this rate for several hours. Recent research suggests that these bomb cyclones occur on average two to three times per winter season, over the coastal waters of the Mid-Atlantic and New England. Central pressures in bomb cyclones can drop into the 950–970 mb range.

The transfer of the ocean's heat and moisture into the overlying cold, dry air is very efficient, causing these storms to draw a significant fraction of their heat from the ocean (similar to the dynamics of a hurricane). Strong storm winds fanning the ocean extract even more heat, and this shot of energy causes Nor'easters to deepen explosively. For these reasons, Nor'easters are classified as a hybrid type of cyclone. While energized by the jet stream and a strong thermal gradient, they also derive a fraction of their energy from the warm ocean, as hurricanes do.

The number of Nor'easters in a given winter varies from year to year, typically on the order of 10–20. Not all storms are equally destructive, and their exact tracks differ; some remain out at sea, others track along the coast, and some have more of an inland trajectory. If the inland air mass is cold enough, very substantial snowfall is generated, impacting major cities along the Northeastern megalopolis. Not all coastal regions of the Mid-Atlantic and New England are affected by the same storm. But the cumulative effect of a season's worth of storms can be tremendous economic and property damage, disrupted transportation networks, and loss of life. Indeed, the seasonal loss of beach sand, destruction of beach property, and flooding can be greater than that caused by a single, strong hurricane making landfall. While hurricanes often garner more media attention, Nor'easters relentlessly hammer away at the Northeast shoreline every year.

Special impact scales have been devised to categorize Nor'easters. The Dolan–Davis Nor'easter Scale is a five-category intensity ranking based on ocean wave energy and the damage that water inflicts to coastlines. Strong wave action causes overwash and inundation, dune erosion, breech of barrier islands, and structural damage. Wave energy develops as a result of both high wind speed and long wind duration. In fact, a Nor'easter's duration factor is just as important as its maximum winds.

Winds blowing from the same direction for many hours create large fetch (the total distance traveled over water). The longer the fetch, the more wind energy gets transferred into the water. For a given wind speed, doubling the fetch will nearly double the wave height. And the wind circulation around Nor'easters tends to be much larger than that of hurricanes. As Figure 6.30 shows, the result is a long fetch of uniform wind direction, which leads to a fully developed wave train. Around hurricanes, which are smaller, wave trains from different directions are more likely to cross and undergo destructive interference, which reduces overall wave heights (although this reduction can be somewhat offset by the much stronger winds in a hurricane).

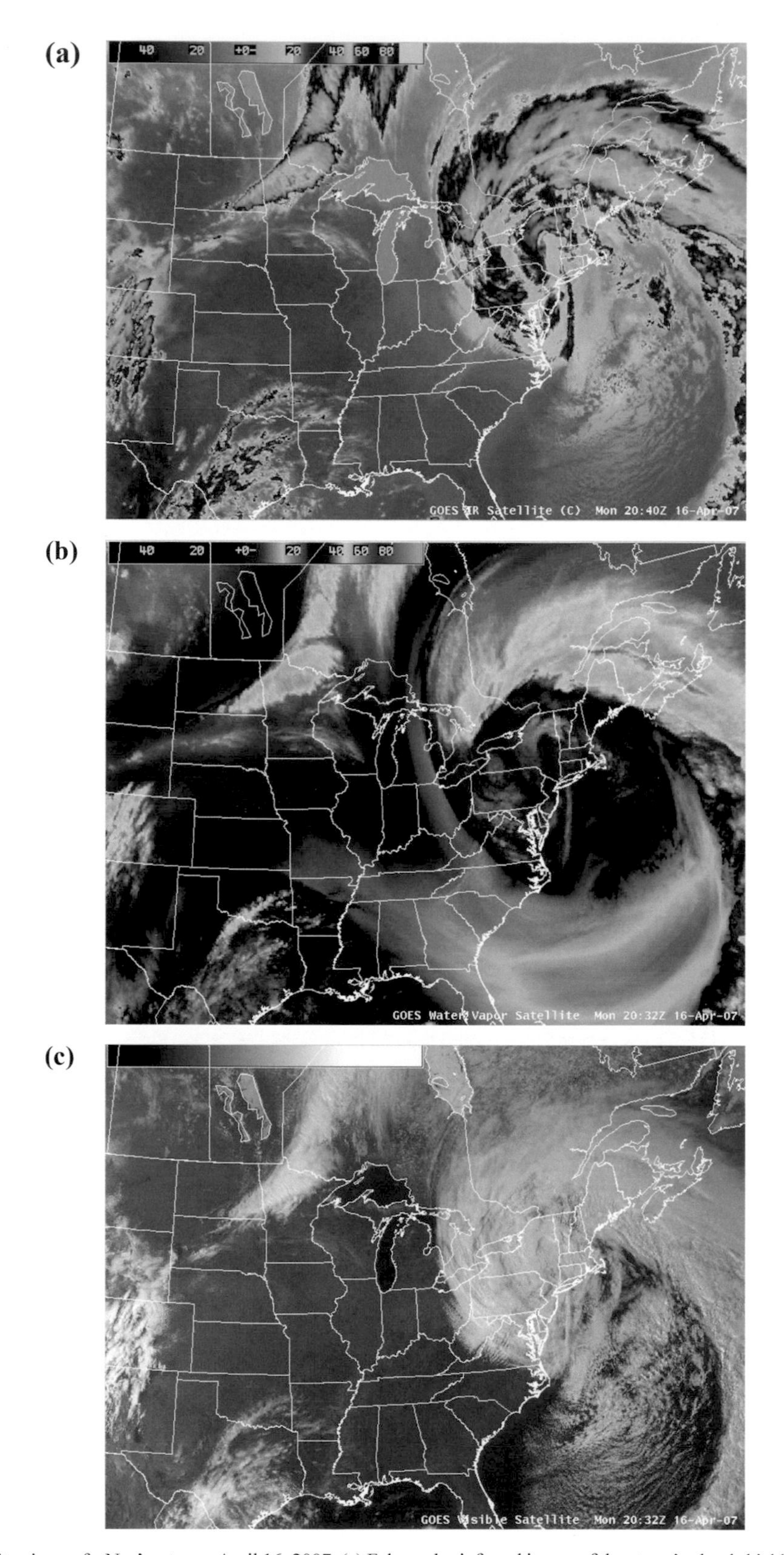

Figure 6.29 Multiple satellite views of a Nor'easter on April 16, 2007. (a) False-color infrared image of the storm's cloud shield; (b) false-color water vapor view of the same storm (blue = humid air mass, orange = dry air mass); and (c) visible satellite image of cloud tops, rendered in black and white. (NOAA.)

Ocean Wave Heights In Nor'easters Often Grow Higher Than In Hurricanes

Hurricane

Compact wind field. Wave sets generated in many different locations simultaneously around compact vortex collide, with interference between some wave trains. This limits wave height growth, even in very high wind states

Wind Fetch

Storm Wind Field

Figure 6.30 Schematic of winds blowing around a hurricane and a Nor'easter. The winds have a significantly smaller fetch around the smaller hurricane, compared to the Nor'easter.

Figure 6.31 shows the surface weather chart for a destructive Nor'easter that hit the East Coast on December 11, 1992. The green arrow shows the direction of strongest winds, blowing from the northeast, along coastal New England (contrast this with the prevailing southwesterly direction of strongest winds in a PNW coastal storm). Solid lines are isobars; note their compact spacing north of the low-pressure center, where the pressure gradient is largest, due to two effects: (1) the surface low itself (which may be rapidly deepening) and (2) very strong high pressure located to the north. The cold, dense dome of high pressure contributes as much to the strong wind field blowing off the Atlantic as does the vortex itself. High pressure over New England also funnels subfreezing air southward, creating an extensive region of heavy snow falling along the western side of the storm.

Nor'easters tend to be long-duration events, a consequence of both their large size and slow movement. Nor'easters form when high-amplitude jet stream patterns (a tightly curving trough and ridge) establish a blocking pattern, causing the surface low pressure to migrate slowly northward or even stall. In terms of coastal inundation and beach erosion, Nor'easters can hammer the coast over multiple high-tide cycles. During each high tide, the flooding and erosion worsen for the same wave height because the waves and tidal influences are additive. Thus, when forecasters assess the damage potential of Nor'easters, the duration of the storm is nearly as important as its actual intensity and the timing of the storm's worst phase relative to astronomical high and low tide.

What might happen to the frequency and intensity of Nor'easters in a global warming world? A few scientific studies have attempted to address this, based on running mathematical models in which climate is allowed to warm. These models show an increase in the number of intense cyclone events over the eastern United States, with more undergoing rapid deepening. As atmospheric water vapor content increases due to warming oceans, Nor'easters are expected to produce heavier precipitation. The increased rate of condensation and attendant increase in latent heat release provide the "energy boost" that strengthens these cyclones.

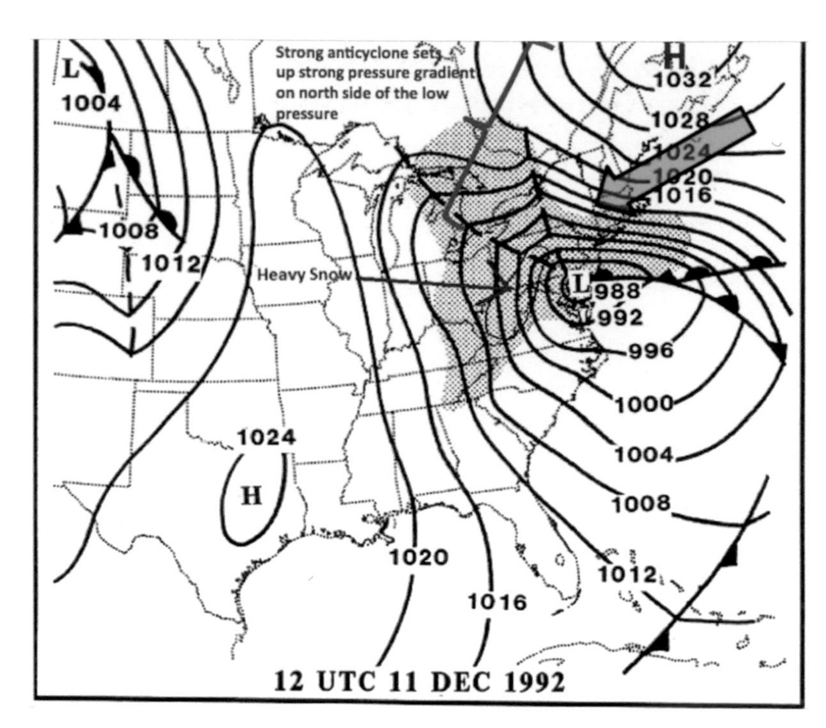

Figure 6.31 Nor'easter, December 1992. This surface weather map shows an intense Nor'easter in early December 1992. Tightly spaced isobars are due to the storm's low-pressure center, combined with strong high pressure located to the north. The pressure gradient generated powerful northeasterly winds that blow off the North Atlantic. (Adapted from Kocin and Ucellini, 1990.)

Wintertime Storms Over the North Atlantic: "Fish Storms"

The term "fish storms" is a colorful term that refers to the exceptionally powerful extratropical cyclones that develop over the North Atlantic Ocean (hence impacting mainly fish, not people). Like Nor'easters, they are most intense and frequent during the winter, and some of them derive from Nor'easters. Each year, several fish storms bomb out, intensifying at dizzying rates. While they possess a cold front and warm front, recent studies have shown that they differ somewhat from classic, land-based extratropical cyclones. They are sometimes called S-K storms after the names of the meteorologists who pioneered their study (Melvin Shapiro and Daniel Kaiser). They are a hybrid type of cyclone, with a cold core aloft but also a pool of warm ocean air close to their center. Because these storms develop so rapidly, the cold front near the inner core "fractures," and a piece of the warm-sector air mass effectively becomes secluded, or trapped, near the storm center. Strong thunderstorms can erupt out of this warm, unstable air mass, close to the storm's inner core. The term "warm seclusion" is used to describe the peculiar structure of S-K storms.

Much less would be known about this type of ocean storm if it weren't for weather satellites. In recent years, satellites containing microwave scatterometers have been able to determine the speed and direction of winds just above the ocean surface. The scatterometer emits microwave pulses, which pass through deep, precipitating cloud layers. The amount of energy reflected back to the sensor depends on the sea state. The more turbulent the ocean (due to breaking waves), the stronger the winds. A fairly accurate map of the wind field beneath the cyclone emerges; an example is shown in Figure 6.32. By examining many of these satellite overpasses, meteorologists have learned that S-K storms can pack a mighty punch of high surface wind. From advances in satellite technology, the true incidence of hurricane-force, sustained wind (> 65 kts [> 74 MPH]) near the core of these storms has surprised meteorologists. This region of intense wind is often found in the storm's southwest quadrant, close to the center, as Figure 6.32 shows.

The occurrence of sustained winds in the 65–70 kt (75–80 MPH) range exemplify the upper limit of sustained wind speed in an oceanic, wintertime cyclone. Another term for this type of extreme, wintertime, oceanic storm is a hurricane-force low (HFL). Hurricane-force lows have been observed in the Pacific as well as the Atlantic. It's important to recognize that these hurricane-force winds are limited to a small area and do not constitute a ring or "eyewall"-type feature typical of tropical cyclones. The winds are of variable duration: Some are as brief as 3–6 hours and, in other instances, last more than 24 hours.

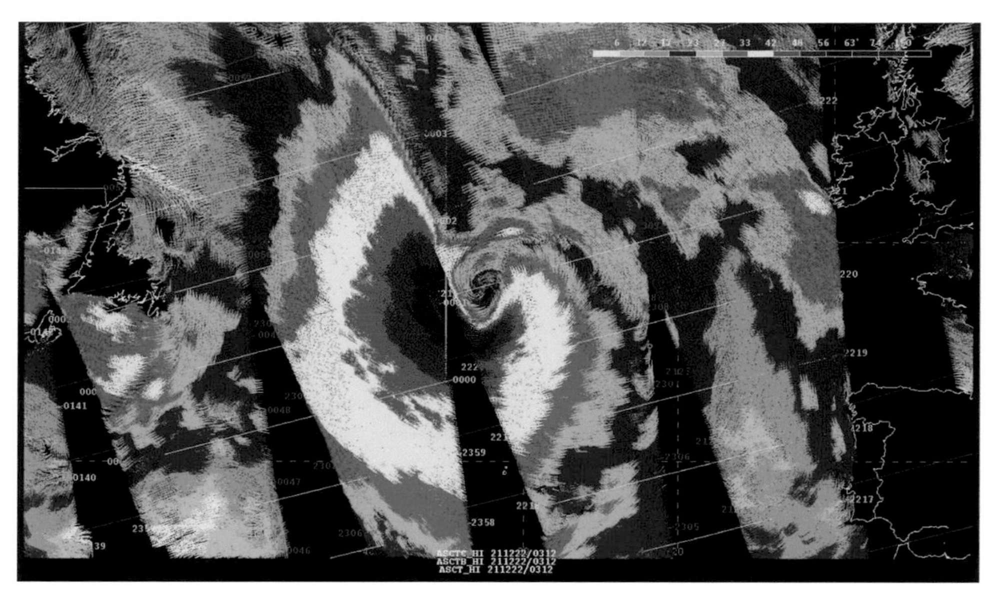

Figure 6.32 Winds beneath an oceanic wintertime cyclone. This satellite image, based on microwave scatterometry, shows exceptionally strong winds swirling around an oceanic wintertime cyclone over the North Atlantic. The strongest pocket of hurricane-force sustained wind (red region, 65 kts [74 MPH]) is located in the storm's southwestern quadrant. (NOAA.)

The huge wind field and extreme winds whirling around S-K storms create ocean fetches that routinely exceed 1600 km (1000 miles). The wind heaves up truly mountainous wave trains not uncommonly in the 12–15 m (39–49 feet) range. These wave trains disperse into a damaging swell that relentlessly pounds the European coastline. During some winter seasons, the parade of S-K storms creates immense damage across the British Isles and Iberian Peninsula. S-K storms are also a constant threat to commercial fishing and shipping vessels that ply the North Atlantic. Shipping lanes between Europe, Africa, and North America are numerous, with ships hauling vast amounts of goods around the western hemisphere.

Additionally, there has been an increasing number of high-profile ship encounters with large ocean cyclones in recent years (likely inevitable, to some extent, due to the great and increasing number of ships on the ocean). The cruise ship *Anthem of the Seas* was crippled by a rapidly deepening, hurricane-force low off North Carolina during February 2016, having encountered 10 m (30 feet) waves and wind gusts over 100 kts (125 MPH). During January 2018, another cruise ship – the *Norwegian Breakaway* – got entangled in a hurricane-force low off Cape Hatteras, North Carolina. Years ago, NOAA's Ocean Prediction Center (OPC) was created to protect these vital shipping lanes, providing detailed weather analyses and marine forecasts across the North Atlantic.

Digging Deeper: Jet Streaks and Bomb Cyclones

Winter storms along the Eastern Seaboard sometimes intensify very rapidly. Recall the definition of a bomb cyclone: a storm in which the central pressure drops at least 24 mb in 24 hours. Coastal storms such as Nor'easters receive an energy boost from the warm ocean surface, in the form of heat and water vapor (the water vapor condenses in rising air, releasing latent heat). This large influx of energy does contribute to rapid storm intensification. But dynamical processes in the jet stream, overlying the developing storm, can also contribute to what some meteorologists refer to as "bombogenesis."

Figure 6.33 shows an intense wintertime Nor'easter (from early February 2010) developing off the Mid-Atlantic coast. The lower panel illustrates the surface features and airflows interacting to generate heavy snow inland of the storm. The center of the storm is shown by the red "L" just offshore, labeled "989 mb Coastal Low." The top panel of this Figure shows airflow patterns at jet stream level (9 km [30,000 feet]). There, a broad jet stream flow surges across the southern United States. Embedded in this flow, on the northern edge, is a closed low-pressure region, or vortex (red "L").

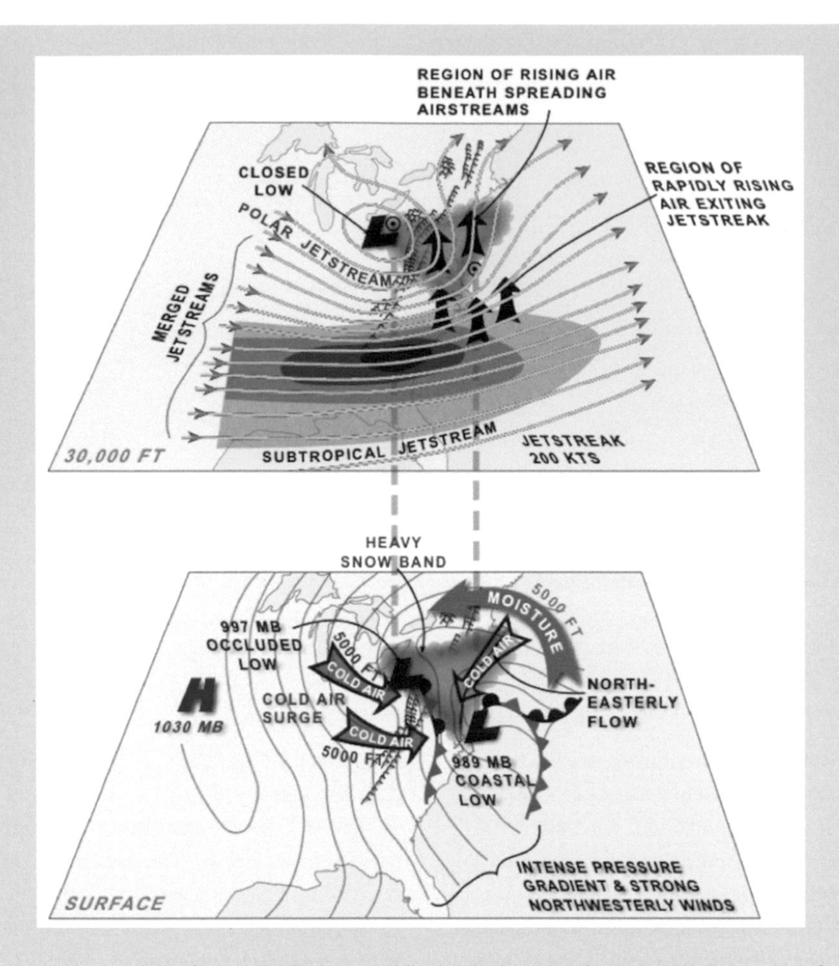

Figure 6.33 Surface and upper atmospheric features leading to a rapidly developing coastal storm or Nor'easter. Key features at the jet stream level (top panel) leading to rapidly rising air (red, vertical arrows) over the core of the cyclone are shown.

Embedded in the general jet stream flow is an even faster pocket of winds, shaded in pink, surging to 200 kts (230 MPH) in the very core (dark magenta shade). This region is termed a jet streak. Forecasters use radiosonde observations, data from commercial aircraft, and weather satellites to detect and analyze jet streaks. Jet streaks are crucial in accurately predicting coastal storm intensity, because a principle of dynamic meteorology states that air exiting the left, front quadrant of a jet streak is forced to ascend vigorously (shown by the cluster of red arrows). Like air exiting the top of a chimney, air is drawn up from below. (Air also is forced to rise vigorously in the right, rear quadrant – but that part of the jet streak is not shown in the graphic.) As air is removed from the surface, above the developing coastal storm, the surface pressure must fall. Thus, when the left exit region of a jet streak develops or migrates over surface low pressure, a rapid reduction in pressure and resulting rapid intensification can take place.

If you study the top panel of Figure 6.33 carefully, you will also note another mechanism that leads to rapidly rising air over the surface cyclone. As air curves sharply northward around the region of upper level, closed-off low pressure, streams of air spread apart. Like a spreading Asian hand fan, the divergence of air draws up air from below (pair of red vertical arrows in the diagram), contributing to further sea level pressure reductions. As it turns out, these spreading airflows and jet streaks are common jet stream features attending intense Nor'easters.

There are rare examples of extratropical cyclones that have bombed out while inland, over North America. Two of these include the "Cleveland Superbomb" of January 1978 (955 mb) and the October 2010 extratropical cyclone that dropped to 955 mb over Minnesota. In the case of the Cleveland storm, some additional energy input contributing to rapid deepening likely derived from heat and moisture off the Great Lakes, but in the Minnesota storm, extreme jet stream dynamics were likely the single cause.

Summary

LO1 Explain how hurricanes develop, intensify, and weaken.

1. Necessary but insufficient large-scale conditions across the tropics for hurricane formation include (1) ocean surface temperature warmer than 26° C (79° F), (2) latitudes 10–30° poleward of the equator, and (3) weak wind shear. The storm's initial spin derives from the merger of multiple, small-scale vortices generated by individual thunderstorms and tropical cloud clusters.
2. The Saharan Air Layer (SAL) is a hot, dry, and dusty air layer flowing westward off Africa, toward the Americas, on the tropical trade winds. Some meteorologists suggest that this dry current impedes the development and intensification of Atlantic hurricanes, although the hypothesis is controversial.
3. Hurricanes intensify through a type of positive feedback process, whereby warm air in the storm's core (generated by latent heat release in tropical thunderstorms) lowers the surface pressure. As winds above the ocean blow more strongly into the low pressure, they extract additional heat and water vapor from the ocean. More vapor leads to more condensation in the thunderstorms, which further warms the core through latent heat release.
4. Periods of rapid hurricane intensification often ensue when the storm moves over pockets of exceptionally warm ocean water, a process that fuels long-lived clusters of thunderstorms called convective bursts. Strong hurricanes can unexpectedly weaken when they undergo eyewall replacement, whereby a secondary ring of wind forms outside the primary ring, causing the storm to weaken in accordance with the conservation of angular momentum. Especially small, intense storms seem to experience the greatest rates of intensity change.

LO2 Discuss several measures of hurricane strength and energy, including the Saffir–Simpson Hurricane Wind Scale and accumulated cyclone energy (ACE).

1. Maximum sustained wind (MSW) and minimum central pressure (MCP) are two key parameters used to rate the intensity of hurricanes. The Saffir–Simpson Hurricane Wind Scale divides hurricanes into five intensity categories based on MSW.
2. Hurricane size can be assessed using the radius of outer closed isobar (ROCI). Atlantic hurricanes have an average storm diameter of 587 km (365 miles) and there is no relationship between storm size and MSW. Storm size is a key determinant of storm surge height, wind damage, and the duration of severe weather. Several new indices such as the Hurricane Severity Index (HIS) and Cyclone Damage Potential (CDP) combine MSW and storm size into a single storm-impact parameter.
3. Accumulated cyclone energy (ACE) measures the kinetic energy contained in each storm. It is based on an integrated measure of the storm's wind intensity and duration. The ACE values of all hurricanes in a given season can be summed to provide ocean-basin-specific, seasonal ACE values. A time series plot of seasonal ACE for many years reveals multi-decadal periods of persistent low hurricane activity, alternating with hyperactive storm eras.

LO3 Summarize the controversy regarding the effects of global warming on changes in hurricane frequency and intensity.

1. Two pioneering research studies released in 2005 demonstrated that strong hurricanes have become more frequent, and hurricane power has rapidly increased, in the decades since the 1970s. These upward trends correlate with a rise in ocean temperature, perhaps due to global warming.
2. The Atlantic Multidecadal Oscillation (AMO) is a feedback between Atlantic ocean temperature and atmospheric wind patterns, causing the North Atlantic surface waters to warm and then cool on a multidecadal cycle. The AMO is tied to cyclic variations of cloud cover, African dust, monsoon rainfall patterns, and wind shear intensity across the tropical and subtropical Atlantic. An active era with above-average hurricane frequency began in 1995 and has extended through at least 2012.
3. The El Niño Southern Oscillation (ENSO), a feedback involving tropical ocean temperature and

wind patterns in the Pacific, can strengthen wind shear across the tropical Atlantic, causing Atlantic hurricane activity to decline during strong El Niño years.

LO4 Discuss the progress made in forecasting hurricane track and intensity.

1. NOAA's National Hurricane Center (NHC), located near Miami, predicts hurricane intensity and movement for the Atlantic and East Pacific Ocean basins. The NHC uses a workhorse fleet of four-engine J-130 aircraft that fly through the developing storm, dropping dropsondes (probes that measure air temperature, moisture, pressure, and winds) during their missions.
2. An important hurricane forecast product is the graphical cone of uncertainty, which projects the average track error from the past five years on either side of the most probable storm track. Hurricane effects often extend beyond the boundaries of the uncertainty cone.
3. During the past several decades, hurricane track errors have been cut in half. However, our skill in predicting hurricane intensity change has improved only slightly. The threat posed by hurricanes is now greater than ever, given the massive increases in coastal property and lives at risk, and hurricanes remain the costliest and deadliest type of natural disaster in the United States.

LO5 Explain how extratropical cyclones develop.

1. Extratropical cyclones intensify through a type of positive feedback process, whereby the jet stream trough and surface low pressure begin to reinforce and self-amplify. The intensification, however, is limited by the process of occlusion, in which cold air wraps around the storm, eliminating the temperature gradient that powers the system. Occlusion is thus a type of self-destructive mechanism.
2. Extratropical cyclones are larger vortices than most hurricanes because the jet stream processes that trigger their formation extend over larger geographical regions. Their larger size translates into weaker winds, even for the same minimum central pressure as a hurricane, because the pressure gradient that generates those winds extends over a much larger area.
3. Some meteorologists hypothesize that extratropical cyclones may become more frequent and intense in coming years, leading to more frequent extreme weather events. The argument is based on the fact that the Arctic is warming faster than the equator, which weakens the polar jet stream. A weaker jet tends to develop stronger meanders (curved regions) which in turn generate more intense cyclones.

LO6 Explain the origins and impacts of strong extratropical cyclones in four susceptible U.S. regions: the Pacific Northwest, Great Lakes, East Coast, and North Atlantic.

1. A full examination of extratropical cyclones reveals narrow air currents, called conveyor belts, that feed into the storm from different locations around it. The warm and cold conveyors provide moisture from different source regions, while the descending dry current may channel high winds to near the surface behind the cold front.
2. Pacific Northwest (PNW) storms are among the most powerful, wintertime extratropical cyclones to strike the United States, both in terms of damaging wind and extreme amounts of precipitation. Both the winds and the precipitation are amplified by the high terrain along coastal Washington, Oregon, and northern California.
3. November Witches are powerful extratropical cyclones that feed off the warm waters of the Great Lakes during autumn. Their extreme winds and high water waves have been responsible for numerous maritime tragedies across the Great Lakes.
4. Nor'easters are a hybrid type of extratropical cyclone that develop along the East Coast during the winter, tracking from North Carolina's Outer Banks to New England. They derive a significant amount of energy from the warmth and high humidity air over the Atlantic's Gulf Stream current. In addition to strong winds and high seas, Nor'easters bring heavy snows and ice to the big inland cities along the Northeastern megalopolis.
5. Fish storms, also called Shapiro–Kaiser (S-K) storms, are powerful wintertime extratropical cyclones that develop beneath the jet stream over the North Atlantic. They feature a secluded core of warm air at low levels and threaten commercial fishing, oceanic shipping, and leisure cruises.

References

Bell, G., 2014. Impacts of El Niño and La Niña on the hurricane season. *Climate.gov*. www.climate.gov/news-features/blogs/enso/impacts-el-niño-and-la-niña-hurricane-season.

Bosart, L.F., W.E. Bracken and J. Molinari, 2000. Environmental influences on the rapid intensification of Hurricane Opal (1995) over the Gulf of Mexico. *Monthly Weather Review*, 128:322–352.

Brueske, K.F., 2001. Creighton Preparatory School, personal communication.

Emanuel, K., 2005. Increasing destructiveness of tropical cyclones over the past 30 years. *Nature*, 436:686–688.

Gray, W.M., 1984. Atlantic seasonal hurricane frequency. Part 1: El Nino and 30 MB quasi-biennial oscillation influences. *Monthly Weather Review*, 112:1649–1668.

Houze, R.A., W.-C. Lee and M.M. Bell, 2009. Convective contribution to the genesis of Hurricane Ophelia, 2005. *Monthly Weather Review*, 137:2778–2800.

Kocin, P. and L. Ucellini, 1990. *Snowstorms Along the Northeastern Coast of the United States: 1955 to 1985*. Boston: American Meteorological Society.

Kocin, P.J. and L.W. Ucellini, 2004. *Northeast Snowstorms. Volume 2: The Cases*. Boston, MA: American Meteorological Society.

Mass, C., 2008. *The Weather of the Pacific Northwest*. University of Washington Press.

Webster, P.J., G.J. Holland, J.A. Curry and H.-R. Change, 2005. Changes in tropical cyclone number and intensity in a warming environment. *Science*, 309:1844–1846.

Williams, F., 2004. *Seasonal U.S. Accumulated Cyclone Energy (ACE) for the Atlantic*. http://la.climatologie.free.fr/.

CHAPTER 7

Winter Weather Hazards: Arctic Air Outbreaks, Nor'easters, Blizzards, Lake Effect Snow, and Ice Storms

Learning Objectives

1. Describe the characteristic types of winter weather hazards that characterize the five U.S. geographical regions (West Coast, Rockies, Midwest/Intermountain Region, Midwest and Great Plains, Southeast and Gulf Coast, New England and Mid-Atlantic).
2. Discuss how heavy snow and ice develop within wintertime storm clouds, including the sources of subfreezing air, the effect of layered air temperature, and the importance of the rain-snow line.
3. Explain how extratropical cyclones generate regions of heavy snow and the manner in which geography modifies snow patterns along the Rockies and East Coast.
4. Discuss the seasonal and geographical distribution of lake effect snow and the key processes giving rise to these localized, heavy snows.
5. Describe how severe ice storms form and the types of damage they cause.
6. Explain the challenges of forecasting heavy snow and ice storms.

Introduction

Few events get more people talking about the weather than an impending snowstorm. "How many inches are they saying?" is a common question at the water cooler, at the dinner table, and on the phone. School kids become excited at the prospect of missed school days. Facebook pages and weather forecasting blogs register millions of hits. The storm becomes the lead story on TV news broadcasts. And for many, there is a sense of anxiety, even dread, because big snowstorms frequently snarl transportation, cripple utilities, maroon people in their homes for days, and require hours – perhaps even days – of backbreaking snow removal.

But snowstorms are just the beginning. Winter across North America offers a wide variety of other high-impact weather events, including ice storms, blizzards, and arctic air outbreaks. The science of these winter events is rich and detailed, and their socioeconomic impacts are numerous and far-reaching. We start this chapter by examining the geographical distribution of heavy winter snow, ice, and associated cold weather hazards. Next, we delve into several important principles behind the formation of snow and ice, including the critical role of temperature as a determinant of frozen precipitation. We then explore the crippling, widespread snowstorms produced by wintertime extratropical cyclones. We describe one of nature's most awesome snow-making machines along the Great Lakes snowbelt, as well as heavy ice storms and the damage they cause. We close the chapter by exploring the great challenge of predicting snow and ice storms, and by examining a relatively new type of snowfall impact scale based on human population density.

U.S. Distribution of Snow, Ice, and Cold Hazards

The wintertime distribution of heavy snow is controlled by two major factors: (1) latitude and (2) geography. The map of wintertime snow accumulation across the United States (Figure 7.1) suggests that latitude is king. For weather systems to generate frozen precipitation, subfreezing air must be present. East of the Mississippi River, the average seasonal snowfall across the continental United States varies from zero along the Gulf Coast to well in excess of 180 cm (71 inches) across New England and

DOI: 10.4324/9781003344988-9

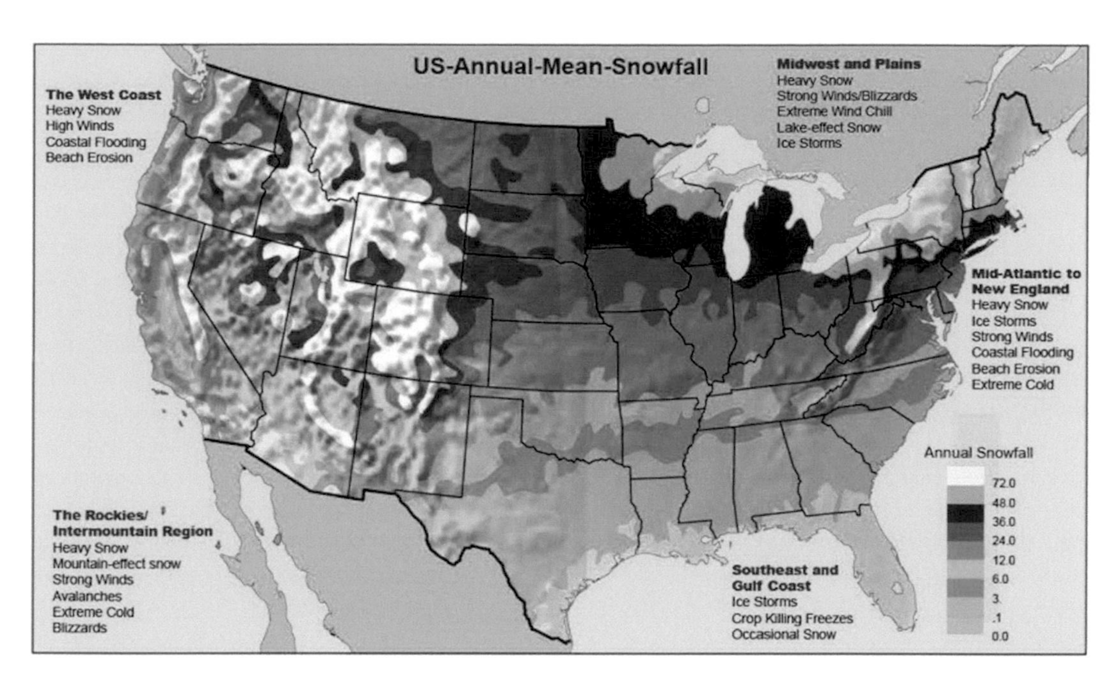

Figure 7.1 U.S. national map of average seasonal snow accumulation. This map shows the long-term (climatological) average snowfall over the course of the year.

the Great Lakes. In the West, altitude combines with latitude to establish heavy snow regions, including the Rocky, Wasatch, Cascade, and Sierra Nevada Mountains. Note also the narrow corridor of enhanced snow along the crest of the Appalachian Mountains.

Even at lower eastern U.S. latitudes, seasonal snow totals can far exceed the annual average. For instance, Baltimore, Maryland, frequently experiences modest seasonal accumulations on the order of 30–45 cm (12–18 inches). During the winter of 2009–2010, however, the region received a whopping 193 cm (77 inches) of snow. Why? This region abuts a major source of atmospheric moisture, the Atlantic Ocean. An intense Nor'easter tracking up the Eastern Seaboard can manufacture snow totals on the order of 60–90 cm (23–35 inches) per storm – more than one season's worth. The problem with seasonal average totals is that they do not reflect the episodic nature of exceptionally heavy snowfalls in the Mid-Atlantic. That is, a truly heavy snow-making Nor'easter may occur only once every three to five years. Two or three such storms in one season (a very infrequent occurrence) yield seasonal totals two or three times the long-term average.

Figure 7.1 divides the United States into five characteristic winter weather impact regions. Each geographical zone has a unique "style" of winter weather hazards. We describe each of the five zones in detail.

Zone 1: West Coast

The greatest snowfalls in the United States are caused by large-scale storm systems, called Pacific coastal lows, in this zone. Along the Cascades, moisture-laden air is drawn into low-pressure centers, then forced to ascend tall mountains, greatly enhancing snow production at high altitude. Sopping-wet snow, called Sierra cement, piles up by the foot along the higher ridge tops from Washington to California. The strong winds at high altitude and streaming through mountain gaps create blizzard (whiteout) conditions. So much snow drifts onto roadways that conventional snowplows are useless; special rotary devices are needed to cut through the deep snow and eject it beyond the road's shoulder. Residents of the mountainous western states are accustomed to the Chain Law, which mandates the use of metal chains on tires when traveling on interstate highways. However, massive snow-making storms do not occur every winter; due to short-term climate oscillations such as El Niño/La Niña (described later in this chapter), some winters accumulate much more snow than others.

Zone 2: Rockies and Intermountain Region

The various mountain ranges between the Sierras/Cascades and the Rocky Mountain Front Range define a broad geographic region including the Wasatch Mountains near Salt Lake City, the Tetons of Wyoming, and the Mogollon Rim across northern

Arizona. Eastward-moving cyclonic storms off the Pacific bring frequent disturbances to the Intermountain Zone. However, these systems arrive moisture-depleted, having precipitated their water along the windward slopes of the Cascades and Sierra Nevada. Heavy snow accumulations still occur, due to orographic enhancement (increase in precipitation over mountains, when moist air is forced to rise), but the snow tends to be drier and fluffier (the so-called champagne powder highly prized by skiers and snowboarders). High winds at altitude lead to blizzard conditions.

Along the Front Range of the Rockies, and particularly along the I-25 corridor through Denver and Fort Collins, upslope snowstorms are common during the winter months. To produce an upslope storm, an extratropical cyclone circulates moisture-laden air from the Gulf of the Mexico into the eastward slopes of the Rockies. The air rises abruptly and cools below freezing, generating surprisingly heavy snow localized to the Denver-Boulder-Fort Collins region and high mountains peaks immediately to the west.

The winter storms of the Rockies and Intermountain Zone provide a key natural service function in this generally arid region: The many feet of snowpack that accumulate each winter provide fresh water during spring and summer, as meltwater flows downstream and fills the reservoirs. This water is crucial for agricultural and urban use. Additionally, the meltwater impounded at altitude provides the potential energy for hydroelectric power generation. Between 20 and 80% of the electricity across a broad region extending from California to Montana to Washington is generated in this manner. Moderate to heavy snowfalls are also critical for the ski industry that is so important to tourism in the scenic western mountains.

Zone 3: Midwest and Great Plains

Heavy snowfalls across the northern and Central Plains accompany migrating extratropical cyclones, developing along the lee (east slopes) of the Rockies. Another common storm type is the Alberta Clipper, which dives south and east of the Rockies (Figure 7.2). Alberta Clippers are extratropical cyclones that originate deep in the Canadian interior and are moisture-starved. Because of their fast movement and lack of significant moisture, snow accumulation is light to moderate but covers a long swath of the northern United States.

Alberta Clippers are frequently followed by intense anticyclones (high-pressure domes) containing brutally cold and dry air. Channeled to the south and east by the Rockies, the dense arctic air blasts across the central and southern Plains and then across the southeast and Mid-Atlantic. Daily high temperatures fail to climb above the teens or single digits. These arctic air outbreaks are preceded by an arctic front (Figure 7.3). Sometimes multiple arctic fronts attend more than one surge of cold air, and a brief cloud band of heavy snow and blowing wind – called a snow squall – sometimes accompanies an arctic front (Figure 7.4[a] and [b]). A snow squall can be likened to a winter thunderstorm – with sudden, strong wind gusts, snow falling at the rate of several inches per hour, even lightning! Hazardous, low visibility conditions – perhaps a whiteout – can occur during a snow squall. It might be clear one minute, then quite suddenly the visibility drops to near-zero. On a fast-moving interstate, the immediate loss of visibility can result in multiple-vehicle accidents. Accordingly, some National Weather Service offices now issue a Snow Squall Warning when localized, intense snow showers are expected to impact interstate or busy highway corridors. This is a short-fused warning typically valid for 30 or so minutes, covering a small geographical region. Warning messages are broadcast on roadside, digital sign boards near the impacted region.

Record low temperatures and dangerously cold wind chill values accompany strong winds behind an arctic front. Wind chill accelerates heat loss from exposed skin surfaces, and it increases with wind speed. A table of wind chill values, which is the apparent temperature due to the cooling effects of the wind, is shown in Figure 7.4(c). Frozen water pipes, road damage, cars and buses that won't start, and various health threats (frostbite, hypothermia, dry skin) are common consequences of arctic air outbreaks. The National Weather Service may issue a wind chill advisory or wind chill warning when the combination of wind and cold may become life-threatening if action is not taken; the specific criteria vary from region to region.

Over the northern and central Plains, high winds created by the strong anticyclone generate blizzard conditions. A blizzard is a period of blowing snow leading to whiteout conditions (less than 400 m or a quarter mile visibility), with winds exceeding 30 kts (35 MPH) for 3 hours or more.

The Great Lakes region sits at a crossroads, influenced by many different wintertime weather systems. Heavy snow-producing extratropical cyclones moving east of the Rockies cross the Ohio Valley, as do Alberta Clippers and arctic air outbreaks. As cyclones cross the relatively warm lake water (provided the lakes have not frozen over), additional moisture increases the rate of snow production, a process called lake effect snow.

Lake effect snow refers to the localized generation of snow bands, often in the form of snow squalls, along the downwind (lee side) of unfrozen lakes. The relatively warm water adds heat and moisture to cold air streaming across the lakes. Because the prevailing surge of cold arctic air is from the northwest or west,

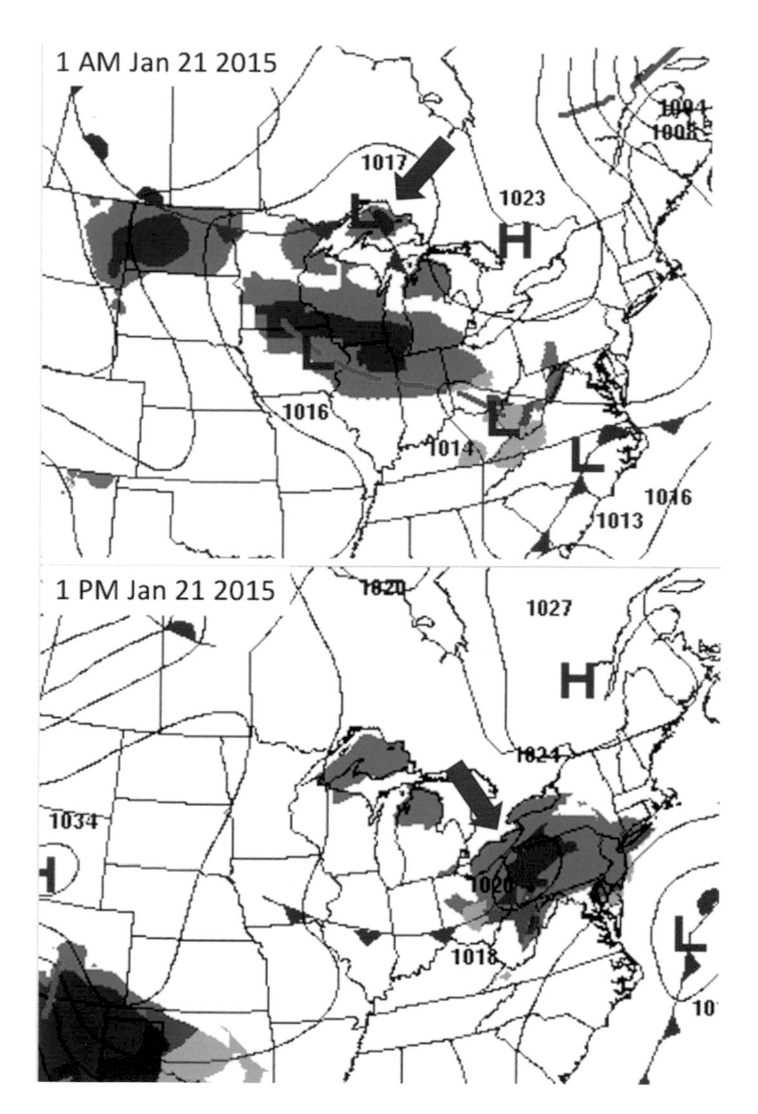

Figure 7.2 Alberta Clipper. This surface weather map depicts an Alberta Clipper, a low-pressure system crossing the northern Plains and Ohio Valley (magenta arrows and red "L"). This weak surface cyclone is attended by a region of light to moderate accumulating snow (blue shades). (Adapted from NOAA.)

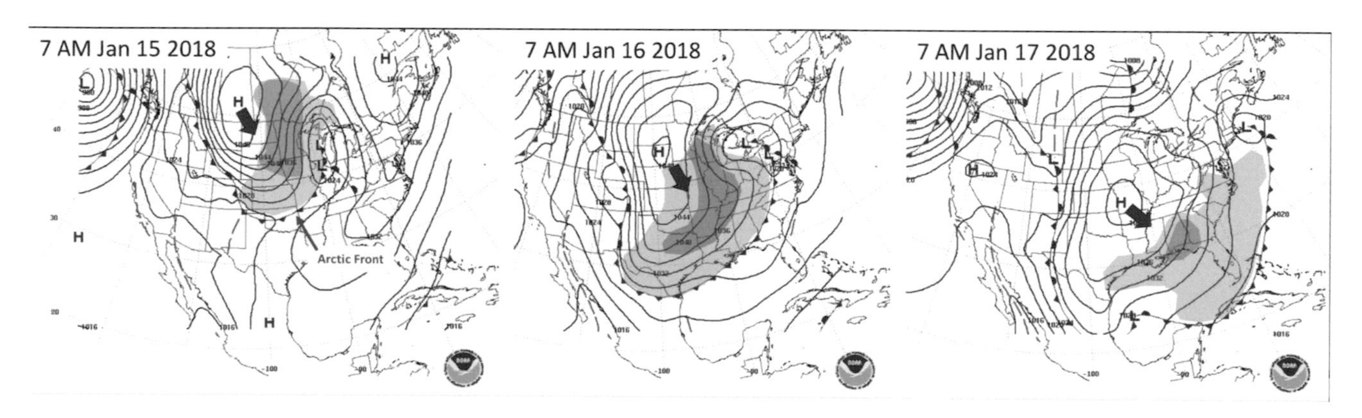

Figure 7.3 Arctic air outbreak behind an Alberta Clipper. This surface weather map shows an arctic air outbreak advancing behind an Alberta Clipper (red "L"). Magenta arrows track the southeastward movement of the 1048 mb anticyclone (blue "H"). The core of the coldest elements are shown by the blue shades. The large pressure gradient between the clipper and the anticyclone implies gusts in the 26–35 kts (30–40 MPH) range across Minnesota, from the northwest, exacerbating the wind chill there. (Adapted from NOAA.)

(a)

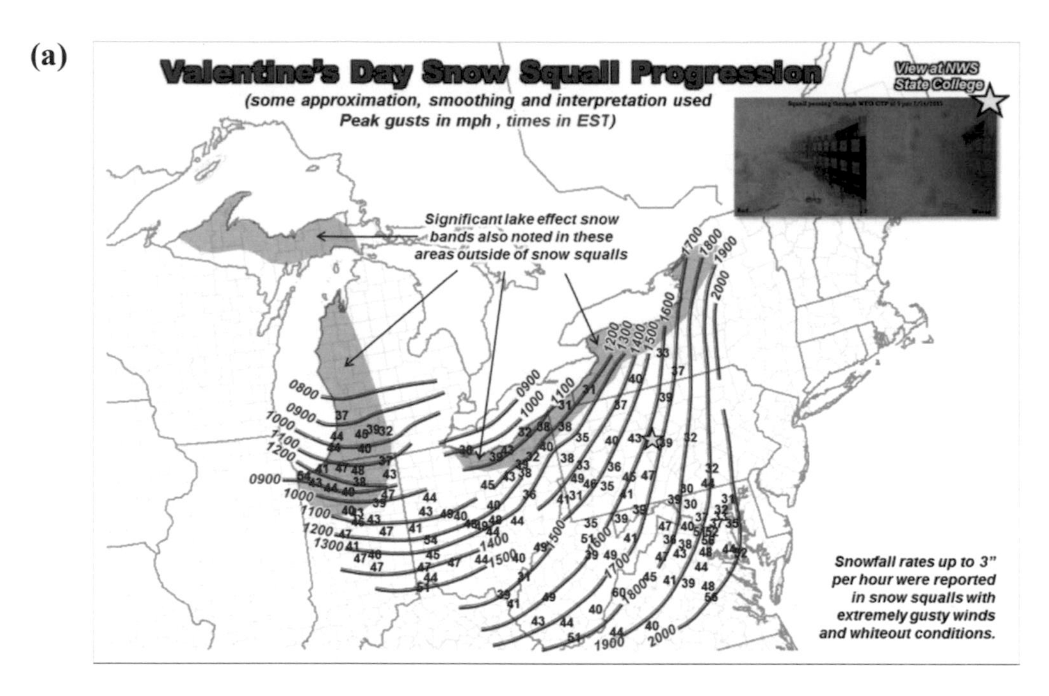

(b)

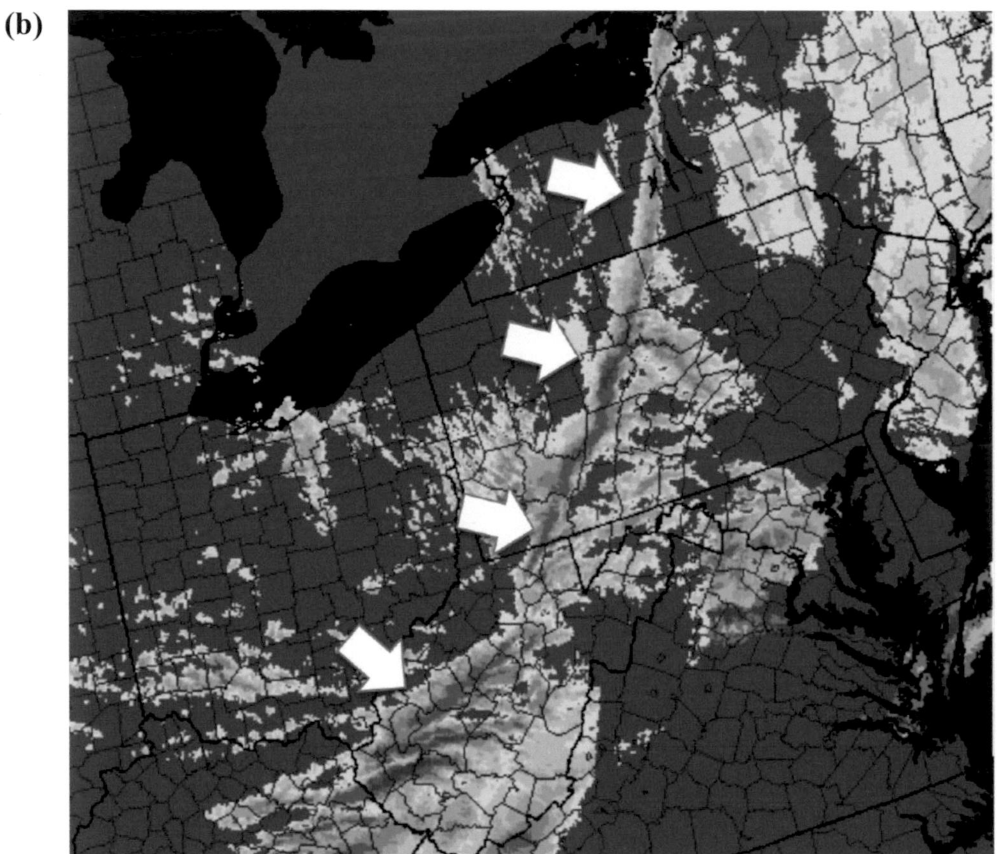

Figure 7.4 Arctic front and snow squall, February 2014. (a) This surface weather map shows the rapid progress of an arctic front and snow squall across the eastern United States on Valentine's Day, 2014. Pink lines (isochrones) show the hourly positions of the arctic front; blue numbers indicate peak wind gusts (MPH) (NWS). (b) Weather radar depiction of the snow squall crossing the Ohio Valley (NWS). (c) The apparent air temperature is shown in the shaded regions of this wind chill, as a function of air temperature and wind speed. Various color shadings indicate time to frostbite on exposed skin. (NOAA.)

(c)

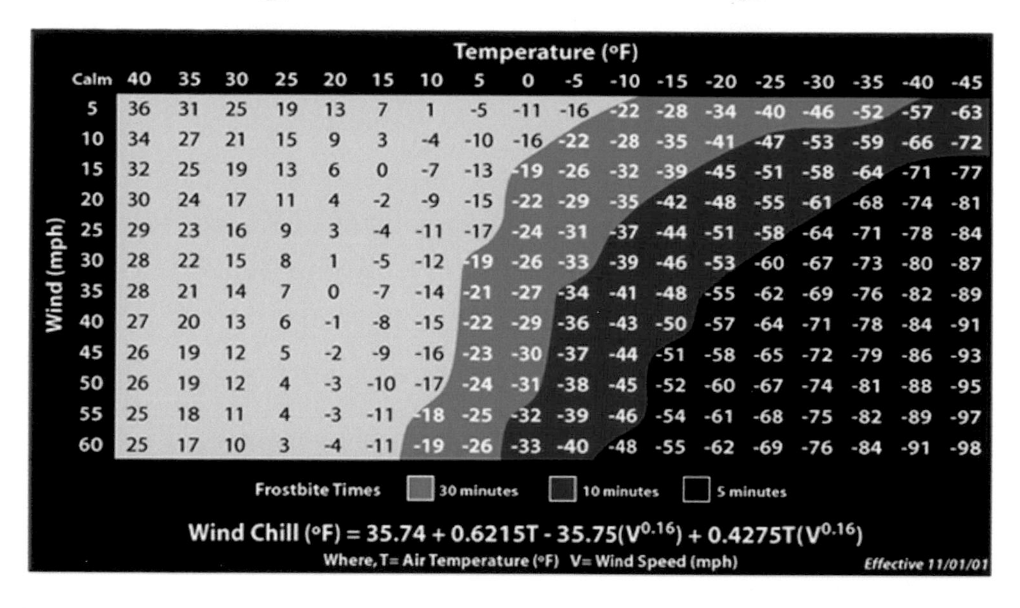

Wind Chill Chart

Wind (mph) \ Temperature (°F): Calm	40	35	30	25	20	15	10	5	0	-5	-10	-15	-20	-25	-30	-35	-40	-45
5	36	31	25	19	13	7	1	-5	-11	-16	-22	-28	-34	-40	-46	-52	-57	-63
10	34	27	21	15	9	3	-4	-10	-16	-22	-28	-35	-41	-47	-53	-59	-66	-72
15	32	25	19	13	6	0	-7	-13	-19	-26	-32	-39	-45	-51	-58	-64	-71	-77
20	30	24	17	11	4	-2	-9	-15	-22	-29	-35	-42	-48	-55	-61	-68	-74	-81
25	29	23	16	9	3	-4	-11	-17	-24	-31	-37	-44	-51	-58	-64	-71	-78	-84
30	28	22	15	8	1	-5	-12	-19	-26	-33	-39	-46	-53	-60	-67	-73	-80	-87
35	28	21	14	7	0	-7	-14	-21	-27	-34	-41	-48	-55	-62	-69	-76	-82	-89
40	27	20	13	6	-1	-8	-15	-22	-29	-36	-43	-50	-57	-64	-71	-78	-84	-91
45	26	19	12	5	-2	-9	-16	-23	-30	-37	-44	-51	-58	-65	-72	-79	-86	-93
50	26	19	12	4	-3	-10	-17	-24	-31	-38	-45	-52	-60	-67	-74	-81	-88	-95
55	25	18	11	4	-3	-11	-18	-25	-32	-39	-46	-54	-61	-68	-75	-82	-89	-97
60	25	17	10	3	-4	-11	-19	-26	-33	-40	-48	-55	-62	-69	-76	-84	-91	-98

Frostbite Times: 30 minutes; 10 minutes; 5 minutes

$$\text{Wind Chill (°F)} = 35.74 + 0.6215T - 35.75(V^{0.16}) + 0.4275T(V^{0.16})$$

Where, T= Air Temperature (°F) V= Wind Speed (mph)

Effective 11/01/01

Figure 7.4 (Continued)

the zones of heavy snow accumulation lie along the immediate eastern and southern margins of the Great Lakes, regularly affecting parts of Wisconsin, Ohio, Pennsylvania, and New York. These snow bands tend to develop after a cyclonic storm has passed to the east, with an anticyclone moving in, such that the "weather clears up stormy." (We discuss lake effect snow in more detail later in this chapter.)

Zone 4: Southeast and Gulf Coast

This region is spared the worst ravages of winter by its proximity to the warm Gulf of Mexico. Arctic air masses streaming southward have often modified (warmed and moistened) as they migrate into subtropical latitudes. However, exceptionally strong arctic air outbreaks can lead to periods of frozen precipitation, including ice storms and light snowfalls. Because several years can pass between events, municipalities often lack the experience and preparedness for coping effectively. Hard freezes can damage crops and fruit trees. Several southern cities (including Atlanta, Georgia, and Raleigh, North Carolina) have discovered just how vicious an inch or two of snow can be, when it falls on frozen roads.

Zone 5: New England and Mid-Atlantic

This region is home to the Northeastern Megalopolis, a sprawling urban corridor extending from Tidewater, Virginia, to Boston, Massachusetts, and inhabited by 52 million residents (based on the 2010 census). Extratropical cyclones approaching from the west bring moderate snow to the region, but none bring more snow than Nor'easters or coastal storms (discussed in Chapter 6). These "snowmakers" are large, slow-moving storm systems that bring a foot or more of snow to several cities within the Megalopolis. Their impact on dense transportation systems (interstate highways, airways, railways) is legendary.

The frequency of snow-producing Nor'easters increases up the coastline from south to north, with Nor'easters most common in the vicinity of Boston. Massachusetts and Maine average at least one heavy snow event (defined as a 25+ cm [10 inches+] accumulation) per year. For comparison, the Washington, D.C.–Baltimore region experiences a heavy snow event only once every three years, and New York City once every two years. The history of the Megalopolis has been shaped in part by heavy snowfalls, bearing names such as "The Knickerbocker Storm" (1922), "The President's Day Storm" (1979), and "Snowmageddon" (2010).

Ice storms are also common across the Mid-Atlantic, a consequence of geography: The northeast-trending Appalachians trap cold, dense air layers on their east side, across the Piedmont region of North Carolina, Virginia, and Maryland. This cold air damming tends to enhance the formation of wintery precipitation (both ice and heavy snow) as humid air flows up and over the cold air off the Atlantic Ocean. Cold air damming also occurs elsewhere in the United States, contributing to upslope snow generation along the Rocky Mountain Front Range.

Formation of Heavy Snow and Ice

Heavy rates of precipitation require the sustained uplift of air along fronts and within the rising air inside low-pressure systems. There must be a steady supply of water vapor in the upflowing air. For heavy snow to be manufactured, the temperature of the air mass must also be below freezing, through a deep layer (1.5–3 km [5,000–10,000 feet]) in contact with the ground. When these three key ingredients – uplift, moisture, and subfreezing air – are all present, the probability of heavy snow becomes more likely, as Figure 7.5 shows.

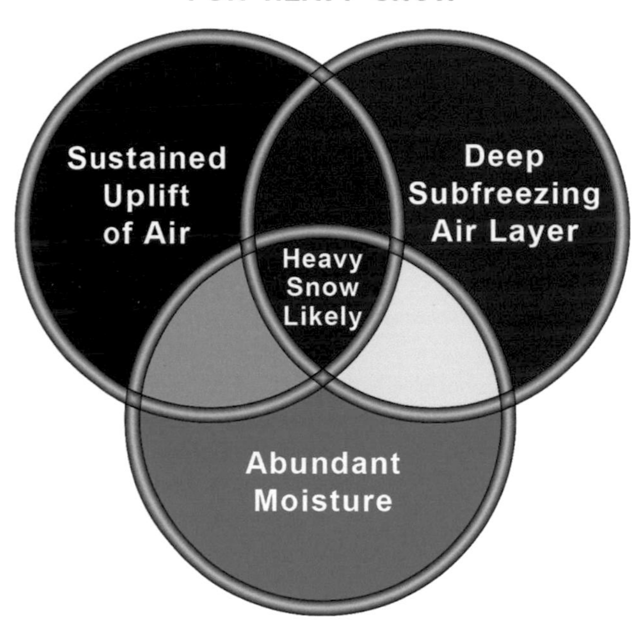

Figure 7.5 Three key ingredients needed for a heavy snowfall. This Venn diagram shows the three ingredients that must be present for heavy snowfall.

Foreseeing where and when all three factors will blend creates a challenging forecast problem. During the winter, large storm systems such as extratropical cyclones generate widespread uplift and process large quantities of humid air. But without the frigid air, rain (not snow) becomes the dominant precipitation type.

Sources of Subfreezing Air Required for Snow Formation

To create a snowfall, the atmosphere must have a means of lifting air to saturation, and the layer of air must be subfreezing (below 0° C [32° F]). A deep, cold air layer is the most critical ingredient in the snowmaking equation. As Figure 7.6 shows, several processes can refrigerate the air, and sometimes these processes act concurrently.

The first panel of Figure 7.6 illustrates the common pairing of an extratropical cyclone (red "L") and a large anticyclone (blue "H") to its west and north. The anticyclone or high-pressure cell circulates cold air toward the south and east, feeding into the back edge of the storm system. The cold air is sourced from Canada (at times as far north as the Arctic). As it becomes entrained into the low-pressure system's winds, the storm's cool conveyor belt mixes into and lifts cold air into an ascending spiral, generating a wraparound cloud shield where heavy snow is manufactured. The shield is often quite extensive, extending for hundreds of miles toward the north and west of the low's center. The process by which winds transport cold air into the interior of an extratropical cyclone, or across the surface of the Great Lakes (leading to lake effect snow) is termed cold air advection. When forecasting heavy snow, it's easy to fixate on the importance of the low-pressure system; however, one must

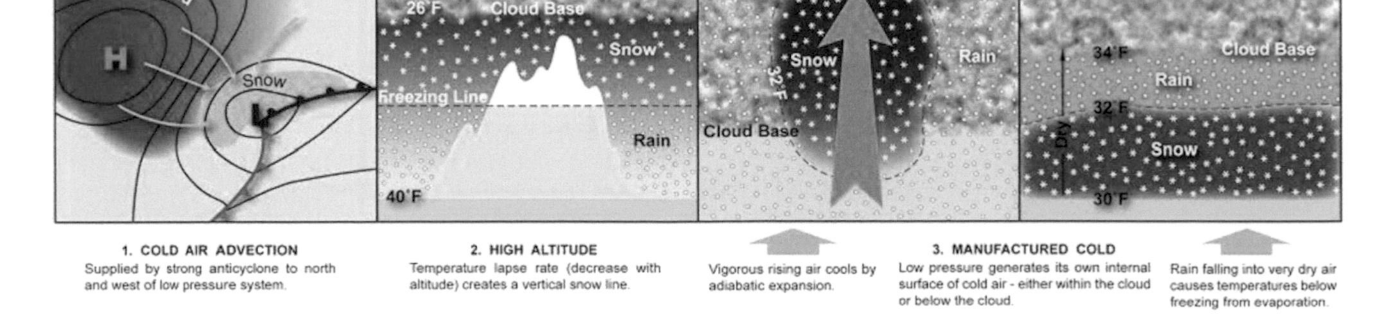

Figure 7.6 Cooling the atmosphere below freezing. This figure shows several ways that the lower atmosphere can cool below freezing, in order to support a heavy snowfall.

not discount the critical role of a cold anticyclone positioned north and/or west of the storm (or Great Lakes).

The second panel in Figure 7.6 shows the effect of high altitude on snow manufacture. High altitude is a major reason why tall, orographic belts across the United States have some of the highest snow totals during winter. The temperature decreases at an average rate of −6.5° C/km (−19° F/mile). Just a few thousand feet up the mountain lies the snow line. Clouds above this altitude are almost always subfreezing (except for 2–3 months in summer) and create snow. Snow accumulates in the upper elevations, turning to plain rain on the lower slopes and valley floor. Changing temperature advection patterns during the storm can raise or lower the snow line by several thousand feet. Snow that accumulates at higher, perpetually colder altitudes is termed orographic snow.

The third and fourth panels of Figure 7.6 illustrate how a storm system can manufacture its own source of subfreezing air when it lacks a source of cold air advection or orographic enhancement. Inside a low, thick cloud mass that is marginally cold enough to produce snow, vigorously rising air can cool to below freezing. Most often, this type of uplift is generated beneath the mid-latitude jet stream. Such uplift creates its own dynamic cooling as rising air cools because of adiabatic expansion.

Finally, a layer of cold air hundreds to a few thousand feet deep can be chilled by a phase change of water called sublimation. To understand this process, suppose that snow is falling from a deep cloud layer. The air beneath the cloud base is a few degrees above freezing but very dry (low humidity). Snowflakes falling into this layer begin to sublimate (transform to vapor). Sublimation extracts significant heat from the surrounding air. Over several hours, the air layer cools and may freeze all the way to the ground. Instead of melting into raindrops, the snowflakes survive and begin accumulating on the surface.

Vertical Temperature Layering and Wintry Precipitation Type

The generation of snow involves multiple processes. Meteorologists must study the microphysical nature of the cloud layer, where processes operate on microscopic scales. This "hidden world" of how snowflakes develop is beyond the scope of this book. Suffice it to say that there are many processes by which microscopic ice crystals within clouds aggregate into larger, precipitating particles, which we call snowflakes. This section assumes we start with a cloud-bearing layer that manufactures snow aloft. Several subsequent processes influence the type of accumulating snow, how rapidly it accumulates, and how it transforms into other types of precipitation, including rain, sleet, and freezing rain.

A special type of microscopic ice crystal, called a dendrite, assumes a pointy, six-sided hexagonal form. Dendrites are one of many crystal shapes that develop under specific ranges of cloud water content and air temperature. Snowflakes are the macroscopic expression of large aggregations of microscopic crystals. Dendritic crystals are particularly adept at interlocking together, like Velcro, efficiently forming visible snowflakes. When the cloud-bearing layer has a specific temperature range favoring dendritic growth (−13° to −17° C [9° to 1° F]), snow formation will be particularly rapid. Forecasted snow accumulation totals are adjusted upward based on this information.

Let's discuss how the vertical change in temperature with altitude, or layering of temperature, is an important determinant of frozen precipitation type (cold rain, wet snow, dry snow, sleet, or freezing rain). Accurate forecasts of a winter storm hinge critically on forecasts of temperature variations in the lower 1.5 km (5000 feet) of atmosphere. Figure 7.7 shows the characteristic temperature layering for several different wintry precipitation types. In all panels, we start with a cloud layer that generates snow several thousand feet above the ground. The rightmost panel shows what happens when the air layer beneath the cloud quickly warms to above freezing temperature, and remains above freezing when it gets to the surface. Snowflakes emerging from the base of the cloud deck immediately melt into raindrops. This scenario describes the miserable "cold rain" outcome – that is, a soaking rain that falls through 1–2° C (34–36° F) air.

The far left panel (a) of Figure 7.7 shows a deep subfreezing air layer, from cloud base to the ground. Snow falling from the cloud layer remains as snow to the surface, accumulating one or more inches. When subcloud air temperatures lie in the −2 to −1° C (28–30° F) range, it's very common to observe snow with a 10:1 liquid equivalent ratio or snow ratio. That is, 25 cm (10 inches) of snow accumulates for every inch of precipitated water (that is, if that snow was melted in a rain gauge). The 10:1 ratio is a good starting point for meteorologists when forecasting snow accumulation.

The two middle panels (b) and (c) of Figure 7.7 represent more complex temperature profiles, leading to the formation of precipitating ice. In Panel (b), snow emerges from the cloud base and immediately melts in a shallow air layer a few degrees above freezing. Such a layer commonly forms along the East and Gulf Coasts, where an elevated mass of relatively warm oceanic air streams into and is lifted upward by the storm (warm

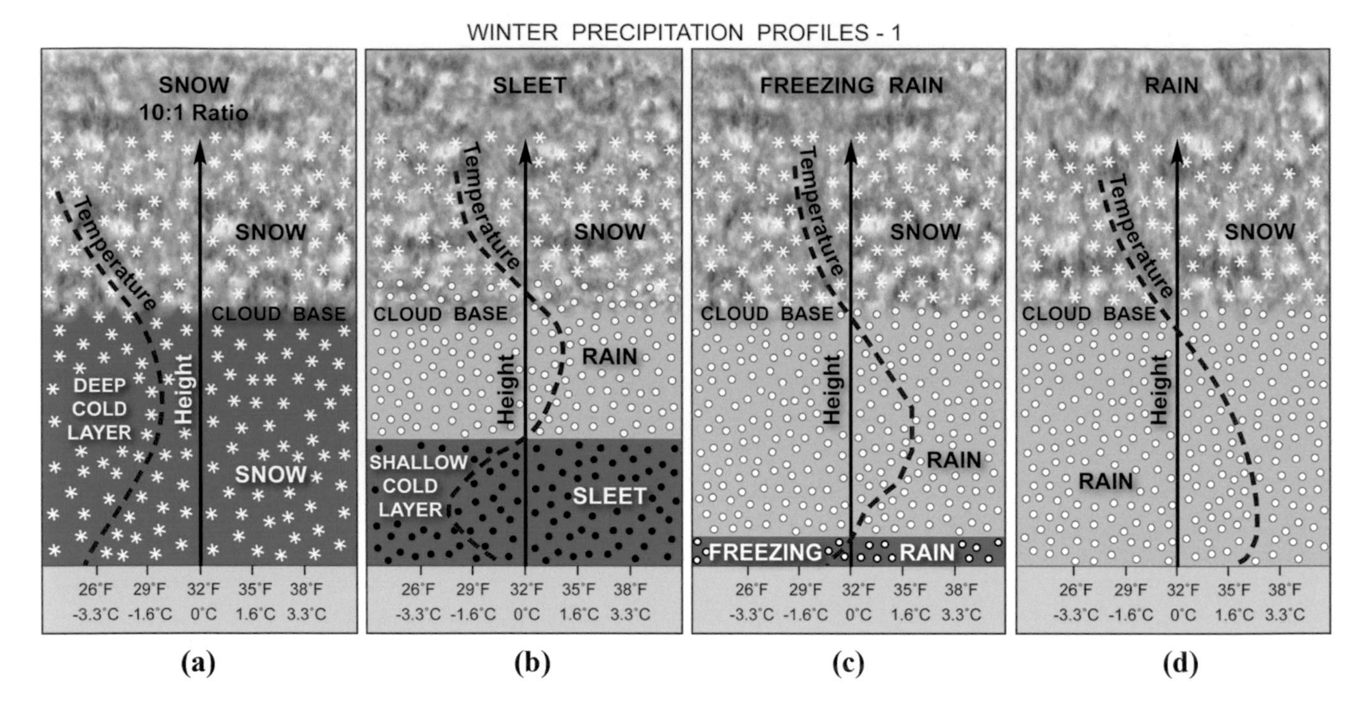

Figure 7.7 Temperature layers and frozen precipitation. This diagram of the lower atmosphere (cloud layer to surface) shows how different temperature layers give rise to various forms of frozen precipitation.

air advection). Beneath the elevated warm layer lies a shallow layer of subfreezing air, several hundred to a thousand or more feet deep. Such a layer commonly forms east of the Appalachians or Rockies, due to cold air damming. Raindrops falling into the frozen air instantly freeze into small grains of clear ice called sleet. The sleet accumulates much like snow and in extreme cases can form a frozen, crusty ground layer several inches thick.

If the subfreezing layer is very shallow close to the ground – or if just the ground surface is frozen – the raindrops falling out of the mild air layer remain liquid to the surface, then freeze immediately on contact (Panel [c]). Here, the "surface" means the soil or grass, along with any objects on the ground, including cars, trees, and utility lines. The resulting freezing rain is the dreaded glaze ice (or black ice on road surfaces) that leads to car accidents on slick roads and to utility outages from the weight of sagging tree limbs and power lines.

As Figure 7.8 shows, liquid equivalent ratios can vary considerably from storm to storm and even change by the hour within individual storms. Ratios on the order of 5:1 describe a heavy, wet, dense snow that makes great snowballs and snowmen. Unfortunately, this type of snow clings as heavy slush to exposed tree limbs and power lines, leading to utility outages. It is also very difficult to shovel, leading to back injuries and heart attacks. The weight of this snow can damage roofs. A liquid equivalent ratio of 20:1, in contrast, is exceptionally dry and fluffy, like talcum powder, easily swept from walkways. However, it is exceptionally prone to becoming windborne, which can lead to snow drifting and whiteout (zero visibility conditions). Both the air temperature and the liquid water content of clouds help determine a snowfall's liquid equivalent. (In Figure 7.8, we consider just the effect of air temperature.)

Panel (c) of Figure 7.8 shows the setup for 5:1 snow. Temperature in the cloud layer is below freezing, but it warms to nearly freezing below the cloud base, down to ground level. This isothermal layer often hovers right at the 0° C (32° F) mark. Falling snowflakes partially melt as they descend, forming a microscopic, watery film that makes them efficiently aggregate (stick together), again like Velcro. If you have ever seen enormous flakes the size of tea saucers fluttering to the surface, you have witnessed 5:1 snow.

Panel (b) of Figure 7.8 (center) shows a subcloud air layer chilled below freezing, down to the low to mid-20s° F. Ice grains in this air aggregate into tiny, compact flakes that do not stick together. In fact, the water equivalent may fall to below 20:1, in which case the snow is feathery light, like fairy dust, the type highly prized by skiers.

Typical snow ratios (average annual values) vary considerably across the continental United States. They exceed 10:1 in snowstorms across the northern tier of the country, commonly 12:1 and higher,

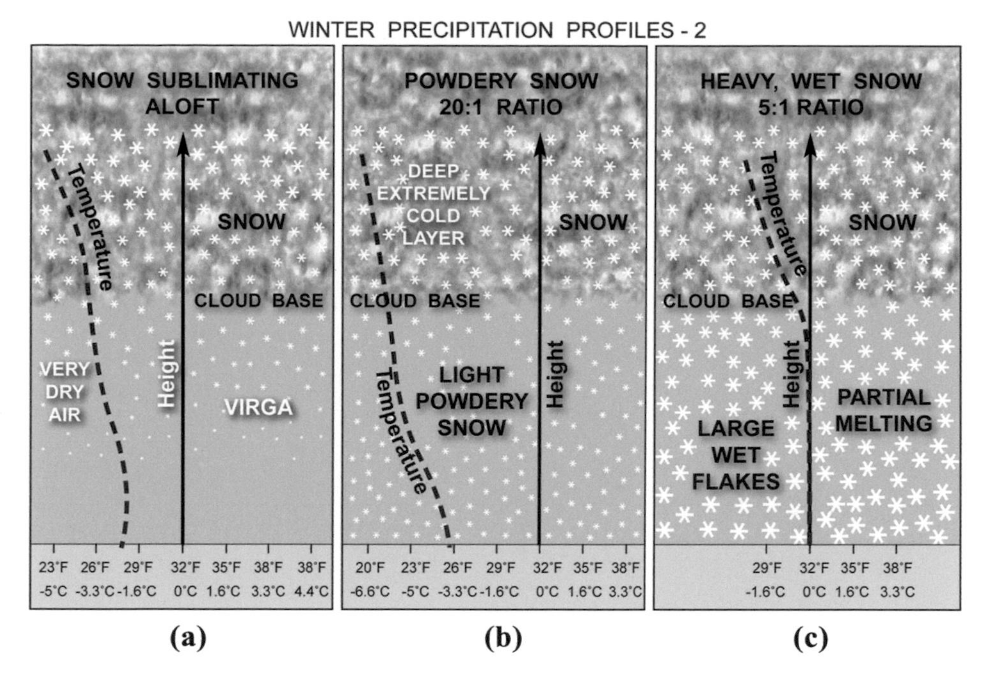

Figure 7.8 Liquid equivalent ratios. This diagram of the lower atmosphere (cloud layer to surface) shows how different temperature and moisture layers give rise to snow with varying liquid equivalent ratios.

where air masses tend to be well below freezing temperature. Across the southern states, ratios are closer to 8:1 because air masses are often close to freezing temperature and the humidity content of air is high, in close proximity with the Gulf of Mexico.

Panel (a) of Figure 7.8 (left side) shows subfreezing that is also exceptionally dry – that is, its humidity (water vapor content) is quite low. This condition often prevails in the intermountain region of the western United States and across the High Plains or when snow just starts to fall from a new cloud layer. The snowflakes sublimate into dry air, transforming to vapor and shrinking, creating virga or fall streaks that are sometimes visible as cottony wisps tapering below the cloud base. No snowflakes survive to accumulate on the surface. A deep, dry subcloud layer can take many hours to saturate or "moisten up"; trillions of snowflakes must be sacrificed before accumulation on the ground can begin. Weather radar shows that snow is reaching the ground, when in fact it's only snowing aloft.

Rain-Snow Line: What a Difference Just 10 Kilometers Can Make!

Snow forecasting is challenging partly because of the phase transitions between rain, freezing rain, sleet, snow, and virga. In one common scenario, an elevated layer of above-freezing air wedges above a colder surface air layer. This wedge occurs when winds from different directions enter an extratropical cyclone at different altitudes, with warm advection aloft (winds from the south and east) and cold air advection near the surface (winds from the north and west). As wind intensifies and temperature patterns change, multiple transitions among snow-ice-rain occur throughout the storm. For instance, precipitation may start as snow, then transition to ice, then to rain. Or the opposite sequence may occur. These so-called "multiple choice days" are the nemesis of weather forecasters. To the public, the forecasts sound wishy-washy, when in fact a lot of complexity must be anticipated.

Figure 7.9 shows that these phase transitions can occur abruptly and over surprisingly narrow zones. Just a few tens of kilometers (miles) can make all the difference between a cold, soaking rain and a crippling, heavy snowfall. Much of winter weather forecasting is based on predicting the exact location of the rain-snow line at the surface. Numerical prediction models with fine grid resolution (a measure of how finely spaced the forecast points are arrayed) are now in use, and they can help to identify subtle shifts in the rain-snow line. The best forecasting technique is to combine these hourly model predictions with weather radar. Polarimetric radar can discriminate between liquid water drops, sleet, wet snow, and dry snow. The meteorologist monitors trends in the rain-snow line's width, location, and movement, as corroborated by reports from Skywarn spotters. Forecasts are adjusted hourly as required (such forecasts are often termed nowcasts).

Figure 7.10 shows a real-world example of the rain-snow line for a Nor'easter moving up the East Coast, as observed

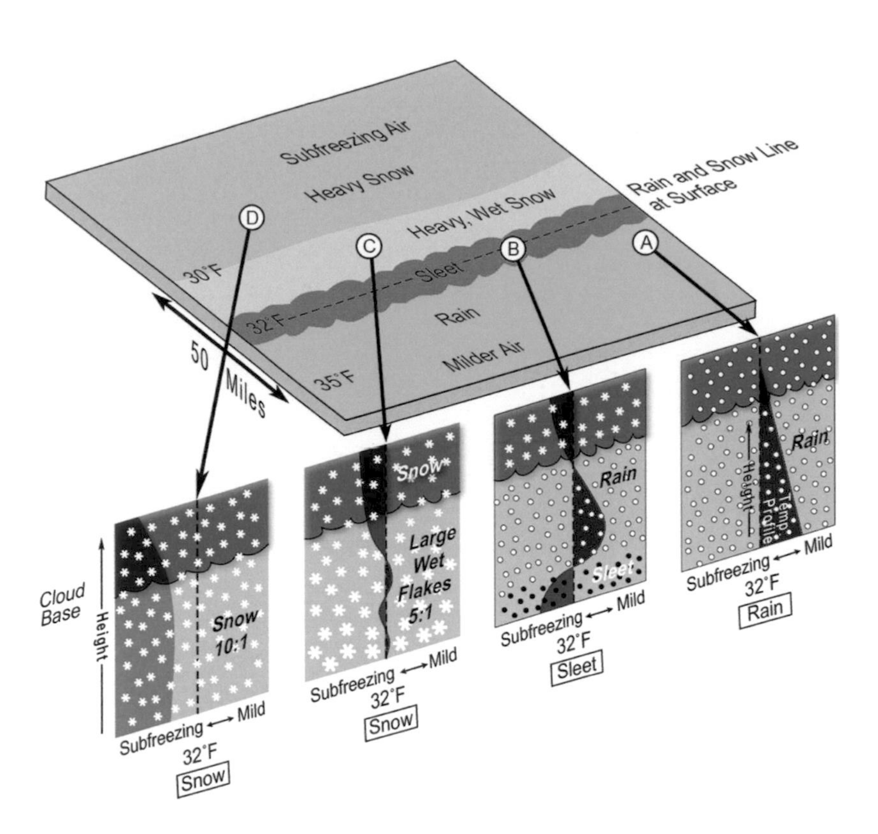

Figure 7.9 Rain–snow line in a winter storm. This diagram shows a three-dimensional depiction of the rain–snow line in a hypothetical winter storm.

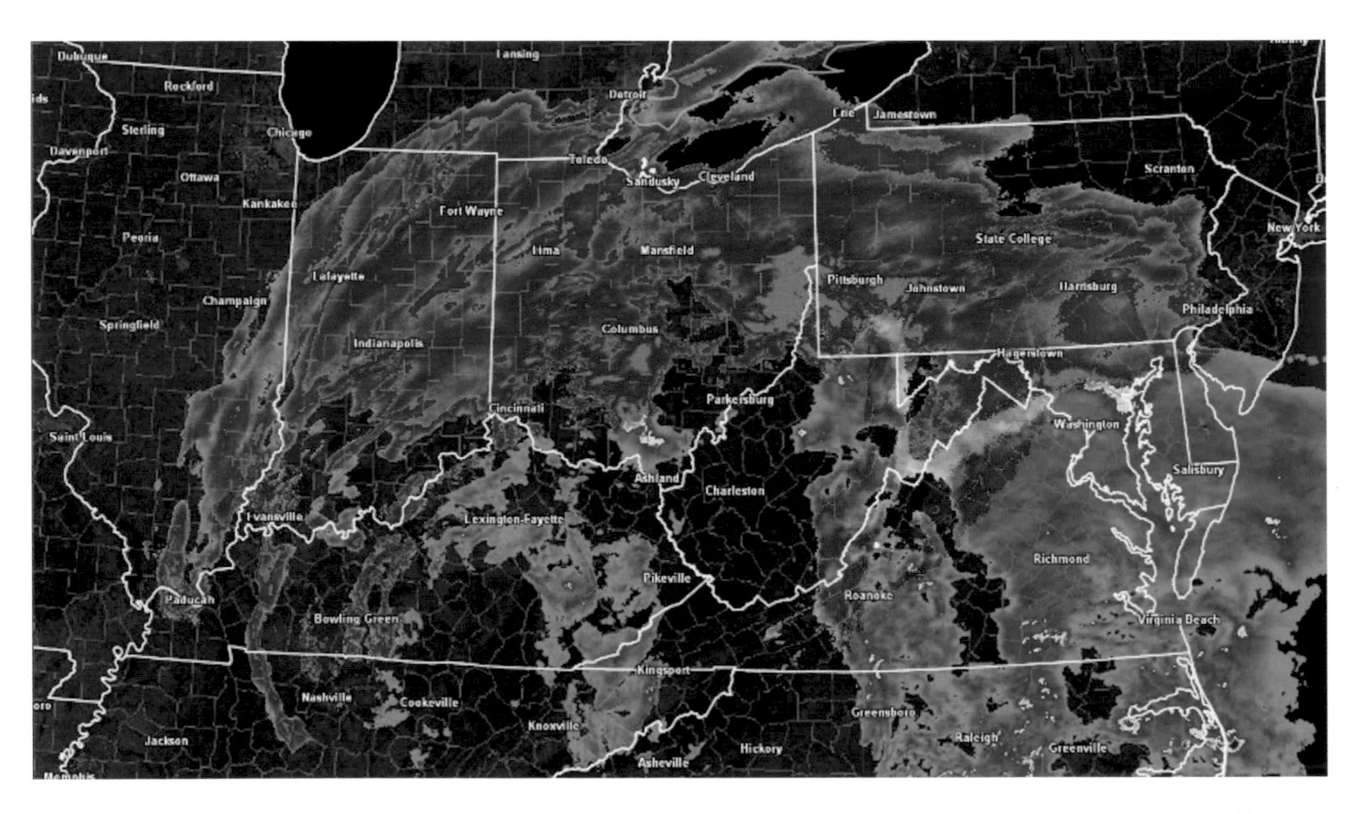

Figure 7.10 Weather radar depiction of heavy winter precipitation across the Mid-Atlantic and Northeast, winter of 2013. The storm pictured here was a coastal low (Nor'easter); blue/purple shades indicate snow, pink shows ice, and green/yellow tones indicate plain rain. (NWS.)

by weather radar. Note the sharp demarcation between rain (green), freezing rain and sleet (pink), and snow (blue). Such sharp transitions are very commonplace in these large winter storm systems.

Intense Snowstorms Created by Extratropical Cyclones

In this section we examine heavy snow that is generated by migrating low-pressure systems across the United States. These include Pacific coastal lows that impinge upon the Cascades and the Sierra Nevada. In many regions, the winter weather impact is also tied to geography, such as high terrain. We focus on two regions: (1) the Front Range of the Colorado Rockies and (2) the eastern slopes of the Appalachians. Historically significant heavy snows are a hallmark of both regions.

Front Range Snowstorms

The densely populated I-25 corridor just east of the Rocky Mountain Front Range frequently experiences heavy snowstorms during the late fall, winter, and early spring. The development of heavy snow in this location is uniquely terrain-dependent, and the area of heavy snow is often quite compact, as Figure 7.11 shows. Snow totals from a single storm can reach three to four feet, with a truly astounding snowfall gradient (the spatial change in snow depth over distance). These heavy snow events fall on a densely populated metropolitan corridor, encompassing Colorado Springs, Denver, Boulder, and Fort Collins. They lead to frequent interstate closures, widespread power outages, and serious disruption of air travel. At the height of a storm, Denver International Airport experiences hundreds of delays or cancellations, and it sometimes closes completely, stranding thousands of passengers for one or more days. These shutdowns ripple across the nation's air traffic corridor because Denver is a strategic hub for transcontinental flights.

Figure 7.11 illustrates an example of a Denver snowstorm on February 2–4, 2012. It dropped 1 m (3 feet) of snow over the foothills of the Denver-Boulder metropolitan area. A separate arm of heavy snow projected to the east (over Douglas, Elbert, and Arapahoe counties) southeast of Denver. This east-west prong is a fairly common pattern.

The large-scale setup for upslope snow requires low-level easterly winds – that is, a flow with an upslope component from the east, toward the mountain crest. High-pressure, low-pressure, or both types of systems can generate this pattern, as Figure 7.12(a) shows. A high-pressure cell, or anticyclone, positioned over Montana can circulate cool, dry air westward. The dense air becomes wedged up against the Front Range as a cold wedge, another form of cold air damming. Simultaneously, an

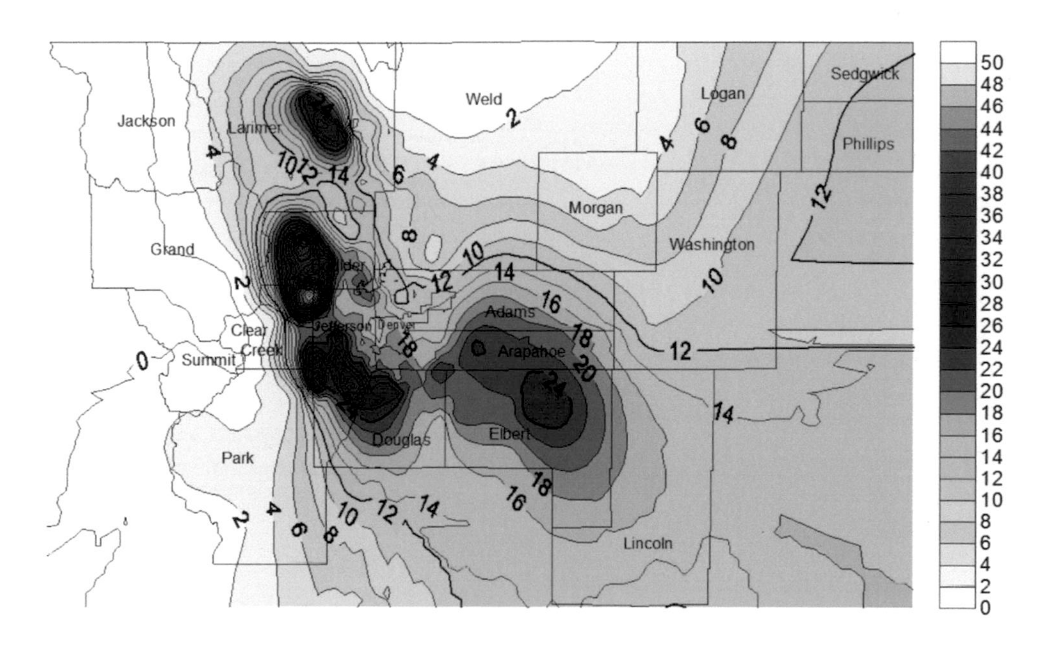

Figure 7.11 Snow accumulation maps from a heavy, upslope snowstorm along Colorado's Front Range on February 2–4, 2012. (NWS.)

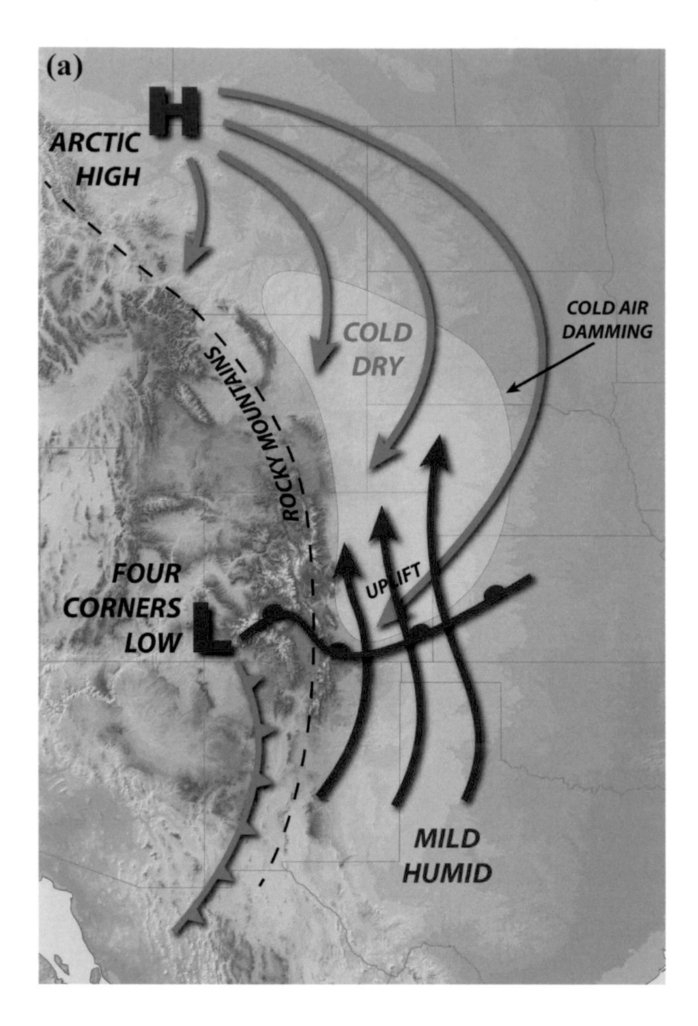

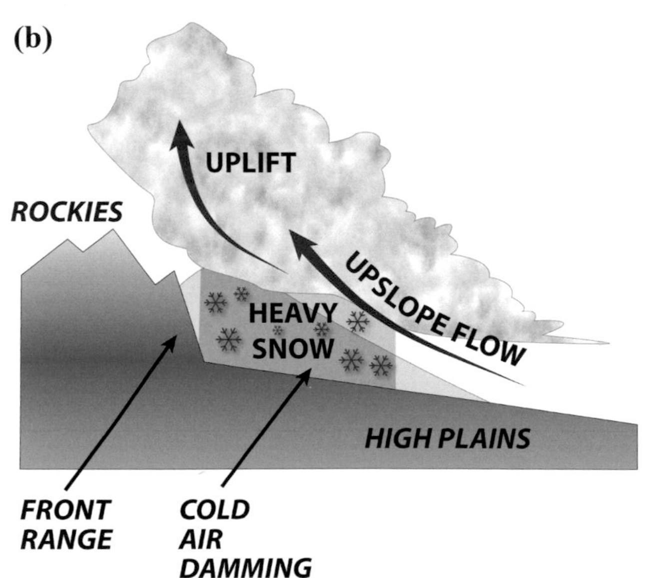

Figure 7.12 Airflow patterns and cold air damming. (a) Typical airflow patterns and cold air damming that lead to upslope snow events along the Colorado Front Range. (b) Overrunning and cold air damming processes.

extratropical cyclone moves into the vicinity of the Four Corners (junction of Colorado, Utah, Arizona, and New Mexico). The storm's winds draw a mild, humid airflow northward from the Gulf of Mexico, then circulate this moisture up and over the cold air wedge. As the air is lifted, it cools to condensation and then freezes, forming an extensive cloud layer that precipitates heavy snow. The situation in which mild, humid air rides up and over cold air is termed overrunning.

The southeasterly flow is the essence of an upslope storm: The airflow gradually rises between the Gulf Coast and Rockies, from sea level to nearly 1.5 km (5000 feet), cooling and becoming nearly saturated along the way. The final, abrupt ascent of this air along the east-facing rampart of the Front Range (an additional 1–1.5 km [3200–4900 feet]) of uplift generates an intense, narrow band of heavy snowfall. Figure 7.12(b) illustrates the overrunning and cold air damming processes.

At times, smaller-scale processes help focus and anchor the region of heavy snowfall against the eastern mountain slopes. The sharp contrast in temperature between the cold northern air mass and warm Gulf air generates a sharp frontal boundary. The front aligns with the eastern mountain slopes, and air is forced upward into a heavy snow band along the front. On the cold-air side of the front, the mountains help channel airflow from the north into a narrow, jet-like current. This barrier jet is often associated with the most intense snow rates. Barrier jets are common to heavy zones of precipitation in other mountainous regions, such as the east-facing slopes of the Appalachians and the western slopes of the Cascades.

The eastward prong of heavy snow south of Denver in Figure 7.11 is a type of orographic effect due to the Palmer Ridge. Figure 7.13 shows the topographic relief of western Colorado. The Palmer Ridge is a smooth, hilly range that broadly rises 0.5 km (1600 feet) above Denver and is oriented at a right angle to the Front Range. Denver and Boulder lie within the corner between the divide and Front Range. Given the geographical setup shown in Figure 7.12(a), surface winds with a northeasterly component are directed upslope – up the Palmer Ridge – toward Denver. As Figure 7.13(b) shows, the heaviest snow falls along Palmer Ridge, over and just south of Denver. In contrast, snow over Fort Collins, 100 km (62 miles) north of Denver, is minimal. This wind–terrain interaction produces a pronounced gradient of accumulation as noted in Figure 7.11 – specifically, inches of snow compared to feet of snow.

Fort Collins, in turn, sits along the crest of the Cheyenne Ridge, another broad ridgeline that projects eastward from the Front Range. As Figure 7.13(a) shows, a strong southeasterly flow at the surface shifts the upslope flow and heavy snow north

(a)

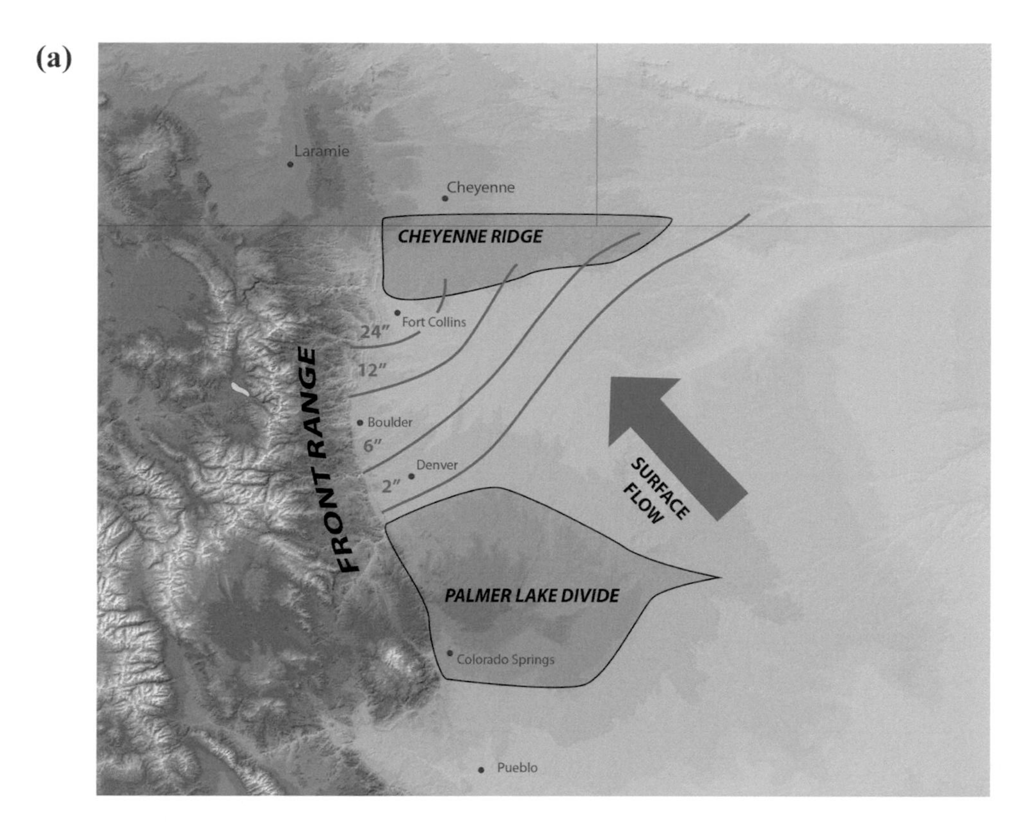

(b)

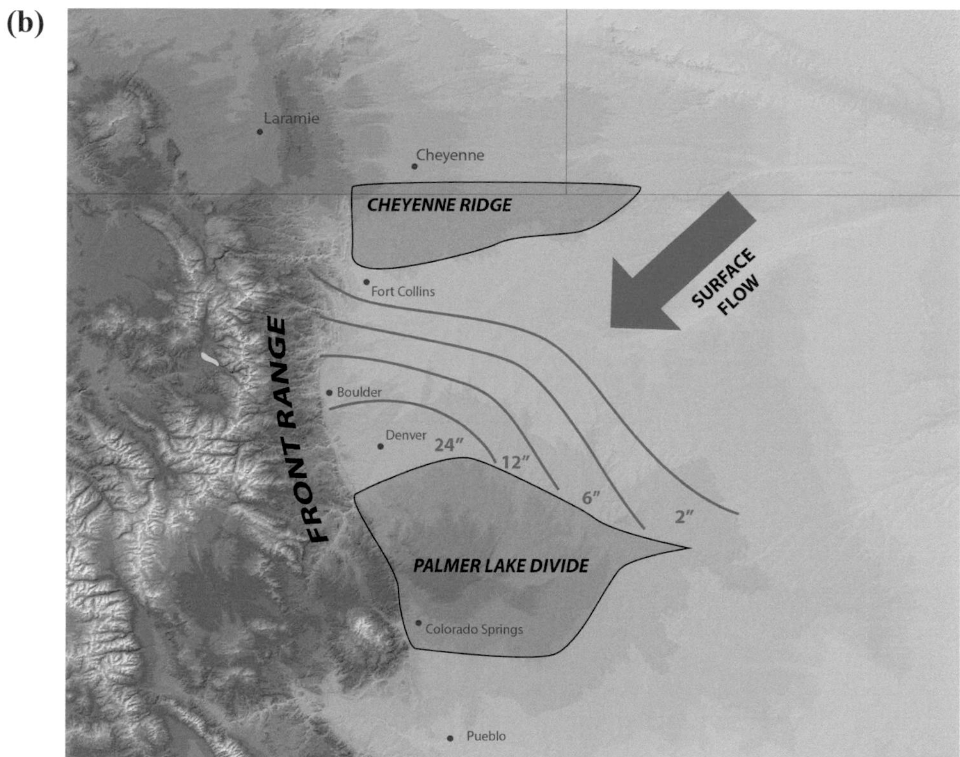

Figure 7.13 Snow accumulation scenarios for low-level winds impinging along the Palmer Divide and Cheyenne Ridge, from different directions, during upslope snowstorms. (a) Heavy accumulation north of the Palmer Divide, from northeasterly surface winds. (b) Major accumulation south of the Cheyenne Ridge, from southeasterly winds.

of Denver, now dumping the snow on Fort Collins. Again, an extreme gradient in accumulating snow ensues – this time, inches in Denver but several feet over Fort Collins and the adjacent foothills.

East Coast Snowstorms

The Northeast Corridor is a vast megalopolis stretching along the Eastern Seaboard, from Tidewater, Virginia, to southeast Maine and encompassing nearly all of the coastal Mid-Atlantic and New England. The region is home to a population of 55 million and several major cities, including Washington, D.C., Baltimore, Philadelphia, New York City, and Boston. It is also home to dense rail networks, interstate highways, and air traffic corridors. During the late fall, winter, and early spring, Nor'easters develop along the coast and hug the Eastern Seaboard (for a review of Nor'easters, see Chapter 6).

Heavy precipitation is a hallmark of these storms because part of their circulation lies over the western Atlantic Ocean. When a subfreezing air mass is also in place, heavy snowfalls deliver a crippling blow to one or more cities within the megalopolis (Figure 7.14). Snow accumulations of around 1 m (2–3 feet) are not uncommon, on an annual basis, over New England. Storms of this magnitude can shut down school districts for an entire week, close major airports, and maroon people in their homes for days before street plows make a first pass.

Figure 7.15 shows the two basic storm patterns along the East Coast. These are named after James Miller of New York University, who wrote the first scientific paper on these storms in 1946. Figure 7.15 (first three panels, January 1966) illustrates the Miller Type A pattern, spanning 24 hours of the storm's characteristic evolution. Recall from Chapter 6 that coastal storms draw their energy from the temperature contrast between cold air over the continental interior and the relatively warm ocean. First, an intense anticyclone over the Plains pushes an arctic air mass south and east to the East Coast. When a wavelike disturbance in the upper-atmospheric jet stream passes over the coastal temperature gradient, an extratropical cyclone is born. The storm intensifies and tracks toward the northeast, along the boundary separating cold and warm air. Once the storm passes over Cape Hatteras, North Carolina, it often undergoes rapid intensification because the system passes over the warm Gulf Stream current. The air overlying the ocean current is exceptionally warm and moist even by wintertime standards, and it feeds additional energy

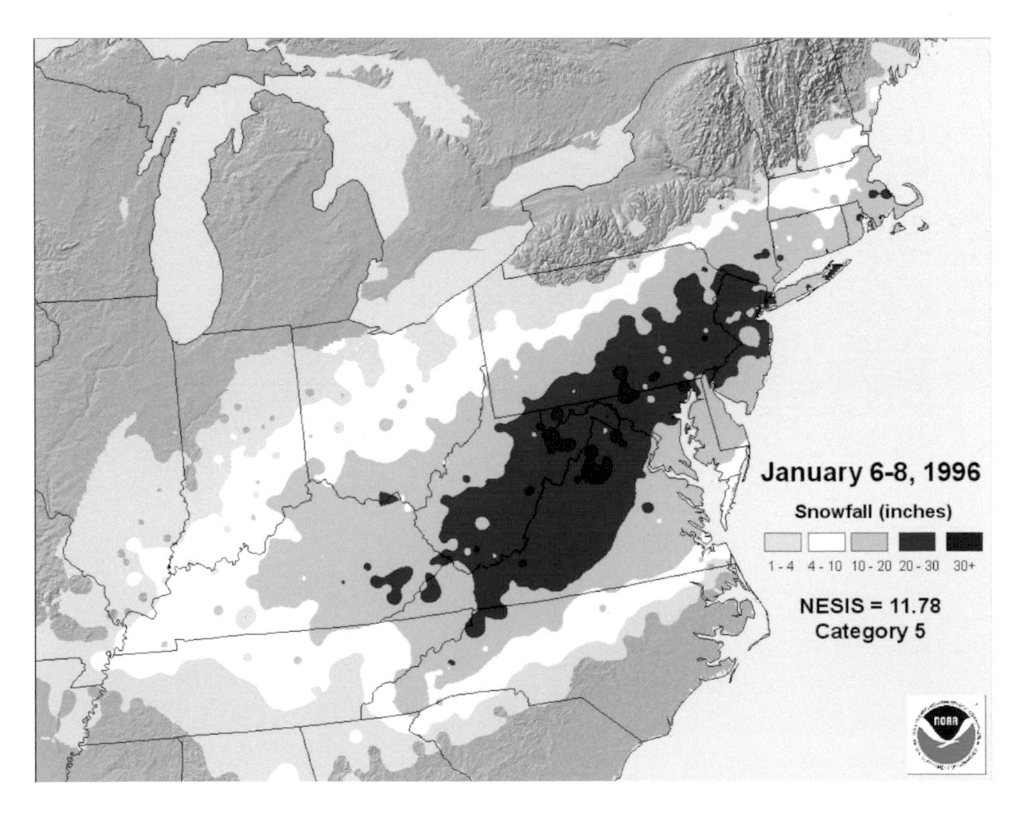

Figure 7.14 Snow accumulation map for a crippling Nor'easter over the Eastern Seaboard, January 2016. The storm dumped 165 cm (65 inches) of snow in North Carolina and brought sustained winds of 50 kts (58 MPH) to the Virginia coastline. (NOAA.)

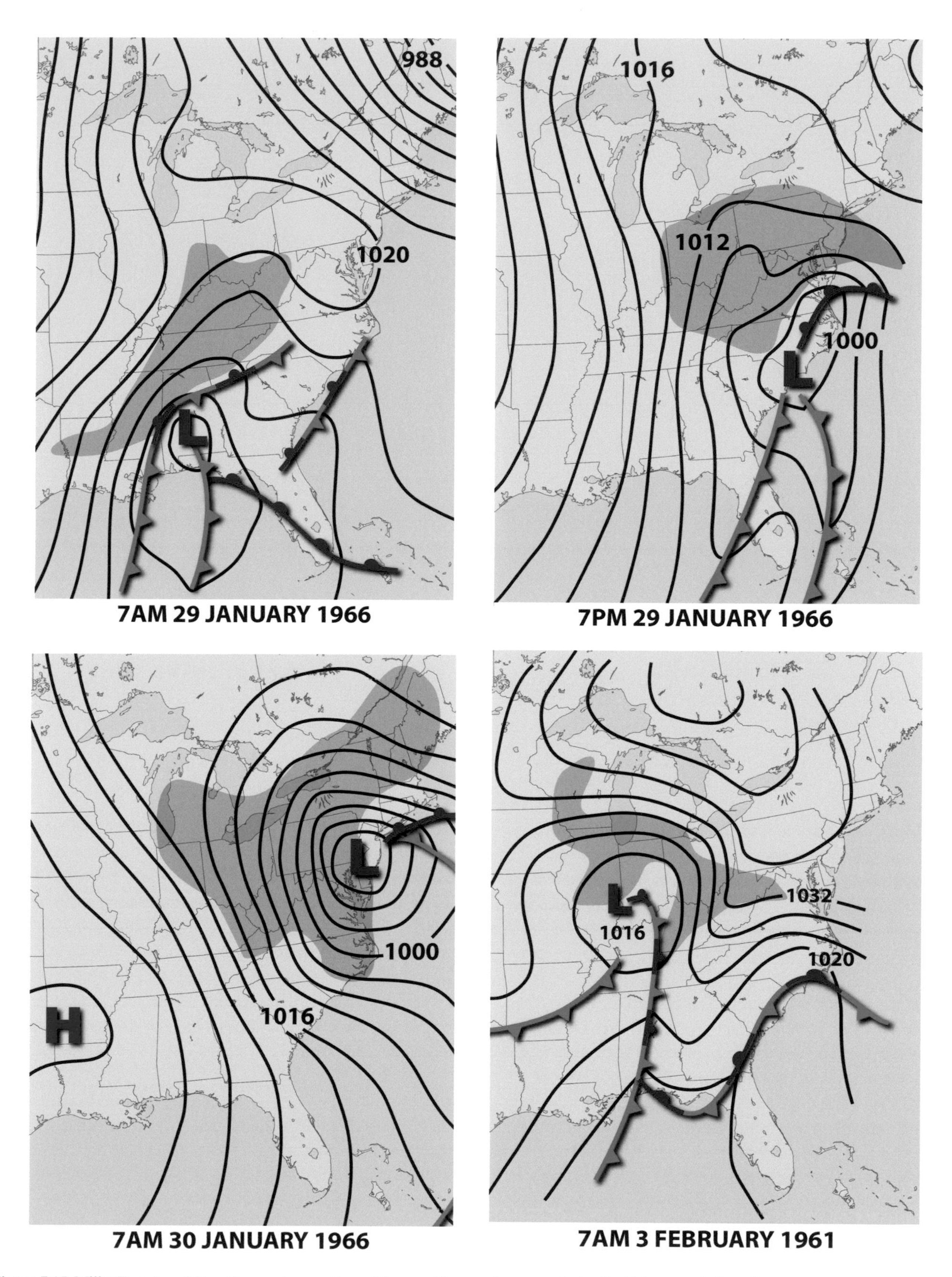

Figure 7.15 Miller Type A and Type B snowstorm patterns. These surface weather maps depict the 24-hour evolution of two types of classic East Coast snowstorms: the Miller Type A (first three panels) and Type B (last three panels) patterns. (Adapted from Kocin and Uccellini, 2004a.)

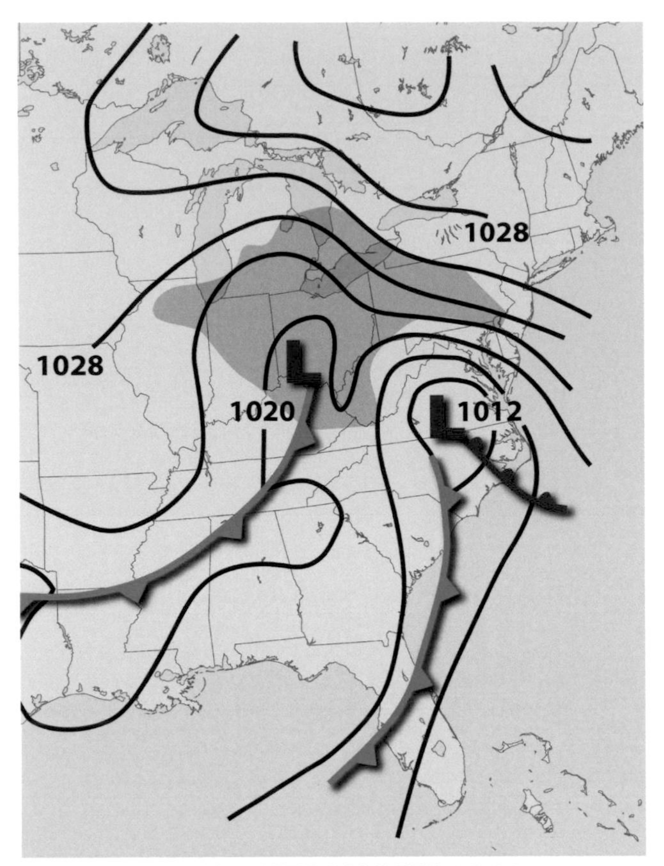

7PM 3 FEBRUARY 1961

7AM 4 FEBRUARY 1961

Figure 7.15 (Continued)

into the storm. The storm, in its full fury, dumps one or more feet of snow over portions of the Megalopolis, along its cold, northwestern periphery. High winds develop in response to a deepening pressure gradient. This situation creates considerable blowing and drifting of snow, sometimes culminating in a blizzard.

Figure 7.15 (bottom three panels, February 1961) shows a 24-hour sequence in the life of a typical Miller Type B storm. This system has a more complex evolution. The primary low pressure first develops over the Ohio Valley and tracks east. Meanwhile, a strong anticyclone over eastern Canada and New England circulates a nose of subfreezing air south, east of the Appalachians. The dense, shallow air layer "locks in" east of the mountains as cold air damming. Overrunning of mild, moist air off the Atlantic causes frozen precipitation to break out across the Mid-Atlantic. Meanwhile, a new low-pressure center begins developing east of the Appalachians, along the coast, in the vicinity of Cape Hatteras. Within 6–12 hours, the coastal low becomes the new, primary storm center and begins tracking up the coastline as a Nor'easter. The Ohio Valley low dissipates along the western Appalachians (this process is termed a center jump). The coastal low rapidly deepens, spreading a wide band of heavy snow inland, blanketing portions of the Megalopolis. Cold air damming is critical to the evolution of this pattern, as the wedge of freezing air – side-by-side with the warm Atlantic air – greatly enhances the temperature gradient from which the storm derives its energy. It also ensures that plenty of cold air is in place for heavy snow.

Because of the narrow rain-snow line, the track of the coastal low is a key determinant of precipitation type. As Figure 7.16 illustrates, a 50–100 km (31–62 mile) shift of the track east or west can make all the difference in the type of precipitation falling over the Megalopolis. An inside track, with the low running up the Chesapeake Bay, places big cities from Washington, D.C., to Philadelphia on the warm air side of the rain-snow line. A cold, heavy rain falls, perhaps mixing

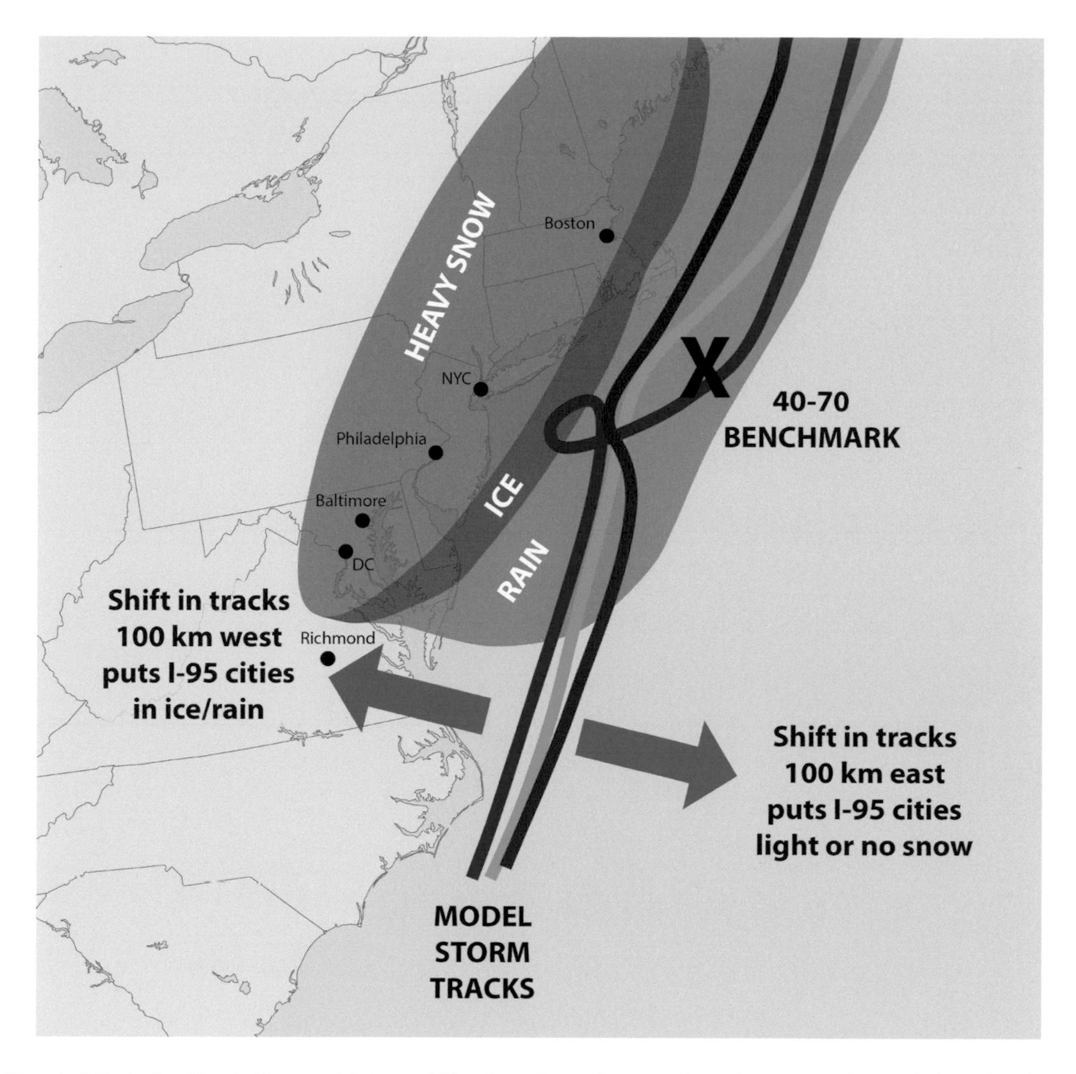

Figure 7.16 Characteristic tracks of wintertime coastal storms. This schematic weather map shows the range of characteristic tracks taken by wintertime coastal storms, along with their associated precipitation impacts, and the importance of the 40–70 benchmark to heavy snow generation over the Northeastern Megalopolis.

intermittently with ice. Farther west, along the crest of the Appalachians, an ice storm ensues. Heavy snow falls farthest west along the precipitation shield, over the mountains of the West Virginia panhandle, western Maryland, and central Pennsylvania. An outside track occurs when the low's center moves well offshore, remaining 150–200 km (93–124 miles) from the coastline. In this scenario, the Delmarva Peninsula and southern New Jersey may experience accumulating ice, while heavy snow falls just east of the major cities from Washington northward (but these cities often get grazed with a few inches of snow). Finally, there is a type of track that spells a snowy "bullseye" for the big cities, which makes snow lovers rejoice and snow haters cringe. This track occurs when the storm tracks slightly offshore, crossing the "40–70 benchmark" – that is, 40° N latitude and 70° W longitude. This track often ensures that a heavy, even crippling snow will blanket several major cities, especially those in New England.

Surface weather maps such as those in Figure 7.15 describe the evolution of these snow-making storms, but there are also important processes in the upper atmosphere. Figure 7.17 depicts a common upper-atmospheric pattern leading to rapid deepening of the coastal low. Here, a single, wavelike trough in the jet stream amplifies (grows larger) over time along the East Coast, as Figure 17(a) shows. Recall from Chapter 6 that the upper-level wave and surface storm are coupled, with one intensifying the other. Other times, the polar jet stream splits into two branches: the northern,

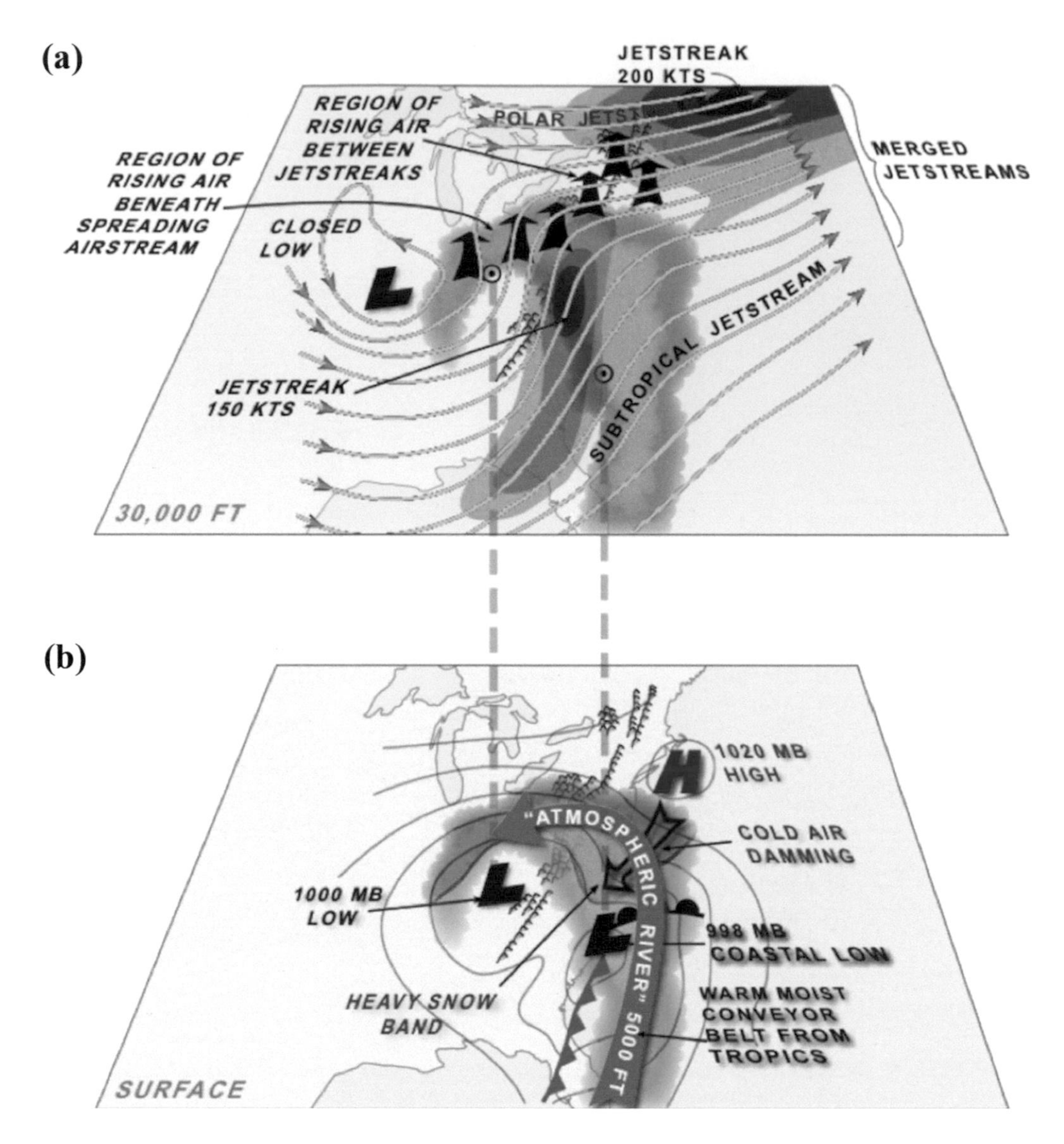

Figure 7.17 East Coast heavy snowstorm, February 2010. These surface and upper atmospheric weather maps depict the major jet stream flow patterns for two heavy snowstorms during February 2010. In the upper panel, troughs in the polar and subtropical jet streams have phased or merged together.

polar jet branch and the southern, subtropical jet branch. Figure 7.17(b) shows this split, but here the waves from both branches merge, a process called wave phasing. Phasing creates one large, intense trough that amplifies and rapidly deepens the surface storm. The surface low may bomb out, as described in Chapter 6. It is during the most rapid deepening phase that the heaviest snow falls, at rates of 5–10 cm (2–4 inches) per hour, occasionally producing a bizarre event called thundersnow. Thundersnow is a brief, heavy burst of snowfall (sometimes with zero visibility) accompanied by lightning and thunder. Small pockets of thundersnow account for some of the most extreme gradients in snow accumulation, as Figure 7.18 shows.

Another important upper-atmospheric pattern along the East Coast is a blocking high pattern (Figure 7.27, left panel), in which a massive ridge of high pressure (anticyclone) develops over Greenland, with a deep trough to its southwest. The surface storm lies close to the center of the trough, and it begins to occlude (as described in Chapter 6). However, the storm's northward progress along the East Coast is impeded or blocked by the high-pressure ridge. Consequently, the storm moves very

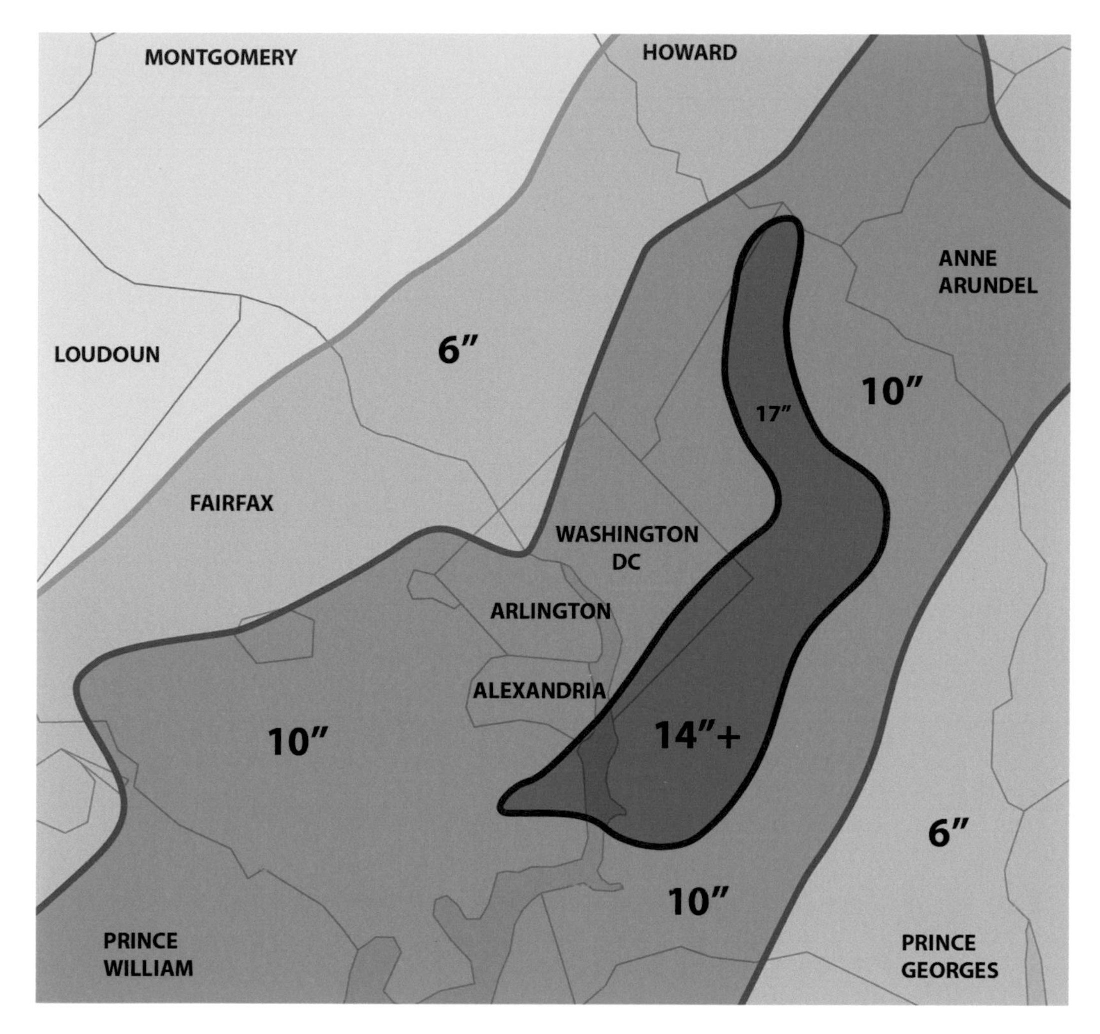

Figure 7.18 Veterans' Day Snowstorm, November 1987. The figure illustrates the extreme snowfall gradient characteristic of some Nor'easters. (Adapted from Kocin and Uccellini, 2004b.)

slowly or becomes stationary. The result is a prolonged period of heavy snow accumulation, perhaps 48 hours or more. This Greenland Block is more likely to become established during the negative phase of a climate oscillation termed the North Atlantic Oscillation, which we discuss in more detail later in the chapter.

Forecasters know that the "devil is in the details" when it comes to heavy snow prediction. A common important feature is a single, heavy snow band that often develops 150–200 km (93–124 miles) to the north-northwest of the storm's center (Figure 7.19[a]). This band forms in a region where airflow characteristics create a front (enhancement of the temperature gradient) in the lower atmosphere, between 1 and 4 km (3300–13,200 feet); along this front, the uplift of cold, saturated air becomes maximized, creating an elongated band of heavy snow. Frequently this band is so slow-moving that it remains nearly stationary for hours. Meteorologists monitor weather radar to track the snow band's movement, coverage, and intensity, issuing heavy snow statements every few hours (another form of nowcasting). These bands generate a long swath of heavy snow, often several hundred kilometers long by 75–100 km (46–62 miles) wide. Exceptionally large snow accumulation gradients occur along the edges of a heavy snow band. An example of a stationary snow band as observed by weather radar is shown in Figure 7.19(b) for a Nor'easter during February 12, 2006. And pockets of unstable air within a snow band can even generate strong convective currents and thundersnow, with snow rates temporarily reaching 3–4 inches per hour (a true whiteout condition).

(a)

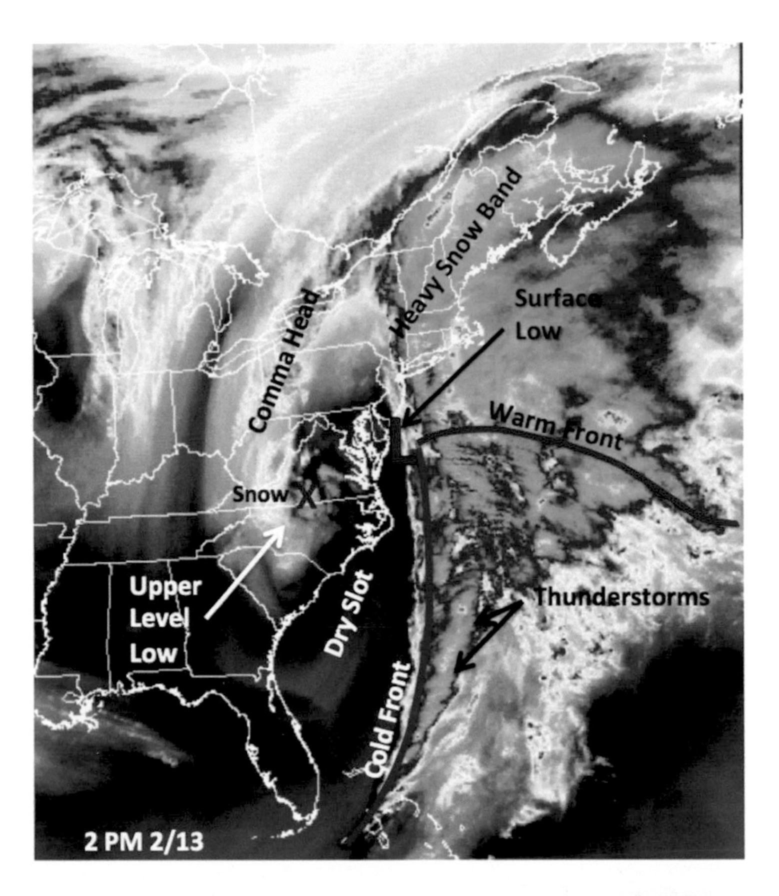

(b)

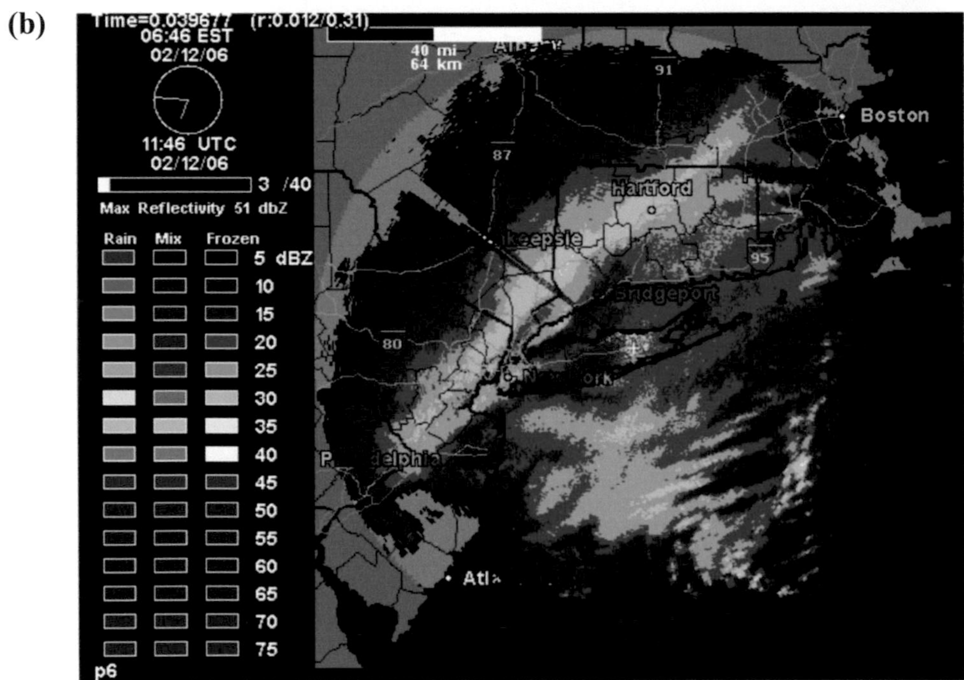

Figure 7.19 Example of a nearly stationary heavy snow band over New England, February 2014. (a) A Nor'easter (coastal) storm ([a] adapted from NOAA) – popularly called "Snochi" because it occurred during the same period as the winter Olympics in Sochi, Japan. (Adapted from NOAA.) (b) A characteristic narrow, stationary snow band from a 2006 Nor'easter. (NWS.)

Lake Effect Snows

Among the most dramatic of all snow regions is the infamous belt across the United States Great Lakes. As Figure 7.20 shows, the annual average snow totals along the eastern shores of the lakes lie in the 250–300+ cm (98–118+ inch) range. All five lakes show this pattern, which is created by westerly to northwesterly cold air surges that cross the lake water, gain heat and moisture, and then precipitate snow on the lee (downwind) shores. The snow typically falls from narrow, streamer-like bands that can extend over 100 km (62 miles), maintaining the same orientation as long as the wind direction does not shift.

The narrow cloud bands are composed of a "train" of shallow, convective cells, sometimes producing lightning and thunder. The snow tends to be squall-like, coming in a sudden burst, with steady snow falling for hours. When a band parks over the same region for 6 or more hours, the heavy snow accumulation becomes highly localized, producing extreme gradients of snow accumulation. Often visually there is a knife-edge demarcation between blindingly heavy snow and clear air; literally, one side of a neighborhood may see a dusting, while the other side receives a foot or more.

There is a well-defined lake effect snow season from late November through January. As Figure 7.21 shows, the season is based on air temperature that is not only below freezing but also significantly colder than the underlying lake water. The difference between water temperature and air temperature is crucial; the greater the difference, the greater the flux (extraction) of heat and moisture from water to air. The lowest thousand meters of cold air become unstable when warmed and moistened. Convection bubbles up over the lake water, reaching heights sufficient to form heavy snow while approaching, then crossing the lee shore. A typical lake effect season may experience half a dozen snow events. The lake effect "machine" shuts down in late winter (specifically February) when the water surface becomes covered with ice. The ice acts like a thermal lid, preventing evaporation and the flux of warm air from water to air.

Figure 7.22 shows an example of a cold air outbreak leading to a lake effect snowstorm. In the typical scenario, a surge of

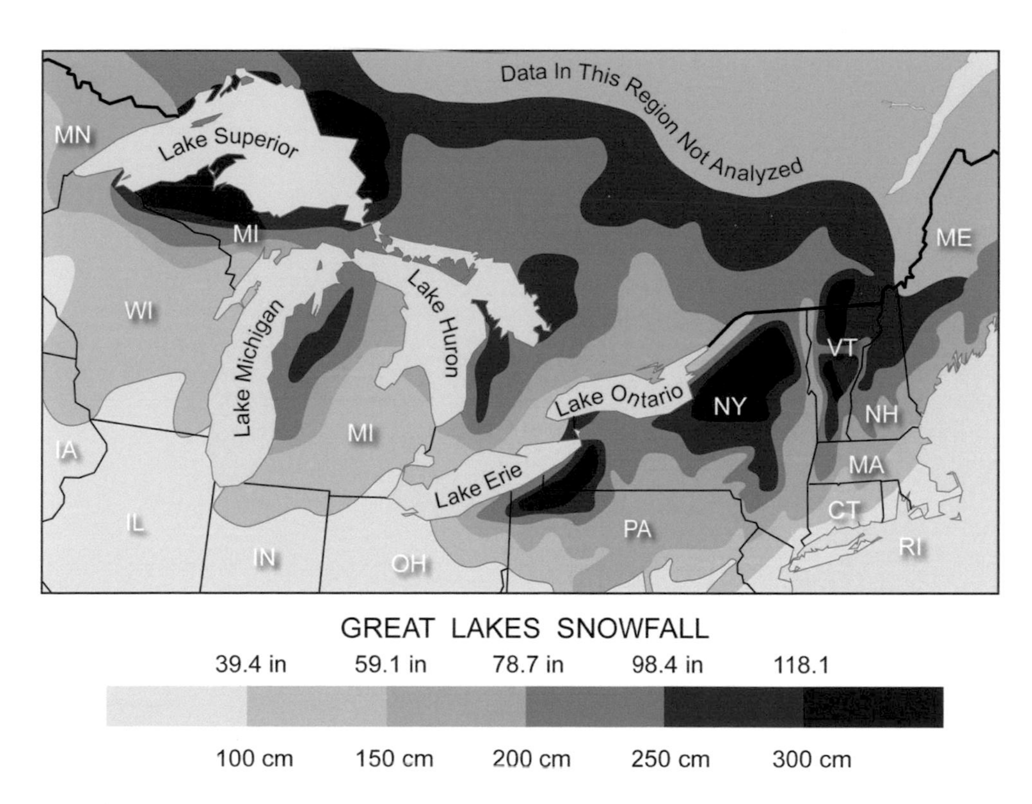

Figure 7.20 Snow accumulation map, Great Lakes. This snow accumulation map for the average winter season across the Great Lakes Region shows tremendous snow totals along the lee shores of each lake.

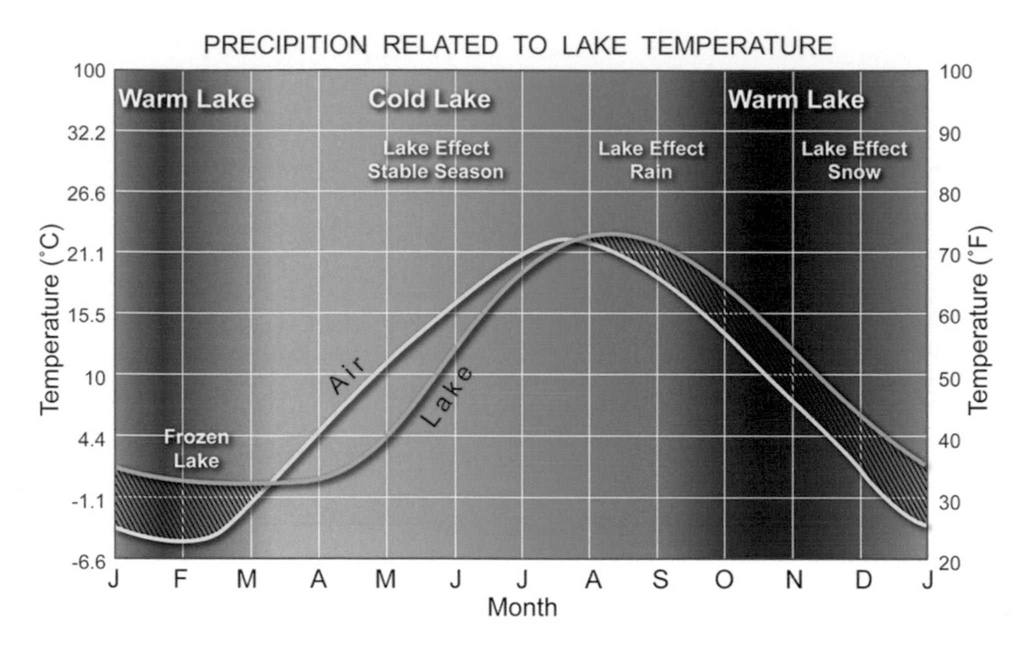

Figure 7.21 Seasonal temperature variation, Great Lakes. Average, seasonal temperature variation for air and lake water. The lake effect snow season occurs when air temperature dips below the water temperature, from late fall through midwinter. (Adapted from NWS.)

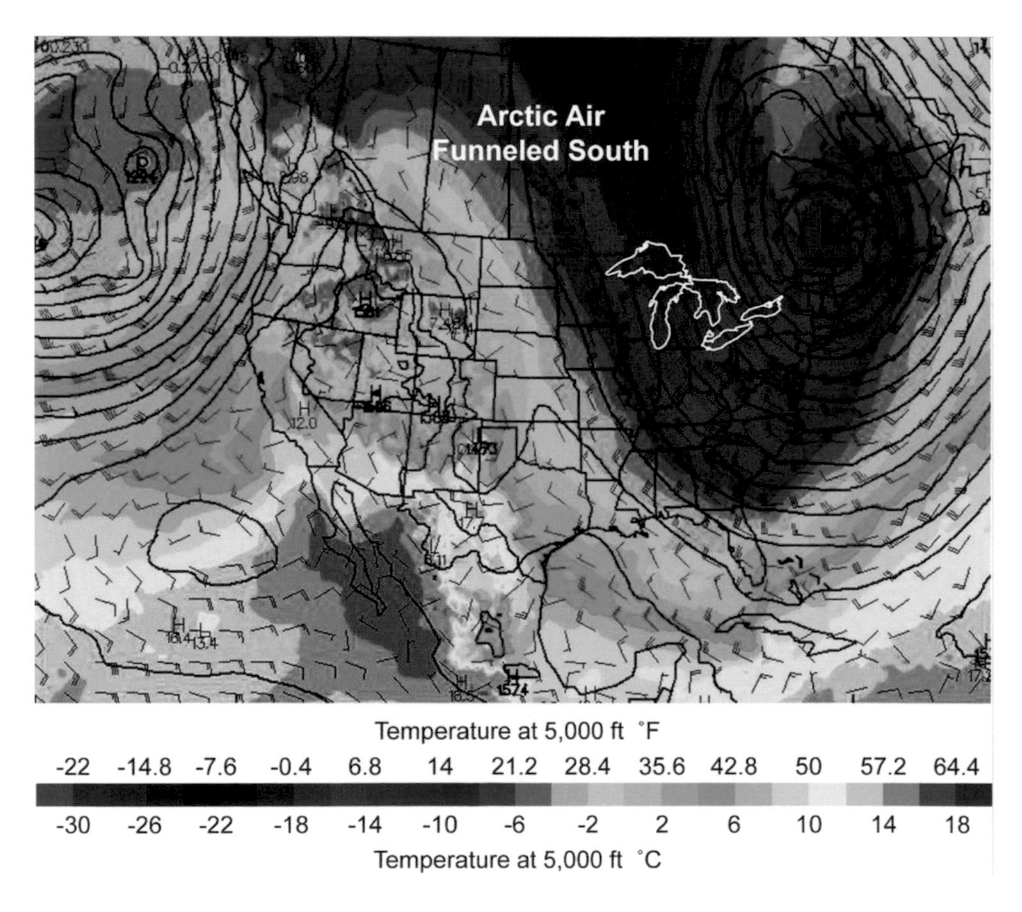

Figure 7.22 Arctic air outbreak. This weather map at 1.5 km (4900 feet) shows an arctic air outbreak being funneled over the Great Lakes, behind a departing extratropical cyclone.

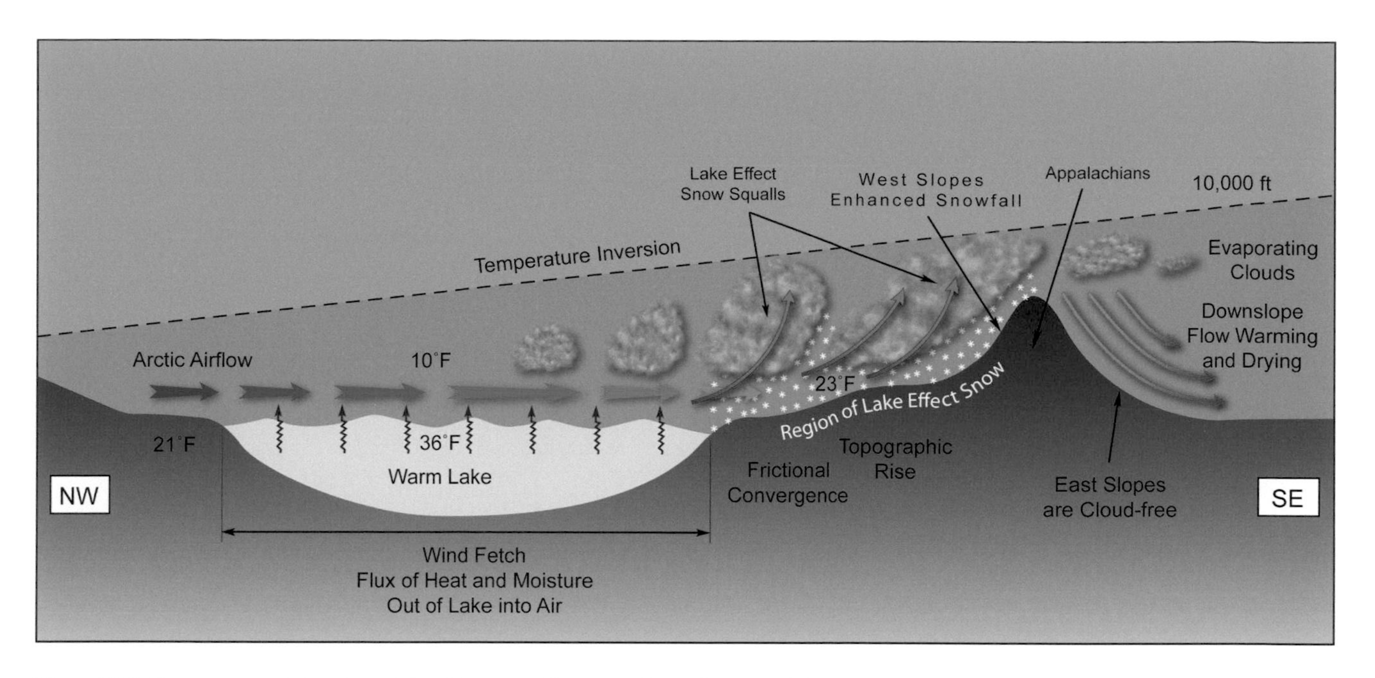

Figure 7.23 Key processes giving rise to lake effect snow. The figure shows the role of lake water, surrounding terrain, and atmosphere in bringing about lake effect snow.

arctic air pushes across the lakes from the northwest, behind a departing low-pressure system. High pressure building across the northern Plains helps accelerate airflow from the northwest. The departing low, which is an extratropical cyclone, may have brought a period of storminess (including widespread snowfall and high wind) to the Great Lakes. Several hours after the low moves away and high pressure moves in, the weather "clears up stormy" as the lake effect machine starts up.

Figure 7.23 provides more detail about the lake effect process. Note the large temperature difference between arctic air and water from northwest to southeast. The air crossing the lake has warmed and moistened considerably while crossing, a process that is maximized by the long fetch over water (see Chapter 6). The longer the transit time over water, the more heat and moisture are extracted from the lake water. In other words, when winds blow parallel to the long axis of the lake, the fetch is longest, hastening destabilization and heavy snow production. Two additional processes enhance the snowfall along the lee shore: (1) frictional convergence, as the air slows abruptly when it comes off the water, causing it to pile up and rise, and (2) topography. Even slight hills can cause orographic lifting, forcing the air to ascend. Long-distance snow streamers, hundreds of kilometers long, can sometimes survive their journey southeastward, all the way to the western slopes of the Appalachians. The west-facing slopes of the Appalachians sometimes give orographic uplift to the moist, streaming air.

The depth of the lake effect clouds is limited by a temperature inversion (air layer in which temperature increases with altitude). Inversions are stable air layers that suppress upward motion. In Figure 7.25, the base of the inversion rises steadily from northwest to southeast. The inversion is created by air sinking down the core of the high-pressure cell that builds from the northwest. Any snow clouds that move beyond the crest of the Appalachians disappear as air subsides, warms, and dries on the lee (eastern) side.

Both the wind direction and air-water temperature difference are critical parameters in forecasts of lake effect snow. Large air-water temperature differences and long fetch both maximize the snow potential. Figure 7.24(a) shows a satellite image of a lake effect snowstorm. Dozens of narrow, cloudy streamers blow off the lakes, and the larger and taller among them generate the snow. Weather radar, shown in Figure 7.24(b), indicates numerous short snow bands crossing Michigan, Indiana, and Ohio. A particularly long snow streamer cuts diagonally across much of Pennsylvania, having developed over the eastern shore of Lake Erie.

(a)

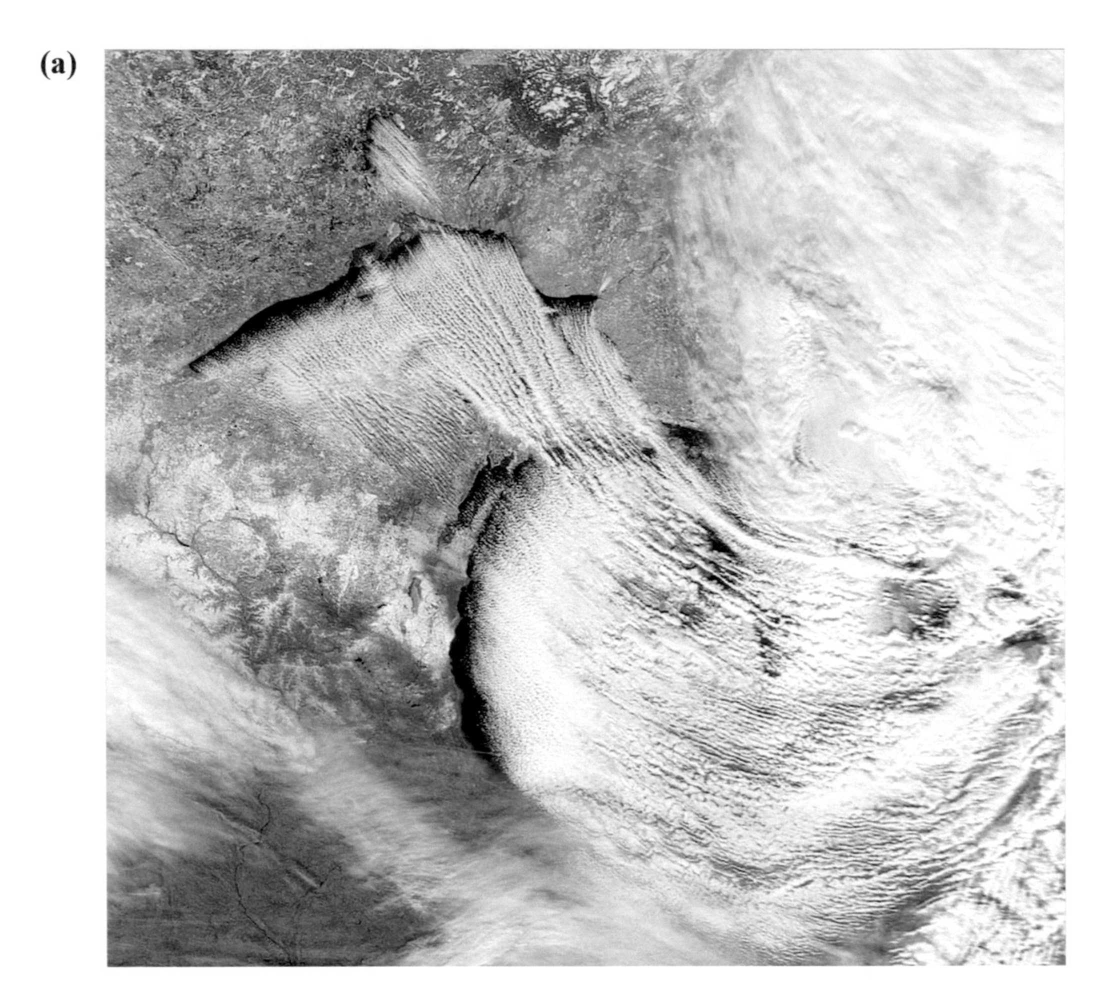

(b)

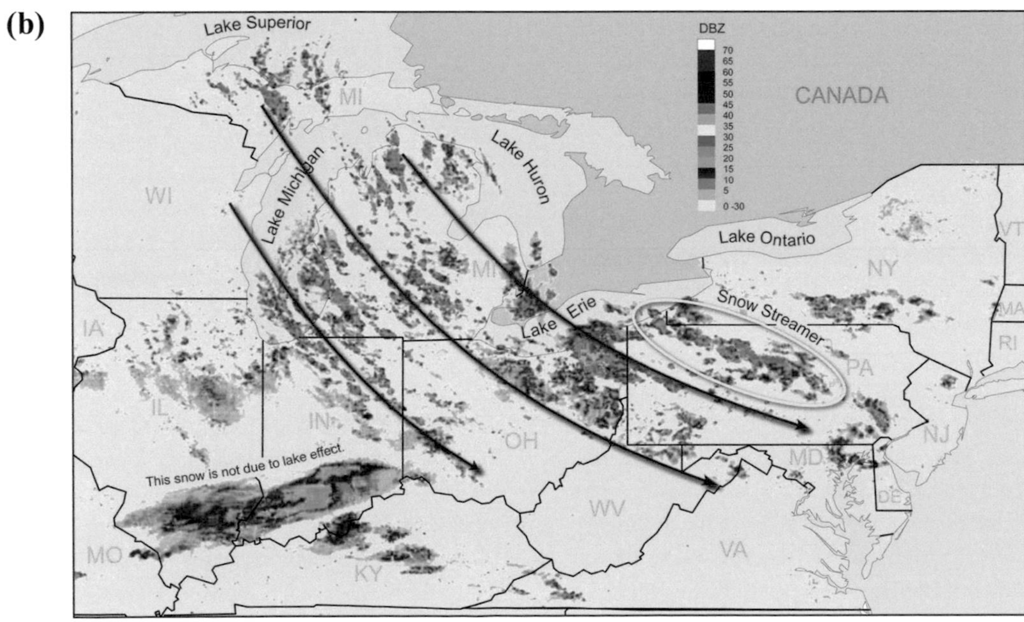

Figure 7.24 Lake effect snowstorm. (a) Satellite view of a lake effect snow event, illustrating multiple cloud streamers developing in a cold, northwesterly flow (NASA). (b) Many narrow snow bands as identified by weather radar. (Based on data provided by NWS.)

Ice Storms

Many people rejoice in a heavy snow (especially kids), but no one hopes for an ice storm. Ice storms – whether an accumulation of sleet, freezing rain, or both – are among the most dreaded and hazardous of all winter weather events. Forecasting the difference between heavy snow and ice sometimes involves very subtle clues. Indeed, many winter precipitation forecasts point to "multiple-choice days" in which people must expect a mixture of snow, sleet, freezing rain, and plain rain. Most ice storms involve some type of phase transition between liquid and frozen precipitation along the rain-snow line. The region of icefall may be narrow (just a few tens of kilometers) or broad (spanning entire states).

Glaze ice accumulations from freezing rain, between 0.6 and 1.3 cm (0.25–0.5 inch) thick, become problematic for trees and power lines, as the weight of ice accretion weighs heavily on limbs and wires. Given the fractal branching pattern of trees, the total volume of ice that can be "stored" within a bare tree canopy, in the form of ice accreted on every branch (large and small) is on the order of several hundreds, if not thousands, of pounds. This ice mass grows exponentially as branch size decreases. As ice accumulations increase to 0.5–1 inch, widespread tree damage and utility outages become a near-certainty. Wind is a factor in utility damage as well; in fact, predictive algorithms consider both wind speed and the amount of ice accumulation in order to assess the probability of localized vs widespread outages. Small amounts of ice accretion combined with high winds or thick ice accretion in a calm wind both create widespread utility damage. The time spent clearing roads of fallen limbs, wires, and poles adds considerably to the time spent reconstructing the electrical grid, which may take several days to a week.

Among the worst ice storms ever experienced in North America was a multiday ice event, lasting from January 5 to 9, 1998, across upstate New York and Ontario. Pockets of up to 10 cm (4 inches) ice accretion left 5 million without power and caused severe damage to massive, high-voltage utility towers. In addition, 35 deaths and nearly 1000 injuries were reported. Over 1.8 million hectares of forest were destroyed.

During an ice storm, black ice (freezing rain on road surfaces) becomes so treacherous that any type of driving is done with great peril. The slightest highway grades, including overpasses, become absolutely impassable. Car accidents are common, as are emergency rooms crowded with people who have fallen and injured themselves. Commercial airlines must engage in extensive wing and airframe deicing, adding to departure delays. Railroads struggle to keep automatic switches operable across main lines, particularly across the densely traveled Northeast Corridor.

Figure 7.25 shows one particularly widespread and highly disruptive ice storm, a classic cold air damming situation from January 1994. The surface weather map shows an extensive wedge of subfreezing air across the entire Eastern Seaboard from Maine to Florida. Here, a shallow but cold and dense air layer had become dammed up against the eastern slopes of the Appalachians. A strong anticyclone (1040 mb) over Maine pumped the frigid air southward, creating the cold dam. Milder air was drawn westward off the Atlantic, over the cold wedge, by a southeasterly current at 1.5 km (5000 feet). This overrunning situation caused widespread cloud and wintry precipitation to break out over the entire East Coast. The pink shaded region denotes heavy ice accumulation (sleet and freezing rain), most concentrated over the Piedmont of North Carolina, Virginia, Maryland, and southeastern Pennsylvania.

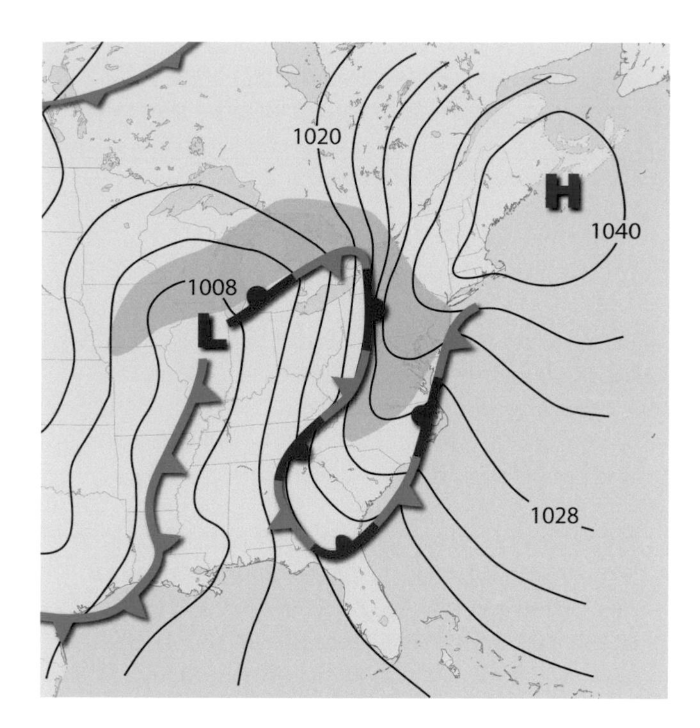

Figure 7.25 Ice storm, January 1994. This surface weather map depicts the key factors that created a crippling ice storm during January 1994 across the Mid-Atlantic states. (Adapted from Kocin and Uccellini, 2004b.)

Forecasting the duration of an ice storm can be challenging because forecasters must predict the breakdown (erosion) of the cold air dam. The dense, cold wedge is stubborn, and it often persists several hours longer than the best predictive models suggest. Strong southerly winds, bearing mild air, are needed to mix out or "scour" the dense air from valleys. Strong winds in the lowest several thousand feet can also mix down warmer air from aloft, from the turbulent airflow these winds generate. Many ice storms are self-limiting for two reasons: (1) When rain freezes into ice on surfaces, latent heat is released, which warms the surface and air layer just above; and (2) rain falling at moderate to heavy rates runs/drips off frozen surfaces faster than it can freeze and accrete. Thus the truly catastrophic, multi-inch ice storms do not arise from heavy rain but rather from lighter and more prolonged rain events, with a cold northerly wind that continuously removes the latent heat.

Forecasting Heavy Snow and Ice

Among the forecaster's greatest challenges is the accurate prediction of heavy snow, ice, high winds, and bitterly cold air. The most common question from the public is, "How much?" But there are many other factors to consider, in addition to the number of accumulated inches of snow. These include the following:

1 *Timing and duration of snow/ice (start and end times)*. A few centimeters (an inch or two) of snow, falling when pavement temperatures are below freezing, can create intense traffic havoc during evening rush hour in major cities, as several recent storms have proved. Raleigh, North Carolina (2005) Atlanta, Georgia (2014), and Washington, D.C. (2016), have all experienced this effect. Traffic on congested roads melted the 2–3 cm (inch or so) of snow (from the pressure of wheels on pavement), and then the liquid rapidly froze into black ice – a type of "flash freeze" effect. In each case, thousands of people, including busloads of students, were stranded on roads for many hours, in some cases overnight. The National Weather Service now issues a special "Potential Winter Commuting Hazard" alert when these types of low snow accumulation but high impact events are anticipated.

2 *Liquid water equivalent (for example, 5:1, 10:1, or 20:1 snow, or anything in between)*. Large snow ratios yield more accumulation for a given amount of liquid equivalent rainfall. Heavy, wet snow requires tremendous exertion to shovel (leading to back injuries, sprains, and heart attacks), and 12–15 cm (5–6 inches or more) can damage trees and utilities.

3 *Wintry precipitation transitions (e.g., snow to rain or rain to freezing rain) and their timing*. Sleet mixing into snow for several hours cuts down greatly on snow accumulation, but it creates additional hazards associated with icy surfaces.

4 *"Stickage" factor*. Will snow accumulate on road surfaces? The answer depends critically on both road temperature and air temperature. When road surfaces are close to freezing, heavy snow rates can quickly chill road surfaces to below freezing, and snow will start to accumulate rapidly. Stickage is also more likely during nighttime hours. Pretreating roads with salty water (brine) or beet sugar solution can effectively reduce the freezing temperature of road surfaces, preventing ice formation and light snow accumulation.

5 *Snow rate,* Will snow falling at several inches an hour overwhelm plows' efforts to keep roads clear? Will heavy snow and/or blowing snow significantly impede highway visibility? Sometimes it's not the accumulation that creates the hazard but rather the near-zero visibility (whiteout) that creates the road hazard, particularly on interstate highways. And if road surfaces are within a degree or so of freezing, a very heavy snow rate can overwhelm melting, allowing accumulation to begin.

6 *Blowing and drifting snow*. Will winds and a powdery snow develop high drifts that cover roadways, block access to structures, and promote low-visibility travel conditions?

7 *Wind chill.* Will the combination of high wind and cold temperatures make conditions perilous for road crews, emergency workers, and those shoveling snow, increasing the risk of frostbite and hypothermia?

Research by Professor Walker Ashley at Northern Illinois University suggests that minor winter snow accumulations can prove more deadly than heavy snowstorms (for data collected in the Midwest). There may be a perception that light snow events are less dangerous and more easily navigated than heavy snowstorms, and so more traffic is on the roads during marginal (or incipient storm) conditions.

A number of winter weather forecast issues are specific to the commercial aviation industry, including airframe and wing deicing; keeping ramps, taxiways, and runways free of snow and slush; and increasing the spacing between arriving aircraft due to low visibility and slick runways. A number of aviation incidents in the United States were caused by winter precipitation, including an Air Florida 737 accident in Washington, D.C., in January 1982, which involved engine icing; an American Eagle turboprop that crashed over Indiana in October 1994 due to severe wing icing and a Southwest 737 that slid off the end of an icy runway in December 2005 while landing at Chicago Midway Airport.

Forecast Models

We discussed numerical weather prediction models, also called forecast models, in Chapter 4. The medium-range (7–10 day) models such as the ECMWF (European model), CMC (Canadian model), and GFS (American models) provide a suite of forecast guidance, including ensemble predictions of storm track and intensity, as well as precipitation amount and type. These models have different strengths and weaknesses (or biases). Forecasters usually "get to know" individual model biases, based on years of experience forecasting in a specific region. Each model's performance must also be carefully validated against real-world observations, to assess the model's typical degree of error. As noted earlier, precipitation type and amount are quite

Table 7.1 National Weather Service Winter Weather Alerts (Mid-Atlantic Region)

Product	Criteria				
	Snow/Sleet	**Glaze Ice**	**Wind**	**Temperature**	**Visibility**
Winter weather advisory	< 12.7 cm/5 inches	< 0.6 cm/0.25 inch			
Freezing rain advisory		< 0.6 cm/0.25 inch			
Wind chill advisory				Variable – region specific	
Wind advisory			S > 27 kts/31 MPH, G > 39 kts/45 MPH		
Winter weather commuting hazard	*See Note 1.*	*See Note 1.*			
Winter storm watch	> 12.7 cm/5 inches*	> 0.6 cm/0.25 inch*			
Winter storm warning	> 12.7 cm/5 inch*	> 0.6 cm/0.25 inch*			
Ice storm warning		> 0.6 cm/0.25 inch			
Blizzard warning	Variable		S or G > 30 kts/35 mph		< 0.4 km/0.25 mile for 3+ hours
Wind chill warning	N/A			Variable – region specific	
High wind warning	N/A		S > 34 kts/39 MPH, G > 50 kts/57 MPH		
Snow squall warning	Up to 5 cm/2 inches/30 min		Wind > 26 kts/30 mph		< 0.4 km/0.25 mile

Notes: in. = inches; S = sustained wind, in miles per hour (MPH); G = wind gust, in MPH

*Either snow or glaze ice combined with snow; exact amounts can vary per region.

Note 1: Generally light accumulation, freezing on contact to glaze ice during rush hour.

sensitive to storm track, and the models often demonstrate large shifts in track from run to run, up to about 72–48 hours from the event. To prevent public confusion, forecasters try to downplay the frequent vacillation, but social media often gets the upper hand. Because modeling software is publicly available, nonmeteorologists often post dire predictions of a major, disruptive snowstorm a week or more in advance, even though the forecast track and intensity are often in significant error that far out. Because of the hype and even panic created by such inaccurate predictions, forecasters are forced to contain the damage, which becomes very distracting and time-consuming.

Indeed, forecasters walk a tightrope when issuing high-impact forecasts for a major metropolitan area. Preparing and shutting down such a region is a costly enterprise, but the costs must be balanced against the need to protect lives and property. Some snowstorms are forecast with great success, with remarkably spot-on, early guidance. One case in point was the March 1993 superstorm that clobbered most of the East Coast, from Georgia to Maine, with heavy snow. Models accurately predicted the track and impacts nearly a week in advance – a feat unprecedented at the time (accurate, medium-range prediction was in its infancy). But sometimes predictions are a spectacular failure. In January 2015, a major Nor'easter was poised to strike New York City with over a foot of snow predicted. Right up to the event, the storm's forecast track differed significantly between the European and American model runs. The local forecast office went with the European model, which statistically was outperforming the American model. Blizzard warnings were issued, and the mayor shut down the city's subway system, an action never before taken in advance of a snowstorm. While the European model placed the snowy bullseye over New York City, the actual storm track was closer to that predicted by the American model – over 160 km (100 miles) east of the European model's prediction. The heaviest snow was thus shifted away from New York City.

Over the past decade, the National Weather Service has worked to increase its public discussions of forecast uncertainty. Instead of issuing a single snow accumulation map, forecasters now use a suite of products that convey the most likely minimum and maximum snow amounts. Probabilities are assigned to different snow-accumulation ranges, for various towns and cities. Such advance uncertainty analysis is important for the planning phase of a high-impact event (for instance, whether to put snowplows on the salt trucks). Airlines may use this information to decide on early cancellations. Once the certainty of a winter event is high enough, the local forecast office issues one or more products (summarized in Table 7.1) in the form of a Watch or Advisory. These may later be upgraded to Warnings when the event is imminent or unfolding.

El Niño/La Niña and the North Atlantic Oscillation (NAO)

For several decades, forecasters have sought to establish connections between the frequent occurrence of heavy snowstorms and short-term climate patterns, including the El Niño/La Niña cycle and the North Atlantic Oscillation (NAO). El Niño/La Niña are disturbances of the tropical Pacific. They involve an amplifying interaction between the upper ocean and atmosphere. These impacts "ripple" into the higher latitudes, shifting the track of jet streams across the United States during winter (Figure 7.26). During an El Niño winter, the subtropical jet stream intensifies across the Gulf Coast and southeastern United States, shifting the track of extratropical cyclones farther south and triggering more coastal storms. When these storms track up the Eastern Seaboard, the air mass temperature determines whether it snows heavily. Statistical studies imply a snowier winter across the mountains of the southwest United States, a large reduction in seasonal snow across the Ohio Valley, and enhanced snow across Maine. During La Niña winters, the phasing of the subtropical and polar jet streams over the northwest United States greatly enhances snow production there, as well as across the far northern Plains, while reducing seasonal snow across the Appalachians and Mid-Atlantic.

The North Atlantic Oscillation is not just an El Niño-like phenomena in the Atlantic; rather, it reflects the difference in air pressure between the subpolar (Icelandic) low and Bermuda High, north–south across the North Atlantic basin. Figure 7.27 shows the characteristic phases of the NAO. During the negative phase of the NAO, the Icelandic Low is replaced by a strong high-pressure ridge over Greenland, while the Bermuda High near Africa weakens. The polar jet stream takes on a significant distortion, with an intense ridge locking in over Greenland (called a blocking high) and a deep trough digging into the southeastern United States. This pattern helps create

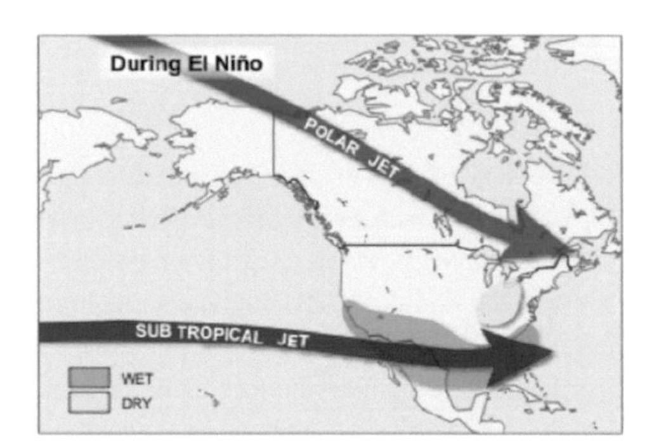

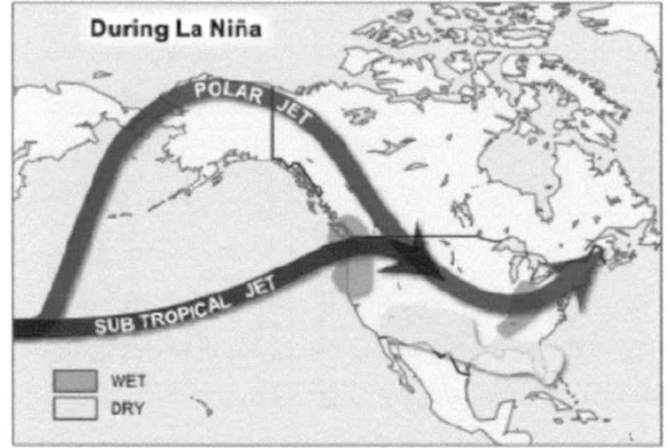

Figure 7.26 Airflow patterns for a typical El Niño and La Niña episodes. The track and intensity of the principal jet streams across North America are altered.

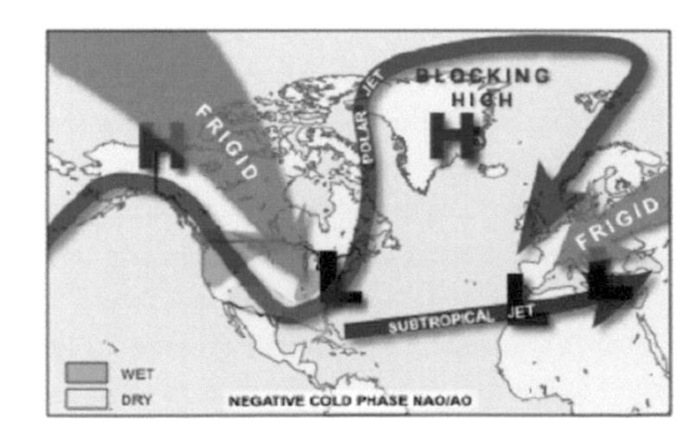

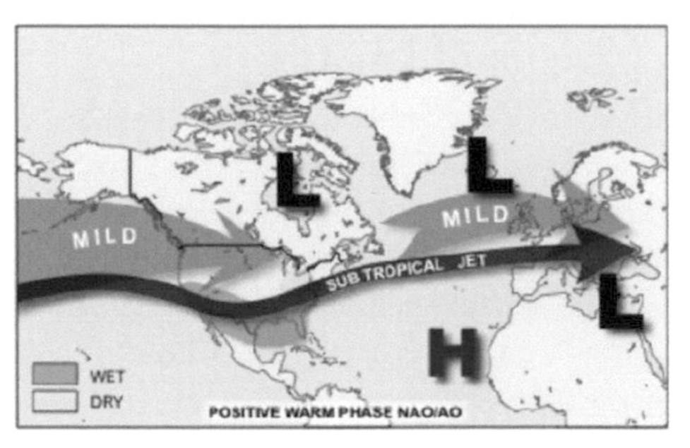

Figure 7.27 Airflow patterns for the North Atlantic Oscillation (NAO). The figure shows the NAO's negative and positive phases, in which the track and intensity of the principal jet streams across the East Coast and Europe are altered.

big snowstorms along the East Coast, in two ways: (1) The trough ensures that arctic air masses invade the East Coast, such that precipitation will be frozen, and (2) coastal storms such as Nor'easters track slowly up the Mid-Atlantic and New England or even become stationary, giving the heavy snow more hours to accumulate. The long-term statistics on seasonal snow accumulation reveal that snowier winters prevail over the Mid-Atlantic, Appalachia, and New England when the NAO locks into an extended negative phase.

Climate models are now being used to anticipate the phase of the El Niño/La Niña pattern, as well as the NAO, weeks to months in advance. An El Niño or La Niña tends to persist for many months, while the NAO is highly variable, changing from week to week. Some regions of the United States, such as the Mid-Atlantic, seem quite vulnerable to these atmospheric regimes. For instance, the record-setting winter of 2009–2010 (all-time seasonal snow accumulation records were broken in Washington, D.C., Baltimore, and Philadelphia) featured an extremely favorable alignment of El Niño with a persistently negative NAO phase.

Snowfall Impact Scales

When forecasting extratropical cyclones, the usual practice is to gauge the storm's overall intensity by measuring the central barometric pressure. However, a storm's very low pressure is not the only metric that can be used to assess the storm's societal impact. Heavy snowfall is certainly a major aspect of these storms. Snowfall impact parameters that map areal snow depth to population density have been developed. Two of these are the Regional Snowfall Index (RSI) and Northeastern Snow Impact Scale (NESIS), both developed at NOAA. Presently these parameters are computed after the fact, but they are useful for comparing and ranking storms in terms of historical significance and for identifying regions that are vulnerable to heavy snowfall. Whereas the RSI is computed for specific geographical regions across the continental United States (Figure 7.28), the NESIS scale was developed to highlight only the populous Northeastern Megalopolis spanning Washington, D.C., to Boston, Massachusetts (home to an estimated 55 million inhabitants).

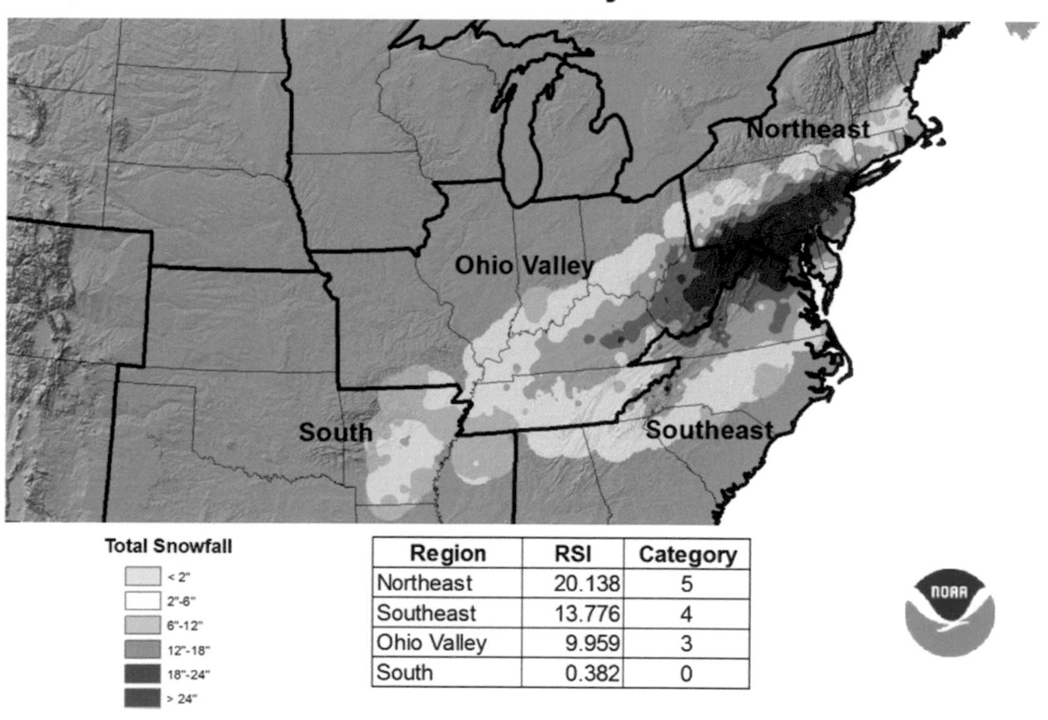

Region	RSI	Category
Northeast	20.138	5
Southeast	13.776	4
Ohio Valley	9.959	3
South	0.382	0

Figure 7.28 Example of Regional Snowfall Index (RSI) calculations for a severe East Coast snowstorm in January 2016. RSI rankings scale from zero (lowest impact) to five (highest impact), and calculations for four different RSI regions are presented in the table. Snow accumulation (inches) is shown by the color figure. (NOAA.)

The NESIS uses a geographical information system (GIS) framework to map a storm's snowfall footprint with population distribution. The goal is to gauge the socioeconomic disruption caused by the storm's heavy snow. This information necessarily decouples snow hazard from storm intensity; not all extremely intense cyclones produce heavy snow (some produce rain instead), and heavy snowfalls are generated by some moderate-intensity storms. The NESIS was developed from a population of 70 snowstorms spanning the years 1950–2000 across the Northeastern Megalopolitan Corridor. Five impact categories are used to rank each storm from Notable (Cat 1) through Extreme (Cat 5).

When looking at RSI and NESIS-rated storms, keep in mind that both the number of people impacted and the snow depth must be large for a storm to achieve a high NESIS value. A narrow swath of heavy snow, even 1 m (3–4 feet), falling across the sparsely populated Appalachian Mountains will not score high; neither will a large swath of 10–15 cm (4–6 inches) snow falling on many cities along the Northeast Corridor. Like so many natural hazards, the typical distribution of NESIS scores follows the law of rare events; that is, the most extreme values are the least common. From 1950 to 2000, there were 45 Cat 1 and Cat 2 NESIS storms, but only 2 Cat 5s. The law of rare events also applies when ranking tornadoes according to peak wind speed or when ranking hurricanes in terms of sustained wind.

Amazing Storms: The North American Blizzard of March, 2017

A recent, significant snowstorm impacted portions of the Mid-Atlantic and New England on March 15–17, 2017. As shown in Figure 7.29, the area of low pressure was a coastal storm (Nor'easter) that tracked from the Delmarva Peninsula to Maine, with pressure dropping rapidly while over the coastal Mid-Atlantic region. A heavy swath of snow, ranging from around a half to one meter (1 to 3 feet) of accumulation, extended from central Pennsylvania through Maine. The bullseye of heaviest accumulation was located over central and upstate New York, with amounts exceeding 1 m (3 feet)

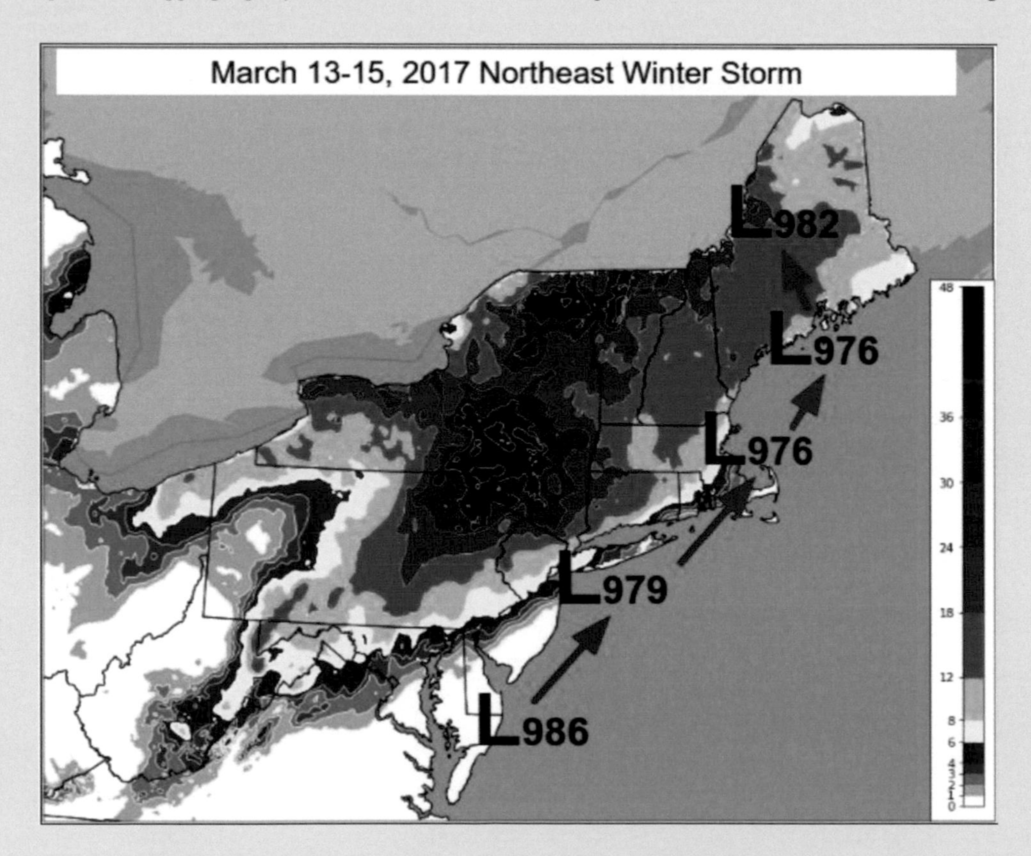

Figure 7.29 Footprint of heavy snowfall left behind by the March 13–15, 2017, blizzard. The track and intensity of the responsible coastal low-pressure system ("L") are shown at various times. (NWS.)

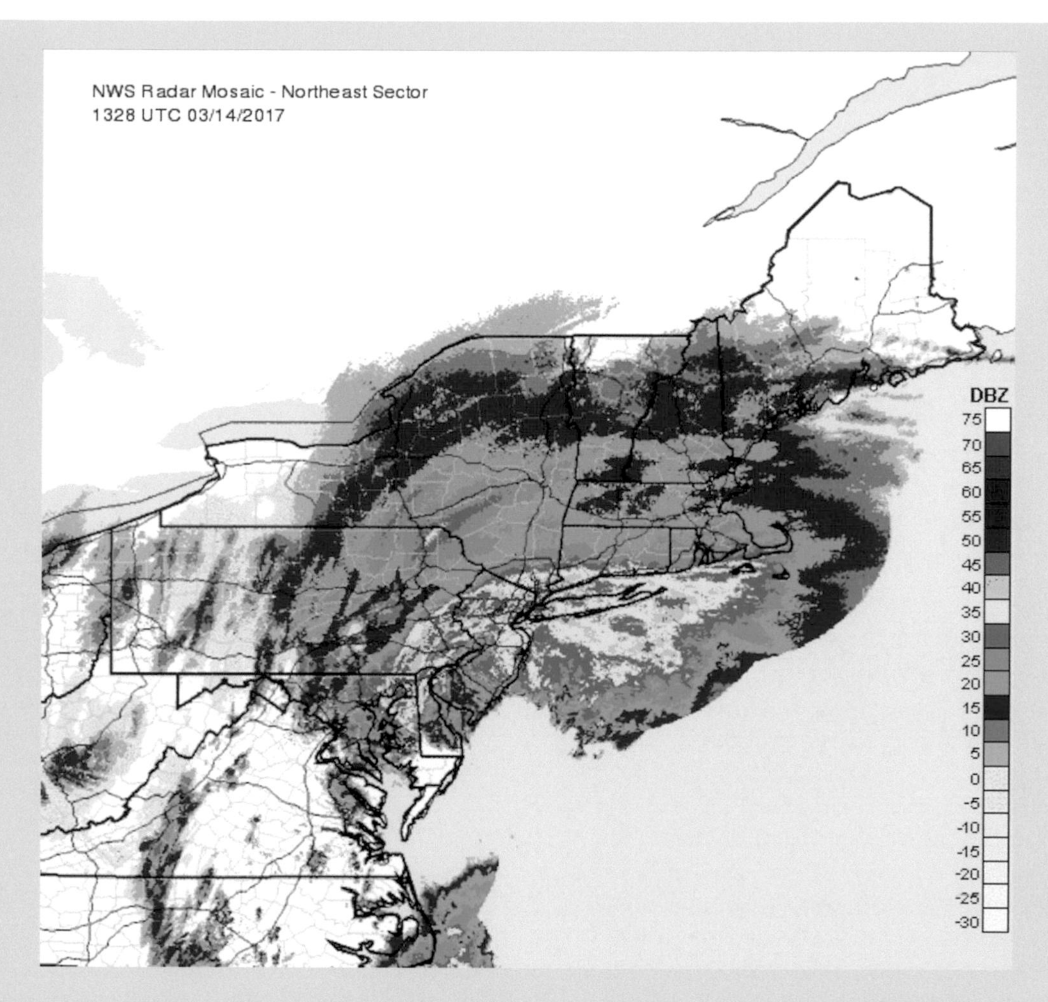

Figure 7.30 Band of heavy snow positioned over southeast Pennsylvania and northern New Jersey. Yellow, orange, and red colors indicate the core of heaviest precipitation as detected by weather radars across the region. (NWS.)

in spots. The storm's maximum snow accumulation was an astounding 107 cm (42 inches) over West Winfield, New York. The storm's track was far enough inland such that the heavy snow swath just missed all the major cities along the I-95 corridor (Washington, Baltimore, Philadelphia, New York City, and Boston). Heavy precipitation still fell over these cities, but enough warm air worked in from the Atlantic Ocean to change an early period of snow to sleet and plain rain.

The storm's meteorological evolution was a classic, textbook example by all accounts, featuring all the dynamical elements discussed in the foregoing chapter. The heaviest snow fell while the storm's central pressure rapidly decreased, and moist Atlantic air streaming into the storm's core ensured plenty of moisture to manufacture several feet of snow. Jet stream processes played a key role in intensifying the upward air motion over the storm's core. First, multiple pieces of jet stream "energy" phased together over the East Coast, as a trough of low pressure in the polar jet stream – over the northern latitudes – merged with a trough in the subtropical jet stream over the southern latitudes. Second, two jet streaks embedded within the merged airstream concentrated the strongest ascent of moist air over the northern Mid-Atlantic. Third, several intense snow bands developed and remained stationary across parts of the affected region – resulting in exceptionally heavy bursts of snow falling over the same locations for many hours. A radar snapshot of a heavy snow band is depicted in Figure 7.30.

This was a very impactful storm, scoring a 3 out of 5 on the NESIS scale (a "major" category snowstorm). Sixty million people felt varying effects from heavy snow, to a wintery mix of snow, ice and rain, to strong wind gusts, coastal flooding and beach erosion. Land and air travel were severely impacted, with cancelation of nearly 8000 commercial airline flights. Blizzard conditions (from a combination of strong winds and heavy snow) were experienced from southeastern New York to southern Maine. Wind gusts in the 61–70 kts (70–80 MPH) range were recorded along coastal Massachusetts, and gusts to 120 kts (138 MPH) razed the high altitude Mount Washington weather observatory in New Hampshire's White Mountains. There were multiple fatalities, including three people who died shoveling snow, and several dying as a result of motor vehicle accidents.

Summary

LO1 Describe the characteristic types of winter weather hazards that characterize the five U.S. geographical regions (West Coast, Rockies, Midwest/Intermountain Region, Midwest and Great Plains, Southeast and Gulf Coasts, New England, and Mid-Atlantic).

1. The Pacific Northwest frequently experiences intense snowstorms from strong extratropical cyclones emerging off the Pacific Ocean, with heavy, wet snow falling in the Cascades and Sierra Nevada.
2. The Intermountain Region (including the Wasatch Range and Tetons) receives occasional, heavy snowstorms from Pacific cyclones, but the moisture-depleted air produces a much drier, fluffier snow.
3. The Rocky Mountain Front Range experiences periodically intense snowfalls, localized to the Denver-Boulder-Fort Collins area, due to the effects of elevated terrain.
4. The Rocky Mountains tend to funnel arctic air masses originating over Canada and created beneath strong anticyclones, southward over the Northern Plains. High winds generated by the anticyclone and associated low-pressure system (Arctic Clipper) create a corridor of blizzard activity over the northern Plains states.
5. Lake effect snow develops in localized regions on the lee (southeastern) shores of the Great Lakes, while Nor'easters (coastal lows) deliver occasionally crippling snowfall to the Northeastern Megalopolis.

LO2 Discuss how heavy snow and ice develop within wintertime storm clouds, including the sources of subfreezing air, the effect of layered air temperature, and the importance of the rain-snow line.

1. Air cold enough to manufacture snow is the most common limiting factor in wintertime storm systems. A cold air mass can be pulled into a low-pressure region, generated by vigorous uplift within the cloud layer (due to jet stream processes), or manufactured by sublimation (phase change of snowflakes to water vapor, in marginally cold air) beneath the cloud layer.
2. Alternating cold and warm air layers between the cloud base and ground create different types of wintry precipitation, including snow, sleet, freezing rain, and plain rain.
3. The rain-snow line is a characteristically narrow zone over which precipitation type changes, due to vertical temperature layering. The position and evolution of this line pose key challenges to winter weather forecasts.

LO3 Explain how extratropical cyclones generate regions of heavy snow and the manner in which geography modifies snow patterns along the Rockies and East Coast.

1. Heavy snowstorms along the Rocky Mountain Front Range are generated by two processes: (1) an anticyclone over the Northern Plains that traps a shallow, cold, dense air layer against the mountains (cold air damming) and (2) a so-called Four Corners Low that circulates milder, moist Gulf air north and west, overrunning the cold air layer.
2. East Coast snowstorms producing heavy snow are created by coastal lows or Nor'easters, with myriad interacting processes that enhance snow rates, including cold air damming; phasing jet streams; formation of nearly stationary, heavy snow bands; and coastal fronts.

LO4 Discuss the seasonal and geographical distribution of lake effect snow and the key processes giving rise to these localized, heavy snows.

1. Highly localized and extreme snow amounts characterize the lee shores of the Great Lakes during late fall through midwinter.

2 Lake effect snow bands develop when an arctic air mass advances across the relatively warm water surface. There is a large transfer of heat and moisture into the overlying cold air, which destabilizes the air, causing heavy snow squalls.

3 Lake effect snow is sensitive to fetch (the total distance that cold air travels over warm water), wind direction, and the temperature difference between the water and overlying air.

LO5 Describe how severe ice storms form and the types of damage they cause.

1 Ice storms including the production of heavy amounts of sleet and/or freezing rain. Freezing rain that accretes onto tree branches and utility lines is a highly destructive type of winter weather hazard. The severity of damage scales directly with thickness of ice accumulation and wind speed.

2 Highly damaging ice storms typify the Mid-Atlantic region, where cold air damming against the Appalachians is overrun by mild Atlantic air, although ice storms can develop anywhere that a stationary front causes mild air to ascend over a freezing air mass.

3 Many ice storms are limited-duration events because, when liquid water freezes, latent heat begins to warm the ground surface.

LO6 Explain the challenges of forecasting heavy snow and ice storms.

1 Numerical weather prediction models are the mainstay of winter weather forecasting, with the emphasis on storm track. Different models have characteristic biases that must be learned through years of forecasting experience in particular regions.

2 While the public tends to focus on snow accumulation potential, other important forecast elements include phase transitions among snow, ice, and rain; the liquid equivalent ratio of snow (10:1 is the most common, but the ratio can vary from 5:1 to more than 20:1); and corollary hazards that maximize the severity of a winter storm, including high wind, wind chill, and whiteout conditions.

3 The Northeast Snow Intensity Scale (NESIS) was developed at NOAA to rate snowstorms in terms of their societal impact. NESIS assigns an intensity category (1–5) to each storm, based on the geographical overlap between snow depth and population density.

References

Kocin, P.J. and L.W. Uccellini, 2004a. *Northeast Snowstorms. Volume 1: Overview*. Boston, MA: American Meteorological Society.

Kocin, P.J. and L.W. Uccellini, 2004b. *Northeast Snowstorms. Volume 2: The Cases*. Boston, MA: American Meteorological Society.

CHAPTER 8

Landfalling Hurricanes: Coastal and Inland Devastation

Learning Objectives

1. Understand the processes by which hurricanes weaken over land. What set of hazards attend hurricanes making landfall along the immediate coast? What hazards unfold once the hurricane is deep inland?
2. Describe the formation of the storm surge, its asymmetry, and the manner in which it is compounded by tidal variations.
3. Distinguish between maximum sustained wind and gust factor. Describe how structures succumb to aerodynamic forces in high wind. Explain the rate at which hurricane winds decay once the storm is inland, and how the areal coverage of damaging wind changes with time. Identify several factors leading to concentrated pockets of extreme wind damage.
4. Explain why hurricanes are capable of exceptionally heavy rainfall, in terms of storm motion, cell training, interactions with mid-latitude weather systems, and orography. What two scenarios explain how heavy rain is distributed with respect to storm track?
5. Define and give examples of the following: post-tropical cyclone, tropical remnant, extratropical transition, hybrid cyclone, rejuvenated storm. Discuss how inland flooding and tornadic activity is sometimes enhanced inland of the coast.

Introduction

It was a storm of epic proportions, striking very late during 2012's Atlantic hurricane season. On October 29, Superstorm Sandy began to ravage the mid-Atlantic, Northeast, and Ohio Valley (Figure 8.1). Lower Manhattan was submerged by a 4 m (14-foot) storm tide, and a massive power outage darkened much of New York City. Jumbled heaps of debris, sodden with mud and mold, were all that remained of beachfront homes in dozens of coastal towns along the New Jersey and Long Island shores. Fires sprouted out of dark flood waters, fueled by ruptured natural gas mains. Nearly 9 million power outages enveloped a massive region bounded by Virginia, Ontario, Ohio, and Michigan. Hundreds of displaced residents remained homeless as winter arrived. In all, 24 states experienced direct impacts from this single, incredibly destructive hurricane (Figure 8.2).

In the warm waters south of Cuba, Sandy quickly incubated into a strong Category 2 hurricane. The storm moved north, toward the East Coast of the United States, and underwent a remarkable metamorphosis. It became part Nor'easter, part tropical cyclone. The system expanded to three times its initial diameter, becoming the largest Atlantic cyclone ever observed. Approaching land, Sandy began to tap an additional energy source, a strong vortex in the mid-latitude jet stream. The vortex propelled an unseasonably cold air mass into Sandy, which wrapped around the storm's tropical core. Even more remarkably, this massive hybrid-type storm made an unprecedented turn toward the west, plowing straight into the New Jersey shore.

We'll revisit Hurricane Sandy's impacts in the "Amazing Storms" box at the end of this chapter. In the meantime, our primary objective is to describe the physical impacts of tropical cyclones during landfall. In this chapter, we also describe why some tropical cyclones remain capable of producing violent weather days after landfall, even those remnants that have traveled hundreds of miles inland. Some of the greatest U.S. weather disasters in history have arisen from this special category of inland storm that transitions from a purely tropical to an extratropical system.

Hurricanes, an intense type of tropical cyclone, are the deadliest and costliest natural disaster facing the United States. Hurricanes Katrina (2005) and Sandy (2012) are the two costliest natural disasters to ever strike the United States. And the nation may be more vulnerable than ever to future hurricane catastrophes. The U.S. coastline, from Texas to Maine, has undergone phenomenal

DOI: 10.4324/9781003344988-10

Figure 8.1 Hurricane Sandy viewed from satellite, October 28, 2012. Note the enormous size of the vortex. The hurricane was in the process of morphing into an extratropical cyclone prior to landfall over New Jersey. (NASA.)

growth of infrastructure, property, and population. We have entered the era of hurricanes that do $100 billion in damage. Hurricane Katrina's (2005) death toll exceeded 1800, a loss of life not seen since the Galveston Hurricane more than a century earlier. More than ever, it is important to cultivate a scientifically accurate understanding of hurricane impacts. By reading this chapter, you will take a giant step toward gaining this crucial understanding.

Hurricanes Weaken in a Characteristic Manner Moving Into Higher Latitudes and Over Land, While Producing Several Mechanisms of Coastal and Inland Damage

We have discussed hurricanes at several points in this book, but before proceeding let's provide a quick review of the processes and terminology related to hurricanes. Tropical cyclones are large, cyclonic vortices containing a warm, inner core. Winds converge into a tight, counterclockwise spiral as air streams into the central region of low pressure. Two common types of tropical cyclones are tropical storms (winds > 35 kts [40 MPH]) and hurricanes (winds > 65 kts [74 MPH]). A circular eyewall is comprises of a central ring of intense thunderstorms, 10–50 km (6–31 miles) in diameter. The eyewall coincides with an annular (ring-shaped) radius of maximum wind. Heavy bands of rain are

Figure 8.2 Catastrophe visits the East Coast. This areal view graphically portrays the immense devastation wrought by Hurricane Sandy along the New Jersey shoreline. The storm surge completely flooded this township and deposited thick layers of sand throughout streets and yards. (USAF.)

arranged in spiral arms surrounding the eyewall. The strongest winds are found in the storm's right hemisphere or semicircle (with respect to storm storm track). As the storm approaches land, winds on the right side push a massive mound of seawater toward shore, called the storm surge.

In the United States, hurricanes are ranked according to the Saffir–Simpson Intensity Scale, which is based on the maximum sustained wind (MSW, a 1 minute averaged wind near the surface). Category 1 hurricanes have winds exceeding 65 kts (74 MPH), while Cat 5 storms have winds that exceed 135 kts (155 MPH).

Hurricanes Weaken Over Land Not Because of Friction but Loss of Their Oceanic Heat Source

Hurricanes extract huge amounts of heat energy from warm ocean water, through the process of evaporation and the transfer of sensible heat. Evaporation is critical since the formation of water extracts ocean heat.

When the vapor later condenses in eyewall thunderclouds, the heat is released into the atmosphere – powering the hurricane.

As a hurricane moves onto land, these energy sources are left behind. Many people incorrectly assume that hurricanes weaken as a result of increased land friction, which slows the wind. While this phenomenon does occur in the lowest few thousand feet above the ground, there is a more important reason why hurricanes weaken. The storm's deeper vortex, which extends to 50,000 or more feet – its "heat engine" – begins running low on fuel. Without continued supply of latent and sensible heat derived from the ocean, the giant convective clouds in the eyewall diminish. Essentially, the storm's deep warm core rapidly cools.

Additionally, as the hurricane approaches land, dry air masses from the interior continent get swept into the storm. Dry air evaporates thunderstorm clouds in the eyewall and rain bands, creating cold downdrafts near the storm center. These downdrafts suppress upward motions that sustain cloud updrafts. A hurricane moving northward into the mid-latitudes also begins to encounter increasingly stiff westerly winds. This wind shear, or change in wind direction and/or speed with altitude, disrupts the vertical integrity of the eyewall and vortex. The top portion of the storm may literally become displaced (sheared) from the bottom half, leading to rapid weakening.

Finally, a hurricane moving north toward the U.S. mainland may encounter cooler ocean temperatures, a feature of higher latitudes. In these cooler waters, the transfer of latent and sensible heat is greatly diminished, which can reduce the hurricane's intensity by one or two categories in the hours just before landfall.

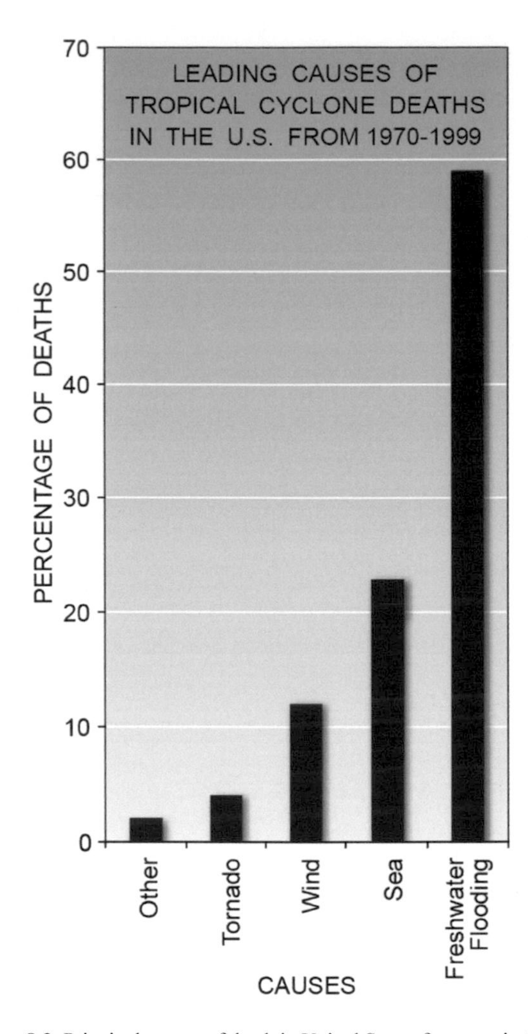

Figure 8.3 Principal causes of death in United States from tropical cyclones (both tropical storms and hurricanes). The overwhelming majority of these deaths come about from freshwater flooding, produced by heavy rain. Outside the United States, storm surge is the predominant cause of death from tropical cyclones.

A hurricane unleashes an arsenal of destructive elements along the immediate coastline and sometimes farther inland.

Hurricanes unleash a powerful triad of damaging effects (sometimes referred to as the "deadly triad") along the immediate shoreline: storm surge, sustained high winds, and heavy rainfall. Figure 8.3 compares the causes of U.S. hurricane fatalities between 1970 and 1999, both along the coast and farther inland. The number one killer is water, and in many storms, this is often freshwater flooding caused by heavy rains. In many parts of the world, storm surge is often the number one cause of death in these storms. For instance, individual storms in the Indian Ocean have drowned nearly 450,000 people in the space of a few hours.

Farther inland, the remnants of hurricanes can remain strong and devastating, sometimes days after the storm makes landfall.

Most often, the lingering effects occur because the remnant vortex combines with a mid-latitude weather system, such as a front, a low-pressure center, or a disturbance in the jet stream.

In all of the cases, the storm acquires a new source of energy. While storm winds typically weaken by 50% or more in the first 24 hours after landfall, on rare occasions hurricane-force winds have been maintained nearly 800 km (500 miles) inland. More commonly, freshwater flooding continues to be a major concern and can even intensify when tropical remnants interact with mountainous terrain such as the Appalachians. Flash floods, extensive river flooding, mudslides, and debris flows are all consequences of this excessive, persistent rain. Finally, swarms of tornadoes sometimes break out ahead of and to the right of the storm's center, even as the tropical system weakens over land! In the sections that follow, we explore the manner in which hurricanes create all of these devastating effects.

One of the Most Deadly and Destructive Elements in a Landfalling Hurricane Is the Storm Surge

The effects of a landfalling hurricane are all too familiar, particularly in the past decade when the United States was dealt a severe double-blow (no pun intended) by Hurricanes Katrina (2005) and Sandy (2012). A massive storm tide (a more precise term describing the surge) over 7.5 m (25 feet) high swamped the Louisiana and Mississippi coasts during Katrina, and a tide of up to 4 m (14 feet) inundated densely populated regions of New York City and northern New Jersey during Sandy. In the section that follows, we describe how the storm tide develops and the manner in which it becomes amplified by a region's specific geography.

Meteorologists and Hydrologists Make a Careful Distinction Between Storm Surge and Storm Tide

Historically, the Galveston Hurricane's (1900) storm tide led to the greatest loss of life of any U.S. hurricane. Estimates range from 6000 to 12,000 deaths. It was likely a Cat 4 (126 kts [145 MPH] sustained winds) at landfall. In fact, the Galveston hurricane was the deadliest natural disaster ever to strike the United States. The city of Galveston, built on a sandbar in shallow Gulf of Mexico water, was extremely susceptible to inundation by storm surge. Lives were lost when people drowned in saltwater, were crushed by a wave of debris swept inland, or trapped under mounds of debris left in the wake of the surge.

Figure 8.4 illustrates the manner in which the sea inundates land during a hurricane. The bottom of the diagram shows mean (average) sea level and the normal tidal range. In the hours before landfall, wind-whipped waves relentlessly pound the shore, gradually increasing in height. But as the hurricane's eyewall moves onshore, a much greater mound of water, the storm tide, swiftly overwhelms the shoreline. The tide does not crash onto the

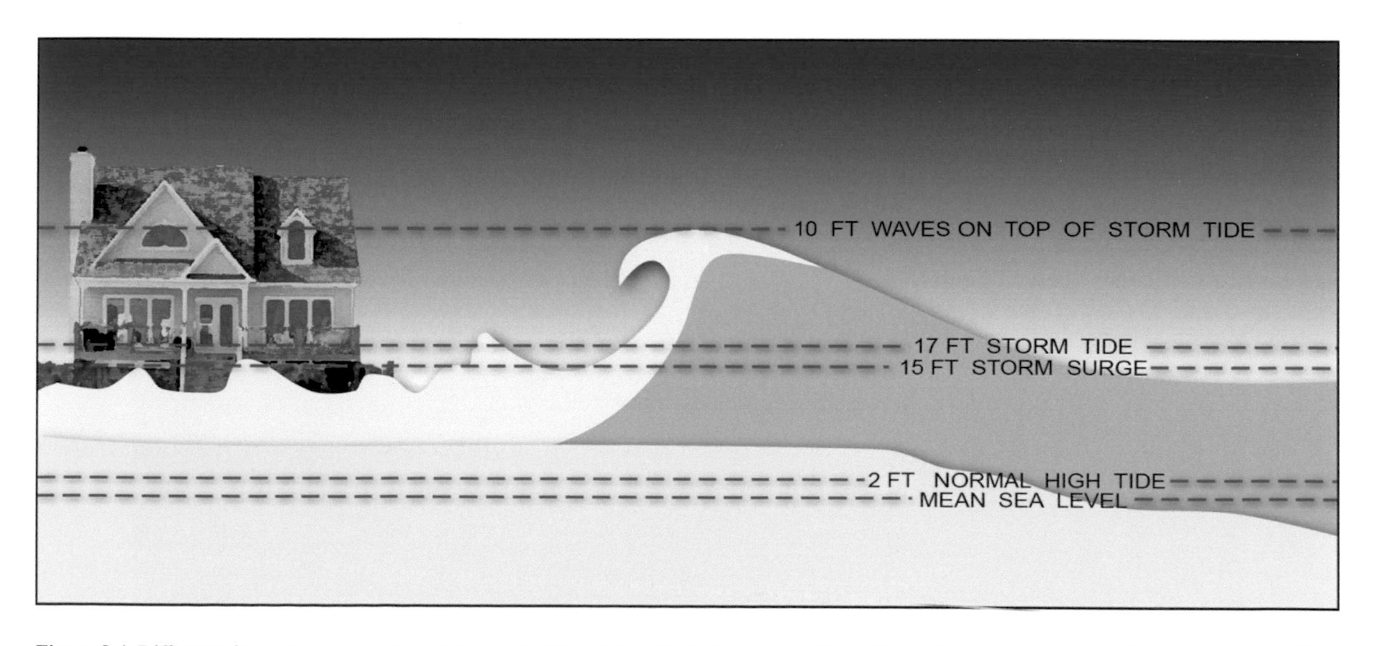

Figure 8.4 Difference between storm surge and storm tide. Storm tide reflects the superposition of lunar tide and storm surge. These two agents build a mound of seawater above mean sea level. The storm tide maintains its elevated form over many hours as a hurricane approaches shore. Individual, transient waves – driven by wind gusts – add further height to the water level.

shore in the manner of an enormous wave; rather, think of it as a swiftly flowing current backed by many miles of high water.

A storm surge will coincide with a phase of astronomical low or high tide. If the surge arrives during high tide, the water flowing inland rises further. It's bad luck when this happens. The sum of the storm surge and astronomical low or high tide is termed the storm tide. The distinction between storm tide and storm surge is not often made in media reports, leading to confusion. For instance, during Hurricane Sandy's assault on New York City, a total storm tide of 4 m (14 feet), recorded at The Battery, reflected a 1.5 m (5-foot) high tide combined with a 9-foot storm surge. On top of the storm tide, which is sustained for several hours, individual wind-generated waves form and dissipate on short timescales. The inland reach of these waves is significantly extended by the elevated ocean water surface.

The sea has great destructive potential. Seawater has a density close to 1000 kg/m^3 (roughly 2200 pounds/yard^3). As the storm tide flows inland at tens of knots, think of the enormous energy involved. Very few structures, short of a steel-reinforced concrete bunker, can stand up to this type of tremendous force. Deep vertical pilings used to anchor structures, such as piers and homes, are undermined by the swirling force of rapidly flowing water. When pilings lose their sandy support, all but the most deeply anchored structures fail.

It's not just water that causes damage. The debris transported by the water compounds the surge's battering effect. Pieces of homes and buildings, including foundation materials, lumber, tree trunks, cars, and boats, mix into and are flushed inland in turbulent storm tide waters. At times, the storm tide builds into an enormous battering ram of churning water, beach sand, and solid objects. Few structures survive this assault. People caught in the surging water experience great bodily trauma, are dragged and pummeled by turbulent waters, become pinned and entrapped by debris, and invariably succumb to saltwater drowning.

Barrier Islands and Beach Heads Are Particularly Susceptible to the Storm Tide

Barrier islands are quite vulnerable to the effects of storm tide. Storm tides can breach these low-lying, narrow strips of sand and vegetation completely in one or more locations. New inlets or channels can be created where a barrier island is close to the mainland. Sinepuxent Inlet, south of Ocean City, Maryland, was created in this manner by the Chesapeake and Potomac hurricane of 1933 (Figure 8.5). Additionally, barrier islands can

Figure 8.5 Hurricanes have the power to completely alter landforms. A powerful hurricane in 1933 created this inlet on the south end of Ocean City, Maryland. (Jane Thomas, Integration and Application Network, University of Maryland Center for Environmental Science.)

be very difficult to evacuate swiftly, particularly when they are choked by large numbers of summertime tourists. Often only one or two causeways connect these densely populated strips with the mainland. If evacuation is not done with sufficient lead time and in stages, pandemonium and gridlock ensue.

Along mainland beaches, wave action removes sand from dunes protecting inland areas from inundation. The effective narrowing of beaches by the cumulative action of intense storms is an ongoing challenge faced by municipalities wishing to preserve their shoreline. Countless millions of dollars are spent each year to replenish the sand lost to storms, either by importing sand or dredging it from offshore. Human alteration of the coastline has exacerbated the damage caused by storm tide. In some locations, dunes are removed or their vegetation removed to facilitate beach access or shoreline view and to build beachfront property. Dunes and coastal marshlands provide a natural barrier against the onslaught of storm tide and greatly diminish wave energy.

Storm Surge Is Created as the Swirling Mound of Water Beneath a Hurricane's Eye Encounters Shallow Coastal Water

How is a storm surge created? The surge is highly asymmetric, occurring just in the hurricane's right-front quadrant (Figure 8.6). On this side of the storm, sustained winds constantly push water toward the shore (curve C). On the storm's left side, winds push water away from the shoreline, out to sea, causing a coastal recession. These effects are strongest at the storm's radius of maximum wind (RMW), as shown on the diagram. Two other important factors contribute to the total surge height (D). The first is the oft-quoted inverted barometer effect or upward suction from low pressure in the storm's eye (A). This effect does cause the sea surface to dome slightly upward. The effect is symmetric across the eye and largely within the RMW. However, it does not account for more than about 10–20% of total surge height; in the worst-case scenario of a 9 m (30 foot) surge, about 1–2 m (3–6 feet) can be attributed to this inverted barometer effect.

The second process, shown by letter B, is a dynamic process that occurs only when the hurricane moves into shallow water. The details are shown in Figure 8.7. In the top panel, a hurricane over deep ocean generates an inward spiral of wind. This wind energy sets the underlying water in motion, creating an inward spiraling water column perhaps 100 m (330 feet) deep. As the water converges inward, it is forced to descend. In deep water, the sinking current can flow away at depth, spreading away from the storm.

In the bottom panel, suppose now the hurricane moves over shallow seafloor. The descending, swirling water column can no longer flow away because it is restricted by friction along the sea floor. But winds continuously push the water inward toward the storm's center. Consequently, a mound of water rapidly builds upward. This dynamic process only operates as the storm nears land, as the seafloor rapidly rises toward shore. It is also symmetric and largely confined to the RMW.

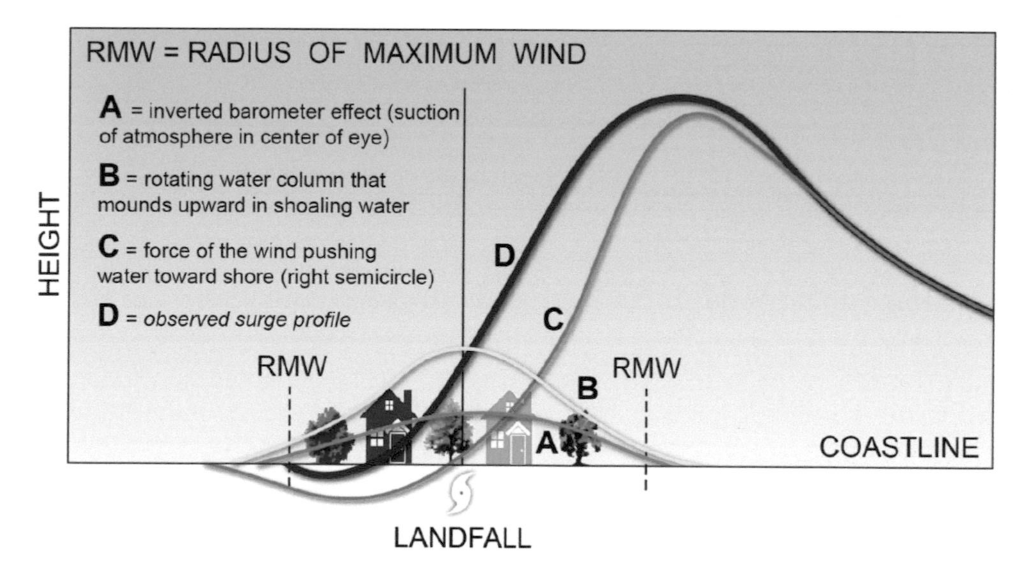

Figure 8.6 A strongly asymmetric storm surge. A large surge is raised to the right of the storm's center (the hurricane is shown approaching the shoreline, as if moving from the reader toward the page). Multiple processes contribute to the total surge height (D) – including the force of the wind (C), the "inverted barometer" effect (A), and a dynamic process due to the swirling column of ocean water as it enters shallow water (B). (Adapted from Simpson and Riehl, 1981.)

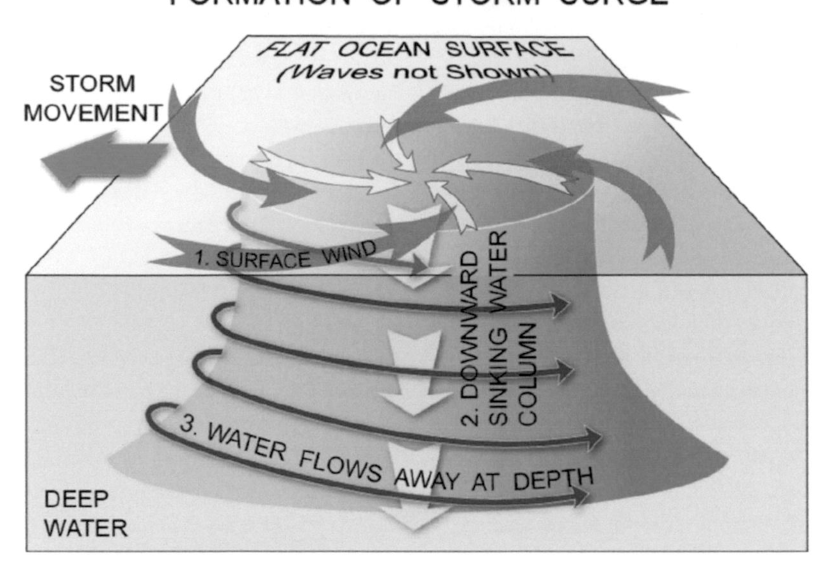

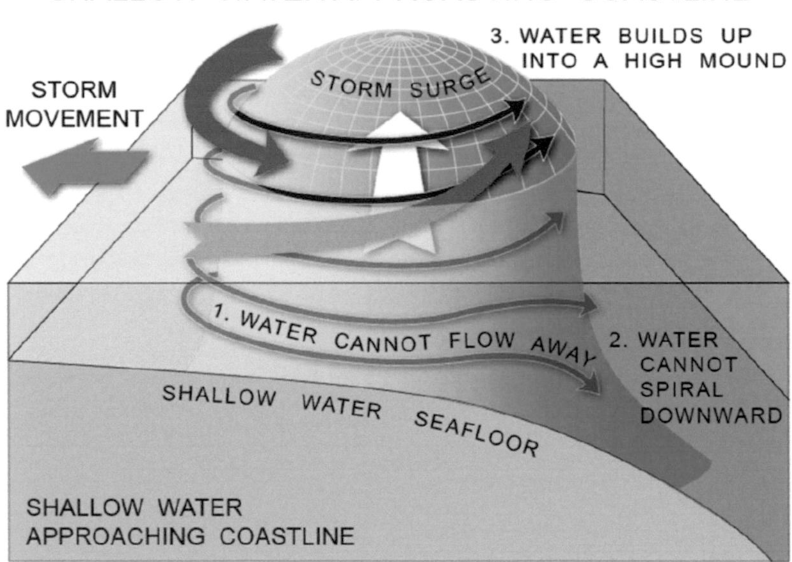

Figure 8.7 Dynamic process giving rise to storm surge. This involves a rotating water column beneath the eye. In open ocean, at great depth, the water can spiral away. When water becomes shallow, outflow at its bottom is restricted, causing the water spiral to build upward. (Adapted from Simpson and Riehl, 1981.)

Inlets and Bays Locally Amplify the Storm Tide

Hurricane Carla was a Category 5 storm that made landfall along the Texas coast in September 1961 (Figure 8.8[a]). The coastal geometry, including the location and size of inlets, bays, and sounds, greatly influenced local surge heights. As Figure 8.8[b] shows, the storm came inland near Corpus Christi, in the vicinity of Porto Lavaca. Surge heights to the left of track were in the 1–3 m (4–8 feet) range, while those on the right side exceeded 3 m (10 feet). This is a general reflection of the storm surge's cross-track asymmetry. But note that the very highest surge developed within deep coastal indentations, including Porto Lavaca's bay (surge = 6–7 m [20–feet]) and Galveston Bay (surge = 5–6 m [16–18 feet]). When the surge enters these narrow channels and bays, wave energy becomes greatly amplified, causing water levels to rise further, concentrating the worst damage over small regions.

(a)

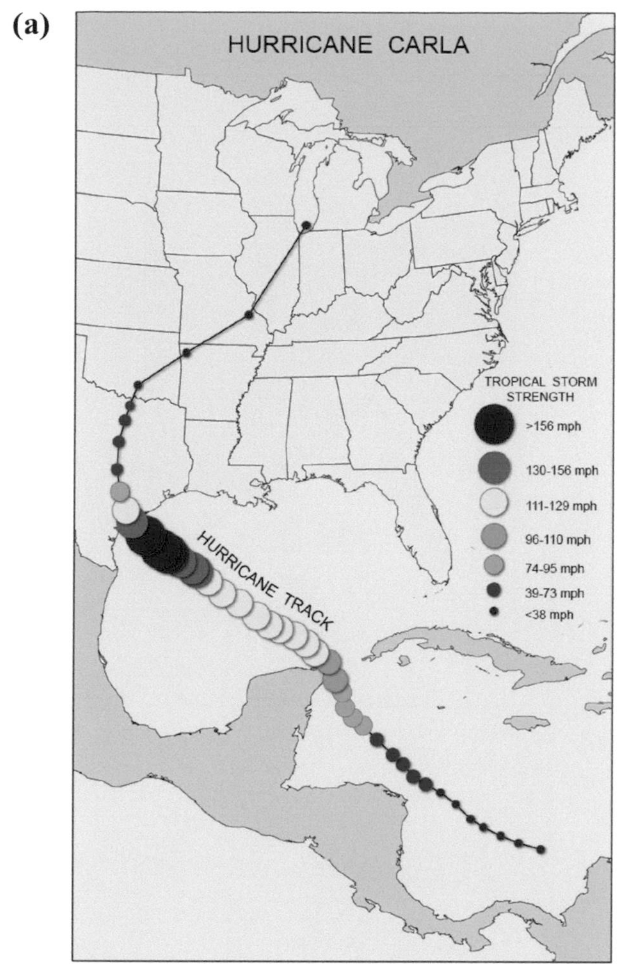

(b)

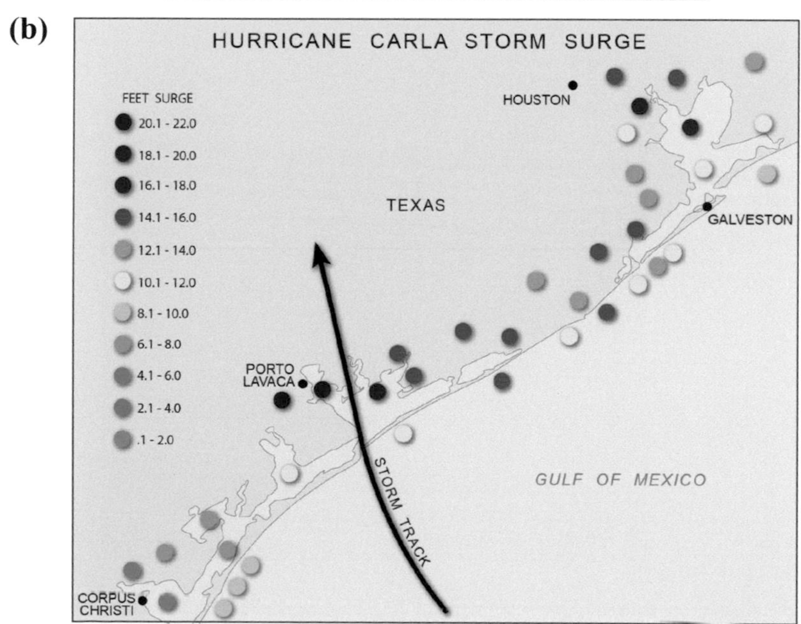

Figure 8.8 (a) Local variations in Hurricane Carla's storm tide. This hurricane, a Category 5 storm in the Gulf of Mexico, struck the Texas coastline as a Category 4 storm. (b) Hurricane Carla raised a tremendously high tide to the right of its track. The highest water levels, up to 6 m (18 feet), developed as the tide became amplified within narrow embayments and inlets along the Texas shoreline. (Adapted from Simpson and Riehl, 1981.)

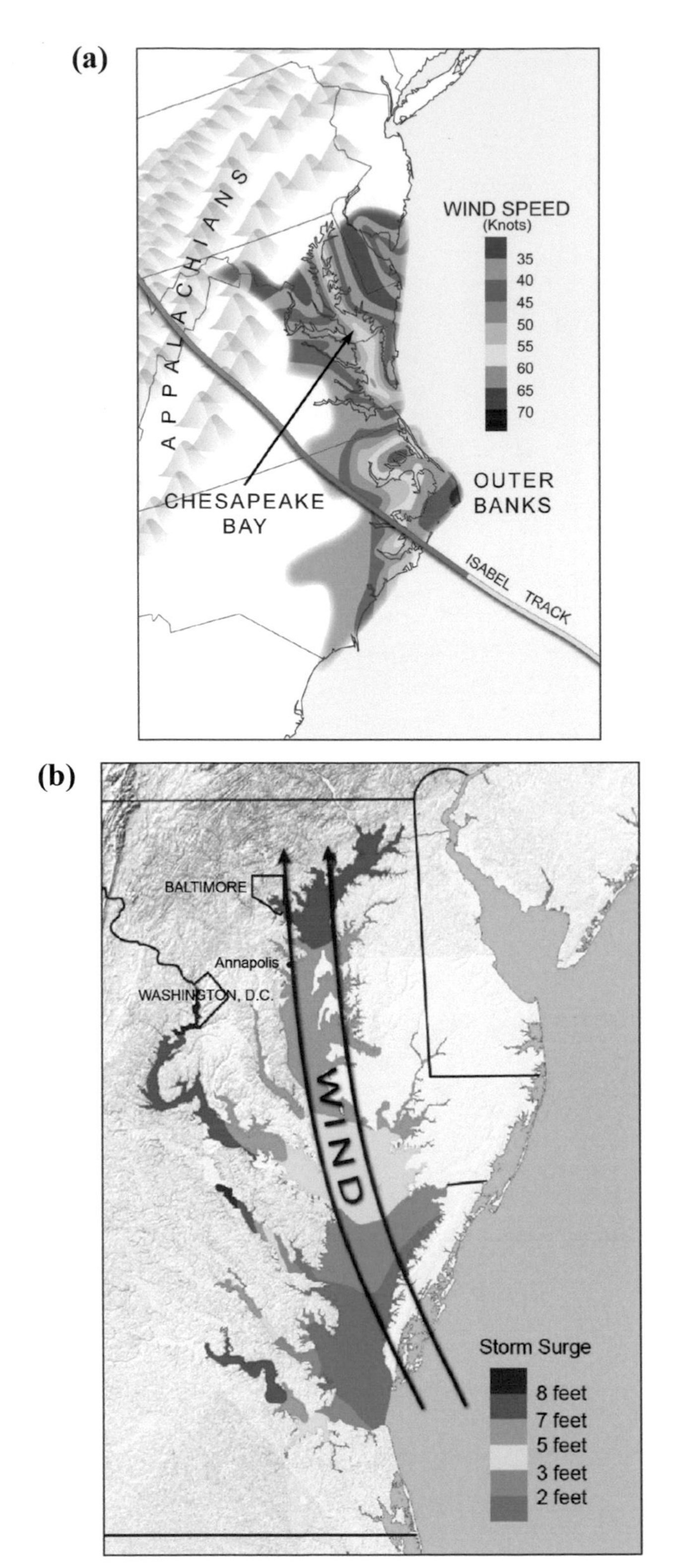

Figure 8.9 (a) Hurricane Isabel's impressive storm tide on the Chesapeake Bay. Hurricane Isabel (2003) took an unusual "inside" track into the Appalachian Mountains instead of recurving along the coastline. As a result, strong winds from the south blew along the long axis of the Chesapeake Bay, pushing water into a surge that moved up the bay toward its northern terminus. (b) An unusual "Bay Surge" in which the highest storm tide developed at the Chesapeake Bay's headwaters near Annapolis and Baltimore, Maryland.

One of the most dramatic examples of surge amplification in a bay occurred in September 2003 after Hurricane Isabel made landfall in the Mid-Atlantic. The Chesapeake Bay is an exceptionally long, straight, and shallow water body extending north from Norfolk, Virginia, to the Susquehanna River, near the Maryland–Pennsylvania border. Isabel followed a highly atypical track; instead of recurving offshore, away from the coastline, the storm barreled straight inland toward the northwest, crossing the Appalachian Mountains into Ohio (Figure 8.9[a]).

The storm reached the North Carolina Outer Banks as a Category 2 storm, then began weakening as it moved inland. The core of the system's highest wind remained south of the Chesapeake Bay, but the winds blew from the east-southeast for several hours, at a sustained speed of 35–43 kts (40–50 MPH). The sustained winds pushed a mound of water straight up the Bay. As the storm tide entered the Bay's narrow northern portion, the tide became amplified, such that the deepest inundation occurred in the Bay's upper reaches (Figure 8.9[b)]. The flood hit Baltimore's Inner Harbor and Annapolis, Maryland, with a storm tide exceeding 2 m (7 feet). Given the proper wind direction, the Chesapeake Bay can act as a giant funnel, concentrating the storm tide in its upper reaches, and even relatively protected regions such as Baltimore and Annapolis – far inland from the Atlantic shoreline – experienced a devastating flood.

This scenario underscores the importance of wind fetch – the distance over which wind blows across the water. The longer the fetch, the greater the duration over which the wind's energy is imparted into the ocean, increasing the height of the waves and storm surge. This duration factor leads to what's termed a fully developed sea. This crucial concept is now embodied in a better understanding of why weak but large hurricanes can develop uncharacteristically large storm tides. This effect is explained in the next section.

Integrated Kinetic Energy (IKE) Is a Better Predictor of Surge Height Than Wind Speed Alone

Recent experience with several landfalling hurricanes suggests that a storm's intensity (maximum sustained wind) is not the only factor in establishing surge height. Historically, large-diameter storms (compared to historical averages) such as Ike (2008) have created storm surge that is significantly higher than one would predict from wind speed alone. Figure 8.10(a) compares the surge raised by Ike at landfall, as a Category 2 storm, vs Hurricane Charley (2004), which struck Florida as a Category 4 system. Charley was a much more compact storm than Ike (Figure 8.10[b]).

Why was Ike's surge so much higher than Charley's surge? An important factor is the total water area over which high winds blow. As more wind energy is transferred into the ocean, the greater the water mound set into motion. The combined effects of wind speed and wind area are captured in a metric called integrated kinetic energy (IKE). In this metric, kinetic energy (proportional to wind speed squared [V^2]) is summed over the total area over which winds are sustained at a threshold speed. The larger the storm's diameter, the greater the fetch; the larger fetch translates into more complete energy transfer between wind and sea, leading to a fully developed sea state – manifest by a higher surge. The IKE of several hurricanes from the past 50 years has been computed over the storm lifetime. By this measure, Hurricane Ike was a much more potent surge generator than Hurricane Charley. Ike's IKE (which is calculated in terajoules) was seven times larger than Charley's, even though Ike's maximum winds at landfall were 25% weaker. Storm diameter, as it turns out, is an important and (until recently) undervalued determinant of a storm's surge-generating potential.

Hurricane Wind Damage Has Several Aspects, Including a Sustained Wind Component and Periodic Gusts to Higher Wind Speed

In this section, we examine damage created by a hurricane's wind. These winds are generated by a large spiral inflow of air, in response to the extreme pressure gradient between the storm's center and its surroundings. In addition to these large-scale winds, various smaller-scale wind processes lead to localized pockets of more extreme wind damage – such that the extent of wind damage reflects considerable horizontal variation.

Major Determinants of Widespread Damage Include Wind Asymmetry Across the Vortex, and the Size of the Vortex at Landfall

Sustained wind speeds in the hurricane vortex increase from the storm's periphery to the radius of maximum wind, just outside the eye. The wind speed increases more rapidly with radial distance as we approach the storm's core. As Figure 8.11 shows, the ring of high-speed wind, which we know as the

(a)

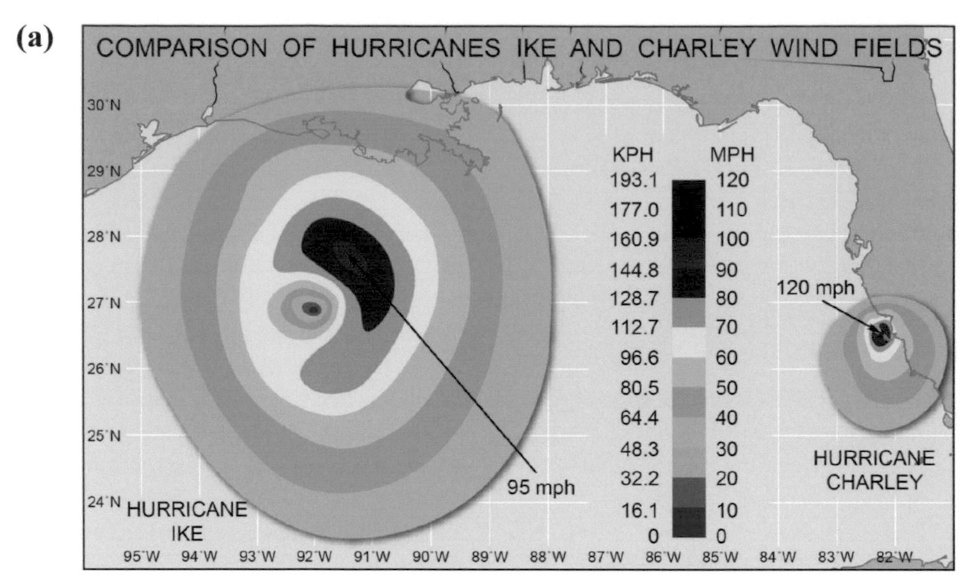

(b)

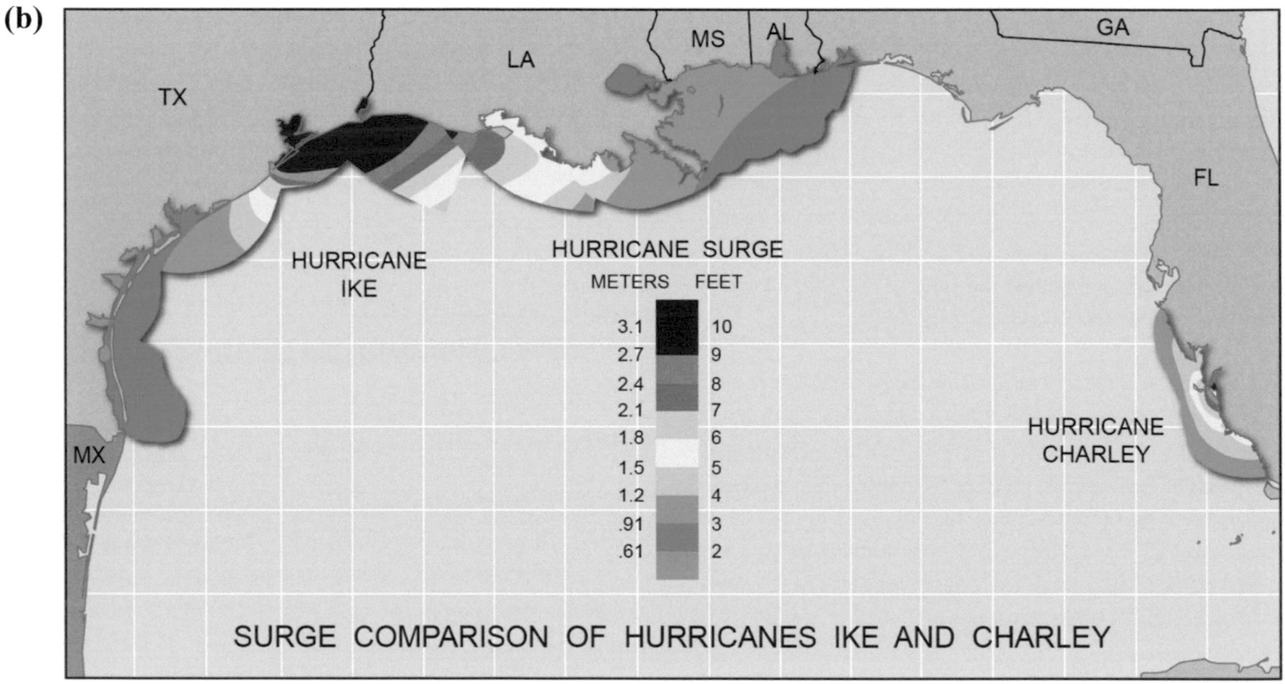

Figure 8.10 (a) Comparison of the wind field between Hurricanes Ike and Charley. Hurricane Ike's vortex dwarfed that of Charley, but Ike was a significantly weaker storm. (b) Not only was the total coastal extent of Ike's storm surge much greater than Charley's, this weaker storm (Ike) paradoxically generated a significantly higher storm surge than Charley did.

radius of maximum wind (RMW), coincides with the eyewall. The graph in Figure 7.11 is called a time series; this type of graph is created when an anemometer records wind speed every few seconds as the storm moves through. Wind speed is plotted in red, wind direction in green. You will note a great deal of gustiness, with large changes in wind speed from minute to minute. But in the background is the maximum sustained wind, which is a kind of average wind speed, shown by the solid black curve. Winds become nearly calm in the center of the hurricane eye. Note how the wind speed increases in inverse relation to the surface pressure (blue curve) – and how rapidly both pressure and wind change very close to the eye.

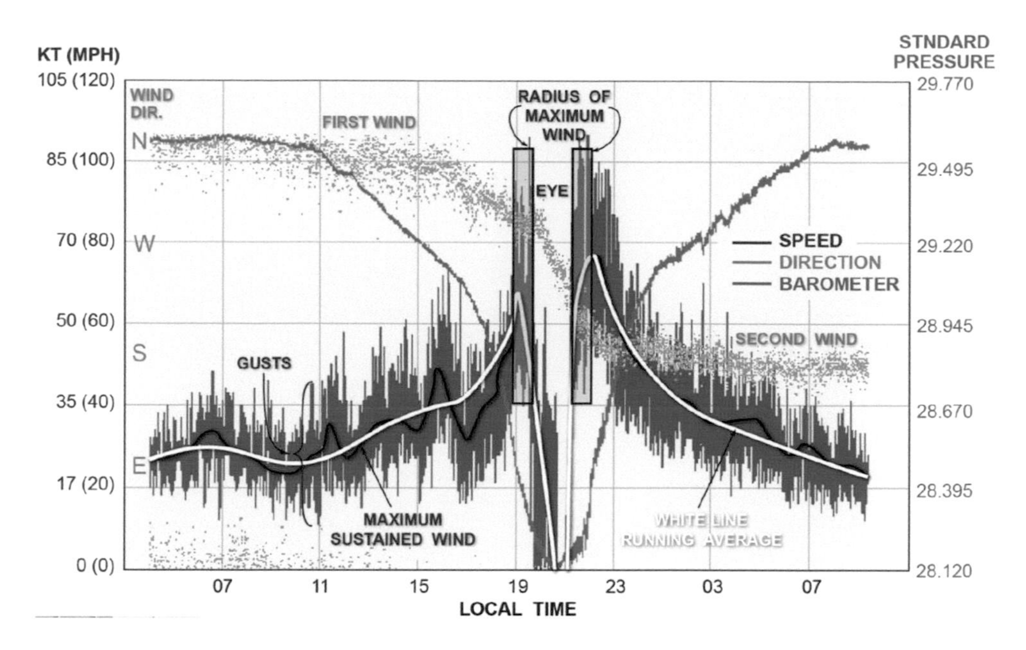

Figure 8.11 Time series tracing of the minute-by-minute variation of a hurricane's wind and pressure, as it moves over a fixed observing site. (Adapted from U.S. Navy Joint Typhoon Warning Center [JTWC].)

The wind field surrounding a hurricane is highly asymmetric, due to the storm's forward motion. Figure 8.12(a) shows a schematic of the wind distribution for a hypothetical hurricane moving northward at 26 kts (30 MPH), with a sustained wind of 87 kts (100 MPH) blowing around the inner core (and generated by the storm's pressure gradient). In a moving storm, the sustained wind is stronger in the right semicircle. This is where the storm's swirling wind and direction of travel are in the same direction and are therefore additive. Adding 87 kts (100 MPH) wind to 26 kts (30 MPH) storm motion (translation) yields 113 kts (130 MPH) in the right semicircle. In the left semicircle, one subtracts the storm motion from the wind, because the storm motion opposes the wind direction. The net wind is 61 kts (70 MPH). The difference in net wind speed, across the storm's core, is 52 kts (60 MPH)! Figure 8.12(b) shows a real-world example for Hurricane Danielle (2000). Across the storm, there is often a substantial difference in the speed between right and left sides (with respect to the storm track), and this difference increases as the storm moves faster. The difference can amount to 43–52 kts (50–60 MPH). Because the storm surge is primarily generated by wind, the surge is also maximized to the right of track.

During landfall, the extent, duration, and severity of wind damage are controlled by three factors: maximum sustained wind (a function of a storm's minimum central pressure), vortex size, and the storm's speed of movement. Hurricane Andrew (1992) was an extremely intense system (Category 5) but also very compact and moving very rapidly. It created a small region of near-total devastation, along its northern eyewall, in Dade County, Florida. Analysis of barometric pressure revealed an amazingly large pressure gradient, approaching 4–5 mb/km, along its northern fringe. Damage was confined to a small area, and the duration of extreme wind was on the order of only 2–3 hours. In this situation, the storm's rapid movement and small size led to brief duration of intense winds over a small region. But the storm's rapid movement increased the cross-vortex wind asymmetry, further increasing the wind speeds to right of track (in the direction north of center). Peak wind gusts exceeded 174 kts (200 MPH), and even well-engineered structures were heavily damaged. These gusts are comparable to a strong tornado! Because of the extreme wind damage, Hurricane Andrew in 1992 became the costliest U.S. natural disaster at that time.

Contrast Andrew with a very large and slower-moving storm such as Hurricane Sandy (2012). Maximum wind gusts at landfall during Sandy were in the 70–77 kt (80–90 MPH) range – half that of Hurricane Andrew. However, the massive size of Sandy's wind field meant that damaging winds were sustained for many hours. Property devastation is also a function of population density and the prevailing type of dwelling. Like Andrew, Sandy struck a very densely populated region with exceedingly high property values. The price tags for Andrew and Sandy (considering the effect of both the wind and surge) ran into to tens of billions of dollars.

(a)

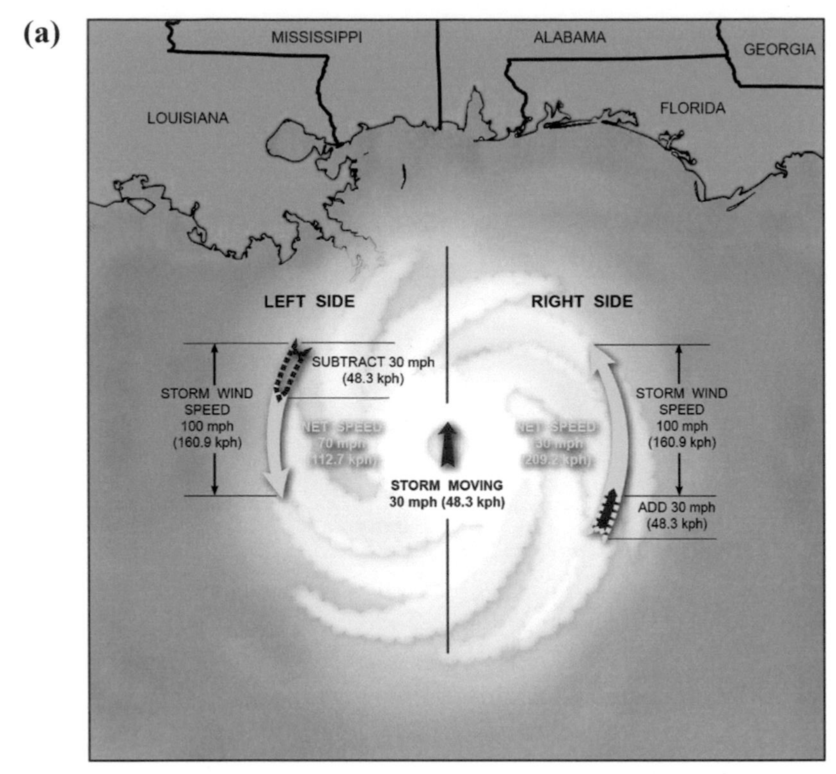

(b)

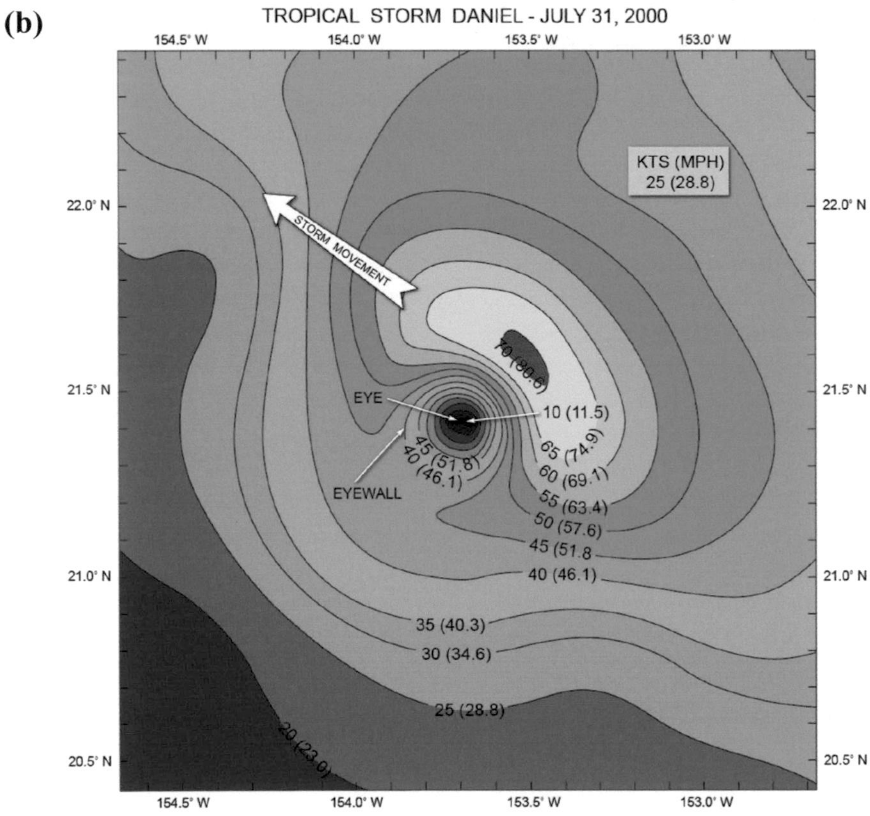

Figure 8.12 (a) How asymmetric sustained winds arise on either side of the hurricane eyewall, due to storm translation. (b) Unequal distribution of sustained surface winds in Hurricane Danielle. This type of wind asymmetry, in which the stronger wind is aligned to the right of the storm center, is typical of a moving hurricane. (NOAA.)

It Is Important to Distinguish Between a Hurricane's Maximum Sustained Wind (MSW) and Gust Factor

Figure 8.11 makes a critical distinction between maximum sustained wind (MSW – solid black line) and hurricane wind gusts (irregular red tracing). The MSW is a 1-minute average wind near the surface; it is the standard by which hurricane intensity is assessed in the United States. Superimposed on the sustained wind are instantaneous wind gusts, or brief excursions of wind speed above and below the MSW. Wind gusts are created by two processes: (1) the turbulent motion of the airflow as it interacts with the surface and (2) internal instability within the airflow. Think of eddies and ripples in a flowing stream – these are analogous to gusts, while the steady current is analogous to the maximum sustained wind.

While the MSW is an averaged value, the gust factor describes the instantaneous variation of wind, with a timescale of seconds to tens of seconds, expressed as a percentage wind speed beyond the MSW. Air turbulence and gust factor are closely related to the degree of surface roughness. Over open ocean, gusts commonly exceed the MSW by 20–30%. While we commonly regard the ocean surface as smooth, bear in mind that it becomes deeply undulating in the presence of storm waves. The waves disrupt the wind field, breaking it into numerous swirls and eddies. Land surface, in contrast, features forest cover, suburban sprawl, urban landscape, and uneven terrain. This type of surface creates greater turbulence than over disturbed ocean, and gust factor rises to 40–50% in excess of MSW. In mountainous terrain, gust factors approaching 100% have been demonstrated. The next section explains why knowledge of the expected gust factor, in addition to the MSW, are critical design elements in building structures to withstand high wind.

Wind Interacts With Structures in a Complex Manner, Setting up Powerful Aerodynamic Forces

The sustained winds exert a force on structures, such as the walls of dwellings, proportional to wind speed cubed (V^2) multiplied by surface area. For the same area, if sustained wind doubles (as it does between a Category 1 and Category 5 hurricane), the damage forces increase by a factor of 4! That's why seemingly small increases in hurricane wind speed can have a disproportionate impact on the amount of wind force and thus damage delivered.

Wind-tunnel studies have demonstrated that homes with gabled (pitched) roofs generate aerodynamic lifting forces as high winds flow over them (Figure 8.13), similar to way that lift is generated when air flows over a curved airplane wing. If the roof is not securely anchored to the walls, then sections of the roof or the entire roof may peel away. With the roof gone, walls lose structural support and collapse outward. This process is enhanced by regions of low pressure that form along the sides of the house as wind streams by – a type of suction force acting to pull the side walls and downwind wall outward. The integrity of walls is also compromised by airborne debris – essentially airborne missiles. Walls are more easily breached in their weaker areas, including large panel windows and garage doors. Eaves or roof overhangs tend to "catch" the wind, causing high pressure to build beneath the roof edges. These forces act in concert with the upward sucking force along the roof, leading to its destruction.

Structural failure is greatly enhanced by transient, buffeting forces that arise from wind gusts. The repeated, on-off "hammering" of a structure by the wind can create a process

Figure 8.13 Schematic of a structure's vulnerability to the wind and resulting aerodynamic forces that lift off the roof and pull the sidewalls outward.

called resonance. During resonance, the structure's natural frequency of vibration is excited by rapid, cyclic variations in wind speed. The structural vibrations rapidly amplify, to the point where the structure literally shakes itself apart. (A simple analogy to resonance occurs when an adult pushes a child on a swing, with each push precisely timed to the arrival of the child.)

Another important damage process is related to the shift in wind direction, as shown in Figure 8.11. The initial hurricane wind to be experienced, as the eye approaches, is called the first wind. As the eye crosses, the wind direction must shift by 180°. The resurgence in wind speed, on the other side of the eye, is termed the second wind. This simple distinction is more than academic. The first wind places high wind loads on structures for several hours. Once the eye passes, extreme forces ramp up for several more hours, but from the opposite direction. The back-and-forth wind loading contributes to structural failure.

Finally, it is worth noting the fate of trees in a sustained high wind. Much of the damage to dwellings and vehicles comes from falling limbs or uprooted trees. Most utility damage, which can lead to multiday power outages, arises from limbs and trees falling on power lines. Removing tree debris from roads delays initial utility access and repair. Soils anchor tree root systems, but these root systems often become saturated and loosened in the heavy rain that accompanies hurricanes, making trees more susceptible to windfall. Furthermore, many hurricanes strike during the late summer and early fall, when trees are in full foliage. Each individual leaf acts as a tiny "sail" that catches the wind, exponentially multiplying the total wind force in the tree crown. For all these reasons, widespread loss of tree cover and extensive power disruptions are common outcomes of hurricane winds at landfall.

As a Hurricane Weakens, Its Winds Decay at an Exponential Rate, But the Wind Field Often Expands

As a hurricane circulation crosses from open ocean onto land, it is tempting to conclude that the surface wind decreases due to increased surface friction. While this is true in lowest few thousand feet of the vortex, surface friction is not "felt" by the vortex winds at higher altitudes. (A useful rule of thumb is that the maximum surface gust on the inland side of a landfalling hurricane approximately equals the maximum sustained wind in that part of the storm still over the ocean.) But the wholesale weakening of the deep vortex is a different matter. As the hurricane "engine" begins running out of fuel at landfall, the sustained wind through the deep vortex spins down in a predictable manner. Figure 8.14 shows the rate at which winds

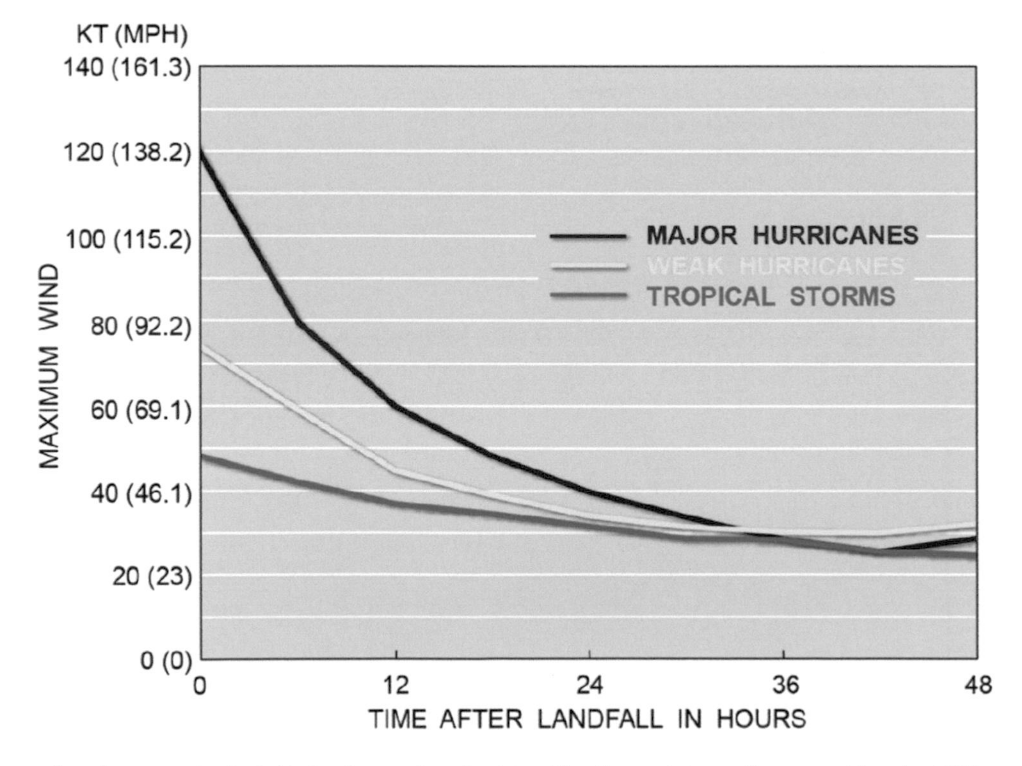

Figure 8.14 Decay of maximum sustained wind in hurricanes of varying intensities. For most storms, the process takes about 24 hours for winds to diminish below tropical storm strength. (Adapted from Emanuel, 2005.)

(a)

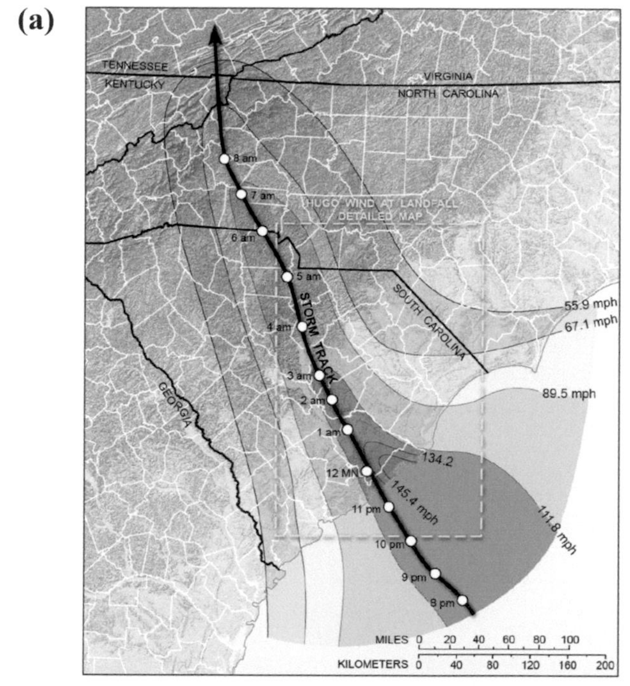

(b)

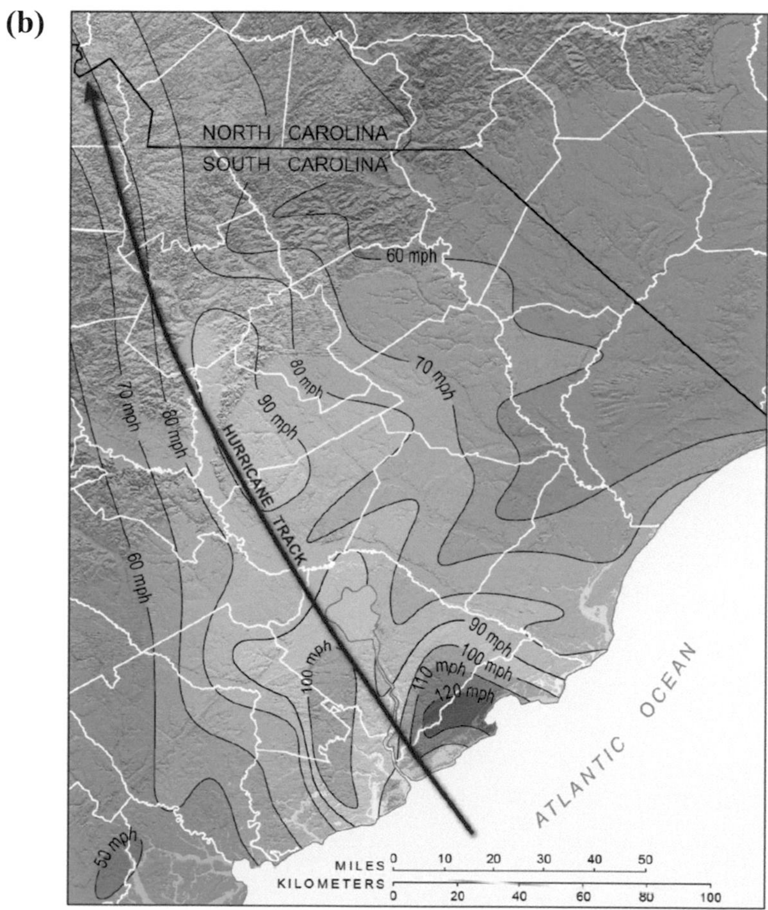

Figure 8.15 (a) Map showing general weakening of Hurricane Hugo's winds at landfall. Weakening became especially rapid once the storm vortex began interacting with elevated terrain. (Adapted from Powell et al., 1991.) (b) Detailed wind analysis during Hurricane Hugo's landfall over South Carolina. The strongest winds, as expected, are found to the right of track, but quickly decay inland. (Adapted from Fujita, 1992.)

decay, for storms of varying intensity. This decay is often exponential, meaning the initial rate of decay is very rapid, followed by a progressively slower weakening. In the first 12 hours, stronger hurricanes experience larger wind reductions than weaker systems. At the end of the first 12 hours, all categories have weakened to 50% of their pre-landfall value. By 36 hours, weakening is nearly complete.

An example of Hurricane Hugo's (1989) wind decay is shown in Figures 8.15(a) and (b) as the storm crossed into South Carolina. Hugo made landfall as a powerful, compact Category 4 storm near Charleston. Shown in the diagram are contours of maximum sustained wind. Following the track inland, note the small, intense region of 104+ kts (120+ MPH) winds to the right of track, along the immediate coast. Winds rapidly diminished as the storm moved inland. Some 80 km (50 miles) inland, sustained winds in the right semicircle were down to 78 kts (90 MPH), with minimal hurricane intensity (65 kts [75 MPH]) on the left side. After another 80 km (50 miles), the post-tropical system encountered the Appalachian Mountains, which disrupted and weakened the winds further.

The rapid "spin-down" of the vortex occurs as the entire wind field expands. Hurricane Irene (2011) illustrates this common behavior (Figure 8.16). Irene was a strong Category 3 system in the Caribbean. The storm began steadily weakening as it

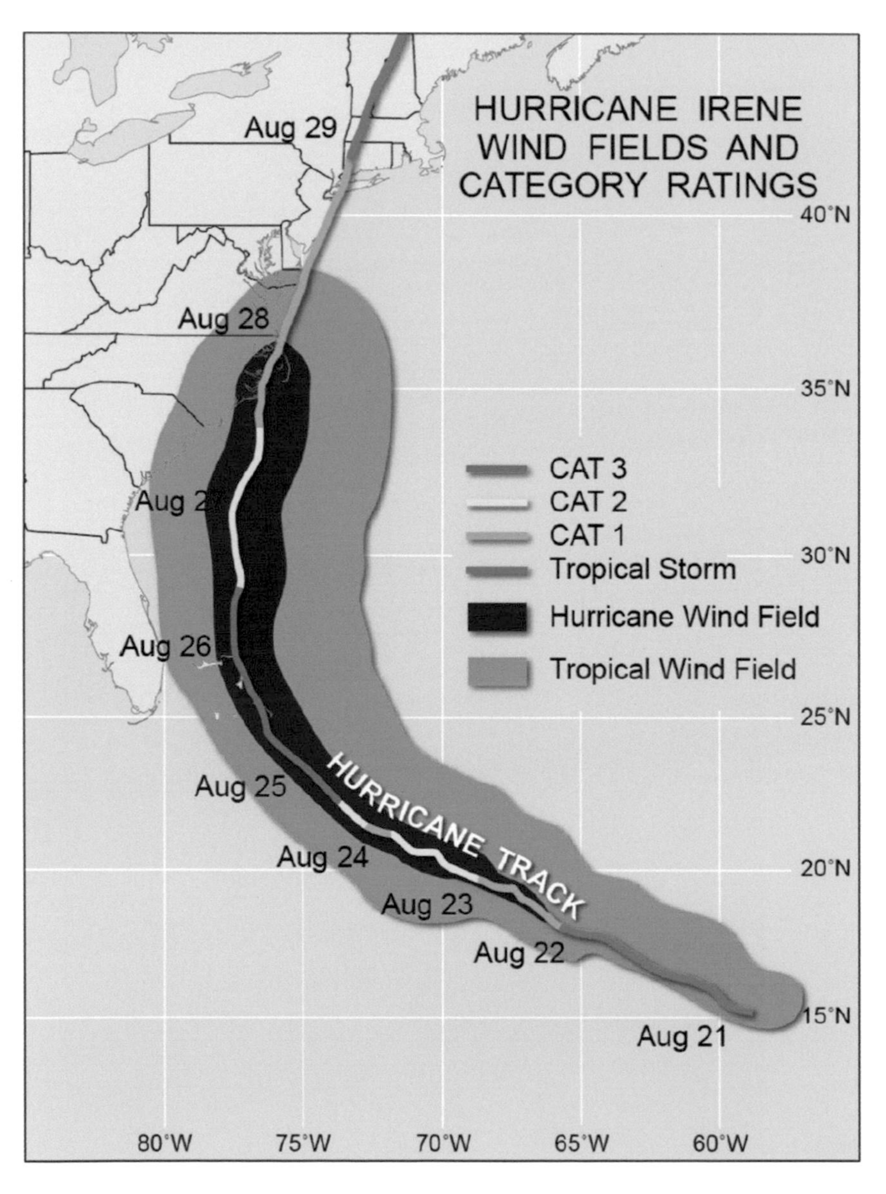

Figure 8.16 Changes in the wind distribution as a hurricane weakens. As is commonly observed, the footprints of both tropical-storm and hurricane-force winds expand as the storm moves into higher latitudes and weakens. (NOAA.)

approached the United States and was down to Category 1 at landfall in the Outer Banks. But note how the width of both the hurricane-force and tropical-storm-force winds expanded, covering more area with time (red and orange shaded regions, respectively).

It is tempting to ascribe this expansion solely to conservation of angular momentum. This principle of physics states that when the winds in a vortex weaken, the vortex expands in size (conversely, an initially large vortex, when contracting, experiences an increase in its wind speed). However, detailed studies reveal that the process is more complicated than this; for instance, expansion often follows eyewall replacement cycles (discussed in Chapter 7). Also, some hurricanes expand during the process of intensification. Figure 8.17 shows how the hurricane's radial wind profile (the variation of wind speed with storm radius) changes during weakening. In this example, a Category 1 hurricane weakens to a tropical storm. The winds along the RMW have died down. But the RMW itself has expanded by nearly 100 km (62 miles). Additionally, the ring of tropical-storm-force winds has enlarged by nearly 300 km (186 miles). A simple analogy of hurricane winds during weakening is to envision a steep mound of rotating sand that is collapsing: the pointed peak flattens, while the diameter of the sandpile increases.

Sometimes during landfall a storm interacts with a nearby high-pressure cell, causing the pressure gradient to intensify on one side of the storm. This can maintain strong, sustained winds for many hours after landfall, while the storm itself undergoes weakening. Such an event occurred when Hurricane Isabel (2003) moved into the Mid-Atlantic, as shown in Figure 8.18.

Prior to Isabel's arrival, a strong fair-weather anticyclone became established over New England. Isabel moved inland to the south of this weather system. The pressure gradient between the storm and the anticyclone strengthened. This interaction enhanced the winds on the northern side of the storm, over the densely populated Northern Virginia-Washington-Baltimore region. Several million power outages resulted from Isabel's passage, even while the hurricane was rapidly weakening to a tropical storm.

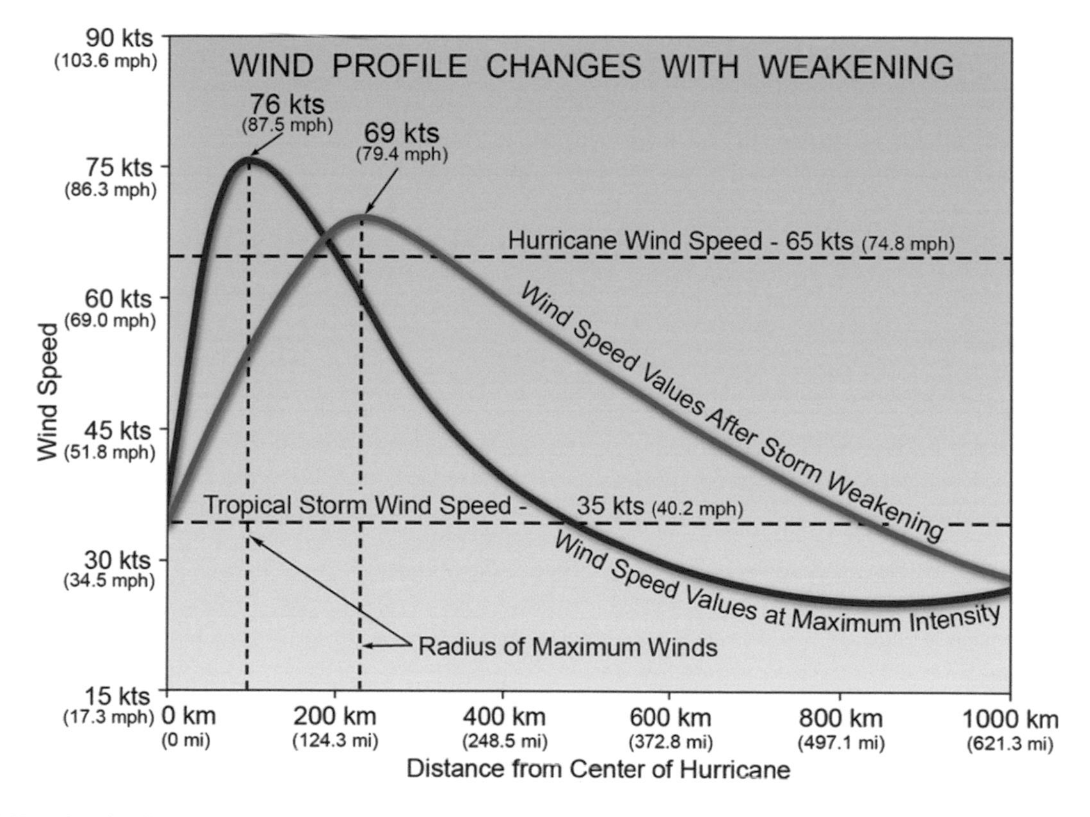

Figure 8.17 How the wind distribution across a hurricane changes during weakening. The wind profile, or strength of the storm's wind as a function of radius, undergoes significant change in a weakening hurricane. Notably, the radius of maximum wind near the center weakens and expands, while wind speeds actually increase at greater distances from the storm center.

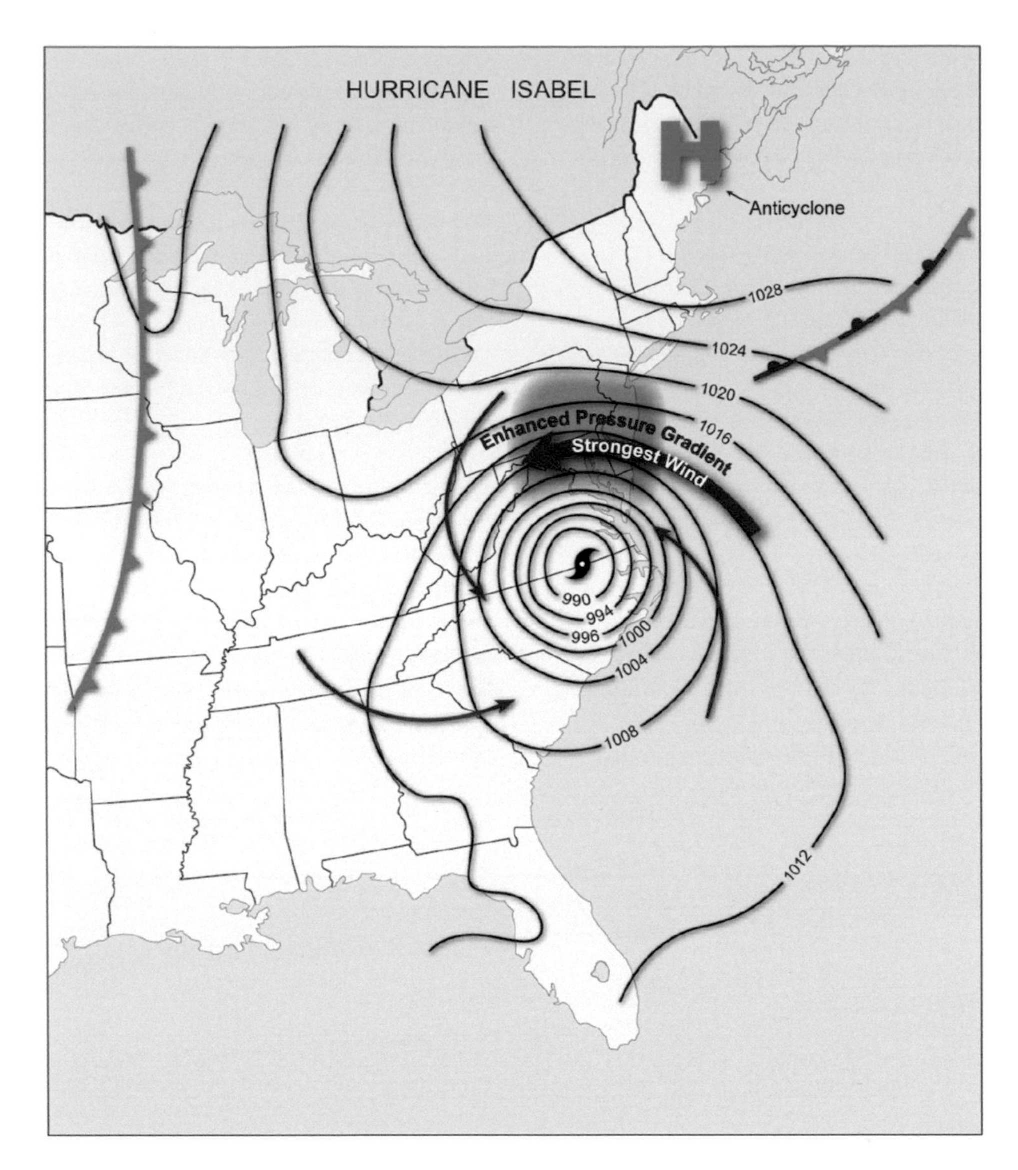

Figure 8.18 Hurricane Isabel's interaction with a strong ridge of high pressure over New England. The inland ridge of high pressure over New England enhanced the pressure gradient along the north side of the storm, maintaining strong winds over the Mid-Atlantic even after landfall.

Downbursts, Mini-Swirls, and Wind Streaks Are All Forms of Localized, Destructive Winds Spawned by Hurricanes

When conducting post-landfall hurricane damage surveys, researchers have long noted small, intense pockets of wind damage. This was especially true over south Florida in the aftermath of Hurricane Andrew. Numerous processes, including those common to severe thunderstorms, are responsible for generating the most extreme winds.

Hurricane Andrew was an exceptionally compact Category 5 storm that struck in 1992 (Figure 8.19). Tropical-storm-force winds extended less than 145 km (90 miles) from the storm's center, making this system among the smallest ever to strike the United States. But for several hours, sustained winds of 130–136 kts (150–157 MPH) hammered the coastline along Biscayne Bay and Dade County, just 40 km (25 miles) south of Miami. Andrew received a pre-landfall energy boost from the warm Gulf Stream, and the storm weakened only slightly as it rapidly crossed the Florida peninsula. Arrested weakening was the result of south Florida's unique land surface – which is actually quite waterlogged, with a large inland body of warm water (the swampy Everglades) and myriad lakes.

In post-storm damage surveys, investigators discovered a pocket of extreme winds approaching 183 kts (210 MPH) along the

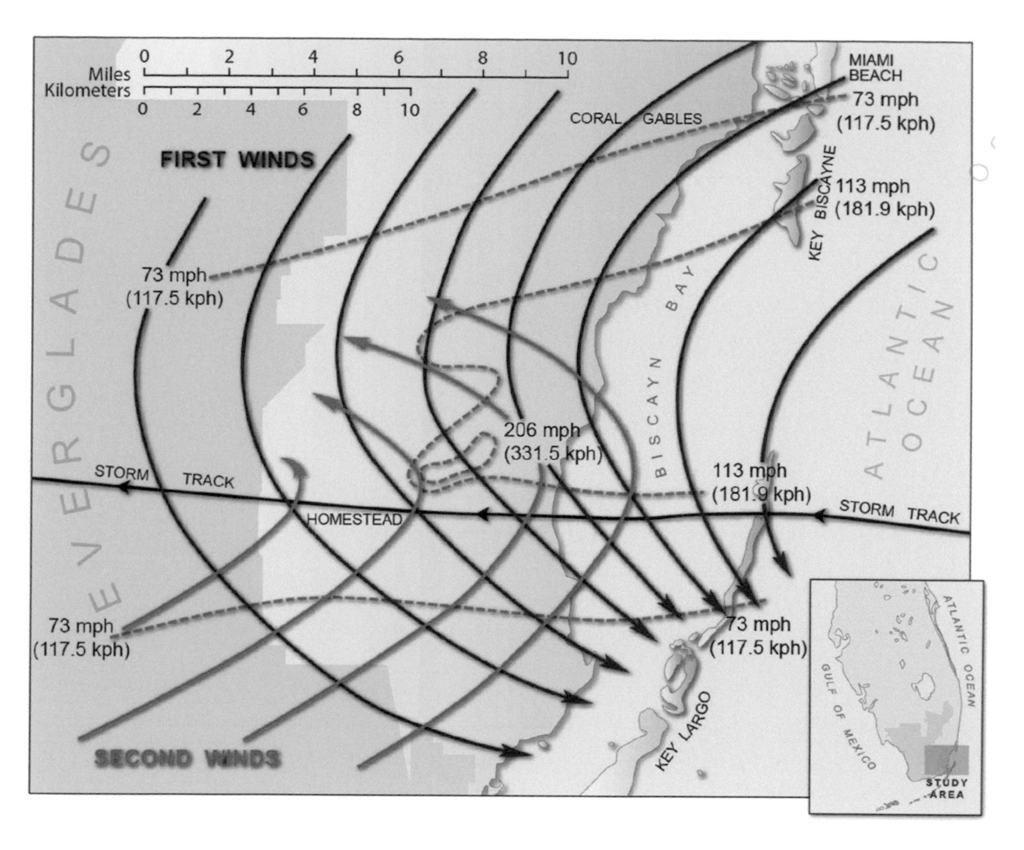

Figure 8.19 Wind analysis of Hurricane Andrew (1992) over southern Florida. The wind field in this storm was unusually intense and compact. The strongest winds were confined to a 16 km (10 mi) wide strip just north of the eye. (Adapted from Wakimoto and Black, 1994.)

northern eyewall, in the vicinity of Homestead, Florida. Damage was near-total (Figure 8.20), with loss of 63,000 homes, 101,000 damaged homes, and a $27 billion price tag (1992 dollars). This single storm became the costliest U.S. natural disaster up to that time.

A narrow, 5 m (16 feet) storm tide impacted Florida's east coast, and up to 36 cm (14 in) of rain fell over the Everglades. But Andrew, metaphorically speaking, seemed to behave more like a monstrous tornado, tens of miles in diameter, sweeping across the southern tip of Florida. As Figure 8.19 shows, the highest winds were contained within 10 km (6 miles) of the storm's center, on the storm's right side.

Figure 8.21 shows an example of downburst winds during the landfall of Hurricane Carla over the Texas coast (September 1961). A downburst is a small-scale (3–5 km [5–8 miles] wide), rapidly descending air mass that blasts outward upon striking the ground. They are the most common agent of wind damage in severe thunderstorms everywhere in the United States. When downbursts occur in hurricanes, they are generated by severe thunderstorm cells embedded in the storm's eyewall or rain bands.

Mini-swirls are a small type of intense vortex just 30–60 m (99–197 feet) in diameter within the hurricane eyewall. They are generated by rapidly moving wind currents that slip past one another, creating turbulent flow (eddies). This slippage is caused by horizontal wind shear – that is, wind speeds flowing in the same direction but at different speeds in parts of the eyewall. Imagine holding your hands vertically in front of you, with a pencil upright in between. Move one hand past the other. This shearing motion causes the pencil to spin, in the manner of a mini-swirl. Mini-swirls create corridors of extreme wind damage within and added to the overall high-speed flow of the sustained wind. They are *not* tornadoes but a type of instability that develops in a strongly varying flow. Their existence became known during a NOAA hurricane hunter flight inside Hurricane Hugo (1985) – when the aircraft with its science crew inadvertently flew right through a mini-swirl (the aircraft managed to limp its way out and land safely)!

Figure 8.20 Landscape of devastation across south Florida in the wake of Hurricane Andrew. Viewing the total destruction of this Florida subdivision from helicopter, one might surmise such damage arose from a strong tornado rather than Category 5 Hurricane Andrew. (FEMA.)

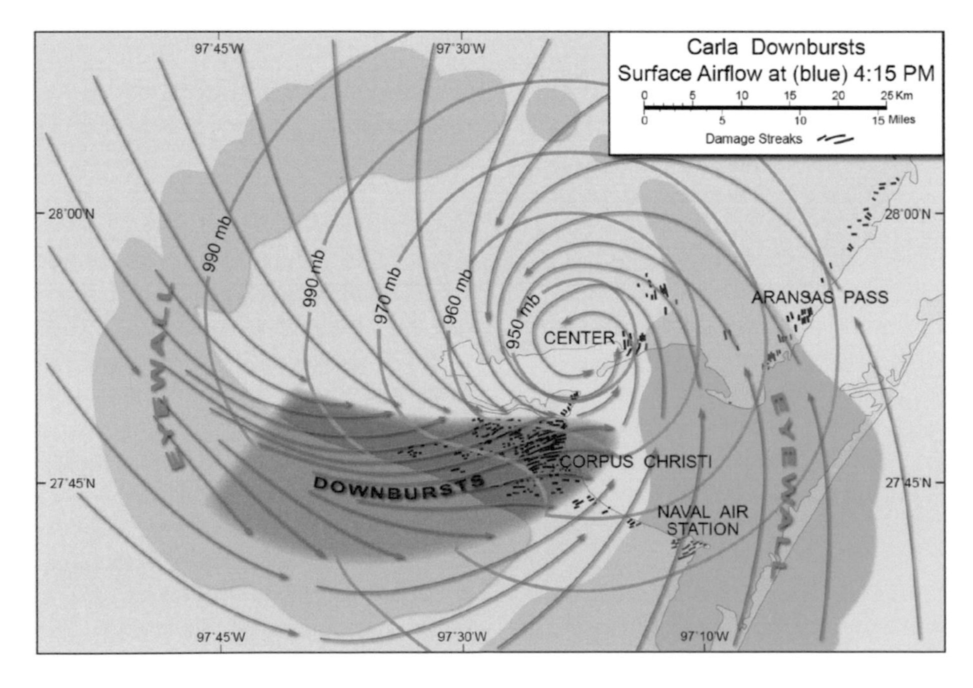

Figure 8.21 Wind damage map in the aftermath of Hurricane Carla, revealing embedded downburst damage. Most commonly observed in severe thunderstorms, downbursts are occasionally generated by hurricanes within the eyewall and rain band regions. These violent outbursts of wind create localized pockets of intense wind damage. (Adapted from Fujita, 1978.)

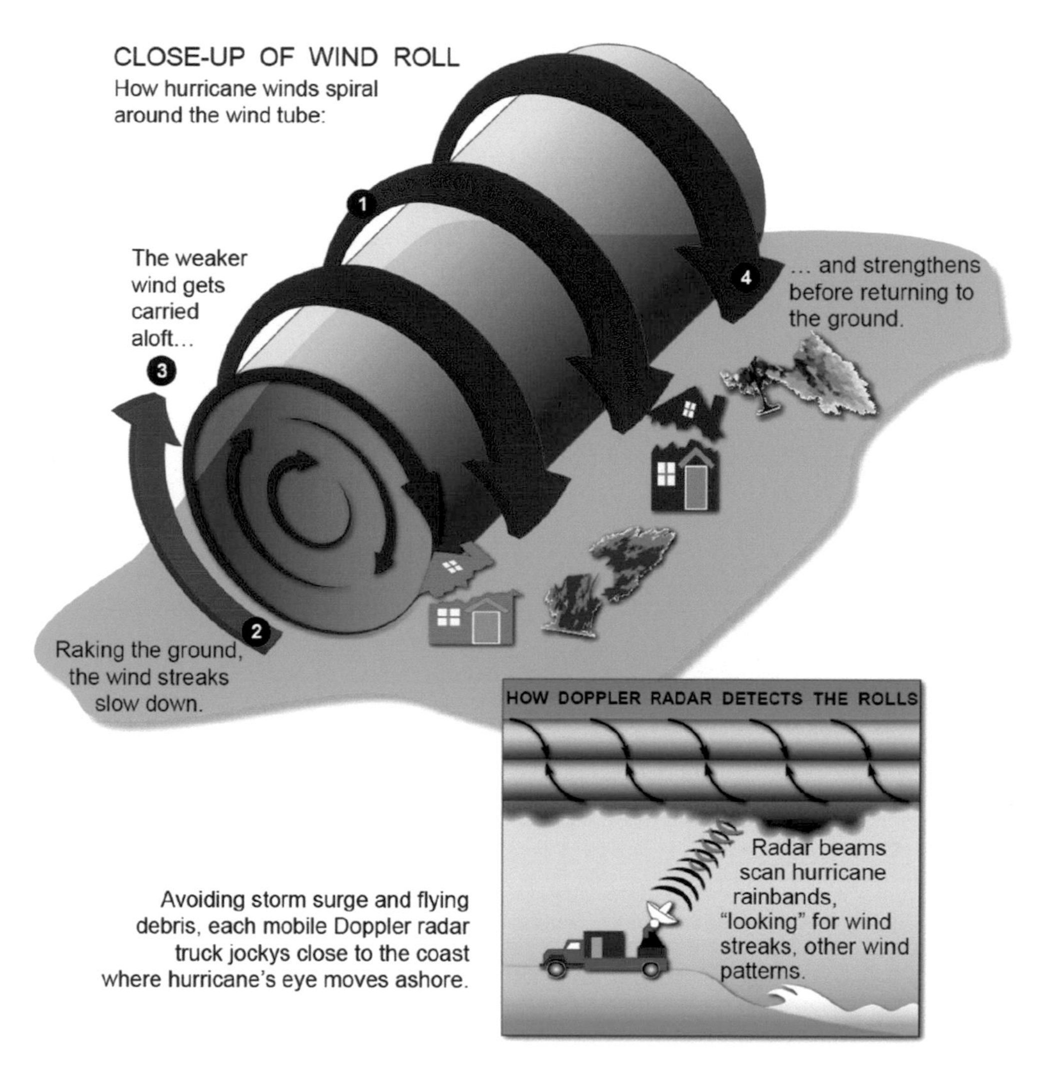

Figure 8.22 Schematic showing the formation of wind streaks above the surface in a landfalling hurricane. Wind streaks are aligned parallel to horizontal, rolling tubes of air above the surface. One of the tubes shown here, by dragging faster winds aloft (freed from surface friction) down to the ground, creates a narrow strip of ground damage parallel to the vortex tube.

Finally, Doppler radar studies in landfalling hurricanes have identified yet another type of wind damage pattern, called wind streaks (Figure 8.22). Wind streaks are narrow, linear swaths or "fingers" of wind devastation that run parallel to one another on the ground, separated by a few hundred meters (yards). Wind streaks are caused by vertical wind shear above the surface. When winds rapidly increase with altitude, the result is a vertical wind shear. When faster airflow aloft slips over slower air at the surface, the wind slips over itself and breaks down into horizontal vortex tubes or "rollers." Back to the hands and pencil, imagine your hands now horizontal, with the pencil lying in between. Move the top hand only and observe the pencil spin. The tubes develop as closely arrayed, parallel rolls, oriented in the same direction as the airflow. Between rolls that have opposing spin, high-velocity air is dragged downward to the surface. The result is a narrow blast of high momentum air striking along a narrow corridor at the surface. These narrow strips of high wind, tens of meters wide and spaced a few hundred meters apart, create long, parallel strips of intense wind damage.

Finally, we note that tornadoes are yet another hallmark of hurricanes that make landfall. Strong vertical shear of airflow in the lowest kilometer (0.6 mile) above the surface is critical for their formation. Tornadoes embedded in hurricanes create narrow swaths of enhanced wind damage, within the larger envelope of the hurricane's sustained winds. However, tornadoes are more likely to form a distance inland from the coast.

Accordingly, we discuss them in the section on inland hazards of tropical cyclones.

Table 8.1 summarizes the various types of small-scale wind damage processes in landfalling hurricanes (downbursts, mini-swirls, wind streaks, and tornadoes). These features are very transient, highly localized, and generally create small zones of most intense wind damage in the eyewall and rain bands.

Table 8.1 Hurricane Wind Damage Mechanisms – Small-Scale Processes

Phenomenon	Location in Storm	Mechanism	Max Winds
Downburst	Spiral bands	Intense downdraft	109–130 kts (125–150 MPH)
Mini-swirl	Eyewall	Horizontal wind shear	*
Wind streak	Spiral bands	Vertical wind shear	*
Tornado	Right front quadrant	Tilting and stretching	57–117 kts (65–135 MPH)

Notes: MPH = miles per hour.

*Maximum wind speeds in these phenomenon are dependent on wind shear, which varies in proportion to the storm's maximum sustained wind and other factors.

Freshwater Flooding Is One of the Most Deadly Aspects of a Landfalling Tropical Cyclone

Flooding from heavy rain is the number one cause of hurricane-related deaths in the United States. Like surge and high wind, the mechanism depends on many details – meteorological, hydrological, and geographical. Landfalling hurricanes commonly produce 15–30 cm (6–12 inches) of heavy rain along the coast, sometimes much more, leading to flash flooding and river floods.

Heavy Rains From Tropical Cyclones at Landfall Ranks as Some of the Most Devastating Impacts Across Broad Regions of the United States

Tropical cyclones, including both hurricanes and tropical storms, are among Earth's most efficient rain producers. They process copious amounts of water vapor into heavy showers. The heaviest rain falls from intense convective cells in the eyewall spiral bands. Between these cells, thick layers of sheet-like cloud (stratiform cloud) generate more widespread, steady rain.

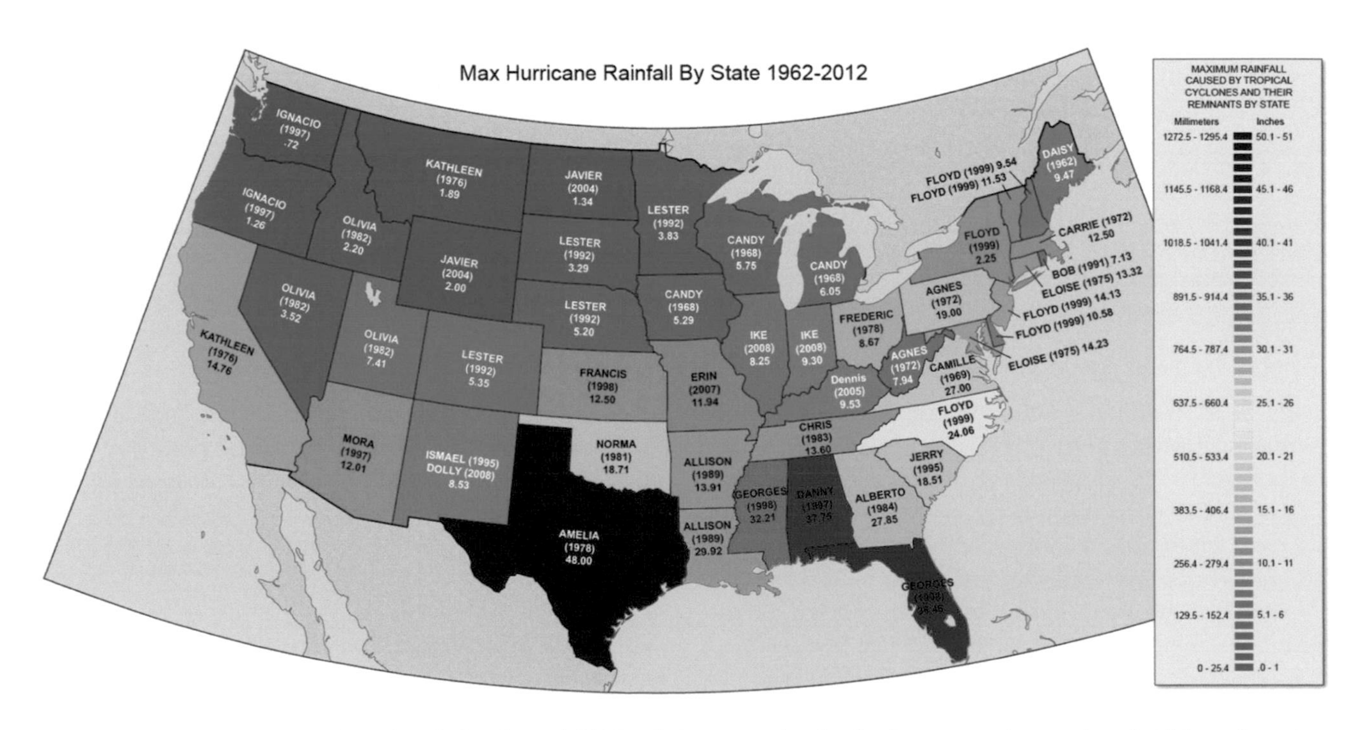

Figure 8.23 State-by-state tally of record tropical cyclone rainfall. Tropical storms, rather than hurricanes, sometimes produce the highest rain totals. (NOAA.)

Wind blows the rain horizontally. Salty ocean spray and heavy rain mix above the ocean. At times, it is nearly impossible to separate ocean from sky.

Figure 8.23 shows the remarkable rain totals created by tropical cyclones and their post-tropical remnants in each of the states. Every U.S. state, at some point since the start of records (1962), has received some amount of tropical cyclone rainfall. As you might expect, the Gulf Coast and Mid-Atlantic states are the rainfall heavy hitters, led by Texas (122 cm [48 inches] from Tropical Storm Amelia in 1978), followed by Florida (98 cm [38.5 inches] from Hurricane Georges in 1998) and Alabama (96 cm [37.8 inches] from Tropical Storm Danny in 1997).

Surprisingly, the amount of rain delivered by a tropical cyclone correlates poorly with its intensity as measured by the MSW. A humble tropical storm can deliver heavier rain than a Category 2 hurricane. Rain accumulation depends on several factors besides vortex strength. One of the most important is the speed at which the storm moves. A slow-moving or stationary hurricane localizes all its rain to one spot. Tropical cyclones, striking the United States in late summer and early fall, often encounter quiescent (inactive) steering winds. That is, there is no fast jet stream over the southern United States; the jet, if present at all, is located over Canada.

Many a Tropical Cyclone Has Languished for Days Along the Gulf Coast States, Resulting in Rain Totals Measured in Feet, Not Inches!

Another reason for heavy rain is the repeated passage of individual rain-producing thunderstorm cells over the same location. This process is called cell training and is illustrated in Figure 8.24. The diagram shows a tropical cyclone tracking slowly W-NW across the Gulf Coast. Large amounts of Gulf moisture enter the storm. Heavy rain showers form in rain band cells (dark green circular regions). One of these bands is oriented east-west across Location A. Winds entering the storm from the east blow parallel to this band. The rain cells, moving with the wind, pass repeatedly over the same location. As long as the spiral band remains stationary, each cell delivers its "cargo" of heavy rain, analogous to a freight train passing overhead. The smaller inset diagram shows how quickly the rain adds up. A sequence of 5–6 rain cells can easily deliver 13–26 cm (5–10 inches) of rain in just a few hours.

Finally, tropical cyclones import a tremendous amount of water vapor over land. This moisture converges near the storm's center as the airflow spirals inward, where it is forced to ascend. Even though the wind vortex may be weakening, the sheer amount of water vapor available to condense into rain is vast; rain production remains a very efficient and widespread process, sometimes persisting over several days of the weakening process.

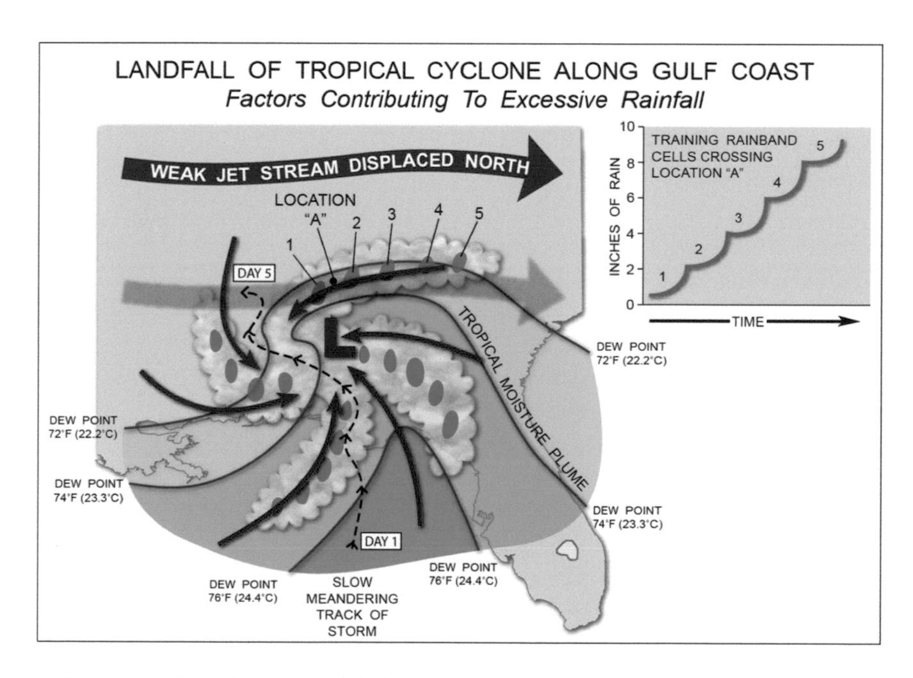

Figure 8.24 The various mechanisms leading to heavy rainfall during a landfalling tropical cyclone: slow storm motion, high moisture content in the air, and cell training within rain bands.

Meteorologists use a simple equation to help predict total rain along the coastline, based on (1) the storm's average rain rate (the one factor that varies with storm intensity), (2) storm size, and (3) speed of storm movement:

$$\text{Total Rain} = (\text{Rain Rate} \times \text{Storm Diameter})/\text{Storm Speed}$$

Consider two quick examples. Storm A produces rain at 5 cm (2 inches) per hour. It is compact, only 200 km (124 miles) in diameter. It also moves rapidly, at 30 kts (35 MPH). Total rain for any location beneath the storm works out to 18 cm (7 inches). Storm B is larger but weaker, moves at 15 kts (17 MPH), with an average rain rate of 3 cm (1 inch) per hour. These numbers yield 68 cm (27 inches) of rain. This is an extreme value but not without precedent along the Gulf Coast.

Let's look at some examples of coastal flooding during tropical cyclone landfalls. Figure 8.25 shows rain accumulations for Tropical Storm Allison (2001), Hurricane David (1979), and Tropical Storm Fay (2008). Allison struck Texas as a tropical storm, weakening to a tropical depression. The system meandered for days along the Gulf Coast, even executing a 360° loop! A total of 51–64 cm (20–25 inches) of rain fell across southern Louisiana. Note the pocket of 89 cm (35 inches) of rain over extreme southeastern Texas. That is almost 1 m (3 feet) of rain.

Hurricane David's rain footprint was more extensive, stretching from Florida to Maine. The storm hit the United States as a Category 2, having weakened considerably from Category 5 status over the Caribbean. Compared to Allison, this system was on the move. Up to 18–25 cm (7–10 inches) of rain fell over the eastern Carolinas. There were isolated pockets of up to 38–51 cm (15–20 inches) of rain, probably from cell training in spiral rain bands.

Finally, Tropical Storm Fay produced the most widespread hurricane flood in Florida's history. The storm traced a path along Florida's peninsula, then the panhandle, blanketing the entire state with heavy rain. During its slow zigzag, Fay made four consecutive landfalls over Florida! As much as 50% of the state received 25–38 cm (10–15 inches) of rain with extreme values exceeding 64 cm (25 inches). Note that Fay's moisture extended as far north as the Smoky Mountains, where the uplift of air along mountain slopes wrung out an additional 25–38 cm (10–15 inches) across high terrain. The enhancement of rain over steep mountain slopes is termed orographic rainfall and is discussed in the next section.

An interesting type of rain-generating process, called a Predecessor Rain Event (PRE), can sometimes occur ahead of tropical cyclones making landfall. A PRE is a coherent, small

(a)

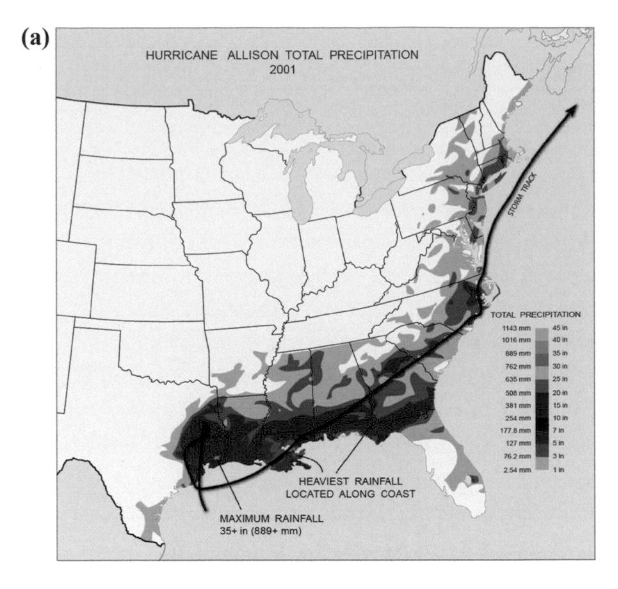

(b)

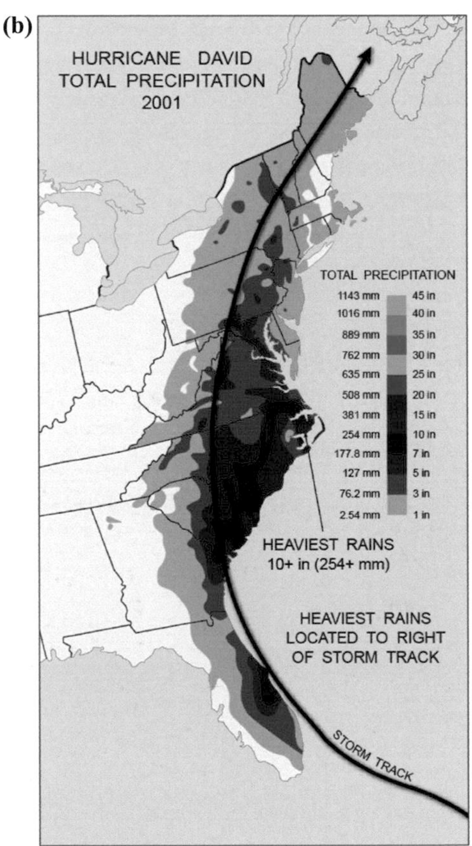

Figure 8.25 (a) Rain accumulation maps from several notable United States landfalling tropical cyclones. Heavy rains fell along the Gulf Coast from Tropical Storm Allison, due mainly to the storm's slow motion and looping track. (b) Hurricane David underwent a typical pattern of recurvature along the East Coast, producing a long arc of heavy rain from Florida to New England. (c) Tropical Storm Fay inundated the entire state of Florida before tracking into the Ohio Valley. The storm was also responsible for heavy rainfall in the southern Appalachians. (NOAA.)

(c)

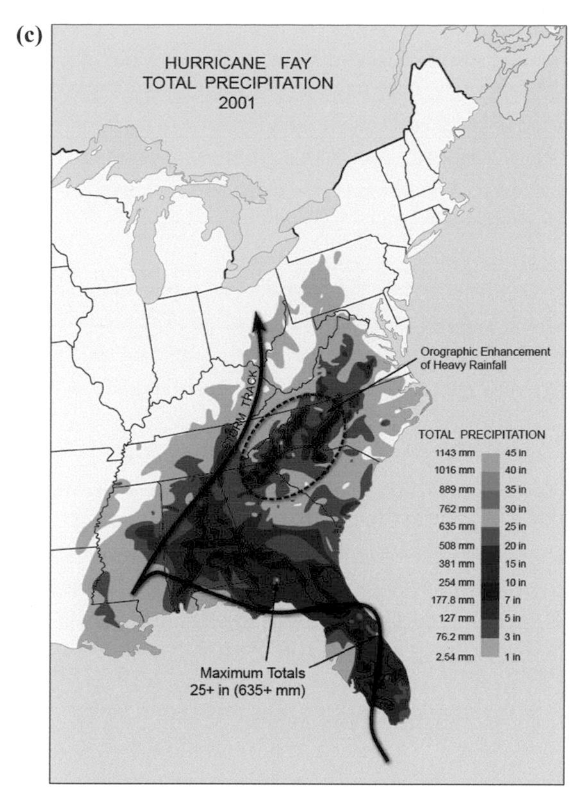

Figure 8.25 (Continued)

region of rainfall, detached from the main storm, often hundreds of miles poleward of the tropical system. They occur when a plume of deep tropical moisture exiting the tropical cyclone interacts with a disturbance in the jet stream, north of the tropical system. At times, PREs have created significant flooding one to two days in advance of the cyclone's main rain shield. Figure 8.26 shows a PRE event that generated surprise flooding during the morning rush hour in New York City on September 8, 2004 – well in advance of Hurricane Frances.

Tropical Cyclones Entering Their Post-Tropical Phase Can Still Create Devastating Inland Impacts

Post-tropical remnants of a tropical cyclone can remain destructive far inland when there is an additional source of energy fueling the vortex. The same energy sources that power extratropical cyclones – strong temperature gradients and dynamic waves in the jet stream – can combine with and sustain remnant tropical systems. These types of hybrid vortex feed off both tropical and extratropical energy sources.

The post-tropical system undergoes significant changes, appearing more and more extratropical over time. For a

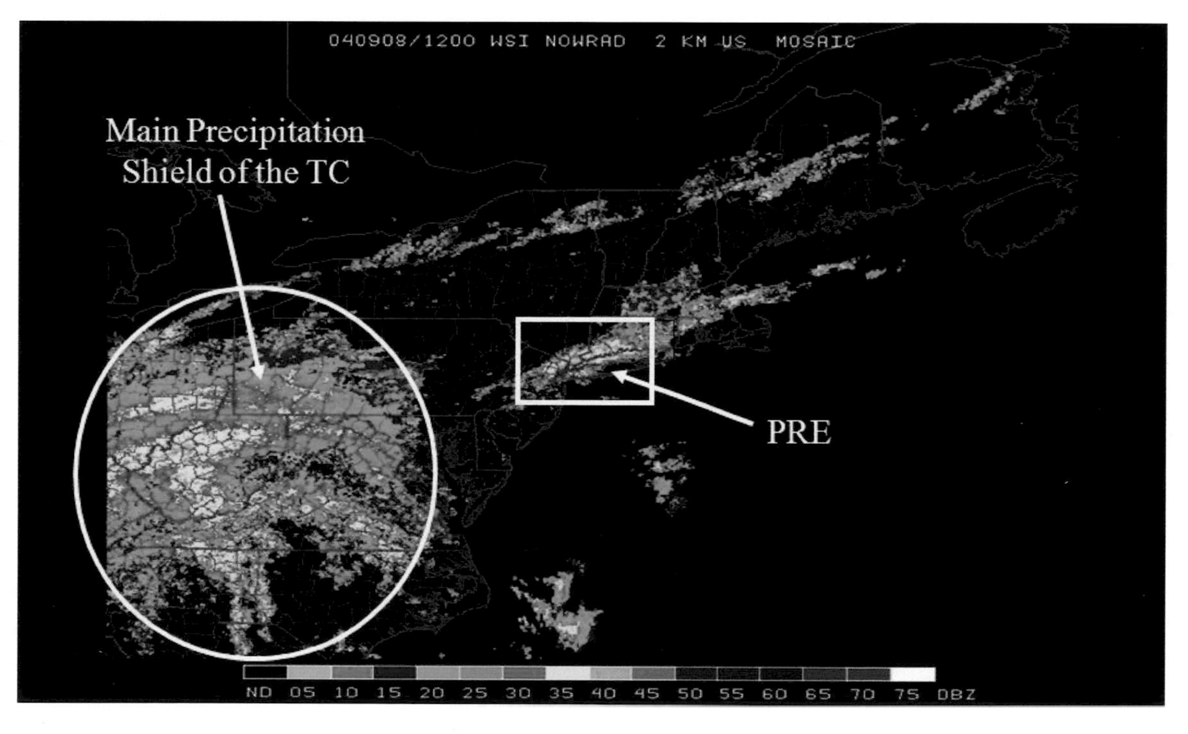

Figure 8.26 A unique flash flood, called a Predecessor Rain Event, generated hundreds of miles ahead Hurricane Frances after landfall. Weather radar showing the small, heavy rain region over New York City on September 8, 2004 – clearly detached from the main, rain-bearing portion of the approaching tropical cyclone. (NWS.)

while, it may become a type of "hybrid" storm – possessing characteristics of both a tropical and extratropical vortex.

This section is about tropical cyclones that move inland, then catch their "second wind," both figuratively and literally. These hurricane remnants continue to produce hazardous weather, sometimes days after landfall. Historically this phenomenon has happened in many locations east of the Rockies but most commonly in the Mid-Atlantic and Northeast. The biggest concern of post-tropical cyclone remnants is heavy rain, leading to flash floods, river floods, mudslides, and debris flows. Tornado outbreaks are also fairly common. Much less frequently, intense hurricane-force winds can be maintained far inland. Even heavy snowfall has occurred – for instance, after the landfall of Hurricane Sandy (2012). Several U.S. weather catastrophes have unfolded during the inland phase; examples are shown in Table 8.2.

In 2010, the National Hurricane Center introduced new terminology to describe tropical cyclone remnants – the so-called post-tropical cyclone phase. Post-tropical cyclones are hurricanes or tropical cyclones that have weakened over land or ocean and that lose key structural attributes including the eye and eyewall. The storm's central warm core will be in the process of cooling. But a weakened vortex of wind remains, typically of a larger diameter than that of a purely tropical system, and one or more spiral rain bands still show up on radar and satellite imagery. Moreover, these post-tropical cyclones may continue to produce heavy rain and high wind.

Table 8.2 Selected U.S. Hurricanes Creating Inland Weather Disasters, 1950–2023

Storm Name	Date	Region Impacted	Inland Hazard(s)
Hazel	October 1954	Mid-Atlantic to Canada	High winds, flooding
Camille	August 1969	Central Virginia	Severe flooding
Agnes	June 1972	Mid-Atlantic	Severe flooding
Eloise	September 1975	Mid-Atlantic	Severe flooding
Floyd	September 1999	Mid-Atlantic, Northeast	Severe flooding
Gaston	September 2004	Central Virginia	Severe flooding
Ivan	September 2004	Mid-Atlantic, Southeast	Tornado outbreaks
Irene	August 2011	Mid-Atlantic, Northeast	Severe flooding
Sandy	October 2012	Mid-Atlantic, Northeast	High winds, flooding, snow
Ida	September 2021	Mid-Atlantic	Severe flooding, tornadoes

Post-tropical cyclones that weaken further – to the point where winds have dropped below 34 kts (39 MPH) – are referred to as remnant lows. Aside from continued moderate-heavy rain, their severe effects are typically limited.

A post-tropical cyclone that morphs into a mid-latitude cyclone, is referred to as a hybrid cyclone. The process by which a tropical cyclone fully transforms into an extratropical cyclone is termed extratropical transition (ET). Just like ordinary extratropical cyclones, these systems have acquired fronts and derive their principal energy from thermal gradients. They become fully coupled to the westerly jet stream aloft. The former warm core of the tropical cyclone transforms into a cold core vortex, characteristic of extratropical cyclones. Research has shown that 46% of Atlantic tropical cyclones undergo extratropical transition, whether over ocean or over land.

This process is illustrated in Figure 8.27.

As a tropical cyclone moves poleward, it encounters a number of changes in its atmospheric and oceanic environment. The Coriolis effect (Earth's spin) becomes stronger, changing the tropical cyclone's wind balance.

Cooler ocean temperatures diminish the supply of sensible and latent heat streaming into the tropical cyclone. At the same time, the storm begins to interact with the strong temperature and humidity gradients typical of the mid-latitude atmosphere. Vertical wind shear increases as the hurricane moves into the domain of the westerly jet stream. A trough in the jet stream may merge with the tropical cyclone. One or more fronts near the surface may become incorporated into the storm's circulation. Both of these interactions increase the uplift of air, escalating the storm's rainfall.

In response to these effects, the post-tropical storm undergoes significant changes. The storm accelerates once it becomes embedded in the faster flow. As a storm's forward speed increases, winds around the vortex become highly asymmetric, blowing more strongly to right of center. Meantime, the heavy rain often shifts to left of track. The storm's cloud and precipitation shield become elongated in the downstream direction. What results is a highly asymmetric, distorted, and often expanded weather system that, while weaker, is still capable of producing destructive weather for several days. Figure 8.28 shows an example, from the satellite perspective, how the physical attributes of a tropical cyclone change during ET. During ET, the hybrid system can remain steady-state for days or slowly decay.

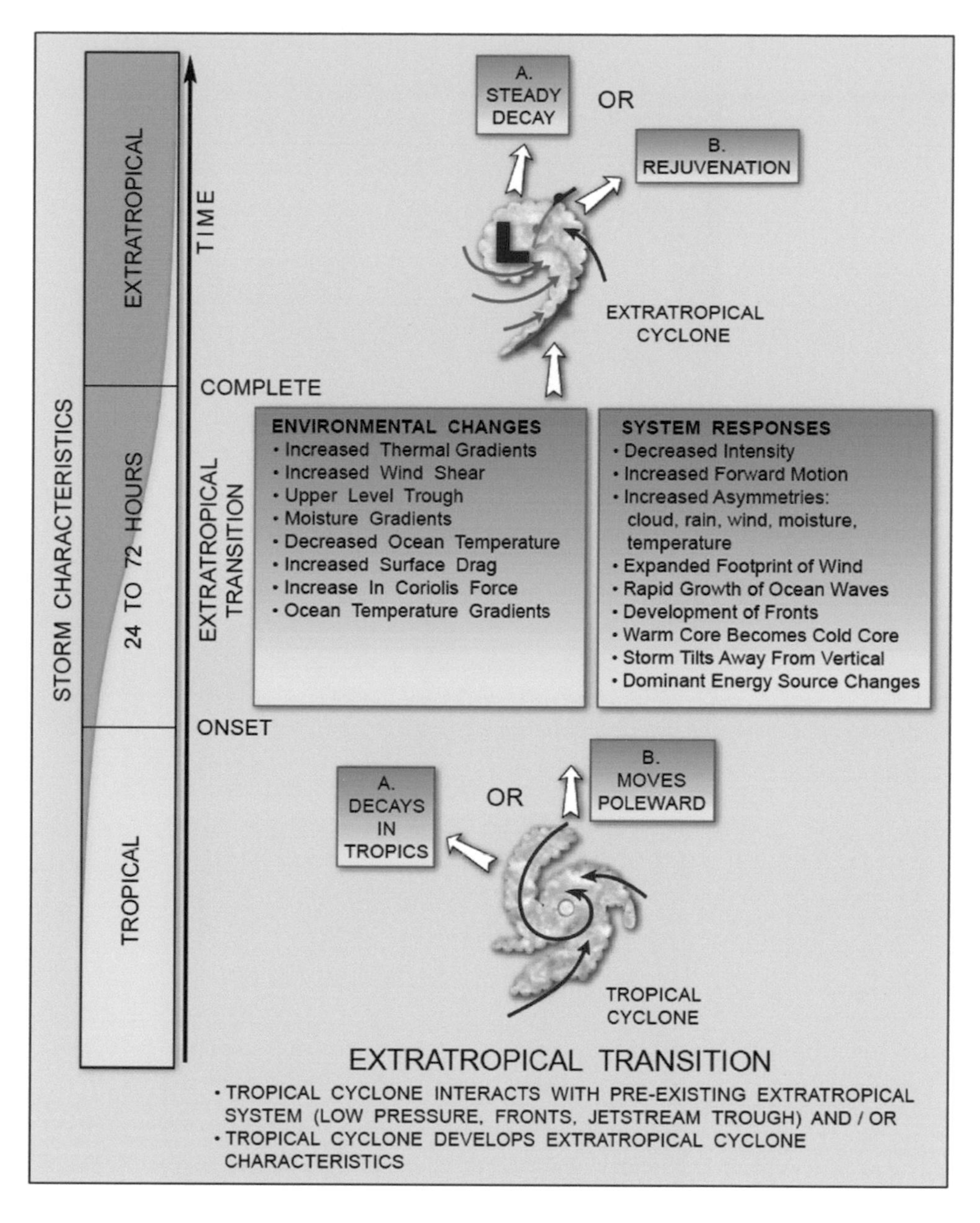

Figure 8.27 The different fates of hurricanes in their post-tropical phase, including the complex process of extratropical transition. (Adapted from Evans et al., 2017.)

Sometimes an inland tropical cyclone can undergo a brief period of re-intensification or rejuvenation. Recent research has identified a class of inland tropical cyclones termed brown ocean storms. Nearly a fifth of inland tropical cyclones (including those over North America, China, and Australia) either maintain intensity at landfall or strengthen while over land. For this to occur, (1) the atmosphere must have minimal temperature variation, (2) there must be ample soil moisture present from recent heavy rains (or the presence of swamps or wetlands), and (3) the evaporation rate from the ground must provide enough energy to the atmosphere – in the form of latent heat – to mimic the ocean. In 2007, the remnants of Tropical Storm Erin moved inland off the Gulf of Mexico and developed an eye over Oklahoma. Tremendous rains in the weeks before landfall had saturated soils over the southern Great Plains. The storm actually became more intense while over land, producing wind gusts to 70 kts (80 MPH) and dropping 25 cm (10 inches) of rain (Figure 8.29).

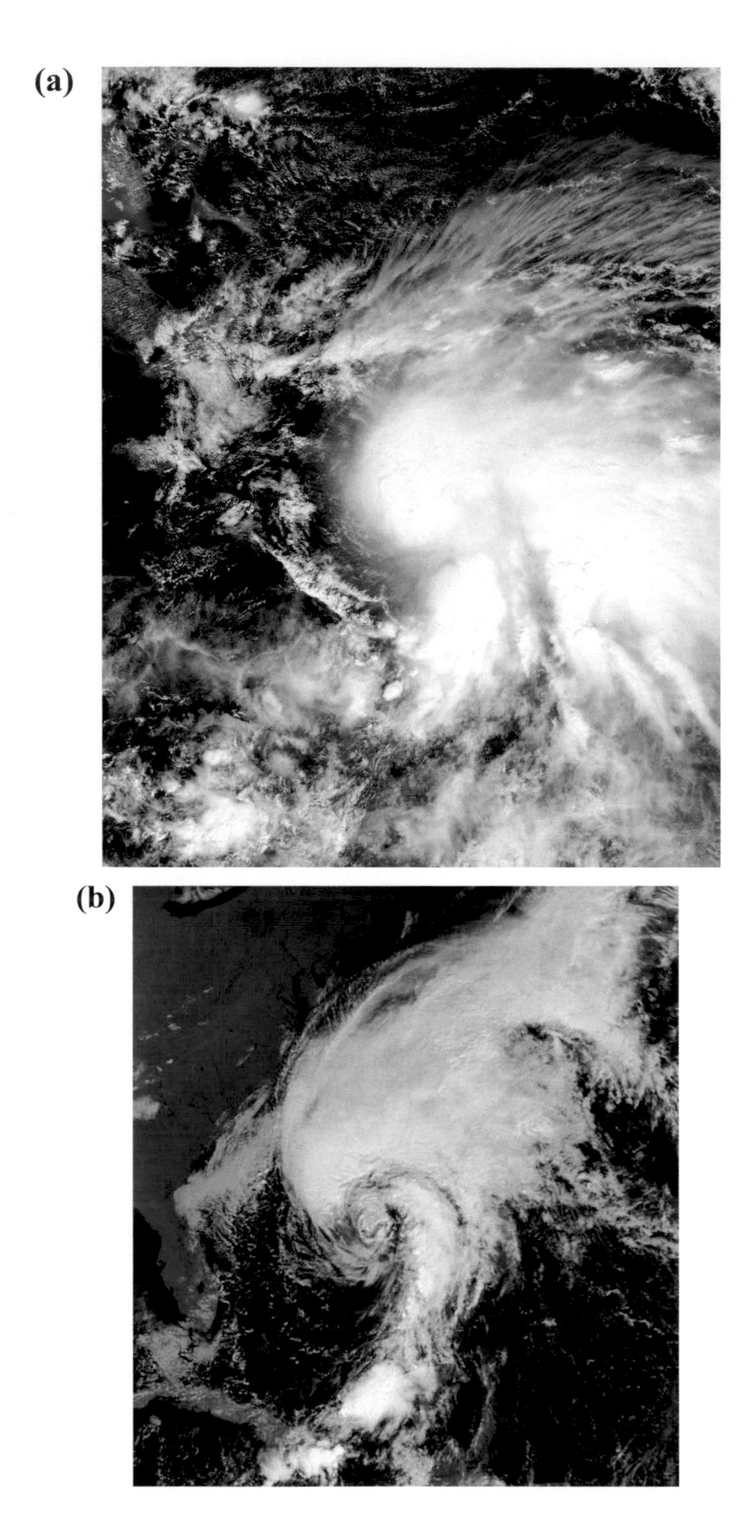

Figure 8.28 (a) Satellite view showing Hurricane Noel (2007) before extratropical transition. Note the symmetric, near-circular mass of tropical thunderstorms clustered near the storm's center. A visible eye, however, remains obscured by high cloud. (b) Hurricane Noel during extratropical transition. The storm takes on the appearance of a classic mid-latitude cyclone, with its distinct comma shape. Clear, dry air intrudes from the southwest, and the cloud shield acquires an asymmetric shape as jet stream winds spread the storm's moisture toward the northeast. (NOAA.)

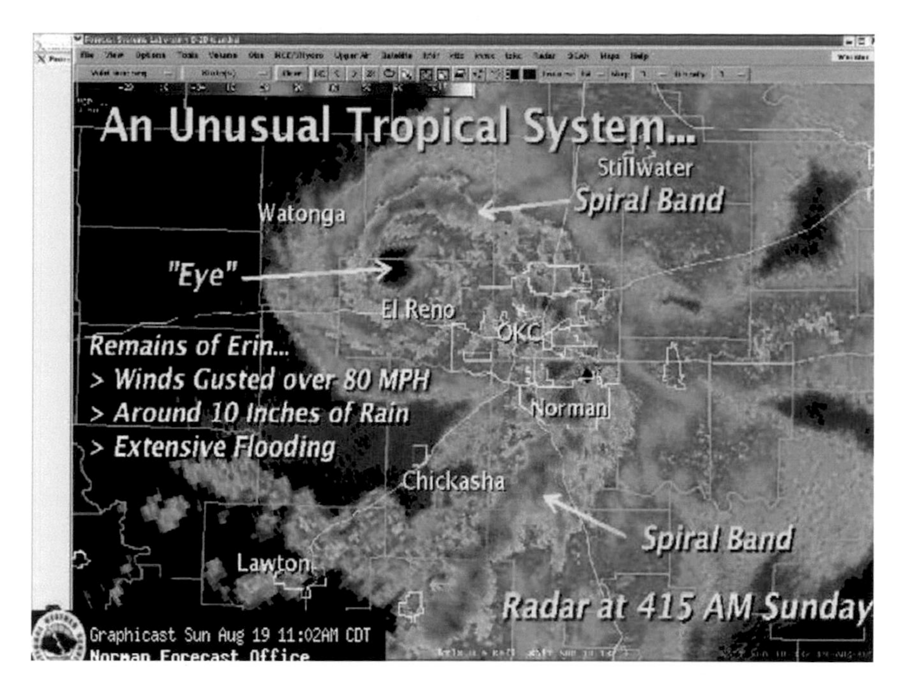

Figure 8.29 Rare example of a tropical cyclone intensifying over land. This shows a system named Erin, which emerged off the Gulf of Mexico less than fully developed, then took on the characteristic eye and spiral tropical cyclone structure over Oklahoma. (NWS.)

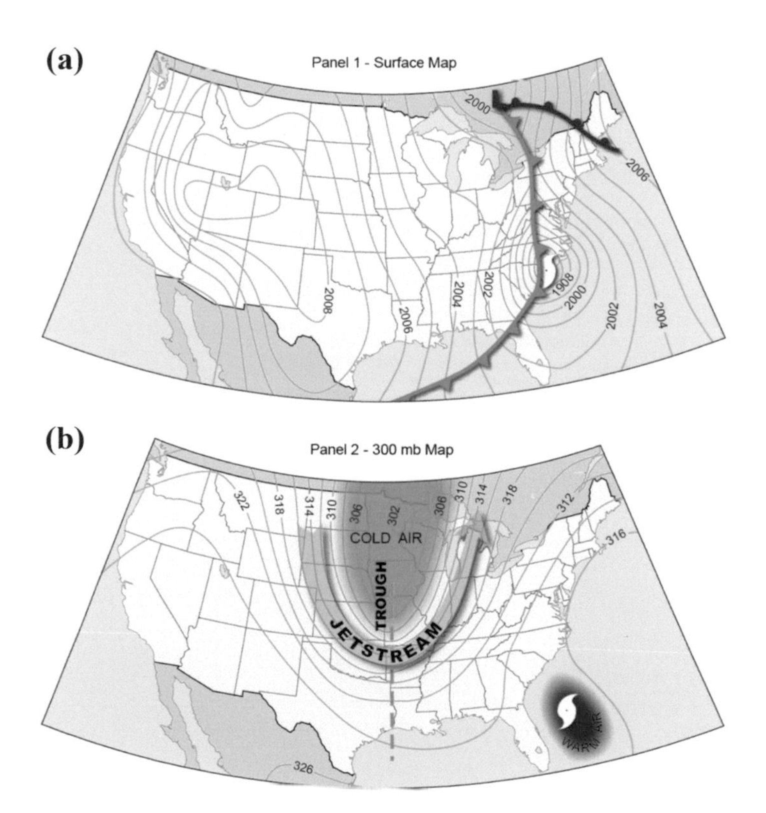

Figure 8.30 (a) Interaction of Hurricane Hazel with a strong cold front and upper-level jet stream trough set the stage for rejuvenation over land. This is the initial phase of the interaction on October 15. (b) Interaction of Hurricane Hazel with a strong cold front and upper-level jet stream trough. This is the initial phase of the interaction. (Adapted from Palmen, 1958.)

Hurricane Hazel (1954) Morphed Into a Powerful Hybrid Cyclone, Remaining a Remarkably Intense Storm Over the Northeastern United States and Canada

Let's now look at a remarkable example of extratropical transition: Hurricane Hazel, during October 1954. In this case, the tropical remnant circulation became coupled with an extraordinarily intense trough in the jet stream over the eastern United States

As shown in Figure 8.30(a), Hazel made landfall along the Carolina coast as a Category 4 hurricane on the evening of October 15, 1954. As the storm moved toward the north, an unseasonably cold and vigorous trough in the jet stream approached the East Coast (Figure 8.30[b]).

The trough helped generate an ordinary extratropical cyclone north of the Great Lakes. A cold front trailed southward from this storm. Hazel rapidly merged with this cold front over the Mid-Atlantic states.

By midnight on the 16th, Hazel's tropical vortex combined with the Great Lakes cyclone (Figure 8.31[a]). The combined system moved over Toronto, Canada, where it led to considerable damage and loss of life. You will note from the appearance of isobars that this hybrid system was just as intense as the hurricane phase in the Carolinas! The strong jet stream trough (Figure 8.31[b]) played a significant role in maintaining Hazel's intensity. In fact, an enormous reservoir of potential energy was generated as cold, arctic air became juxtaposed against warm, tropical air. This generated an intense thermal gradient and reservoir of potential energy. As cold air sank and warm air

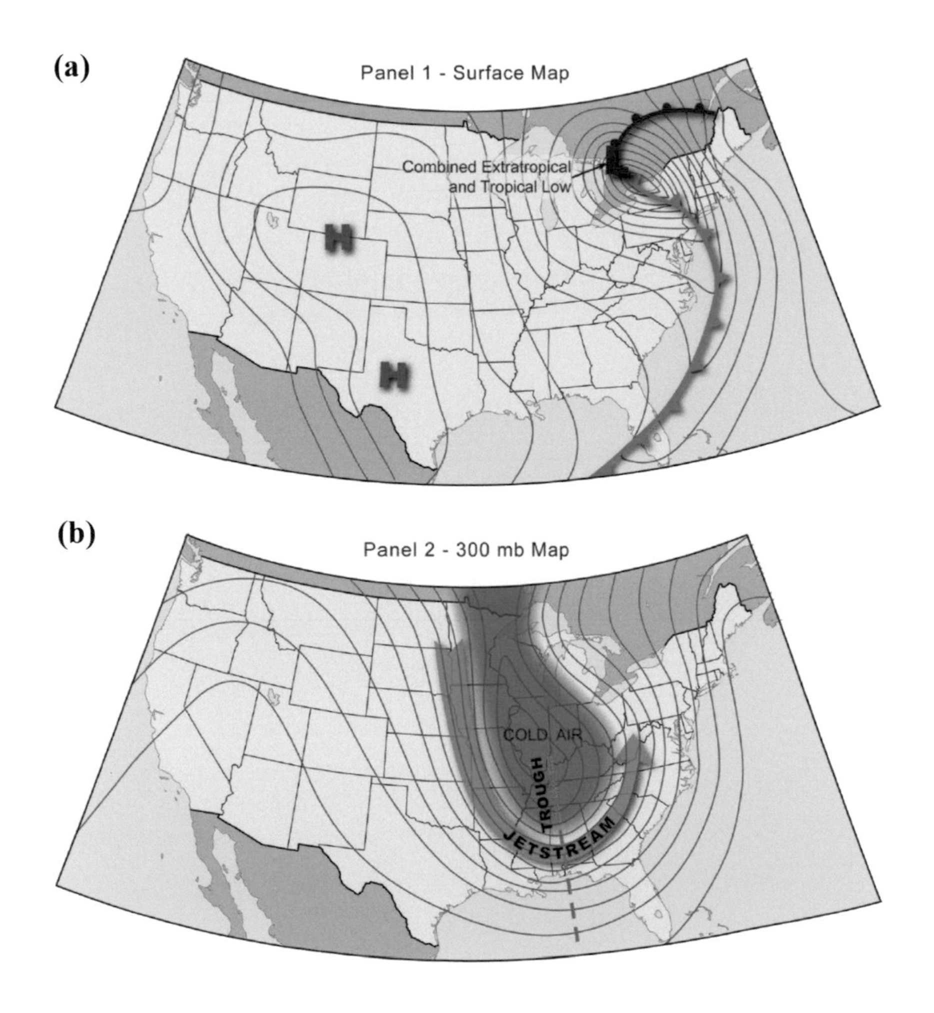

Figure 8.31 (a) Interaction of Hurricane Hazel with a strong cold front and upper-level jet stream trough. The post-tropical cyclone and extratropical cyclone have merged into a giant hybrid vortex. (b) The intensified trough at jet stream level, with extremely cold air circulating far to the south, and powerful post-tropical vortex over Canada. (Adapted from Palmen, 1958.)

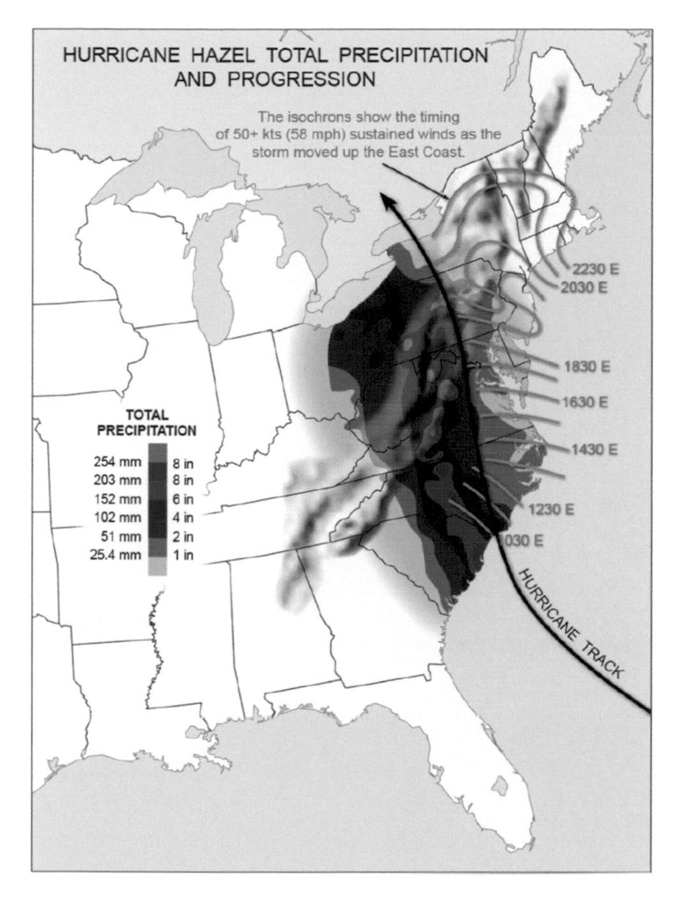

Figure 8.32 Devastating inland effects of Hurricane Hazel's extratropical transition and rejuvenation. A swath of heavy rain developed to the left of a track along the Appalachians, while sustained 50+ kt (58+ MPH) winds rapidly swept northward up the East Coast, on the right side of the system. (Adapted from NOAA and Palmen, 1958.)

ascended, a tremendous amount of kinetic energy was created, on the order of 1 petajoule (1 quadrillion joules). This energy is equivalent to the detonation of 16 Hiroshima atomic bombs. This extra energy counteracted spin-down of the tropical vortex as it left behind its oceanic heat source. As a result, hurricane-force winds were maintained for hundreds of miles inland.

Because the storm was embedded in fast jet stream winds, it zipped northward at speeds of 43–48 kts (50–55 MPH). Within 24 hours of landfall in the Carolinas, Hazel traveled all the way to southeastern Canada. Sustained winds near 70 kts (81 MPH) were experienced in Norfolk, Virginia, and Washington, D.C., with gusts approaching 87 kts (100 MPH). Winds gusted to 78 kts (90 MPH) in New York. As Figure 8.32 shows, these high winds were generated along the right side of the storm's track. Note the steady progression of the 50 kt (58 MPH) isotach up the East Coast. The storm's fast movement contributed to these exceptionally strong inland winds.

A swath of heavy rain fell to the left of track, along the Appalachians. Overall, 10–15 cm (4–6 inches) of rain fell, with isolated pockets exceeding 15 cm (6 inches). The storm's rapid movement in this case limited the total rain accumulation.

Extremely Heavy Rainfall Far Inland Is Commonly Generated by Hybrid Cyclones, Particularly Over the Appalachians

Some of the greatest flood disasters in the history of the Mid-Atlantic and New England have unfolded when post-tropical cyclones and their hybrids interact with mountainous terrain and preexisting fronts. Disastrous floods ensue, including flash floods and river floods. Mudslides and debris flows along mountain slopes may also result.

Hurricane Floyd (1999; Figure 8.33) generated exceptionally heavy rains during its extratropical transition. Floyd struck the Carolinas on September 16 as a strong Category 2 hurricane. As Figure 8.34 shows, the storm tracked northeastward along the Outer Banks and Delmarva Peninsula. An elongated swath of exceptionally heavy rain, 18–25 cm (7–10 inches), fell from North Carolina through Maine, impacting 13 states. Maximum amounts of rainfall (38–51 cm [15–20 inches]) accumulated over portions of eastern Virginia and North Carolina. These rains generated widespread river flooding across eastern North Carolina, where several major rivers exceeded 500-year flood levels, cresting 6–7 m (20–23 feet) above flood stage. Floodwaters were heavily contaminated with pesticides, fertilizer, and hog waste across this largely agricultural region. Damage to dwellings from the storm's effects was extensive – nearly 75,000 homes – forcing 10,000 people into temporary shelter, with total economic losses approaching $5 billion. We should note that Floyd's flood was "preconditioned" by the arrival of Tropical Storm Dennis, which only a week earlier deposited 18–25 cm (7–10 inches) of rain across eastern North Carolina.

As in the case of Hazel, the location of inland, heavy rain was along Floyd's left side. Nearly half of all landfalling tropical cyclones have a left-of-center (LOC) rain distribution. Figure 8.35(a) shows how LOC heavy rain is produced. The key aspect is interaction between the tropical cyclone and a jet stream trough, located upwind of the storm. Timing is everything. When the trough captures a tropical cyclone, copious water vapor condenses in the trough's vigorously ascending air. A steady supply of low-level tropical moisture enters the storm from the south and east

Figure 8.33 False-color infrared satellite image of Hurricane Floyd (1999) during its extratropical transition phase. (NOAA.)

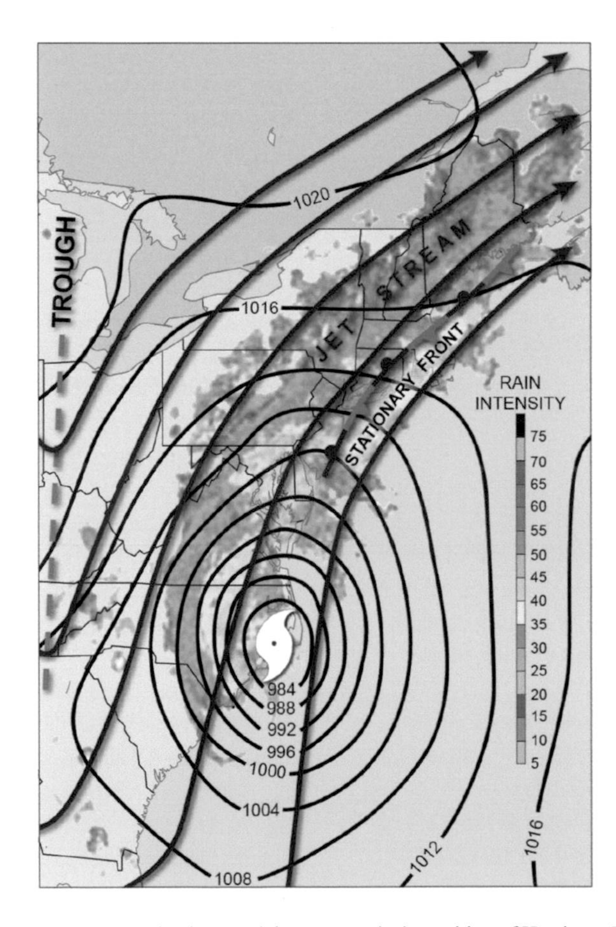

Figure 8.34 The various mid-latitude weather elements that hastened the extratropical transition of Hurricane Floyd over the Mid-Atlantic. Floyd became embedded in an upper-level trough and strong jet stream flow. Note the extension of heavy rain up the East Coast, hundreds of miles ahead of the main storm, along a type of coastal front, established by the strong, onshore flow of warm, humid air off the Atlantic. (Adapted from Atallah and Bosart, 2003.)

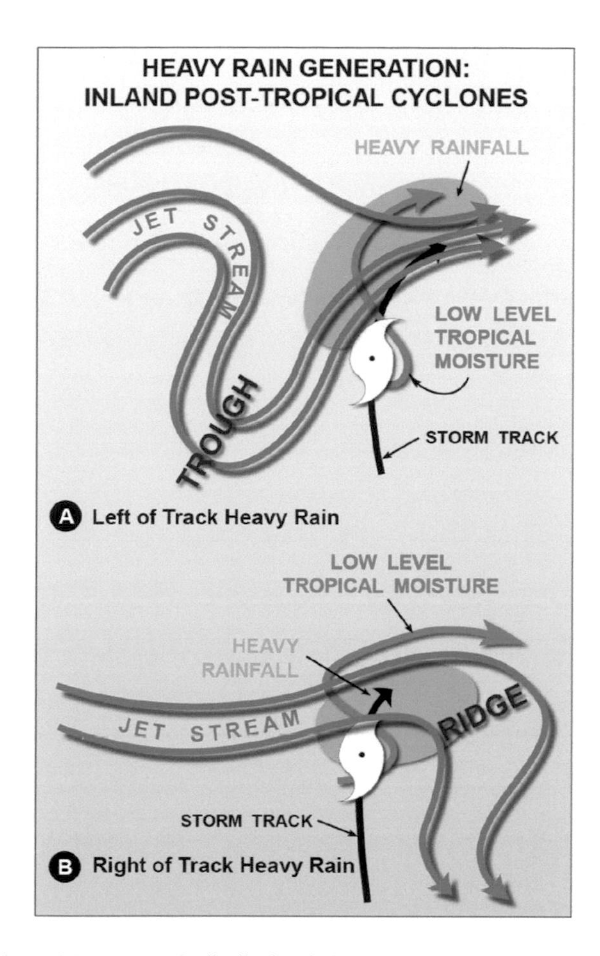

Figure 8.35 Heavy rain distribution during the hybrid phase of an inland, post-tropical cyclone. These models illustrate how the region of heavy rain formation shifts either left of the storm's track (a) or right of track (b), depending on the nature of the jet stream interaction. Both types occur in roughly equal proportion. (Adapted from Atallah et al., 2007.)

(green arrow). These weather systems are optimized, heavy-rain-generating machines. If a decaying tropical cyclone interacts with a ridge downwind in the jet stream, heavy rain generation shifts to right-of-center (ROC; see Figure 8.35[b]).

Finally, heavy rains can be generated when post-tropical remnants interact with mountainous terrain. In this situation, there is enhanced uplift of moisture-laden air along mountain slopes. This interaction can also involve the capture of the remnant by a mid-latitude weather system, in the manner of Floyd. Figure 8.38 shows examples of orographic rain enhancement in Hurricanes Agnes (1972) and Eloise (1975). These two storms produced the floods of record in Pennsylvania and Maryland, respectively. The flood of record is the greatest historical flood in a given state. Agnes struck the Florida panhandle as a Category 1 storm, Eloise as a Category 3. Curiously, both systems followed an identical inland track and produced the heaviest rain over northern Virginia, Maryland, and Pennsylvania, far from landfall.

In Agnes (Figure 8.36[a]), the forced ascent of humid, tropical air up east- and south-facing slopes of the Appalachians produced incredibly heavy rainfall over several days. The heavy rain was also left-of-center, implying (correctly) that a mid-latitude trough in the jet stream enhanced the uplift of moist air. A weather front parallel to the Appalachians and along the storm's northward track further enhanced rain production over Virginia and Maryland. When Agnes' post-tropical vortex moved off the North Carolina Outer Banks, a second vortex developed over the Appalachians and tracked northward, creating two rain-generating systems operating in tandem! The inland vortex eventually slowed, becoming nearly stationary over western Pennsylvania. The slow movement contributed to very heavy rain accumulations across Pennsylvania.

Hurricane Eloise (Figure 8.36[b]) took a track very similar to Agnes, but Eloise struck Florida as a Category 3. Major flooding once again ensued in the central and northern Appalachians, days after landfall, producing Maryland's flood of record.

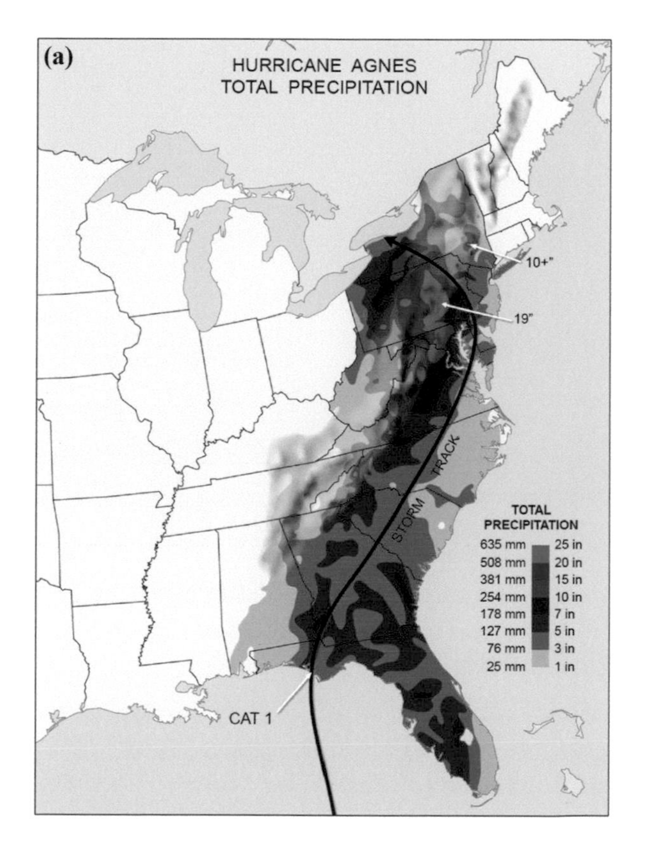

Figure 8.36 (a) Although making landfall in Florida, the remnants of Hurricane Agnes did not produce devastating rain until hundreds of miles north. There, rich tropical moisture combined with mid-latitude weather systems over steep terrain to produce Pennsylvania's historical flood of record. (b) Bearing an uncanny resemblance to Hurricane Agnes, Hurricane Eloise made landfall over Florida and disgorged heavy rainfall several days later over the Maryland-Pennsylvania mountains. (Adopted from NOAA.)

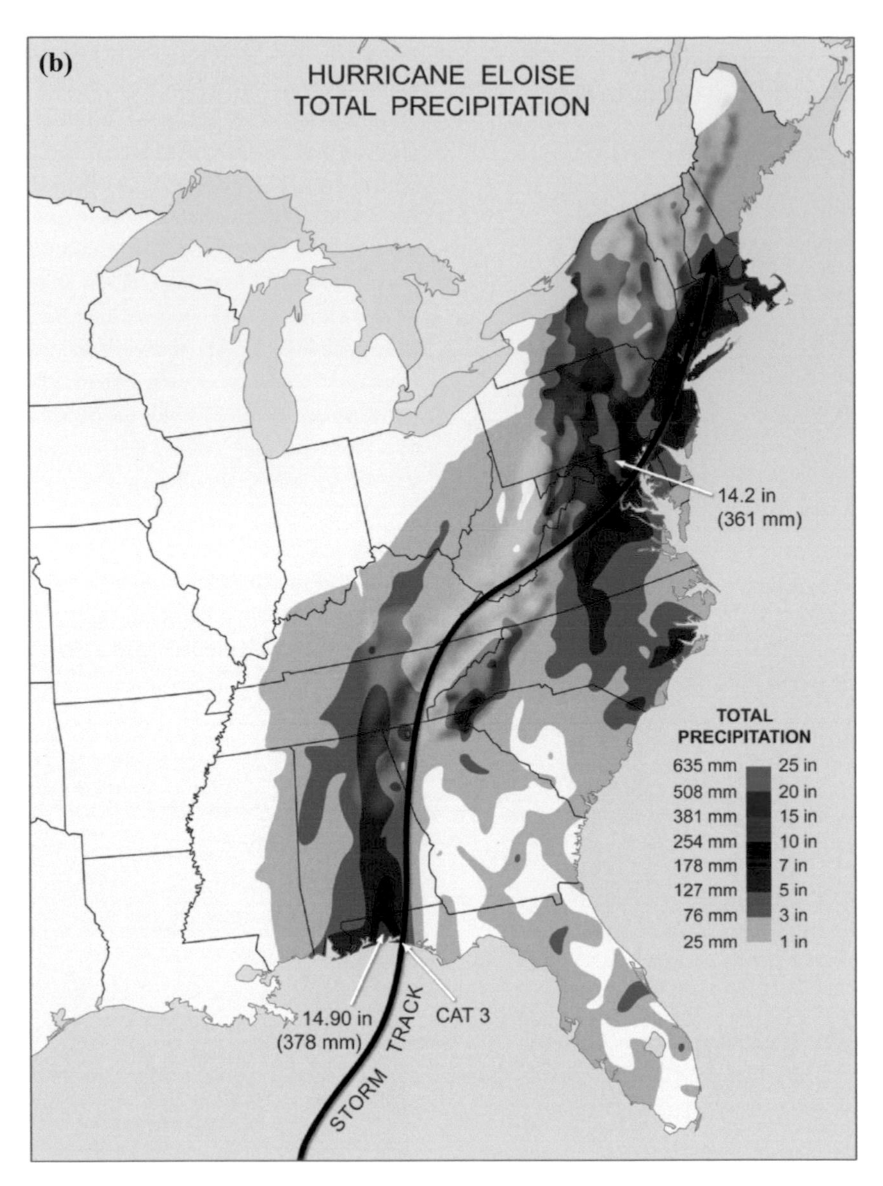

Figure 8.36 (Continued)

Following Landfall of a Weakening Tropical Cyclone, Numerous Tornadoes May Develop Inland

In addition to heavy rain, hurricanes sometimes play a final destructive card after landfall. When a tropical cyclone forms a tornado, the parent storm's spin is focused down to very small time and space scales. A tornado is a type of severe local storm. However, there are a few "twists" involved in the process. First, the tornadoes are most likely to occur only when a tropical cyclone weakens over land. Second, the spin actually arises from horizontal twisting of air. Third, tropical-cyclone-spawned tornadoes are among the weakest of all tornado intensity categories, with winds that rarely exceed 117 kts (135 MPH) – and more commonly 57–97 kts (65–110 MPH).

To understand the process that creates these tornadoes, examine Figure 8.37. Surface friction causes winds in the lowest mile of the hurricane to slow down. As in the case of wind streak formation, friction creates a vertical wind shear. Air within this layer begins to "tumble" over itself, forming long, horizontal "rollers" above and parallel to the ground, and perpendicular to the wind direction. Rain bands in tropical cyclones contain convective cells with updrafts. (Updrafts are narrow currents of rapidly rising air in convective cloud cells, as described in Chapter 9.) As vortex loops are drawn into an updraft, they tilt from horizontal to vertical. The updraft then stretches the vortex tubes vertically, such that they contract, intensifying their spin (think of the ice skater effect, recalling conservation of angular

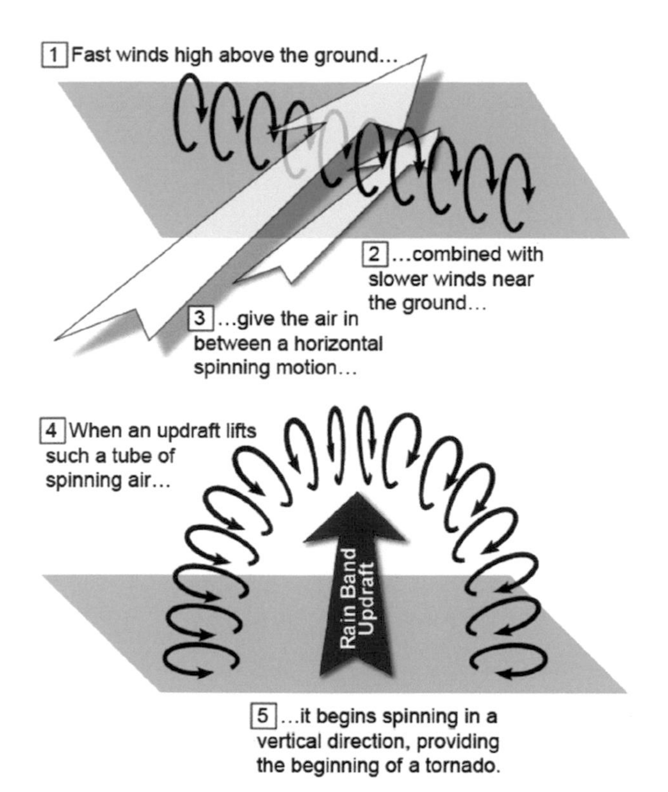

Figure 8.37 Tornadoes in tropical cyclone rain bands acquire their "twist." The spin starts in the horizontal plane, as fast winds aloft overtake air slowed at the surface by friction. This process imparts a rolling motion to the air. Later, updrafts in rain band convective cells draw up loops of spin, tilting horizontally rotating air to the vertical. Stretching of air in the updraft amplifies the vortex into a weak tornado.

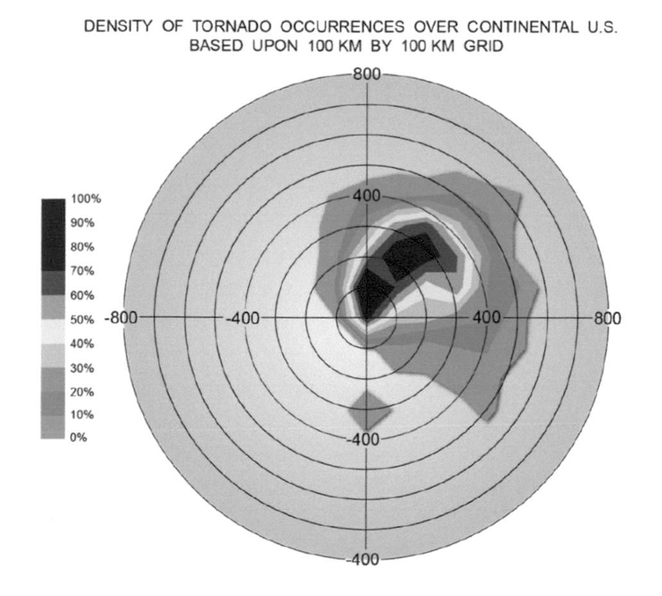

Figure 8.38 Location of tropical cyclone tornadoes after landfall. Most of these tornadoes, while characteristically weak, tend to cluster in the northeast quadrant of post-landfalling storms. (Adapted from Schultz and Cecil, 2009.)

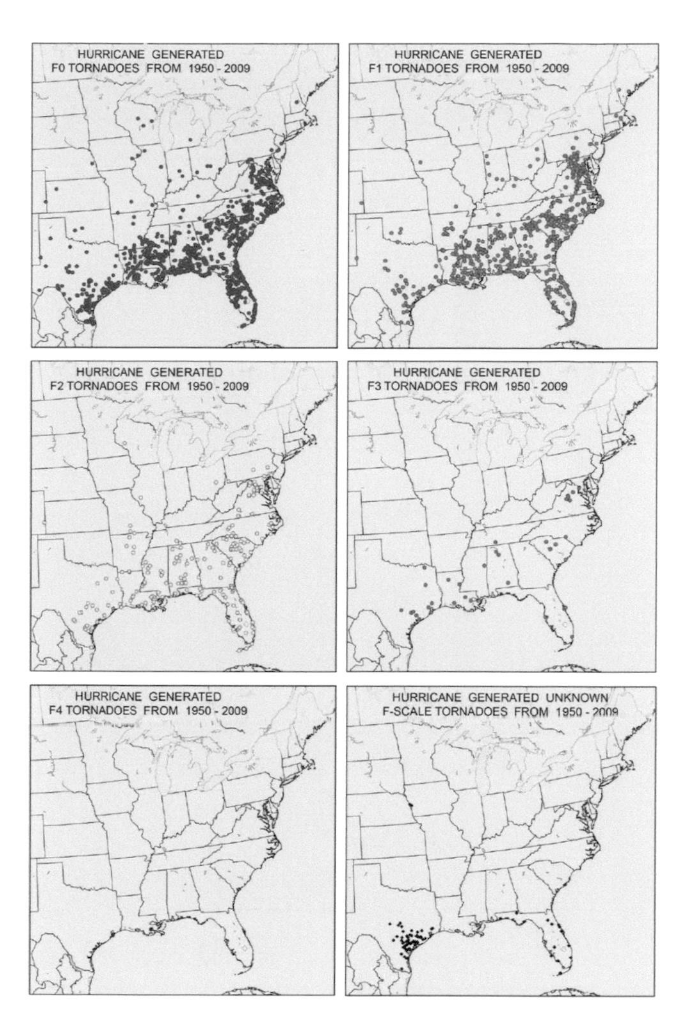

Figure 8.39 Breakdown of tropical cyclone tornadoes from 60 years of data, by intensity (colors) and location relative to the coastline. The great majority of these tornadoes are considered weak and short-lived, and most occur within 322 km (200 miles) of the coast. (Data from Schultz and Cecil, 2009.)

momentum). In this manner, one or more tornadoes are created within the spiral rain bands of a landfalling tropical cyclone. But over open water, there is not enough surface friction to slow down the air and get it to "tumble."

Figure 8.38 shows the typical distribution of tornadoes within a landfalling tropical cyclone. The distribution is asymmetric, with most tornadoes developing in the storm's right front quadrant and within a 400 km (249 miles) radius of the storm's center. This quadrant is where the fastest winds flowing around the right of track (in part due to storm motion) are slowed by surface friction.

Figure 8.39 shows the location of all hurricane-produced tornadoes over the United States for a 60-year period. These data

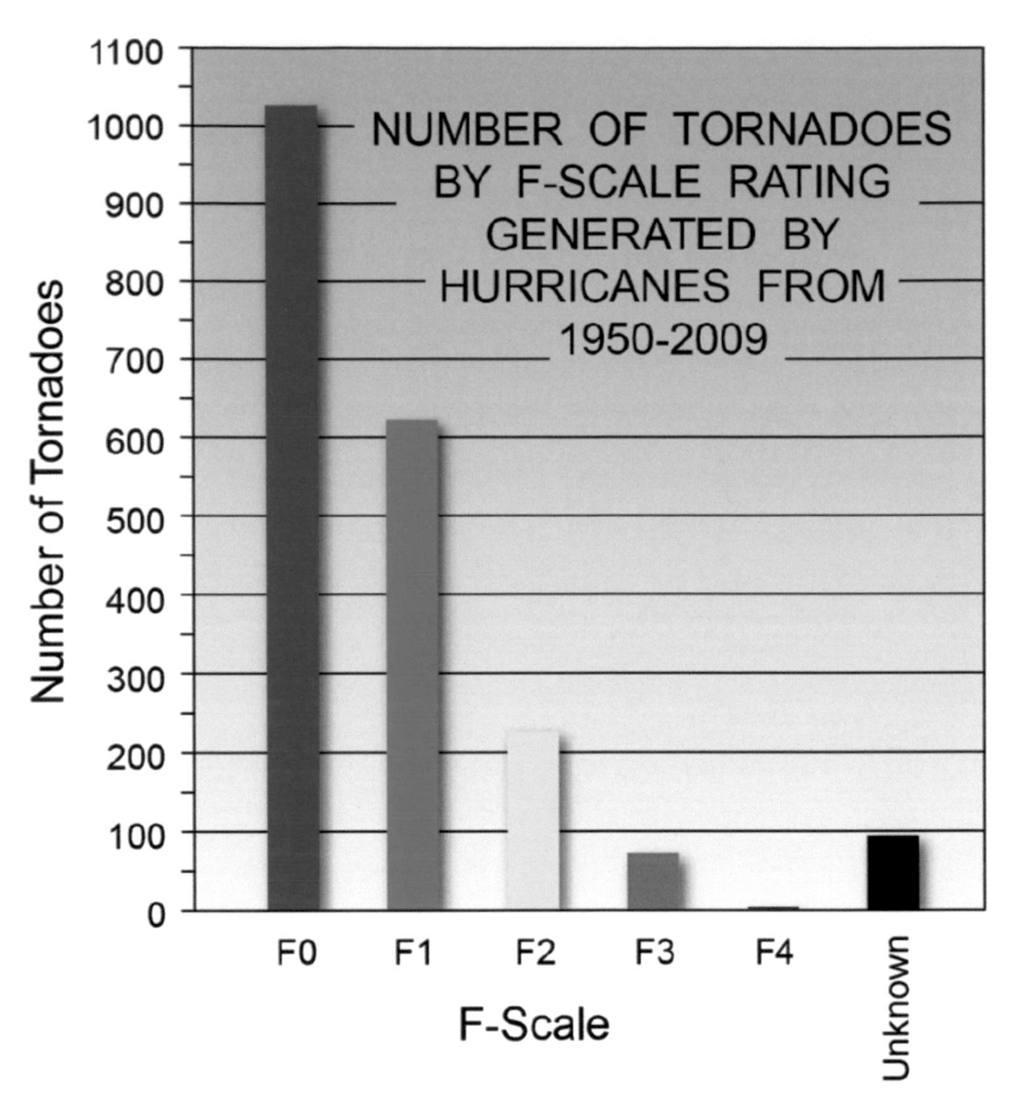

Figure 8.40 Frequency distribution of tropical-cyclone-spawned tornadoes in the 60-year study shown in Figure 8.39. (Data from Schultz and Cecil, 2009.)

are stratified according to Fujita or F-scale wind intensity, which ranks tornadic wind speed on a scale from F0 (weak tornado) to F4 (violent tornado). Most tropical cyclone-spawned tornadoes are weak F0–F1 category (winds to 96 kts [110 MPH]), and occur within a few hundred kilometers of the coastline; 63% of all these tornadoes form within 161 km (100 miles) of the coast, and an additional 25% within 322 km (200 miles) of the coastline.

Figure 8.40 presents a histogram (frequency distribution) of tropical cyclone tornado intensity, based on this 60-year climatology. Note that about only 15% of these tornadoes exceed F1 intensity.

Until 2004, the greatest number of tornadoes for any U.S. hurricane stood at 115, spawned by Hurricane Beulah (1967) across Texas. Beulah's record number of tornadoes was beaten by Hurricane Ivan in 2004 as it tracked across the Southeast and Mid-Atlantic. Ivan created pockets of numerous tornadoes (called tornado swarms) over a 4-day period, as Figure 8.41 shows. Ivan came ashore as Category 3 in Alabama. The storm generated 117 tornadoes over land, including a swarm of 40 tornadoes over northern Virginia on September 17. A small number of these tornadoes were rated as F2–F3 (winds up to 143 kts [165 MPH]). On each day, tornadoes tended to cluster during the afternoon – the time of maximum solar heating, when the atmosphere is most unstable (i.e., warm surface air will rise through a deep layer, forming thunderstorms). This makes sense, since strong convective updrafts are needed to tilt horizontal vortex tubes into the vertically spinning tornadoes.

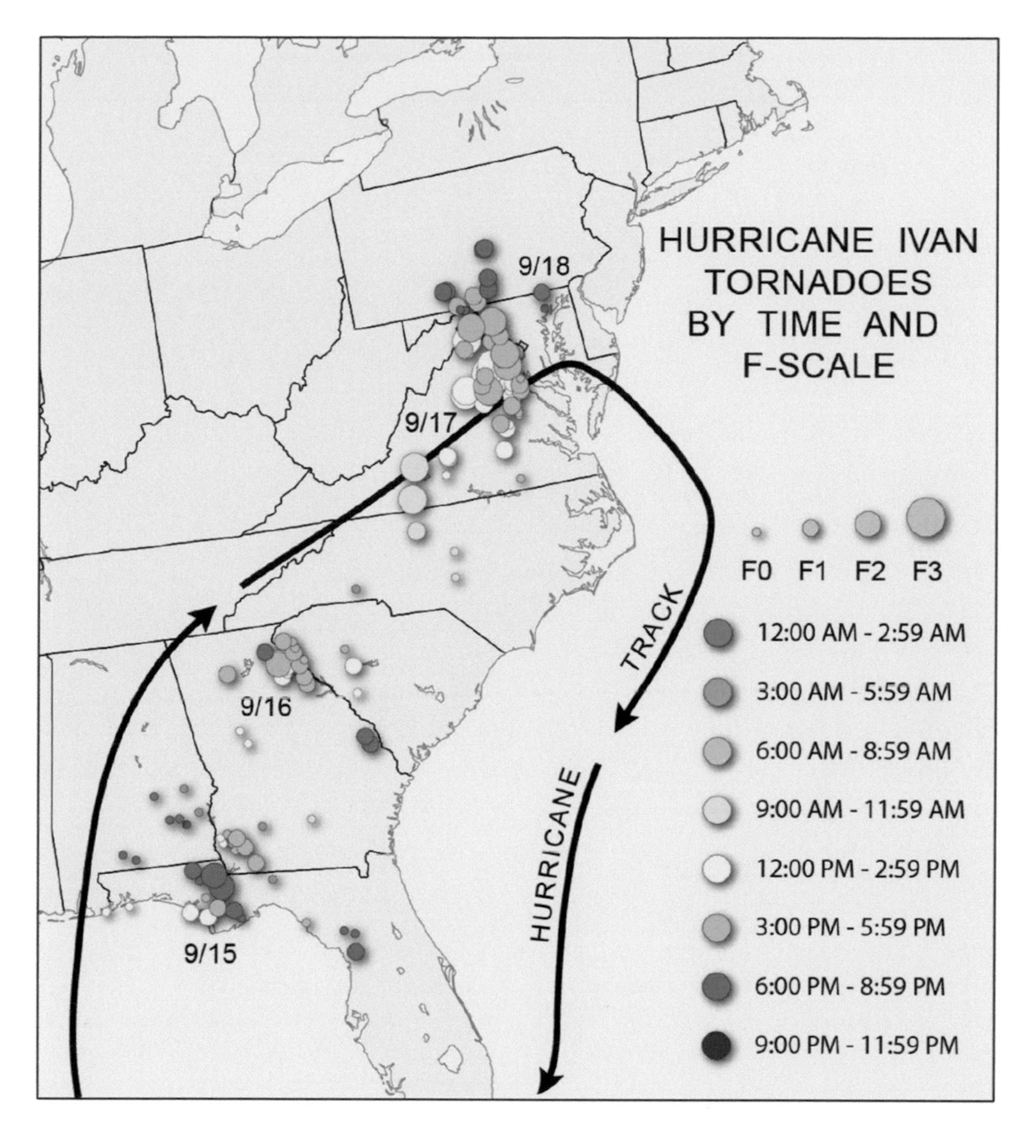

Figure 8.41 Hurricane Ivan (2004) was the most prolific inland-tornado-producing hurricane in U.S. history. Several tornado outbreaks were generated along Ivan's multiday path through the Southeast and Mid-Atlantic. Outbreaks clustered in the afternoon hours, when the atmosphere was most unstable from solar heating.

Amazing Storms: Hurricane Katrina (2005) – Anatomy of the Great Flood

Hurricane Katrina marked the first time the United States crossed the $100 billion disaster threshold. It was also the first time in 77 years that fatalities from a single storm (1833 deaths) surpassed the 1000-death mark. In many ways, we have made remarkable progress in terms of predicting storm track and landfall location, in some cases up to 7–10 days in advance. But our nation's exponential growth of population and property in low-lying, extremely vulnerable locations such as the Gulf Coast (Figure 8.42) and New York Harbor have rendered our nation more susceptible than ever to catastrophic hurricane death tolls, once thought to be a relic of centuries past.

Katrina attained rare Category 5 status in the Gulf of Mexico and maintained this intensity up to 24 hours from landfall. As Figure 8.43 shows, the center passed very close to downtown

Figure 8.42 Doomsday in New Orleans, early September 2005 – a great flood at the hands of Hurricane Katrina and a failed defense system. (NOAA.)

New Orleans, but the city was spared the highest winds and surge, which were located in the storm's right semicircle. A massive storm surge, in the range of 7–8 m (23–26 feet) obliterated sections of the Mississippi coastline. Portions of the coastline were inundated to a distance of 10–19 km (6–12 miles) inland. New Orleans and surrounding parishes, which sit 1–2 m (3–7 feet) below mean sea level, were 80% submerged from a combination of ocean and lake water. For years, it was thought that an extensive system of levees and pumps would protect the city from a storm of up to Category 3 intensity. As Figure 8.43 shows, Katrina was rapidly weakening as it moved north across Louisiana and could be properly characterized as a system transitioning from Category 3 to Category 2 as it passed the city. In spite of this weakening, 53 levees in and around New Orleans were breached or overtopped during a 6–8 hour period. The failure of the city's defenses – which can be partly traced to inadequate design and construction – is the worst civil engineering disaster in U.S. history.

Figure 8.44 shows that Katrina's great flood unfolded in two parts. The first came in the early hours of August 29 as the storm approached Lake Borgne. While still a Category 3 storm, high winds from the east – the storm's First Wind – built a high storm surge in the progressively narrowing Mississippi Sound. This surge flowed westward through Lake Borgne and a portion entered the Industrial Canal, a straight channel created as a shortcut connecting New Orleans to the Gulf of Mexico. The Canal funneled and concentrated the surge further, amplifying it by 20% and propelling it rapidly westward at high speed into the city. Figure 8.46 shows the arrival of this surge at the Industrial Canal's mouth around 7 a.m. Levees along the Canal protecting the city's western boroughs failed, and Gulf saltwater began pouring into the city.

Throughout the morning, Katrina continued to move rapidly northward. By noon, the center passed to the northeast of New Orleans and the storm weakened to Category 2. A strong, sustained wind from the north-northwest – the storm's second wind – set up a southward-flowing surge across Lake Pontchartrain. Surge and storm waves exerted tremendous pressure along the southern wall of the lake, blowing out several levees along canals connecting the lake with the city's northern boroughs. These included the 17th Street and London Avenue canals. These levees succumbed between 10 and 11 a.m.

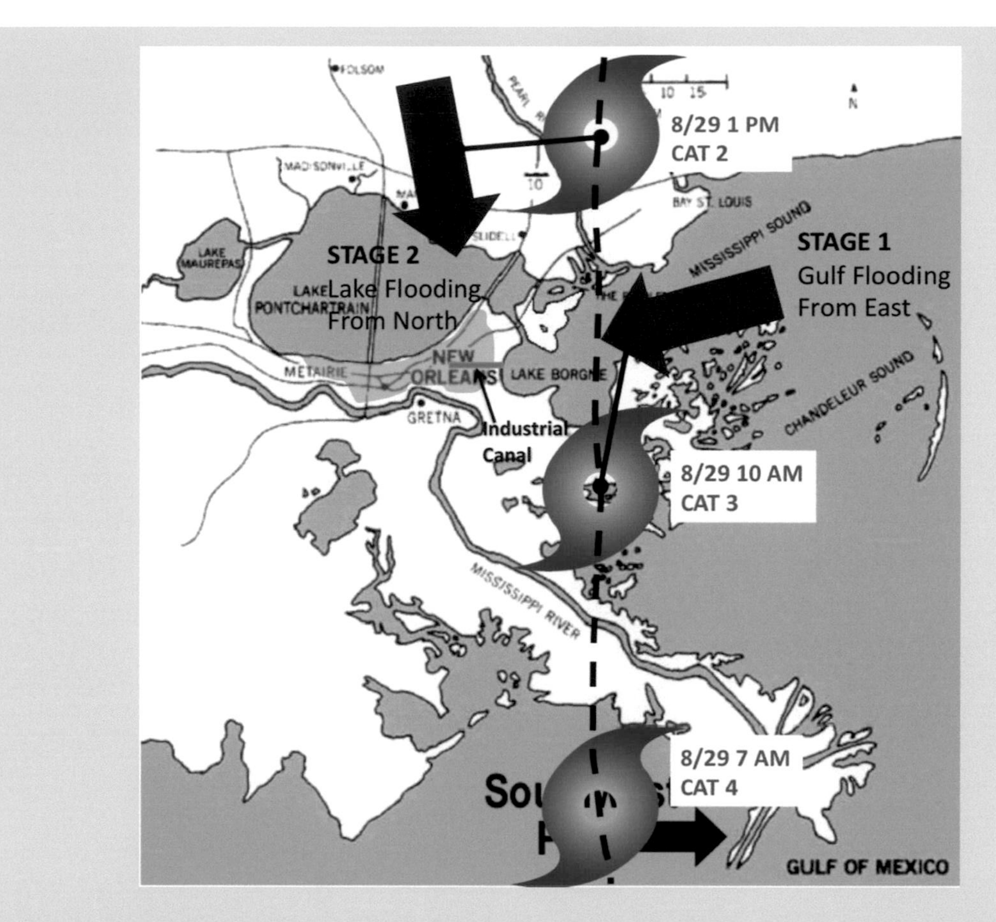

Figure 8.43 Schematic showing the timing and direction of Katrina's assault on New Orleans, compared to the track and intensity of the storm. The Great Flood unfolded in several stages and from multiple directions.

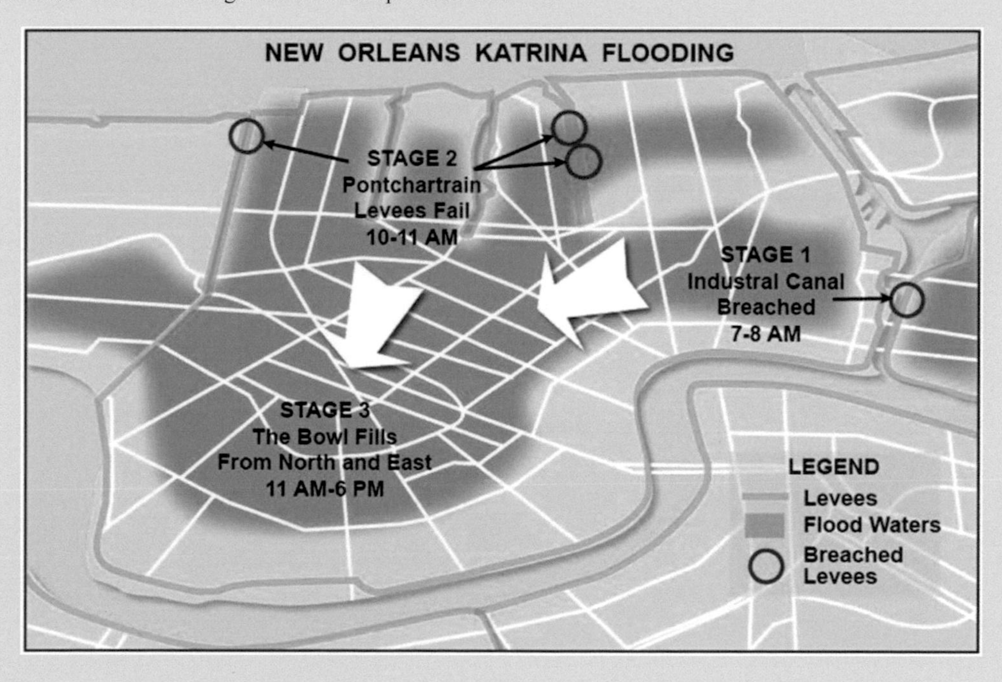

Figure 8.44 Sequence of the Great Flood's three stages and locations of principal levee failures.

Amidst the chaos of the storm – including the high wind and heavy rain, power outages, impassable roads, and communication system failures – authorities were slow to realize when and where the breaches occurred. But throughout the afternoon, the central city's vast bowl was filling with water. This third stage continued into the evening of August 29. The floodwaters were derived from the Gulf via the Industrial Canal breach and Lake Pontchartrain. It took nearly two months to pump out the city; the city's defenses were optimized to prevent water from entering the city, not removing a massive volume of water. The nightmare, worst-case scenario that scientists and engineers had predicted for decades had finally unfolded. And once again, this grand, historical city – dry and in the midst of recovery – lies vulnerable to another Great Flood.

Amazing Storms: Hurricane Sandy (2012) – Diverse Impacts From a Hybrid Super Storm

Hurricane Sandy was a meteorological surprise with so many bizarre twists that many scientists (including the authors of this text) were left shaking their heads in disbelief. Like Katrina, this system garnered "superstorm" status in terms of its size, power, impacts, and price tag approaching $100 billion. Like many of the large storms before it, Sandy underscores the vulnerability of low-lying, densely populated regions including New York Harbor. It also foretells a new era of extreme devastation from natural hazards such as hurricanes. This hazard exists despite the nation's technological prowess and the fact that the United States has arguably done the most of any nation to advance the science of hurricane behavior and prediction.

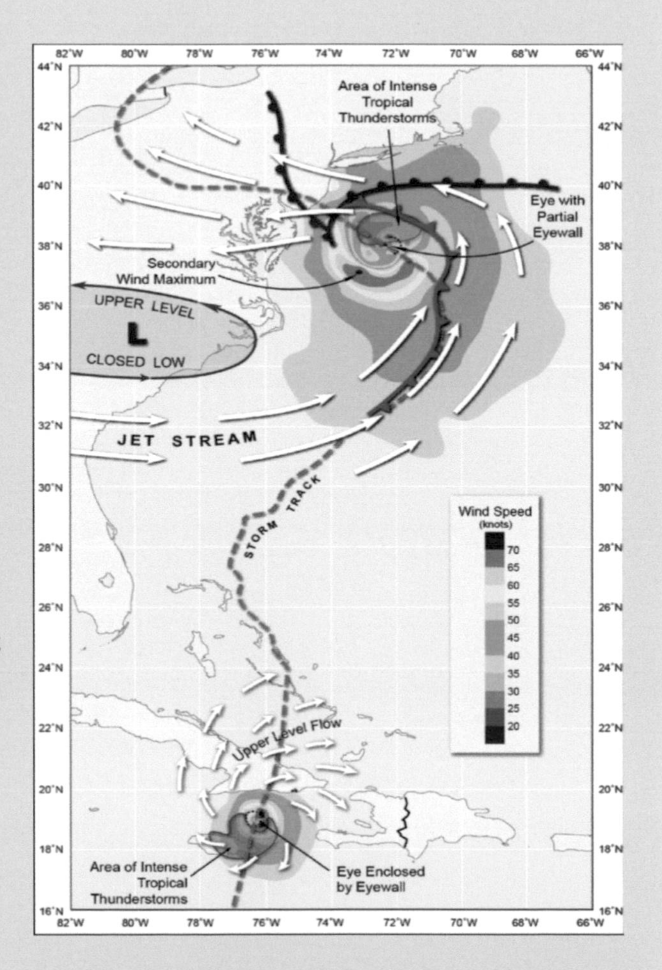

Figure 8.45 The life and times of Superstorm Sandy. Shown here is the complex evolution of a cyclonic vortex through several different configurations, including hurricane, hybrid vortex, and massive Nor'easter. Track is shown by the dotted line.

Medium-range (7–10 day) weather prediction has improved markedly in the past 10–15 years. Some of the forecast models – namely the ECMWF – were portraying an ominous (and largely accurate) scenario as many as eight days in advance of Sandy's landfall. The models were also suggesting two very uncommon characteristics for a hurricane off the East Coast. First: The tropical cyclone would combine with a deep, cold mid-latitude trough in the jet stream, creating an enormous hybrid cyclone – the largest ever documented in the western Atlantic – with a blend of hurricane and Nor'easter characteristics. Second: The huge vortex would turn due westward, straight into the New Jersey shoreline. Such a sharp, left-hand turn is unprecedented in all hurricane tracks documented since the mid-1800s and presented a forecast dilemma, as not all of the forecast models predicted this "left hook" (Figure 8.45). The bizarre turn was due to a highly anomalous jet stream track, with large distortions (troughs and ridges) across the eastern United States, Canada,

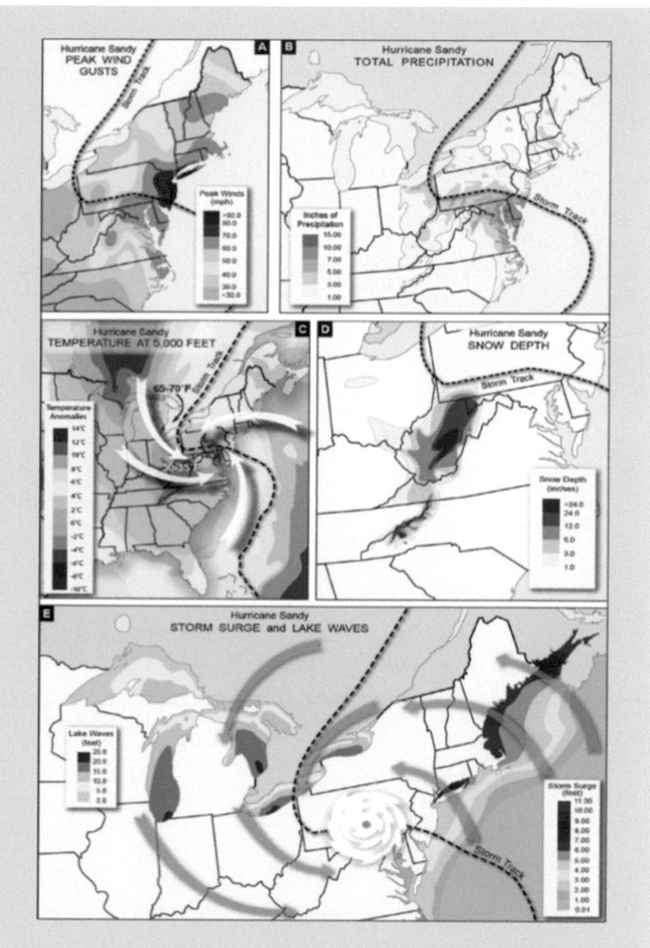

Figure 8.46 Superstorm Sandy was a multifaceted storm in terms of weather and water impacts – combining high wind, heavy rain, bizarre temperature fluctuations, heavy snow, destructive coastal storm surge, and high waves across the Great Lakes.

and Greenland. A massive ridge of high pressure sat south of Greenland, blocking northward progress of the storm. Meanwhile, an unseasonably intense trough of low pressure over the eastern United States would act to pull the storm westward, directly into the East Coast.

Figure 8.45 also traces the evolution of this superstorm from its purely tropical beginnings to a massive hybrid system off the Mid-Atlantic coastline. The hurricane developed unusually late in the Atlantic hurricane season (the last week of October) and reached near-Category 3 status before weakening over Cuba. Once north of the islands of the Caribbean, the storm encountered a series of upper-level troughs in the westerly flow. A particularly potent vortex lay over the southeastern United States and contained a core of very cold Canadian air. This vortex entrained Sandy into its periphery and began circulating a tongue of cold air around the system's warm tropical core. The strong thermal gradient thus created enhanced the storm's kinetic energy and caused the vortex to expand. A large-diameter ring of high wind developed on the storm's southwest side, at a great distance from the center and to the left of track. In essence, this tossed the textbook notion that a hurricane's strongest winds occur in the right semicircle "to the wind"! The inner core still retained tropical characteristics, including a warm central eye and ring of eyewall thunderstorms.

Just a few hundred miles off the coast of New Jersey and New York, Sandy was morphing from this hybrid configuration into a full-scale Nor'easter, complete with warm and cold fronts and a massive precipitation shield. In fact, the storm made landfall as an extratropical cyclone. This created a novel dilemma for forecasters: Should they use hurricane warnings, which convey a heightened level of seriousness for emergency planners, or should they instead issue separate gale warnings, flood warnings, and coastal flood warnings under the guise of the more technically correct (and powerful) Nor'easter? This was not an easy decision. Ultimately, the hurricane warnings were dropped, and this action did create confusion among those making preparations for Sandy's impacts. Since then, however, the National Weather Service has made a policy decision to stick with hurricane warnings should a powerful hybrid system again threaten the coast.

Figure 8.46 illustrates the great diversity of oceanic and sensible weather impacts during Sandy's transit from New Jersey to Ontario. Highest sustained winds with gusts to 70–78 kts (81–90 MPH) occurred to the right of track, along the New Jersey shoreline and Long Island (a). Heavy rain was concentrated left of track, consistent with our model of extratropical transition involving a jet-stream trough located upstream (b). Panel (e) shows the height of the storm surge. Note the high values over a large stretch of coastline to the right of track. Surge was particularly focused in narrow inlets including New York Harbor, Long Island Sound, and the Gulf of Maine. For decades, hurricane scientists and emergency planners had noted the unique vulnerability of New York Harbor to hurricane storm surge, given its funnel-shaped geometry that causes the surge to concentrate and build in amplitude. Population density is very high in the lowlands surrounding the Harbor. Note also the unusual occurrence of lake effect surge. So massive was the strong wind field of this storm, that waves of up to 8 m (26 feet) were raised along the lee shores of the Great Lakes, hundreds of miles from the seaboard.

Panel (c) reveals another bizarre impact: an inverted temperature gradient across the Northeast and Mid-Atlantic. Instead of warm temperatures to the south and colder conditions to the north, Sandy's circulation drew unseasonably cold air from Canada across the central Appalachians. Meanwhile, an influx of warm air from the

Atlantic's Gulf Stream was circulated over New England. While New York and Maine enjoyed high temps to near 21° C (70° F), portions of West Virginia were plunged below freezing. This set up the Central Appalachians for perhaps Sandy's most bizarre twist of all: heavy, wet snow, accumulating to nearly 1 m (3 feet) in spots (d), from moisture of hurricane origin!

Compared to Katrina, Sandy's death toll was much lower (around 150 direct fatalities in the United States), but the property damage was exceptionally large, given the high density of beachfront properties and seaside communities and the infrastructure and population of New York City. Both of these storms were essentially "worst-case scenarios" for two of the most hurricane-vulnerable regions in the United States.

Amazing Storms: Hurricane Harvey (2017) – A New U.S. Flood Record

On August 25, 2017, a monster Cat 4 hurricane named Harvey slammed onshore along the Texas coastline. All of the expected catastrophes attending such a powerful storm lashed at Texas, including sustained winds of 113 kts (130 MPH), a destructive storm surge, and torrential rains. The winds within the 32 km (20-mile) diameter eyewall were notably destructive, wherein smaller, embedded vortices of tornadic intensity completely leveled blocks of homes in Rockport, Texas. Within hours of landfall, Harvey's wind intensity rapidly slackened. But this was just the storm's opening act. In spite of weakening to a tropical storm by the next day, a far more significant flood disaster would unfold along the Texas coast over the next several days, producing the greatest rain accumulation on record in the United States from a tropical cyclone.

The storm's rain totals are portrayed in Figure 8.47, over a 5–6 day period, as the storm slowed to a crawl and stalled just offshore of the Houston region. Rain totals of 51–76 cm

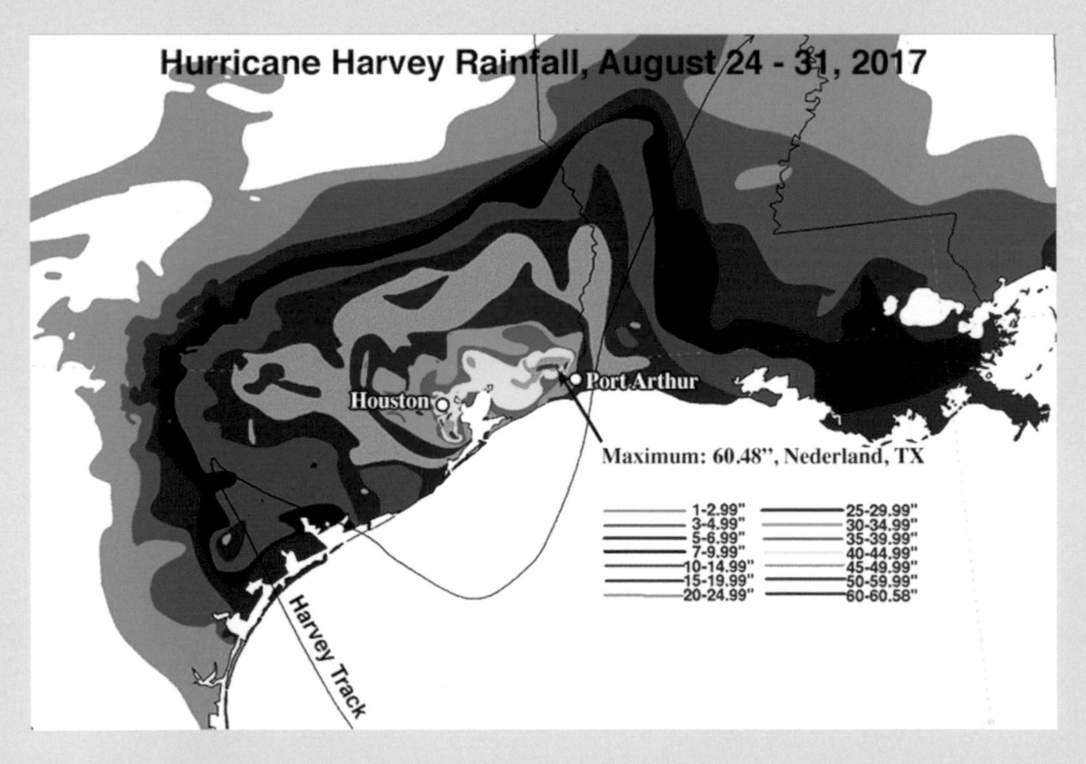

Figure 8.47 Rain accumulation map for Hurricane Harvey along the U.S. Gulf Coast. Storm track (generally from west to east) is indicated by the black line. (NOAA.)

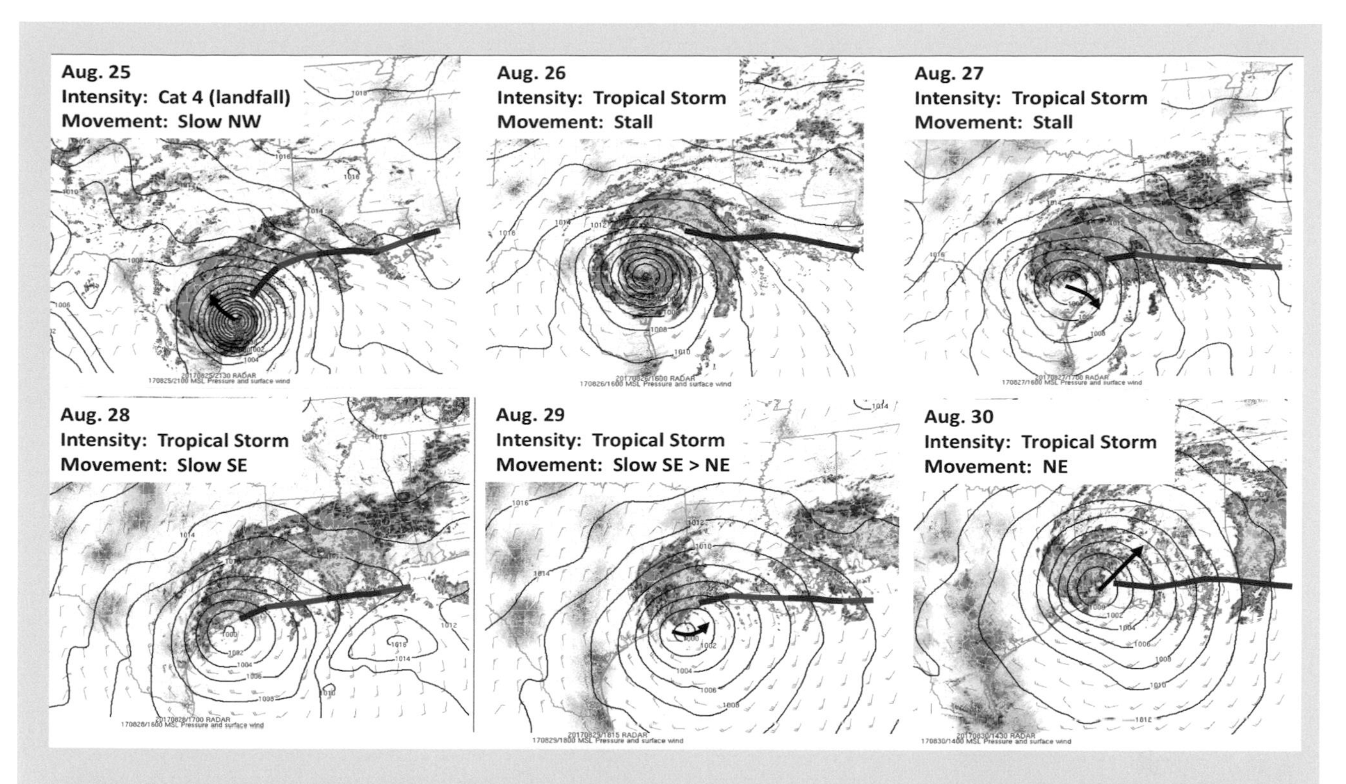

Figure 8.48 Daily surface weather charts and radar imagery during Harvey's rain-laden trek across the Gulf Coast. Radar snapshots are shown by shaded color regions (orange = heaviest rain rates). Circular black lines denote isobars. Heavy black arrows indicate direction of storm movement from one day to the next. Stalled weather front is showed by red and blue heavy lines. (Adapted from NWS.)

(20–30 inches) were widespread across the central and northern Texas coast, but a more extreme zone of 100–113 cm (40–45 inches) encompassed greater Houston and points to the east, including Port Arthur. Several smaller pockets of up to 150 cm (60 inches) occurred just south of Houston and north of Port Arthur, with an absolute point maximum of 151 cm (60.5 inches) reported at Nederland, Texas. Calculations suggest that the total volume of rainwater dumped across the Texas coast would completely fill a cube 4.8 km (3 miles) wide on each side, containing a total of 25 trillion gallons of fresh water.

The Houston urban region, with all of its pancake-flat, suburban sprawl, is the fourth most populous spot in the United States. Much of what used to be a permeable landscape with tall, dense prairie grass, has been subsumed by impervious blacktop, concrete and roofing material. These factors, and proximity to a rich moisture source (the Gulf of Mexico) create a "perfect storm" of conditions for catastrophic flooding.

Figure 8.48 shows a six-day analysis of weather maps during Harvey's slow transit across Texas and Louisiana. Notice, for instance, how rapidly the overall storm weakens between August 25 and August 27 (the isobars became much less numerous and more spread apart). The great irony is that the formerly fierce windstorm at landfall transformed into a much feebler wind event – while its highly efficient, rain-producing apparatus ramped up. Much of the storm's circulation moved back over the warm Gulf waters, where winds helped evaporate vast amounts of water vapor. A stalled frontal boundary along the Gulf Coast helped converge and lift that moisture to the east of Harvey's center. Corridors of rain cells blossomed and moved off the Gulf and over the same regions in and around Houston repeatedly, for days. This effect, called training (described further in Chapter 14), generated a pulsing of rains with rates at times exceeding 5–7.5 cm/h (2–3 inches/h).

Summary

LO1 Understand the processes by which hurricanes weaken over land. What set of hazards attend hurricanes making landfall along the immediate coast? What hazards unfold once the hurricane is deep inland?

1. Hurricanes moving over land weaken primarily through loss of sensible and latent heat release, as the circulation moves away from the ocean heat source. Additional factors include wind shear, which disrupts the vortex; ingestion of dry air; and movement over cool, coastal shelf water.
2. Along the immediate coast, the principal hurricane hazards are heavy rain, high wind and storm surge. Freshwater flooding accounts for most U.S. hurricane fatalities.
3. Inland hazards often continue in the hurricane's post-tropical phase, sometimes hundreds of miles inland. These effects include flooding from heavy rainfall and swarms of tornadoes.

LO2 Describe the formation of the storm surge, its asymmetry, and the manner in which it is compounded by tidal variations.

1. Storm surge refers to the wind-induced rise of ocean water, building in height as the storm moves over shallow water. Storm surge height is greatest in the storm's right semicircle.
2. Storm tide combines the storm surge with tidal variations along the coast. Surge is enhanced during high tide, and its impact is reduced during low tide.
3. Inlets and embayments along the coast enhance the height of storm surge by funneling and concentrating the wave energy.
4. Integrated kinetic energy (IKE) combines the effects of the storm's maximum sustained wind and areal footprint of wind on the ocean. The total water area set into motion by storm wind is an important factor in determining surge height, independent of wind speed.

LO3 Distinguish between maximum sustained wind and gust factor. Describe how structures succumb to aerodynamic forces in high wind. Explain the rate at which hurricane winds decay once the storm is inland and how the areal coverage of damaging wind changes with time. Identify several factors leading to concentrated pockets of extreme wind damage.

1. A hurricane's maximum sustained wind (MSW) is the 1 minute averaged wind just above the surface. MSW varies inversely with the minimum central pressure. MSW ramps up exponentially between the outer edges of the storm and the eyewall.
2. MSW is not evenly distributed around the storm but is maximized in the right semicircle, where the swirling wind and storm motion act in the same direction.
3. Gust factor relates to the degree of turbulence in the airflow, with gusts exceeding MSW values by 20–30% over open water and 40–50% over land.
4. Aerodynamic forces due to wind impinging on structures are substantial, varying with the square of wind speed. Repeated buffeting of structures by gusts can lead to failure due to resonance effects. Shift in wind direction between the storm's first and second wind (crossing the eye) adds a further damage mechanism.
5. The storm's MSW typically decays to 50% of the over-ocean value within 12 hours of landfall, with continued exponential decay through 24 hours. The decay is often associated with an expansion of the storm's wind field. The expansion shifts the radius of maximum winds outward and may result in higher wind speeds at large distances from the center.
6. Downbursts within a hurricane's eyewall and rain bands are created by small pockets of rapidly descending air (intense downdrafts). They create localized (3–5 km [5–8 mile] wide) pockets of fan-shaped damage at the surface.
7. Mini-swirls are intense vortices embedded in the eyewall winds at the surface, only a few tens of meters across, creating streaks or swaths of enhanced wind damage.
8. Wind-streaks arise from the abrupt decrease in wind speed at the surface over land due to friction. Horizontal vortices or "rollers" created by

this wind shear transport high-velocity airflow to the surface, creating narrow swaths of enhanced wind damage just tens of meters wide.

LO4 Explain why hurricanes are capable of exceptionally heavy rainfall, in terms of storm motion, cell training, interactions with mid-latitude weather systems, and orography. What two scenarios explain how heavy rain is distributed with respect to storm track?

1. Tropical cyclones at landfall are very efficient rain producers, characteristically producing 15–30 cm (6–12 inches) of rain, sometimes much more. Rain efficiency arises from slow movement of the storm, training of convective rain cells within rain bands, and high moisture content of tropical air feeding the storm.
2. Total rain production in a tropical cyclone along the coast is proportional to the storm's average rain rate and storm diameter and inversely related to storm speed. Because so many factors influence rain production, total rain is often not well correlated with a tropical cyclone's wind intensity at landfall.

LO5 Define and give examples of the following: Post-tropical cyclone, tropical remnant, extratropical transition, hybrid cyclone, rejuvenated storm. Discuss how inland flooding and tornadic activity is sometimes enhanced inland of the coast.

1. A weakening tropical cyclone over land becomes post-tropical when it loses key defining attributes such as its eyewall and warm core, and it may further degrade to a remnant low (winds below 34 kts [39 MPH]).
2. Post-tropical systems may combine with one or more mid-latitude weather systems and begin to acquire extratropical characteristics. During this process, called extratropical transition, the system becomes a hybrid cyclone. If the hybrid acquires a cold central core and fronts, it completes its transition into an extratropical cyclone. Additional energy acquired from mid-latitude weather systems may enable the hybrid storm to briefly rejuvenate, with continued production of severe weather – including high, sustained winds – far inland.
3. During extratropical transition, extremely heavy rainfall – leading to flash floods, river floods, mudslide and debris flows – can continue inland over mountainous terrain. Flood catastrophes of this sort have included Camille (1969), Agnes (1972), Eloise (1975), and Lee (2011).
4. Generation of heavy rain during extratropical transition shifts to left-of-center when the post-tropical vortex interacts with a jet stream trough located upstream. A right-of-center distribution unfolds when the system interacts with a jet stream ridge downwind.
5. Hurricane tornadoes develop over land, paradoxically during the period when the parent storm is weakening. Land friction slows winds in the lowest 1 km (3300 feet), creating horizontal vortex tubes. Updrafts in rain band convective cells draw up vortex tubes, tilting them into the vertical, and amplifying their spin through vertical stretching. Tornadoes created in this manner are typically weak and short-lived.
6. Hurricane tornadoes tend to concentrate in the storm's right front quadrant and cluster in the afternoon hours, when solar heating and thus air mass instability is strongest.

References

Atallah, E. and L.F. Bosart, 2003. The extratropical transition and precipitation distribution of Hurricane Floyd (1999). *Monthly Weather Review*, 131:1063–1081.

Atallah, E., L.F. Bosart and A.R. Aiyyer, 2007. Precipitation distribution associate with landfalling tropical cyclones over the eastern United States. *Monthly Weather Review*, 135:2185–2206.

Emanuel, K., 2005. *Divine Wind: The History and Science of Hurricanes*. Oxford: Oxford University Press, 285 pp.

Evans, C., et al., 2017. The extratropical transition of tropical cyclones, Part I: Cyclone evolution and direct impacts. *Monthly Weather Review*, 145:4317–4344.

Fujita, T. Theodore, 1978. *Manual of downburst identification for Project Nimrod*. Satellite and Mesometeorology Research Project Paper No. 156, University of Chicago, Chicago.

Fujita, T. Theodore, 1992. *Memoirs of an Effort to Unlock Mystery of Severe Storms*. Chicago: University of Chicago Press.

Palmen, E., 1958. Vertical circulation and release of kinetic energy during the development of Hurricane Hazel into an extratropical storm. *Tellus*, 10:1–13, Sweden.

Powell, M.D., P.P. Dodge and M.L. Black, 1991. The landfall of Hurricane Hugo in the Carolinas: Surface wind distribution. *Weather and Forecasting*, 6:379–399.

Schultz, L.A. and D.J. Cecil, 2009. Tropical cyclone tornadoes, 1950–2007. *Monthly Weather Review*, 137:3471–3484.

Simpson, R.H. and H. Riehl, 1981. *The Hurricane and Its Impact*. Baton Rouge, LA: Louisiana State University Press.

Wakimoto, R.M. and P.G. Black, 1994. Damage survey of Hurricane Andrew and its relationship to the eyewall. *Bulletin of the American Meteorological Society*, 75:189–202.

PART III

Severe Local Storms and Their Weather Hazards

CHAPTER 9

Structure and Evolution of Ordinary Thunderstorms

Learning Objectives

1. Describe the three principal types of ordinary thunderstorms: single cells, multicells, and squall-line systems.
2. Explain the process of atmospheric convection, which is driven by buoyancy and develops in order to transport excess surface heat upward.
3. Discuss how wind shear gives rise to different types of thunderstorms.
4. Describe the characteristic life cycle of the thunderstorm cell, including development of the updraft, downdraft, and cloud electrification.
5. Understand how multicell thunderstorm systems propagate and move relative to the prevailing airflow.
6. Describe how squall lines develop and the hazards they bring.
7. Explain the geographical distribution of thunderstorms across the United States and how thunderstorms vary with time of day.

Introduction

This chapter examines thunderstorms, a common type of local storm. Severe local storms include not only thunderstorms but also their progeny – tornadoes, damaging straight-line wind, hail, intense lightning, and flash floods. But what exactly does "local" mean? In most cases, one or more thunderstorms are responsible for producing severe weather on time and space scales much smaller than that of an extratropical cyclone – hence the term "local," referring to regions that are tens to a few hundreds of kilometers in scale. Many times, these local storms are embedded within and move with extratropical cyclones. But occasionally severe thunderstorms may develop into massive systems, impacting more than one state and lasting 24 or more hours. In this case, the distinction between local scales (individual towns, cities, or counties) and synoptic scales (large, contiguous regions of the United Sates) becomes blurry.

Not all thunderstorms produce lightning and thunder. In fact, the more general term for any localized, deep storm cloud is atmospheric convection. Convection refers to the concentrated, vigorous, upward transport of warm air and compensating downward movement of cold air. Deep convective clouds over tropical oceans produce very little lightning compared to their land-based counterparts. In the United States, deep convective clouds may generate flash floods and even tornadoes, without appreciable lightning.

There Are Three Types of Ordinary Thunderstorms: Single Cells, Multicells, and Squall Lines

Thunderstorms develop in a great many ways and assume myriad forms. The simplest type is an isolated single cell. Larger and longer-lived clusters of single cells are called multicell storms. Extensive, fast-moving lines of cells can be organized into a squall line. In each case, the fundamental building block of the storm is a deep convective cloud cell containing an updraft and a downdraft. All three storm types can produce various types of severe weather, including torrential rains and flash flooding, intense lightning, large hail, damaging winds, and tornadoes.

But these three archetypes are part of a spectrum of thunderstorms; there are many "shades of gray." Storms frequently morph from one type to another, and large thunderstorm systems often contain more than one type. So in a sense this chapter oversimplifies the wide variety of storms found in nature.

DOI: 10.4324/9781003344988-12

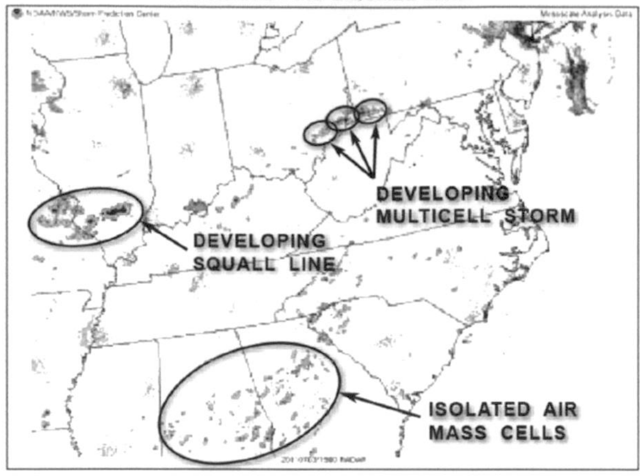

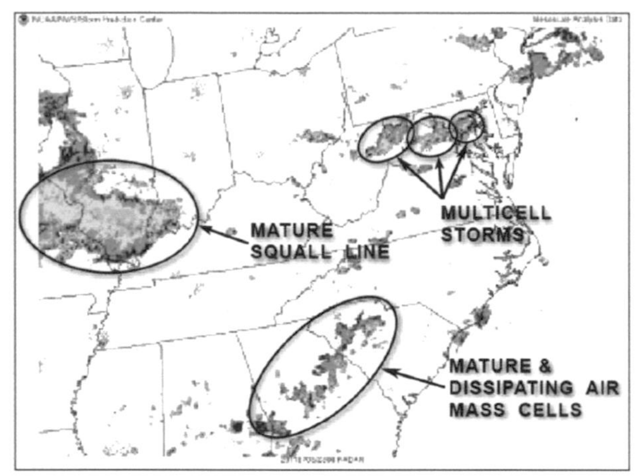

Figure 9.1 Visual depictions of thunderstorm types. The first radar image (top) illustrates isolated air mass thunderstorm cells, a developing multicell storm, and a developing squall line. As time progresses, these entities mature into more dominant storm features (bottom).

Figure 9.1 illustrates these three basic storm types. This is a weather radar snapshot of thunderstorms on a summer day in 2011. You will note the simultaneous development of all three storm types: isolated single thunderstorm cells over the southern Appalachians, a group of multicells growing along a cold front in the Ohio Valley, and a squall line developing along a stationary front in the upper Mississippi Valley.

New Technologies Help Us Detect Thunderstorms

Technology – including weather radar and satellites – has helped revolutionize our understanding of thunderstorms. It now permits much more timely and geographically specific warnings of severe weather hazards. Figure 9.2 shows four different views of the same thunderstorm complex over Pennsylvania's Allegheny Plateau. The areas impacted by these thunderstorms appear very different, depending on the type of detection equipment being used. Panel (a) shows a visible satellite image of the storms, presenting the cloud top structure. Panel (b) shows the same cloud tops using infrared (in false colors, with the coldest and tallest clouds colored orange). From these two images, one might conclude that potentially severe weather is impacting much of western Pennsylvania. However, NEXRAD radar reflectivity (Panel [d]) reveals that only narrow corridors within these clouds were producing heavy rainfall (the most intense rain cells are denoted by red and magenta). However, these rain regions also contained intense lightning, as shown in Panel (c).

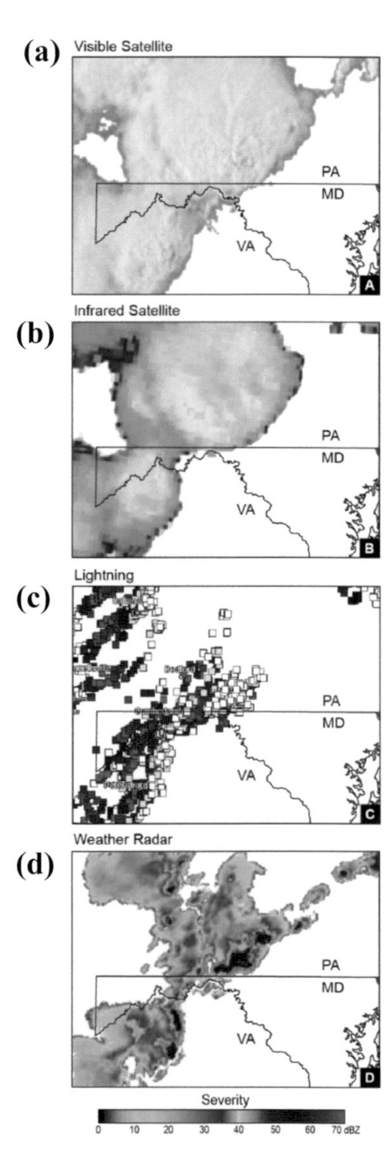

Figure 9.2 Four remotely sensed images of the same thunderstorm. Each image reveals different information that can be used to analyze the storm's impact.

Atmospheric Convection Moves Warm Air Upward, Sometimes Producing Hazardous Weather as a Result

Before we delve into the various types of severe convective weather, we must first study the fundamentals of ordinary thunderstorms. This boils down to understanding buoyant forces and wind shear (the change in wind speed and/or direction with height). Why do thunderstorms develop in the first place? What is their characteristic size and lifetime? How do they become electrified? Why is vertical wind shear the most important determinant of thunderstorm type? How does the annual distribution of thunderstorms vary geographically across the United States and throughout the day?

Thermal Instability in the Atmosphere Drives Thunderstorms

Whenever air above the surface warms to excessive levels, warm air will be moved upward by buoyant, ascending air parcels. This is the essence of atmospheric convection. Thunderstorm cells can be likened to atmospheric heat pumps. Updrafts remove heat from the surface, and downdrafts bring cold air downward. A thermally unstable situation (cold air overlying warm air) is rendered stable, and convection ceases. Every afternoon, hundreds of convective clouds erupt over Earth's heated landmasses. An example over the Amazon rain forest, as viewed from the Space Shuttle, is shown in Figure 9.3. When the upward heat transfer becomes especially vigorous, severe local weather develops.

Air parcels moving heat upward cool adiabatically as they ascend. An unsaturated parcel cools at the dry adiabatic lapse rate of 10° C per km (18° F/3300 feet), while a saturated (cloud-containing) parcel cools at a much slower rate, on the order of 5° C per km (9° F/3300 feet). The saturated parcel cools at a slower rate because condensation releases latent heat; the added warmth partially counteracts the adiabatic cooling.

A conditionally unstable atmosphere is predicated on the release of latent heat once a rising parcel achieves saturation. The initially rising parcel may not have enough energy to

Figure 9.3 The Amazon rain forest is a massive thunderstorm-generating region. Each day in the Amazon huge amounts of water vapor are driven into the atmosphere to form clouds, many of which become thunderstorms. (NASA.)

ascend through a deep layer. But if the parcel can be lifted to its condensation level (perhaps by a cold front or by air flowing up a mountain), latent heat warms the parcel. The additional buoyancy then propels it upward on its own; if the parcel ascends to tropopause level, a thunderstorm is born.

Thunderstorm Cells Contain Updrafts Rising Through a Deep Conditionally Unstable Air Mass

Most deep convective clouds, including thunderstorms, are driven by the ascent of air parcels in a conditionally unstable atmosphere. What determines how tall and how intense thunderstorm clouds become? The depth of the conditionally unstable layer determines their height, and the magnitude of the parcel's buoyancy determines the updraft strength. A basic thunderstorm requires that the unstable layer extend high enough so that the upper reaches of the cloud become frozen. When parcels exceed the temperature of the surrounding air by a few degrees, ascent becomes vigorous. Thunderstorms also require sufficient moisture in the low-level air (if the air is extremely dry, condensation may be insufficient to enhance cloud buoyancy) and a mechanism to nudge parcels to their condensation level.

Thunderstorm clouds are called cumulonimbus (literally, "raining cumulus") clouds. Cumulonimbus clouds typically reach heights of 13–17 km (8–10 miles) in the mid-latitudes, depending on the season and the synoptic weather setting. Moist, humid air parcels ascend through a cumulonimbus cell in a more or less continuous current termed the updraft, which may reach speeds of 20 kts (23 MPH) or more.

Formation of Rainfall, Lightning, and the Downdraft Are Related Processes

Within the updraft, water vapor condenses into cloud droplets. Cloud droplets grow to raindrop size by collision-coalescence. When the mass of rain suspended aloft becomes too heavy, the precipitation falls out as an intense, sudden shower. In the upper reaches of the cumulonimbus cloud, where temperature falls below 0° C (32° F), a fraction of the liquid water freezes into various forms of ice (including snow and graupel [small, soft ice particles]). The process of freezing releases additional latent heat, further enhancing updraft buoyancy and enabling the cloud to grow to great heights.

Small ice crystals accrete into large snowflakes, and graupel particles grow by riming (recall that riming occurs when supercooled water in a cloud freezes on contact with an ice crystal). As the mass of heavy ice particles descends through the cloud, it melts, contributing to the volume of rain.

In the middle regions of the cumulonimbus cloud, both liquid and frozen forms of water coexist. The mixture promotes a process called charge induction, which is the generation of static electrical charge. Charge induction is the first step in cloud electrification and the subsequent electrical discharge we view as lightning.

Additionally, the large downward cascade of precipitation starts a core of air moving downward. Drier environmental air is drawn across the cloud edges in the wake of this descending current. Some of the cloud water and rain evaporates into the drier air. Evaporation extracts heat from the air, cooling the descending mass of air. The rapidly moving downrush is termed the downdraft, which can exceed 10 m/s (33 feet/second).

A thunderstorm cell, with a lifetime on the order of 30–45 minutes, contains a warm, buoyant updraft, a cool descending downdraft, heavy precipitation, and lightning. The essence of deep convection is this: Warm air is moved upward, and surface air is cooled. The unstable atmosphere is stabilized in the storm's wake.

There Is Variation in the Type of Thunderstorm Updrafts

Most types of ordinary thunderstorm cells draw their unstable air from just above the warm surface. Their updrafts are rooted in the surface. Other types of storms can tap an elevated layer of instability, beginning their upward growth 2–3 km (1–2 miles) above the surface. And still other types of storms can develop in an environment that is surprisingly dry above the surface. Let's look a bit closer at the unique situations that lead to these different types of thunderstorm updrafts.

Figure 9.4 illustrates the three different types of thunderstorm cells. The diagrams are designed to show the process by which buoyant air in these updrafts rises. As you follow the path of a rising updraft, you can compare its temperature with that of the environment surrounding the cloud. As long as the rising air is warmer than its environment, the air is buoyant and will continue to ascend. The temperature and dew point temperature variation with height, in the environment of the thunderstorm cloud, is shown by the red and green curves, respectively. The blue line with arrows illustrates the temperature along the updraft's path of ascent.

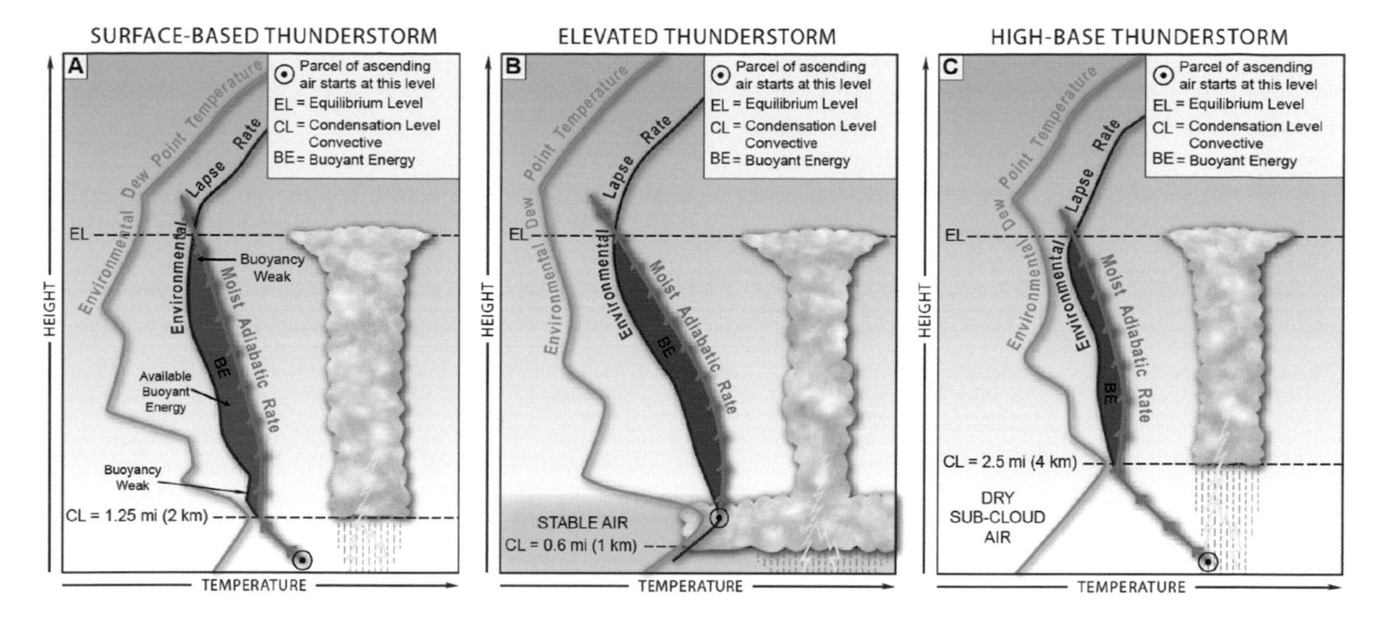

Figure 9.4 Three different types of thunderstorm updraft configurations. (a) Air is buoyant from the surface to the equilibrium level (EL). (b) A temperature inversion (cool air beneath warm air) from the surface to about 2 km (1.2 miles). The temperature inversion produces a stratus layer (shallow, layer clouds) below the more typical vertical structure of an elevated thunderstorm. (c) A high-base thunderstorm where condensation is delayed due to lower dew points associated within arid or semiarid regions.

Surface-Based Thunderstorms Are Most Common During the Late Afternoon Hours

Figure 9.4(a) shows the development of a surface-based thunderstorm. Initially unsaturated but humid surface air is lifted. It cools and dries adiabatically in the lowest 2 km (1.2 miles), then becomes saturated at the condensation level. Thereafter, it rises at the moist adiabatic rate. Above the condensation level, the updraft remains warmer than its environment. Near the tropopause, the parcel becomes cooler than the environment, just above a point called the equilibrium level (where air temperature inside and outside the updraft become equal). The deep layer between the condensation and equilibrium levels is conditionally unstable and supports a thunderstorm cloud of great depth.

The layer of buoyant air can be thought of as a reservoir of potential energy: Where there is buoyancy, the updraft accelerates upward and gains kinetic energy (energy of motion). The amount of updraft potential energy can be quantified and is called available buoyant energy.[1] The region of buoyant energy is shaded red between the environment and updraft temperature curves in Figure 9.4(a). The buoyant energy lets us quantify the degree of atmospheric instability and is a critical forecasting parameter for thunderstorms and severe weather. The larger the buoyant energy, the more unstable the air mass, and/or the deeper the unstable layer. The units of buoyant energy are measured in joules of energy per kilogram of air (J/kg). Ordinary, nonsevere thunderstorms require a buoyant energy of at least a few hundred J/kg. When buoyant energy values exceed 2500 J/kg, severe thunderstorms containing especially vigorous updrafts are likely. Extreme values of over 5000 J/kg can precede a violent tornado outbreak in the Great Plains.

During the afternoon, as the Sun warms the surface layer of air, the air mass becomes more unstable as buoyant energy increases. Maximum buoyant energy values occur in the late afternoon or early evening. This process explains why thunderstorms are typically a late afternoon phenomenon over much of North America during the warm season.

Elevated-Type Thunderstorms Often Occur Overnight

Figure 9.4(b) illustrates a different type of thunderstorm, the elevated thunderstorm. Air rising in this storm's updraft is rooted not at the surface but at some higher air layer. There are situations in which a cool, stable air mass lies close to the surface. The region north of a warm front (recall that warm air spreads over cooler surface-based air ahead of a warm front) is one common situation. Another situation occurs at nighttime, when the air layer above the surface cools after sunset and becomes stable. In these situations, there may be a reservoir of buoyant energy, several kilometers above the surface. The reservoir of available buoyant energy, in other words, does not extend from the surface upward. Ahead of a vigorous front, air parcels may be nudged upward from a higher starting point (the diagram shows this happening from the 2 km [1.2 mile] level), above which they become freely buoyant. This elevated, unstable layer contains substantial buoyant energy, through a deep enough layer to sustain a thunderstorm.

Elevated thunderstorms, while capable of producing locally heavy rain and lightning, are relatively benign. They rarely produce damaging wind, large hail, or tornadoes.

High-Based Thunderstorms Are Common in the High Plains and Can Trigger Wildfires

The final type of thunderstorm is the high-base thunderstorm, illustrated in Figure 9.4(c). These local storms are found in semiarid environments, such as the High Plains and southwest desert regions of North America. Here, the moisture content in the lowest several kilometers is relatively low, but, because of strong solar heating, the air mass becomes unstable through a deep layer. However, updraft air parcels must rise a great vertical distance in order to achieve condensation – typically 3–4 km (2.0–2.5 miles). But at the condensation level, there is enough moisture to promote cloud formation, and the updraft realizes available buoyant energy through a deep layer. Intense thunderstorms can still develop. High-based thunderstorms are notorious for starting forest fires by virtue of their lightning, which occurs in the absence of rain. Without the quenching effect of heavy rain, the wildfire grows unabated.

Atmospheric Instability Often Changes Throughout the Day Due to Numerous Processes

Measures of updraft buoyancy such as available buoyant energy are rarely static and nonvarying. The atmosphere is constantly evolving. The severe thunderstorm potential may be nil in the early morning, but increase during the afternoon hours. Infrequently, the potential for severe thunderstorms will actually increase overnight. Much of the potential depends on changes in the amount of available buoyant energy. Figures 9.5 and 9.6 illustrate common ways that available buoyant energy changes.

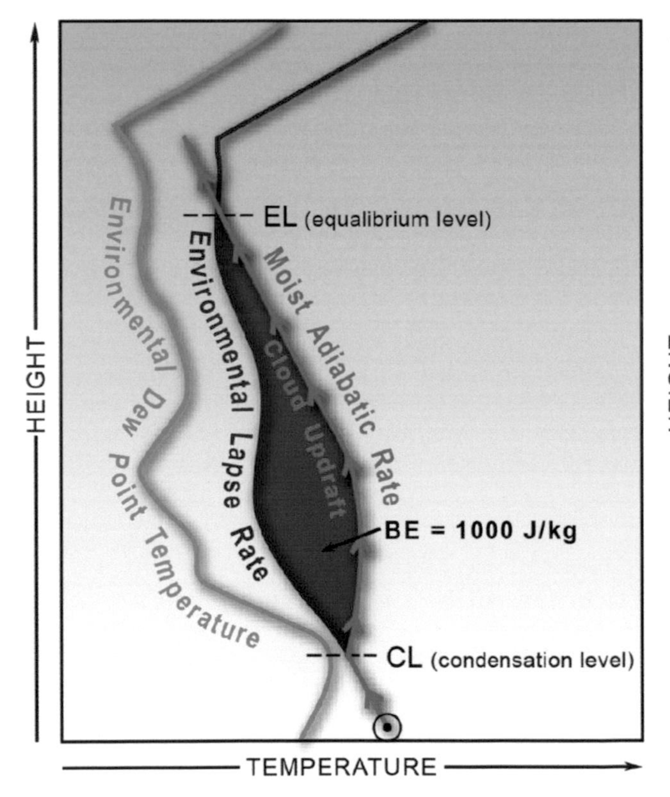

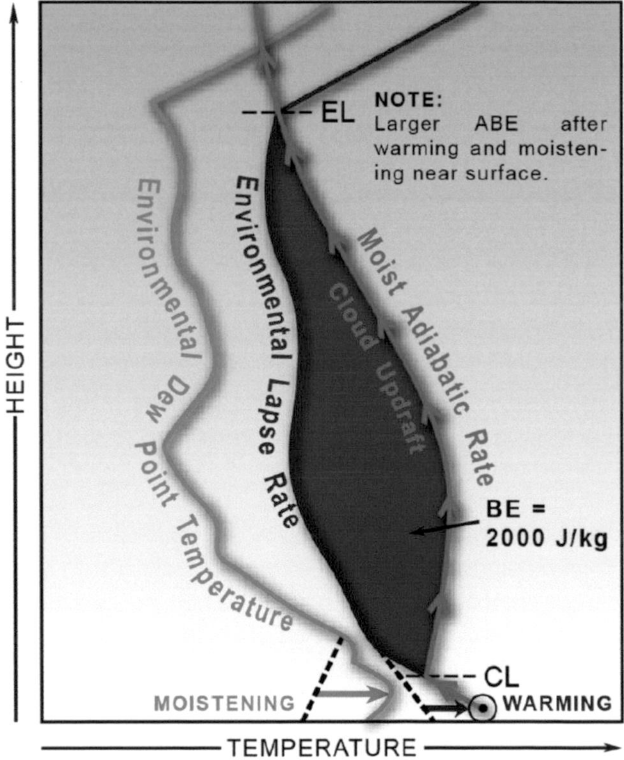

Figure 9.5 How afternoon heating and moistening of the surface air layer leads to increased buoyancy in a cloud updraft. Only the lowest levels experience a temperature and humidity increase.

Figure 9.5 shows the process by which available buoyant energy increases from afternoon solar heating. If the upper tropospheric temperature remains unchanged, but the surface air warms and moistens, available buoyant energy increases. At Time 1, in the late morning (left panel), the available buoyant energy is 1000 J/kg. This energy in itself is sufficient to support thunderstorm development. But the potential for severe thunderstorms may arise with continued afternoon heating and moistening. Warming might occur because of the strong summertime sun or because southerly winds usher in moist air from the Gulf of Mexico. The air can moisten because the previous day's rain evaporates from soil and vegetation. By Time 2, mid-afternoon (right panel), the available buoyant energy has jumped to 2000 J/kg. The shaded red region in the diagram has become fatter, meaning the amount of updraft buoyancy has increased. Clearly, we would anticipate more intense thunderstorm growth during the late afternoon.

Figure 9.6 shows a different scenario. What if surface temperature and humidity remain unchanged, but the upper layers of the troposphere cool down? This happens when a colder air mass invades the region aloft, usually when a trough in the jet stream passes through (recall that troughs contain cores of cold air). At Time 1, the available buoyant energy is a modest 1000 J/kg. But after several hours of cooling aloft (Panel B, Time 2), a buoyant updraft rising from the surface finds itself in a cooler environment. Because of the parcel's temperature excess, it can rise freely through a deeper layer. Available buoyant energy becomes larger, 1500 J/kg. Again, one would expect more vigorous thunderstorms at Time 2.

There are frequently situations where the processes illustrated in Figures 9.5 and 9.6 occur simultaneously. In the warm sector of extratropical cyclones, particularly during spring, southerly winds usher in warm and humid air at low levels, and strong northwesterly winds aloft bring in cold air. This process causes available buoyant energy to become extreme, resulting in widespread, severe thunderstorms – including tornado outbreaks.

This situation is illustrated in Figure 9.7. Within the extratropical cyclone's warm sector, a ribbon of warm, humid air streams northward below 3 km (2 miles), while in the

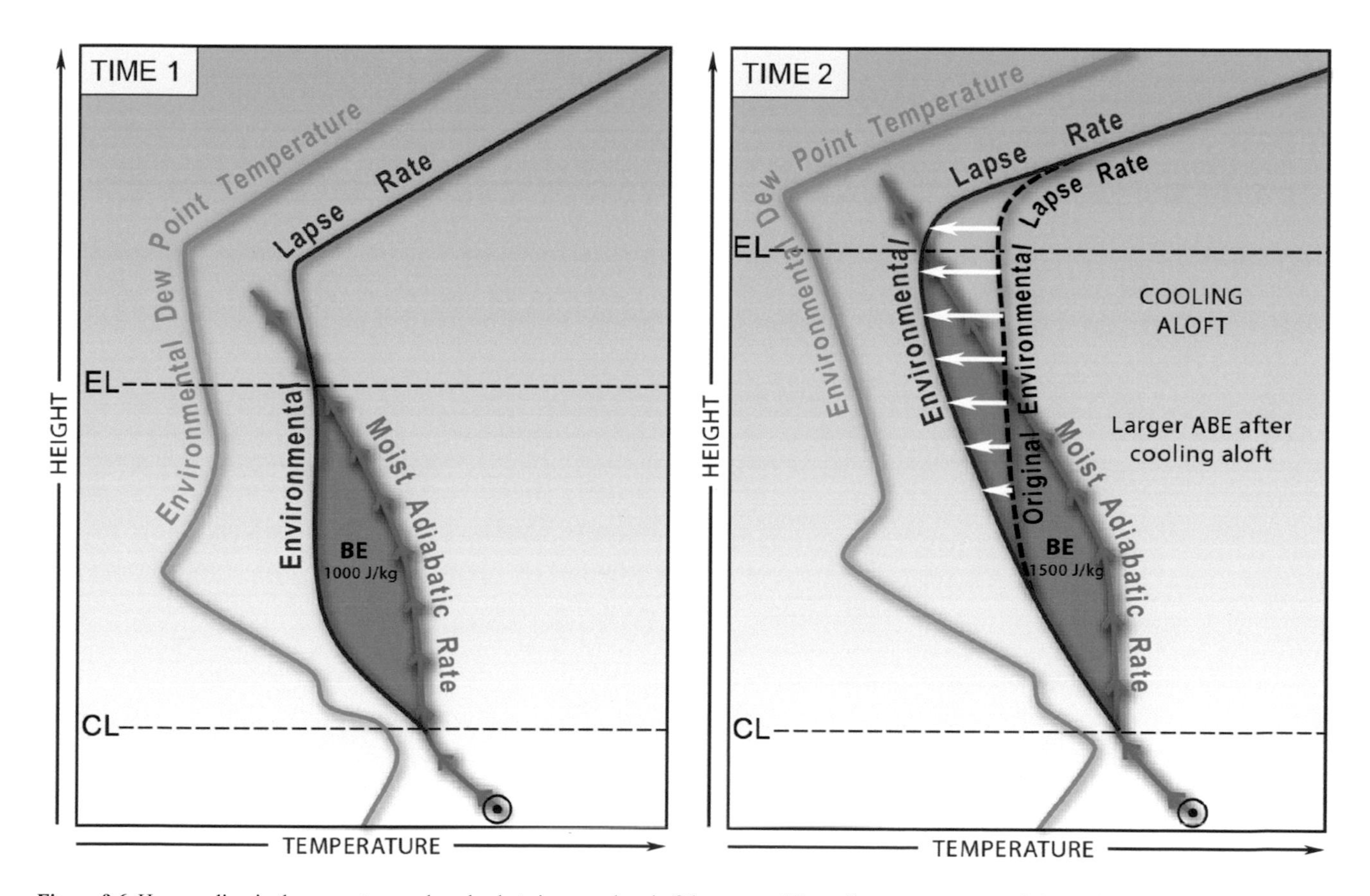

Figure 9.6 How cooling in the upper troposphere leads to increased updraft buoyancy. The surface temperature and dew point temperature have not changed.

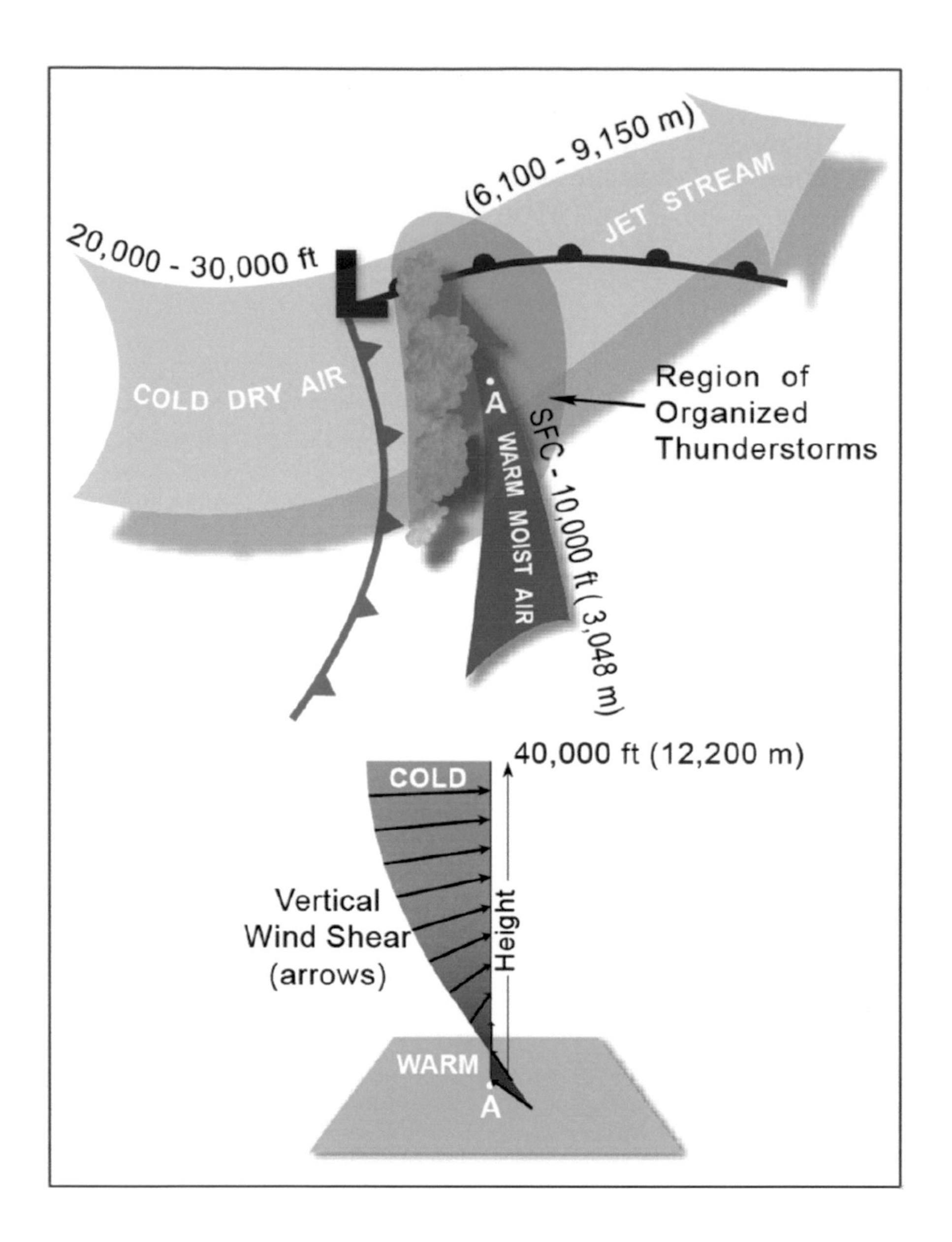

Figure 9.7 In extratropical cyclones, warm air flowing from the south at low levels interacts with cold air from the west at high levels. Strong wind shear is also associated with this temperature pattern in the storm's warm sector.

upper troposphere, cold air arrives from the west. Within the green shaded region of the cyclone, intense thunderstorms erupt – particularly along and ahead of the cold front. The cold front is often the trigger for thunderstorms, lifting unstable air to condensation. Elevated-type thunderstorms may develop north of the warm front. Springtime extratropical cyclones have a propensity for breeding episodes of severe local storms, partly due to winds at different levels that bring in different air temperatures. But extratropical cyclones also increase the vertical wind shear, the change in wind speed and/or direction with height. The next section explains why wind shear is the single most important factor that determines how thunderstorms become organized and which of these produces severe weather.

Wind Shear – The Change in Winds With Altitude – Helps Organize and Strengthen Thunderstorms

You may have heard the term "wind shear" in the context of aviation, especially as a cause of airplane crashes. When wind speed and direction change abruptly just above the surface, an aircraft can experience significant changes in airspeed, pitch, and lift, and experience dangerous turbulence. However, in the context of thunderstorm formation, we are most interested in the wind shear through a deep atmospheric layer, particularly the lowest 6 km (4 miles).

In the mid-latitudes, the winds usually blow from the west through much of the troposphere, and they increase speed with increasing height. Because the westerly jet stream is strongest in the winter (November–March), wind shear also tends to be largest during this time. Depending on the shear, thunderstorm cells can acquire vertical rotation, which serves as a precursor to violent tornadoes and damaging hail. The shear helps organize storm cells into long-lived squall lines, parts of which can bow outward, generating violent straight-line winds. As a general rule (with some exceptions), the stronger the wind shear, the more severe the thunderstorm, given sufficient moisture and thermal instability.

The concept of wind shear is complex because there are so many possible ways that winds vary with altitude. For now, let's focus on how the winds in the environment of the thunderstorm lead to three ordinary (nonsevere) types of storm that we introduced earlier in this chapter: single cells, multicells, and squall lines.

The Extratropical Cyclone's Warm Sector Often Contains Significant Wind Shear

Once again let's examine Figure 9.7, which shows a common scenario in which organized clusters or lines of thunderstorms develop. The warm sector of an extratropical cyclone provides thermal instability, a source of low-level moisture, lifting of air (along the cold front and warm front), and vertical wind shear. All of these elements come together in a "sweet spot" within the warm sector (shown by the green shaded region in Figure 9.7). The bottom part of the diagram shows how the winds turn clockwise and increase in speed with height. Low-level southeasterly flow gradually changes to strong westerly winds aloft. The lengthening arrows show that wind speeds increase with height.

Weak Wind Shear Promotes Short-Lived, Single Thunderstorm Cells

For the sake of simplicity, let's ignore significant directional changes and focus just on changes in wind speed with height and on how these wind profiles influence thunderstorm formation (Figure 9.8). Panel (a) shows very weak winds throughout the depth of the troposphere, everywhere less than 5 kts (6 MPH). This example typifies a "weak wind shear" type of thunderstorm. The thunderstorm cell is vertically erect; no part leans due to the wind.

In the early stage of the cell's formation, the warm updraft surges upward. When precipitation develops, it drags down a core of chilly downdraft air with it. The downdraft emerges

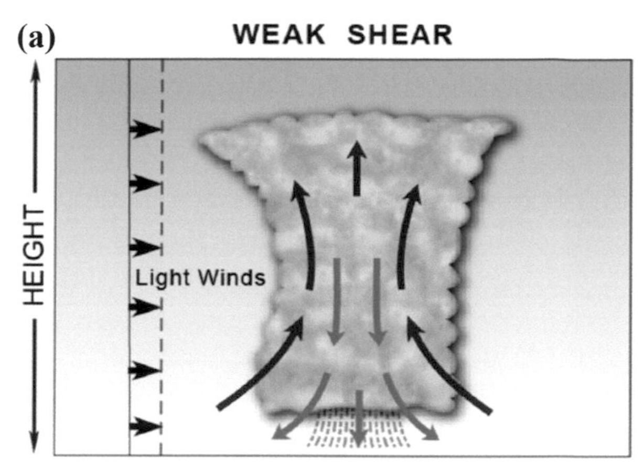

This cell features a vertical updraft. The downdraft develops within the updraft, reducing its buoyancy. This leads to a short-lived storm.

(b) DESTRUCTIVE SHEAR

HEIGHT

60mph (97kph)

This cell features a severely inclined updraft. There is little vertical growth. Turbulent mixing destroys buoyancy and no storm develops.

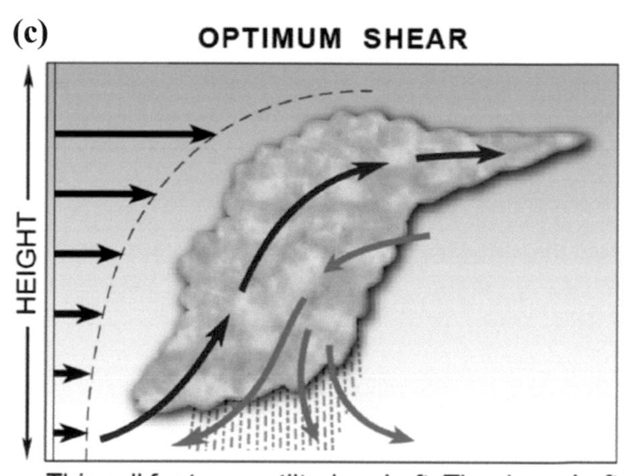

This cell features a tilted updraft. The downdraft develops beneath the updraft.

Figure 9.8 Different intensities of vertical wind shear (wind speed that increases with height) greatly influence the formation of thunderstorms. Shown are weak, moderate, and strong shear and their impacts.

from the base of the cloud and spreads outward along the surface. But when the downdraft develops from within the updraft, it smothers the updraft. The downrush of air and its cold temperature destroy buoyancy. Without buoyancy, continuous upward motion inside the cloud is not possible. Thus this type of single-storm cell has a "short fuse" and dissipates after about 30–40 minutes. The storm rains itself out, and any remaining clouds aloft slowly evaporate.

Extreme Wind Shear Prevents Thunderstorms From Developing

Panel (b) of Figure 9.8 illustrates the opposite extreme, very strong wind shear. Within the 0–6 km (0–3.7 mile) layer, the winds increase rapidly with height, attaining 60 kts (69 MPH) near the 6 km (3.7 mile) level. As the cloud updraft initially pushes to greater heights, the plume of rising air is bent over, elongating in the downwind direction. Buoyancy tries to make vertical headway, but the winds blow the current sideways. The cloud cannot attain depths great enough to produce a thunderstorm. Additionally, strong turbulence in the airstream mixes cooler and drier environmental air into the updraft, robbing the updraft of its buoyancy. In excessively large shear – sometimes called destructive shear – only a shallow, highly inclined convective cloud develops, then quickly dissipates.

Moderate Amounts of Wind Shear Promote a Long-Lived Thunderstorm

Given these two extremes of shear – one weak, one destructive – is there an intermediate configuration that promotes long-lived thunderstorms? Indeed, Panel (c) of Figure 9.8 illustrates a "goldilocks" situation. Wind speeds increase with height but not drastically, reaching about 40 kts (46 MPH) in the upper troposphere. With such a gradual increase and without excessive winds at any one level, the cloud updraft attains a modest tilt or incline. Buoyancy can push the cloud to great heights, but its precipitation falls out from beneath the updraft, not through the updraft (as in Panel [a]). The mass of falling ice and rain does not "weigh down" the updraft.

Cooler, drier environmental air enters the storm at mid-levels along its forward side.[2] As precipitation falls through this air, some of it evaporates and chills the air further. A strong downdraft develops and spreads out along the surface. In fact, as the leading edge of cold air advances along the surface, it acts as a miniature cold front, helping to lift buoyant air into the updraft. This type of synergy between the drafts can sustain a thunderstorm for many hours.

The Life Cycle of a Single, Ordinary Thunderstorm Cell Is Well-Defined, With a Series of Stages

Short-lived thunderstorms dotting the landscape on a hot summer day are a single-cell, or "pop-up," type of storm. They are especially prevalent along the spine of the Appalachians or Rockies when winds are weak. Many tens to perhaps a hundred single storm cells may develop within a large, unstable air mass. In this section, we examine the anatomy and life cycle of the simplest type of thunderstorm cell, the single-cell thunderstorm.

Single cells experience maximum growth around the time of peak afternoon heating, when the air mass is most unstable, and then dissipate by sunset. The storm cells are usually 10 km (6 miles) or less in diameter, last 30–45 minutes, and contain a single, erect updraft and downdraft. Because the updraft does not become very intense or long-lived, these cells rarely produce severe weather. However, if the instability is extreme (buoyant energy in excess of 3000–3500 J/kg), air mass thunderstorms may briefly produce a localized episode of damaging winds, small hail, and intense lightning.

Visual Cloud Features of Single-Cell Thunderstorms

Figure 9.9 is a photograph of a typical air mass thunderstorm. Much of the visible cloud mass takes the shape of an anvil, as updraft air spreads laterally at the stable tropopause. The upper reaches of the cloud are composed entirely of ice crystals and small snowflakes, while the bottom half contains a mixture of supercooled cloud water and rain. The symmetric appearance of the anvil cloud shows that there is very little wind shear; if winds were blowing strongly at high levels, the anvil cloud would elongate into a narrow streamer aimed downwind (see Figure 9.35).

The Life Cycle of a Single-Cell Thunderstorm Begins With the Growth Stage

A weather experiment in the 1940s conducted over Ohio and Florida allowed meteorologists to define a characteristic life cycle for ordinary single thunderstorm cells. There are three stages in the life cycle, each spanning a 10–20 minute duration. In the first step, the growth stage (Figure 9.10), the cloud contains a central updraft of buoyant air. Warm and humid air converges into the cloud base, feeding the updraft. Within

Figure 9.9 Visual appearance of an air mass thunderstorm. An air mass thunderstorm clearly displays the interface between strong rising air and stable stratosphere, resulting in a large spreading anvil cloud composed of ice crystals and small snowflakes. This very unique vantage point is provided by the NASA ER-2 aircraft flying at 65,000 feet. (NASA.)

the updraft, a core of precipitation first develops at 2–4 km (1.0–2.5 mile) altitude. Cloud droplets grow to raindrop size by frequent collisions among smaller droplets. Above the freezing level, drops are supercooled. During the growth stage, the cloud is dominated by updraft, and raindrops are growing large enough to begin falling out. The cloud is termed a towering cumulus.

During Its Life Cycle, a Thunderstorm Reaches Peak Intensity During the Mature Stage

Stage 2, the mature stage (Figure 9.11), is the actual thunderstorm stage. The massive cloud, called a cumulonimbus, grows to tropopause height (13.6–16.7 km [44,600–54,800 feet]) and the updraft may overshoot into the lower stratosphere by a kilometer or so. The icy cloud top becomes sharply defined, assuming the shape of a blacksmith's anvil. This occurs when updraft air, having lost its buoyancy in the warm (stable) tropopause layer, is forced to diverge laterally, spreading the cloud top sideways. A heavy rain shower now reaches the surface. All three phases of water coexist during the mature stage: water vapor, converging into the cloud base updraft; liquid water (cloud water, raindrops, and supercooled water drops) in the middle regions of the cloud; and ice (cloud ice crystals and snowflakes) in the frozen upper portion. The cloud layer between the pure ice and pure liquid is called the mixed phase region, and it is critical for both cloud electrification and the formation of hailstones (hail is discussed in Chapter 10). "Mixed phase" refers to the simultaneous occurrence of supercooled water drops, ice crystals, and small, soft ice particles (called graupel).

Formation of Lightning During the Mature Stage

Lightning is a significant thunderstorm hazard, responsible for several fatalities and injuries each year in the United States. Additionally, lightning kills livestock, damages trees, knocks out utilities, damages electrical equipment, and sparks structural fires. Lightning is a very intense, electrostatic discharge – a narrow, elongated channel of plasma (an ionic form of matter) with a current of several tens of thousands of amperes (A), up to 300 million volts (V), and a temperature of nearly 50,000° C (90,000° F).

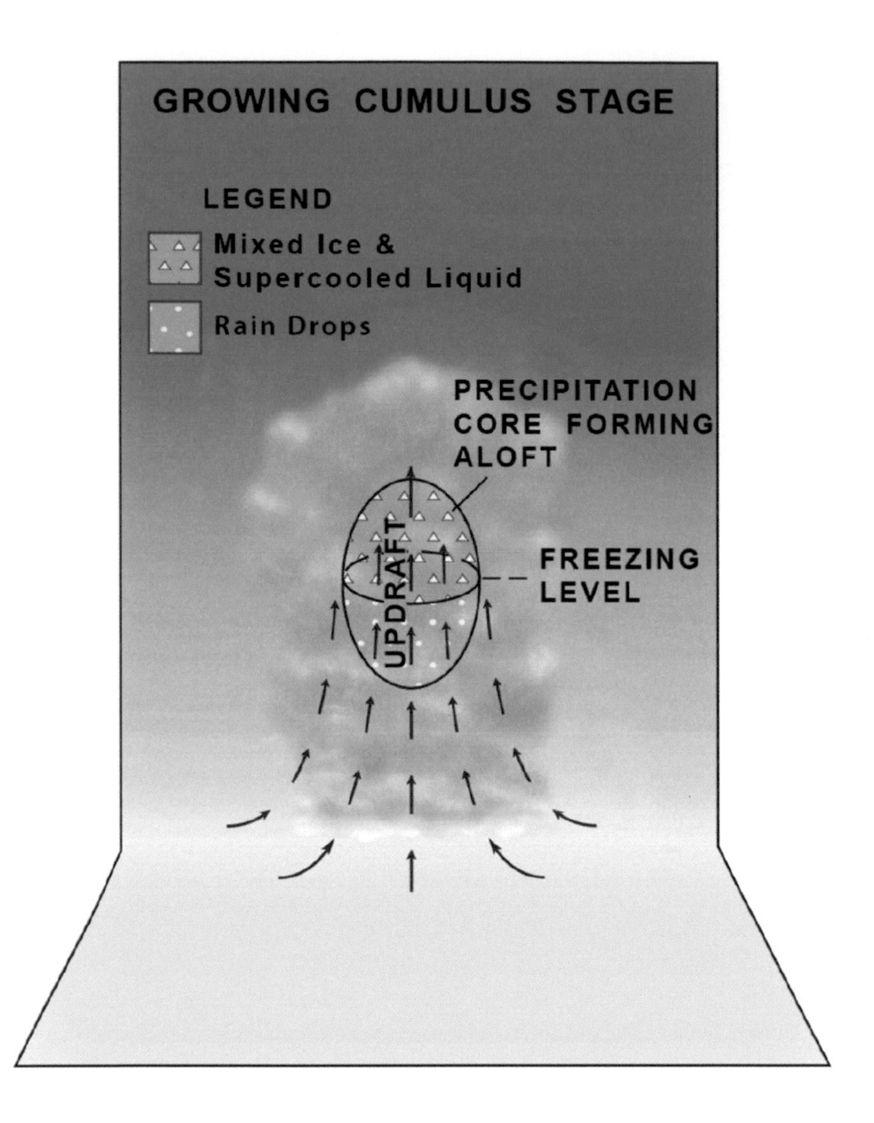

Figure 9.10 Growth phase of an air mass thunderstorm. In this phase, warm moist air feeds in from the surface in all directions. This buoyant air produces strong updrafts that generate towering cumulus clouds.

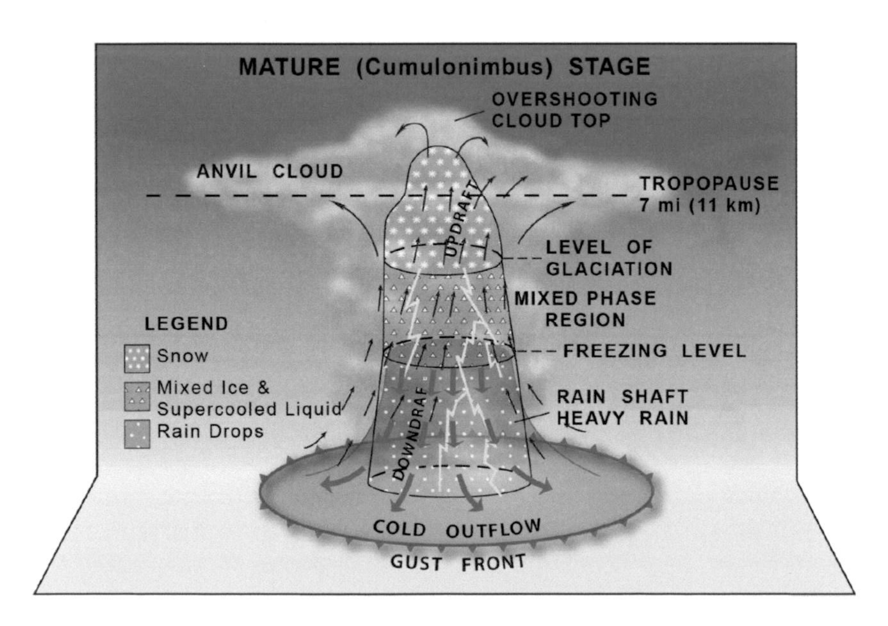

Figure 9.11 Mature stage of thunderstorm development. Thunderstorms are structured with a heavy rain shower in the lower portion of the storm, a mixed phase region immediately above the freezing level, and a frozen phase of ice particles in the upper reaches of the storm. The convective movement of frozen particles produces the enormous static charges that result in lightning.

The formation of lightning requires a deep convective cloud containing a strong updraft and a mixed phase region. A two-step process, termed charge induction and separation, is hypothesized to generate lightning. In a vigorous updraft, ice crystals and graupel particles collide with one another. During collision, ice crystals acquire a positive static charge, and graupel particles acquire a negative static charge. This is the charge induction portion of the process.

The small, lightweight ice crystals are swept upward in the updraft, building a concentration of positive charge near the top of the storm cloud. Graupel particles, which are heavier, settle out as part of the cloud downdraft, depositing a region of negative charge in the cloud's middle section and near the cloud base. This sorting of charge, based on the mass of the ice particle, is termed charge separation. The intense region of negative charge at cloud base can induce a charge of the opposite sign (positive) to build along the ground surface, beneath the storm cloud. The charge can become particularly concentrated around tall, pointed ground objects such as big trees, towers, buildings, etc. Figure 9.12 summarizes the process of charge induction and separation. This is a simplified version of the actual process, as recent studies using special weather balloons released inside thunderclouds reveal a more complex, layered charge structure, with as many as four opposing charge layers within the updraft and downdraft regions.

So far, we have described the formation of very strong electrical fields within the cloud and along the surface but not the actual lightning. The potential difference or voltage grows to very great levels. However, air possesses a strong resistance to current flow. In order for lightning to discharge these charge centers, a conductive path through the air must first develop. This process occurs very rapidly but is invisible. It starts with the propagation of leaders, or conductive (ionized) paths, simultaneously building downward from the cloud base and up from the ground level. Quickly, the leaders connect. With the sudden establishment of a conductive channel between cloud base and ground (or within the cloud), a surge

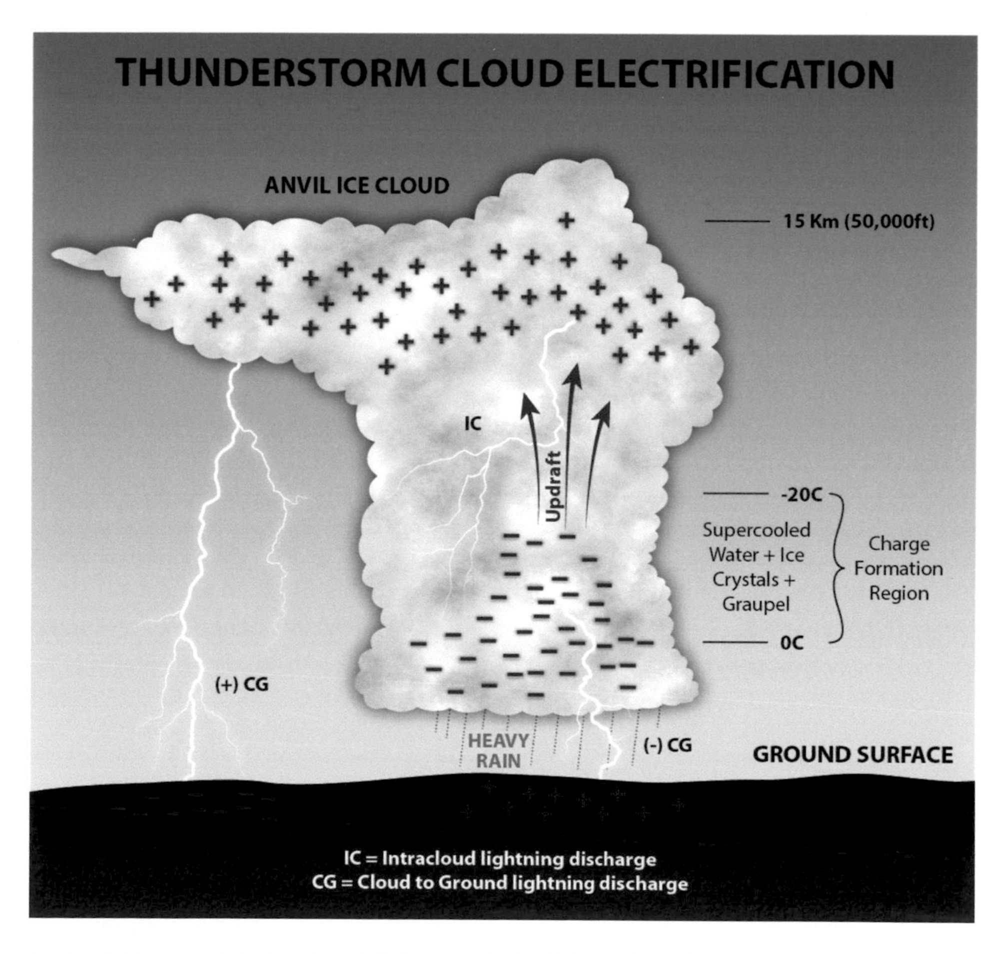

Figure 9.12 Schematic showing how cloud drafts and precipitation types lead to lightning formation in a thunderstorm.

Figure 9.13 Vivid example of a cloud-to-ground lightning stroke.

of current passes through the channel. This surge, the return stroke, is the visible part of the lightning flash. Oftentimes, there are repeat surges through the same conductive channel. The entire stroke, though, is exceedingly brief, i.e., 50–200 microseconds.

Most cloud-to-ground (CG) lightning strikes (Figure 9.13) are of negative polarity – meaning a negative electrical charge flows from the cloud base toward the ground. Fewer than 5–10% of all CG discharges are positive polarity, which originates from the top of cloud (where the positive charge is concentrated) and may strike the ground many miles from the thunderstorm cloud (up to 40 km [25 miles distant]!). Because of the longer distance from cloud top to ground, positive CGs convey significantly more current than negative CGs, last longer, and cause greater heating of structures on the ground. The lightning may reach 1 billion V and convey 300,000 A. Positive CGs are thus more lethal and more likely to start ground fires, including wildfires. "Anvil crawlers" refer to elevated, rapidly branching lightning emanating high up in the anvil (stratiform cloud layers) of a mature thunderstorm (Figure 9.14).

Thunder is an acoustic (sound wave) generated by the explosive expansion of superheated air within the lightning channel. At sea level, these sound waves travel approximately 661 kts (760 MPH). Over an elapsed time of 5 seconds, the sound wave has traveled approximately 1 mile. This is the basis for estimating your distance to the lightning strike.

Discharges between charge centers within a thunderstorm cloud create intracloud (IC) lightning (Figure 9.15). The brilliant light emitted by this discharge tends to be very diffuse, as the light is scattered internally by trillions of water

Figure 9.14 Anvil crawlers – a web of lightning that propagates at high speed horizontally, beneath the base of elevated ice clouds in the thunderstorm's anvil cloud region. (Axel Rouvin.)

droplets and ice crystals; the resulting flash is often termed sheet lightning. At night, this can produce a mesmerizing, stroboscopic display as a distant thunderstorm seemingly pulses with blue-green, internal flashes. Lightning detection networks reveal that there are many more IC flashes, relative to CG discharges, in most thunderstorms; upward of 80–90% of all lightning generated by thunderstorms, in fact, may be in the form of IC lighting.

Total lightning refers to the sum total of all IC and CG flashes for a given thunderstorm. The discharge rate of CG or total lightning is a useful indicator of a storm cell's intensity; generally, the higher the flash rate (per minute), the more vigorous the updrafts and downdrafts – and the more severe the potential of the storm cell. Meteorologists use flash rate, in conjunction with weather radar, to assess the intensity trends of thunderstorms – whether in isolated cells, lines, or large clusters of cells. It is now common practice to combine lightning and radar data into single images or short movies, as a means to indicate the electrical threat posed by an approaching system of thunderstorm cells. Lightning stroke density can identify the most intense (and potentially severe) storm cells, as well as trends in cell intensity. Alternatively, fine-scale lightning detection networks being tested across several regions of the United States now create highly detailed maps of individual lightning strokes, combining cloud-to-ground and intracloud detection, with as much detail as weather radar alone (Figure 9.16).

Figure 9.17 shows global lightning incidence. Shown is flash density, here defined as the number of flashes per square kilometer of ground per year. Over North America, there is

Figure 9.15 Striking photo of an intracloud lightning discharge, with branches of the flash emerging from the cloud interior. (UK Met Office.)

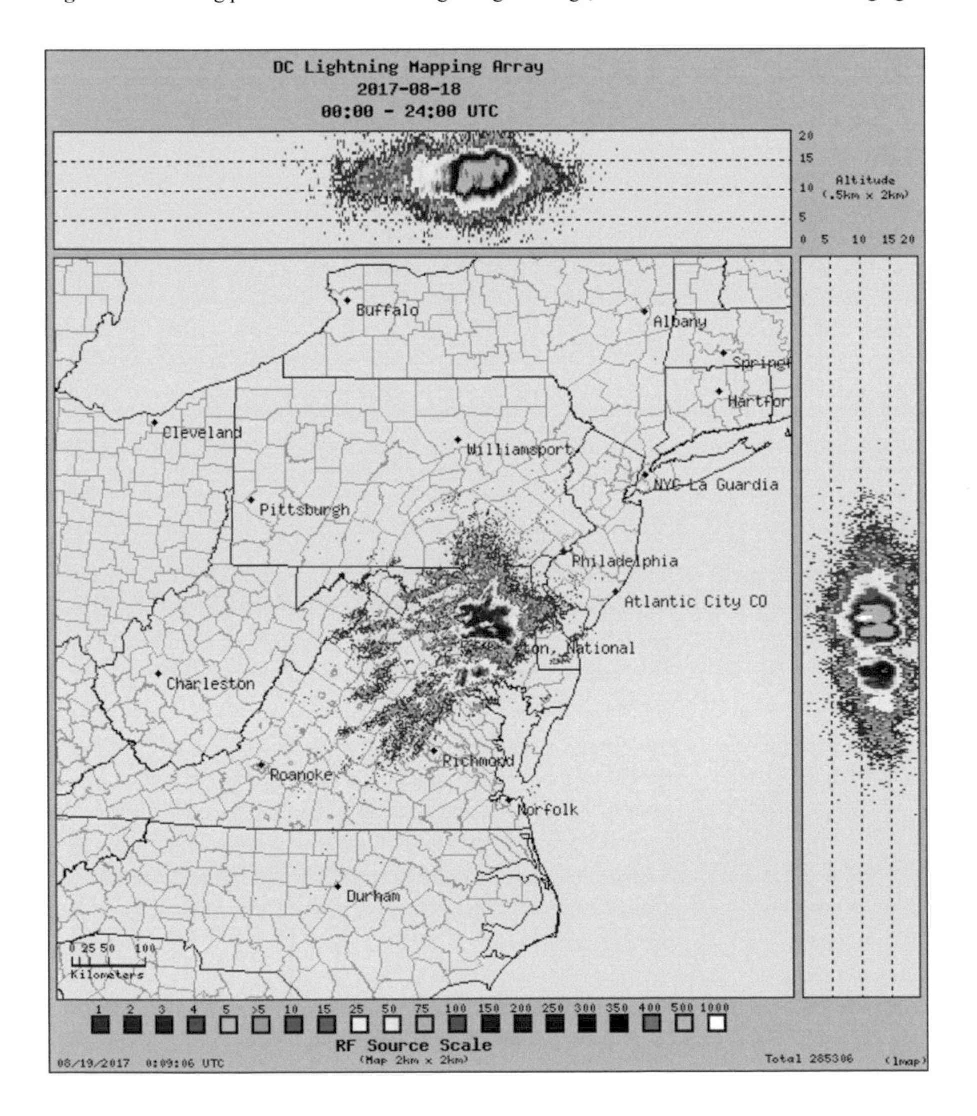

Figure 9.16 High-resolution lightning detection network (the D.C. Lightning Mapping Array, or DC-LMA), which portrays highly detailed maps of lightning evolution in storm clouds, such as these cells on August 19, 2017. The altitude scales on the margins show the vertical height of lightning stroke origins. (NASA.)

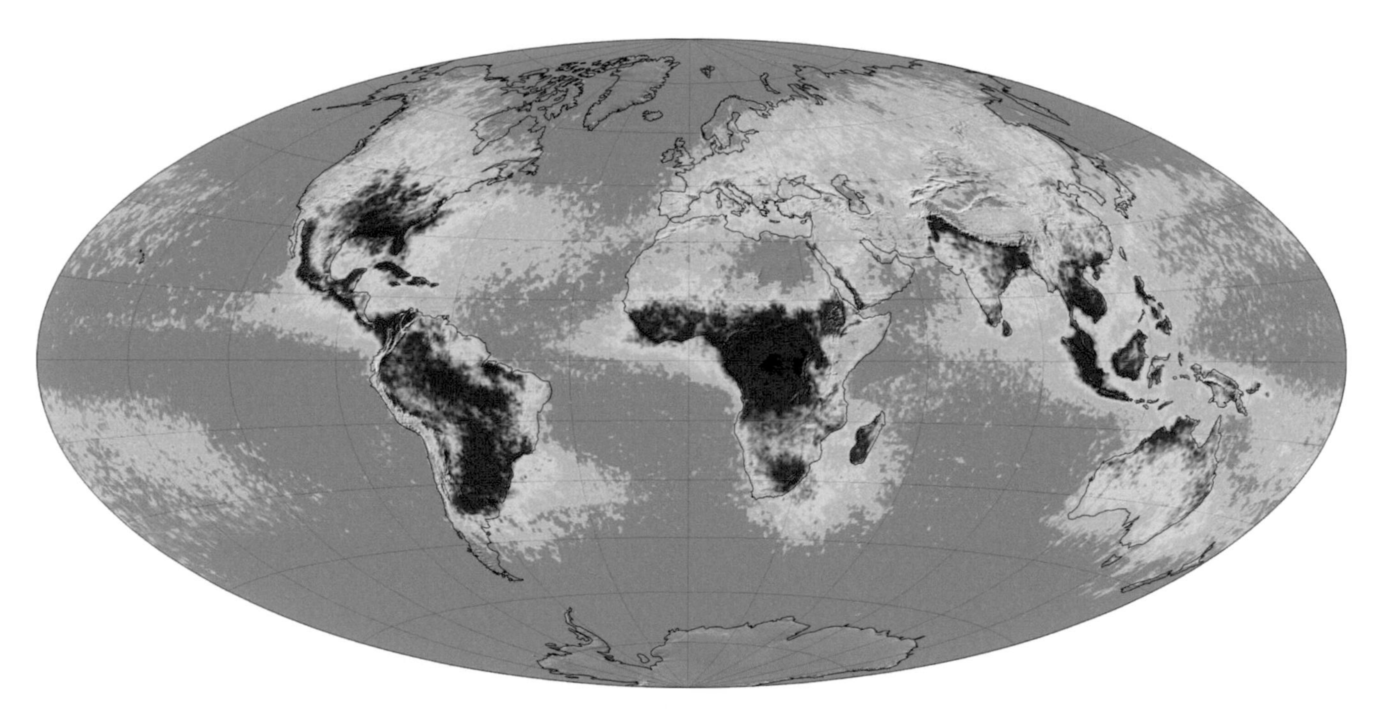

Figure 9.17 Global climatology of lightning flashes per square kilometer by geographical location, based on NASA satellites for the years 1995–2002. Dark red tones indicate up to 70 flashes/km²/yr. (NASA.)

a significant annual number of flashes over the eastern half of the United States. The large maximum over the southern Mississippi Valley and Southeastern United States reflects northward penetration of warm, humid, unstable air off the Gulf of Mexico – drawn into extratropical cyclones as they transit the country from west to east. Additionally, during summer, a strong southerly flow off the Gulf (generated by the Bermuda High over the western Atlantic) draws this unstable air mass inland. Within the United States, Central Florida is the "lightning capital." In addition to the constant invasion by Gulf and Caribbean air, this region experiences a convergence of sea breeze circulations that lead to almost daily thunderstorm growth during the warm season (see the Amazing Storms box at the end of this chapter for more on Florida thunderstorms).

The Downdraft, Creating Surges of Strong Wind, Forms During the Thunderstorm's Mature Stage

During thunderstorm maturity, a downdraft develops within the lower portion of the storm cloud. The downdraft is caused when the rapidly sinking mass of precipitation drags cloud air downward. The cloud air is chilled by evaporation of rainwater as drier environmental air mixes in. When the downdraft strikes the surface, the downflow is directed sideways as a windy blast of cool, nearly saturated air. The lateral spread of downdraft air is called the storm's outflow. The outflow is a cool, gusty layer about 0.5–1 km (0.3–0.6 mile) deep. As it passes through, we experience a sudden drop in temperature by 10–15 C degrees (18–27 F). Wind gusts may briefly reach 30–50 kts (35–58 MPH). The air pressure briefly rises a few mb because the mass of cool air is denser than surrounding air and thus weighs slightly more.

It's possible to observe the spread of an outflow using satellites and weather radar. Figure 9.18 shows a view from the GOES satellite of an outflow. A group of thunderstorm cells developed over the Texas Panhandle and produced extensive outflows at the surface that merged together. Outflows are relatively cloud-free except for a ring of new convective clouds that develop along their gust front. Here, unstable air is forced to converge and rise upward.

Figure 9.19 shows a dramatic example of an outflow obtained by a weather radar. Here, a cluster of thunderstorm cells over Nebraska produced an enormous, symmetric, expanding

Figure 9.18 A circular outflow pattern around a multicell thunderstorm complex as viewed by satellite. (Adapted from NOAA.)

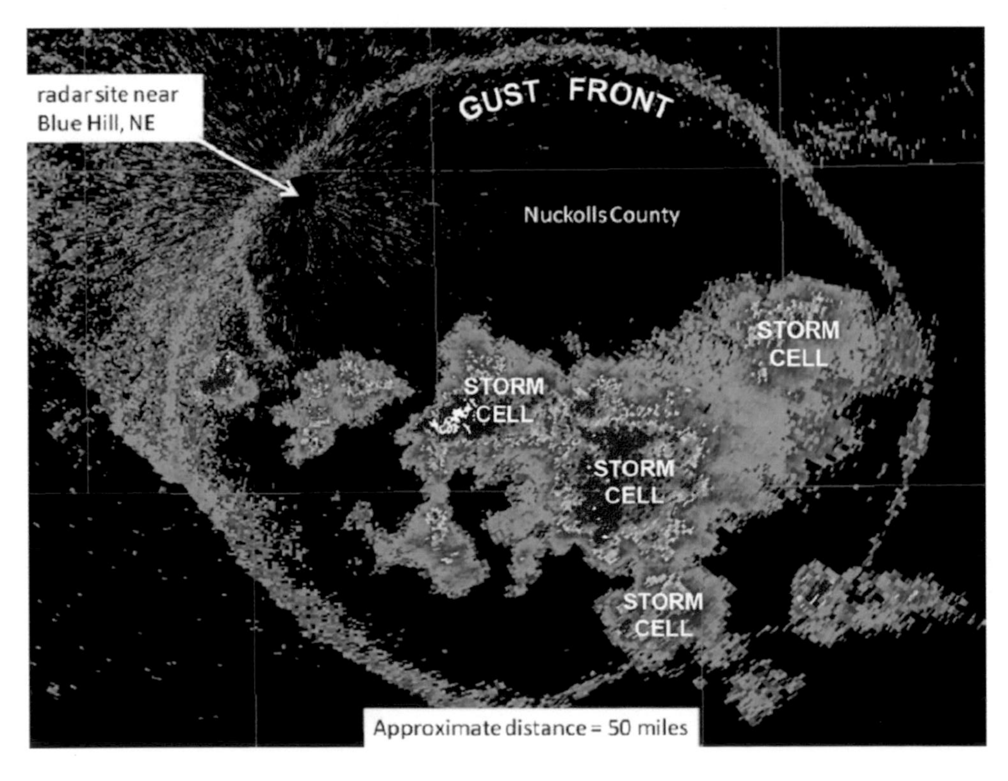

Figure 9.19 A circular outflow pattern around a multicell thunderstorm complex as viewed by weather radar. (NWS.)

Figure 9.20 Photo of a shelf cloud resulting from strong downdrafts lifting chilled, moist air to condensation, ahead of the rain shaft. (Jason Weingart, NASA.)

outflow. The narrow ring is generated by the gust front, where dust and insects were lofted by strong winds. These small targets mimicked raindrops, making the gust front boundary readily visible.

Figure 9.20 illustrates a dramatic cloud feature often visible along the leading edge of an approaching thunderstorm, called the shelf cloud. In this figure, the storm is advancing from left to right. The shelf cloud develops as the cool outflow lifts warm, humid air just ahead of the storm, like a miniature cold front. Because the outflow air is very moist, condensation occurs below the main cloud base, making the shelf cloud appear low hanging, dark, and ominous. The shelf cloud often tumbles over itself like a rolling pin, testifying to turbulent airflow. The shelf cloud marks the leading edge of the outflow, a boundary called the gust front. The gust front usually is invisible unless outlined by blowing dust.

The Final Act of a Single-Cell Thunderstorm Is Its Dissipating Stage

We've now arrived at the dissipating stage of our single storm cell (Figure 9.21). The strongest wind gusts, heaviest rain, and most intense lightning occur near the end of the mature stage. A weak updraft may still be present in the upper reaches of the anvil cloud, and intracloud lightning may continue for some time. But the lower half of the updraft has been essentially eliminated by the downdraft's cold downrush. Moderate rain continues to fall, depleting the cloud's store of liquid and ice. The lower portion of the cloud essentially "rains itself out," and its winds become gentle. The anvil cloud may slowly drift away from the region where the storm first formed or may vanish in place. We say that the anvil cloud sublimates (its cloud ice changes phase directly to water vapor) in the dry upper-level air.

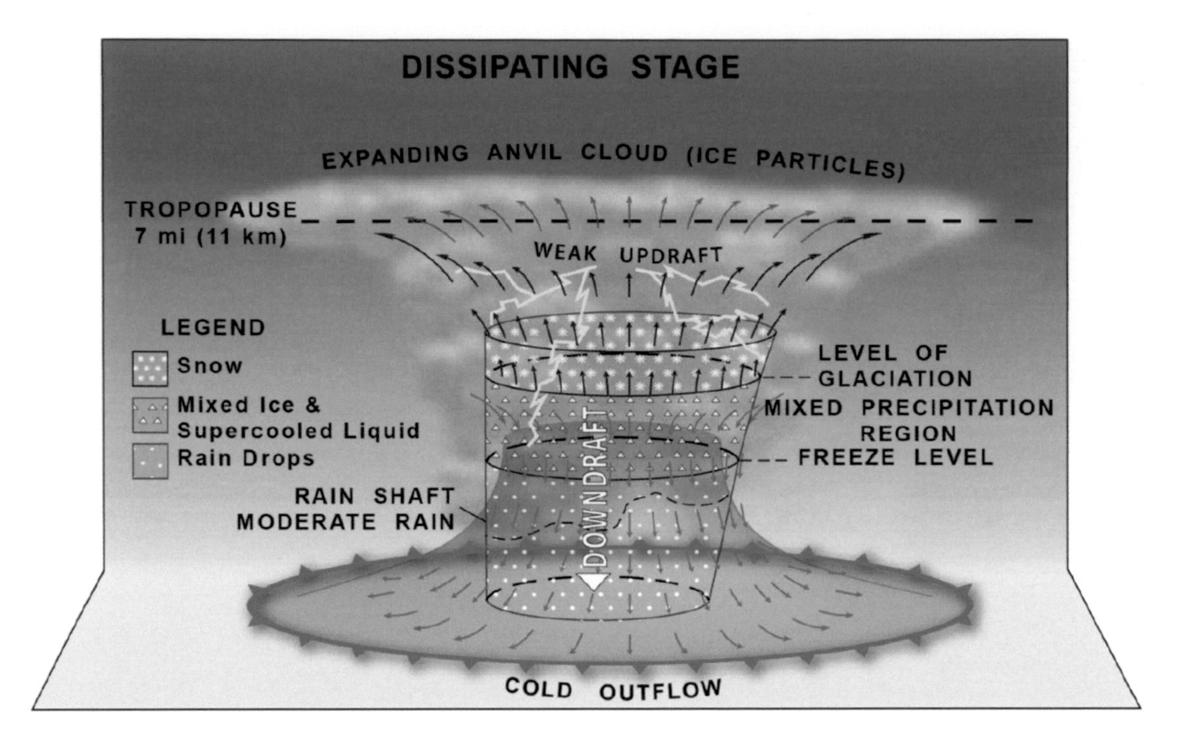

Figure 9.21 The dissipating stage of thunderstorm development is dominated by downdrafts with strong outflow. This process effectively cuts off the energy necessary to perpetuate the storm, and it dies.

Multicell Thunderstorms Are Large, Long-Lived Ensembles of Single Thunderstorm Cells Evolving as a System

As wind shear increases, multicell thunderstorms are likely. The multicell complex consists of multiple cells clumped as an aggregate. On weather radar, often multiple, embedded rain cores can be discerned; at other times, the large mass appears amorphous. Each core or cell is in a different stage of its life cycle. The lifetime of the multicell thunderstorm can last several hours. Cells continually form, mature, and decay within the larger system. Individual cells move with a speed and direction equal to the average wind through the deep layer. However, the movement of the entire storm complex may deviate from this motion.

Figure 9.22 is a photo of a multicell thunderstorm. The cloud system is extensive, composed of several adjacent convective cells. Young, growing cells appear on one side of the complex. A larger, more mature cell is next to it, to the right. A dissipating cell has an anvil of spreading ice, towering behind the other cells, to the right of the mature cell.

In a Multicell Thunderstorm, Individual Storm Cells Continuously Regenerate

Figure 9.23 shows a model of a multicell thunderstorm and its relationship to the deep wind shear. Westerly winds increase in speed with height. The multicell is approximately 80 km (50 miles) in diameter and composed of four cells, each in a different phase of development. Cells move with the prevailing westerly wind.

A broad outflow of cool air, fed by several downdrafts, spreads outward along the surface. At the surface, along the west side of the storm, this dense airflow meets the prevailing westerly winds head-on. Here, the airflow converges; the air is forced to ascend. In this zone, new cloud updrafts develop. They deepen and mature, moving off toward the east, and dissipate. The cycle repeats itself. This process shows how the larger storm complex is created and why it outlives an individual cell.

The important feature to note is the manner in which the ensemble of cells hangs together as a single, large entity. Yet within this ensemble, individual cells are forming and decaying, each going through its own life cycle. New cells form continuously on the upwind side of the complex, and old cells eventually decay on the downwind side. While each individual

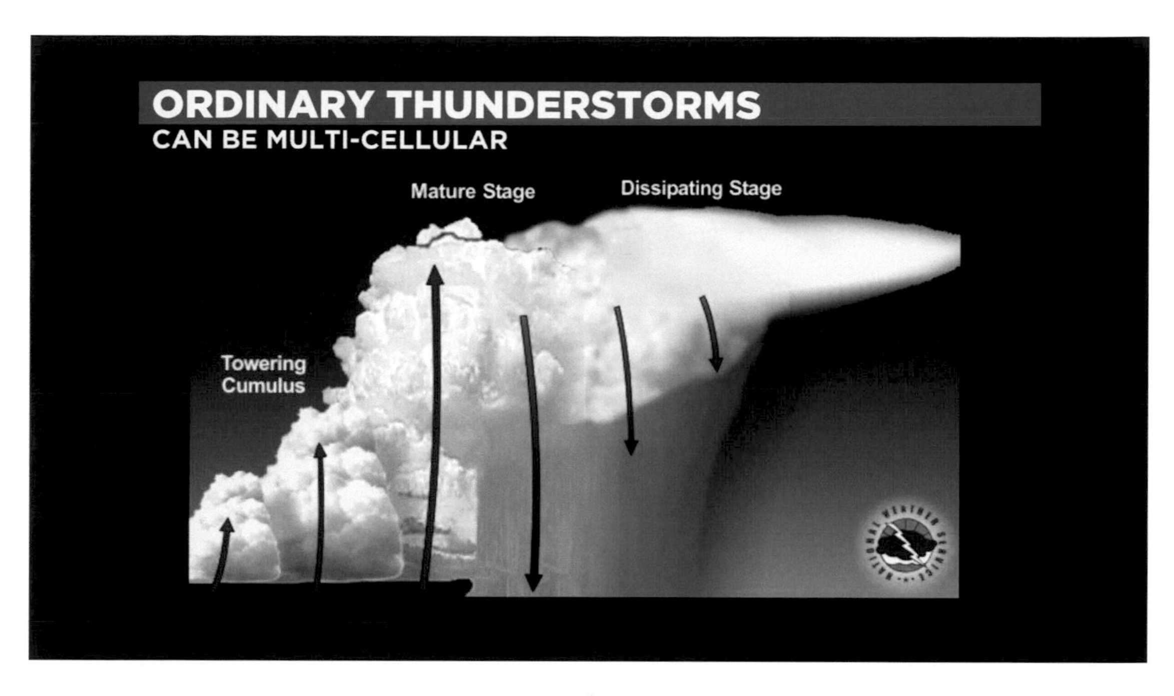

Figure 9.22 A multicell thunderstorm shows several distinct "towers" of clouds, each representing a convective cell in a different stage of the life cycle. (NOAA.)

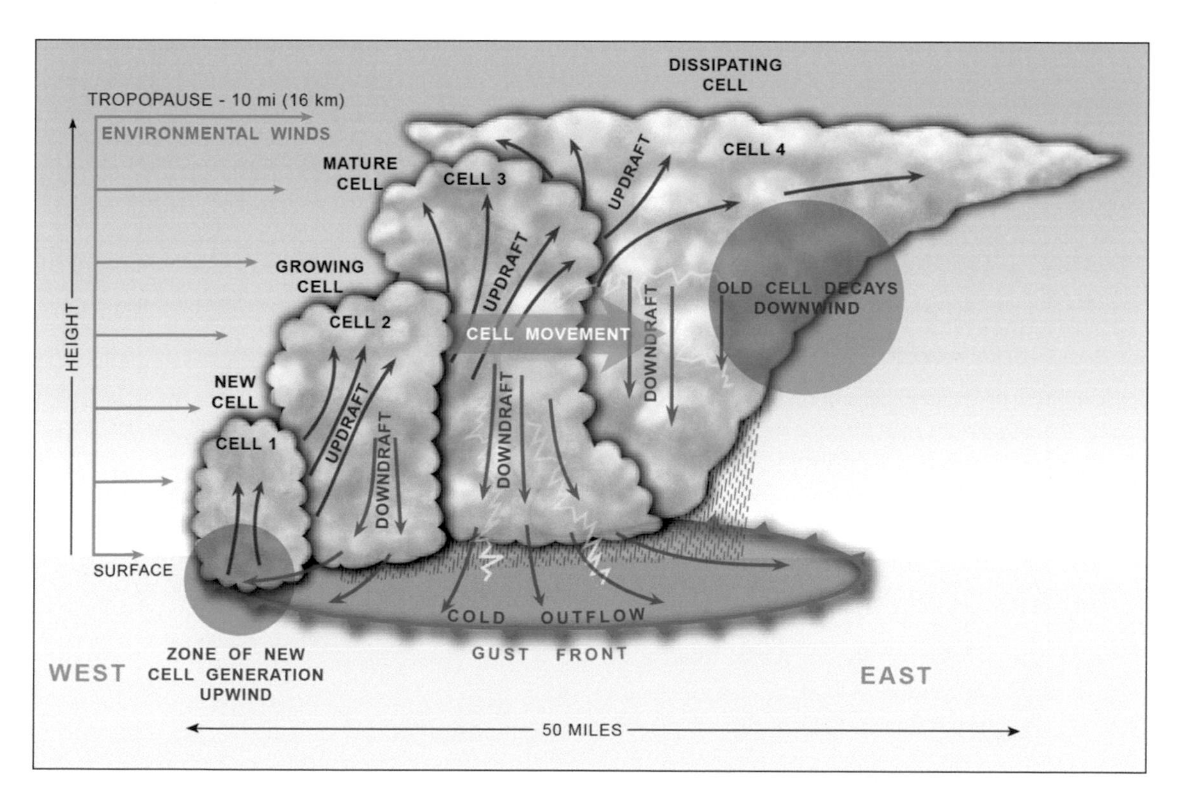

Figure 9.23 Schematic of a multicell thunderstorm. These systems are a product of cold outflow colliding with warm moist air flowing into the cells near the surface. Wind shear pushes newly formed cells off to the east. This dynamic provides the lift necessary to regenerate cells on the west side of the complex.

cell may last less than an hour, the lifetime of the larger ensemble may stretch to 3 or more hours.

The wind shear has another important effect. Because winds increase with height, the updraft within each cell is tilted downwind. Moderately tilted updrafts tend to persist. The storm outflow undercuts warm, humid air on the west side, lifting it and regenerating cells in the system. Updraft and downdraft cooperate to maintain the system. And with more intense, longer-lived drafts, there is an increased likelihood that the multicell storm will produce severe weather, including large hail, torrential downpours, and downbursts.

Multicell Thunderstorms Frequently Move More Slowly Than the Prevailing Winds

When discussing multicell storms, it is important to distinguish between cell movement and cell propagation. Propagation refers specifically to new cell growth on the upwind side of the system. You can think of propagation as regeneration. There are many possible ways that winds can interact with a multicell, and cell propagation need not occur on the upwind side of the storm, as depicted in Figure 9.23. Other processes can set up a zone of cell regeneration in other parts of the system.

Why does a multicell frequently move more slowly than the prevailing winds? We can think of the problem in terms of vectors (arrows that graphically represent wind speed and direction of movement). In Figure 9.23, individual cells move from west to east. The cell movement vector is thus oriented from west to east. But cell propagation occurs on the upwind side of the system, and cell decay occurs on the downwind side. We can represent the process of cell regeneration by its own vector, which points from east to west, opposite the direction of cell movement. Simple vector addition gives us the net movement of the multicell as a whole.

Figure 9.24 summarizes this process. New cells develop on the west side of the storm, where the winds intersect the storm

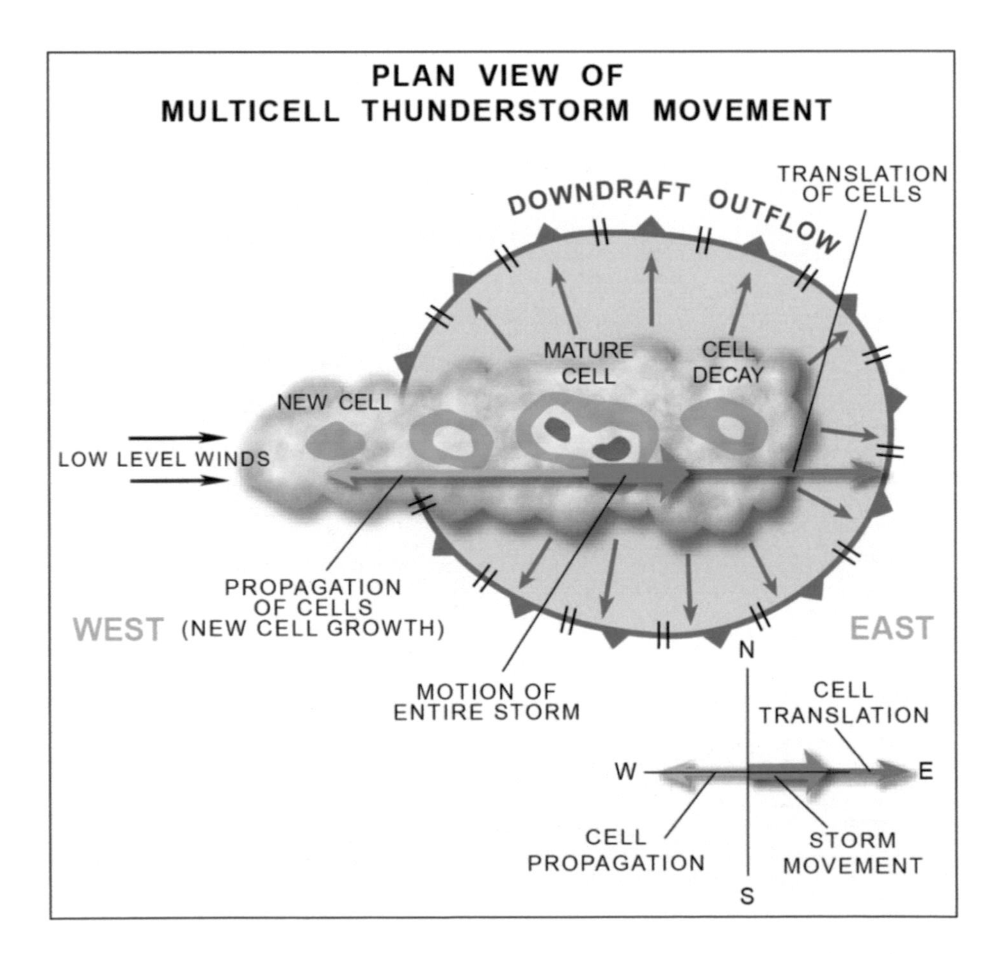

Figure 9.24 Movement of a multicell thunderstorm. The entire *system* moves as a result of the growth of new cells and the movement of mature cells. Here, cells grow, or propagate, toward the west while prevailing winds move mature cells toward the east. The sum of these vectors determines the overall storm movement, which may be slower than the prevailing winds.

outflow, and air gets an upward lift. Cells move toward the east and decay on the east side of the storm. The cell motion is shown by the orange vector. Cell propagation, in yellow, is toward the west – opposite the direction of cell motion, but the magnitude of this vector is not as large. Nevertheless, when the cell translation and propagation vectors are added, the resultant vector (green) gives a net multicell motion that is slower than that of individual cells. The upshot is that when the multicell slows down but individual cells continue to move through the complex, the potential for flash flooding increases. In any one location, the rain accumulates repeatedly and over longer duration.

Of all the types of ordinary thunderstorm cells, multicells are the most ubiquitous, and they occur in all types of environmental settings. But the most extensive and longest-lived form of ordinary thunderstorm system is the squall line, which we discuss next.

Squall-Line Thunderstorms Can Be Thought of as "Rolling Thunder"

Squall lines are among the largest and longest-lived thunderstorm-bearing cloud systems. Lines or arcs of thunderstorms may extend more than 500 km (310 miles) in length and 100 km (62 miles) in width. Figure 9.25 shows a squall line viewed from space. The line developed ahead of a cold front, part of an extratropical cyclone that moved across the Great Plains. The line consists of a dozen or more individual thunderstorm cells, and the anvil clouds of each cell have merged into an extensive, continuous plume of ice cloud.

Squall Lines Develop From Large-Scale Lifting of Air and Can Sustain Themselves for 6 or More Hours

How does a continuous line of thunderstorms develop? The basic ingredients include an unstable air mass, abundant low-level moisture, vertical wind shear, and a large-scale mechanism to force the air upward – most commonly in the form of a cold front. Figure 9.26 shows how the squall line often begins as isolated cells along a cold front, which merge into a solid line over time. The first cells develop outflow boundaries, and these boundaries begin colliding. New storm cells erupt where low-level air converges and rises between boundaries. Over time, cells fill in the gaps. Along the leading edge of the cells, adjacent outflows merge into a continuous gust front. As the cool, dense outflow plows forward, it lifts warm, humid air ahead of the

Figure 9.25 Satellite view of a typical squall line. This image shows a number of individual thunderstorm cells embedded in the extensive anvil cloud of a squall line, which has developed in advance of a cold front. (NOAA.)

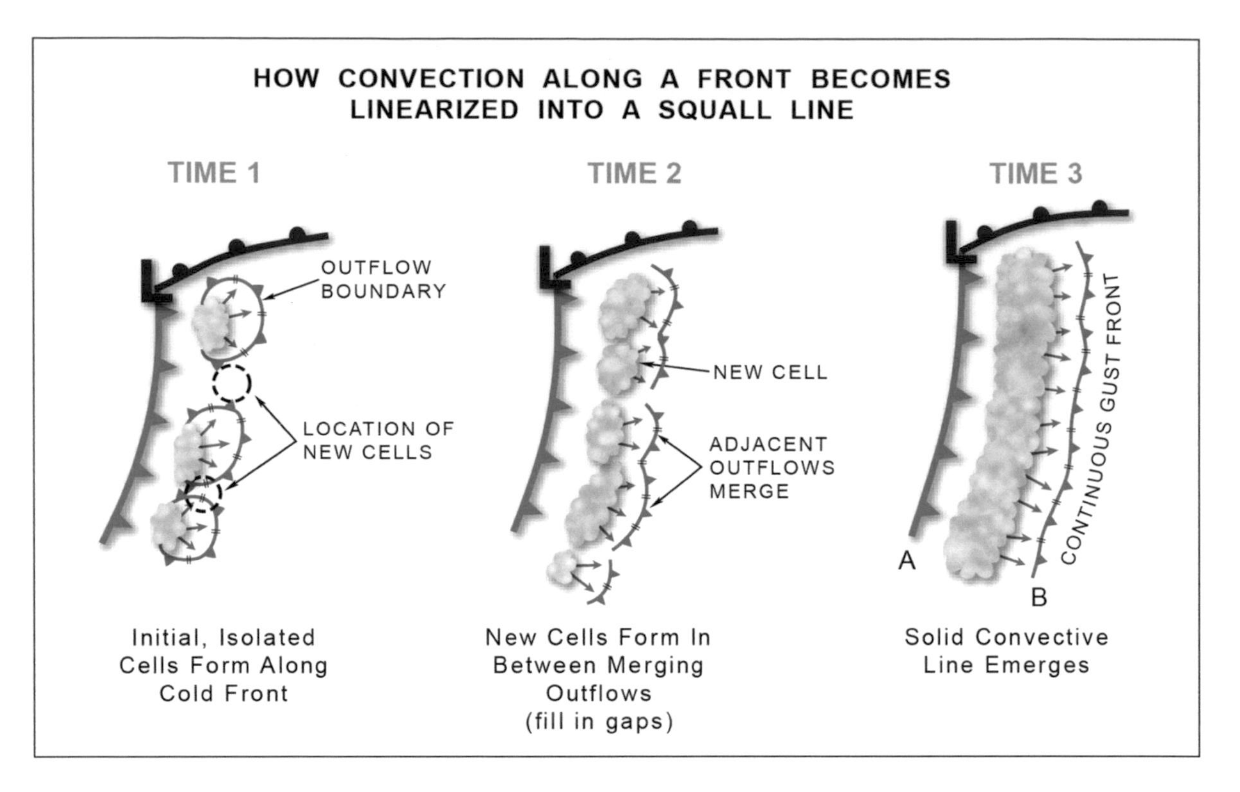

Figure 9.26 How individual thunderstorm cells multiply along a frontal boundary and merge into a squall line.

line. New storm cells develop along the leading edge of the gust front. In this manner, the squall line propagates itself.

Squall lines are very efficient in sweeping through the warm sector of a cyclone and processing enormous volumes of low-level, unstable air. Large amounts of water vapor lifted to the condensation level also make squall lines heavy rain producers. While a squall line may last several hours, perhaps 6 or more, eventually it weakens and dissipates. Dissipation happens when the gust front and its outflow advance many miles out ahead of the convective line. New cells no longer redevelop along the edge, where they would otherwise become incorporated into the main system. Sometimes the squall line sweeps across the entire warm sector of a cyclone, then dissipates when it reaches cooler, stable air north of the warm front.

Squall Lines Typically Develop in the Warm Sector of an Extratropical Cyclone

Squall lines frequently develop along cold fronts, but other times they develop hundreds of miles ahead of a cold front. To understand some of the reasons why, we turn to Figure 9.27(a), which shows a squall line developing ahead of a cold front, where westerly low-level air behind the cold front converges with a southerly flow ahead of the front. In the upper atmosphere, a trough of low pressure is also present. Sustained, large-scale ascent of air ahead of the trough cools and saturates the troposphere, promoting widespread cloud growth within the squall line.

Figures 9.27(b) and (c) illustrate situations in which the squall line may develop 100–200 km (62–124 miles) or so ahead of the cold front. Such growth might be triggered by air converging into a prefrontal surface trough or the dryline (a boundary between very humid and very dry air). Squall lines also start up along an elevated cold front, sometimes called a cold front aloft (CFA). In this case, an especially cold, dry surge of air sweeps around the cyclone above and ahead of the surface cold front. As this air mass plows into the warm sector, it can lift a deep layer of warm, humid air.

Inside the Squall Line Are Multiple, Highly Organized Airflows

We now peer inside the squall line to study its airflow and the way that it manufactures rain. To simplify our analysis, we can look at the squall line as a two-dimensional (2D) cloud system. That is, if we slice through the storm perpendicular to its motion, we find a similar structure no matter where we make the cut.

Figure 9.28 shows an east-west slice across a north-/south-oriented squall line. The system's updraft and downdraft are

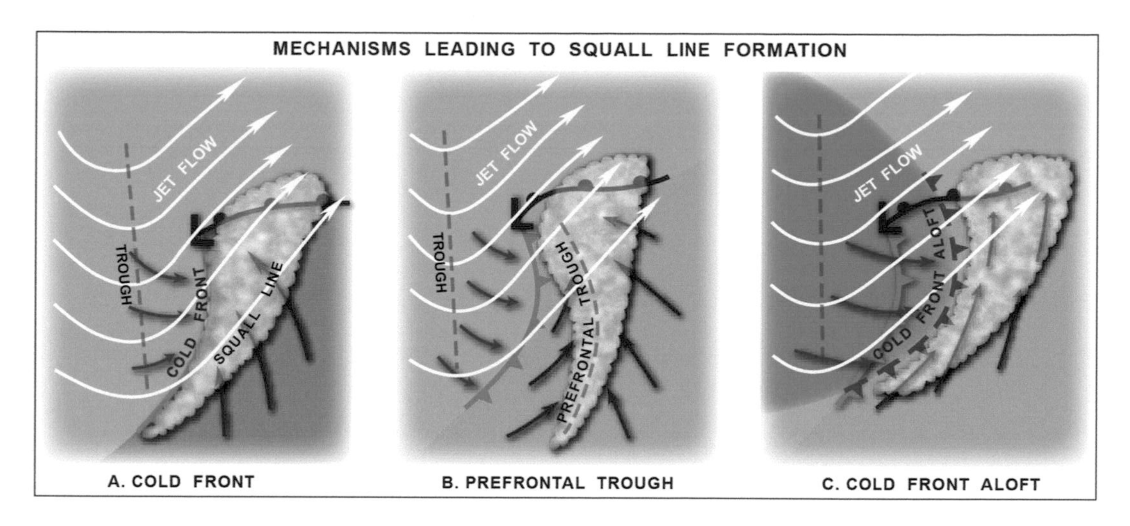

Figure 9.27 Various large-scale factors leading to squall line development. Squall-line development can be influenced by the position of the cold front and associated troughs. Squall lines can form along a cold front (a) or in advance of the cold front if a prefrontal surface trough is present (b). Sometimes cold air aloft extends ahead of the surface front, causing the squall line to form ahead of the surface cold front (c).

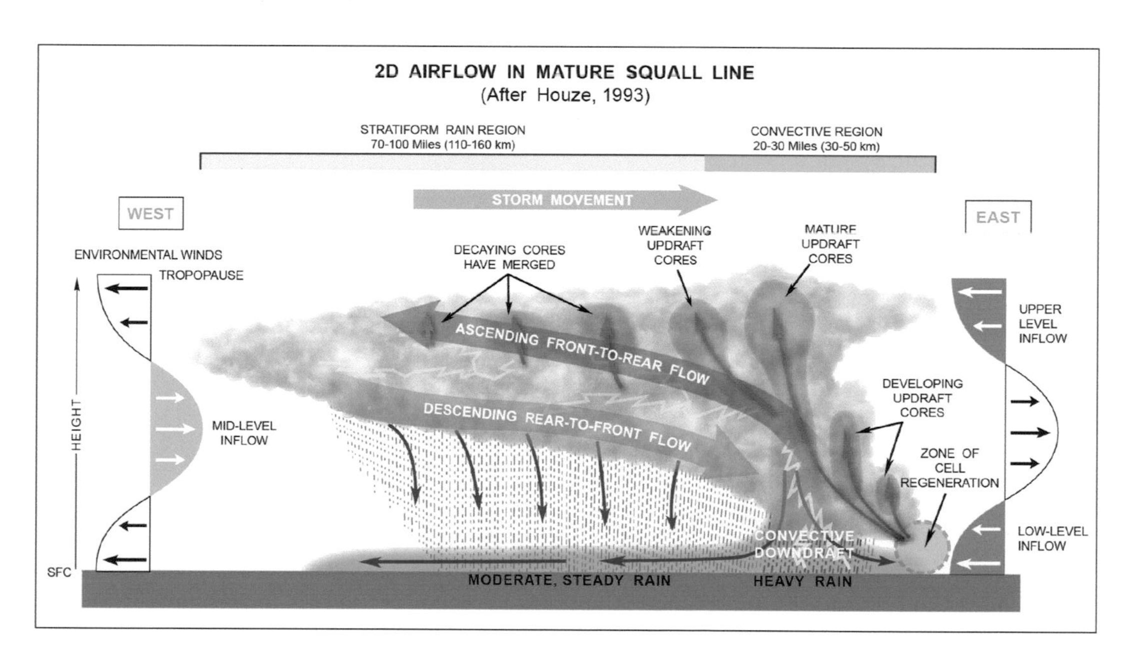

Figure 9.28 Cross section through a squall line illustrating airflow structure and cloud features. A broad, sloping updraft ascends from front-to-rear across the system, overlying a sinking ribbon of air (the downdraft) streaming from rear to front. These airflows represent a highly organized and persistent configuration. (Adapted from Houze, 1993.)

similar to that in a multicell storm but mirror-imaged (compare with Figure 9.23). In fact, one can think of a squall line as a row of forward-propagating multicells. Individual cells are pushed eastward by the prevailing westerly flow. Regeneration of new cells (propagation) occurs along the squall's leading (eastern) edge. The storm's propagation vector, which is directed from west to east (from the back edge of the storm toward the front), is in the same direction as cell movement. These vectors are added, and the resulting squall line movement is faster than the average westerly airflow (Figure 9.29). While squall lines tend to move briskly, they often contain very broad cloud shields that take hours to cross a location. Rain accumulations on the order of 3–7 cm (1–3 inches) are not uncommon as one of these lines passes through.

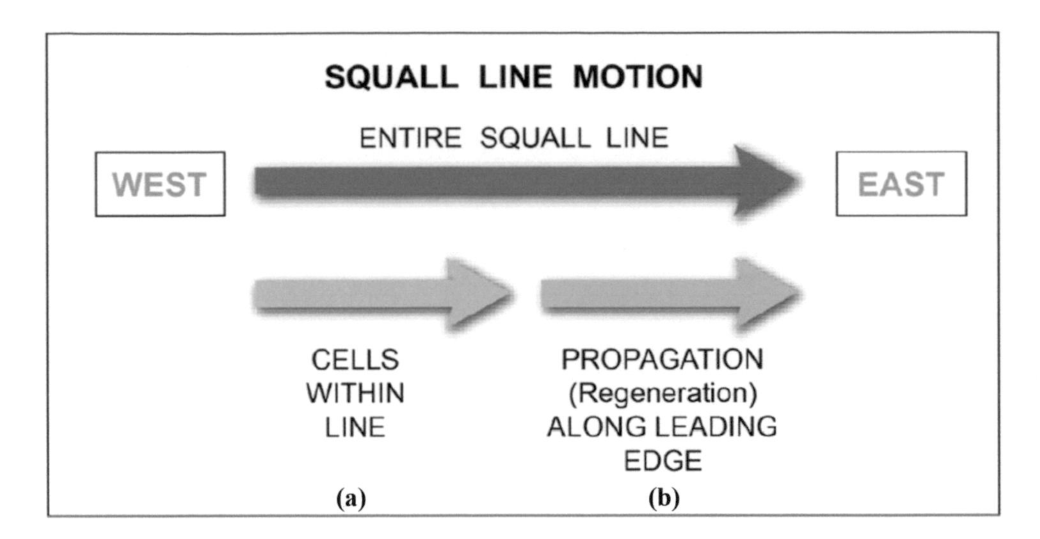

Figure 9.29 How a squall line moves. Squall line motion is the vector sum of two kinds of movement: (a) cells within the squall line moving with the prevailing wind and (b) propagation (regeneration) of cells along the leading edge. The sum of these two vectors can produce a squall system that moves faster than the prevailing wind.

Squall Line's Convective Region Contains the Most Extreme Weather, Including Torrential Downpours

Now let's go back to Figure 9.28 to examine some peculiar types of air motion not found in multicells or single cells. Low-level inflow of warm, humid air enters the squall system from the east. This unstable air ascends in updrafts along the leading edge of the system. The leading region consists of a number of rapidly growing cells. Rain cores develop within these cells aloft. The mature convective core is the deepest part of the system. A heavy rain shower falls from this cell, accompanied by lightning, and a convective downdraft spreads cool, saturated air along the surface. Where the leading edge of this outflow pushes ahead of the convective edge, air is lifted into the developing cells. The tall rampart of developing and mature convective towers is typically 30–50 km (19–31 miles) wide. When squall-line weather exceeds severe thresholds, most of the damaging elements (hail, torrential rain, and strong straight-line wind) come from this region.

Stratiform Rain Region in a Squall Line Produces Widespread Rain That Can Linger for Hours

In the upper levels ahead of the convective portion of a squall line, winds blow through the system, from east to west. This current is termed the front-to-rear flow. Such a flow is a consequence of the wind shear and the storm system's movement. The current pushes updraft cores toward the rear of the cloud system. As they move rearward (relative to the larger system, which is advancing toward the east), the updraft cores weaken and coalesce into a broad region of gently rising air. These remnants help build a thick, extensive layer of cloud that trails behind the convection like the smoke plume of a steam train. From these clouds, light to moderate rain falls. Because the clouds are more layered and stratified (as opposed to deep, vertical cloud towers), they are collectively referred to as the stratiform rain region.

The trailing stratiform region is variable in width (and sometimes fails to develop at all), but it is typically 100–150 km (62–93 miles) wide. Upward air motions are less than 1 kt (1 MPH), quite gentle compared to the intense convective updrafts that commonly exceed 20 kts (23 MPH). While the updrafts in deep convective towers are concentrated and intense, ascent in the stratiform layer is much more diffuse and widespread. It may occupy 70–80% of the squall line's total cloudy area. Accordingly, these different regions produce their rainfall in different ways (see Digging Deeper: How Squall Lines Produce Heavy Rainfall).

Along the back edge of the squall line, a different air current impinges on the storm from the west. Mid-level cold and dry air is pulled into the interior of the stratiform clouds. Rain falling into this dry air chills the air by evaporation. This chilling causes the broad ribbon of rear-to-front flow to descend. When this current plunges rapidly to the surface, exceptionally violent windstorms can develop, creating a fast-moving, arc-shaped squall line called a bow echo (discussed in Chapter 13).

Digging Deeper: How Squall Lines Produce Heavy Rainfall

Figure 9.30 shows a time series of rain intensity and accumulation for a squall line over a 2 hour duration. The figure shows both the instantaneous rain rate (white curve) and the cumulative totals, as might be recorded in a rain gauge, at 20-minute increments (blue bars). Notice the brief period of heavy rains, up to 3–5 cm/h (1–2 inches/hour), caused when the convective region passes through. Then there is a brief lull in precipitation followed by resumption of rain at a moderate rate (0.5–1.0 cm/h [0.2–0.4 inches/hour]). Depending on the speed of the squall line and the dimensions of the stratiform rain region, flash floods sometimes arise.

To see how squall lines manufacture heavy rain in different ways, refer to Figure 9.31. Within the convective towers, vigorous condensation occurs within updraft cores. Below the freezing level, large raindrops develop via collision and coalescence and fall out as localized, heavy showers. Above the freezing level, supercooled water drops undergo riming – they freeze on contact with ice crystals, producing frozen graupel particles. Graupel are small, white, mushy pellets just 0.1–0.3 cm (< 2/10 inch) diameter. The smaller particles (light enough to remain suspended), including cloud ice crystals, get pushed out of the updraft core by the impinging front-to-rear flow. But as they exit, they also descend. The resulting trajectories create a fountain-like pattern. The stratiform cloud region is saturated with water vapor. This vapor condenses directly onto ice particles (a process called vapor deposition), fattening them into snowflakes.

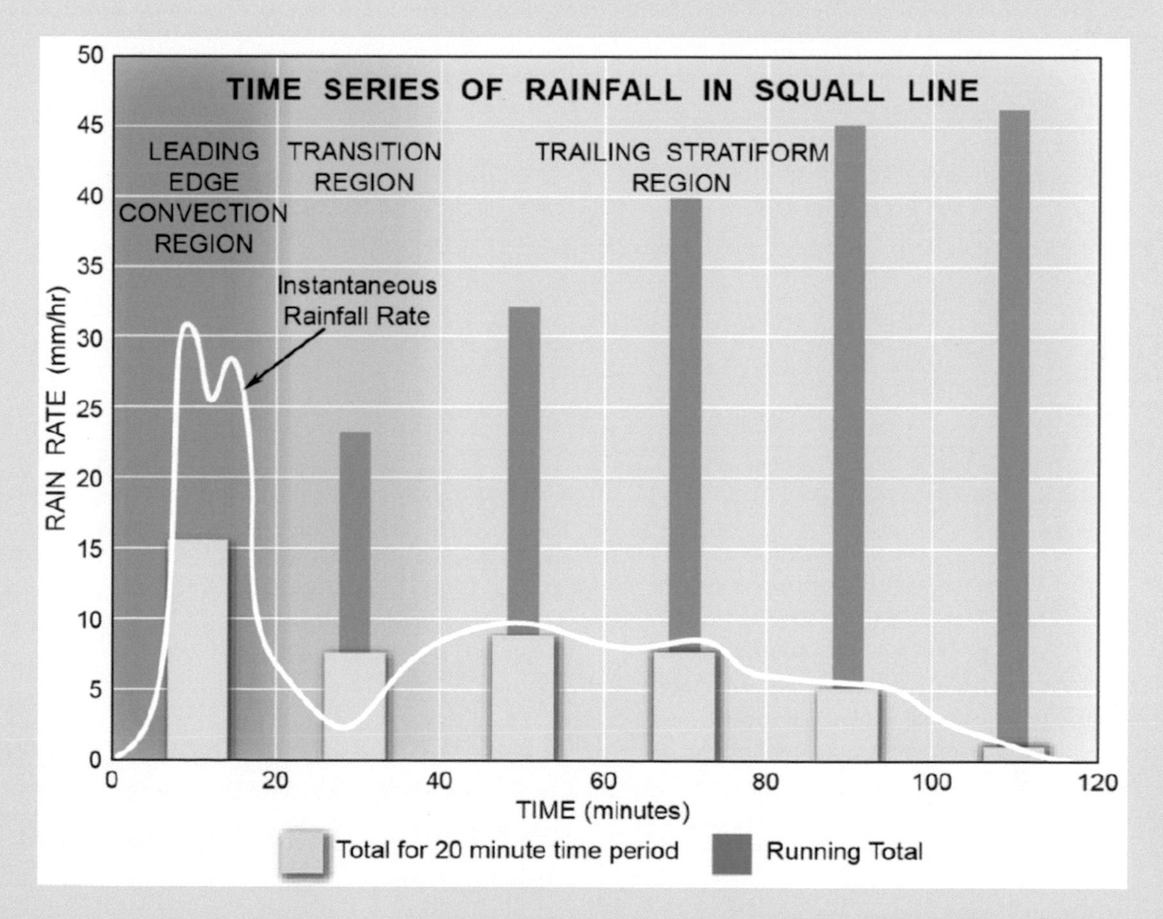

Figure 9.30 Time history of rain accumulation in a typical squall line. Rainfall is not uniform throughout a squall line, nor is the rate of rainfall consistent, as illustrated in this graph. Brief heavy rains first fall from the convective line, followed by up to several hours of moderate-intensity rainfall from the stratiform rain region.

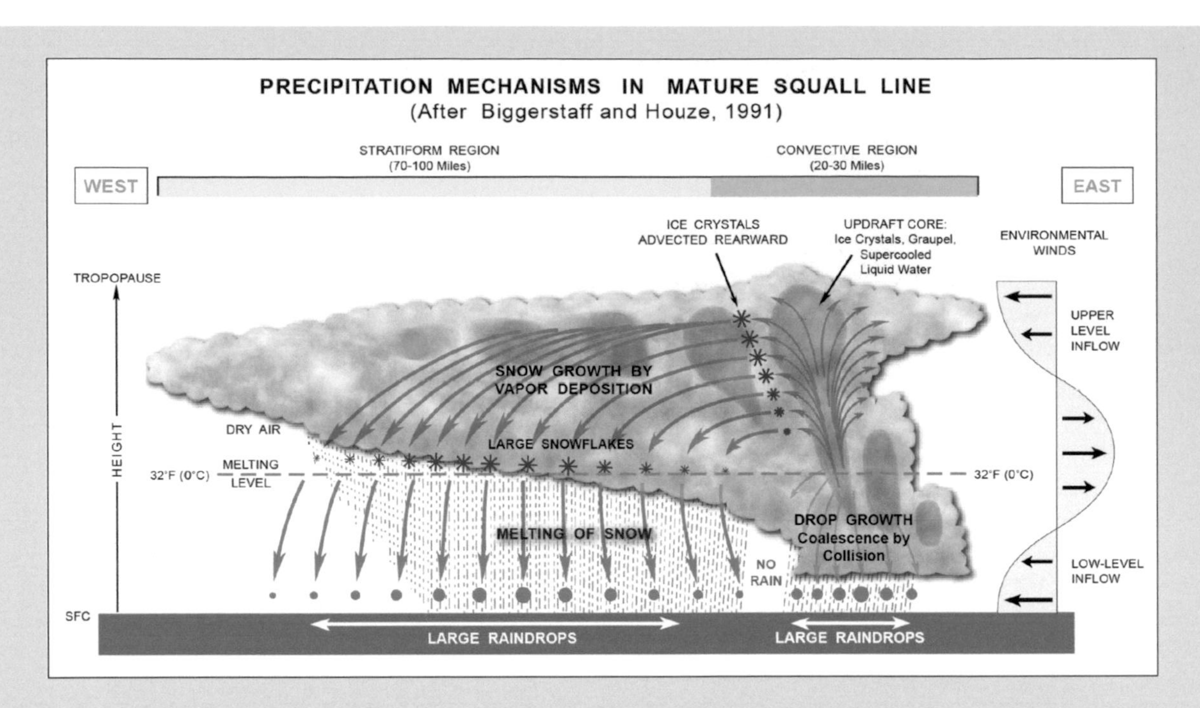

Figure 9.31 Mechanisms generating heavy rainfall in a squall line. Depicted in this diagram are the various in-cloud processes that generate rainfall within a squall line, including liquid and frozen particles of various sizes, types and distributions. (Adapted from Biggerstaff and Houze, 1991.)

These flakes are heavy and fall out of the stratiform cloud layers. As they settle through the melting (0° C or 32° F) layer, they become water-coated and "sticky," clumping together to form even larger snowflakes. At a temperature of about 5° C (41° F), the flakes completely melt, forming large drops of rain that reach the surface.

While the leading convection often appears solid and unbroken in radar images, there are times when this is not the case. Sometimes, given special configurations of wind shear that include strong directional wind change, the leading convective "line" consists of isolated and very intense cells. These contain rotating updrafts and are termed supercells and may produce violent tornadoes.

Squall Lines Are Readily Monitored on Weather Radar

Figures 9.32 and 9.33 show how squall lines appear from the vantage of weather radar. Figure 9.32 shows the same squall line as in the satellite image in Figure 9.25. Note the solid line of very intense convective rain cores (colored red and magenta), from which rain is falling at a rate of several inches per hour. The line is sweeping across Kansas from west to east. In this case, an extensive stratiform region has not yet developed. Most stations would measure only a brief (30–45 minute) episode of intense rain as the convective line sweeps through.

Figure 9.33 illustrates a massive squall line with a classic stratiform rain region. Again, the line moves from west to east. The leading convective portion is a solid, continuous zone of heavy rain (red colors). Behind this line, widespread moderate (and even locally intense) rains cover a broad region. Whether or not a squall line develops a well-defined stratiform rain region depends on many factors: Moist upper levels, air forced to ascend over a wide area beneath the jet stream trough, and strong shear all help generate an area of stratiform rain.

The Climatology of Thunderstorms in the United States Is Characterized by Strongly Contrasting, Regional Variations

We complete this chapter by briefly discussing the geographical distribution of all thunderstorms (ordinary and severe) across the continental United States. This climatology

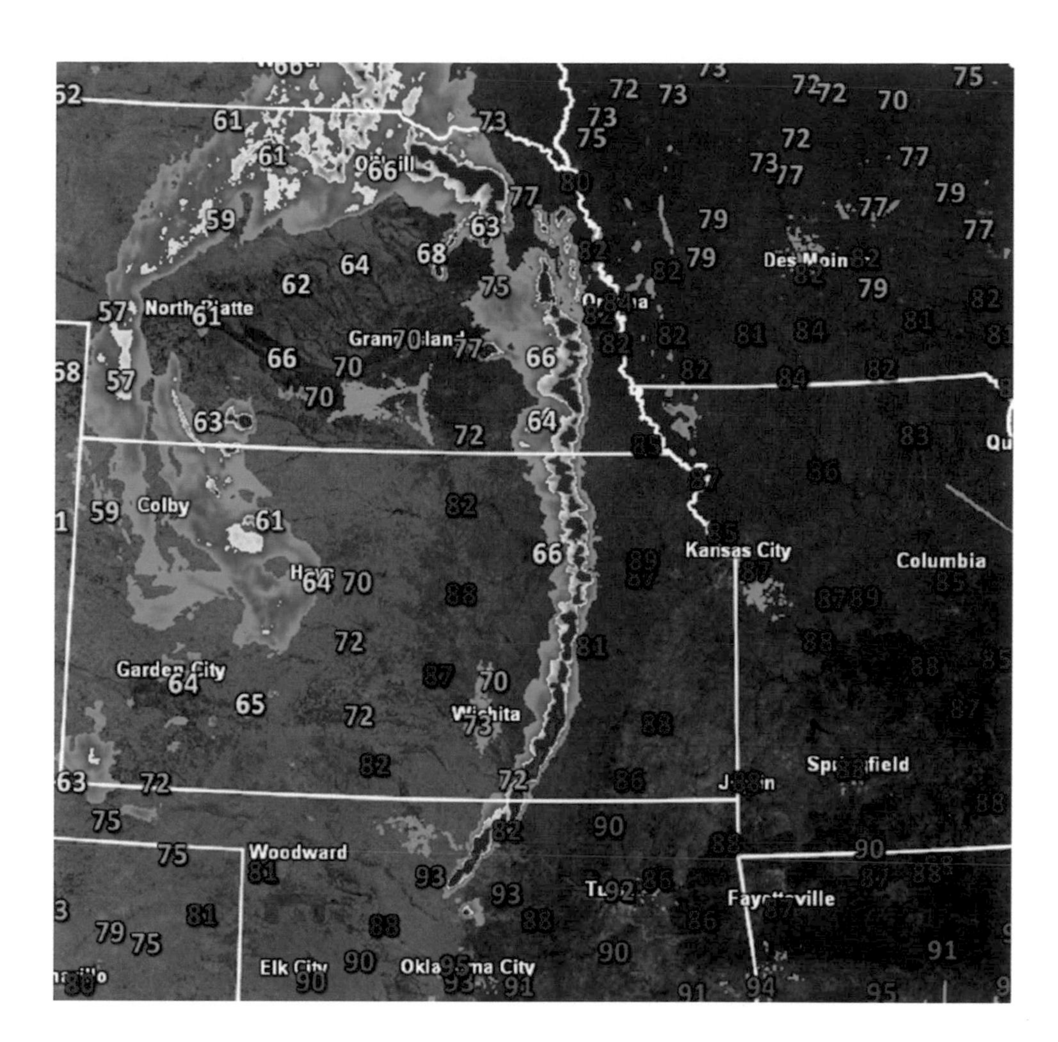

Figure 9.32 Radar image of a narrow squall line lacking a well-defined stratiform rain region. Very heavy rainfall (illustrated in reds and oranges) is common to squall lines. (NWS.)

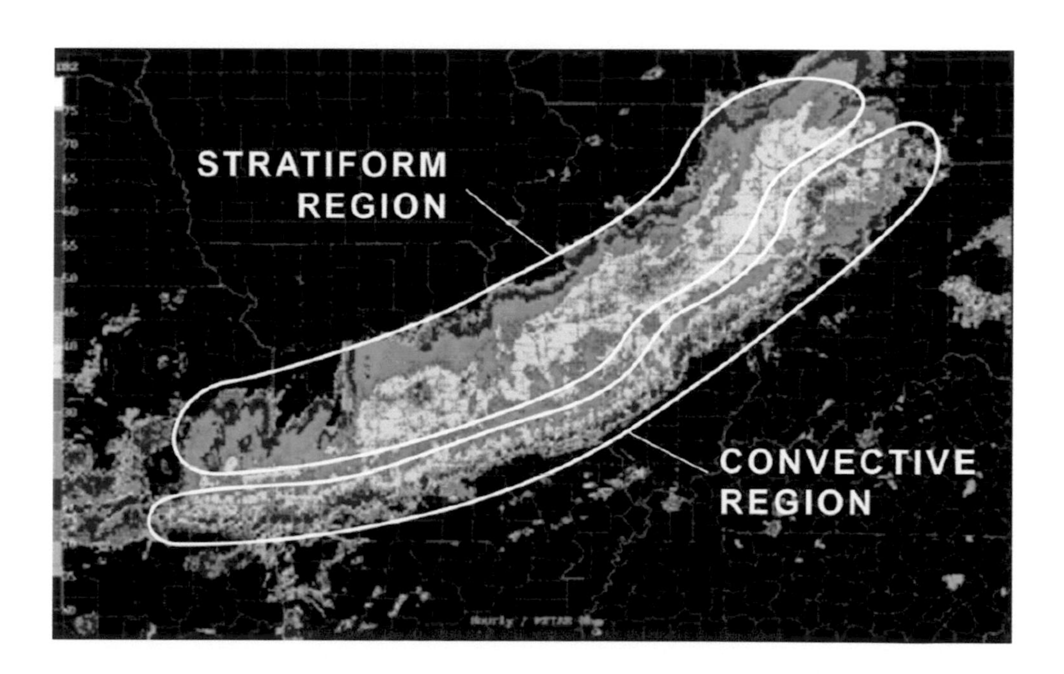

Figure 9.33 Radar image of a squall line with a well-developed trailing stratiform rain region. A sharply defined leading convective line (red) is followed by a broad, trailing stratiform rain region in this radar image. (NWS.)

presents the regional variations, as well as how thunderstorm activity varies according to time of day. Figure 9.34 is a map of the summertime (June–August) storm distribution. Cool colors (green) indicate low frequency; warm colors (orange) indicate high frequency. When studying this map, bear in mind that thunderstorms require three ingredients to develop: unstable air, relatively high moisture content, and a lifting mechanism. Quite commonly, fronts associated with extratropical cyclones provide the uplift. But the map shows that there are also sources of uplift tied to geography and regional wind circulations.

Thunderstorm "Hot Spot" Is Located Over the United States Southwest During the Summer Monsoon Season

Thunderstorms, in general, are rare over the extreme western United States, from Washington through coastal California. While Pacific moisture is often abundant, air masses over the eastern Pacific tend to be cold and stable, thanks to the cold California Current (a chilly ribbon of ocean water flowing southward from the Gulf of Alaska). But paradoxically, thunderstorms frequently bubble up over Arizona, New Mexico, Utah, and Colorado during the summer months. The cause is an area of surface low pressure (called a thermal low) that develops over the hot, elevated terrain. The low persists for many weeks and pulls in a stream of humid air from the Gulf of California. This circulation pattern is part of a regional-scale system of winds called the North American Monsoon. During summer monsoon season, intense thunderstorms erupt in the late afternoon and evening hours across the U.S. Southwest.

Another Thunderstorm "Hot Spot" Is Located Along the Rocky Mountains and High Plains

Summertime thunderstorms also concentrate east of the Four Corners region, along the eastern slopes of the Colorado

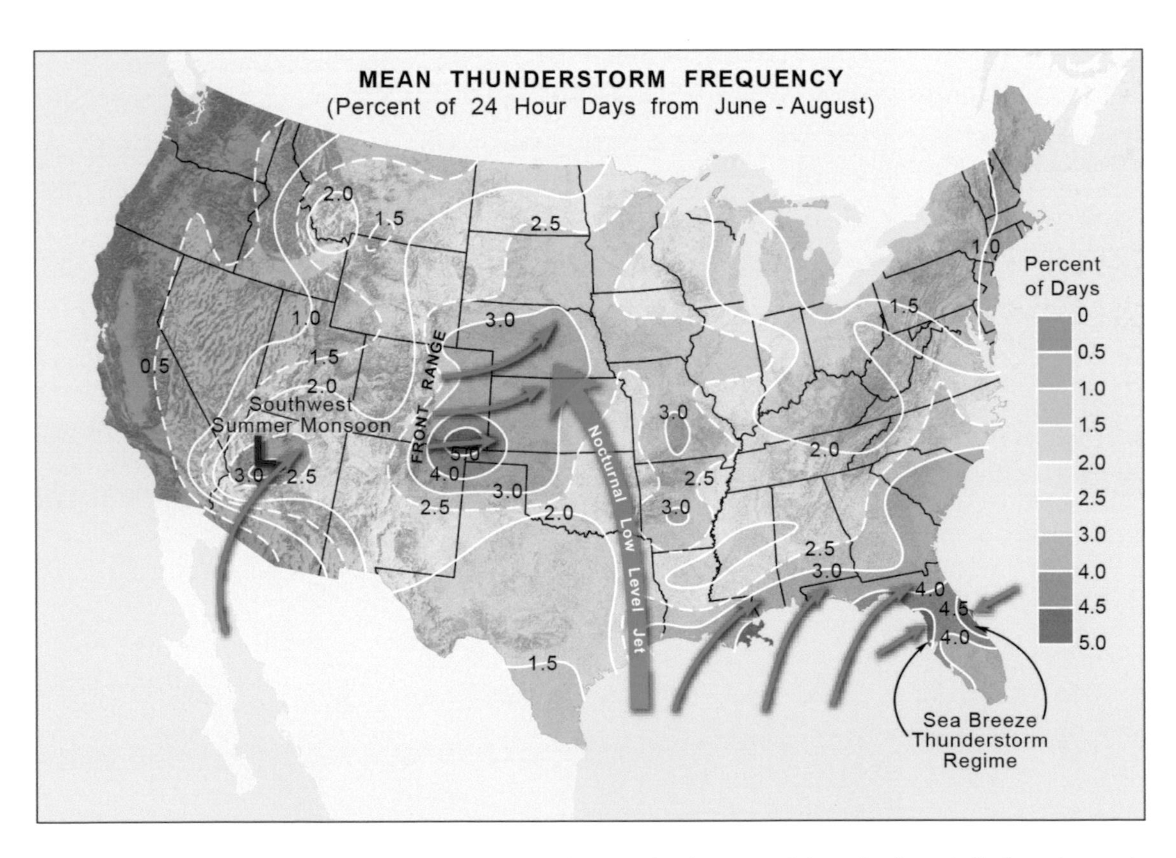

Figure 9.34 Frequency of occurrence of thunderstorms across the United States. Thunderstorm activity varies dramatically from place to place. Low occurrences are found in the Pacific Northwest and New England, where cool air does not provide adequate lifting for storms to develop. In stark contrast, in Arizona a strong thermal low develops during the summer and acts as a firing mechanism for numerous strong thunderstorms. The Great Plains region (especially the Four Corners area) produces a huge number of thunderstorms due to strong thermal heating, a good source of moisture from the Gulf of Mexico, and the added lifting offered by the front range of the Rocky Mountains. Finally, the area along the East Coast spawns many thunderstorms in the summer months due to strong thermal heating and a good supply of moisture moved inland by the Atlantic Bermuda High.

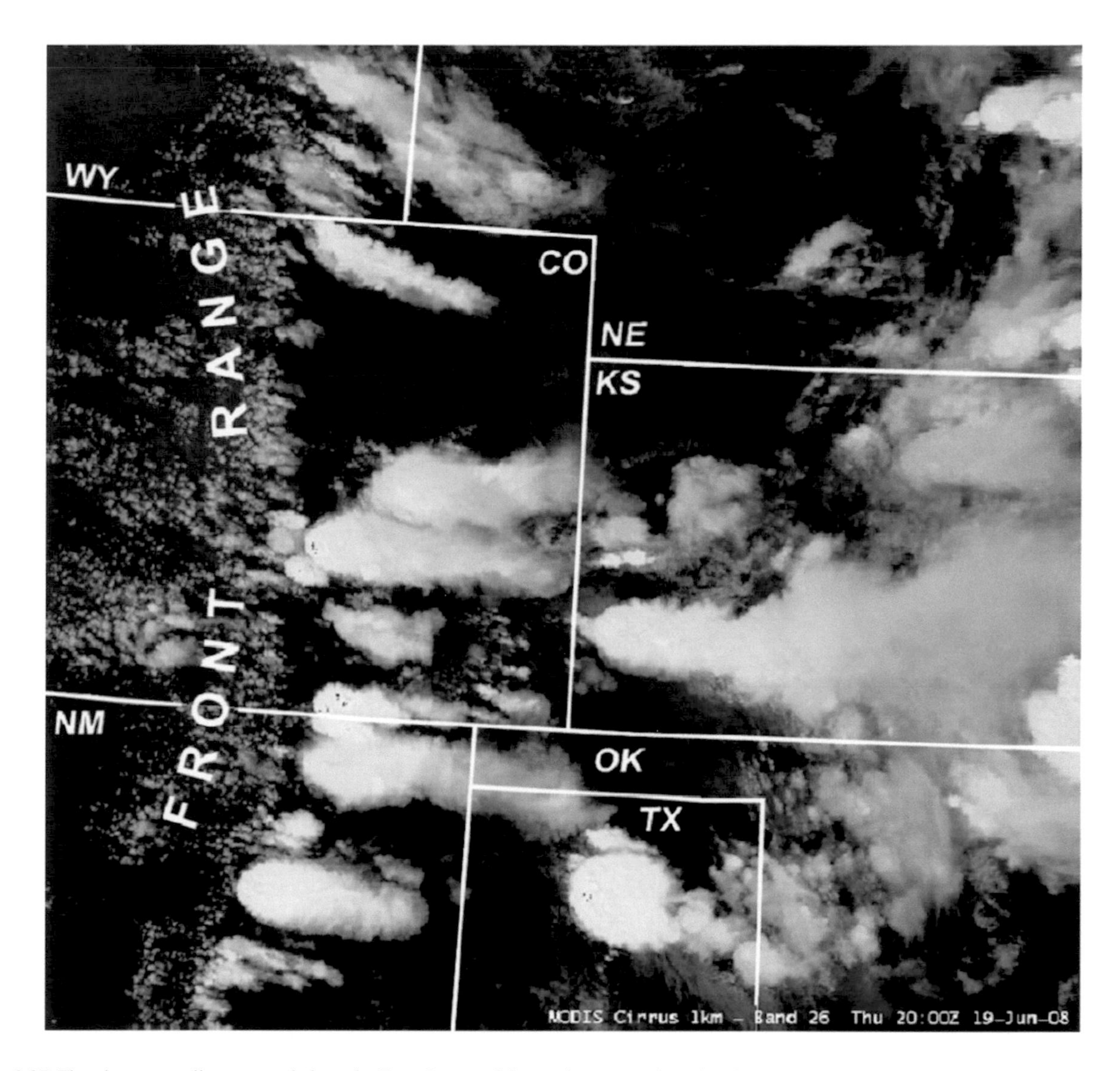

Figure 9.35 Thunderstorm cells generated along the Front Range of the Rocky Mountains. Thunderstorms triggered over the Rockies often become energized late at night by the arrival of very moist air. In fact, the heaviest and most widespread summertime rains to fall over the High Plains take place during the overnight hours. (NOAA.)

Plateau, and over the High Plains. This fact might seem odd at first, given the great distance of these mountains from the Pacific. But the air mass becomes quite unstable and low-level moisture frequently arrives from the Gulf of Mexico. Recall that mountain ranges act as “elevated heat sources.” During the high afternoon sun, a daytime valley breeze develops, causing air to ascend from valleys up mountain slopes. This type of regional circulation pattern was discussed in Chapter 2. The ascent of warm, humid air in the afternoon triggers widespread thunderstorm formation along the eastern slopes and spine of the Rockies. Once formed, the storms often drift eastward on prevailing westerly winds over the High Plains. During the evening and overnight hours, these High Plains storms remain active. The satellite image in Figure 9.35 shows thunderstorm cells dotting the crest of the Front Range and foothills on a hot afternoon. Prevailing westerly winds (wind shear) blow off the anvil ice clouds toward the east.

Florida Is the Thunderstorm Capital of North America

We come now to a third high-frequency thunderstorm zone along the U.S. southeastern coast, particularly over the Florida peninsula. Very warm and humid air streams across the region from the Gulf of Mexico and western tropical Atlantic are pumped inland by a persistent summertime weather system in the North Atlantic called the Bermuda High. But Florida is unique in that dual sea breezes converge along the axis of the peninsula from both coasts. Air converging between the breezes is forced upward, generating afternoon and evening thunderstorms across Florida. While this is a general explanation, the reality is a bit more complicated; for more details, read the special box titled "Digging Deeper: Florida is the U.S. Thunderstorm Capitol!" Florida is also the number one spot in the United States in terms of cloud-to-ground lightning frequency.

The Time of Day for Peak Thunderstorm Activity Depends Critically on Your Location

Figure 9.36 shows the local daily time of peak thunderstorm activity. There is tremendous variation across the United States. The figures use hot colors for afternoon thunderstorm maxima and cool colors to represent nocturnal storms. Since summertime air masses become the most unstable during the afternoon sun, it's not surprising to find that much of the country experiences a late afternoon-early evening incidence of thunderstorms.

However, the Rockies and the nocturnal low-level jet conspire to produce a great deal of overnight storminess across the High Plains. The regional wind system called the nocturnal low-level jet was discussed in Chapter 2. This is a low-altitude "river" of high wind that develops during the night over the southern Plains. This southerly current transports large amounts of moisture northward from the Gulf of Mexico. This may come as a surprise to those who do not live in the area, and one can only imagine how these storms hampered settlers pushing across the Plains, with nothing more than leather hats and wagon tarps for protection!

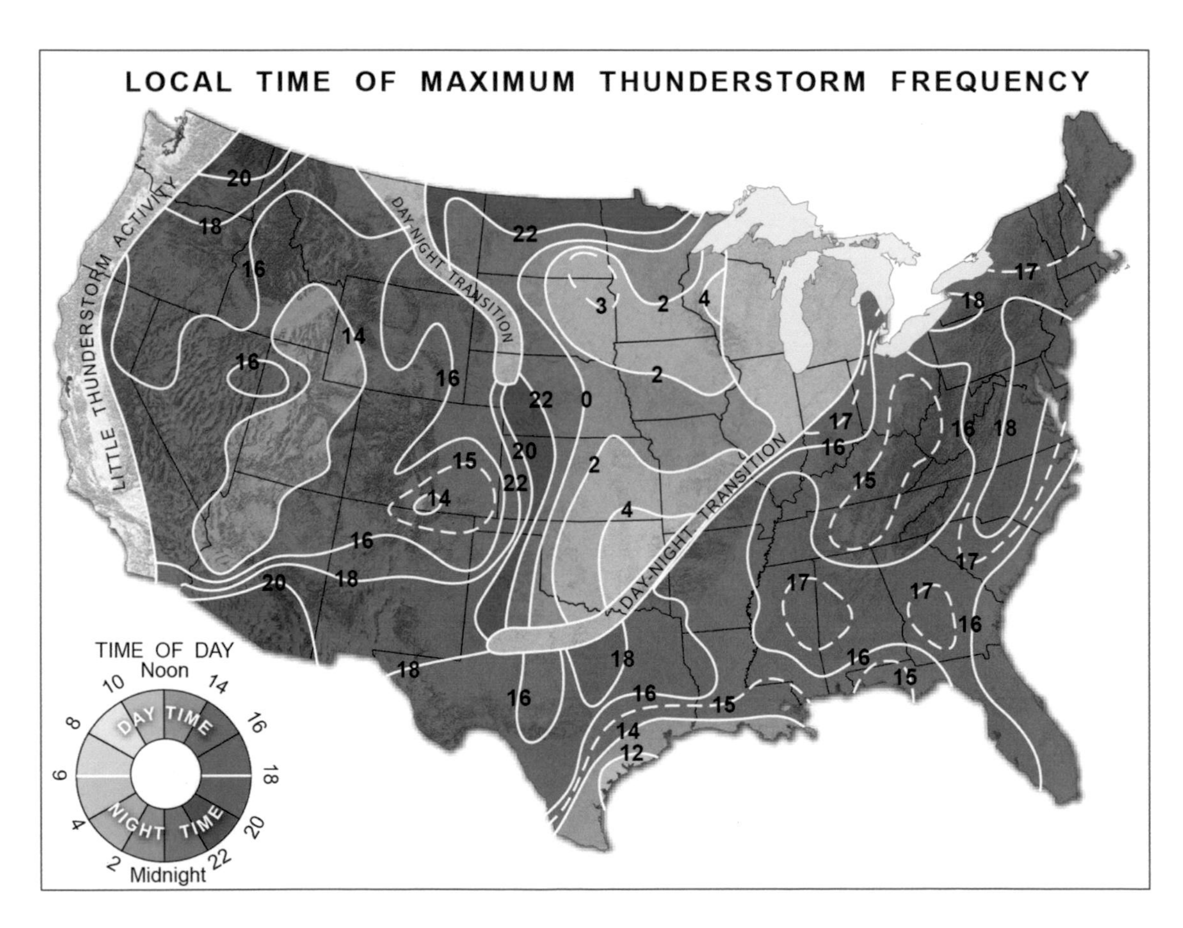

Figure 9.36 Time of occurrence of thunderstorm activity across the United States, which varies markedly from region to region. There is a general absence of thunderstorms along the West Coast of the United States. The western and eastern regions observe maximum thunder activity from about noon to midnight, while the central portion of the country experiences maximum activity after midnight and into the early morning hours. (Adapted from Kessler, 1986.)

Amazing Storms: Florida Is the U.S. Thunderstorm Capitol!

In this chapter, we discussed many of the ways that unstable air is lifted to create thunderstorms, including lifting along the fronts of extratropical cyclones, up mountain slopes, and ahead of coastal sea breeze boundaries. Along both coasts of Florida's peninsula, sea breeze boundaries move inland on many days. But to say that the unstable air is simply "squeezed upward" between these boundaries is an oversimplification. Let's examine a sequence of illustrations that capture the evolution of thunderstorms on a typical summer day over Florida (Figure 9.37). On most summer days, a prevailing wind blows across the peninsula. The wind is most often from the east, due to tropical trade winds or the clockwise flow around the Bermuda High. On other days, winds blow from the west, off the Gulf of Mexico. Depending on the wind direction, usually one of the sea breeze boundaries moves inland farther than the other.

In Figure 9.37, the day starts with prevailing easterly winds. The east coast sea breeze (ECSB) gets pushed inland because the onshore flow and prevailing winds are in the same direction and reinforce each other. The west coast sea breeze (WCSB), on the other hand, remains anchored along the coastline; here, the prevailing winds oppose the breeze's onshore flow.

As the morning unfolds and the land surface heats up, the air mass over Florida becomes unstable. Thunderstorms erupt along both sea breeze boundaries, but only the ECSB moves inland. It generates strong convergence and uplift of air. A line of thunderstorms develops along the ECSB and sweeps across the peninsula. Thunderstorms along the

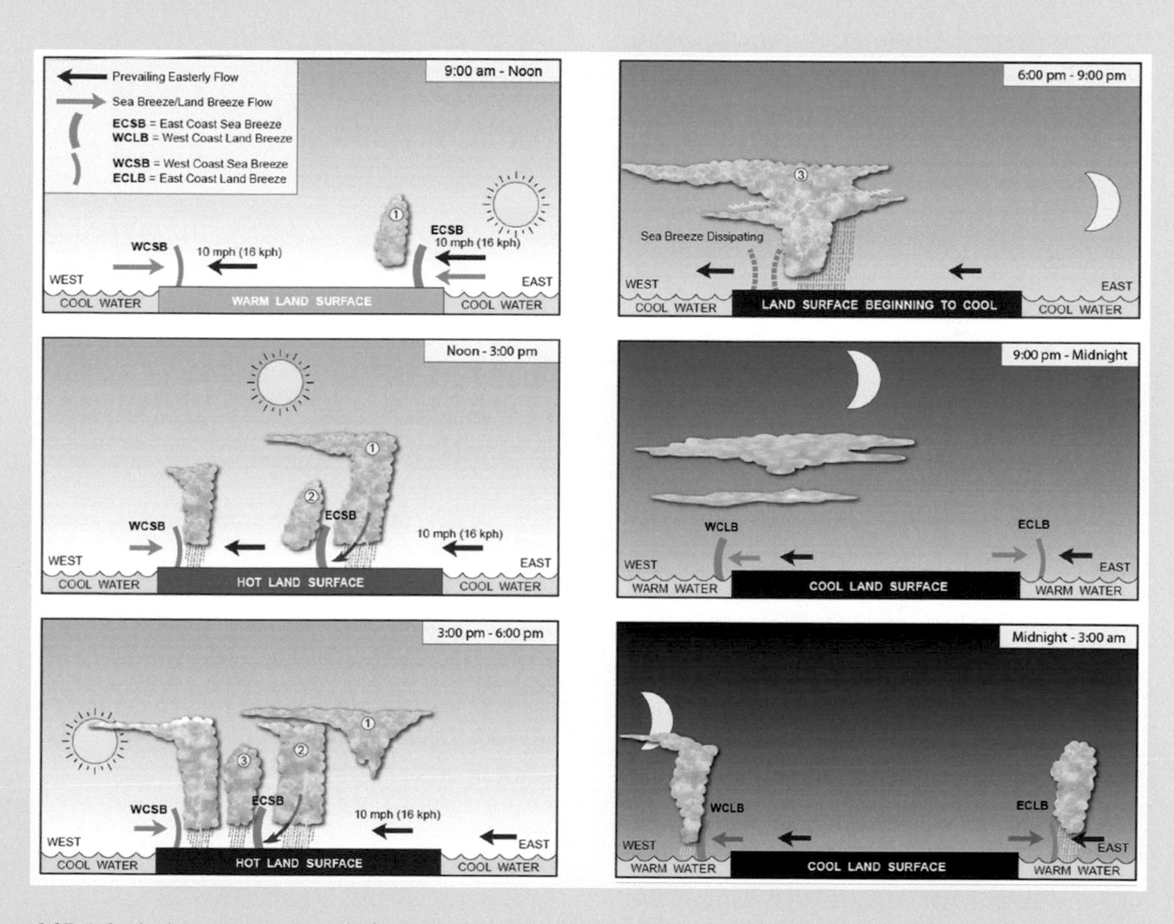

Figure 9.37 A day in the evolution of Florida thunderstorms. The prevailing wind is from the east. Intense afternoon heating of the peninsula causes the sea breeze to move inland, firing off thunderstorms along its leading edge. These thunderstorms move across the peninsula and decay by evening. Nocturnal cooling of the land generates an offshore land breeze circulation that triggers overnight thunderstorm activity over adjacent oceans. The cycle repeats itself on the next day.

west coast remain close to the coastline. By late afternoon, thunderstorms from the east coast push across Florida, merging with west coast storms. These mergers create heavy rains and frequent cloud-to-ground lightning.

Notice that thunderstorms regenerate along the ECSB boundary (storm cells labeled 1, 2, and 3) as that boundary moves toward the west. Successive generations of storms effectively propagate the line of heavy weather across the peninsula. Cold downdrafts from these clouds reinforce the ECSB boundary, intensifying it and helping to propel it across Florida.

By nightfall, the land surface cools and the air stabilizes. Remnants of dissipating thunderstorms (anvil clouds) slowly drift off the peninsula. Late at night, cool air flows off the peninsula as east and west coast land breezes. Warm, humid ocean air converges and ascends along these boundaries, triggering a nocturnal round of thunderstorms several miles off both coasts that persist into the early morning hours.

We can also view the evolution of Florida thunderstorms from space (Figure 9.38), using a series of snapshots recorded every hour during the daytime. On this particular day, the prevailing wind blew from the west, and the WCSB pushed deeper into the peninsula. See if you can follow the maze of sea breeze boundaries, a lake breeze boundary (LB), and thunderstorm outflow boundaries (OB) as they trigger at least three sequences of thunderstorms (TS).

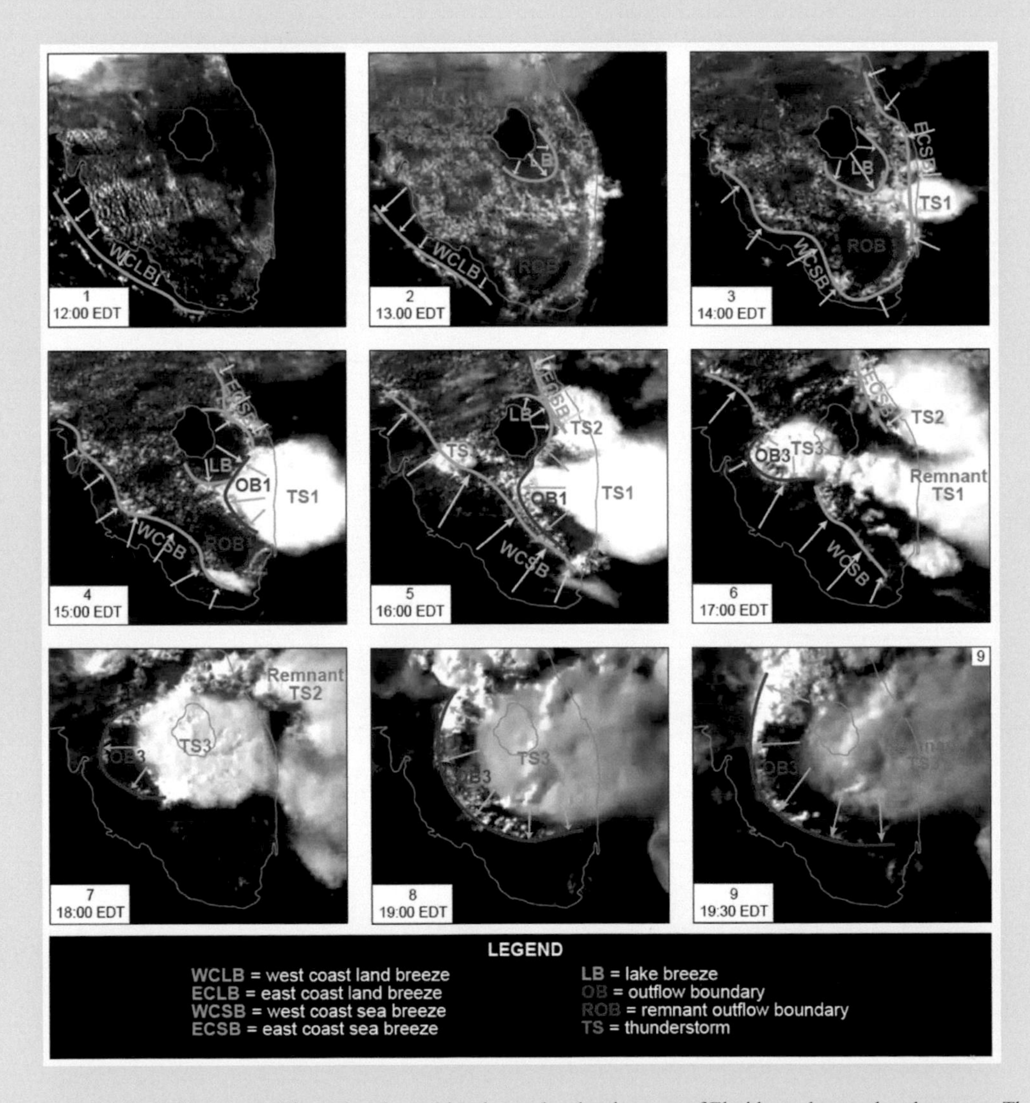

Figure 9.38 Sequence of satellite images illustrating the rapid and complex development of Florida sea breeze thunderstorms. The prevailing wind is from the west. Not only sea breezes but also outflow boundaries generated by thunderstorms and even the Lake Okeechobee breeze influence where and when individual thunderstorm cells develop. (Adapted from NOAA.)

Summary

LO1 Describe the three principal types of ordinary thunderstorms: single cells, multicells, and squall-line systems.

1. Thunderstorms are responsible for much of the hazardous weather produced by severe local storms. Thunderstorms can occur in isolation, aggregate into larger complexes, and are often embedded in larger-scale traveling disturbances such as extratropical cyclones.
2. A more generalized term describing deep, localized thunderstorm clouds is atmospheric convection, as not all intense convective clouds produce lightning.
3. The basic building block of any thunderstorm is the convective cell, which contains an updraft and a downdraft.
4. Deep convective clouds develop in an unstable atmosphere, and these clouds act as heat pumps to move excess surface heat (both sensible and latent heat) upwards. The various forms of severe weather (torrential rain, lightning, damaging winds, hail, tornadoes) are essentially by-products of this necessary heat transfer.

LO2 Explain the process of atmospheric convection, which is driven by buoyancy and develops in order to transport excess surface heat upward.

1. Thunderstorms are described as cumulonimbus clouds composed of liquid water in their lower parts, ice particles in their upper (anvil) regions, and a mixed phase region of both supercooled liquid and ice in the middle. The mixed phase portion is critical for lightning formation.
2. Surface-based thunderstorms develop as warm surface air heated by the afternoon Sun and buoyed upward through a deep, unstable layer.
3. The buoyant ascent in elevated thunderstorms begins when an unstable layer a few thousand meters above the surface is lifted. These storms develop along warm fronts and during the overnight hours and rarely produce severe weather.
4. High-based thunderstorms develop in an unstable layer that is semiarid, common to the western United States. These storms produce little rain, and their lightning is known to ignite wildfires.
5. The degree and depth of the atmosphere's unstable layer is a predictor of thunderstorm severity and is characterized by calculating buoyant energy (J/kg). The buoyant energy of a cloud updraft can be increased by warming and moistening the surface layer and/or by cooling the upper troposphere.

LO3 Discuss how wind shear gives rise to different types of thunderstorms.

1. Vertical wind shear over a deep layer (0–6 km) is an important determinant of thunderstorm organization and severity. Modest increases in wind speed with height cause the updraft to tilt and unload its precipitation into a separate downdraft, leading to a longer-lived storm cell. In the absence of shear, the downdraft quickly forms within the updraft and eliminates its buoyancy.
2. The three types of ordinary thunderstorms are single cells, multicells, and squall line systems.

LO4 Describe the characteristic life cycle of the thunderstorm cell, including development of the updraft, downdraft, and cloud electrification.

1. The ordinary single cell thunderstorm forms in an environment of weak or no wind shear. It undergoes a life cycle consisting of three stages that spans 30–45 minutes. In the developing stage, the convective cell rapidly deepens as its updraft ascends, and a pocket of rain develops aloft.
2. In the mature stage of thunderstorm evolution, the downdraft develops when falling rain drags a mass of air toward the surface. Dry, cool air entering the sides of the cloud causes rain to evaporate, chilling the air, which contributes further to its descent. The downdraft creates a blast of cool air that spreads along the surface, called the outflow. The gust front is the leading edge of the outflow.
3. During the thunderstorm's mature stage, regions of static charge build within the cloud as supercooled water droplets and ice particles interact within the cloud. Electrical discharge (lightning) occurs both within the cloud and between cloud and ground.
4. During a thunderstorm's dissipating stage, the cloud becomes downdraft-dominated and much of the remaining cloud water is gradually converted to rainfall of diminishing intensity.

LO5 Understand how multicell thunderstorm systems propagate and move relative to the prevailing airflow.

1. Multicell thunderstorms, the most ubiquitous type of thunderstorm, consist of an ensemble of individual cells that coevolve. The merged outflow of these cells lifts warm, moist air, forming new cells repeatedly in the same location, often on the west side of the storm. Wind shear pushes newly formed cells off to the east. In this manner, the overall system perpetuates itself for many hours and can become very intense.
2. Multicell storm systems often move more slowly than the prevailing winds. This is because the regeneration or propagation of cells occurs on the upwind side (back edge), while at the same time cells are moved toward the downwind side (front edge).

LO6 Describe how squall lines develop and the hazards they bring.

1. Squall-line thunderstorms are extensive convective cloud systems, long-lived and often very intense. They typically develop ahead of an extratropical cyclone's cold front in conjunction with a trough in the jet stream.
2. The extensive gust front along the leading edge of a squall line lifts an enormous volume of unstable air and moisture. New cells develop repeatedly along the gust front and weaken as they move rearward. Squall lines often move faster than the prevailing winds because cell regeneration occurs out ahead of the system, as opposed to the back edge (as in multicells). Squall lines eventually weaken once the gust front outruns the leading edge by many miles; newly formed cells no longer feed directly into the main cloud system.
3. Squall lines generate heavy rains via different cloud mechanisms. Brief, heavy showers are concentrated along the front of the line, produced by deep convective cells. A more horizontally extensive layer of cloud develops as remnants of convective cells merge on the backside of the system. Within this stratiform rain region, moderate-intensity rain may fall for several hours.
4. The deep convective clouds of a squall line often merge into a continuous line. At other times, the leading convection consists of discrete cells called supercells. The rotating updrafts in these supercells often produce violent tornadoes. At other times, the leading convective line bows outward and produces a long-lasting sequence of violent downdraft winds.

LO7 Explain the geographical distribution of thunderstorms across the United States and how thunderstorms vary with time of day.

1. Thunderstorms across the United States vary dramatically from place to place. Summertime hot spots of activity occur along the Rockies, the semiarid southwest, the Central Plains, and along the Gulf Coast, including the Florida peninsula. These geographical settings create regional wind circulations that favor the uplift of humid, unstable air during the summer.
2. Across much of the United States, the time of day most frequented by thunderstorms is late afternoon and evening, when the air mass is most unstable (from solar heating). However, broad regions of the Central Plains experience a nocturnal maximum in thunderstorms. This is partly due to a nighttime wind circulation that imports moisture northward from the Gulf of Mexico and late afternoon thunderstorms that drift eastward off the Rockies.

Notes

1 Also called convective available potential energy (CAPE).
2 This process is called entrainment.

References

Biggerstaff, M.I. and R.A. Houze, 1991. Kinemetic and precipitation structure of the 10–11 June, 1985 squall line. *Monthly Weather Review*, 119:3034–3065.

Kessler, E., 1986. *Thunderstorm Morphology and Dynamics*, 2nd ed. Norman, OK: University of Oklahoma Press.

Houze, Robert A, 1993. *Cloud Dynamics*, 1st ed. Academic Press, Boston, MA.

CHAPTER 10

Severe Thunderstorms, Emphasizing Supercells and Damaging Hail

Learning Objectives

1 How do instability and wind shear act to generate severe local storms? Explain what distinguishes a severe local storm from an ordinary thunderstorm, and the types of weather hazards not explicitly covered in the standard definition of a severe thunderstorm.
2 Discuss in general terms the different categories or classes of severe local storms and their organizing features.
3 What characteristic types of severe weather hazards does each main type of severe local storm generate?
4 Discuss the two prevalent, seasonal types of weather patterns (cool vs warm months) giving rise to severe weather outbreaks.
5 Explain how the distribution of severe thunderstorm hazards shifts through the course of the year and the interannual variability of hazards.
6 Describe the basic properties of supercell thunderstorms. What is a mesocyclone? Distinguish between classic, low-precipitation, high-precipitation, and mini-types of supercells.
7 How do supercells develop? Trace in particular the origins of the mesocyclone's rotation from genesis through storm maturity.
8 Discuss the formation of damaging hail, and discuss the societal impacts of this severe thunderstorm hazard.
9 How do severe thunderstorms impact agriculture? The power grid? Commercial aviation? How is thunderstorm vulnerability shaped by patterns of human behavior?

Introduction

Isolated thunderstorm cells are localized and short-lived (about 30 minutes). Multicells and squall lines cover more area and may persist 2–3 hours. All of these thunderstorms generate heavy rain, lightning, strong wind gusts, and sometimes small hail. Fortunately, the vast majority of thunderstorms in North America remain below severe limits. But what exactly is a severe thunderstorm?

According to the National Weather Service (NWS), a severe thunderstorm must produce one or more of the following:

1 wind gusts exceeding 50 kts (58 MPH);
2 hail size greater than 2.5 cm (1 inch) diameter;
3 one or more tornadoes.

Some of these criteria are arbitrary; for example, a 50 kt (55 MPH) gust might very well fell a large, diseased tree limb and is therefore considered a severe wind gust. Hailstones only a half inch in diameter can devastate crops, yet such small stones are not considered "severe."

The three criteria for severity do not capture all the destructive facets of a thunderstorm. A flash flood can kill and cause extensive property damage, but the NWS issues a separate set of flash flood watches and warnings, not necessarily inclusive of thunderstorms. Some thunderstorm cells produce intense lightning, leading to home fires, downed trees, and livestock and human fatalities – but the NWS does not issue Severe Lightning warnings. Nonetheless, the imperfect set of criteria used to identify severe thunderstorms works for the majority of cases.

This chapter lays out the fundamentals of severe thunderstorms, including the types of organized severe storm systems, the requirements for their formation, the geographical and seasonal distribution of severe local storm hazards, and socioeconomic impacts.

How a Thunderstorm Becomes Severe

Thunderstorms become severe for a variety of reasons, but the two most important are:

1 an extremely unstable air mass, and
2 strong vertical wind shear.

DOI: 10.4324/9781003344988-13

Recall that an unstable atmosphere is one in which temperature decreases rapidly with altitude, as the surface layer warms, and/or the atmosphere aloft cools off. Cloud updrafts experience buoyancy in this type of environment, rising to great height. Deep cumulonimbus clouds contain large amounts of liquid water and ice, generating heavy rain showers, lightning, and strong wind gusts. The downdraft, which is driven by the weight of falling precipitation and evaporative cooling, generates powerful wind gusts at the surface. Pockets of updraft and downdraft comprise a single thunderstorm cell.

The most extreme type of instability develops during the summer months, when solar heating is strongest. During spring, the ground is warming and cold air masses from Canada invade the United States at high altitudes. Both processes destabilize the atmosphere. One measure of an atmosphere's instability is available buoyant energy (ABE), which, for a moderately unstable atmosphere, is about 1500–2000 joules per kilogram (J/kg). But during late spring, the most unstable air masses have ABE in excess of 3000 J/kg. In the most extreme cases, ABE approaches 5500 J/kg.

When instability increases to excessive levels, the updraft becomes very intense and often punches into the lower stratosphere. As humid air is literally sucked into the base of the storm cloud from miles around, the rates of condensation and freezing of moisture increase. A rapidly rising updraft levitates a tremendous mass of liquid water and ice, which facilitates the growth of large hailstones (some of which may weigh several pounds). The mixture of ice and water roiling within the turbulent updraft induces static electrical charge. Lightning flash rates, an indicator of updraft strength, may approach 100 discharges per minute. Cloudbursts lead to flash flooding. The torrent of water also drags pockets of air downward, the most extreme of which become localized downbursts (isolated pockets of intense, outflowing winds with speeds up to 130 kts [150 MPH]). Through the formation of precipitation, a vigorous updraft thus leads to an intense downdraft.

Wind shear promotes organized, long-lived thunderstorm systems. Without shear, the updraft remains vertically erect. The downdraft, created by rain, descends straight through the updraft, smothering it. Wind shear tilts the updraft so that it leans over. The downdraft then develops beneath the sloped updraft rather than within it. When they occur in isolation, shear-enhanced thunderstorms are called multicells. When moist, unstable air is scooped upward along a cold front, a squall-line system evolves. Wind shear in this setting promotes tilted drafts that fuse along the line.

Note that wind shear refers not only to winds that increase speed with altitude but also to winds that change direction. A certain configuration of strong shear near the surface causes air to rotate. As a storm develops within the sheared air layer, the storm incorporates this spin into its drafts. A storm with a helical or rotating updraft confers a high level of organization and energy efficiency. Helical flows dissipate energy slowly and help stabilize a storm cell, making it long-lived.

Types of Severe Thunderstorms

Figure 10.1 shows the various types of thunderstorms, ranging from ordinary cells (left side of the diagram) to highly organized, severe storm systems (right side). The organization and severity of each storm depend on the levels of air mass instability and wind shear. All three types of severe thunderstorm contain one or more vortices that confer an added degree of intensity and longevity to each type of storm. The vortices all arise from specific configurations of wind shear in a highly unstable atmosphere.

- Supercells, which contain a rotating updraft called a mesocyclone (sometimes called a "meso"), persist in a quasi-steady manner for many hours. The helical updraft also serves as an incubator for strong and violent tornadoes.
- Bow echoes are arc-shaped lines of fast-moving thunderstorms. They contain vortices at both ends of the line. These bookend vortices draw a stream of air into the back of the storm and accelerate it to high velocity. If this current descends to the surface as a downdraft, it generates an explosive outflow of air containing downbursts.
- The mesoscale convective complex (MCC) is the largest type of severe local storm. It contains a large, cyclonic vortex embedded in deep rain clouds. The vortex increases the efficiency with which the storm condenses moisture, making MCCs very prolific flash flood generators. The large vortex can also outlive its parent thunderstorm complex, triggering a new round of thunderstorms the next day.

In reality, nature does not always create storms that are easily classified. The warm sector of a springtime extratropical cyclone often contains multiple types of thunderstorm (for example, a mix of ordinary multicells, supercells, and squall lines). One storm type can also evolve from another. An ordinary multicell

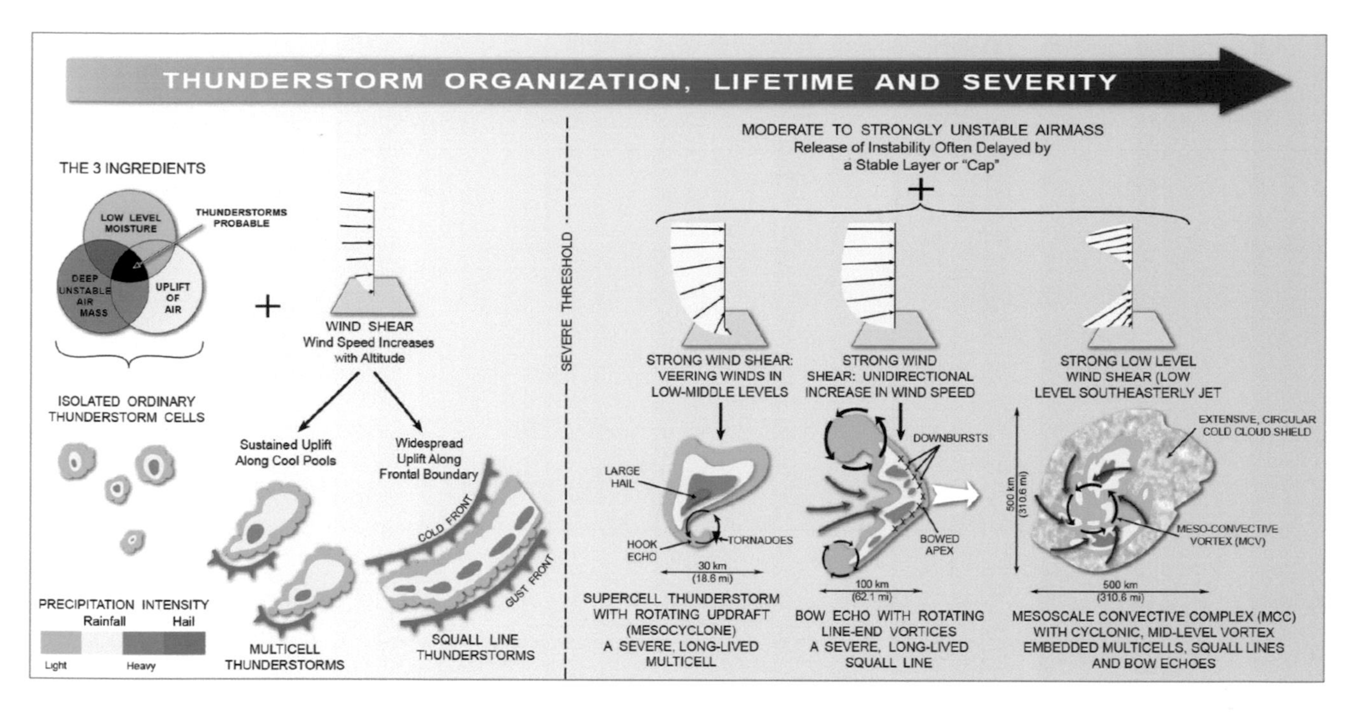

Figure 10.1 Spectrum of severe thunderstorms. Shown is the great variety of thunderstorms, many organized around one or more intense vortices.

may evolve into a long-lived supercell if the wind shear becomes more favorable. A supercell may morph into a bow echo, switching the severe weather from violent tornadoes to downbursts.

Categories of Severe Weather Generated by Thunderstorms

Table 10.1 summarizes the types of severe weather generated by the major categories of thunderstorm. The table includes one additional severe storm type not illustrated in Figure 10.1, the pulse-type severe storm. Pulsing cells are ordinary, isolated cells that produce a brief (10–15 minute) period of severe weather before dissipating. They are neither long-lived nor prolific producers of severe weather. In the absence of wind shear, an extremely unstable air mass allows the updraft and downdraft to strengthen to severe levels, releasing a torrent of heavy rain, hail, and downbursts. Because there is no wind shear, intense drafts cannot be sustained. The entire storm cell develops, matures, and literally collapses in the space of 30–45 minutes.

Figure 10.2 shows the life cycle of an ordinary thunderstorm cell vs a pulse-type severe cell. In the ordinary cell (top panel), the updraft reaches the tropopause in 15–20 minutes. Rain reaches the surface, but the core downdraft smothers the updraft, robbing its buoyancy. Without the continued ascent of humid air, the updraft cannot sustain condensation. The storm rapidly dissipates after 30–40 minutes.

The pulse-type storm, with stronger instability and greater updraft buoyancy, develops a stronger, taller core of precipitation. The cloud top rapidly surges to 15.0–16.5 km (49,000–54,000 feet). Suddenly, the huge mass of suspended precipitation (rain and hail) descends, causing the upper portions of the cell to collapse. An intense cloudburst erupts below the cloud base, accompanied by a powerful downdraft that may contain small hail and one or more downbursts. The downdraft rapidly spreads through the lower portion of the cloud, replacing the updraft. After 40 minutes or so, the cell dissipates.

Pulse-type severe storms generate isolated pockets of flash flooding, hail, and damaging wind. Because the cells cycle so rapidly, their severe stage is too brief to be captured by weather radar – which requires minutes or more to "sweep" the volume of air surrounding it. By the time a Severe Thunderstorm Warning is issued, the storm's severe phase is winding down.

In their mature phase, ordinary isolated cells, multicells, and squall lines may produce sub severe forms of weather, including

Table 10.1 Types of Thunderstorm Organization vs Severe Weather Potential

Storm Type	Hail Size	Strong Straight-Line Winds	Downburst Potential	Tornado Potential	Flash Flood
Isolated ordinary cells	None or small size	Unlikely	None	None	Slight
Ordinary multicell storms	Small to medium size possible	Probable	Low risk	Low risk of weak tornadoes	Moderate
Ordinary squall line	Small to medium-sized possible	Highly likely	Low to moderate risk	Low risk of weak tornadoes	Moderate
Pulse-type isolated cells	Small to medium-sized likely	Highly likely	Moderate risk	Low risk of weak tornadoes	Moderate
Supercell	Medium to large size likely	Highly likely	Moderate to high risk	Possible strong or violent category	High
Bow echo	Small to medium-sized possible	Highly likely	Highly likely	Possible weak to moderate category	Slight
Mesoscale convective complex	Small to medium-sized possible	Probable	Moderate risk	Low risk of weak category	High

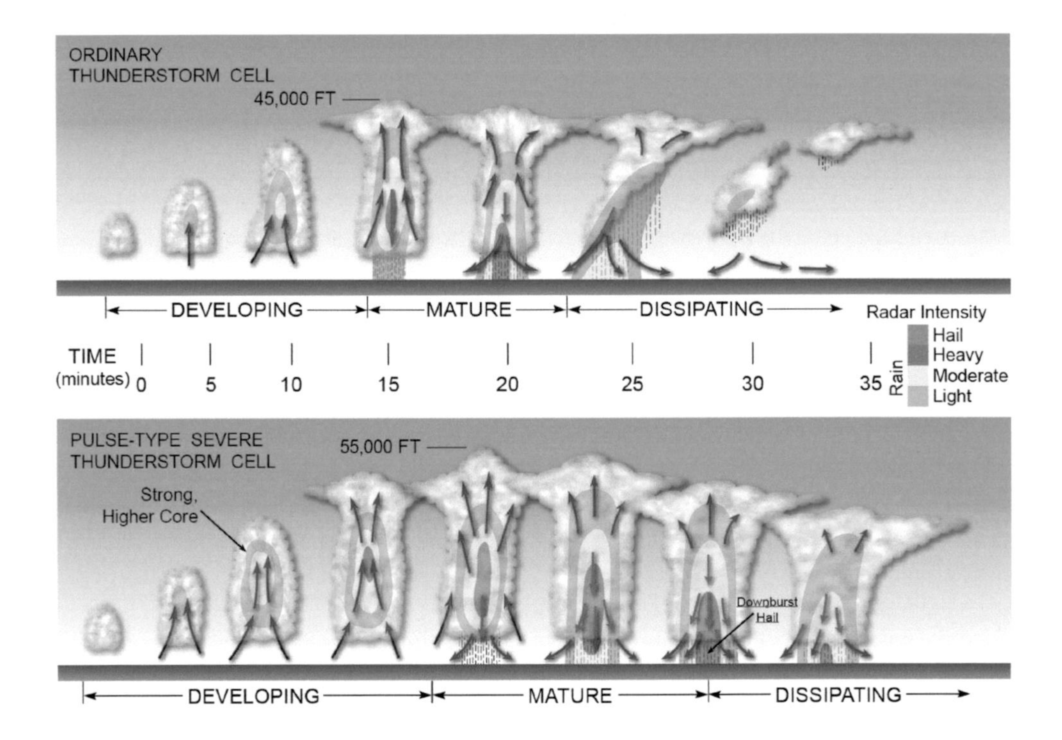

Figure 10.2 Structure of the pulse-type severe thunderstorm. To a large degree, pulse-type thunderstorms owe their existence to strong vertical pulses that lift shafts of moisture upward, leading to strong convectional precipitation. Because these storms have difficulty maintaining vertical instability due to self-generated downdrafts, they tend to have fairly short life cycles.

a cloudburst, small hail, and strong (but not damaging) wind gusts. Supercells are the most common producers of severe weather across the full spectrum, including flash floods and downbursts. Supercells are also champion creators of giant hail and all violent tornadoes. Bow echoes are profligate downburst producers and can also spawn tornadoes of weak-to-moderate intensity. MCCs are efficient producers of flash floods, particularly at night. During their afternoon formative stage, MCCs can breed other forms of severe weather including downbursts, hail, and tornadoes.

The occurrence of one type of severe weather can distract from another type generated by the same storm. For example, the Mayfest Storm in Fort Worth, Texas, was a supercell that occurred on May 5–6, 1995. The supercell was embedded in a larger bow echo thunderstorm complex. The storm struck suddenly during a local outdoor festival called Mayfest, where 10,000 people were gathered. More than 100 persons were injured in this severe hailstorm, and downburst winds propelled hailstones sideways at terrific speeds. Hailstones exceeded 4 inches in diameter, generated more than $1 billion in damage, and commanded national media coverage – cars, windows, and roofs suffered catastrophic damage from the enormous volley of hail. The Mayfest supercell was one of the most destructive hailstorms in U.S. history. However, flash flooding from the same storm was responsible for all of the 13 fatalities on that day.

Seasonal Occurrence of Severe Local Storms

This section examines two broad types of seasonal weather patterns that combine high instability and strong wind shear. These seasonal patterns persist for several months. Many single or multiday outbreaks of severe local storms develop within these large-scale settings. One pattern is characteristic of summer months, while the other is typical of cooler weather.

Cool-Season Pattern

In the warm sector of an extratropical cyclone, warm and humid air streams northward above the surface (Figure 10.3). At the same time, a vigorous trough in the jet stream approaches from the west. Cold air advection at high levels overruns the warm layer, leading to a very unstable atmosphere with significant buoyant energy (Point C). The southerly winds near the surface veer to southwesterly and then to westerly at the jet stream level. The vertical shear at Point C reflects both strong speed and directional shear, the precise configuration leading to supercells. During the afternoon, supercells erupt within the cyclone's warm sector, leading to an outbreak of strong to violent tornadoes and damaging hail.

Closer to the cold front, strong uplift of moderately unstable air occurs near Point D. A squall line develops, sweeping westward through the warm sector. In this setting, winds characterized by strong speed shear (but not directional shear) favor the development of bow echoes. These bows generate long sequences of downburst winds.

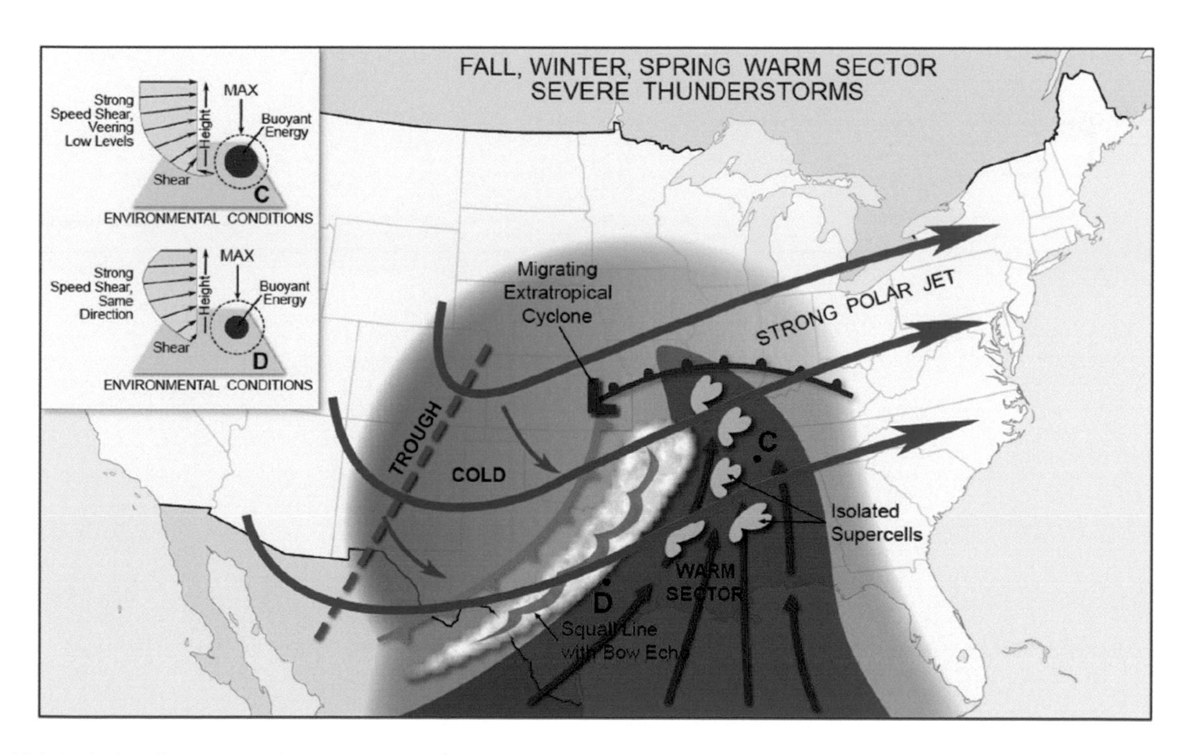

Figure 10.3 Typical cool-season thunderstorm pattern. A warm, moist layer of air flowing in from the SSW encounters a trough that cuts across this stream of energy. The interaction causes an eruption of thunderstorms often accompanied by tornadoes.

As the parent cyclone travels east-northeast, severe local storms follow along. The outbreak commences when daily instability is strongest, in the late afternoon. Wind shear and sustained uplift along the cold front can trigger additional rounds of severe storms overnight. The next day, another outbreak of severe weather may develop farther north and east. This mode of severe weather typifies the cool season in the United States (November–April), when cyclones are strongest and a strong jet stream provides ample wind shear.

Warm-Season Pattern

The prevailing summertime mode of severe local weather in the United States features extreme instability but only moderate wind shear (Figure 10.4). By summer, the polar jet lifts into Canada and weakens, and a large ridge of high pressure (and surface anticyclone) builds east of the Rockies. Shear decreases over the United States. But with intense solar heating, and evaporation of soil and plant moisture, the available buoyant energy content of the lowest atmospheric layers rises to extremely high values (points A and B). Strong southerly flow from the Gulf of Mexico creates a plume of high humidity wrapping around the western limb of the anticyclone. An oppressively hot and humid air mass stagnates along a stationary front across the Great Lakes region. This front separates the heat dome to the south from the cooler air of Canadian origin.

Ring of Fire

The great arc of strongly unstable air defines a so-called atmospheric Ring of Fire along the southeast United States, central Plains, and Great Lakes regions. Isolated late afternoon and evening thunderstorms occur daily within this plume. These storms often take the form of pulse-type severe storms. Thunderstorms are exceptionally numerous along the Gulf Coast and Florida, where sea breezes move inland during the afternoon. But the lack of wind shear prevents the formation of widespread, severe thunderstorms. Few thunderstorms develop in the core of the anticyclone, where air is sinking and thereby stabilizing the air mass. Sinking air warms and dries the middle troposphere, creating an inversion layer that suppresses convection. The exception may be isolated mountain thunderstorms that erupt along the spine of the Appalachians, where air streams up mountain slopes and is effectively forced through the inversion layer.

Northwest Flow Aloft (NWFA)

There is one location where moderately strong wind shear prevails – along the northern rim of the anticyclone, in the vicinity of the polar jet stream (Point B). The winds blow

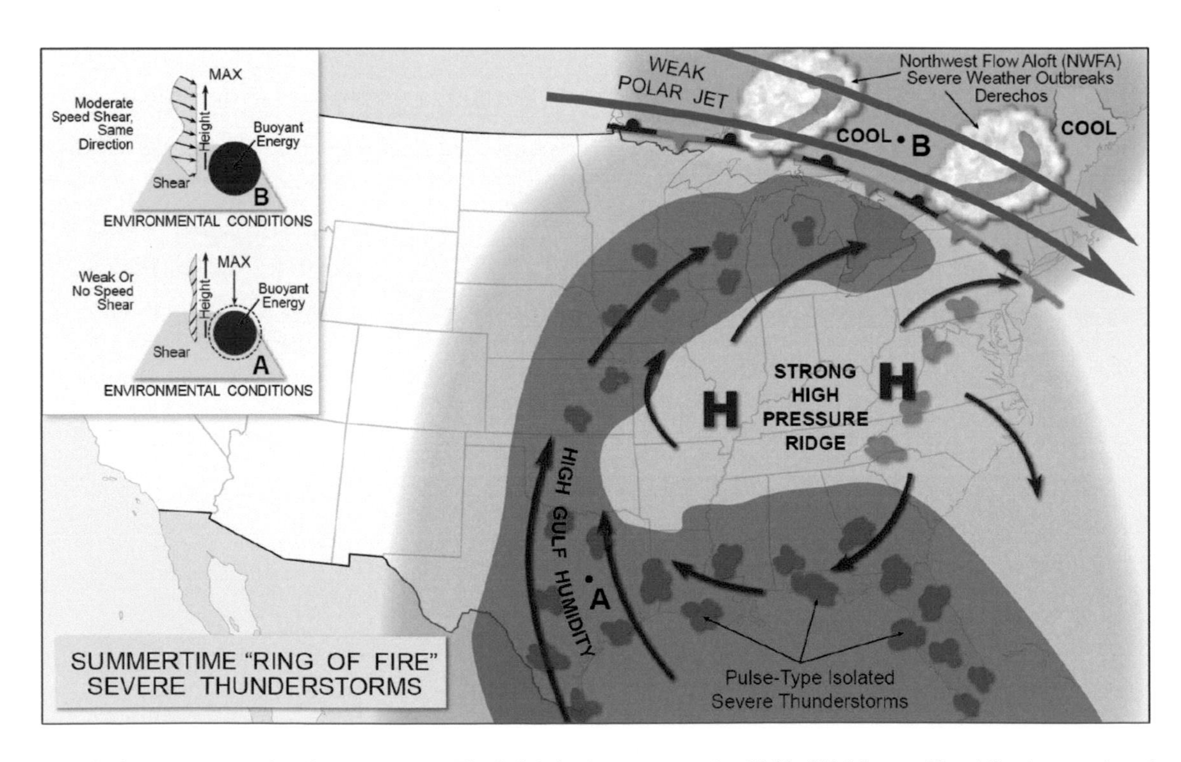

Figure 10.4 Typical summer season thunderstorm pattern. The dark red color represents the shield of high heat and humidity that permits pulses to develop along the frontal boundary.

uniformly from the west-northwest, and wind speed increases significantly with altitude. Outbreaks of severe thunderstorms can develop quite explosively in this region. This weather pattern, most common during June–August, is called a Northwest Flow Aloft (NWFA) weather pattern. The name illustrates the contrast with cool-season extratropical cyclones, which contain strong southwesterly winds aloft in the warm sector. Late spring/early summer NWFA severe weather outbreaks generate strong tornadoes, but the mid-summer NWFA events favor a different type of severe convective windstorm called a derecho.

Derechos

A derecho is a long-lived, convective windstorm generated by one or more bow echoes. The bow echoes create a continuous sequence of downbursts, often extending along corridors hundreds of miles long. Bow echoes track along the stationary front, where instability is concentrated and uplift is sustained. These organized, fast-moving bow echoes create the larger derecho track, which are exceptionally violent.

Derechos can have the same damage impact as a tornado outbreak or a landfalling hurricane. Sometimes two or three derechos follow the same geographical corridor, striking sequentially within the space of a few days. Downbursts progress northwest to southeast across a broad corridor that encompasses the Upper Midwest, Great Lakes, New England, and Mid-Atlantic. Because these storm complexes race along the northern rim of the anticyclone, they are sometimes called ridge rollers. People are particularly vulnerable to NWFA outbreaks because these storms track along a region of high population density (the large cities along the base of the Great Lakes), at a time when people are outdoors engaged in recreational activities.

Geographic Distribution of Thunderstorm Hazards

Many maps portray the climatology, or annually averaged geographic distribution, of severe thunderstorm hazards. For instance, such maps show that tornadoes strike the Great Plains in the spring months, with the bull's eye centered on central Oklahoma. These maps, however, fail to convey the dynamic nature of severe weather. That is, distributions vary considerably from year to year, both in the number of severe weather events and geographic location. Many factors explain this interannual variability, including variations in the intensity and track of extratropical cyclones, the location and strength of the summer heat ridge, and quasi-periodic climate oscillations such as El Niño or La Niña.

To portray the dynamic nature of severe weather, we can show the geographical distribution using a spatial technique called centroid analysis. The distribution of a severe weather hazard can expand or contract, or it can form separate centers or centroids, in different years. Given a scattering of severe weather reports, we can chart the center of the activity over several years. Figure 10.5 shows distributions and centroids for tornadoes over a 6-year period. For each year, the figure show maps for 2 months: December (representative of the low period of tornado activity) and May (the peak of tornado activity).

It is clear that not all years are big tornado producers, and there are significant seasonal shifts in the overall pattern. In Figure 10.5, note that wintertime tornadoes concentrate along the Gulf Coast states, but the centroid can shift hundreds of miles, from southern Georgia (2005) to central Louisiana (2009). In May, tornado activity is quite significant. The central tendency of the tornado distribution moves around a small region composed of Kansas, Oklahoma, Arkansas, and Missouri. In some years, the distribution clusters close to the centroid, while in other years many more states experience tornadoes.

Figure 10.6 shows centroids and distributions of damaging thunderstorm winds, exclusive of tornadoes. These winds include straight-line winds and those produced by downbursts, which generate devastation similar to that of weak- and moderate-intensity tornadoes. Once again, the figure presents six years of data, showing both the winter minimum (January) and summer maximum (July). Annual downburst activity peaks later than tornadoes. Note the large interannual variability captured by January 2008 and January 2009. The southeast United States is the region most affected by damaging thunderstorm winds during winter. In July, damaging winds become very widespread, with far greater geographic coverage than late spring tornadoes. Thus downbursts are a more common thunderstorm hazard. Maximum July activity is found in the Ohio Valley, with very little year-to-year variation in the centroid. The Ohio Valley lies within the principal U.S. corridor of summertime derecho activity.

Finally, Figure 10.7 maps damaging hail reports and corresponding centroids. Once again the figure illustrates contrasting low (December) and high (June) activity months. There is significant seasonal difference in large hail events. Apart from one year (2005), the cool season

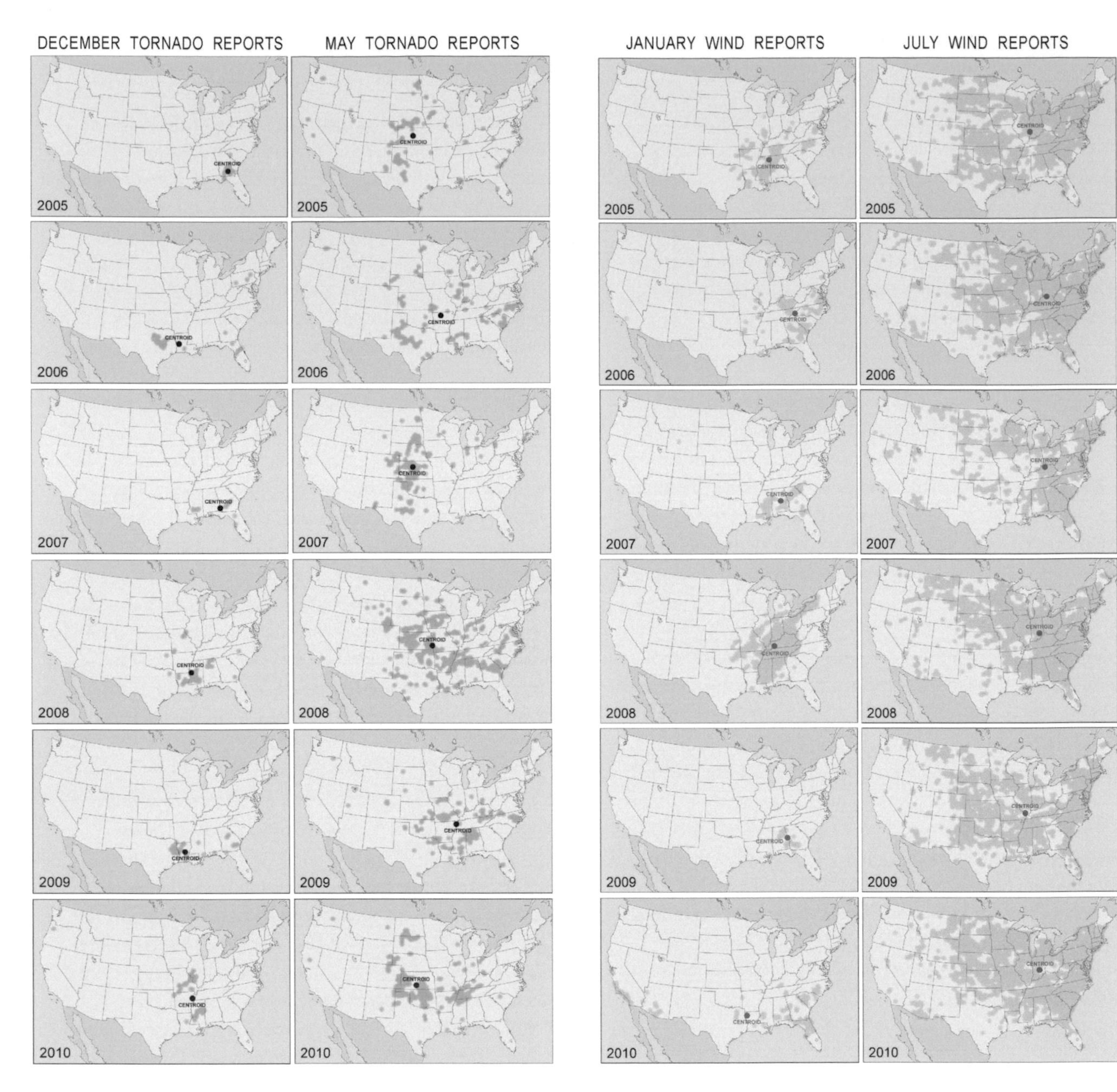

Figure 10.5 U.S. tornado reports, 2005–2010. This collection of maps presents both the low and active months during the tornado season for the 2005–2010 period. Note that there is variation in active periods (May 2008) and low periods (May 2005). Furthermore, tornado activity is greatly compressed into the southern Gulf states during the December period, but the centroids move into the central portion of the country during the active (May) portion of the year.

Figure 10.6 Damaging thunderstorm wind reports, 2005–2010: A collection of maps plotting wind reports for the period. While there are similar geographical distributions between the low (December) and active (July) periods, there are many more damaging wind reports than tornado reports. Also, the centroid locations for the high-activity (July) periods are considerably farther east than for the tornadoes.

is largely devoid of large hail – with just a few reports clustering in the Mid-South. In June, most U.S. regions east of the Rocky Mountains report hail. In some years, hail is curiously absent along the Gulf Coast states. The most consistent year-to-year coverage appears to be the Central Plains region.

What conclusions can we draw from Figures 10.5–10.7? While there are similarities among all of the map collections, there are also clear variations from one year to the next. Therefore, while we can discuss average conditions in broad, general terms, significant differences do occur from year to year. The causes of these differences may be as important to study as the processes that work to establish long-term averages.

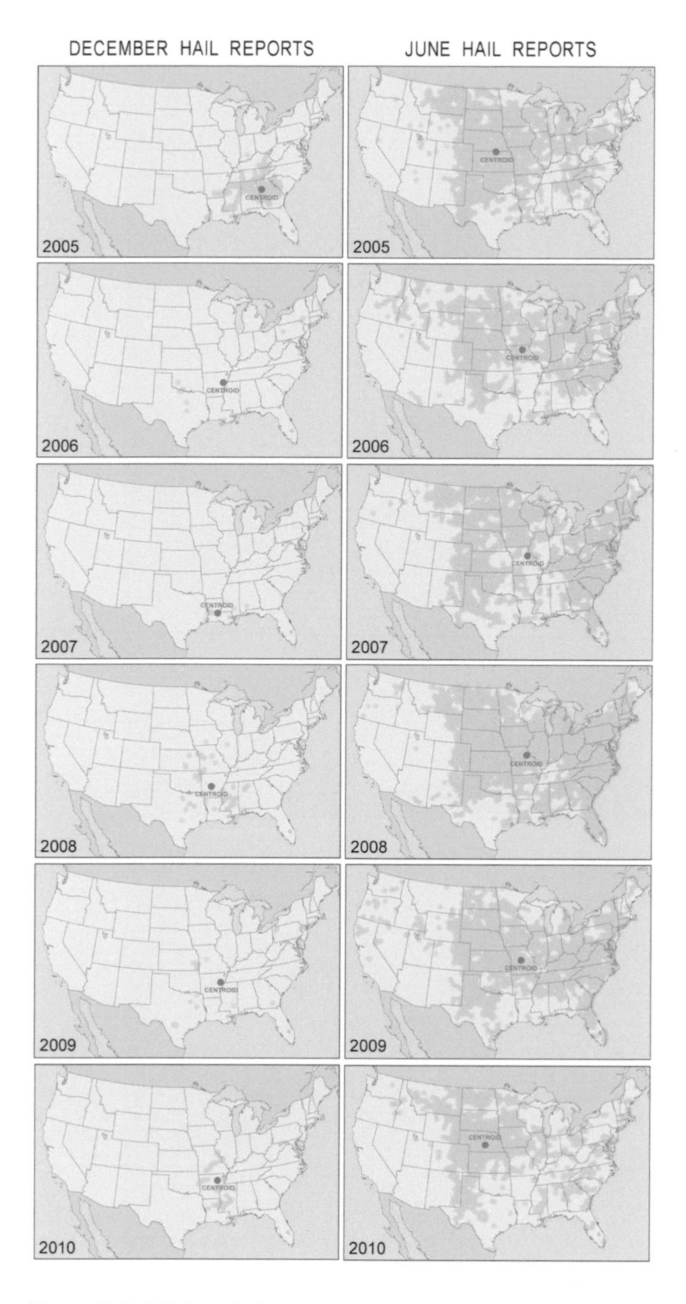

Figure 10.7 U.S. large hail reports, 2005–2010. This figures presents a collection of maps plotting hail reports. Again, variations in the spatial distribution of patterns emerge during both the low (December) and high (June) periods of activity.

Supercells: The Severe Thunderstorm Archetype

Supercells are particularly fearsome severe local storms. They are the most prolific breeders of tornadoes, and they account for the vast majority of strong and violent tornadoes in the United States. In a nutshell, a supercell is an exceptionally long-lived, vigorous multicell thunderstorm – one that contains a helical updraft or mesocyclone. Supercells require significant updraft buoyancy (available buoyant energy) and strong wind shear. Winds in the storm environment must increase significantly in speed and direction in the lowest 5–6 km (16,500–20,000 feet). The winds must also shift clockwise with increasing altitude, a process called veering. Veering is very common in the warm sector of extratropical cyclones. Once fully formed, supercells persist up to 5–6 hours, generating sequences of tornadoes and laying down deep swaths of large hail, downburst winds, and flash flooding.

Panel A of Figure 10.8 depicts an isolated classic (typical) supercell from long distance. These are truly awe-inspiring cloud systems. The mushroom or anvil cloud shape occurs when fast-flowing updraft air is forced to spread outward at the stable lower stratospheric boundary (tropopause), at an altitude of 15.0–18.2 km (49,000–60,000 feet). Individual supercells are typically tens of kilometers in diameter, and are in fact an amalgam of numerous convective cells.

Panel B of Figure 10.8 illustrates a portion of the cloud base in a region of the supercell containing the mesocyclone. This is the supercell's "business end." The vigorous spin of the mesocyclone is readily discernible in these images. The low-hanging, circular, rotating cloud deck is called a wall cloud. Warm, humid air feeds the mesocyclone's updraft as it spirals inward and rises through the base of the wall cloud. In many cases, the wall cloud hangs just a few hundred feet off the surface and may be 5 or more km (3 miles) in diameter. If a strong or violent tornado were to develop, it would do so along the back edge or center of the wall cloud. A swath of heavy rain mixed with hail descends near the back edge of the wall cloud in Panel B. Strong downdraft winds, possibly containing downbursts, blast the surface beneath the rain shaft. If you see something that looks like this, it's time to run for cover!

An additional feature that helps forecasters identify a supercell is termed the V-notch, seen in weather radar (Figure 10.9[a]). This notch is created when strong environmental winds split around the mesocyclone, which behaves like a solid, vertical cylinder of swirling air. The vertical updraft and its meso act like a vertical cylinder. Two streams of heavy precipitation develop downwind of the meso on either side, forming the wings of the "V."

Figure 10.9(b) is the storm's overshooting cloud top, seen from satellite, which casts a dark shadow on the lower cloud anvil. The anvil cloud of ice particles streams outward and away from the updraft in an expanding plume. Fast jet stream flow near 15 km (49,000 feet) can stretch anvils more than a hundred

Figure 10.8 Structure of supercells. (a) A majestic, rotating supercell thunderstorm seen at sunset. Upper portions of the anvil are composed entirely of ice crystals and snow. (NWS.) (b) Wall cloud – the base of a supercell's rotating updraft (mesocyclone). (NOAA.)

(a)

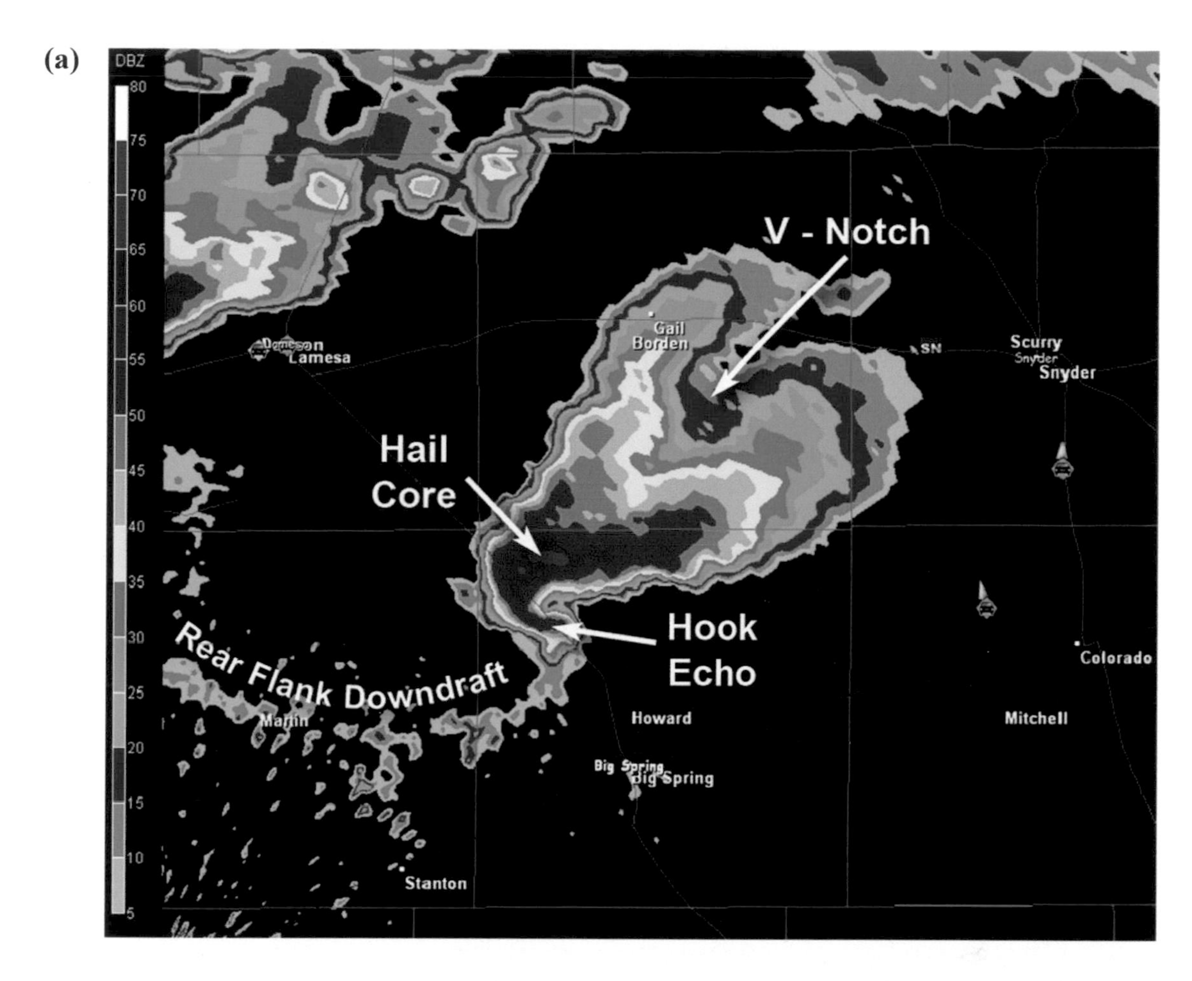

(b)

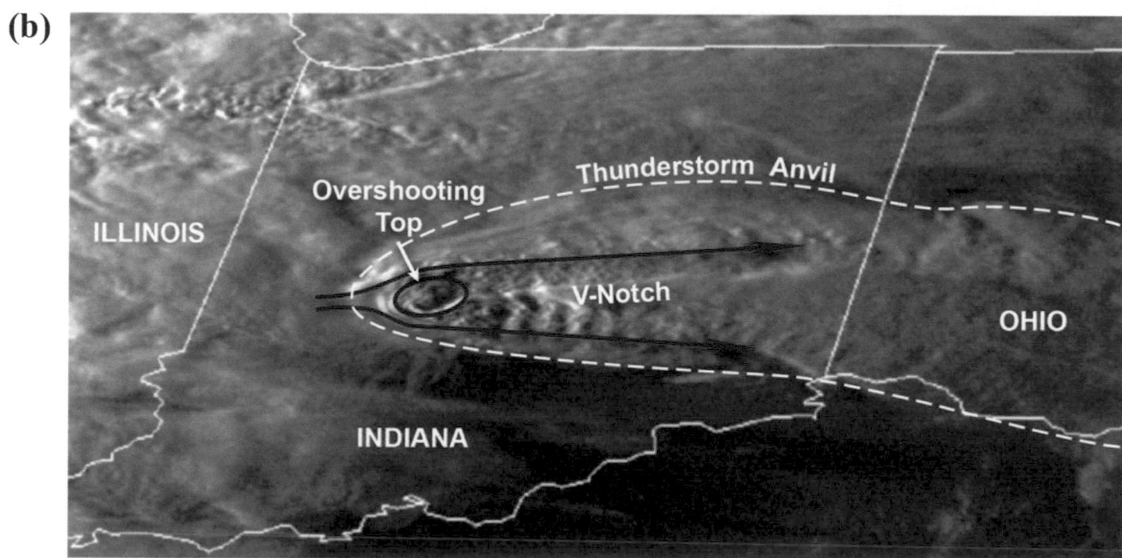

Figure 10.9 V-notch supercell signature. (a) This radar image illustrates a V-notch that is formed as a fast flow passes from southwest to northeast across this supercell, thereby dragging and diverging the currents outward toward the northeast. (b) This satellite image illustrates the same effect shown in (a) but streaming from west to east.

miles downwind. Along the back edge (upwind side) of the anvil is the overshooting cloud top where the strong updraft penetrates a short distance into the stratosphere – essentially coasting to a stop, based on its inertia. This portion of the supercell may appear to "bubble up" 2–3 km (6500–10,000 feet) above the top of the horizontal anvil, and it is a sure sign that the updraft is especially intense (updraft velocities can approach 115 kts [130 MPH] in the most vigorous supercells).

Supercell Air Circulation and Precipitation Pattern

The strong, deep updraft with its mesocyclone is the defining feature of a supercell. This updraft serves to organize the air circulation of the entire storm. Figure 10.10(a) provides a three-dimensional schematic of the essential flow features in a mature supercell. Figure 10.10(b) presents the view of the same supercell from the south, with key cloud and precipitation features identified. The mesocyclone extends through much of the cloud layer, but it is often most intense in the storm's mid-levels. The strong inflow of warm, humid air feeds the meso from the south and east. Vapor condenses into liquid cloud droplets and then freezes into ice particles called hydrometeors, which are moved downwind at high levels (toward the east-northeast) by strong environmental winds impinging on the storm from the west and southwest.

The hydrometeors grow to precipitation size and fall out in a cascade of heavy precipitation on the storm's north and east flanks. A downdraft, called the forward flank downdraft, develops within this cascade, due to the drag of falling precipitation on the surrounding air and evaporative cooling. Downdraft outflow spreads ahead of the storm along the surface as a gust front, accompanied by strong wind. A second downdraft develops during the later stages of the supercell. This spiral downdraft forms on the upwind side of the mesocyclone. As it descends, it spirals around the meso close to the surface. This downdraft develops its own gust front that surges into the supercell's inflow sector, helping to converge and lift warm, humid air into the meso. Powerful outflow winds frequently generate downbursts within this rear flank downdraft (RFD). The rear flank downdraft is intimately linked to the formation of tornadoes. However, not all mesos produce tornadoes, and only about 25% of supercells become tornadic.

The flanking line is a conveyor belt of developing convective cells that form along the RFD. As these cells grow, they merge into the main storm updraft. These cloud elements may contain pockets of developing hailstones, and they supply hail embryos into the main updraft. Once levitated within the powerful main updraft, they can grow to enormous size. Hail is finally released in a torrent along the back edge of the meso, where stones mix with curtains of heavy rainfall.

The hook echo is a distinguishing feature of the supercell observed by weather radar (Figure 10.11). Red shades show areas of strong precipitation. Although the mesocyclone's spin is not directly observed by conventional radar, we can infer a counterclockwise circulation in the storm's southwest corner. Here, an arc of heavy rain and hail curves into the meso in the form of a spiral or hook. A persistent, well-defined hook is the hallmark of a meso, and it increases the probability that one or more tornadoes will develop. If a tornado develops, its location is shown by the black, inverted triangle in Figure 10.10, at the interface of the meso and RFD. In the case of this very intense supercell over Oklahoma, the small region of magenta shading in the ball of the hook is actually debris levitated by the winds of a violent tornado – a feature called the debris ball.

In many ways, a supercell resembles a miniature version of an extratropical cyclone; it contains a central vortex, an RFD that mimics a cold front, and a miniature warm sector along the southeast flank of the storm.

Supercell Types

Supercells span a spectrum, ranging from low-precipitation to high-precipitation varieties. The classic type discussed thus far lies in between. Figure 10.12 illustrates the types of supercells. On the far right side of the spectrum is the HP (high-precipitation) supercell. This is a very large storm with an extensive ice anvil cloud. The area of heavy rain can be very extensive and produce flash floods. The purple region shows a core of large hail along the back edge of the mesocyclone. Because of low visibility in heavy rain, features such as the wall cloud may be obscured or impossible to spot. Tornadoes generated by HP supercells may be "rain-wrapped" or enshrouded in curtains of heavy rain, making advance warning difficult. On radar, the HP supercell often lacks a distinct hook echo and appears "kidney bean" shaped. The mesocyclone is located at the crook of an indented region (the inflow notch) on the southwest side of the storm cell.

Like the classic form, an LP or low-precipitation supercell contains a mesocyclone in the southwest quadrant. However, very little precipitation reaches the surface. The appearance of an LP supercell on weather radar reveals a very weak storm. LP supercells rarely produce tornadoes, but occasionally they generate one or more funnel clouds aloft. They are typical of the Great Plains and may develop along a special type of weather front called a dryline. Visually, the cloud base of LP supercells is quite striking, dominated by a low, rotating wall cloud (Figure 10.13) and very little, if any, precipitation.

(a)

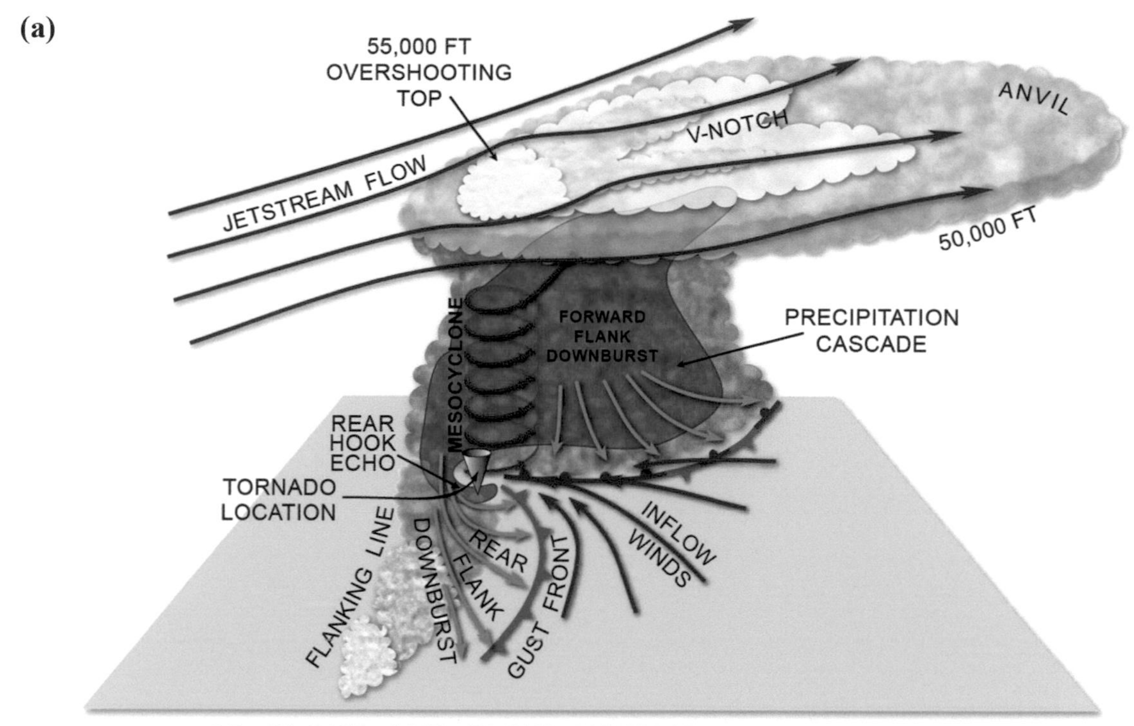

(b)

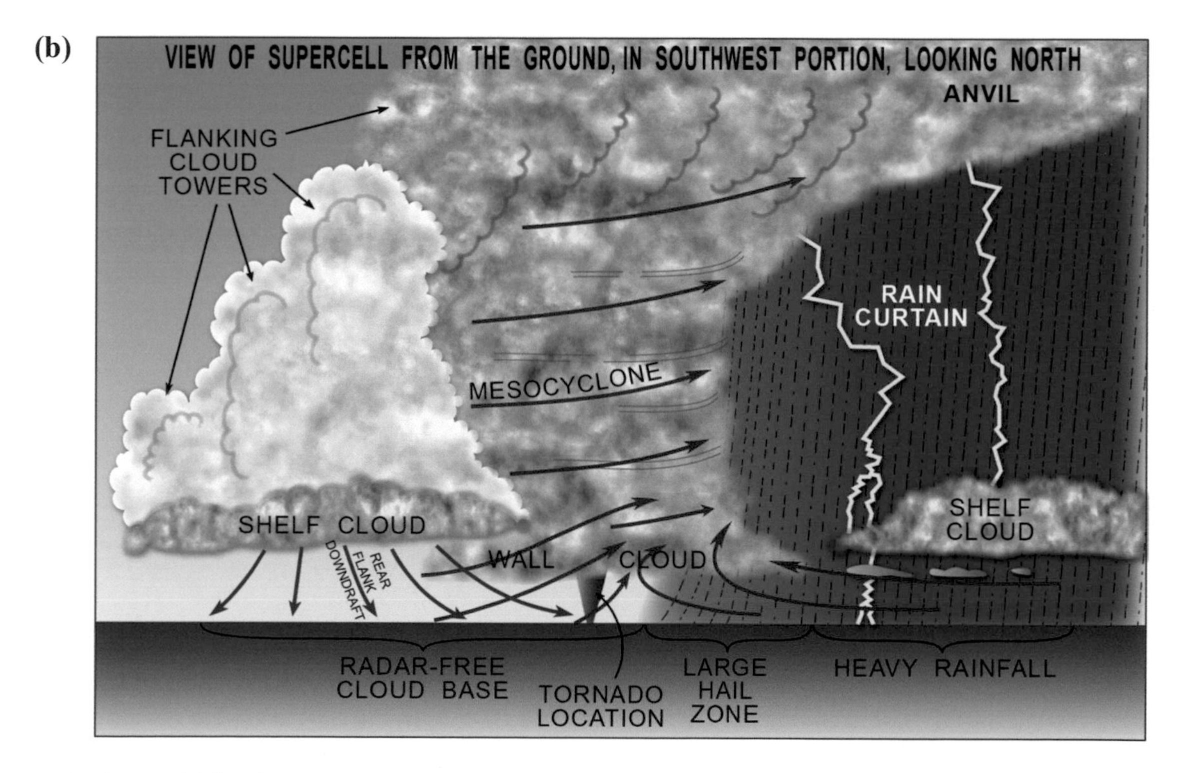

Figure 10.10 (a) The essential flow features in a mature supercell. The dynamics within supercells are complex and tightly integrated. Updrafts add warm, moist air, and downdrafts undercut the inflow of warm air; both of these processes work to accelerate and strengthen the storm. Notice that a strong overshooting top breaks through the top cloud layer and can actually cause a diversion of the jet stream flow. (b) This image presents the view of the same supercell shown in Panel (a) from the south, with key cloud and precipitation features identified, as might be viewed by a storm spotter.

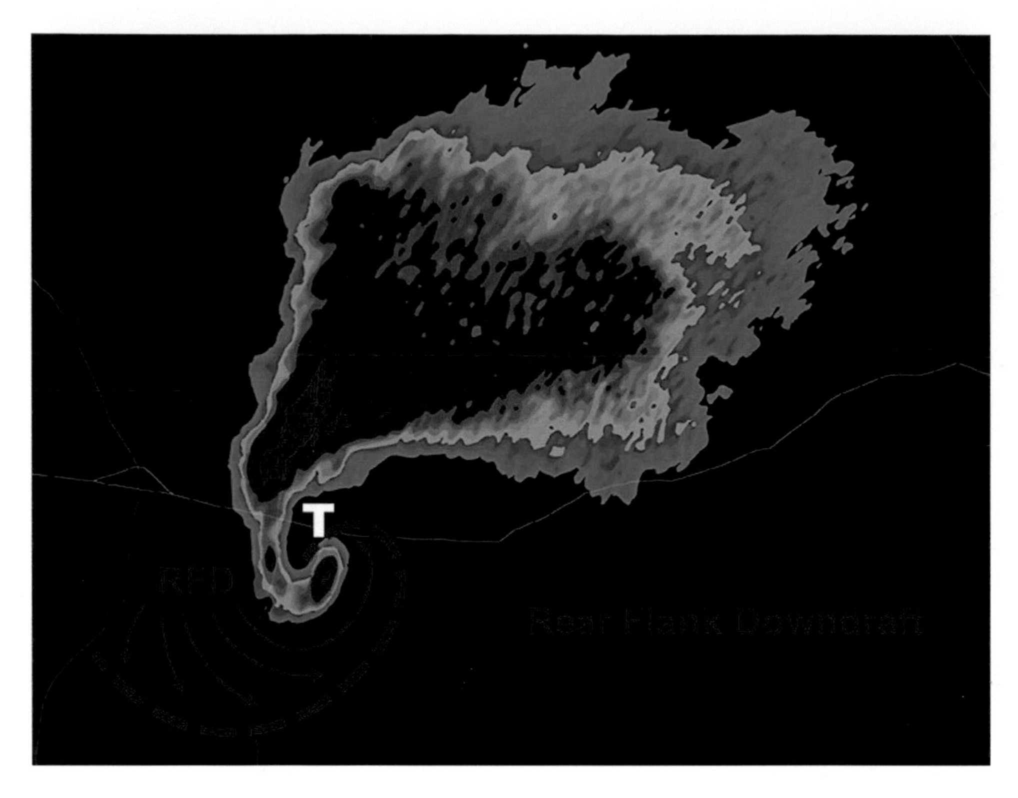

Figure 10.11 Supercell hook echo. This radar image clearly illustrates the hook echo that often develops along the southwest border of the mesocyclone. The location of the tornado is shown by the "T," and the RFD is the rear flank downdraft that wraps around and contributes to tornado formation. (NWS.)

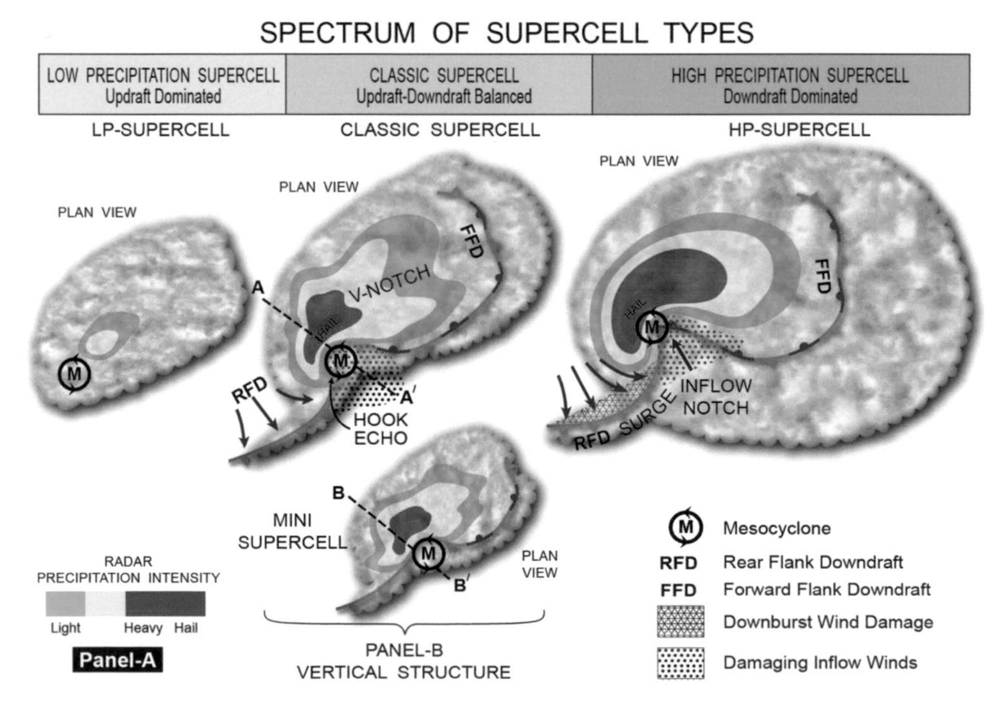

Figure 10.12 Spectrum of supercells. Less severe storms are on the left; more severe storms are on the right. The physical size of the supercell and the magnitude of the precipitation change over this spectrum.

Figure 10.13 Typical LP supercell of the great plains. The cloud base of LP supercells can be quite striking. It is dominated by a low, rotating wall cloud and very little, if any, precipitation.

HP supercells do not breed tornadoes with the frequency of classic supercells. The formation of tornadoes depends on a delicate balance between the strength of the updraft and downdraft. HP cells are downdraft dominated. (Because they produce so little rain, LP supercells never develop a downdraft and thus lack the basic set of air currents needed to generate tornadoes.) The RFD of an HP supercell increases in intensity and coverage over time, surging rapidly around the mesocyclone. The RFD surge is accompanied by downbursts. In fact, HP supercells often morph into a bow echo-type storm.

Figure 10.12 also shows a mini- (low-top) supercell. This storm is essentially a classic supercell, scaled down in depth and size. It contains a mesocyclone, which tends to be shallow and weak. A region of heavy rain develops close to the meso, which may contain hail. Mini-supercells are capable of producing weak to moderate tornadoes.

The structural differences between classic and mini-supercells can be more easily seen in the vertical orientation Figure 10.14 (left panel) shows the structure of a classic supercell. The core of heavy rain is deep and narrow, and it includes a hail shaft (purple). Next to the hail region is an area of weak radar energy, called the vault. The vault contains the updraft and its meso. In a strong updraft, growing cloud water droplets are flung rapidly from its top and sides. There is not enough time for cloud droplets to grow to precipitating size. Hence, this region of the storm reflects very little radar energy.

In the mini-supercell (right panel), the cloud top is much lower. Rain is less intense, and the vault is not as apparent. Because mini-supercells can appear innocuous on weather radar, they require extra vigilance on the part of meteorologists. And because minis are shallow, they often go undetected at long distances (the lowest radar beam just skims the cloud tops). Mini-supercells tend to develop over the Mid-Atlantic and New England, and they are often present in the spiral rain bands of tropical cyclones over land.

Why is there a spectrum of supercell types? Research suggests that the amount of wind shear in the storm environment plays a key role, particularly the presence or absence of strong winds at high levels. When winds are strong, precipitation becomes displaced far downwind of the updraft, where it remains aloft or evaporates before reaching

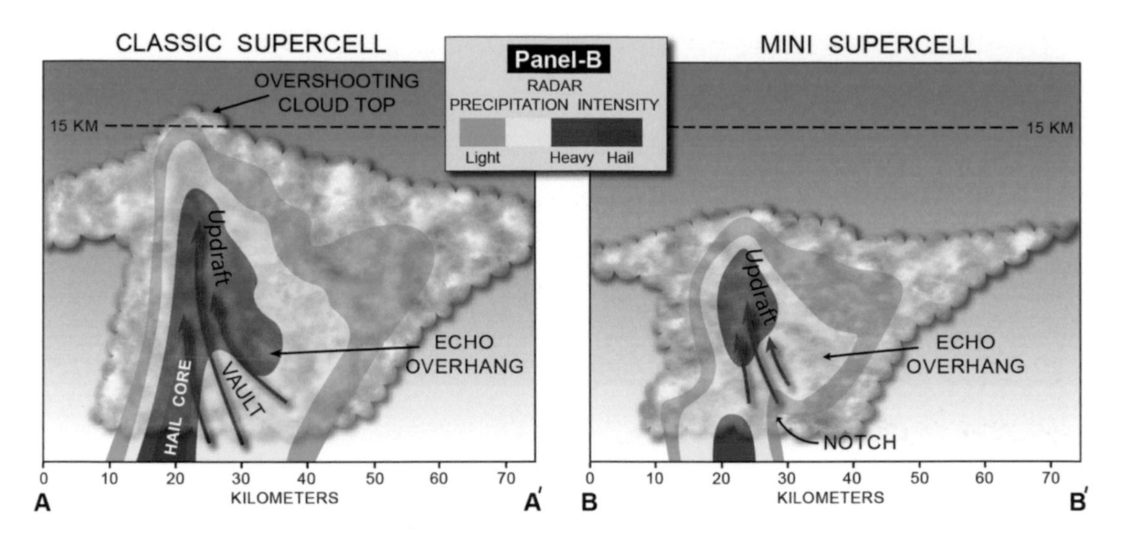

Figure 10.14 Classic and mini-supercell structure. The classic supercell (left panel) extends upward considerably farther and contains a much deeper precipitation core and a vault caused by strong updrafts. The mini-supercell (right panel) appears innocuous on weather radar and often goes undetected at long distances (the lowest radar beam just skims the cloud tops).

the surface. This set of circumstances creates an LP supercell. When winds aloft are weak, the mass of precipitation forms very close to the updraft, encircling the meso and generating an HP supercell. The precipitation collects in one spot, is not spread out by winds, and therefore becomes locally concentrated.

Supercell Formation

Large-Scale Setting

Supercells most often develop in the warm sector of a vigorous extratropical cyclone during the spring months. When supercells become especially numerous in the warm sector (Figure 10.15), a tornado outbreak is likely. The formation of supercells is tied to three key factors: (1) geography, which creates three different types of air masses, (2) development of explosive instability, and (3) a very specific type of vertical wind shear.

Figure 10.16 illustrates a supercell's key large-scale ingredients at different levels of the atmosphere. The bottom panel (surface) shows an extratropical cyclone. Three different air masses come together to create several frontal boundaries. The warm front occurs where warm, moist southerly flow overrides cooler, drier air to the north. Where the cold, dry air undercuts warm air from the northwest, the cold front becomes established. A third type of boundary, the dryline, separates hot, dry air from warm, moist air. The hot, dry air originates over the U.S. Desert Southwest region, while a warm, humid air mass streams north from the Gulf of Mexico. The dryline is located within the cyclone's warm sector, ahead of the cold front. As the dryline advances eastward, it lifts unstable air ahead of it, forming a squall line that may contain supercell thunderstorms.

As we proceed upward from the surface, the airflow veers from south to west. At 1.5 km (5000 feet), a moist flow of air from the Gulf feeds into the cyclone. The flow may be concentrated into a narrow, fast ribbon (the low-level jet). Above this layer, at 3.0 km (10,000 feet), warm and dry winds overspread the moist plume. This warm, dry layer streams off the elevated, arid plateau of Mexico. The dry and hot layer may take the form of a fast-flowing mid-level jet. Finally, at 9.0 km (30,000 feet), there is a strong trough in the jet stream. A fast-moving pocket of air, called a jet streak, may curve around the base of the trough. Upon exiting the jet steak, the airflow fans apart (in a process called diffluence), drawing up air from below, ahead of the trough. The sustained uplift of air helps to intensify thunderstorms to severe levels within a widespread region of the cyclone's warm sector.

Let's examine the vertical structure of the atmosphere at Point A in Figure 10.16 from a radiosonde (Figure 10.17). The left panel shows the variation of temperature and dewpoint with height, while the right panel shows changes in wind speed and direction (i.e., wind shear variation). On the left, the elevated plume of

Figure 10.15 An active supercell day. This radar image illustrates a typical cluster of supercells establishing along an extratropical cyclone's cold front during the warmer season.

hot and dry air is very important. The source of this air layer is often traced to the Rockies over the southern United States and the Mexican Plateau. In this layer, temperature increases with altitude; this layer therefore is a temperature inversion that serves as a stable layer or cap. Below it lies a layer of warm and humid air. During the early afternoon, the low-level air rising in cloud updrafts often has insufficient buoyancy to penetrate the inversion layer. The cap impedes the premature, early-day release of instability. But as the Sun continues to heat the surface, the temperature climbs. The situation is loosely analogous to a pressure cooker – with the inversion layer bottling up the instability beneath it. Eventually, by late afternoon, the air warms sufficiently for updrafts to break the inversion layer. Then large amounts of available buoyant energy are released through a deep layer, almost explosively. This process of delayed release greatly contributes to the strength of a supercell's updraft.

Creating the Mesoscylone

Vertical wind shear profile plays an important role in creating spin within the supercell's updraft. As the wind shear structure in Figure 10.17 shows, environmental winds must impinge on the storm from many directions. But, most importantly, these winds must both veer and greatly accelerate just above the surface.

Figure 10.18 illustrates something paradoxical. The initial source of the mesocyclone's rotation comes from horizontally spinning air! Winds that blow faster with height create a strong shearing effect. The air tumbles over itself, forming invisible, horizontal

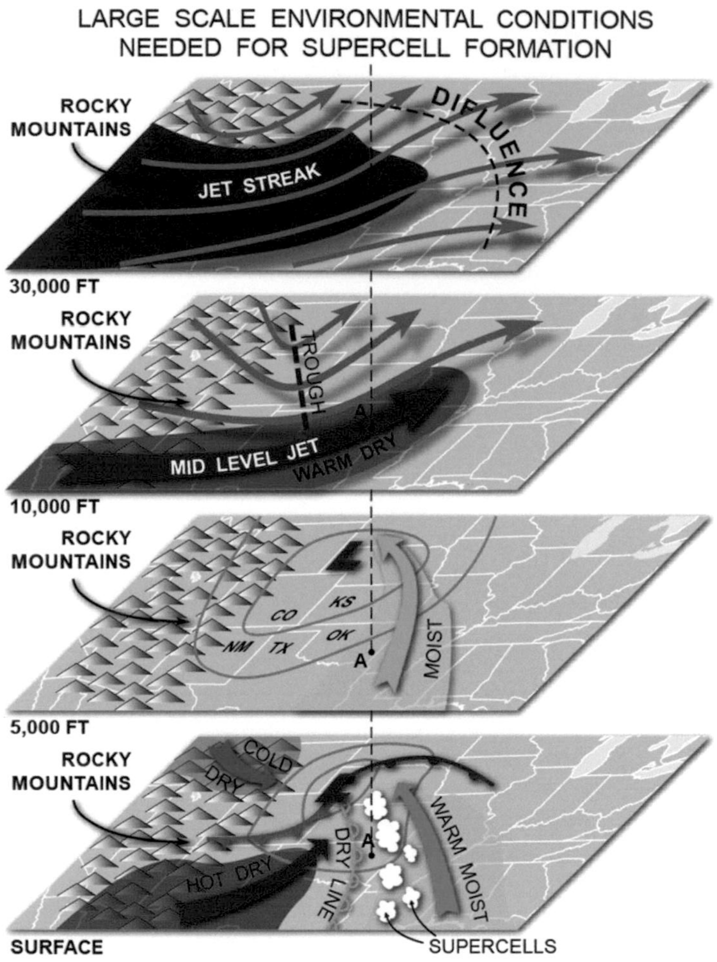

Figure 10.16 Composite atmospheric structure favoring supercell formation. This diagram shows a supercell's large-scale ingredients at different elevations within the extratropical cyclone as it advances from the U.S. Southwest.

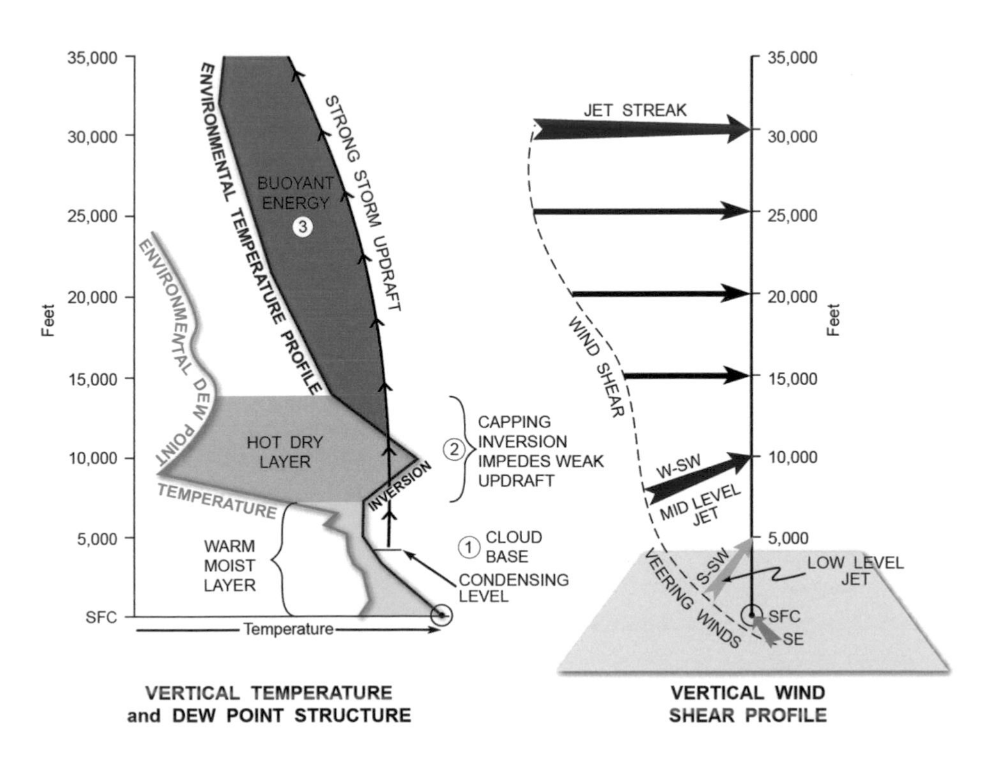

Figure 10.17 How the atmosphere becomes explosively unstable. A cap of hot, dry air acts as a barrier to the upward movement of energy, concentrating low-level heat and moisture to extreme levels. Once the cap is eroded, the pent-up energy and moisture are free to shoot upward. The result is rapid development of one or more supercells.

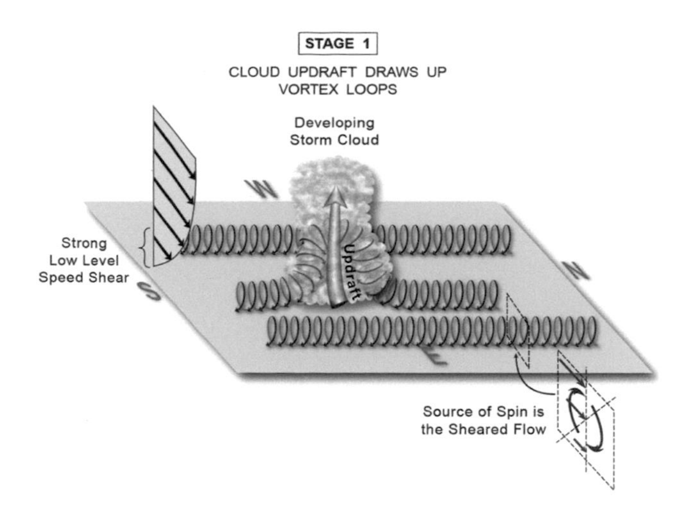

Figure 10.18 Origins of a supercell's spin. The developing supercell contains horizontal rolling vortex tubes of air brought about by the inflow of horizontal air that feeds the updraft. The updraft latches onto a vortex tube and drags it upward into a vertical position. A vertical vortex couplet forms. (Adapted from Klemp, 1987.)

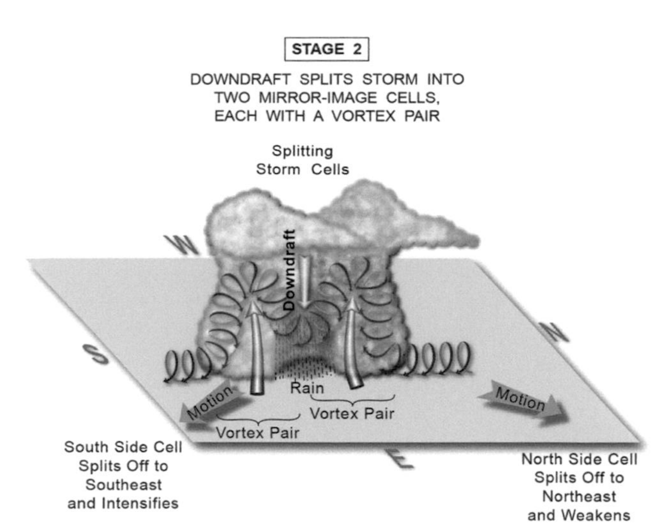

Figure 10.19 Splitting a storm into a vortex pair. After the vertical vortex couplets form (Figure 10.18), a strong vertical downdraft takes control and pushes downward on the section that connects the vortex couplets. The storm cloud is divided into two, and mirror-image storm clouds are created. The southern cell begins moving toward the southeast and intensifies, while the northern cell moves toward the northeast and weakens. (Adapted from Klemp, 1987.)

vortex tubes. Imagine a pencil placed between your two outstretched hands. As your bottom hand stays put, begin sliding the top. The pencil acquires horizontal spin. In a vortex tube, the direction of spin is initially clockwise. Now imagine cloud updrafts rising up out of the wind-shear layer. Updrafts draw the vortex tubes upward, first deforming them, then tilting them to the vertical (try visualizing this process using a slinky). This process creates a vertical vortex couplet, or pair of counter-rotating vortices, inside the growing storm cloud. The southern vortex spins counterclockwise, and the northern vortex rotates clockwise. Note that the direction of horizontal spin became reversed in the southern cloud vortex. The original horizontal vortex tube spun clockwise, but the new vertical vortex spins counterclockwise. Again, you can use a slinky to visualize this phenomenon.

The storm cloud continues to grow, precipitation falls, and a downdraft develops (Figure 10.19). The downdraft effectively splits the storm cloud and its vortex tubes into two identical halves. Each new cloud system contains its own vertical vortex pair and its own updraft. Mirror-image storm clouds are created. An updraft develops within each vortex tube – one on the southern side, one on the northern side – as air is sucked into the low-pressure core of each vortex. The southern cell begins propagating (regenerating) toward the southeast while the northern cell propagates toward the northeast. This process describes the evolution of a splitting cell.

Figure 10.20 shows that the veering (clockwise turning) of low-level wind selectively intensifies the southern storm. In this cell, the counterclockwise vortex becomes co-located with the updraft, forming a helical updraft or mesocyclone. The main downdraft and heavy rain region contain an anticyclonic vortex (a meso-anticyclone). Wind in low levels impinges on the storm from the southeast. This interaction develops a local region of high pressure near the mesocyclone and a local region of low pressure near the meso-anticyclone. Winds in the upper levels impinge on the storm from the west. A region of low pressure forms aloft along the mesocyclone, and a region of high pressure forms along the meso-anticyclone.

These pressure changes aloft and at the surface are critical. They lead to a vertical acceleration, directed from high to low pressure (shown by the arrows), that further intensifies the buoyant updraft (much more so than in an ordinary thunderstorm). Air ascends from high pressure below, toward low pressure above, along the length of the mesocyclone. The ascending air stretches the meso, contracting it width-wise, causing it to spin faster. At the same time, the downdraft within the meso-anticyclone intensifies; air descends from high pressure aloft, toward low pressure below. The descending air flattens out the meso-anticyclone, expanding it and weakening it. Slowly, over a matter of 1–2 hours, these processes create a supercell storm dominated by a single, intense, helical updraft. This mesocyclone is typically 5–10 km (3.1–6.2 miles) wide and several kilometers deep and contains spiral winds exceeding 45 kts (52 MPH). All the while, the supercell slows down and

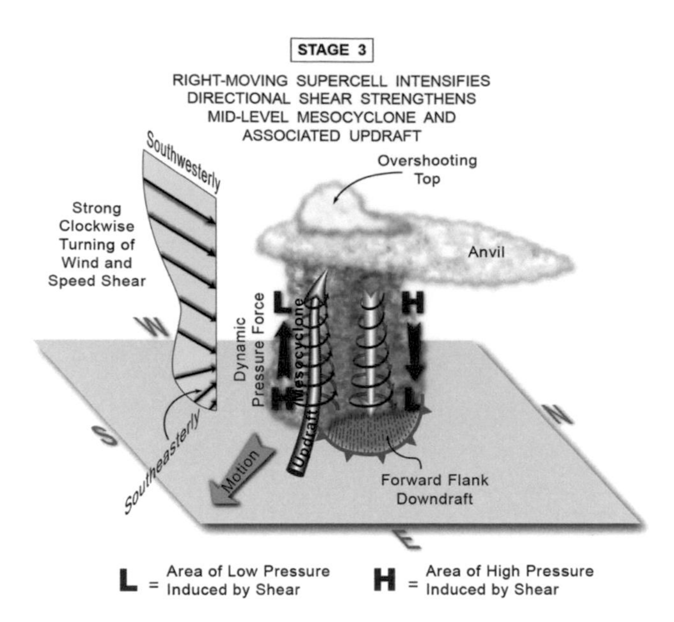

Figure 10.20 Strengthening of the supercell mesocyclone. This illustration shows how the mesocyclone develops from the modified vertical vortex couplet. The mesocyclone is the "business end" of the storm from which much of the violent weather emanates. (Adapted from Klemp, 1987.)

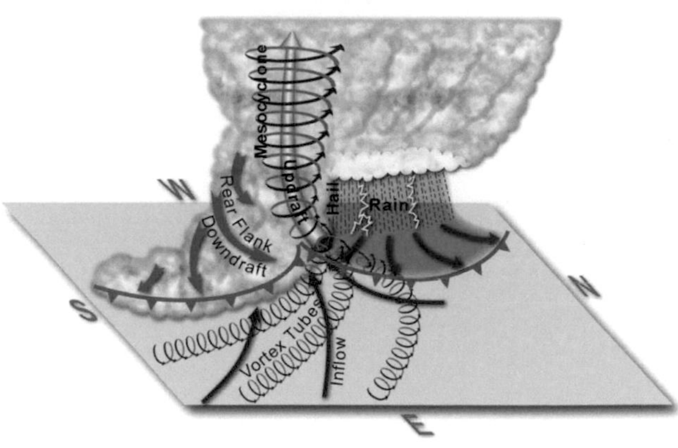

Figure 10.21 Mature, steady-state supercell. The mesocyclone continues to spin as it continually sucks vortex tubes into its lower end. As the rear flank downdraft intensifies, the probability of tornadoes increases. (Adapted from Klemp, 1987.)

begins deviating toward the right, heading southeast. It becomes a right-moving storm with respect to the prevailing jet stream flow, which is generally from the west.

It is important to recognize that the supercell's updraft is partly driven by buoyant forces (rising, warm air) and by strong pressure forces. The pressure force can contribute 50% to the total updraft strength. This set of processes is unique to a supercell's modus operandi. Incredibly strong cloud updrafts can approach 115 kts (132 MPH). These updrafts explain why supercells are prolific producers of large hail. Only the most intense updraft can levitate stones long enough for thick layers of ice to accumulate.

While the direction of a mesocyclone's spin (counterclockwise) is the same as that of its parent cyclone, the origin of the meso's spin is not directly related to Earth's rotation. Why? A vortex as large and long-lived as an extratropical cyclone is influenced by the Coriolis effect. In comparison, the sense of a meso's spin derives from vertical wind shear that is partly created by friction with the land surface, and temperature gradients aloft – both of which are processes that act independently of the Earth's spin.

Figure 10.21 shows the airflow in a mature, steady-state supercell. The storm has taken 1–2 hours to reach this point. The mesocyclone continues to acquire spin from horizontal vortex tubes near the surface. These tubes are literally sucked inward, drawn into the base of the mesocyclone, which acts like a vacuum hose. The vortex tubes turn abruptly upward, and their sense of spin reorients to counterclockwise. By adding new spin to its base, the mesocyclone grows downward from mid-levels, reaching the surface. Once it approaches the surface, the stage is set for possible generation of tornadoes.

Lives of Supercells: Splitting, Merging, and Transforming

Once a supercell reaches its mature phase with a strong, deep meso, it may remain steady for 2–3 hours. The helical nature of the updraft contributes to stability. Supercells are most intense and long-lived when they are isolated, with access to a steady supply of warm, humid air rich in vortex tubes. Otherwise, nearby storm cells compete for available energy and spin. The lone supercell out ahead of a squall line or the southernmost storm in a line of supercells is often the one to watch out for!

Random interactions and mergers between adjacent storm cells frequently occur. This makes the task of predicting severe weather, particularly tornadoes, very difficult. Figure 10.22 illustrates some ways that supercells interact and evolve. Panel (a) illustrates the familiar process of storm splitting, which creates a supercell pair. With the correct wind shear, the right-moving cell becomes the severe, long-lived supercell, while the left cell typically weakens but may still produce a brief period of severe weather.

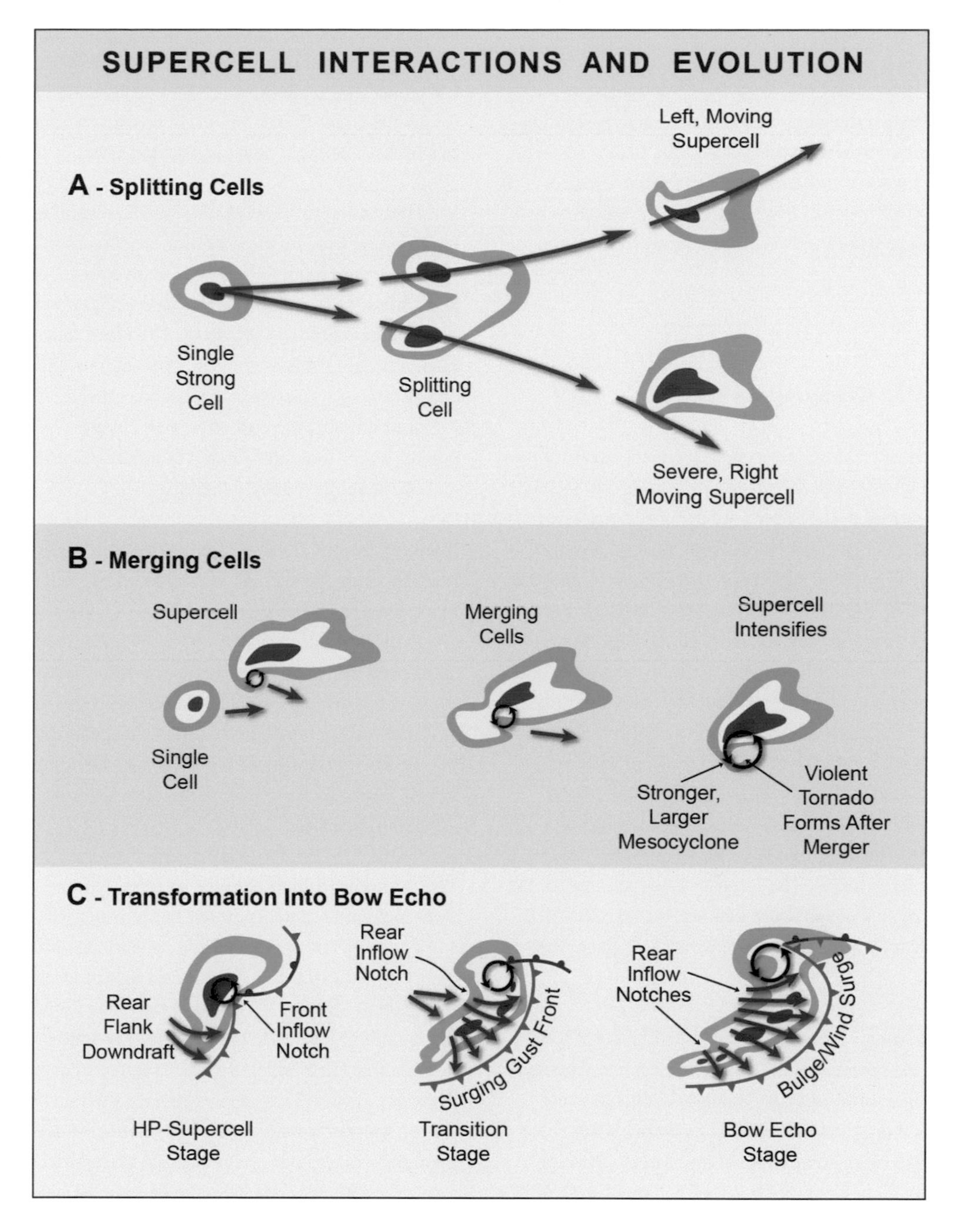

Figure 10.22 Interactions and evolution of supercells. Supercells can take several paths of evolution as they mature. (a) How a cell can split and become two cells. (b) The merging of two cells into one. (c) A cell transforming from an HP supercell into a bow echo.

In Panel (b), a single, fast-moving storm cell overtakes and merges with a slower, right-moving supercell. The merging phase is a ripe time for tornado development. If the storm updrafts combine in an optimal manner (not all mergers are constructive; some weaken the supercell), the mesocyclone expands and strengthens. The merger is immediately followed by a strong or violent tornado.

Panel (c) captures a common type of evolution for an HP supercell. Recall that HP supercells contain a large, intense rear flank downdraft. When the downdraft surges and expands, the kidney-bean-shaped HP supercell becomes an arc-shaped bow echo. The mesocyclone migrates to the northern end of the bow, where it may expand and create a slowly spinning comma head structure (perhaps even generating tornadoes). The morphed storm, now in bow-echo stage, creates a cluster of downbursts along the leading edge of downdraft.

An HP supercell that transitions into a downburst-producing bow echo is called a Pakwash storm, so-named for a prototype event that took place across Northwest Ontario, Canada, in 1991. The intense downbursts in that cell reached 115 kts (132 MPH) and blew down 1500 km^2 of trees in the Pakwash Forest.

Large and Damaging Hail

Among the dangerous triad of severe thunderstorm hazards is large and damaging hail, defined by the NWS as 2.5 cm (1 inch) diameter stones or greater. Hailstones are spherical masses of ice that fall from a thunderstorm cloud. They are composed of rimed ice – that is, droplets of supercooled (unfrozen) water that freeze upon impacting the stone, building up successive layers or spherical shells of ice. Stone sizes exist along a spectrum, ranging from pea size, to marble (nickel) size, to ping pong ball size, to golf ball or egg size, up to baseball and even softball size. As stone size increases, the density of the hailfall tends to decrease; that is, there are fewer large stones per unit area compared to small stones. But the fall velocity of the largest stones can exceed 87 kts (100 MPH) – readily punching holes through car windshields, destroying roofs, solar panels, and deeply denting metal vehicles. The largest stone measured in the United States fell over Vivian, South Dakota, on July 23, 2010; this stone was 20 cm (8 inches) in diameter and weighed nearly 2 pounds!

Tornadoes and hurricanes frequently draw headlines, but hail is responsible for the greatest share of annual property damage, with nearly 70% of insured property damage claims from severe storms. For more than a decade, U.S. homeowners have incurred more than $10 billion in annual hail losses. Damage claims for individual hail-damaged cars can range from $15,000 to $20,000. Homeowners in parts of Colorado, an especially hard-hit hail state, have replaced roofs for the fourth time in seven years. During some of these remediations, roofs were so completely destroyed that occupants of the homes had to live in hotels for several months. Hailstorms costing $2–$3 billion are now becoming more commonplace, in such diverse regions as Phoenix, Arizona; Denver, Colorado; and Minneapolis, Minnesota.

Hailstorms account for tremendous agricultural losses annually in the United States, in terms of both crops and livestock. Farmers in hail-prone regions frequently take out hail insurance; in Illinois alone, the collective, annual price tag of this coverage exceeds $600 million. As shown in Figure 10.23, hail – and particularly large, damaging hail – tends to occur most often across the Great Plains region. In any given year, regions within this broad "hail belt" may experience seven or more hail days.

Why are hailstorm costs escalating? The main reason is that we continue to build more homes in vulnerable locations . . . including regions with large suburban sprawl, such as Denver and Dallas. Homes are also becoming larger, with the average size of a home having grown by 1000 feet^2 since the early 1970s. Bigger roofs mean larger hail targets, and more windows and relatively weak (easily shredded) vinyl siding to be replaced. Finally, the cost of asphalt shingles has risen, due to market forces that have tended to inflate the price of crude oil and decrease the supply of asphalt.

Small hailstones are also capable of inflicting great damage. Marble- or egg-sized hail, when blown sideways by strong thunderstorm gusts, can complete denude trees of their leaves and leave acres of mature crops totally shredded. Hail is also a significant hazard to commercial aircraft – not only parked at the gate but during unexpected encounters when flying too close to thunderstorm cores. Fortunately, such encounters are rare, but the fiberglass nose of the aircraft is destroyed as hail stones impact with relative (closing) velocities approaching 261 kts (300 MPH) or more.

The formation of hail requires an intense thunderstorm updraft. A large volume of supercooled liquid water (SCW) must be suspended by the updraft current. Small ice particles, called embryos, begin growing into larger hailstones as SCW accretes and instantly freezes. The longer the stone remains suspended in the updraft, the more layers of ice are added. Stones growing to large size (golf ball and larger) require updraft speeds on the order of 43–52 kts (50–60 MPH) to levitate the heavy stones. Storms containing the very strongest and longest-lived updrafts – supercells – have updraft speeds that can reach an astounding 87 kts (100 MPH). Truly massive hailstones are levitated by such drafts, growing to baseball and softball size. At some point, the growing stones become too heavy for even the strongest updraft, and they fall out. As Figure 10.24 shows, these massive stones fall earthward along the edge of the updraft (on the northern and western sides of a supercell updraft). What ensues is a narrow curtain or cascade of large stones that pummel to the ground, called a hail shaft.

As the stones descend, they partially melt, reducing their volume, but the melting ice extracts heat from the layer of air in contact with the stone – forming a thin, chilled air layer that can reduce the rate of melting. Large stones are built from

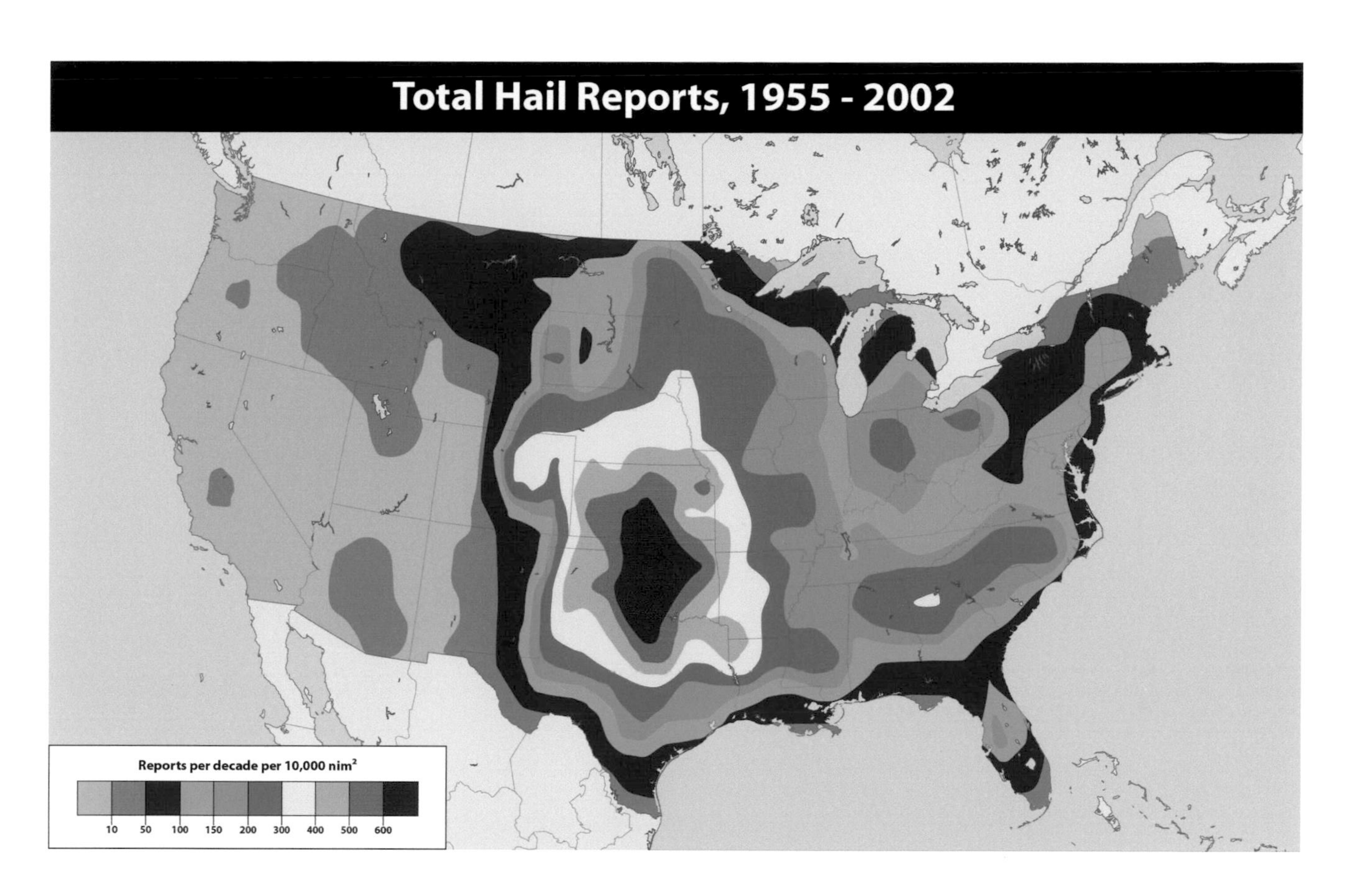

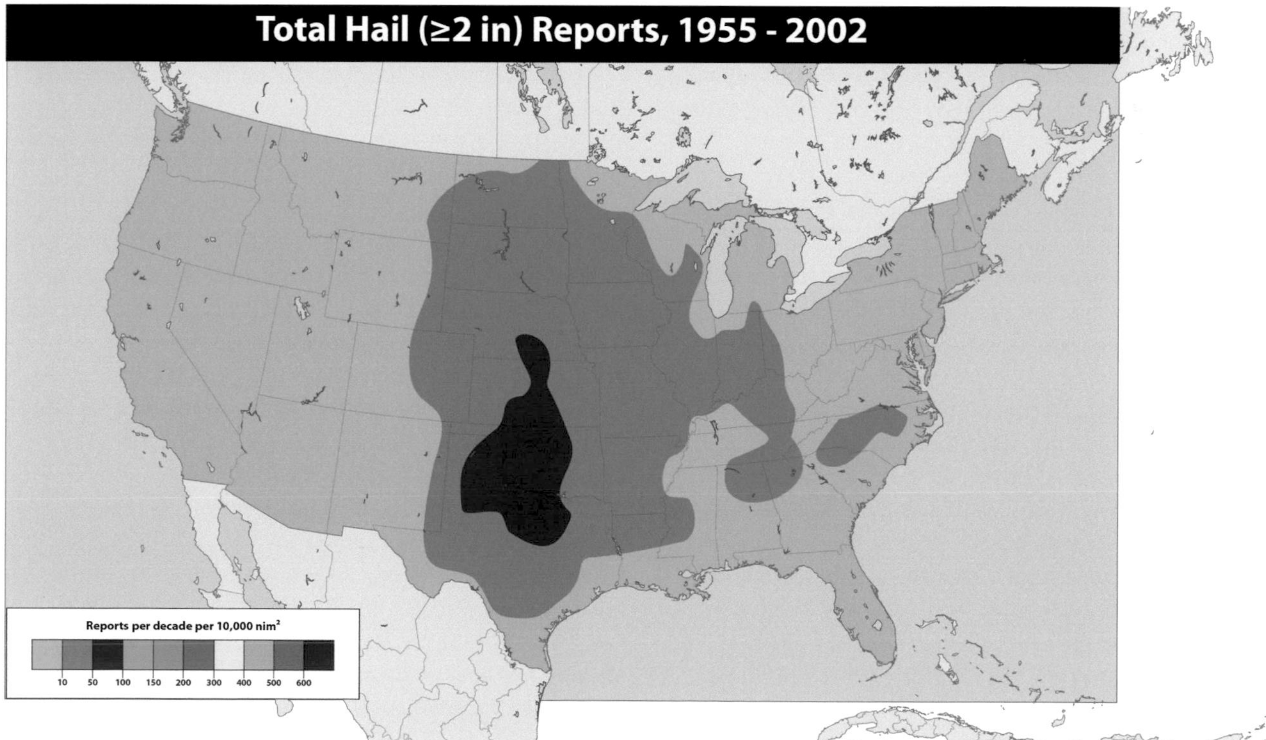

Figure 10.23 Decadal hail incidence across the geographical domain of the continental United States. (Top) The very conspicuous Great Plains maximum. (Bottom) The most common regions of 2 inch and larger hail.

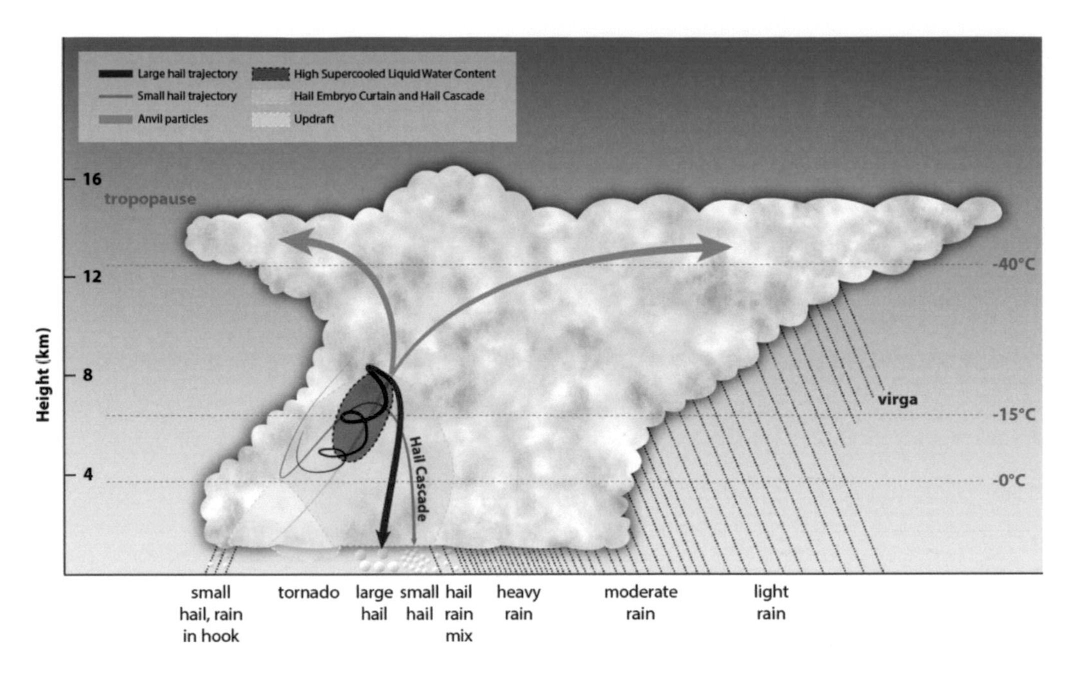

Figure 10.24 Schematic hypothesizing large hail growth inside a supercell thunderstorm. The largest stones build in size during their spiral ascent in the updraft region, then cascade out adjacent to the updraft.

concentric spheres or layers of ice, some layers clear and some opaque. This layering is a result of the stone passing into and out of multiple pockets of SCW. Rapid riming in the presence of large amounts of SCW traps microscopic air bubbles as the water freezes, forming a whitish ice layer. Slower riming allows air to escape, rendering the ice layer clear. Slice an onion in half, and the layering bears resemblance to a hailstone's structure (Figure 10.25[a]). Occasionally, large stones have several lobes or spikes, resembling the end of a mace. These are thought to form when small stones accrete onto the surface of a larger stone, as the stone descends at high velocity and overtakes smaller stones in its path (Figure 10.25[b]).

The lower the freezing level in the atmosphere, the less the falling stones melt, and the larger these stones are when reaching the surface. This is why hail is rare in low latitude (tropical) zones where the air mass remains above freezing through a deep layer above the ground. The height of the freezing level over the Plains (which varies from day to day) can help discriminate those situations that are more likely to favor large hail-bearing thunderstorms. Even in hail-prone areas, people often wonder why the time of year most likely for large chunks of ice (hail) falling out of clouds is also the warmest time – namely, the summer months. It does seem paradoxical! The reason is that in order for large hailstones to grow, extremely strong updrafts are necessary, and thus the air mass must be very unstable. As a rule of thumb, the most unstable air masses occur on those days when the ground surface is heated most strongly, when the strong summer sun is overhead. While supercell storms are the largest hailstone producers, intense multicell storms and even short-lived, pulse-type storms can deliver damaging hailstones in the quarter- to golf-ball-sized range.

Severe thunderstorms can occasionally deliver such a massive volume of hail that it accumulates several inches deep; indeed, there are documented cases where the depth of hail has exceeded 0.3 m (1 foot). These storms have been coined "Storms Producing Large Amounts of Small Hail," or SPLASH, storms. The ensuing hail swath or hail streak can be readily observed from the air and even from space (Figure 10.26). A typical location embedded in a hail swath experiences a deafening – if not outright terrifying – hailfall for 15–20 minutes. Strong thunderstorm gusts may blow the hail into drifts several feet high. Drifting hail can clog storm sewers and small streams, leading to flash flooding from heavy rain that mixes with the hail. These hail swaths can be many tens of miles in length but are often less than a mile wide. Portions of Colorado, east of the Rockies, are notorious for so-called "plowable hailstorms" requiring snowplow crews to clear the streets in the hours after these storms. Structural damage typical of large hail, especially when windblown, is shown in Figure 10.27.

(a)

(b)

Figure 10.25 Characteristics of large hailstones. (a) A large stone sliced in half, revealing "onion-like" layering of clear and opaque ice layers. (NWS). (b) The bizarre shape of spiked hail. (National Center for Atmospheric Research.)

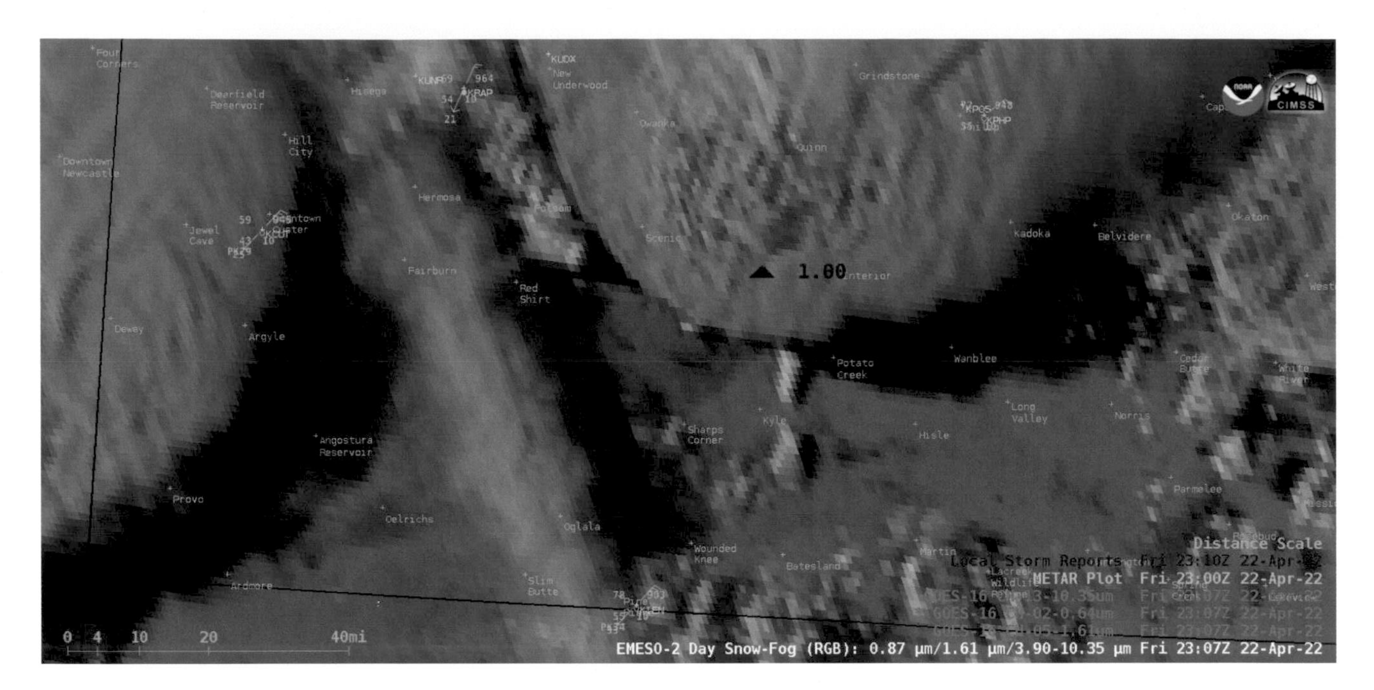

Figure 10.26 Satellite view of a hail swath. The generating storm cell has moved on, but a conspicuous linear tract of ground ice remains for many hours, shown here between adjacent thunderstorm cells. (NOAA.)

Figure 10.27 Severe damage to a mobile home due to large hail. Vinyl siding and bottom skirt were severely damaged, in addition to window glass shattered and interior blinds obliterated. (National Severe Storms Laboratory.)

Could global warming be partly responsible for the escalation of hail damage in the past decade or two? Possibly. The processes that would seem to promote hail growth and frequency may operate in varied and contradictory ways. With warming climate, hotter surface temperatures may very well promote more frequent and more violent thunderstorms, due to an increase in air mass instability and cloud updraft buoyancy. However, the freezing level may also be moving higher above the ground, meaning hailstones have greater opportunity to melt on the way down. Additionally, wind shear (the increase in wind speed with altitude) may become weaker in the mid-latitudes (caused by a general weakening of the polar jet stream), which will reduce the severity of thunderstorm cells.

Societal Impacts of Severe Thunderstorms

Of the several hundred thousand thunderstorms that develop over the United States every year, fewer than 10% are severe, and the peak incidence of severe local storms is May through July. However, the annual property losses from severe storms are large, ranging from $5 billion to $10 billion. Figure 10.28 reveals that the trend in annual losses has been upward since 1980. Losses in 2011 were staggering, approaching $26 billion, largely as a result of numerous record-breaking tornado super outbreaks that year. Total losses in that year, combining property and agricultural losses, hit $47 billion. However, the upward trend is instigated by a factor beyond the hazard itself: large increases in population, insured property, and suburban sprawl. Over the past decade, more people and property have been placed in harm's way.

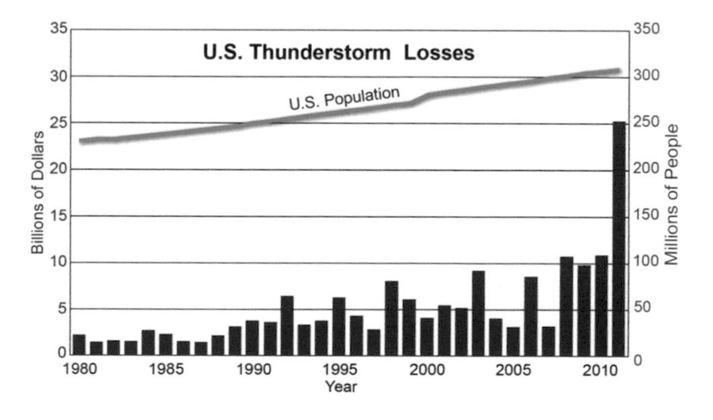

Figure 10.28 Property loss trends due to thunderstorms. Losses grew steadily from 1980 to 2011. There have been considerable fluctuations from one year to the next (compare 2005 and 2006). Comparing the population trend line (blue) with increased losses (red columns), it is evident that much increase in property loss can be attributed to more people being placed in harm's way as populations grow and people "fill in space."

Figure 10.29 shows the distribution of the 30,000 severe thunderstorm reports in 2011, including tornadoes (blue, 1894 reports), large hail (9417 reports), and damaging winds (18,685 reports). The Rocky Mountains sharply delineate dense hazard coverage over the eastern two-thirds of the United States, but parts of the west are not immune. Nontornadic winds outnumber all other severe thunderstorm hazards and are the most widely distributed thunderstorm hazard. In 2011, large hail events clustered over the Great Plains. That same year, large and violent tornadoes were especially concentrated in the Mid-South.

But, as we have seen, severe weather distributions and frequency, as well as storm fatalities, can vary markedly from year to year. In terms of long-standing (30+ year) averages, flash floods remain the number one thunderstorm killer (107 fatalities per year), with lightning ranked second (65 fatalities per year). Annual tornado deaths are more variable; the number depends both on the number of springtime tornado outbreaks and on whether a violent tornado strikes major population centers. While the average annual number of tornado deaths in the United States is 65, the death toll in 2011 was 553 (accompanied by 5400 injuries).

The effects of thunderstorm damage and disruption are diverse and wide-ranging. All modes of transportation are affected, but commercial aviation is particularly vulnerable. Who hasn't experienced a flight delay or cancellation due to thunderstorms? Who hasn't endured a power outage from a nearby lightning strike or tree limbs falling on utility lines? The agricultural sector is hit especially hard. Farmers incur billions in annual thunderstorm losses, not just to crops but also livestock, and in the ability to store and transport produce. But these losses go largely unreported in the news.

Impacts on U.S. Farming

Each year, thunderstorm hazards (flash floods, hail, and high winds) cause massive agricultural loss. Figure 10.30 shows the state-by-state distribution of select agricultural disaster claims due to thunderstorm hazards in 2012. However, remember that any single year is not necessarily representative of long-term trends in thunderstorm losses. Also, some hazards are regional, while others are more widespread. Some areas of the United

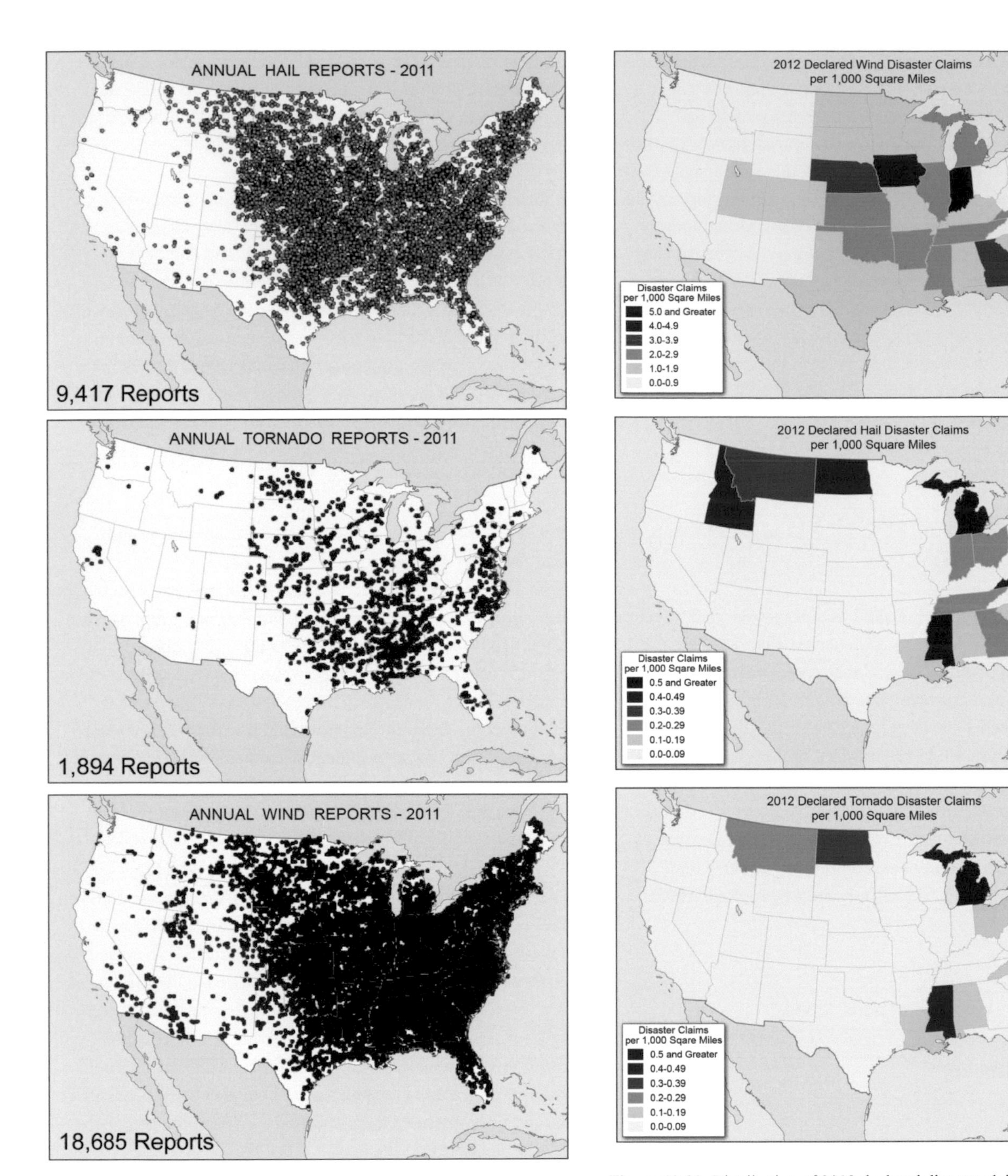

Figure 10.29 Distribution of damaging hail, tornado, and wind reports for 2011. Total reports were about 30,000. The number of tornado reports is quite low compared to hail and wind reports. Wind reports (excluding tornadoes) account for more than 60% of the reports.

Figure 10.30 Distribution of 2012 declared disaster claims for wind, hail, and tornadoes. Since the information is for a single year, it provides only a snapshot of conditions in that year. The areas most affected are widely scattered, and the tornado data appear to be influenced by warm and cold seasons.

States, such as Mississippi, experience triple hazards: hail, damaging wind, and tornadoes.

Hail damage is the long-term driver of annual crop loss. The damaging effects of hail include defoliation (loss of leaf area), broken stalks, and severely bruised leaves. Young crops are particularly vulnerable. Even small stones (less than golfball-sized, the threshold used to define a severe thunderstorm) are capable of tremendous damage if windblown, shredding the leaves off entire fields. A sudden fall of dense, large stones often kills domesticated animals caught outdoors. Severe winds of all types – including tornadoes, straight-line winds, and downbursts – blow down crops. Additionally, dry topsoil and small stones mobilized by high winds can sandblast plant leaves.

Flash floods provide another set of crop-damaging processes. Among the worst is erosion of topsoil and the valuable nutrients that it contains, including nitrogen and phosphorous. A study conducted in Iowa revealed that a single thunderstorm on farm fields mobilizes 100 tons of soil. "Downstream" impacts must also be considered. The rainwater bearing high nutrient loads is flushed into streams and rivers, leading to a process called eutrophication that causes algae to proliferate. Raindrop impact dislodges soil grains, which spatter upwards and coat plant leaves. These mud deposits stress plant leaves and diminish their photosynthetic efficiency. Standing water that partially or totally inundates crops leads to oxygen depletion. Once the water drains, waterlogged soils promote root and stalk rot. Waterlogged grain used to feed livestock must be disposed of because molds and their toxins are highly poisonous to those animals.

Impacts on Utilities

Data from the North American Electric Reliability Council (NERC) from 1984 to 2006 demonstrate that thunderstorm hazards cause 29% of all outages that affect 50,000 customers or more. Thunderstorms are on par with equipment failure as the leading two causes of utility shutdown. A large derecho (severe convective windstorm) traveling from Iowa to the Atlantic Ocean on June 29, 2012, knocked out power to nearly 9 million customers! Restoration times for this particular storm ranged from several days to more than a week. The fast-moving storm struck during the peak of a record-breaking, 38+° C (100+° F) heat wave, exacerbating everyone's misery. The heat wave and its toll can be thought of its own "disaster following a disaster" of the derecho. Such a "one-two punch" is common in areas of the Southeastern United States such as Florida, which endures the ravages of a hurricane, followed by days of the extreme heat and high humidity of a tropical air mass. Largely because of thunderstorm events, blackout frequency shows a pronounced summertime peak. Blackouts are four times more likely during the late afternoon relative to morning, correlating with the afternoon peak in thunderstorm development.

Whether an outage will occur depends not only on storm intensity and duration but also on the density of foliage and above-ground transmission lines. The duration of restoration efforts depends on (1) magnitude of total system damage, (2) degree of preparation by the utility company, and (3) related hazards that impede restoration efforts. Utilities can perform repairs more efficiently if they are able to stage crews and stockpile supplies in advance. Given sufficient early warning of a storm, utilities line up contractors from outside the region. However, these advance efforts require sufficient forecast lead time and reliability. Large cyclonic vortices such as hurricanes provide a much longer lead time than a short-fused, severe local storm. Summertime thunderstorms form and decay in the space of hours and can strike randomly. Utilities cannot be repaired until obstacles such as fallen trees are cleared from roadways and access points. Flooded streets may hinder repair vehicles. Fires created by downed utilities must be extinguished, and all live wires across an outage region must first be eliminated. Flooded substations must be pumped out before repair work can begin.

Utility loss is a high-impact and frequently recurring problem in many suburban and urban regions of the United States. There is an incorrect perception that frequent, violent storms are "the new norm." More likely, outages have gotten worse because the amount of above-ground, built infrastructure has increased dramatically over the past few decades. The utility grid is more stressed than ever. Meanwhile, for decades urban tree growth has been encouraged for aesthetic and practical reasons. Tree crowns have been allowed to expand, and utilities struggle to maintain annual pruning programs. As with so many natural disasters, the causes of damage to utility systems are multifactorial, a result of both the natural hazard and anthropogenic forces.

Impacts on Commercial Aviation

Thunderstorm-related weather is the greatest cause of all aviation delays, which peak during the summer months (Figure 10.31). Staying far from thunderstorm cells is a number one safety priority for airlines, mandating 16 km (10 miles) minimal separation from the strongest storm cores. These cells contain regions of extreme turbulence, lightning, icing, and hail – all of which can easily damage an airplane. Pilots on commercial aircraft receive continual storm structure and intensity information from nose radar, which also may include the location of lightning discharges and turbulence levels. Often they will receive directives from air traffic control to vector around problematic storms, and pilots can request vectors of their own to avoid looming storm cells.

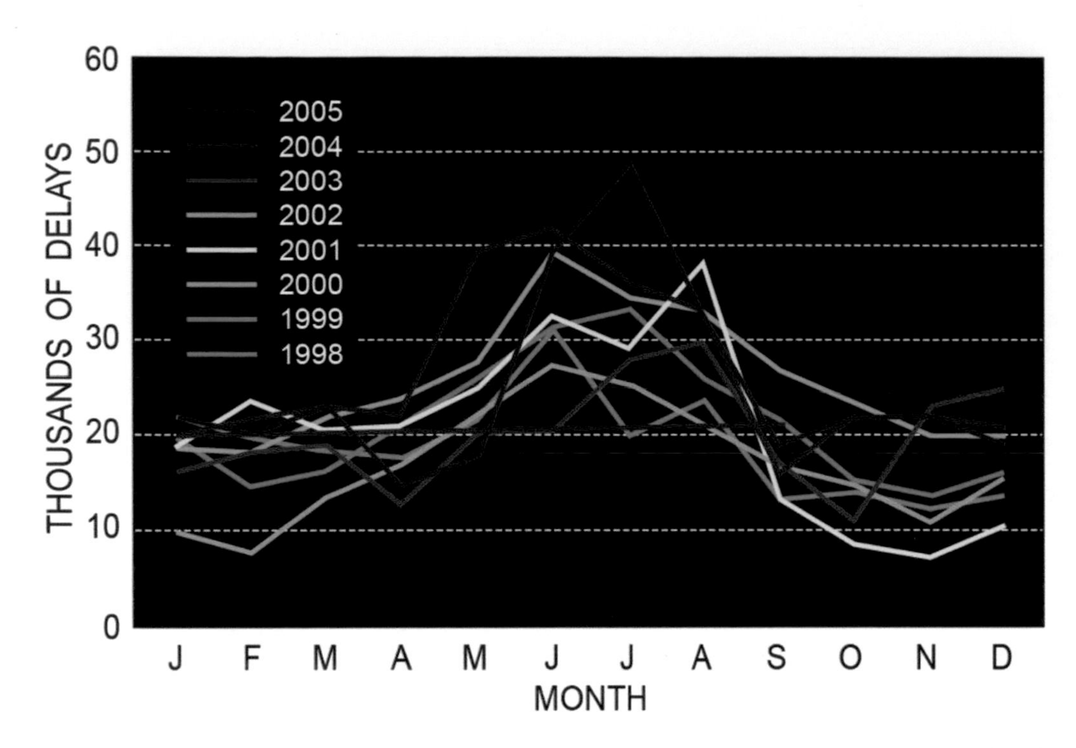

Figure 10.31 Weather-related delays for air traffic, 1998–2005. Delays peak in the warm season. The reasons are a greater number of thunderstorm events and an increase in total volume of air travel during this season.

Close to the surface, aircraft are subject to strong wind shear (downbursts, as described in Chapter 13), and a landing runway may be invisible in heavy rain. The control tower may increase the arrival and departure spacing of aircraft; this slows down operations but is necessary to increase safety margins. Crews that service aircraft on the ground are vulnerable as well, and a "ground stop" may be initiated so that personnel can take shelter when lightning is detected close to the air terminal.

In addition to these in-flight hazards, thunderstorms wreak tremendous havoc on the efficient flow of air traffic across the country. Thunderstorms often create a "ripple effect" that adversely impacts air traffic hundreds, even thousands of miles away from a storm complex. A squall line located along the en route portion of flights may require vectoring of many aircraft onto alternate airways, requiring an increase in time spent airborne, leading to delays and increased fuel consumption. Severe thunderstorms close to or over major air terminals and hubs are even more problematic. A severe storm over the Dallas/Fort Worth International Airport, for instance, may shut down that facility for 30 or more minutes. Aircraft destined for Dallas from other airports, from as far away as the East or West Coast, may be placed on hold until the weather over Dallas clears, or they may be diverted to other cities perhaps 50–100 miles away. Increasingly, one strategy utilizes ground stops, keeping aircraft that have not yet departed on the ground, avoiding bottlenecks when many aircraft converge on a closed terminal, and adding to the cost and time delay of diversions.

The National Airspace System (NAS) is composed of a dense network of jetways (above 5.5 km [18,000 feet]) connected by radio navigation beacons. These beacons and the intersections where jetways cross comprise a series of navigation fixes. The specific routes flown between airports are composed of a sequence of these navigational fixes. These routes are chosen to maximize fuel economy and minimize airspace congestion. A constant horizontal and vertical distance along routes separates aircraft. A small navigation computer on the aircraft flies the prescribed route by coupling with the plane's autopilot. Figure 10.32 shows how routings may be altered "on the fly" to safely navigate airborne traffic around difficult weather regions. For pilots, it's often a simple manner of reprogramming their navigation computers with a few touches of the keypad. But at certain times of day, the NAS is saturated in many regions, such as the Northeast United States, with other regions close to saturation, and so there is less wiggle room for alternate routings and diversions than one might think.

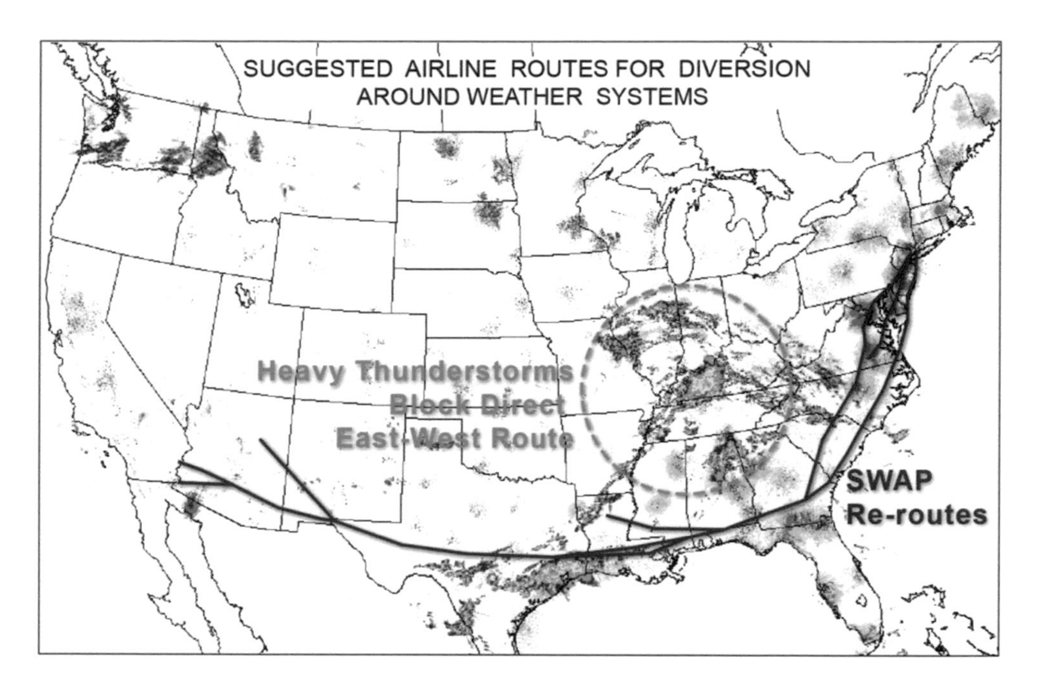

Figure 10.32 Example of a severe weather avoidance plan (SWAP) route. The alternate routing to avoid this storm line, issued by air traffic control, can be quickly implemented on the navigation computer onboard many commercial aircraft.

Amazing Storms: The Denver Cyclone and Severe Thunderstorms Along Colorado's Front Range

Severe thunderstorms strike certain U.S. locations with great frequency. The Great Plains are a hotspot for severe local storms during the warm season. So is a small geographical region nestled in a topographic "bowl" along Colorado's Front Range. This location, shown in Figure 10.33, is centered on the Denver metropolitan region but also includes Boulder and many towns along the densely populated Interstate 25 corridor. To the north of the region lies the east-west Cheyenne Ridge; to the south, the Palmer Divide; and to the west, the steep rampart of the Rockies. Within this topographic depression, a persistent, small-scale vortex often becomes established when the westerly flow crosses the mountains and interacts with southeasterly winds streaming off the High Plains.

The so-called Denver Cyclone is a shallow vortex, about 100 km (62 miles) in diameter, into which spiral winds converge in a counterclockwise manner. Given sufficient low-level humidity and an unstable air mass, the uplift of air within the circulation serves as a focus for early afternoon thunderstorms. These storms commonly initiate over elevated terrain to the southwest of Denver, with a squall line coalescing along the cyclone's southern flank by late afternoon. The squall line may further intensify and migrate toward the east. Sometimes the line of thunderstorms continues to propagate away from the region, through the evening, and evolve into a larger mesoscale convective complex (MCC) over eastern Colorado and western Kansas during the overnight hours.

The Denver Cyclone and its associated region of thunderstorms create local severe weather over the densely populated region, with significant impacts on flights into and out of Denver International Airport (DIA). Denver is a strategic stopover for commercial air traffic traversing the United States. Delays or closures of this airport can create a serious and costly ripple effect throughout the U.S. National Airspace System. Thunderstorms associated with the Denver Cyclone frequently close one or more of DIA's four principal arrival gates during the summer months – leading to holds and diversion of aircraft in DIA's terminal airspace, as well as ground stops and delays at distant airports.

Additionally, the Denver Cyclone contributes to an unusually high frequency of small tornadoes along the I-25 corridor. These tornadoes often develop not from supercells but rather from smaller and shorter lived multicell storms that breed a weak type of tornado called a landspout. The generation of tornadoes is hardly surprising, since the tight spiral of the Denver Cyclone creates a thunderstorm environment rich in spinning air. Individual thunderstorm cells act to focus this background rotation into small-scale vortices, such as landspouts. Other types of severe weather impacts from the Denver Cyclone include downbursts, hail, and flash floods.

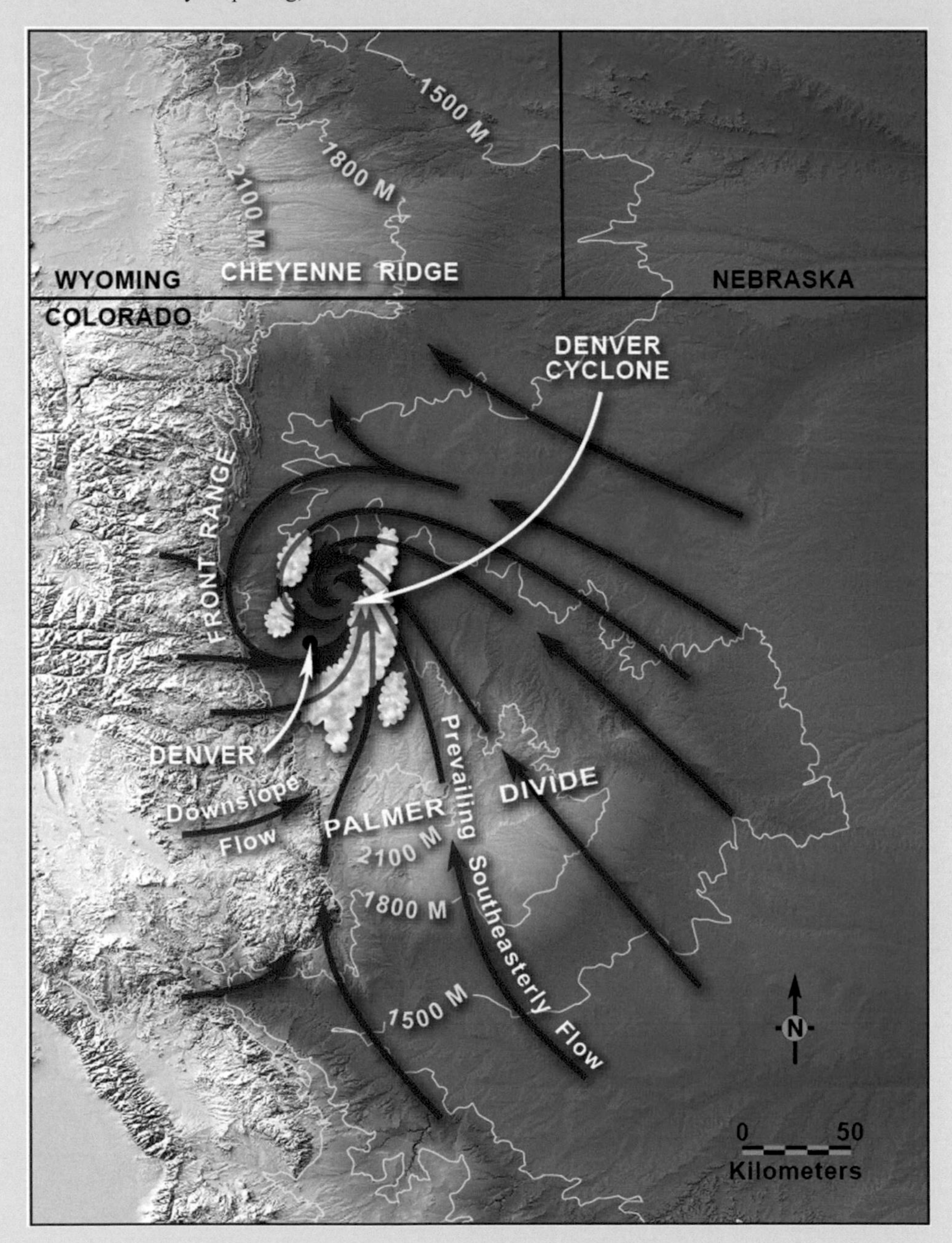

Figure 10.33 Location of the infamous Denver Cyclone. The Denver Cyclone is sandwiched between the Cheyenne Ridge to the north and the Palmer Divide to the south and is bounded on the west by the Front Range of the Rockies. As winds from the west interact with a moist southeast inflow, the cyclone sets up early afternoon thunderstorms. The storms often begin to the southwest of Denver, and, as they mature, they travel northeastward onto the Plains. Occasionally they transition into large mesoscale convective complexes (MCC) that travel into Kansas during the overnight hours.

Summary

LO1 How do instability and wind shear act to generate severe local storms? Explain what distinguishes a severe local storm from an ordinary thunderstorm, and describe the types of weather hazards not explicitly covered in the standard definition of a severe thunderstorm.

1. Thunderstorms become severe when an exceptionally unstable air mass combines with vertical wind shear (changes in wind speed and direction with altitude).
2. Strong instability strengthens the updraft, leading to torrential forms of precipitation and a more intense downdraft. Wind shear separates the updraft and downdraft, so that they do not interfere, and creates a source of spin within the storm.
3. Severe thunderstorms, by definition, contain wind gusts exceeding 50 kts (58 MPH), large hail (> 2.5 cm [1 inch diameter]) and tornadoes.
4. Fewer than 10% of all thunderstorms over the United States become severe, and lightning and thunder do not always occur.
5. A severe local storm is distinguished from a large-scale, intense vortex (i.e., extratropical cyclone, tropical cyclone) because the damaging weather occurs over a smaller area and is of short duration.
6. In addition to damaging winds, hail, and tornadoes, severe local storms can also generate dangerous, frequent lightning and flash floods.

LO2 Discuss in general terms the different categories or classes of severe local storms and their organizing features.

1. The most organized and long-lived types of severe local storm (supercells, bow echoes, and MCCs) each contain one or more vortices. The vortices promote exceptionally vigorous updrafts and confer a "steady state" nature to the storm.

LO3 What characteristic types of severe weather hazards does each main type of severe local storm generate?

1. A category of severe local storm, which arises in situations of very strong instability but weak or nonexistent wind shear, is the pulse-type severe storm: an exceptionally intense but short-lived single storm cell that may generate a short period of torrential rain, intense lightning, small hail, and damaging wind gusts.
2. Supercells, which contain a rotating updraft called a mesocyclone, produce the full spectrum of severe weather and are the most prolific generators of large hail and nearly all strong and violent tornadoes.
3. Bow echoes are fast-moving arcs of thunderstorms, with a pair of vortices that accelerate airflow from the back to the front of the storm. The air current descends to the surface as a violent blast of wind, called a downburst.
4. Exceptionally long-lived bow echoes produce elongated swaths of extreme wind damage, extending hundreds of miles, the product of tens or even hundreds of individual downbursts. This type of severe local storm is termed a derecho.
5. Mesoscale convective complexes (MCCs) are the largest category of severe local storm. On satellite imagery, they are recognized by their oval or round shape, hundreds of miles in diameter. MCCs generate locally heavy rains, leading to flash flooding, and reach peak intensity during the overnight hours.

LO4 Discuss the two prevalent, seasonal types of weather patterns (cool vs warm months) giving rise to severe weather outbreaks.

1. During the cool months, severe local storms frequently develop within the warm sector of an extratropical cyclone. Widespread or unusually intense storms constitute a severe weather outbreak on one or more days, including squall lines and supercells. These outbreaks are most intense and widespread in the late winter–early spring across the Mid-South and Central Plains.
2. During the summer, severe storms typically develop within a hot and humid air mass, along the western margin of a large-scale anticyclone over the eastern two-thirds of the United States. While an extratropical cyclone is often not present, a stationary boundary is, separating the "heat dome" to the south from cool Canadian air. Derechos frequently develop along the northern limb of the anticyclone, tracking along the

boundary, across the upper Midwest, Great Lakes, and Mid-Atlantic states.

LO5 Explain how the distribution of severe thunderstorm hazards shifts through the course of the year and the interannual variability of hazards.

1. Severe thunderstorms across the United States exhibit strong seasonal variations, with different types of hazards peaking at different times of the year. However, there are also significant year-to-year variations both in the geographical distribution and number of severe local storms.

LO6 Describe the basic properties of supercell thunderstorms. What is a mesocyclone? Distinguish between classic, low-precipitation, high-precipitation, and mini-types of supercells.

1. The supercell thunderstorm is often regarded as an "archetype" for the generation of warm season severe weather hazards. The key to its destructiveness lies in the formation of its mesocyclone, a vertical vortex 3–6 km (5–10 miles) in diameter, coincident with the storm updraft.
2. There is a spectrum of supercell types, ranging from low-precipitation (LP) supercells, classic supercells, high-precipitation (HP), and mini-supercell varieties. LP supercells rarely produce tornadoes. Classic supercells have a balanced updraft–downdraft circulation that leads to prolific tornado generation. HP supercells are dominated by heavy rainfall and intense downdrafts, and they occasionally "morph" into bow echoes.

LO7 How do supercells develop? Trace in particular the origins of the mesocyclone's rotation from genesis through storm maturity.

1. Multiple scales of atmospheric motion interact to create supercell thunderstorms. The particular geographical setting of the Great Plains enables an explosively unstable air mass to develop, involving three different types of air mass.
2. The mesocyclone, paradoxically, derives its spin from winds that increase speed with altitude close to the surface (vertical wind shear). The shear generates horizontal vortex tubes, which are drawn upward by the storm updraft and are reoriented into the vertical. Additional processes including clockwise-turning (veering) winds intensify and further organize the mesocyclone.

LO8 Discuss the formation of damaging hail and discuss the societal impacts of this severe thunderstorm hazard.

1. Hailstones form during the riming of ice particles, whereby small ice crystals become coated by layers of supercooled liquid water, and the liquid freezes on contact. In this manner, successive layers of ice build stones to large sizes.
2. Very intense, long-lived updrafts are required to grow hailstones to a large size, since these heavy stones must remain levitated many thousands of feet above the surface. This makes supercell thunderstorms the ideal "factory" of the largest hail.
3. Large hail accounts for 70% of all annual severe thunderstorm losses, in excess of $600 million per year in agricultural losses.
4. An unusual type of hailstorm is the so called SPLASH, or Storm Producing Large Amounts of Small Hail, which often leaves behind a narrow swath of hail that accumulates to a great depth, enough to require plowing the roads.

LO9 How do severe thunderstorms impact agriculture? The power grid? Commercial aviation? How is thunderstorm vulnerability shaped by patterns of human behavior?

1. Severe thunderstorms inflict tens of billions of dollars in damage to U.S. property and agriculture each year. Fatalities exhibit irregular patterns that vary by the number and type of severe weather outbreak occurring in a given year. Three aspects of society that are exceptionally vulnerable to the effects of severe local storms include the agricultural segment, electrical utility infrastructure, and commercial aviation.
2. Hail, high wind, and flash flooding combine to produce a wide variety of crop and livestock loss. Hail stones need not be large to bruise or destroy plant leaves, as small stones blown sideways in downdraft winds can quickly decimate crops. Flash floods lead to irreversible soil erosion, the loss of critical soil nutrients, and the rotting of plant root systems.
3. Several notable power outages, of exceptional duration and widespread impact, have become

more frequent during the summer months, primarily as a result of severe thunderstorms. While the utilities often state that storms are becoming more violent and commonplace, the root causes must include human factors such as increased suburban sprawl, lack of tree pruning efforts, and the unabated growth of tree crowns in and around overhead electrical wires.

4 The National Airspace System has reached the point of saturation within several air travel corridors across the United States. A summertime peak in air traffic delays, which have become problematic in recent years, are due in large part to the increased frequency of severe thunderstorms during these months.

5 Air traffic controllers employ several strategies to cope with lines or clusters of thunderstorms crossing congested air routes, including ground stops, ground delays, severe weather avoidance plan (SWAP) reroutes, and diversion of arriving aircraft to nearby air terminals.

Reference

Klemp, J.B., 1987. Dynamics of tornadic thunderstorms. *Annual Reviews of Fluid Mechanics*, 19:269–402.

CHAPTER 11

Tornadoes: Structure, Evolution, and Genesis

Learning Objectives

1. Identify the different types of tornado-producing storms and the characteristic intensity of tornadoes produced by each.
2. Understand the physical properties of tornadoes and governing forces, including cyclostrophic balance, the conservation of angular momentum, vorticity, circulation, suction vortices, swirl, and asymmetric distribution of winds.
3. Describe how Doppler radars have yielded new insights about tornado structure and evolution, including single- and multi-vortex tornadoes.
4. Define wind load and explain how a tornado's winds create destructive aerodynamic forces that damage dwellings.
5. Explain the core principles associated with tornado formation (tornadogenesis).

Introduction

Tornadoes strike the United States sporadically and violently. No other location in the world experiences more tornadoes than the central and eastern United States. These intense, vertical vortices produce the strongest of all surface winds. In the most violent tornadoes, wind speeds approach 300 kts (345 MPH), far greater than those in a Category 5 hurricane (which can reach 136 kts [157 MPH]). But tornadoes are extremely localized storms; the typical tornado track is about 50 m (165 ft) wide by 2–3 km (1.2–1.8 miles) long. While a single hurricane can adversely impact a 10,000 km^2 (3860 mi^2) coastal region, less than 1% of the U.S. population will find themselves in the path of a tornado during the course of their lifetime. To understand the relative rarity of tornadoes in any one location, consider the statistics: About 10% of all U.S. thunderstorms each year become severe, but only 10% of the severe storms, or 1% of all thunderstorms, spawn a tornado.

In this chapter, we explore all facets of tornado science. We start by introducing tornado classifications and describing the physical properties of tornadoes. We then discuss the forces that give rise to extreme winds and explore key properties that distinguish large, violent tornadoes from smaller, more transient tornadoes. We show how portable Doppler radars have mapped the internal, three-dimensional structure inside tornadoes. We close by examining how tornadoes damage built structures and looking closely at the enigma of tornadogenesis – the mysterious process by which tornadoes develop within the parent thunderstorm. The next chapter places the devastation wrought by tornadoes and tornado outbreaks in a societal context.

Tornado Classifications

By definition, a tornado is a violent, small-scale rotating column of air, or vortex, in contact with the surface, spawned from a thunderstorm cloud. Like hurricanes, tornadoes have a vast range of wind intensities, from 60 to 300 kts (69–345 MPH). The weakest tornadoes are very small (just a few meters wide) and short-lived, on the order of minutes. Tornadoes are usually classified in three broad categories:

- Weak tornadoes have winds below 97 kts (112 MPH).
- Strong tornadoes have winds that range from 97 to 143 kts (112–165 MPH).
- Violent tornadoes – with winds over 143 kts (165 MPH) – are also the largest, up to a kilometer or more wide, with tracks that extend 30–40 km (19–25 miles). A long-track, violent tornado can persist an hour or more, destroying many communities across several counties, even multiple states. Only about 10–12 of these "monster" tornadoes occur in the United States during a typical year. While violent tornadoes account for just 1% of all tornadoes, they cause 70% of all tornado fatalities.

DOI: 10.4324/9781003344988-14

Nature produces tornadoes in many ways. The most severe thunderstorms, the supercells, are prolific tornado breeders; they generate 79% of all reported tornadoes and nearly all violent tornadoes. Weaker tornadoes are frequently spawned by especially violent squall-line thunderstorms called bow echoes. In bow echoes, a swath of damaging winds arises from both downbursts and weak tornadoes. This category of storm may account for as many as 18% of all tornadoes. Tropical cyclone remnants also develop tornadoes after landfall, generating 2–3% of annual tornado counts, usually within 200 miles of the coastline. Landspouts, a type of weak tornado, originate from isolated convective cells that may not be intense enough to qualify as thunderstorms.

Fundamental Tornado Structure, Wind-Generating Forces, and Spin

As we start to look more closely at tornadoes, let's begin by examining some tornado types.

How to Identify a Tornado

Tornadoes take on a variety of appearances.

Tornadoes come in an amazing array of shapes, sizes, colors, and configurations. There is no one "prototype" design that applies to all tornadoes. Figure 11.1 illustrates many of the common variations. Panel (a) shows a classic, funnel-shaped tornado. The main portion of the vortex is rendered visible by water vapor that has condensed into tiny cloud droplets, creating the funnel cloud. The vapor condenses when air rushing into the low-pressure region of the vortex expands adiabatically and cools, lowering the air temperature to its dew point. Soil and debris lofted from the surface whirls around the base of the tornado. If the surrounding air has a low dew point, a funnel cloud may be unable to condense, and only the dust whirl is visible.

Panel (b) shows a large, conical tornado that has become partially rain-wrapped. A tornado can become obscured by dense rain curtains. Rain-wrapped tornadoes present a hazard to community storm spotters, who may fail to notice these "invisible" tornadoes before they strike a town.

Panel (c) displays a classic stovepipe vortex, essentially a wide, straight cylinder between cloud base and surface. Stovepipes tend to be strong tornadoes. They often appear gray or dark blue because they lack direct illumination; the great thickness of the overlying storm cloud absorbs and scatters sunlight.

Panel (d) shows the most dreaded type of tornado, a massive wedge vortex. Often, these violent tornadoes assume a rolling, turbulent, boiling appearance; the classic funnel shape is not present. These most violent of tornadoes, while extremely rare, may be 2–3 km (1.2–1.8 miles) wide (the record for the widest tornado stands at 3.5 km [2.1 miles]).

Panel (e) shows a rope tornado. The rope often signals the final, dissipating stage of a tornado. The rope often takes on a highly constricted, sinuous form, and a segment of it may even lie horizontal. The rope vortex has been deformed and stretched along the dome of rain-chilled air emanating from the storm, the storm's downdraft.

When illuminated directly by sunlight, the true color of a tornado's funnel cloud is white, like any other cloud, as Panel (f) shows. A funnel cloud is a tornadic vortex that has not contacted the surface.

Finally, Panel (g) illustrates a large, violent tornado containing multiple suction vortices. The lower end of violent tornadoes often breaks down into separate suction vortices, becoming a multiple-vortex tornado. The most intense winds on Earth are concentrated at the tip of these miniature vortices, which spin not only about their own axes but also orbit around the main axis of the parent tornado.

The shape, color, and intensity of a tornado can vary considerably throughout its lifetime. For instance, a tornado in its developing stage may have a very narrow funnel. As it matures, the tornado may take on a stovepipe or conical form. Passing over plowed farm fields, the funnel color may change from gray to brown or black, as soil is lofted. Passing over a lake, the funnel may become white or blue as water is ingested. Finally, as a tornado begins dissipating, it will often "rope out."

"Dynamic Suction Pipe"

Tornadic vortices behave like a dynamic suction pipe.

The same set of physical processes governs most types of tornadoes. Like any atmospheric vortex, the swirling wind is established by a balance of wind-generating forces. Curved flow around a large-scale synoptic vortex (such as an extratropical cyclone) is in a state of gradient wind balance (i.e., a balance of the pressure gradient, Coriolis, and centripetal forces). These large vortices include the effect of the Earth's spin via the Coriolis effect. Tornadic vortices are many orders of magnitude smaller. We can ignore the effect of Earth's spin at this tiny scale (see "Digging Deeper: Measures of Intense Vortices . . . and Why Are Anticyclonic Tornadoes Uncommon?").

Figure 11.1 Diversity of tornado appearances. Tornadoes may appear as slender as a rope or as a massive funnel more than 1 km wide. Multiple tornadoes can appear simultaneously, and some are so shrouded in rain that the naked eye cannot detect them. (NOAA.)

Figure 11.1 (Continued)

The winds that swirl around a tornado's low-pressure region are in a state of cyclostrophic balance, as shown in Figure 11.2. The cyclostrophic wind arises when the inward-directed pressure-gradient force is balanced by an outward-directed centrifugal force.

When in cyclostrophic balance, air can flow neither toward the central axis of the vortex nor away from it; all flow is confined to a swirling sheath around the vortex axis. However, cyclostrophic balance breaks down at the surface. When the free end of the vortex contacts the ground, friction slows the swirling wind in a shallow layer. As Figure 11.2 shows, centrifugal force is proportional to the velocity squared (V^2). At the surface, as the wind slows, the centrifugal force must weaken. But the pressure-gradient force remains unaltered. Air can now rush inward toward the central axis of the vortex. We can think of a tornado vortex as a kind of dynamic suction pipe. Air swirls around the pipe, but at its base, it can spiral inward, in response to suction. The swirling air enters the vortex interior and rises upward through the pipe from below.

Figure 11.3 illustrates how a tornado is embedded in a larger mesocyclone. Let's consider a strong or violent tornado, the type that forms in supercell thunderstorms. Recall that supercells have a tall, rotating mesocyclone several kilometers in diameter, embedded inside the main updraft. The base of the mesocyclone, 2–10 km (1.2–6.2 miles) in diameter, extends partially downward below the cloud base, just off the surface. The much smaller

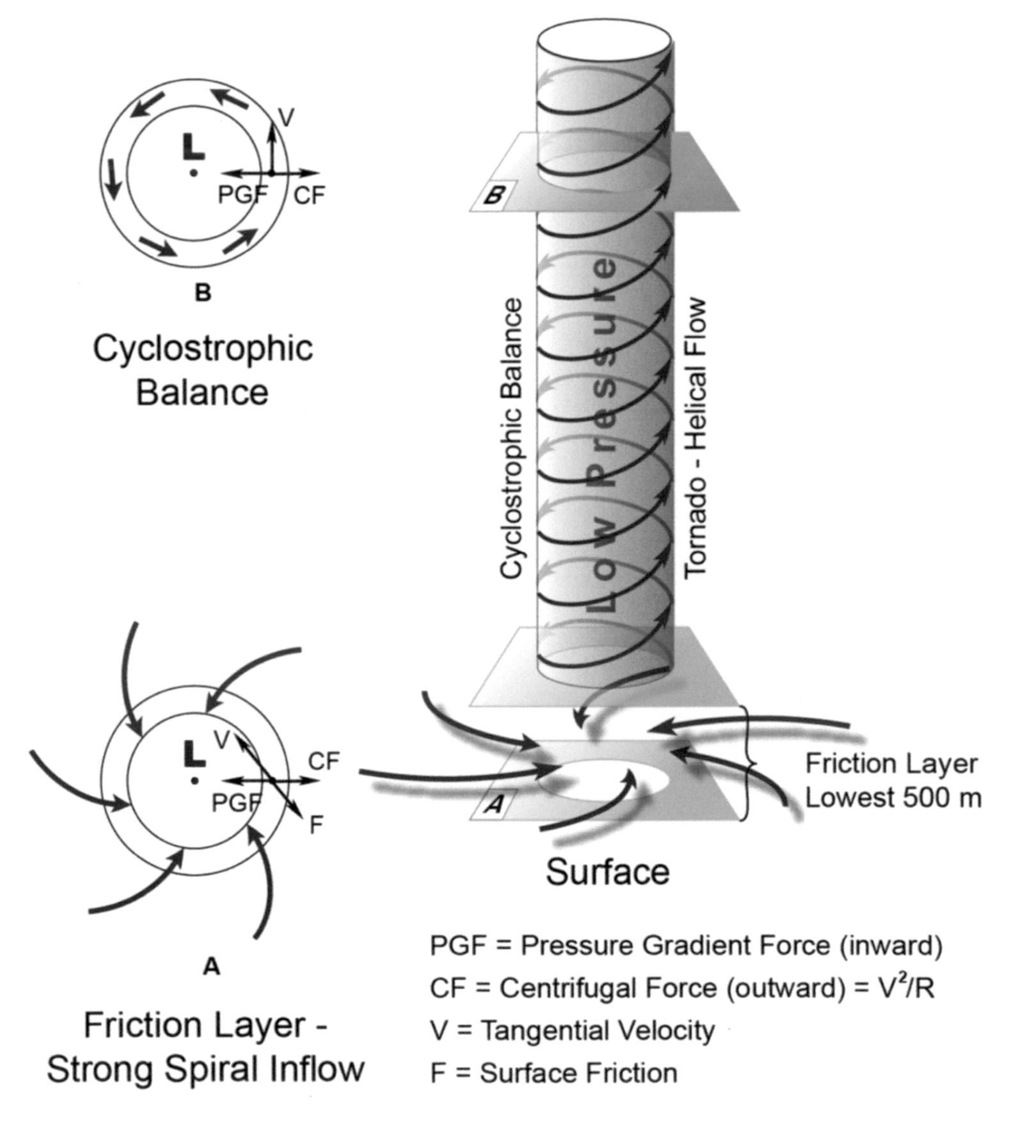

Figure 11.2 Cyclostrophic balance. The inward-directed pressure gradient force is balanced by the outward-directed centrifugal force, producing a swirling cylinder of air around the vortex axis. This balance is disrupted only very close to the surface, when friction with the ground slows the centrifugal force. However, the pressure-gradient force is unaltered. Wind rushes in, creating the equivalent of a giant suction pipe.

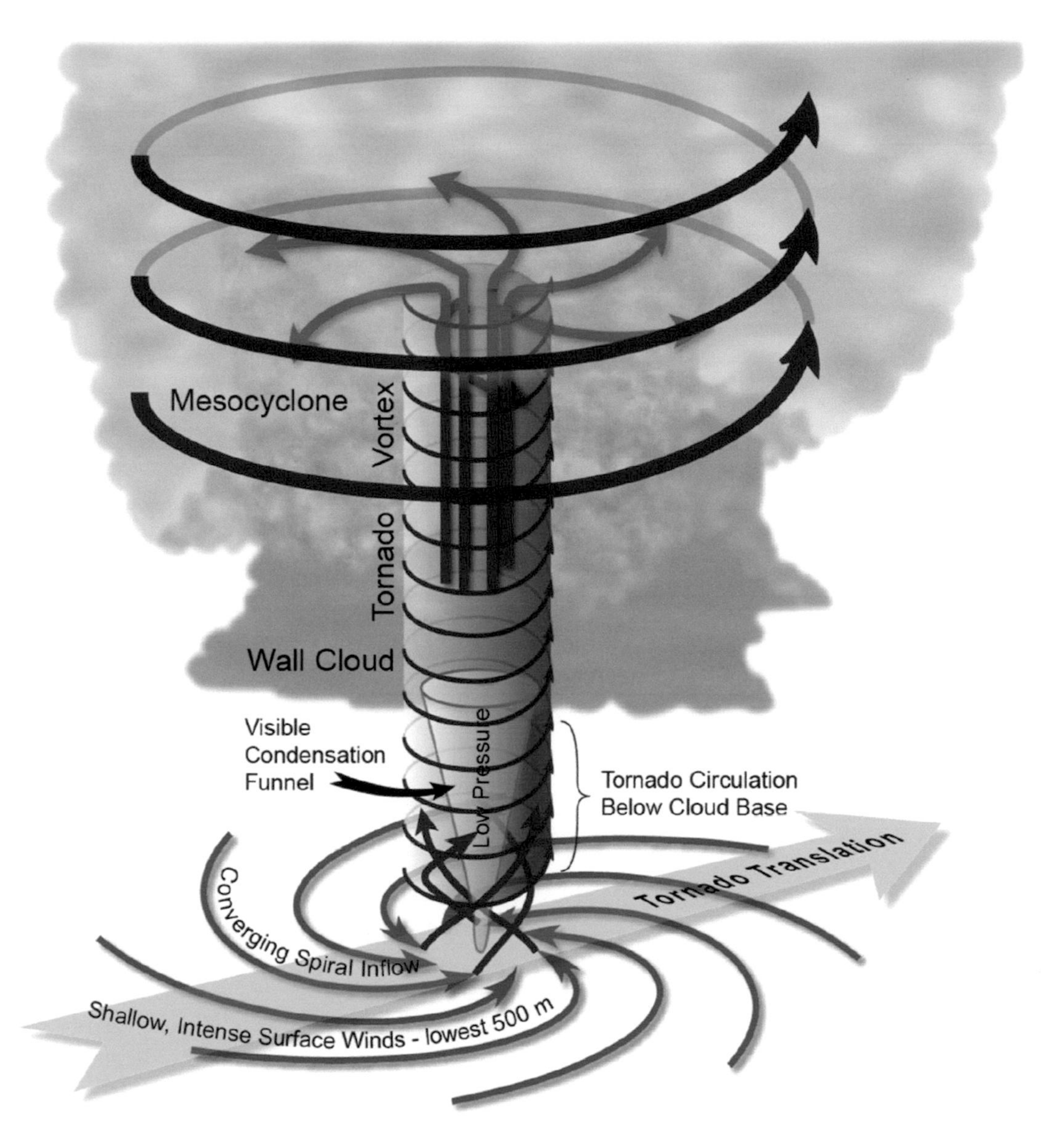

Figure 11.3 Tornado embedded in its parent mesocyclone. The mesocyclone is part of a supercell thunderstorm. The low pressure inside the mesocyclone core acts to evacuate the rising, swirling air in the tornado, thus maintaining a low pressure in the tornado vortex and a "suction" that is fed by air entering the tornado vortex at the surface. As long as this dynamic is maintained, the tornado continues to operate.

tornado (perhaps 500 m [500 yards] in diameter) develops inside the mesocyclone. The rotating winds of the mesocyclone supply angular momentum to the tornado, and the low pressure in its core helps evacuate air from the interior of the tornado vortex. As a dynamic suction pipe, the tornado attempts to fill with air sucked in its base. But as long as the mesocyclone pulls air out of the tornado vortex (at its top), the air never completely fills the tornado, and a low-pressure core is maintained. Surface friction also limits the amount of air that can be sucked into the lower end of the tornado. Significant pressure reductions are found inside tornadoes – typically several tens of millibars. The largest pressure deficits are on the order of 100 mb, or about 10% of mean sea level pressure.

At the base of the tornado, air converges in a tight spiral in a shallow air layer. In fact, powerful inflow or feeder winds can create considerable damage some distance from the tornado vortex. Note that the funnel cloud, which often tapers toward the surface, is smaller in diameter than the tornado vortex, which occupies a near-vertical cylinder down to the surface. The tornado extends upward some distance inside the larger mesocyclone. The mesocyclone and its tornado translate (move in a straight path) along the ground, often at speeds of 30–40 kts (35–46 MPH), creating a narrow path of destruction as the tornado moves across the landscape.

Digging Deeper: Measures of Intense Vortices . . . and Why Are Anticyclonic Tornadoes Uncommon?

Two important measures of tornado intensity are circulation and vorticity. These concepts explain why all tornadoes need not have winds that swirl in a counterclockwise (cyclonic) direction.

Circulation (C) is a general measure of the size and strength of a vortex. It is proportional to tangential velocity (V) and vortex radius (R):

$$C = 2\pi VR$$

Circulation increases when either the wind speed and/or the vortex radius increases. Vortex size exerts a much greater influence on circulation than radius does. As Table 11.1 shows, the circulation of a small, weak tornado is 1000 times less than that of a typical weak hurricane, but both possess the same wind speed. The circulation of a large, violent tornado is 35 times larger than that of a weak tornado.

The circulation does not indicate how rapidly the vortex winds make a single revolution. The rate of spin is commonly expressed as vorticity, ς, which is calculated as tangential velocity divided by radius ($\varsigma = 2V/R$). Vorticity is expressed in units of inverse seconds – that is, a vorticity $\varsigma = 1/s$ is interpreted as "one complete revolution per second." For a given tangential velocity, the smaller the vortex, the larger its vorticity. The table compares the vorticities of a weak tornado, a violent tornado, a weak hurricane, and a strong hurricane. The enormous range in vortex radius is responsible for the wide range of vorticity. The vorticity of a large, strong hurricane is 1000 times smaller than that of the weak tornado. And the vorticity of a dust devil just 10 m (33 feet) across is ten times larger than that of the most violent tornado. You have probably observed dust devils, which are compact swirls of soil, leaves, and debris that form over a hot surface but lack a distinct funnel cloud and last less than a minute.

To summarize: Hurricanes are characterized by very large circulation and small vorticity, while most tornadoes have large vorticity and comparatively small circulation. But what does all this have to do with a tornado's direction of spin? The vast majority of tornadoes – weak, strong, or violent – spin in a counterclockwise direction; they are cyclonic vortices. But anticyclonic (clockwise-spinning) tornadoes do occur; once thought to be extremely rare, they are being discovered in greater numbers (thanks in large part to recreational and professional storm chasers on the Great Plains). In fact, anticyclonic tornadoes are sometimes paired with cyclonic tornadoes beneath the same supercell. However, most anticyclonic tornadoes are comparatively weak and short-lived compared to their cyclonic counterparts.

It is a common misconception that the Earth's spin, which is counterclockwise, directly influences the direction of tornado spin. Many people mistakenly think that the Coriolis effect controls the spin of a tornado and even the direction that water swirls down a sink drain. The bottom line is this: The tornado vortex is far too small and short-lived to be influenced by the Earth's rotation. We can think about a tornado's spin in terms of vorticity. Knowing Earth's radius and rotational velocity, we can easily compute Earth's vorticity at its surface. This works out to 0.00015/s. The vorticity of the smallest tornado is about 3/s. Thus the Earth's vorticity is only one part in 10,000 of a tornado's vorticity – far, far too small to directly influence a tornado's spin.

So why do so many tornadoes spin counterclockwise? Strong and violent supercell-spawned tornadoes are embedded within the mesocyclone, which spins counterclockwise. But the sense of mesocyclone spin is established by airflow that is initially horizontal and characterized by a rapid increase in wind speed with altitude – a property independent of the Earth's spin.

Table 11.1 Properties of Various Storm Vortices

Vortex Type	Core Radius (R)	Tangential Velocity (V)	Circulation (C)
Weak tornado	10 m (330 feet)	35 m/s (68 kts/ 78 MPH)	2198 m^2/s
Violent tornado	100 m (3300 feet)	125 m/s (243 kts/280 mph)	78,500 m^2/s
Weak hurricane core	10 km (6.2 miles)	35 m/s (68 kts/ 78 MPH)	2,200,000 m^2/s
Strong hurricane core	50 km (31 miles)	70 m/s (61 kts/ 156 mph)	3,378,378 m^2/s

Notes: m = meters; m/s = meters per second; m^2/s = square meters per second.

Tornadoes and the Conservation of Angular Momentum

When a whirling vortex contracts, it speeds up; when it expands, it slows down.

A very important concept in meteorology is the conservation of angular momentum, L. Conservation of angular momentum simply states that the product of the vortex radius, mass, and wind speed must remain constant. Angular momentum in any rotating wind system is defined by the following equation:

$$L = VRm$$

where V is the wind's tangential velocity, V (the speed of the swirling wind, or spin), R is the vortex radius, and m is air density (mass). The product of these quantities, $L = VRm$, is conserved. In other words, if radius (R) decreases (i.e., the vortex shrinks or contracts), tangential velocity (V) must increase, to keep angular momentum constant (we ignore variations in mass for the sake of simplicity). If radius increases, spin must decrease. Recent Doppler radar observations show that the radius of tornadoes undergoes pulse-like variations along the track. When the vortex shrinks, wind speeds momentarily increase.

Figure 11.4 illustrates the basic concept of angular momentum. We start with a supercell thunderstorm and its mesocyclone. The mesocyclone has cyclonic (counterclockwise) angular momentum. The embedded tornado has a much smaller radius, perhaps 10–20% the diameter of the mesocyclone. Accordingly, as angular momentum is conserved, cyclonic tornado winds must increase in speed. Some tornadoes break down into two or more suction vortices near the surface. These tiny vortices, perhaps only 10 m across, reflect the final shrinking of the vortex system; the highest tangential winds are thus found within these vortices.

Figure 11.5 quantifies these principles further. From top to bottom, we see the complete hierarchy of vortices that give rise to intense tornadic winds. Tornadoes lie at the base of an energy flow, starting with the large-scale extratropical cyclone (top panel). As the spatial scale shrinks, rotating winds become more sharply focused and intense. Each successive panel shows orders of magnitude reduction or shrinkage in vortex area, with a rise in tangential wind at each step. The extratropical cyclone is a broad vortex, on the order of 1000 km (621 miles). In a mature cyclone, winds spin at 20–30 kts (23–35 MPH) around its center at low levels. Within the cyclone's warm sector, supercell thunderstorms develop. These storm cells each contain a mesocyclone, with a diameter

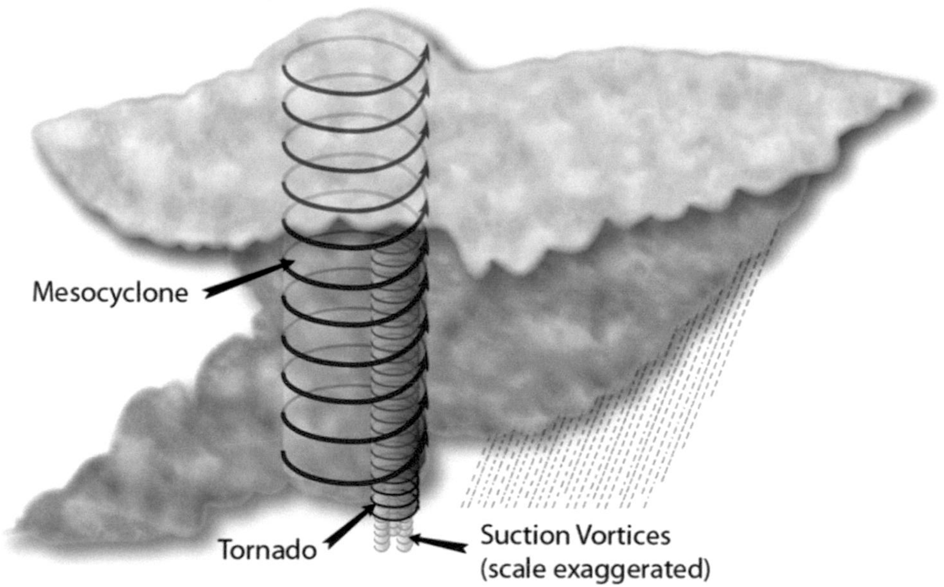

Figure 11.4 Conservation of angular momentum. As we move from the mesocyclone to the tornado, there may be a tenfold or greater reduction in radius. Therefore, all else being equal, the tornado's wind speed must increase to offset the change in size. Likewise, suction vortices are considerably smaller in radius than the tornado, so they must experience a proportional increase in wind speed.

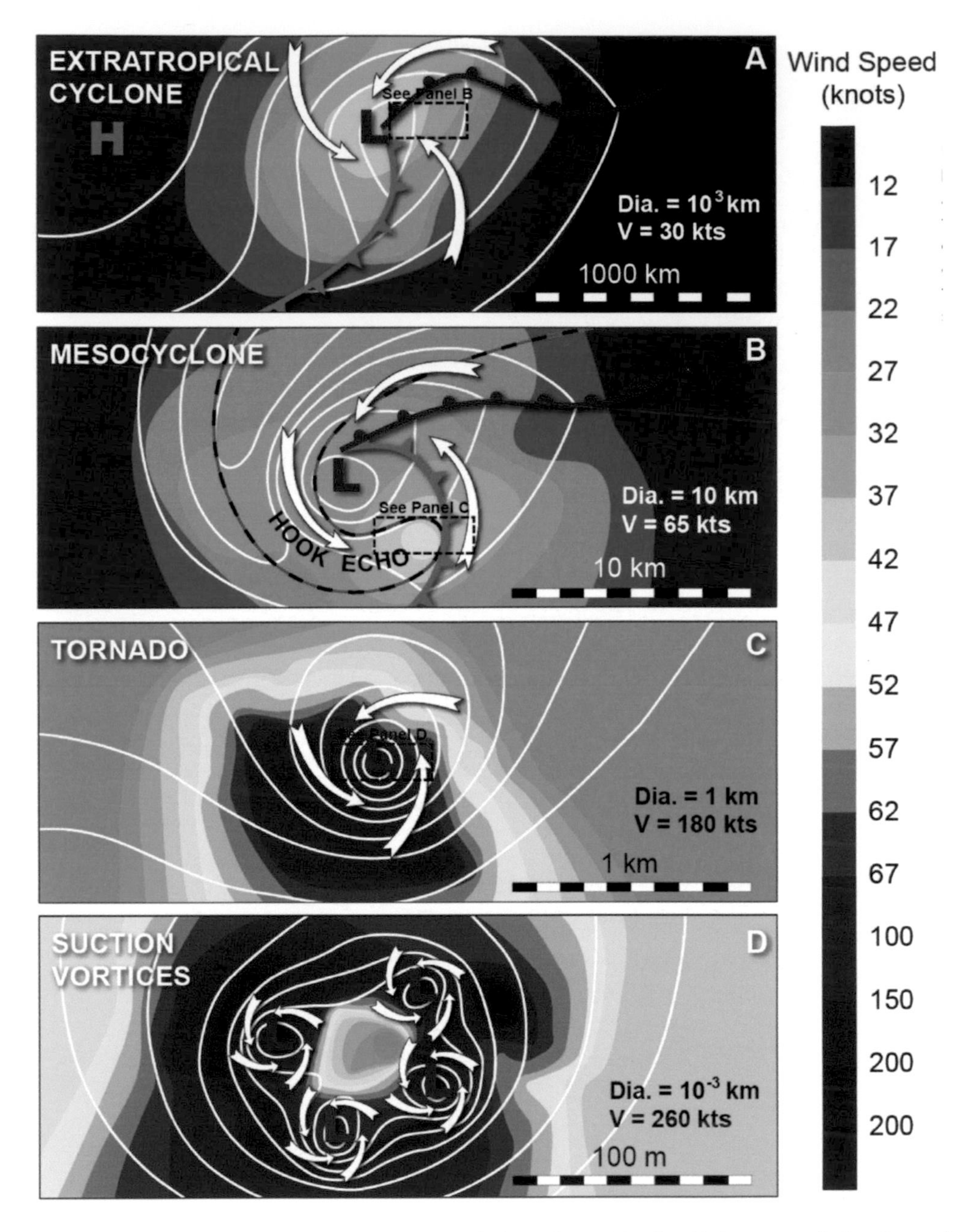

Figure 11.5 Conservation of angular momentum in a hierarchy of atmospheric vortices. We begin with the extratropical cyclone and progress downward in scale through mesocyclone, tornado, and suction vortices. Due to conservation of angular momentum, as the size of the vortex decreases, the speed of the rotating winds in the vortex increases. (Adapted from Fujita, 1981.)

of 10 km (6 miles). While the direction of spin is the same in the cyclone and the mesocyclone (counterclockwise), the winds of the mesocyclone spin at 50–70 kts (58–81 MPH). Stepping down further, we arrive at the tornado vortex with a diameter of 1 km (0.6 mile) embedded within the mesocyclone. Winds in a strong tornado circulate counterclockwise in the 150–200 kt (173–230 MPH) range. As tornadic winds devolve into 10 m diameter suction vortices (lowest panel), winds approach 300 kts (345 MPH). The entire vortical wind system in Figure 11.5 spans six orders of magnitude (a scale factor of 100,000), and the rotating winds have increased from 30 to nearly 300 kts (35–345 MPH) – a grand example of angular momentum's conservation.

Wind Distribution

Tornadoes have a very concentrated and asymmetric wind distribution.

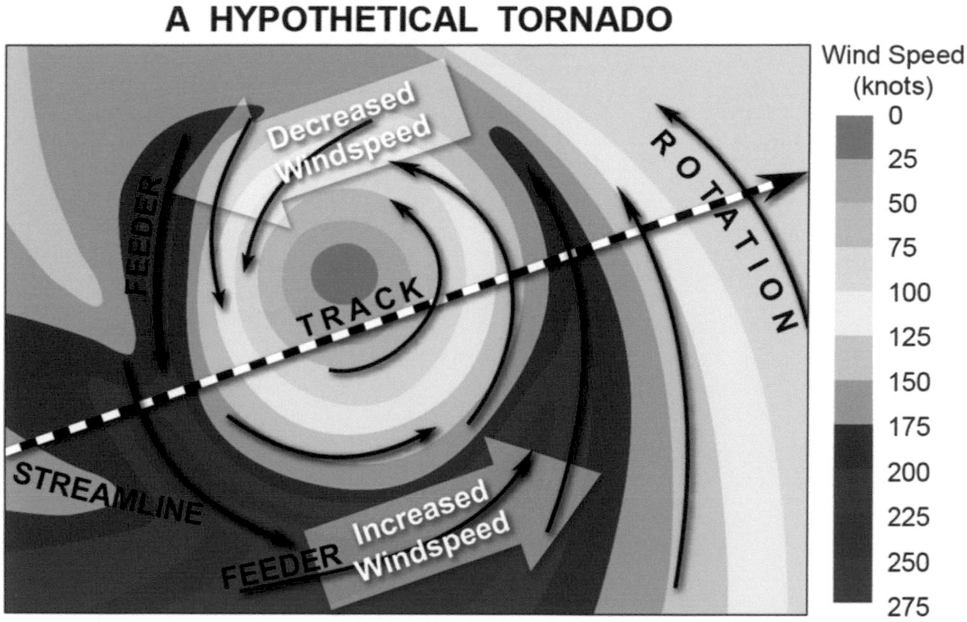

Figure 11.6 Wind asymmetry in a typical tornado. The speed with which a tornado tracks (translational speed) can have significant effect on the wind speeds associated with each side of the tornado. To calculate the differences in wind speed, we add the track speed to the right-side tornado winds and subtract the track speed from the left-side tornado winds.

Generally speaking, tornadoes travel in straight paths because they are embedded in a supercell, which is itself propelled by large-scale winds such as the jet stream. Occasionally, tornado translation slows or becomes erratic, or the tornado becomes stationary. Tornadoes can even make loops; such is the case when the supercell slows or stops moving, and the tornado orbits counterclockwise around its parent mesocyclone.

If a tornado is moving along the ground, the winds that swirl around the vortex (the tangential wind) are not the same on both sides of the tornado track. As the vortex moves in a straight path, the direction and speed of movement must be added to or subtracted from the tangential wind. Let's imagine a tornado moving along a straight path, at a brisk speed of 50 kts (58 MPH). As Figure 11.6 shows, the tornado is tracking from W-SW to E-NE. If the tornado were stationary, it would have a uniform tangential velocity of 200 kts (230 MPH) everywhere around the center. But the entire vortex is moving. On the right side of the tornado path, the speed of the tornado's movement – its translation speed – adds to the tangential velocity; both the wind direction and the tornado movement act in the same direction. In this case, we add 50 kts + 200 kts to get a total wind of 250 kts (288 MPH). To the left of the tornado track, we subtract the translation speed from the tangential wind, because the wind swirls in the opposite direction that the tornado is moving: 200 kts minus 50 kts yields 150 kts (173 MPH) total wind to left of track. Thus the total wind across the tornado varies considerably, amounting to 100 kts (87 MPH) difference (250 [288 MPH] on the right vs 150 kts [173 MPH] on the left), giving rise to an asymmetric wind field. This large difference occurs over very small distances, on the order of just tens to perhaps a few hundred meters.

The asymmetric wind field explains why the worst tornado damage often appears sporadically and is confined to a small area. Homes on one side of a neighborhood street may be devastated, while those on the opposite side may suffer only minor damage. The location of a structure with respect to the tornado track and the distance from the tornado center greatly determines the damage it receives.

Another type of wind asymmetry arises from feeder bands. Feeder bands are narrow jet-like ribbons of air that stream into the tornado, along the surface. As Figure 11.6 shows, the strong inflow of air swirling into the vortex is sometimes concentrated in narrow corridors. These feeder bands may be only a few tens of meters wide, and they may contain locally more intense winds than the surrounding vortex.

Figure 11.7 shows a post-tornado damage assessment following a devastating 1979 tornado that occurred in Wichita Falls, Kansas. The assessment is based on aerial surveys and illustrates many of the important concepts we have discussed so far. First, note the very linear and narrow track of damage caused by the

Figure 11.7 Destruction caused by the April 10, 1979, Wichita Falls tornado. The highest winds occur on the right side of the counterclockwise-rotating tornado. With the more rare clockwise-rotating tornado, winds are stronger on the left side. (Adapted from Fujita, 1981.)

tornado. As the tornado approached the small town of Sikes, it changed direction slightly, executing a shallow "S" turn. The path is generally a uniform 500–700 m (1640–2300 feet) wide. This tornado was rated as a violent tornado, with winds up to 225 kts (259 MPH). There is a very large asymmetry in total wind speed across the path, with the highest winds confined to a very narrow corridor (only about 100–200 m [328–656 feet] wide) to the right of the tornado track. The core of high winds is relatively uniform up to Location A, then becomes sporadic until Location B, at which point the tornado begins to rapidly weaken. To the right of track, particularly in the region of the "S" turn, prominent feeder bands arc inward toward the track's center from the southwest. Additional feeder bands curve inward from the northeast left of track, just east of Fairway Blvd. The key conclusion here: When a particular tornado is rated as violent, the highest winds occupy only a small percentage of the total path area, typically about 10–20%.

Novel Insights Into Tornado Structure and Wind Speed

Until the late 1980s, meteorologists learned about tornadoes' internal structure and maximum wind speeds by direct observation, laboratory experiments, and computer simulations. Instruments placed directly in the path of tornadoes experienced a low success rate of collecting data, usually because (1) the tornado missed the instrument, (2) the tornado was too dangerous to safely approach, or (3) the instrument was destroyed. Few, if any, devices can withstand the winds of a violent tornado. But the quest to understand the tornado's internal airflow continues because such knowledge provides the engineering insights needed to build windproof dwellings.

The late 1980s saw a breakthrough in tornado observation technology. Doppler radars, which measure wind velocity from a distance, were built small enough to be completely portable. A tornado intercept team could transport the device in a car or mount it on a small truck, and position the device within 1–2 km (0.6–1.2 miles) of the tornado. These devices were pioneered by Professor Howard Bluestein and his graduate students at the University of Oklahoma. In the past two decades, Dr. Joshua Wurman (Center for Severe Weather Research) has greatly refined portable Doppler techniques using his Doppler on Wheels (DOW). The DOW uses a large radar dish (Figure 11.8) that increases the beam's resolving power, down to a pixel size of about 60×60 m (60×60 yards). But progress measuring wind speeds in tornadoes has been slow. Tornadoes remain dangerous and difficult to approach, and the scientific field programs devoted to intensive tornado research are costly. On average, only a small handful of useful portable Doppler tornado datasets are gathered in any given year. Nevertheless, several theories proposing the structure and evolution of tornadoes are finally being verified by these Doppler datasets. Incidentally, the highest ever Doppler-measured wind in a tornado stands at 300 kts (345 MPH).

Figure 11.8 Doppler on wheels (DOW). Scientists now use DOW, with its greater resolving power, to better understand the science of tornadoes. However, it is still difficult and dangerous to position DOW effectively. (Josh Wurman, Center for Severe Weather Research.)

Evidence From Doppler Studies

The tornado's innermost "secrets" and peak winds are being revealed.

Figure 11.9, which depicts a strong to violent tornado, synthesizes much of what we know about the anatomy of a tornado, based on portable Doppler research studies. The tornado is sliced in half, revealing its hollow inner core. Intense tangential winds form a vertical cylinder extending from above the cloud base to the surface. The color scale indicates the relative speed of the tangential wind. To understand the figure properly, you must visualize that the tangential wind comes out of the page on the left side, then enters the page on the right side (note that this movement completes the counterclockwise swirl). Now imagine the tornado moving in a straight line away from you. This movement makes the total vortex wind stronger on the right side. Note how the tangential winds also decrease with height. Tangential winds are strongest in a shallow ring at the base of the tornado, just above the surface. The entire vortex also widens gradually with height. The overall tornado structure bears an uncanny resemblance to the inner vortex of a hurricane.

Superimposed on the rotational or tangential wind is a secondary circulation – a strong, radial air flow moving toward the vortex near its base (red arrows). On each side of the vortex, the inflow penetrates close to the vortex center, where it erupts violently upward as a powerful ascending jet – the updraft – within the ring of rotating air. The combination of the tangential velocity and strong updraft describes a counterclockwise helical flow, one in which air spirals inward and upward at very high speeds around the vortex core.

The core of a strong-violent tornado contains a central downdraft. Low pressure at the base of the vortex pulls air toward the surface. Air rising in the helix expands adiabatically and cools, and its vapor condenses, forming a cylindrical funnel cloud. But in the core, descending air compresses and warms the air, evaporating cloud droplets and clearing out the core. Any cloud droplets that remain in the core region are centrifuged outward by vortex rotation.

At the base of the tornado vortex is the debris cloud or dust whirl, consisting of soil, bits of vegetation, dust, and the pulverized remnants of structures such as wood panels, tree limbs, bricks, glass, shingles, and wall insulation. Centrifugal force slings the largest pieces outward beyond the radius of maximum tangential winds. Smaller particles remain embedded in the helical walls of the vortex and may ascend to great heights, forming a debris sheath around the tornado.

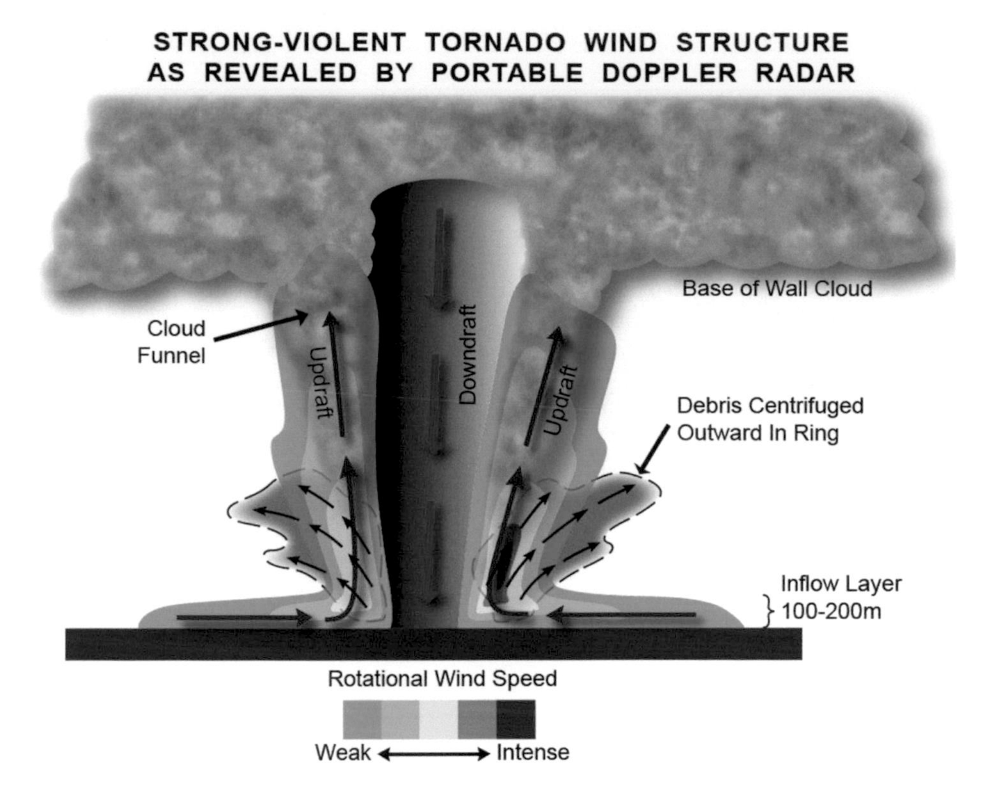

Figure 11.9 Cross section of a tornado. Imagine the tornado moving away from you. Winds are strongest near the surface and decrease with altitude. Furthermore, winds are stronger on the right side than on the left side due to the storm's forward movement. Air feeds in near the surface and then flows upward in a clockwise rotation as the mesocyclone draws air out of the tornado at the top.

Suction Vortices

Suction vortices are the most destructive aspect of large, violent tornadoes, as revealed in Figure 11.5. The parent tornado vortex can be envisioned as a giant ring of counterclockwise-swirling air. Numerous subvortices rapidly form, just tens of meters in diameter, orbit some distance around the parent vortex, and rapidly dissipate. Suction vortices orbit around the parent vortex at about half the speed of its tangential wind. Wind speeds in their cores can be much higher than in the main vortex. Homes and structures subject to the passage of a suction spot experience absolute peak winds that can push 261 kts (300 MPH) and beyond, along with extremely rapid wind accelerations in the horizontal and vertical. These near-explosive accelerations exert tremendously rapid and violent changes in pressure on structures, causing instant failure.

As Figure 11.10 shows, homes unlucky enough to be hit by a suction vortex are often completely demolished. The formation, movement, and dissipation of individual suction vortices account for much of the irregularity in a tornado's damage pattern – forming the perception that a tornado "hops and skips" across the landscape. As a suction vortex orbits the parent tornado while the tornado translates in a straight line, cycloidal etchings (curved damage streaks) are embedded in the main path (Figure 11.11).

There is a simple number that predicts whether a single or multiple vortex will form.

Laboratory simulations of scaled-down tornado vortices and the recent DOW observations confirm that a tornado's structure is determined by a simple mathematical parameter called the swirl ratio, S. Swirl ratio determines whether a tornado contains a simple, rotating ring of wind or a large ring containing multiple suction vortices. The swirl ratio is defined as the speed of the tangential wind, V_t, divided by the vortex updraft, w. For S values less than one, the spin is weak compared to the upward motion in the tornado core. This configuration describes the structure of weak tornadoes, shown in Figure 11.12 at the very top. As tangential winds increase relative to upward motion, the tornado undergoes a transition in structure. A central downdraft begins to develop and divide the vortex into two cells. At very high values of S, from 2 to 6, vortex breakdown occurs. The core downdraft penetrates to the surface. At the interface between air descending

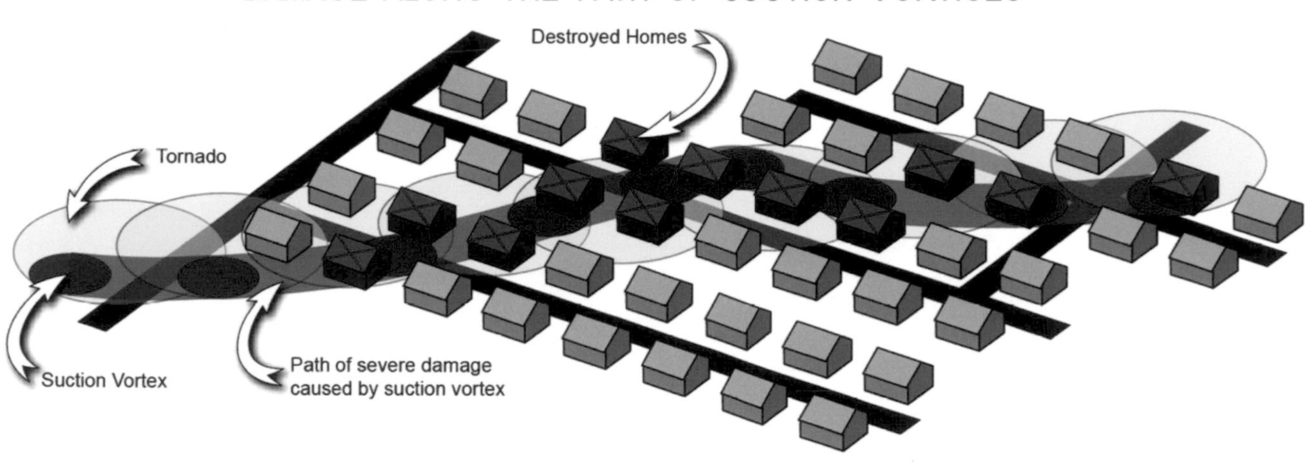

Figure 11.10 Destruction along a suction vortex. Shown is how homes along the track of suction vortices are far more vulnerable to major destruction.

Figure 11.11 Cycloidal marks (damage streaks). This aerial image shows cycloidal marks etched into the landscape as a result of the suction vortex within a tornado. (National Archives and Records Administration.)

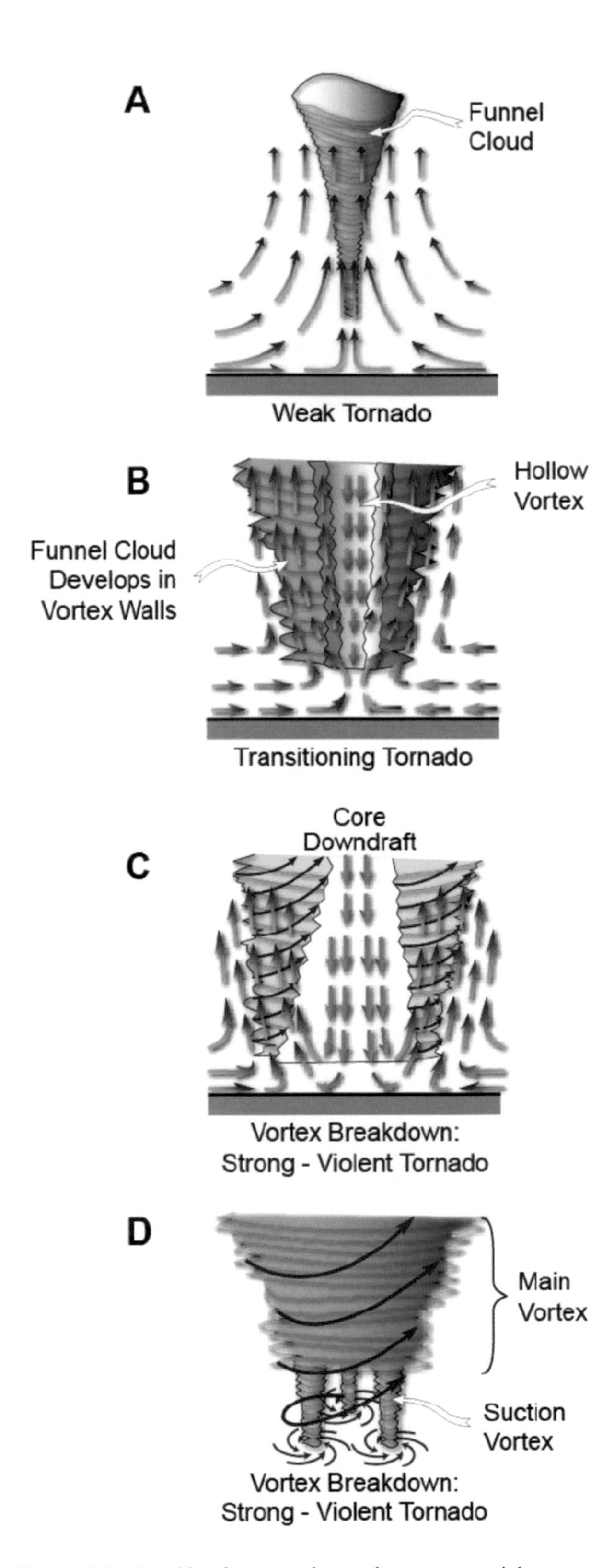

Figure 11.12 Transition from a weak tornado to a strong, violent tornado. (a) A weak tornado. (b) A central downdraft begins to divide the vortex into two parts. (c) Shearing increases both horizontally and vertically as the vortex divides. (d) As the process continues, multiple suction vortices form. (Adapted from Davies-Jones, 1986.)

in the core and the annular updraft, very large horizontal and vertical shears (changes in wind speed and direction) develop. In this region, the flow becomes highly unstable, devolving (breaking down) into two or more suction vortices. Vortex breakdown is characteristic of strong-violent tornadoes that contain very strong tangential winds. Look again at the multiple-vortex tornado in Figure 11.1(g). Each suction vortex is just meters wide and only a few tens of meters tall. But the total vortex winds (suction vortex wind + parent tornado wind) add up to the most extreme wind velocities in a confined space on Earth.

Tornado Wind Loads and Aerodynamic Forces Causing Structural Failure

The most violent tornadoes create widespread destruction of brick and mortar structures, heavy vehicles, and engineered structures such as schools. Witness the aftermath of a violent tornado in Joplin, Missouri, in 2011 (Figure 11.13). Here, near-total destruction except to the most strongly engineered commercial buildings resulted from winds that approached 174 kts (200 MPH) across a 1.6 km (1 mile) wide path through the city center. Trees were completely denuded of all limbs, vegetation, and bark. Sheet metal from roofs was twisted around tree trunks. Pavement was lifted from roadbeds. Entire neighborhoods were rendered unrecognizable. At $3.6 billion (2023), it is the costliest tornado in U.S. history.

Wind Acceleration and Wind Load

Forces exerted on structures depend on wind speed, structural shape, and surface area

Because tornadoes are typically small and travel quickly, structures are exposed to peak winds for only seconds. However, the rate of change of the wind speed, or wind acceleration, is likely to approach 5 Gs (five times the acceleration of gravity) when an intense suction vortex strikes a structure. The force per unit area, or wind load, exerted by tornadic winds varies with the square of wind speed. So a doubling of wind speed (between a weak and strong tornado or between a strong and violent tornado) leads to a quadrupling of force. The force exerted on structures by a high-end violent tornado is 15–20 times larger than the wind load exerted by a weak tornado. Some studies suggest that the relationship may be even larger, with wind pressure forces varying according to the cube of wind speed.

Peak tangential winds are not the only measure of a tornado's ability to destroy. The updraft is also critical. An updraft of

Figure 11.13 Post-tornado Joplin, Missouri, 2011. Even large structures like the hospital in the background (white building) were severely damaged. (Courtesy of Carol Rabenhorst.)

80–90 kts (92–104 MPH) lofts heavy structures or pieces of debris, which tangential winds then readily accelerate. In fact, the simple sum of a 100 kt updraft operating in tandem with 200 kt (230 MPH) tangential wind yields a total (helical) wind of 225 kts (259 MPH). Additionally, the duration of high wind is also important. In very large tornadoes or those that translate slowly, structures are subject to long-duration winds (perhaps a minute or more). These winds can cause structures to vibrate in a manner that contributes to their destruction.

Abrupt changes in wind direction are also significant. Winds from different directions exert stress on structures from multiple angles. Many structures have a weak wall. If that wall experiences direct onslaught from the highest wind, the entire structure succumbs as the weak wall fails. A common example of a weak wall is the front of a house with a large picture window, an entrance door, and a garage door. All of these portals are relatively frail and easily compromised by the wind.

How and Why Tornadoes Destroy Houses

The key to a home's survival is to keep the roof on!

Figure 11.14 shows what typically happens when a wood-frame-and-siding house is subject to high wind loads. If the front wall of the house lies just to the right of the tornado path and is subject to peak winds, glass windows and doors are blown inward. High-speed air rushes inside the house, instantaneously raising the internal pressure. Moreover, wind flowing around the sides and back wall of the house creates a pressure reduction along the external surfaces. Consequently, these remaining walls (sides and back) are pushed outward, along a pressure gradient between inside and outside. A pitched roof exacerbates home damage. Roofs are designed with a pitch so that they easily shed rainwater, but this pitch makes them vulnerable to high wind. As wind flows over the rooftop, an aerodynamic lift force develops, similar to air flowing over the curved upper surface of an airplane wing. Consequently, the roof or large portions thereof may peel away. With the roof detached, freestanding walls lose their support and more easily fall outward. Large eaves or roof overhangs allow the wind to more effectively build wind pressure beneath roof edges, increasing the upward lift force.

As the roof lifts off and external walls fall outward, it appears that the house explodes. For decades, such explosions were wrongly attributed to the pressure drop inside the tornado's core. Homeowners in the path of a tornado were encouraged to open one or more windows to "ventilate" the home prior to vortex passage. However, we now know that this logic is faulty. Tornadoes do indeed have reduced core pressure; however, the shroud of whirling debris that strikes the home before the core arrives more than adequately eliminates pressure differentials, by poking numerous holes through windows, doors, and walls. Much of the damage rendered to homes and buildings comes from the tremendous impact of wind-borne debris: pieces of

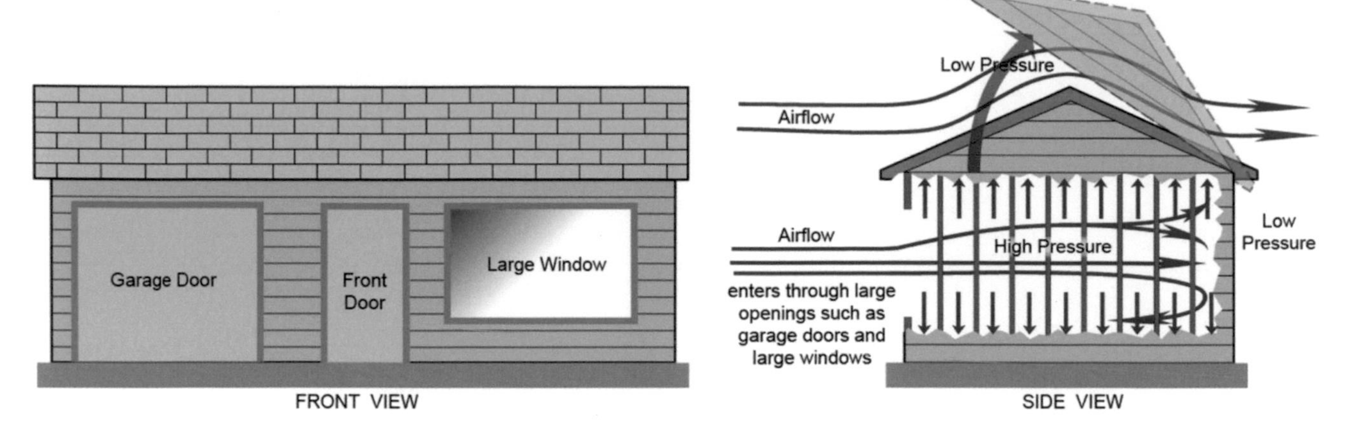

Figure 11.14 Aerodynamic wind forces on a home. When wood-framed homes are exposed to strong tornadic winds, such as those found in the right side of the tornado, the structure is often destroyed. Winds penetrate the structure through weak zones such as large windows and garage doors. Wind going around and over the house creates low pressure above and behind the structure. As a result, the home "explodes" outward, destroying the structure.

2×4 inch lumber, bricks, metal signs, tree limbs, glass shards, plywood, and various other dense objects hurled at a speed of several hundred knots.

Homes not built according to strict code often succumb through loss of the roof. Figure 11.15(a) shows the major design elements that buttress and secure a roof to the walls of a typical wood-frame home. Diagonal roof trusses are attached to the horizontal top plate that overlies vertical wall studs. In Figure 11.15(b), a low-budget method of attachment utilizes three toe nails – that is, nails driven in diagonally, to connect the top plate and truss. This is not a very secure method of attachment. The upward lift force and/or torque (twisting force) exerted between roof and walls by the wind can easily wrench trusses free from the top plate. The preferred method is to more strongly secure the top plate and trusses with preformed metal plates, as shown in Figure 11.15(c). Each precisely shaped plate secures the stud and truss in multiple locations, greatly increasing the attachment surface area. The plate also resists bending and twisting by wind forces. In houses that use this method, the roof is more likely to remain attached to the house – and as the saying goes, "If you save the roof, you save the home."

Figure 11.15(c) also illustrates the importance of securely attaching the house's walls to its foundation. Low-budget methods may rely on friction and gravity (weight of the home) to do the trick. Without metal plates and anchor bolts, the entire home can literally be shoved off its foundation slab by a tornado's winds. In fact, this type of destruction happens so often that damage-assessment teams call these homes "sliders."

Mobile (or manufactured) homes are especially dangerous places during a tornado. The tornado fatality rate is considerably higher in manufactured homes than in permanent homes; in fact, mobile homes dominate the tornado fatality statistics for all structures. According to tornado researchers Kevin Simmons and Daniel Sutter, mobile homes, permanent homes, automobiles, and buildings account for 43%, 31%, 9%, and 5% of national tornado fatalities, respectively. In the Southeast, mobile homes account for 58% of tornado deaths.

Mobile homes are very lightweight and constructed of comparatively flimsy materials. They have a very large ratio of surface area to weight, and they are easily tumbled or rendered airborne. Windborne debris easily penetrates their thin walls and roofs.

Some mobile-home owners secure their homes to the ground using metal tie-down straps, usually one on each end. These straps encircle the home and attach to metal shafts augured into the ground. But these straps provide a false sense of security and are not effective at winds above 60–70 kts (69–81 MPH). They cannot protect a mobile home from the impact of wind-borne debris. Damage surveys reveal that straps are often severely rusted at ground level and therefore snap at low wind speeds.

Tornado Safety

Here is the bottom line: Lacking a basement, put as many walls as possible between you and the tornado.

There are some simple, common-sense rules regarding tornado safety. For permanent homes with a basement or storm cellar,

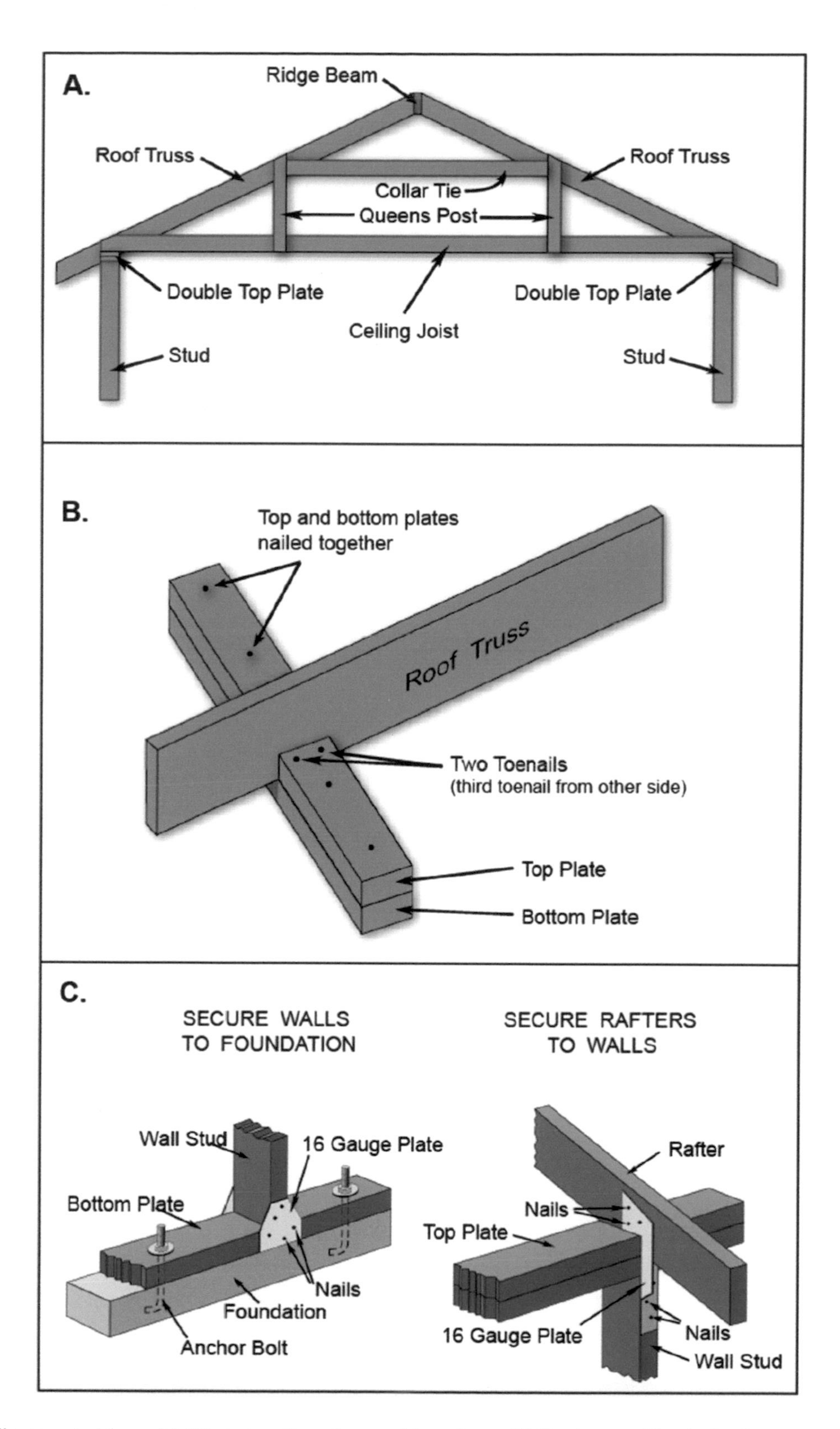

Figure 11.15 Building to protect the roof. (a) Typical roof truss in a wood-frame house. (b) How trusses of low-budget houses are often attached to the top wall plate with toe nailing, a technique that gives the roof little ability to resist being wrenched loose in tornado-strength winds. (c) A much stronger method of attaching both roof trusses and the wall studs. A few extra dollars spent on these construction techniques increases the odds that the roof will stay attached to the house, thereby greatly increasing the likelihood that the house will remain intact.

relocate there, preferably beneath a heavy workbench or beneath the staircase. If there is no basement, put as many walls between you and the tornado as possible. Seek shelter in an interior closet or in a bathroom, climbing into the tub or shower stall. If you live in a tornado-prone area, consider constructing a tornado safe room, which is a small, above-ground, structurally reinforced room in the center of the house.

If you are in a public building, such as a school or airport terminal, move away from plate-glass windows. Seek an interior hallway, crouching low and protecting your head and neck with hands. Move out of large, open spaces such as gymnasiums that have large-span, flat roofs (these are flimsy and readily collapse). If you are in a mobile home, get out – either to a community storm shelter or into a low ditch. Cars or trucks are death traps, easily flipped or rendered airborne; never try to out-drive a tornado. Leave the vehicle and crouch in a low ditch. Finally, never seek shelter under a highway overpass. This has proven to be a fatal mistake; the tornado's winds accelerate substantially while passing through the narrow passage between bridge deck and ground.

Theories of Tornado Formation

For years, the public has been sold a gross oversimplification about how tornadoes form – specifically, "cold and warm air masses clashed over the Great Plains, leading to severe storms and twisters." If the process were this simple, we'd have tornado outbreaks during all hours of the day, every day of the year! Cold and warm masses continuously interact along fronts, but many of these interactions do not produce severe thunderstorms. In fact, the complete process by which tornadoes develop within thunderstorms – termed tornadogenesis – remains one of the most elusive questions in meteorology – but new insights are filling in knowledge. And because tornadoes claim many lives and destroy billions of dollars of property, there is real urgency to "crack the code."

For strong and violent tornadoes, the first step requires formation of a supercell thunderstorm. But tornadogenesis occurs on smaller space and timescales than the larger supercell. While a supercell may take 1–2 hours to mature, encompassing a region of several hundred square kilometers, the tornado develops in just minutes, on scales of 1 km^2 (0.6 mi^2) or less. Tornadogenesis is a very focused, small-scale process. There are subtle differences between supercells that do not create tornadoes and those that do, and in fact the majority of supercells (70–75%) fail to spawn tornadoes. In the past several decades, scientists have identified many ways that tornadoes develop, both within supercells and in other types of storm cells. In this section, we review the important principles, starting with the genesis of strong and violent tornadoes.

Supercell Tornadoes

Supercells generate many of the strong tornadoes and all violent tornadoes across the United States.

A defining characteristic of a supercell and the key to its severity and longevity is its rotating or helical updraft, the mesocyclone ("meso"), located in the storm's mid-levels (2–6 km [1.2–3.7 miles]) and southwest quadrant. Most mesos do not produce tornadoes, and some of the most intense mesos ever documented on Doppler radar failed to create a tornado. A tornado is more likely when the mid-level mesocyclone builds down to the surface. The mechanism that produces the low-level meso differs from genesis of the mid-level meso. To understand how a low-level meso develops, we must first look at a basic mechanism: The storm manufactures its own horizontal spin in the lowest 1 km (0.6 mile) of the storm, which then becomes tilted into the vertical.

As Figure 11.16 shows, horizontal vortex tubes can develop in two ways beneath a supercell. First, the mesocyclone aloft can induce a powerful inflow of air just above the surface by means of its suction effect. This establishes a strong vertical shear, in which winds increase speed rapidly with altitude (this is a different source of wind shear than the deeper, environmental winds that initially create the mesocyclone). This streaming air overturns or tumbles along a horizontal axis. Imagine holding up one palm, placing a pencil on that palm, and then moving the other palm across the pencil, mimicking strong winds blowing above the surface. The pencil rolls over and over. Second, there is often a boundary or front separating cold, rain-cooled air (the forward flank downdraft) from warm inflow air feeding into the base of the updraft. Along this boundary, buoyant warm air rises on one side; meanwhile, dense, cold air sinks on the other side of the boundary. The rising and sinking create an overturning motion along the boundary. A horizontal vortex tube develops along the length of this downdraft boundary, just above the surface. This process can take up to an hour or so to create horizontal spin.

The updraft of the supercell features a region of low pressure near the surface, and air streams inward to fill the low. Some of the air streams parallel to and through the vortex tubes, drawing them inward toward the low. As the inflow turns abruptly upward, entering the base of the updraft, the orientation of the tubes changes from horizontal to vertical. As the vortex tubes ascend, they consolidate, coalescing into a low-level

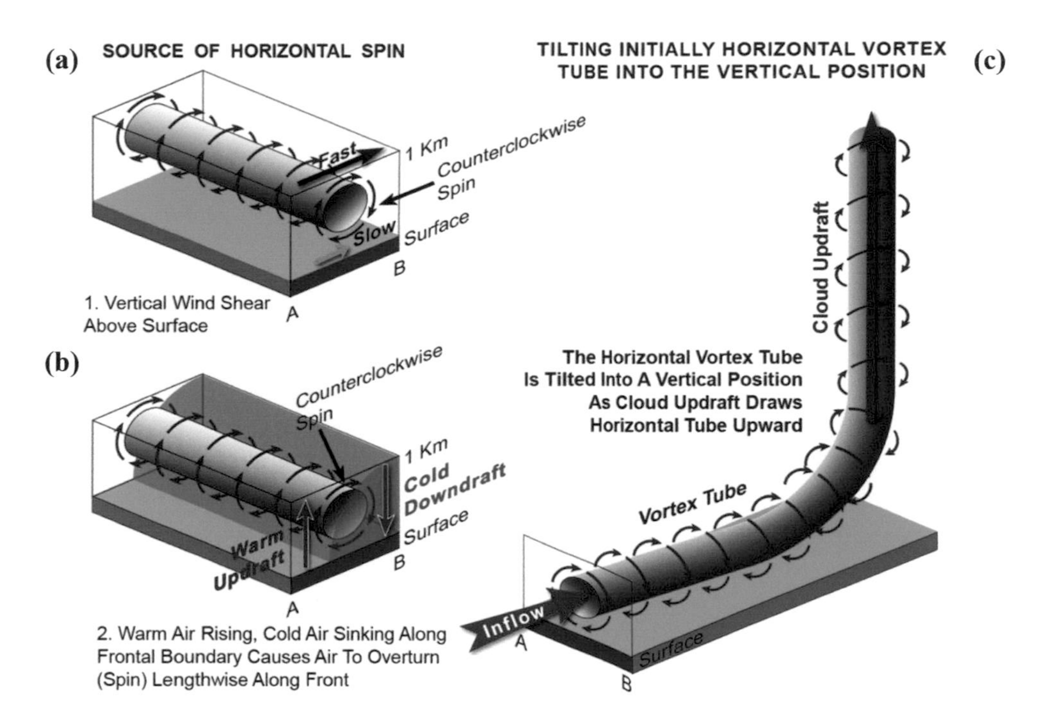

Figure 11.16 Tilting storm-generated horizontal spin to vertical. Horizontal rotating vortices can result from two different processes. (a) First, vertical wind shear is caused by two different wind speeds (slow near the surface and fast at higher levels above the surface). (b) The second method is common along small-scale fronts beneath the supercell, where warm air rising in updrafts meets downdrafts of cold subsiding air. (c) A horizontal vortex tube is tilted upward into a vertical position, resulting in the formation of a low-level mesocyclone in a supercell. Tilting occurs when strong ground flow into the mesocyclone draws the horizontal vortex tube inward toward the storm's center. As the inward flow turns abruptly upward, the horizontal vortex tube is lifted into a vertical position.

mesocyclone. Figure 11.17 is a three-dimensional rendering of the process. You will note that the sense of spin initially present in the horizontal vortex tubes is preserved as they are tilted. As the tubes lift, they spin counterclockwise, the same direction as the supercell's mid-level meso. Formation of a low-level meso is a critical early step that may lead to tornadogenesis.

A second important step comes next. In tornadic supercells, 1–2 hours after the supercell forms, a new downdraft develops. This spiral downdraft, which forms along the southwestern flank of the supercell and wraps around the mesocyclone, is called the rear flank downdraft (Figure 11.18). This downdraft divides the mesocyclone so that it is composed of downdraft and updraft in roughly equal parts (note the horseshoe-shaped warm and cold regions in the figure, both inside the meso). At the boundary between the updraft and downdraft, there are exceptionally strong shears (rapid changes in vertical and horizontal air motion). Air spirals up into the updraft portion of the low-level meso (red arrows) (Figure 11.19). Meanwhile, the rear flank downdraft spirals counterclockwise downward, wrapping around the low-level meso (solid blue arrows). When the tornado does form, it often does so near the center of the divided mesocyclone, indicated by the "T" inside a small circle on the diagram.

Within this highly dynamic region of the supercell, there are two major modes by which the tornado develops. Starting in the 1970s, meteorologists noted that some tornadoes first developed in mid-levels of the mesocyclone, near its center, building downward. The signature of the nascent tornado could be identified aloft on Doppler radar up to 15–20 minutes prior to touchdown. A theory called the dynamic pipe effect explains how these tornadoes build downward. Figures 11.20 and 11.21 illustrate the progressive, downward growth of the tornado vortex, as cyclostrophic balance becomes established at progressively lower and lower levels. As each new layer of spinning pipe is added to the lower end, the pipe extends down to the surface. Then surface friction establishes strong inflow of air into the tornado's base.

But more recently, tornado field programs revealed that not all tornadoes develop in this top-down manner. Many form from

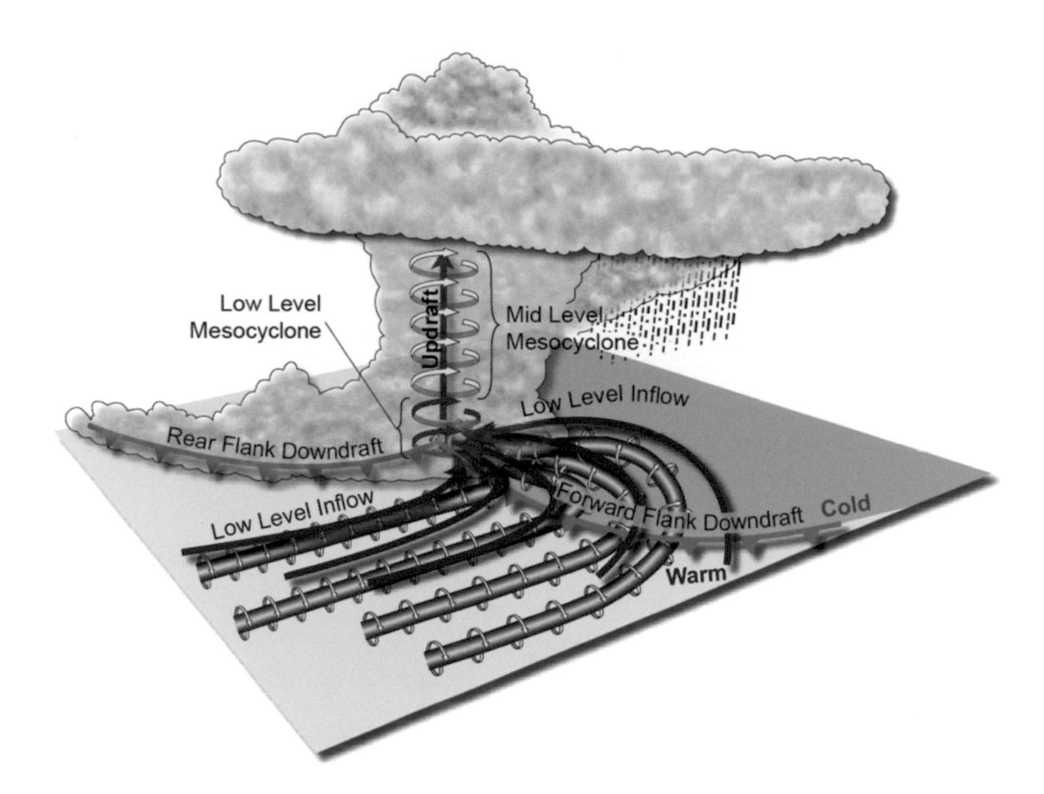

Figure 11.17 Formation of the low-level mesocyclone. This illustration applies the principles in **Figure 11.16** to a 3D supercell. Notice how the horizontal vortex tubes are drawn into the mesocyclone and raised into a vertical position. The strong updraft at the core of the mesocyclone strengthens the vertical vortex tube. (Adapted from Klemp, 1987.)

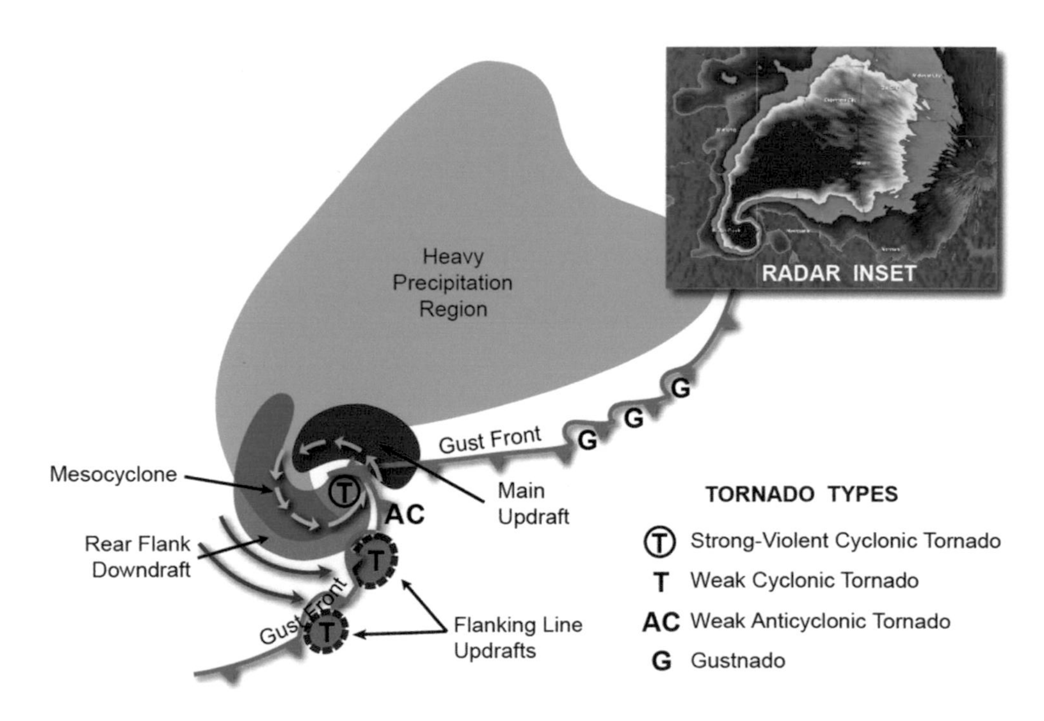

Figure 11.18 Divided, low-level mesocyclone. Within the mesocyclone are strong areas of updrafts and downdrafts that produce violent shearing effects near the center. This is the likely location for the birth of a strong or violent supercell tornado (indicated by the letter "T" inside the circle). (Adapted from Lemon and Doswell, 1979.)

Figure 11.19 Major supercell wind currents giving rise to tornado formation. The storm's rear flank downdraft (blue arrows) wrap around the tightly spiraling mesocyclone updraft (red arrows), giving rise to very strong wind shears.

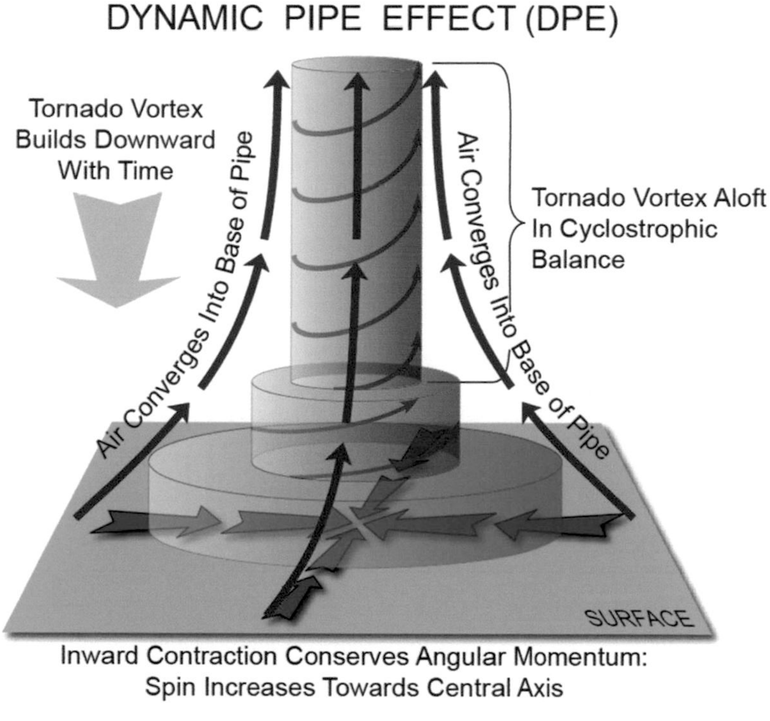

Figure 11.20 How a tornado aloft can progressively extend downward to the surface. As warm air is drawn in at the surface and is caught up in the strong vertical updraft found at the center of the tornado aloft, the flow spirals tighter, drawing the tornado downward toward the surface.

the bottom up, starting very near the surface. More and more data from detailed field studies using portable Doppler radars support the "ground-up" formation theory. So meteorologists have sought a mechanism by which intense vertical spin can be created at the surface. A basic principle, first recognized in the mid-1980s, requires a downdraft for this type of tornadogenesis. As Figure 11.22 illustrates, when horizontal vortex tubes are tipped into the vertical by an updraft, vertical spin cannot develop at the surface because the updraft transports the spinning tube upward, away from the ground (top panel, Times 1–3). However, if a downdraft does the tipping, horizontal spin is transported toward the ground (bottom panel, Times

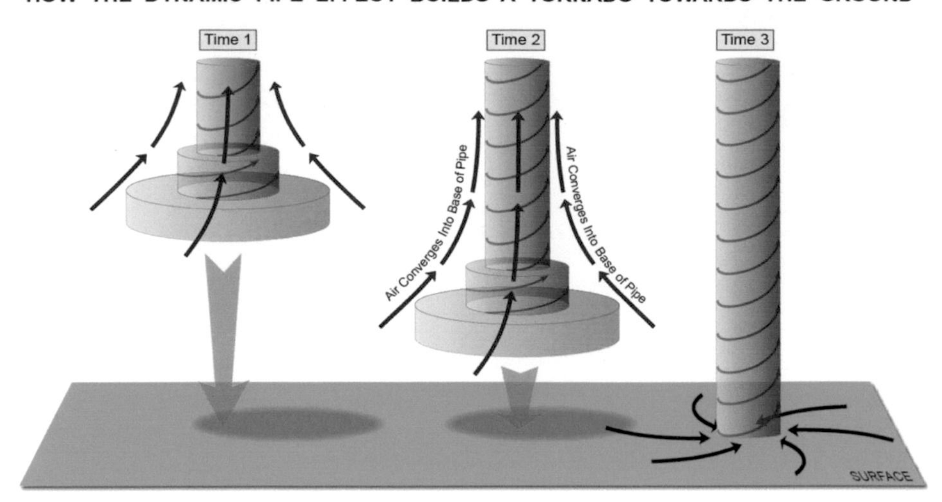

Figure 11.21 Stepwise progression of a tornado being drawn downward toward the surface. This is a hypothesized manner in which concentrated and self-amplifying spin in the tornado column descends to ground level.

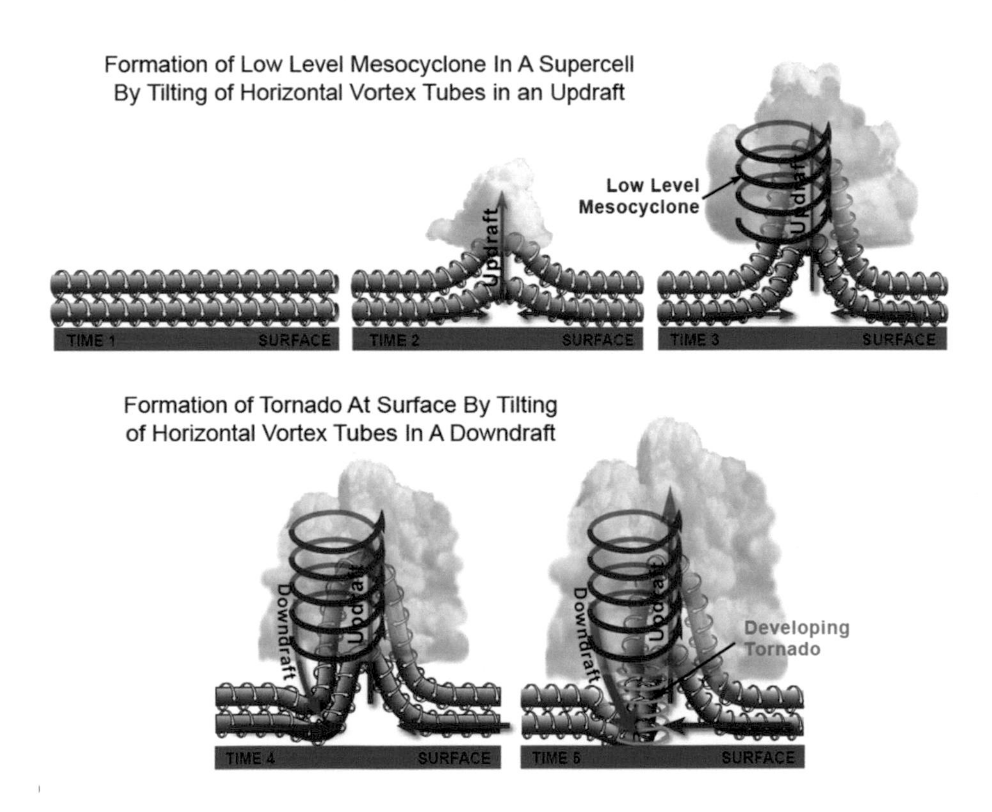

Figure 11.22 Stepwise progression of horizontal vortex tubes being drawn upward into a vertical position. In Times 4 and 5, a vigorous downdraft tilts horizontal spin to the vertical and downward toward the surface. Additionally, stretching of the vortex in the adjacent updraft accelerates the wind speeds. (Adapted from Markowski and Richardson, 2010.)

4 and 5). The rear flank downdraft, which is part of the divided mesocyclone, provides the needed mechanism. Then, as newly tilted vortex tubes are drawn into the adjacent updraft, they are stretched vertically. The stretching is critical, as it contracts these tubes. Conservation of angular momentum requires that they spin up rapidly to tornadic intensity, as Figure 11.23 shows.

Computer simulations and observations both show that air enters the tornado from the supercell's downdraft region. It is

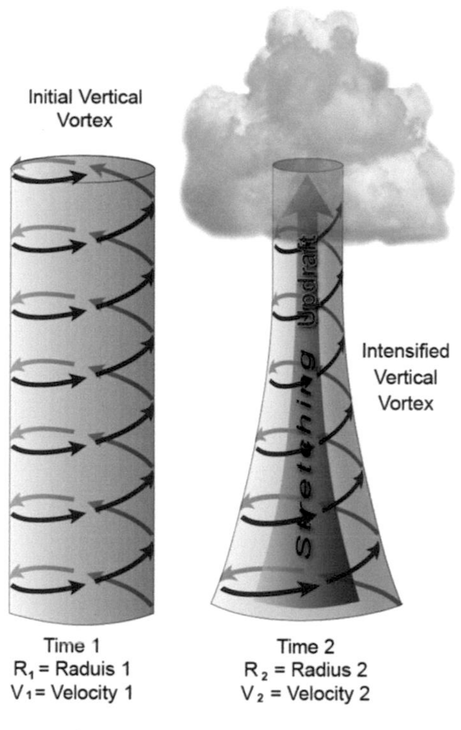

Figure 11.23 Final spin-up of the tornado. Stretching the vortex tube causes the wind speed to accelerate through the conservation of angular momentum.

somewhat ironic that tornadogenesis occurs when the supercell is transitioning from updraft-dominant to downdraft-dominant, which often marks the beginning of the storm's demise.

Ultimately, tornadogenesis involving the downdraft is a self-limiting process. The downdraft air, being relatively cool and dense, wraps around the mesocyclone, much in the way that a cold front eventually occludes an extratropical cyclone. After about 20–30 minutes, the supercell updraft is cut off from its source of warm, buoyant inflow. As the updraft weakens, so does the mesocyclone. The tornado's low-pressure core begins to fill with air, and it dissipates.

In short, many scientists believe that the rear flank downdraft plays a critical role in the formation of strong and violent tornadoes. Studies demonstrate that the majority of high-end, supercell-spawned tornadoes (strong-violent) initiate deep within the rear flank downdraft. Periodic surges in the rear flank downdraft – generated by heavy, descending rain cores – may also be important. Backed up by powerful computer simulations of supercell dynamics, other scientists believe the source of a tornado's spin derives from horizontal vortex tubes forming along the forward flank downdraft; these tubes get drawn into the low-level mesocyclone, tilted and stretched.

Forecasters cannot yet discriminate supercells that become tornadic from those that do not. Supercells often do not form, mature, and decay in isolation, but frequently interact with other storm cells. As they move, supercells pass across localized gradients in air stability and wind shear, encountering various air boundaries. All of these processes influence tornadogenesis in both constructive and destructive ways, but we do not yet have the technology to measure these variations.

Finally, it has been suggested that tornadogenesis is a fragile process, requiring an optimum set of conditions. For instance, the air in the rear flank downdraft must not be too cold. The temperature of the downdraft air is also critical: too cold, and the dense air resists upward lifting and stretching inside the updraft. All of these conditions help explain the relatively scarcity of tornadoes as a severe local weather phenomenon.

Digging Deeper: A Possible Mechanism for Tornadogenesis – The Rear Flank Downdraft

A number of recent scientific investigations of tornadoes involving portable Doppler radars and computer simulations suggest that one possible route to supercell tornado formation lies in the evolution of the storm's rear flank downdraft. As the downdraft descends and begins to wrap around the low-level mesocyclone, it generates one or more horizontal rings or tubes of rapidly spinning air, very close to the ground. One of these is shown in Figure 11.24(a). If the vortex tube advances beneath the end of the mesocyclone, inflowing updraft air rapidly draws the northern vortex tube upward, reorienting it into the vertical (Figure 11.24[b]). Now the ingested tube is stretched, shrinking and intensifying it, and the concentrated, counterclockwise spin is added to the background spin provided by the mesocyclone. In the final stage (Figure 11.24[c]), the vortex tube has intensified into a tornado.

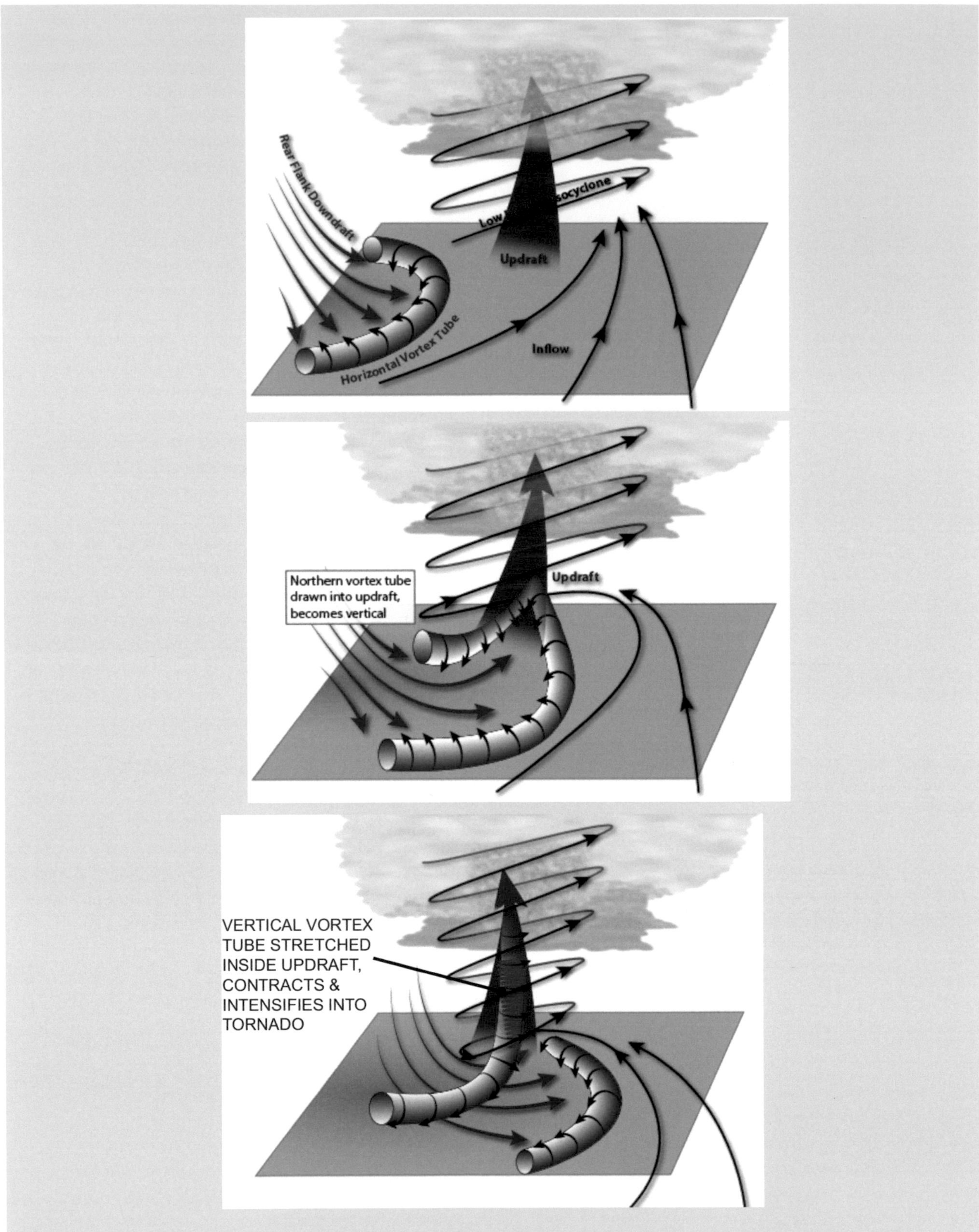

Figure 11.24 Hypothesized steps in tornado formation, from the "ground up," in such a manner that invokes dynamics of the rear flank downdraft. (Adapted from Markowski et al., 2008.)

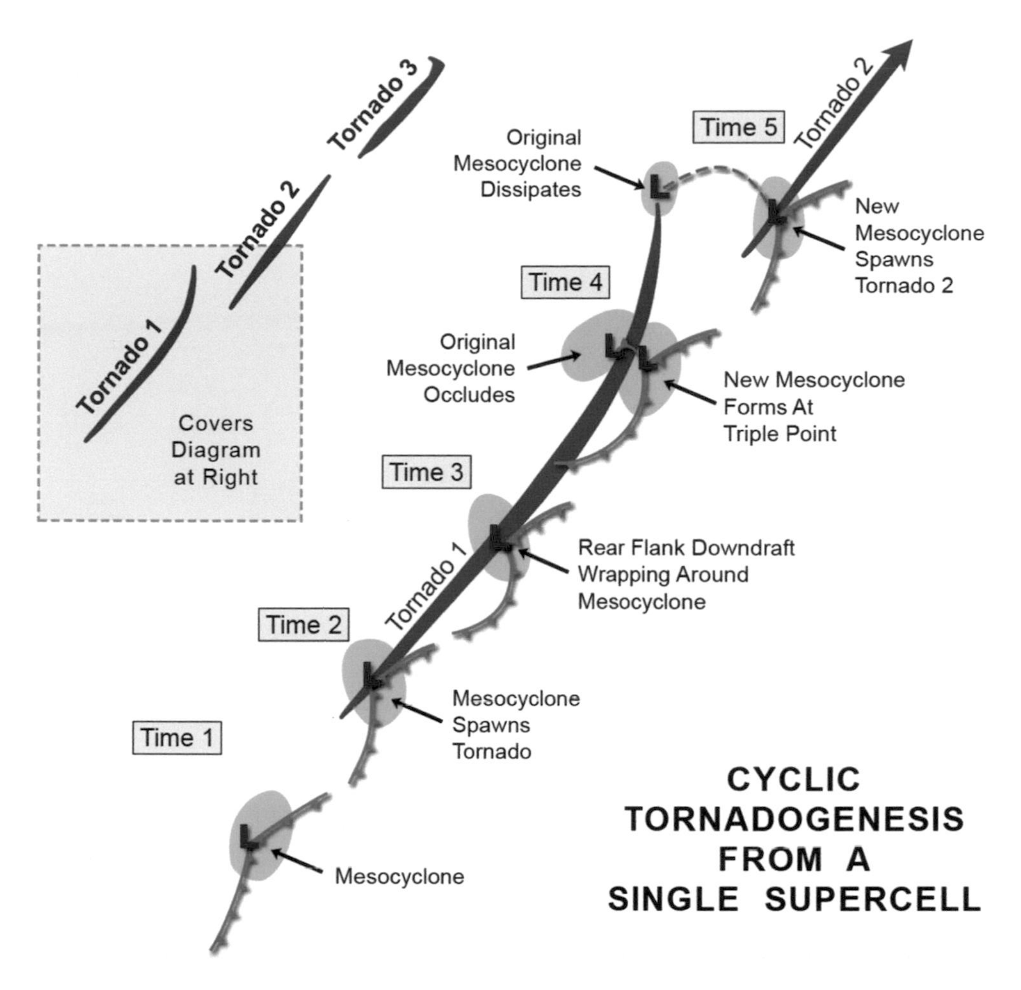

Figure 11.25 Cyclic tornadogenesis. A supercell has the potential to produce a series of successive tornadoes that can span many tens or even hundreds of kilometers. At Time 1, a mesocyclone forms. At Time 2, a tornado is spawned as the meso intensifies. At Time 3, the tornado intensifies. At Time 4, the strong rear flank downdraft occludes the updraft and thereby weakens the tornado. A new mesocyclone forms to the right of the old dissipating meso. At Time 5, a new tornado is generated in the second mesocyclone, and the cycle repeats itself.

Cyclic Tornadogenesis

Some supercells generate a sequence of tornadoes, one after another. Exceptionally long tornado swaths in fact may be created by a succession of tornadoes, with the demise of one tornado overlapping with the genesis of another close by.

Supercells can be very long-lived, producing severe weather for many hours along a track that extends hundreds of kilometers. Long-lived supercells often produce multiple tornadoes, in cycles that last 30 minutes to an hour. The succession (serial generation) of tornadoes can sometimes leave a long path of damage that is mistakenly attributed to a single, long-lived tornado.

Figure 11.25 illustrates the processes of cyclic tornadogenesis, in which a series of tornadoes are spawned. In our example, the same supercell tracking from southwest to northeast produces a series of three tornadoes. There is a brief gap in tornado path coverage between successive tornadoes. At Time 1, the entire supercell is not illustrated, just the mesocyclone and associated small-scale fronts or air mass boundaries associated with the rear flank and forward flank downdrafts; the larger region of heavy precipitation (rain and hail) has been omitted for clarity. By Time 2, the first tornado forms within the mesocyclone. This tornado intensifies and widens as it tracks northeast. At Time 3, the tornado has matured, but the rear flank downdraft begins wrapping around the mesocyclone. At Time 4, the downdraft occludes the updraft, and the tornado weakens. However, intense uplift of air along the leading edge of the rear flank downdraft leads to formation of a new low-level mesocyclone, to the east of the original meso. By Time 5, this new meso has intensified while the original meso dissipates. Henceforth, a new tornado develops within the second meso, and the cycle repeats itself.

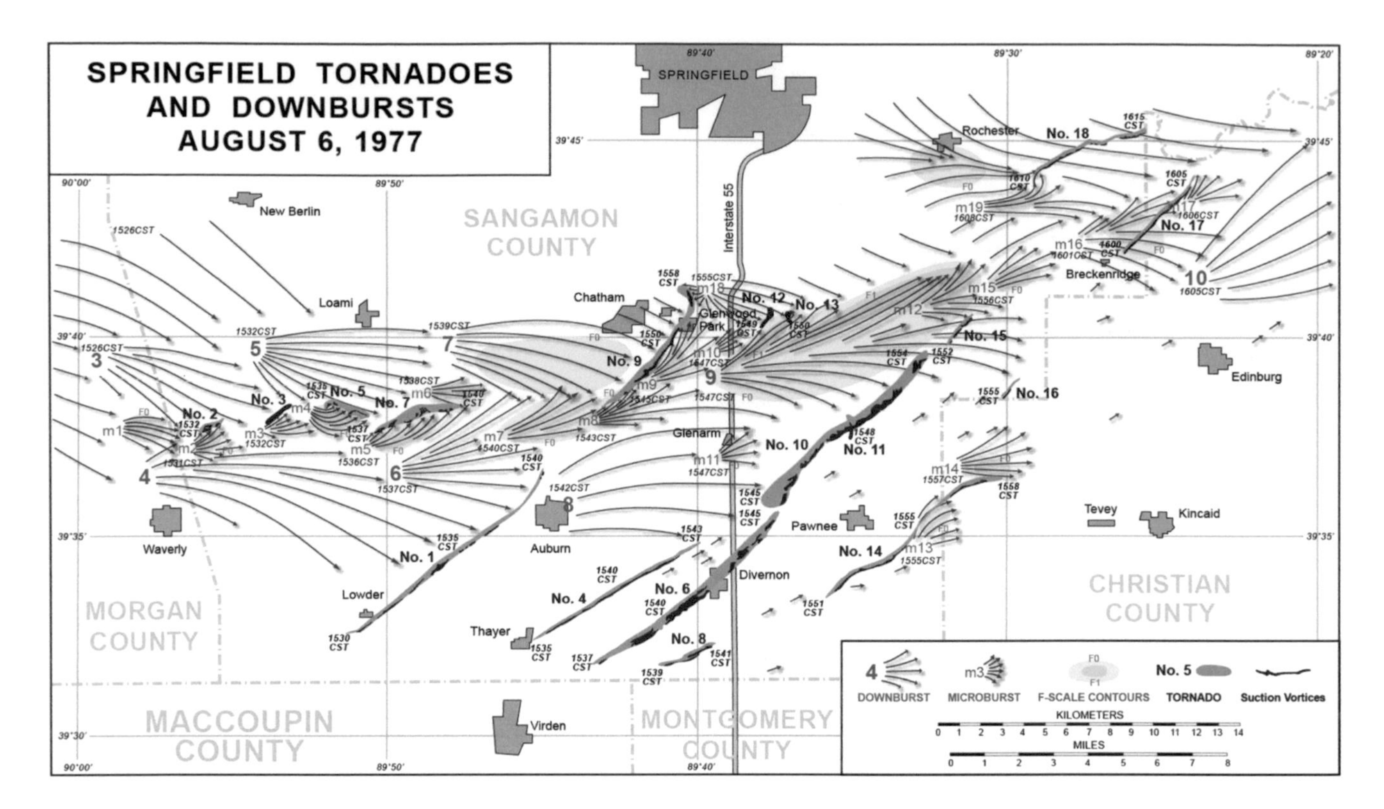

Figure 11.26 Damage from a bow echo, near Springfield, Illinois, August 1977. This bow echo traveled from west to east. The post-damage survey shows how a single bow echo storm complex can generate numerous tornadoes and a series of strong downbursts. (Adapted from Forbes and Wakimoto, 1983.)

Tornadoes Produced by Bow Echoes

Recall that a bow echo is a fast-moving, bow- or arc-shaped line of intense thunderstorms that produces devastating, localized streaks of wind damage called downbursts. Bow echoes also generate tornadoes. Figure 11.26 shows a post-storm damage survey of a bow echo that traveled from west to east near Springfield, Illinois, in August 1977. The burst-like streaks of wind damage are shown by the blue patches and arrows; each fan-shaped region is an individual downburst. Downbursts contain a straight-line or divergent wind that is nonrotary in nature, and the winds can approach or exceed 100 kts (115 MPH), the same intensity as weak tornadoes. Microbursts are small versions of downbursts. In the figure, note that this bow echo generated a series of 10 downbursts and 17 smaller, embedded microbursts. It is not uncommon for a single bow echo to generate a mixture of downbursts and tornadoes.

The damage survey reveals narrow, linear swaths (colored red) marking the tracks of tornadoes. This bow echo produced a total of 18 tornadoes. Many of the tornadoes occurred south of the main downburst corridor (numbers 1, 4, 6, 8, 10, 11, 14, 15, and 16) and were likely generated by small mesocyclones (weaker and shorter-lived than a supercell meso) along the forward edge of the bow echo. The remainder developed within and along the axis of downburst activity, particularly at the interface of downbursts and microbursts. These tornadoes were smaller, weaker, and shorter-lived than the others.

Figure 11.27 shows where tornadoes commonly develop within bow echoes. In this plan view, green shading indicates the region of heavy rain, and the most intense convective cells are located along the storm's leading edge. The system is moving quickly toward the east-southeast. The heavy blue arrows denote the fan-like sweep of downdraft air flowing through the bow echo, creating a blast of strong, rain-cooled wind at the surface. The leading edge of this massive downdraft is the gust front. Behind the gust front are small pockets of intense wind; these are the individual downbursts and microbursts. The small tornadoes develop from very small and shallow vortices, smaller than a supercell's mesocyclone, along the bow echo's leading edge. These tornadoes are indicated by bold "Ts." Their hypothesized formation involves very strong wind shears and tilting of vortex tubes but from mechanisms that may differ from the formation of supercell tornadoes.

Landspouts

Waterspouts are weak vortices, pendant from convective clouds, over warm bodies of water. The clouds need not be

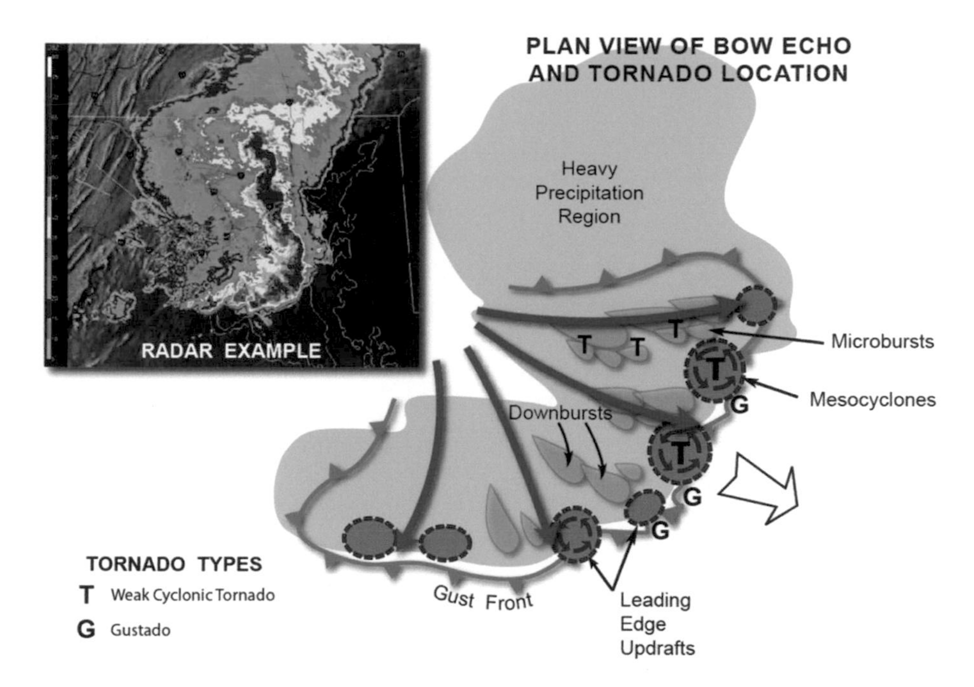

Figure 11.27 Bow echo with embedded tornadoes and downbursts. Bow echoes are often associated with tornadoes and downbursts, which are usually fast-moving features. Cold downdrafts of air spread outward toward the leading edge of the gust front. Downbursts and microbursts are embedded behind the gust front. Tornadoes can form along the edge of the bow echo, where they are embedded in small mesocyclones called misocyclones.

Figure 11.28 Landspouts (weak tornadoes) lack a well-defined funnel. While landspouts can form most anywhere, they are particularly common along the Front Range in the Boulder–Denver region of Colorado, as shown here. (National Severe Storms Laboratory.)

thunderstorms. Storm chasers along the Great Plains have long recognized a weak, short-lived type of tornado that develops from the base of towering cumulus clouds, some of which lack lightning and thunder. Figure 11.28 shows an example. Here we see two of these vortices lofting soil and dust in Washington State. These tornadoes lack well-defined funnel clouds (the air beneath the cloud base is too dry to support condensation), and there is no visible wall cloud, rainfall, or lightning. Often, a cluster or line of these tornadoes develops simultaneously, spaced a few kilometers apart, from a row of towering cumulus clouds. They are most common just east of the Colorado Rockies, in the Boulder–Denver area, along the I-25 corridor. There, a large, terrain-induced wind circulation called the Denver Cyclone develops during the warm season. These tornadoes are called landspouts because they are thought to develop in a similar manner to waterspouts.

Figure 11.29 shows a plan view of typical meteorological features that generate landspouts. The process starts with a wind-shift zone across which horizontal wind shear is present. Air on the southeast side of the boundary has a southerly component, while air on the west side has a northerly component. The sense of horizontal shear imparts small regions of counterclockwise spin along the boundary. These individual regions, often less than 1–2 km (0.6–1.2 miles) across, are called misocyclones, and they lie very close to the ground. When isolated convective cells develop atop misocyclones, cloud updrafts stretch each misocyclone vortex into a tighter, faster spinning vortex, turning the misocyclone into a landspout. This process is identical to that illustrated in Figure 11.23, involving the conservation of angular momentum. Small pockets of heavy rain may fall from individual cloud cells (shown by the green regions).

Figure 11.30 illustrates the three-dimensional evolution of a landspout-bearing cloud line. Panel (a) shows the wind-shift zone, with opposing surface winds on either side. The leading edge of this boundary often becomes distorted into a series of wavelike undulations. Between the lobes and clefts, small

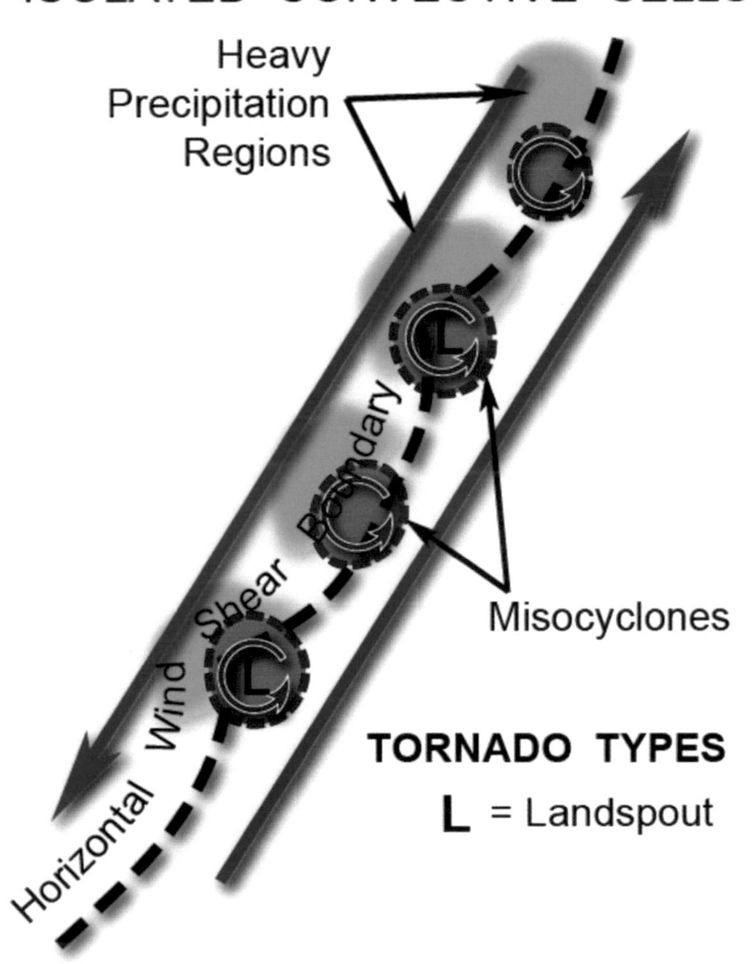

Figure 11.29 Formation of landspouts. Opposing horizontal wind shear forms misocyclones. When isolated convective cells stretch these small vertical vortices, landspouts are formed.

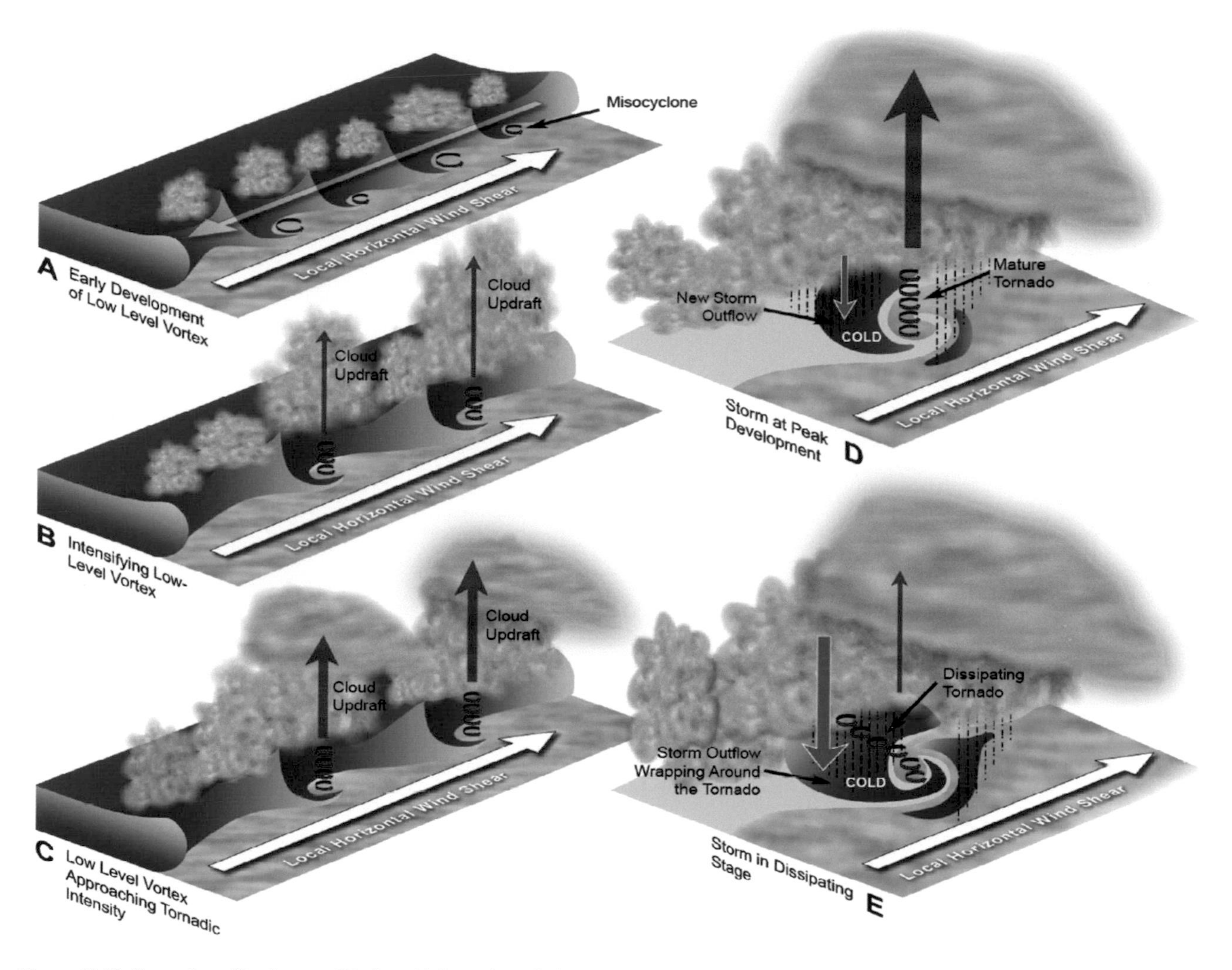

Figure 11.30 Formation of landspouts: 3D view. (a) Opposing winds cause undulations and horizontal shear along the cold/warm air boundary, setting in motion the rotating cylinders of air. (b) Local cumulus cells stretch the vortex cylinders. (c) The vortices continue to be stretched, and their wind speeds accelerate. (d) Cold downdrafts begin to encircle the vortices and cut off the energy inflow. (e) The landspout in its final phase of dissipation. (Adapted from Lee and Wilhelmson, 1997.)

counterclockwise vortices develop. These are the misocyclones. In Panel (b), cumulus clouds containing updrafts begin forming along the zone. Individual misocyclones often merge and coalesce, intensifying, and cloud updrafts amplify spin by vortex stretching. In Panel (c), as convective clouds grow upward, misocyclones spin up to near-tornadic intensity. Further merger or consolidation of vortices may occur at this time. Panel (d) illustrates a mature convective cloud, which deposits a heavy rain shower west of the updraft region. A cool downdraft accompanies the shower. When the downdraft reaches the surface, it wraps around the tornado. The convergence of dense air spiraling inward provides a final, brief period of vortex contraction, reaching maximum intensity. Finally, in Panel (e), downdraft dominates the cloud. It completely wraps around the tornado, often distorting it into a horizontal rope-like shape. This is the tornado's dissipating phase. The total process from updraft formation to development and dissipation of the tornado is often no longer than 15–20 minutes.

Gustnadoes

A final tornado variant is the gustnado, shown in Figure 11.31. A gustnado is a very brief, weak tornado that sometimes develops along gust fronts associated with thunderstorms and heavy rain showers. Often, all that is visible is a tiny dust whirl at the surface, along the edge of the storm cloud, in front of an opaque rain curtain. Sometimes several whirls are observed along a single gust front. There is often no funnel

Figure 11.31 Gustnado. A gustnado is a very weak, short-lived tornado that originates along a gust front. Lacking significant pressure reduction for condensation and thus a funnel cloud, gustnadoes are often marked only by the appearance of a dust whirl. (Jefferson County Kansas Emergency Management.)

cloud. Gustnadoes represent the weakest end of the tornado spectrum. The maximum swirling wind speeds in most gustnadoes usually lies below 50–60 kts (58–69 MPH). Many thousands of gustnadoes probably occur every year in the United States, and many are never observed. Meteorologists disagree on whether to classify them as tornadoes, and they are not officially logged in the U.S. tornado database. They form in a manner analogous to Panel (a) in Figure 11.30. Gustnadoes form within the lobes and clefts of the gust front when part of the outflow becomes distorted or surges ahead. The rain-cooled, chilly downdraft air at the gust front is dense, hugging the ground; it resists vertical lifting and stretching if updrafts are present above. Therefore, gustnadoes rarely amplify to strong tornadic intensity.

Amazing Storms: The 2013 El Reno, Oklahoma, "Mega Tornado"

The National Weather Service has coined this enormous and extremely violent tornado, which struck close to Oklahoma City, Oklahoma, on May 31, 2013, "the most dangerous tornado in storm observing history." It was given this description because, ironically, the tornado did not damage much in the way of structures but rather killed four storm chasers in a single stroke. These direct deaths by the tornado wind itself (and not other factors, such as losing control of the vehicle on rain-slicked roads) are believed to be the first in the many decades of tornado chasing.

The tornado, rated EF5 based on Doppler estimates, is also believed to be the widest tornado on record, having swelled rapidly at the end of its 40 minute long track to 42 km (26 miles) wide. Mobile Doppler radars assessed peak winds of 256 kts (295 MPH) near the ground, making it the second strongest tornado measured by Doppler radar.

Detailed reconstruction of the tornado track from radar, video, and storm damage analysis reveals an oddly shaped and unusually wide swath of wind damage, shown in Figure 11.32. While the overall track length was only 26.1 km (16.2 miles), the unusually wide wind envelope is apparent toward the end of the track. Following what is believed to be the center of the tornadic vortex (dashed black curve), note that the tornado makes several abrupt turns (i.e., 1815 CDT and 1830 CDT) and even loops (1821 CDT and 1825 CDT). As the tornado expanded, numerous sub-vortices (smaller, more intense tornadoes embedded within the main tornadic vortex) began to develop on the southern flank of the parent vortex. Sub-vortices (also called suction vortices), as you recall, are common features of large, intense tornadoes. A peculiar anticyclonic tornado developed at 1820 CDT, also on the southern quadrant of the main tornado. The strongest damage indicated in this diagram is EF3 (red shaded regions). As described in Chapter 12, the EF intensity scale is based on wind damage, not measured wind speeds. This tornado covered sparsely inhabited terrain. While the Doppler radar measured EF5 level winds, it was determined that those peak winds occurred within sub-vortices, and the vortices did not hit engineered structures.

One chase vehicle, belonging to The Weather Channel, was seriously damaged by the tornado, injuring meteorologist Mike Bettes. Another chase vehicle was driven by a team of meteorologists, including Tim Samaras, his son, and colleague Carl Young. All three died when their vehicle was rolled by the tornado's winds. A fourth fatality occurred in another chase vehicle, driven by Richard Henderson. Mobile Doppler analysis after the fact revealed that these vehicles were hit by sub-vortices.

Why were there so many fatalities, including those of veteran storm chasers? As the vortex grew in size and intensity, it became enshrouded in swirling curtains of heavy rain and dust. These conditions obscured the presence of the sub-vortices, which as times were clocked at orbiting the main vortex at a rate of 152 kts (175 MPH!). Escape routes were limited, and the rapid speed of the sub-vortices left precious few seconds of reaction time. Also, the outer edge of the parent tornado itself was nebulous to the eyes of trained chasers. This large tornado is probably best described as a type of "multiple vortex mesocyclone" (MVMC) – where the division between mesocyclone and embedded tornadoes becomes blurred. Was the visible edge of windblown rain and dust a downward extension of the mesocyclone's circulation to the ground (which sometimes can be safely penetrated) or the outer edge of a particularly wide tornado? All of these factors conspired to create a very unusual and hazardous situation for chasers that day, even those with decades of scientific experience.

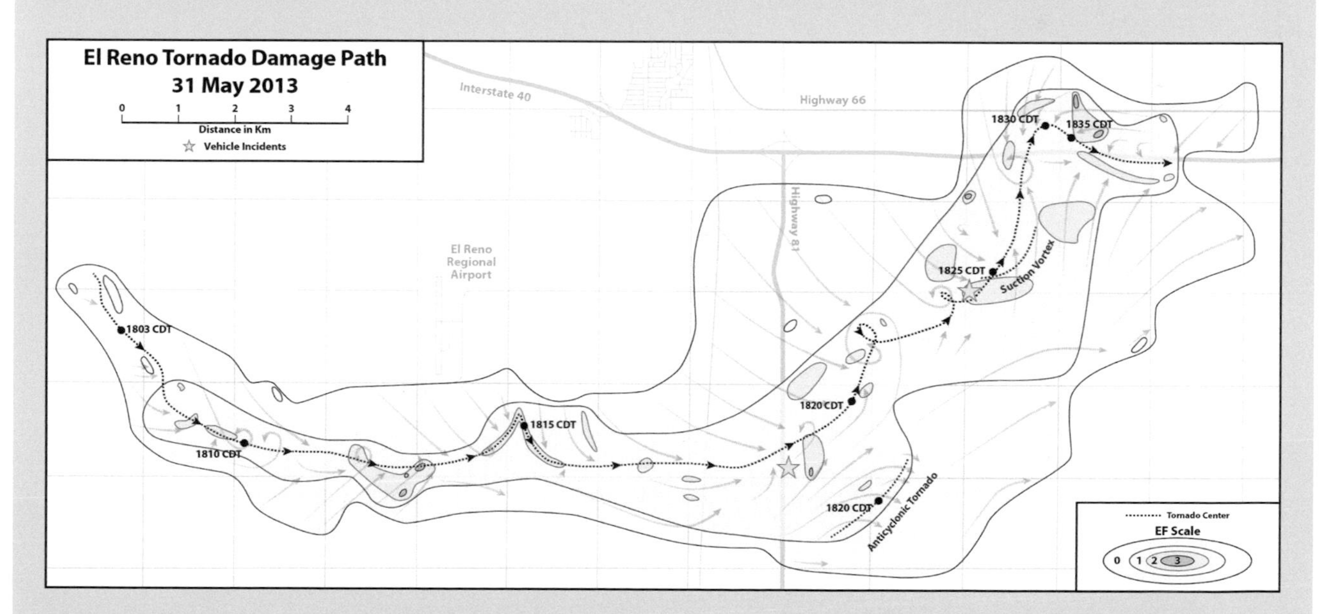

Figure 11.32 Schematic of the El Reno, Oklahoma, tornado. Track reconstruction shows contours of estimated wind intensity based on structural damage (shaded colors), center track of the tornado (dotted black curve), and direction of surface airflow (magenta shaded arrows). Yellow stars mark the location of motor vehicle casualties. (Adapted from Wakimoto et al., 2016.)

Summary

LO1 Identify the different types of tornado-producing storms and the characteristic intensity of tornadoes produced by each.

1. Tornadoes are very rare severe local storms; fewer than 1% of all thunderstorms during the year in the United States spawn a tornado.
2. Tornadoes can be broadly classified as weak (winds < 111 MPH), strong (111–165 MPH), and violent (winds between 166–200+ MPH). Severe supercell thunderstorms spawn most of the strong and violent tornadoes, while weaker tornadoes are generated both by supercells and by bow echoes. While violent tornadoes account for 1% of all U.S. annual tornadoes, they produce 70% of all tornado fatalities.
3. The tornado, which is a tubular, low-pressure vortex of wind in contact with the surface, develops a funnel cloud when humid air expands and cools adiabatically inside the tornado core, to the dew point temperature of the air.

LO2 Understand the physical properties of tornadoes and governing forces, including cyclostrophic balance, the conservation of angular momentum, vorticity, circulation, suction vortices, swirl, and asymmetric distribution of winds.

1. The swirling wind in a tornado vortex arises from cyclostrophic balance, in which the inward-directed pressure gradient force is balanced by the outward-directed centrifugal force.
2. Tornadoes behave like a dynamic suction pipe. Air can only enter the vortex tube at its lower end, at the surface. This generates a shallow layer of intense inflow or feeder winds converging in a spiral beneath the tornado.
3. Strong and violent tornadoes are embedded in a larger vortex contained inside the supercell, called the mesocyclone. Low pressure in the mesocyclone helps remove air that flows into the base of the tornado and supplies angular momentum that helps spin up the tornado.
4. Tornado formation and maintenance are predicated upon the conservation of angular momentum, which is the product of a tangential velocity, vortex radius, and air density. Tangential velocity is the speed of the swirling wind. As a mesocyclone contracts (decreases in radius), tangential winds must increase to maintain constant angular momentum. The base of large, strong tornadoes may devolve into numerous suction vortices; their very small radius requires that tangential velocity increase further, creating the fastest, localized winds inside the tornado.

LO3 Describe how Doppler radars have yielded new insights about tornado structure and evolution, including single and multi-vortex tornadoes.

1. Portable research Doppler radars have enabled scientists to study the 3D airflow inside both weak and violent tornadoes. In addition to the swirling wind, tornados also feature strong inflow of air into their base and a powerful updraft within the vortex ring. Large, strong/violent tornadoes contain rapidly sinking air in a clear "eye."
2. Tornadoes move along the ground, or translate, in a relatively straight path. This translation speed must be added to the tangential wind on the right side of the tornado track and subtracted from the left side. This creates an asymmetrical pattern of wind damage across the tornado. Much higher winds usually prevail on the right side of track, creating an uneven damage pattern as the tornado moves along the ground.
3. Anticyclonic (clockwise spinning tornadoes) do occur in the northern hemisphere, but they are few in number compared to cyclonic (counterclockwise) tornadoes. The direction of a tornado's spin is not determined by the sense of the Earth's rotation, as the Earth's spin represents only 1 part in 10,000 of a tornado's spin.
4. Suction vortices develop in strong/violent tornadoes at high swirl ratio (i.e., the tangential velocity greatly exceeds the updraft speed), at which point the vortex breaks down or devolves into several smaller sub-vortices. As a tornado translates, individual suction vortices may orbit the tornado's axis, leaving cycloidal marks or etchings in the damage swath.

LO4 Define wind load and explain how a tornado's winds create destructive aerodynamic forces that damage dwellings.

1. The wind load in a tornado is the pressure or force exerted by the wind on structures, which varies with the square of the tornado's wind speed. As wind flows over and around a dwelling, aerodynamic forces cause the roof to lift upward and the walls to move outward. Additionally, debris tossed at high velocity by the tornado's winds destroys walls, windows, and doors.
2. Mobile homes are easily damaged by tornado winds, because their thin walls are easily penetrated by debris, and their high surface area-to-weight ratio makes them susceptible to rolling.

LO5 Explain the core principles associated with tornado formation (tornadogenesis).

1. Tornadogenesis describes the process of tornado formation within thunderstorms. The majority of supercell thunderstorms do not produce tornadoes. Those that do form develop within the supercell's mesocyclone, as it interacts with a cool downdraft of air that wraps around the mesocyclone late in the supercell's lifetime.
2. Several dynamic principles help form tornadoes at the surface: (1) horizontal vortex tubes above the surface are tilted into the vertical by the downdraft; (2) once tilted by the downdraft, the vertical tubes are stretched in the adjacent updraft (divided mesocyclone); (3) by conservation of angular momentum, as a vortex tube stretches and its radius decreases, tangential winds increase to tornadic intensity.
3. Cyclic or serial tornadogenesis is the process by which a single supercell, lasting many hours, produces a succession of tornadoes.
4. Landspouts are a type of weak tornado that form in an analogous manner to waterspouts. Small vertical circulations called misocyclones develop close to the surface when horizontal wind shear along an air mass boundary spins up and consolidates. When the updraft of a growing storm cloud becomes colocated with a misocyclone, it is stretched vertically, intensifying to tornadic wind speed. Landspouts are common in High Plains thunderstorms, particularly along Colorado's Front Range.

References

Davies-Jones, R.P., 1986. Tornado dynamics. In *Thunderstorm Morphology and Dynamics*, E. Kessler, Ed., 2nd ed. Norman, OK: University of Oklahoma Press, 197–236.

Forbes, G.S. and R.M. Wakimoto, 1983. A concentrated outbreak of tornadoes, downbursts and microbursts, and implications regarding vortex classification. *Monthly Weather Review*, 111:220–235.

Fujita, T., 1977. Anticyclonic tornadoes. *Weatherwise*, 30:51–64. Washington, DC: Heldref Publications.

Fujita, T., 1981. Tornadoes and downbursts in the context of generalized planetary scales. *Journal of the Atmospheric Sciences*, 38:1511–1534.

Klemp, J.B., 1987. Dynamics of tornadic thunderstorms. *Annual Reviews of Fluid Mechanics*, 19:269–402.

Lee, B.D. and R.B. Wilhelmson, 1997. The numerical simulation of non-supercell tornadogenesis. Part II: Evolution of a family of tornadoes along a weak outflow boundary. *Journal of Atmospheric Sciences*, 54:2387–2415.

Lemon, L.R. and C.A. Doswell, 1979. Severe thunderstorm evolution and mesocyclone structure as related to tornadogenesis. *Monthly Weather Review*, 107:1184–1197.

Markowski, P., E. Rasmussen, J. Straka, R. Davies-Jones, Y. Richardson and R.J. Trapp, 2008. Vortex lines within low-level mesocyclones obtained from pseudo-dual-Doppler radar observations. *Monthly Weather Review*, 136:3513–3535.

Markowski, P. and Y. Richardson, 2010. *Mesoscale Meteorology in Mid Latitudes*. West Sussex, UK: Wiley-Blackwell.

Wakimoto, R.M., N.T. Atkins, K.M. Butler and H. Bluestein, 2016. Aerial damage survey of the 2013 El Reno tornado combined with mobile radar data. *Monthly Weather Review*, 144:1749–1776.

CHAPTER 12
Tornado Outbreaks, Detection, Warning, and Societal Response

Learning Objectives

1. Discuss how tornadoes are rated and ranked according to the Enhanced Fujita wind damage scale, and how meteorologists distinguish between tornado and downburst damage.
2. Discuss how annual tornado incidence varies geographically across the United States and also by time of day, recognizing that there is more than one U.S. "tornado alley."
3. Discuss how the reporting of tornadoes, tornado fatalities, and damage have changed over the past 100 years, including influences by U.S. population growth and short-term climate fluctuations and also short-term climate fluctuations including El Niño and La Niña.
4. Distinguish between a tornado outbreak, outbreak sequence, super outbreak, and tornado family. Describe the damage and fatality characteristics for two exceptionally intense tornado years (1974 and 2011).
5. Illustrate how tornadoes are predicted, beginning three days before a major outbreak, down to the time of actual tornado touchdown, and discuss how the effectiveness of the tornado warning system is evaluated.
6. Describe the ways that the broadcast media and ordinary citizens are using technology to improve tornado detection.

Introduction

In the previous chapter, we examined the various types of tornadoes, the large-scale environment in which they form, tornadogenesis, and the manner in which tornado winds inflict their damage. In this chapter, we focus on the societal dimensions of tornadoes. We start by examining how tornadoes are reported, surveyed, and rated. Next we explore the geographical, seasonal, and diurnal distribution of tornadoes, examining how U.S. property damage and mortality have changed over the years. We describe tornado outbreaks and outbreak sequences, analyzing whether certain climate cycles, such as El Niño, control the number, location, and intensity of tornado outbreaks. Finally, we discuss tornado watches and warnings, detailing how the emerging technologies may increase lead time for tornado forecasts and the accuracy of tornado path predictions.

Human vulnerability to tornadoes (and many other severe storms) is the combined effect of the tornado hazard (the place and time of tornadoes) and the societal risk. Risk includes people's actions that place them in harm's way. We examine this idea in light of the devastating 2011 tornado season, which shattered established records and ended a 40-year period of markedly reduced tornado fatalities. Did an increased hazard, a heightened risk, or both contribute to this devastating year?

U.S. Tornado Database: Surveying and Ranking Tornadoes

It is difficult to accurately gage the intensity of tornadic winds; most instruments designed to measure violent wind speeds do not survive the encounter. Here we discuss a tornado wind intensity scale based on the degree of damage, from which we infer the speed of a tornado's winds. We also describe how meteorologists distinguish among tornadic and other types of damaging thunderstorm wind.

The Fujita (F) and Enhanced Fujita (EF) Scales

Unarguably, the most prolific tornado researcher was Dr. Theodore Fujita (1920–1998) of the University of Chicago. Based on his exhaustive tornado-damage surveys, he introduced a tornado intensity rating scale in 1971. This scale assigned wind speed values based on extent of damage to structures. Fujita devised the scale in collaboration with Allen Pearson, head of the

DOI: 10.4324/9781003344988-15

Table 12.1 Tornado Wind Speed Ratings

Fujita Rating (1971–2007)	Wind Speed 56–74	Enhanced Fujita Rating (post-2007)	Wind Speed (mph)	Subjective Category
F0	35–63 kts; 40–72 MPH	EF0	56–74 kts; 65–85 MPH	Weak
F1	64–97 kts; 73–112 MPH	EF1	75–96 kts; 86–110 MPH	Weak
F2	98–136 kts; 113–157 MPH	EF2	97–117 kts; 111–135 MPH	Strong
F3	137–179 kts; 158–206 MPH	EF3	118–143 kts; 136–165 MPH	Strong
F4	180–226 kts; 207–260 MPH	EF4	144–174 kts; 166–200 MPH	Violent
F5	181–276 kts; 261–318 MPH	EF5	175+ kts; 201+ MPH	Violent

Notes: MPH = miles per hour; F = Fujita; EF = Enhanced Fujita.

National Severe Storms Laboratory. Table 12.1 shows the Fujita (F) scale, which includes six categories, F0–F5. The F-scale is a tornado damage scale; the corresponding wind speeds are educated guesses and not readily verified. Indeed, verifying wind speeds is quite difficult in practice. Anemometers (mechanical devices designed to measure wind speed) cannot survive the direct impact of a strong or violent tornado, are often not in the right place (think of how small a tornado is), and attempting to place one in the direct path of a tornado is dangerous.

In the past decade, engineers and meteorologists concluded that the original F-scale wind speeds are overestimates, particularly at F2 and higher. Also, a particular wind speed can cause different levels of damage, due to differences in building engineering and construction. In 2007, a revised or enhanced Fujita (EF) scale (shown in Table 12.1) was put into practice. EF wind estimates are based on a more precise interpretation of U.S. construction practices, including 28 categories of indicators that assess damage to barns, manufactured homes, high rises, municipal buildings, towers, and trees. Note that strong tornadoes are those rated F2 and F3, while violent tornadoes are rated F4 or F5. Typically, a significant tornado is one that is rated F2 or stronger. Presently, the EF scale recognizes no upper bound to tornado wind speeds, although the highest speed ever directly measured, using portable Doppler radar, is 244–279 kts (281–321 MPH). This wind was measured 100 m (330 feet) above the surface in a 1999 EF5 Oklahoma tornado (note the range of uncertainty in the wind estimate). It is likely that the surface wind speed at the very surface was weaker due to friction.

In this chapter, we refer to both the Fujita (F) and Enhanced Fujita (EF) Scales, depending on the era in which the data were collected and analyzed.

The most common tornadoes, accounting for three-quarters of all tornadoes, are the weak EF0 and EF1 types. Figure 12.1 shows the inverse relationship between tornado intensity and frequency. We have repeatedly encountered this frequency distribution – called the law of rare events – in our survey of meteorological hazards, including hurricanes, flash floods, and so on. Only about one EF5 and ten EF4 tornadoes occur in the United States each year. As Figure 12.2 reveals, the great majority of tornado fatalities arise from violent tornadoes, which account for less than 1% of all tornadoes, and fewer than 3% of all tornadoes cause fatalities.

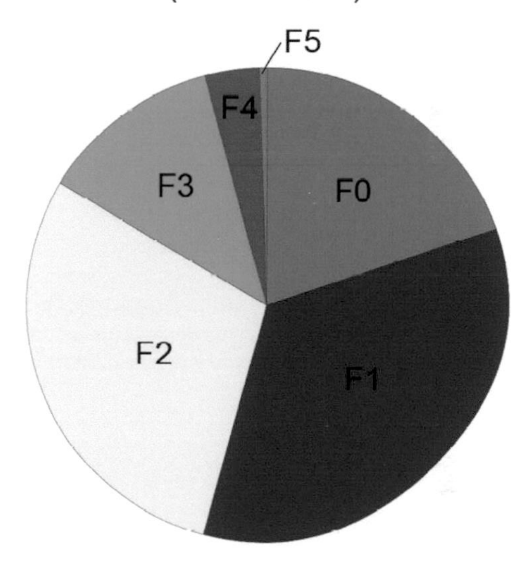

Figure 12.1 Inverse relationship between tornado frequency and F-scale rating. More than half of all tornadoes are rated EF0 or EF1.

Wind-Damage Surveys: Detective Work After the Storm

As Figure 12.3 shows, the damage from a strong or violent tornado is often catastrophic, with complete destruction of permanent homes and other engineered structures. Eyewitnesses

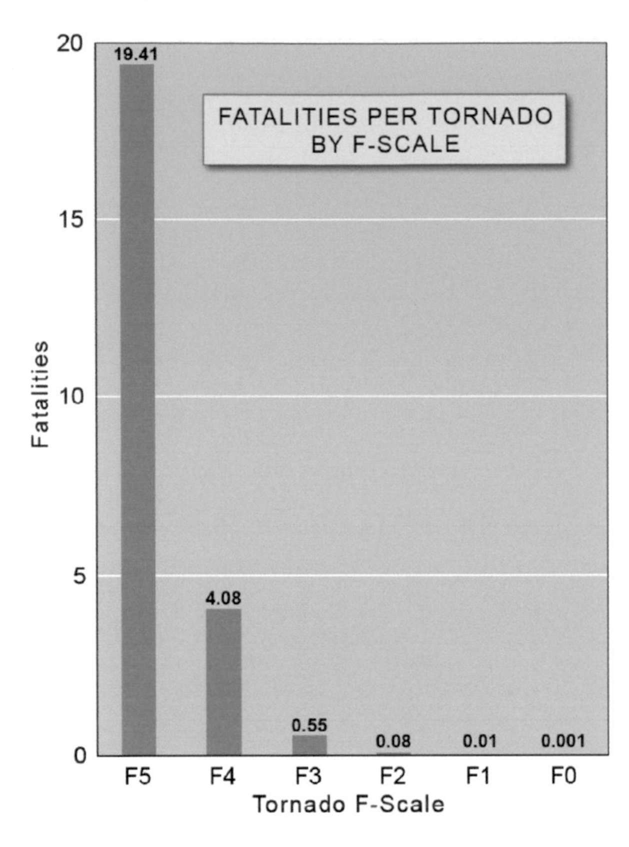

Figure 12.2 The deadly nature of EF4 and EF5 tornadoes. On average there are nearly 20 deaths for every F5 tornado. At the other end of the scale, there is on average one death for every thousand F0 tornadoes.

often record video footage of a massive tornado and post it to YouTube. These tornadoes instantly generate regional and perhaps national media coverage.

Severe thunderstorms create a variety of destructive winds, including powerful, localized blasts of air called downbursts with wind speeds that can approach 130 kts (150 MPH). In a downburst, the winds are straight-line, as opposed to rotary. Distinguishing between a tornado and a downburst is important scientifically, in order to maintain an accurate national database documenting all tornadoes. For this reason, the National Weather Service (NWS) conducts a post-storm damage survey to determine what caused the storm damage. These surveys may include interviews with eyewitnesses, ground surveys, video footage, and Doppler radar images. Most importantly, the NWS collects photographs via an aerial survey (using a light aircraft, drone, or helicopter). These post-damage surveys characterize tornadoes according to severity, path width, path length, and wind intensity.

From the ground, tornado and downburst damage can appear very similar. Often, a walk through a post-storm neighborhood reveals a jumble of fallen trees, limbs and utility lines. Unless viewed from the air, it is difficult to determine the mode of damage – whether by tornadoes or straight-line winds. Figure 12.4 summarizes the differences between tornadoes and downbursts. The tornado is an

Figure 12.3 Extreme damage to well-built frame structures from the violent (EF5) Hackleburg, Mississippi, tornado of April 27, 2011. (NWS.)

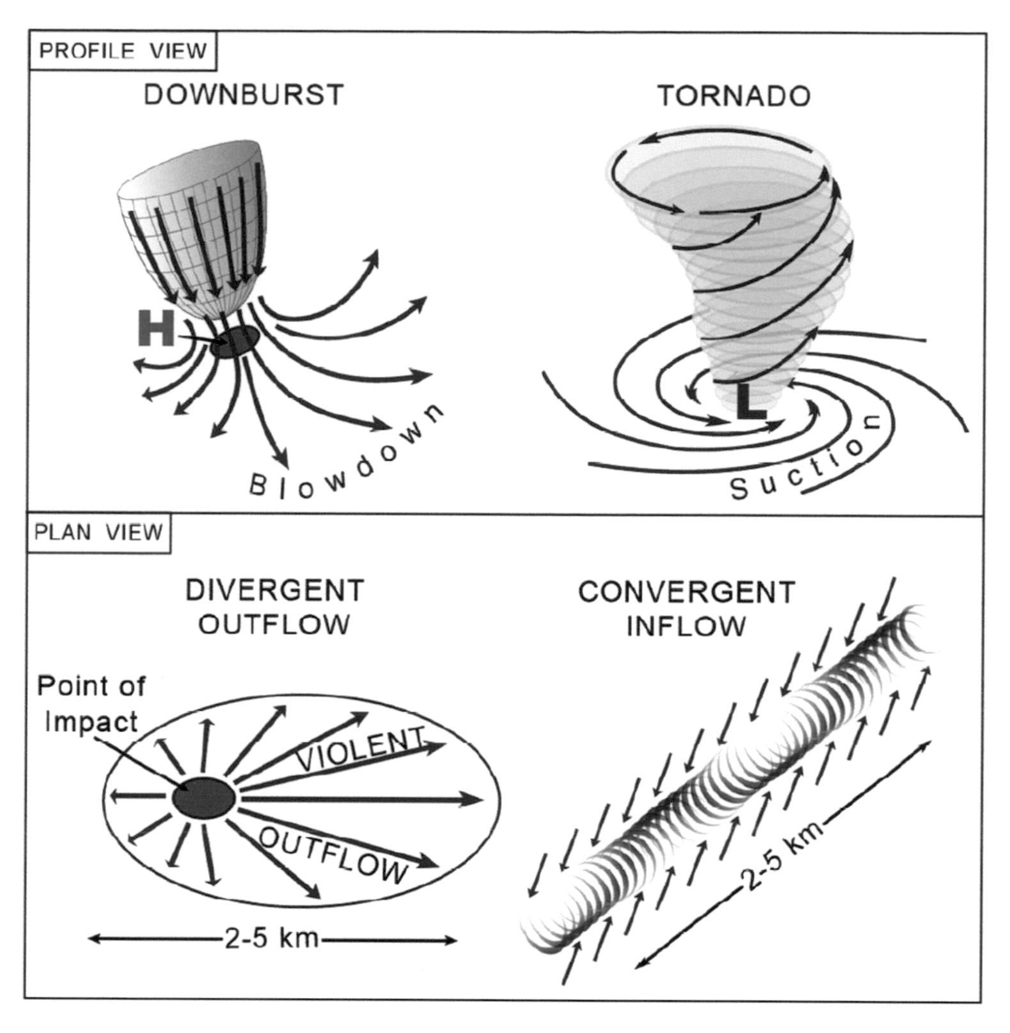

Figure 12.4 Difference in wind patterns associated with downbursts and tornadoes. The plan view depicts a diverging trajectory of winds resulting from downbursts. A tornado produces a pattern of swirls and counter winds on either side of the tornado.

elongated, narrow vortex of intense low pressure (L) in contact with the ground. Its suction generates a powerful, spiral inflow of wind just above the surface. As air spins around the vortex, it also rises upward. From the air, the tornado leaves a narrow, linear damage track. Objects such as trees lean inward toward the axis of the path, converging inward toward a common center. A downburst is a rapidly descending mass of air. When the sinking current strikes the ground, it creates an outward-directed blast, like aiming a garden hose at the ground. The point of impact generates a small region of intense high pressure (H) from which the outflow emanates. The downburst generates a circular, elliptical, or fan-shaped pattern of debris outflow. Felled trees lean outward from the impact point (Figure 12.5). When you think downburst, imagine a blast of air striking the surface; when you think of a tornado, visualize the end of a vacuum hose moving above the surface.

Aerial surveys are especially important when a single thunderstorm produces a mixture of downburst and tornado damage. Figure 12.6 illustrates just how complicated the process gets. This survey was done in the wake of a violent F5 tornado and accompanying downburst winds from a severe thunderstorm over Iowa. Damage generated by tornadoes is shown in red, with darker red shades corresponding to higher F-scale categories. Blue shaded regions denote pockets of downburst damage. Downbursts are also rated using the F-scale, and darker blue shades indicate higher F-scale wind categories.

In Figure 12.6, a supercell produced multiple tornadoes and downbursts. The storm tracked from southwest to northeast across Boone and Story Counties, Iowa. The total swath of wind damage extended nearly 40 km (25 miles), commencing at 1425 Central Standard Time (CST), broadening toward the northeast. The storm's mesocyclone first generated a sequence of tornadoes. They included

Figure 12.5 Classic downburst damage in a forested region. In this aerial view, arrows align with the downed trees. A divergent wind produces a pattern on the ground that radiates out from the point of impact. (Adapted from NWS.)

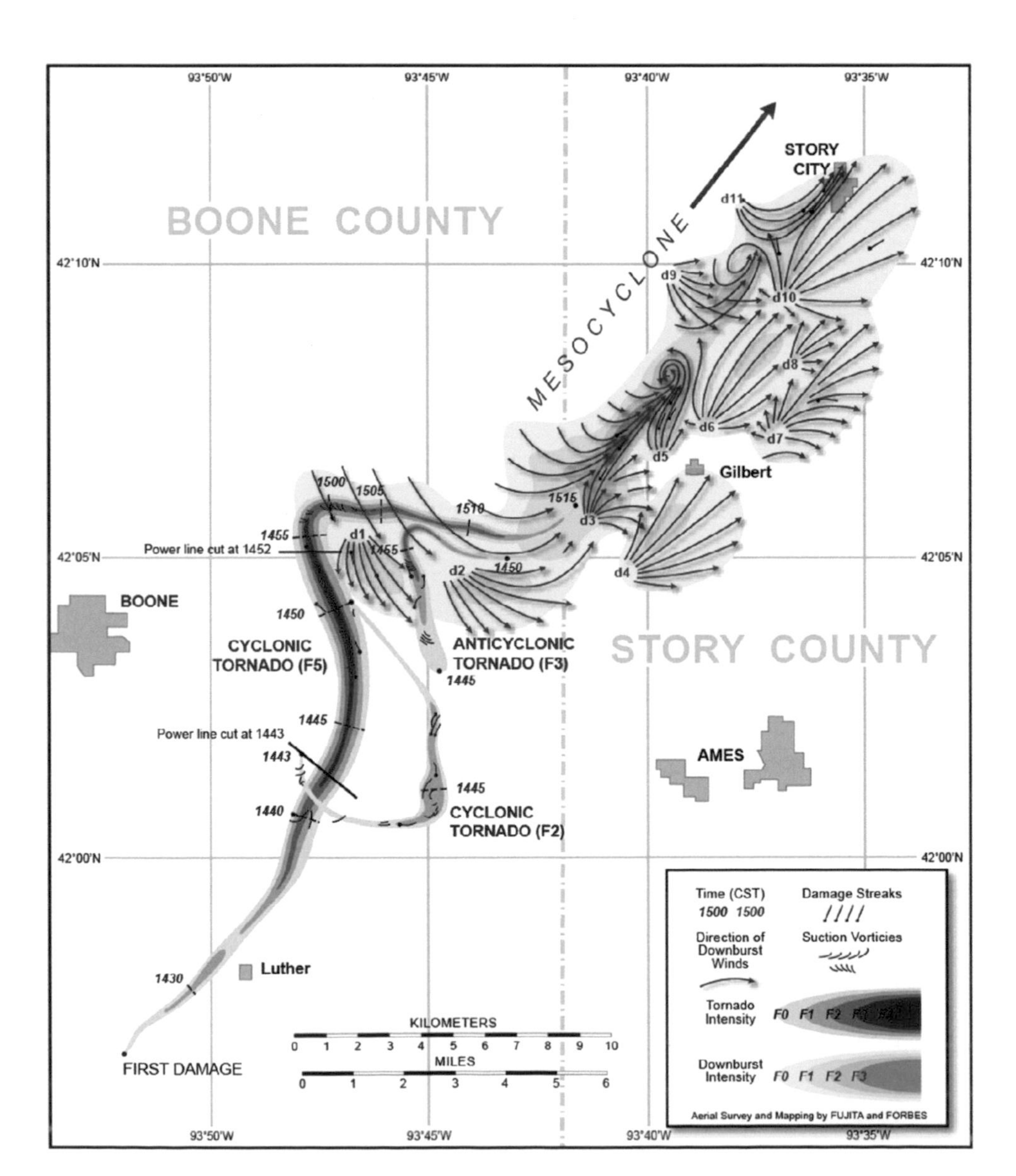

Figure 12.6 Tornadic and downburst wind patterns along the storm track. The supercell produced several tornadoes (red tracks) in its first hour, then produced an extensive family of downbursts (blue regions) as it continued to track toward the northeast. (Adapted from Fujita, 1978.)

a long-track, wide F5 tornado and two smaller tornadoes. At one point (1445 CST), all three tornadoes were rolling across the fields!

The main F5 vortex is typical of violent tornadoes: The wind speeds are extreme, and the track is long (in this case, nearly 30 km [19 miles], lasting almost an hour). The tornado's width also explains why such "monster" tornadoes (F4 and F5) are so destructive, causing a disproportionate number of all tornado fatalities.

While this tornado began as a humble F0, it quickly evolved into an F5 during the first 15 minutes of its life cycle. F5 intensity damage does not span the entire width but rather is confined along a narrow strip just a couple tenths of a kilometer wide.

Most tornado paths strike a southwest to northeast orientation. The interactions between a supercell's mesocyclone and rear flank downdraft may deflect the tornado in a different direction. Both the F3 and F5 in Figure 12.6 hooked abruptly eastward at 1500 CST.

From 1500–1515 CST, tornadoes began weakening while the supercell switched into a downburst-generating mode. Recall that supercells sometimes morph into bow-echo type structures that generate families of downbursts. After about 1455 CST, the supercell produced downbursts for nearly an hour. Some contained winds rated at F3 intensity (> 137 kts [158 MPH]).

Tornado surveys like the one shown in Figure 12.6 are very exhaustive, sometimes requiring months of careful documentation and analysis. Researchers must sort through multiple reports of tornadoes, some of which turn out to be duplicates. Sometimes a single, exceptionally long-track tornado turns out to be multiple tornadoes generated in rapid succession by the same parent supercell.

Careful storm damage surveys often reveal suction vortices within the core of strong and violent tornadoes. The highest winds, approaching 260 kts (300 MPH), are contained in these tiny whirlwinds, which produce the greatest number of fatalities. Figure 12.8 shows a highly detailed damage survey of a tornado that contained suction vortices. Each fatality is identified by a black dot. In all cases, the dots coincide with a suction vortex (each vortex is shown by a magenta-colored arc). Suction vortices are identified

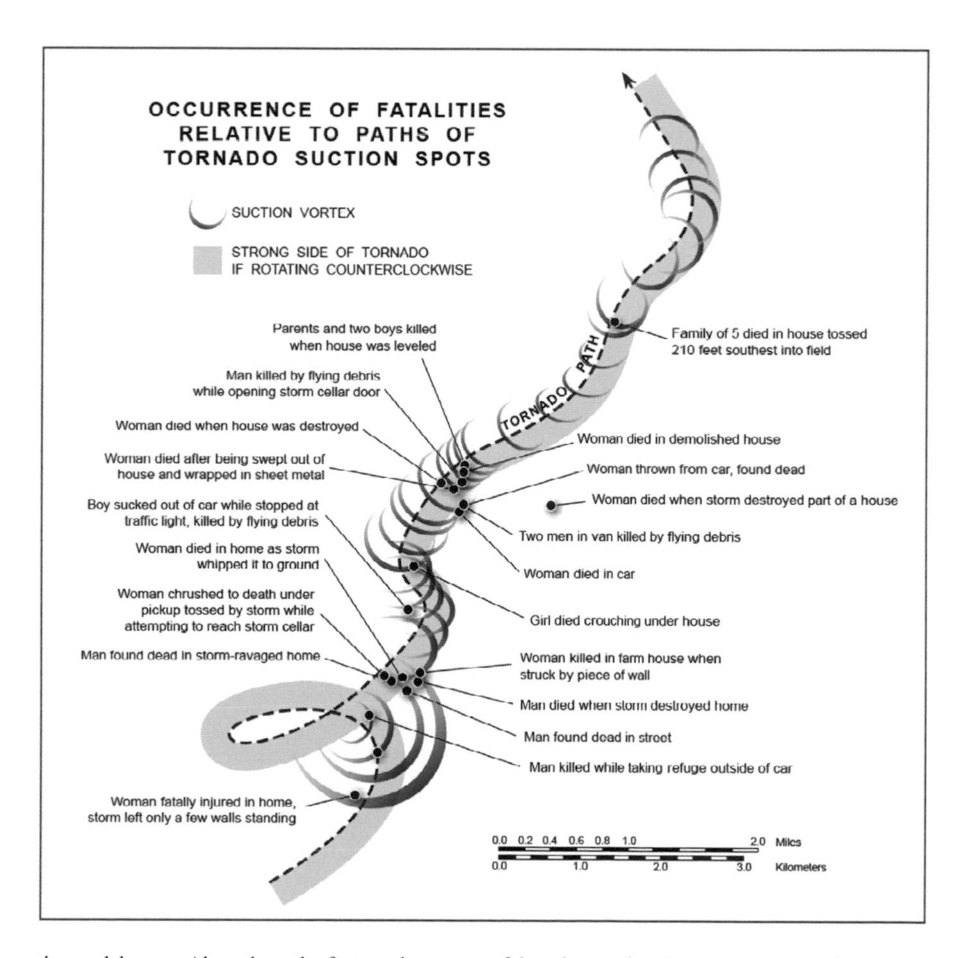

Figure 12.7 Suction vortices and damage. Along the path of a tornado are powerful suction vortices that can cause great damage to structures and fatalities. As this diagram illustrates, most fatalities are close to the center of the tornado path, where suction vortices are generated. (Adapted from Fujita, 1970.)

Figure 12.8 LaPlata, Maryland, tornado damage swath, created by an EF4 tornado on April 28, 2002. (NASA.)

by their cycloidal etchings in the debris field. Also note the complex movement of this tornado, which executed a 360° loop.

It is now possible to delineate the long track of a violent tornado on a single digital image obtained from satellite. Figure 12.8 shows the swath cut by the destructive LaPlata, Maryland, tornado of April 28, 2002. True-color visual imagery (top panel) reveals the linear scar of near-total devastation through the city's core.

The assignment of specific F- and EF-scale wind categories is an imperfect process. Rating a tornado's wind speed is based on subjective interpretations of structural damage. Different interpretations arise because engineering and building standards vary. For example, consider two homes of identical design in two different communities. To save money, one builder fails to secure the roof to the walls with storm clips, relying instead on nails, gravity, and friction. Homes with below-average construction disintegrate more easily, leading to a comparative overestimate of tornadic winds. Identically constructed homes oriented in different directions will experience contrasting damage. For instance, if the garage door faces the tornado, the house will fare worse than if its back side takes the brunt of the wind. A garage door is flimsy compared to a solid back wall.

In addition, some tornadoes are probably not rated close to their maximum intensity because they occur in remote locations and do not damage engineered structures. For of all of these reasons, it is important to recognize that the EF scale is a damage scale first and a wind-intensity scale second.

Finally, recent research suggests that, historically, tornadoes have likely been underrated by at least one intensity category. This is especially true of tornadoes traversing undeveloped or sparsely occupied areas, with few structures to disturb. Portable Doppler radar observations of actual wind speed have allowed researchers to quantify the disparity with EF damage scale estimations.

Tornado Climatology: Geography and Seasonality of U.S. Tornado Distribution and Frequency

When people think of "Tornado Alley," visions of the Great Plains come to mind. It may come as a surprise that the Great Plains is not the only location where a great number of violent, killer tornadoes occur every year in the United States. In this section, we explore the detailed distribution of U.S. tornadoes both in space and through the course of a typical year.

U.S. Tornado Belts and Alleys

Where exactly is the region commonly referred to as Tornado Alley? In broad geographical terms, the region includes the central Plains, particularly northern Texas, Oklahoma, Kansas, Nebraska, and the Dakotas. Scientifically, however, there is no consensus on what precisely constitutes Tornado Alley. However, tornado scientists have known for years that other regions outside the Plains experience formidable tornado activity.

Figure 12.9 presents a map of U.S. tornado frequency, based on all tornado reports from 1880 to 2003, normalized (adjusted) for variable reporting area. The map gives the number of tornadoes per 1000 square miles. While most U.S. tornadoes are

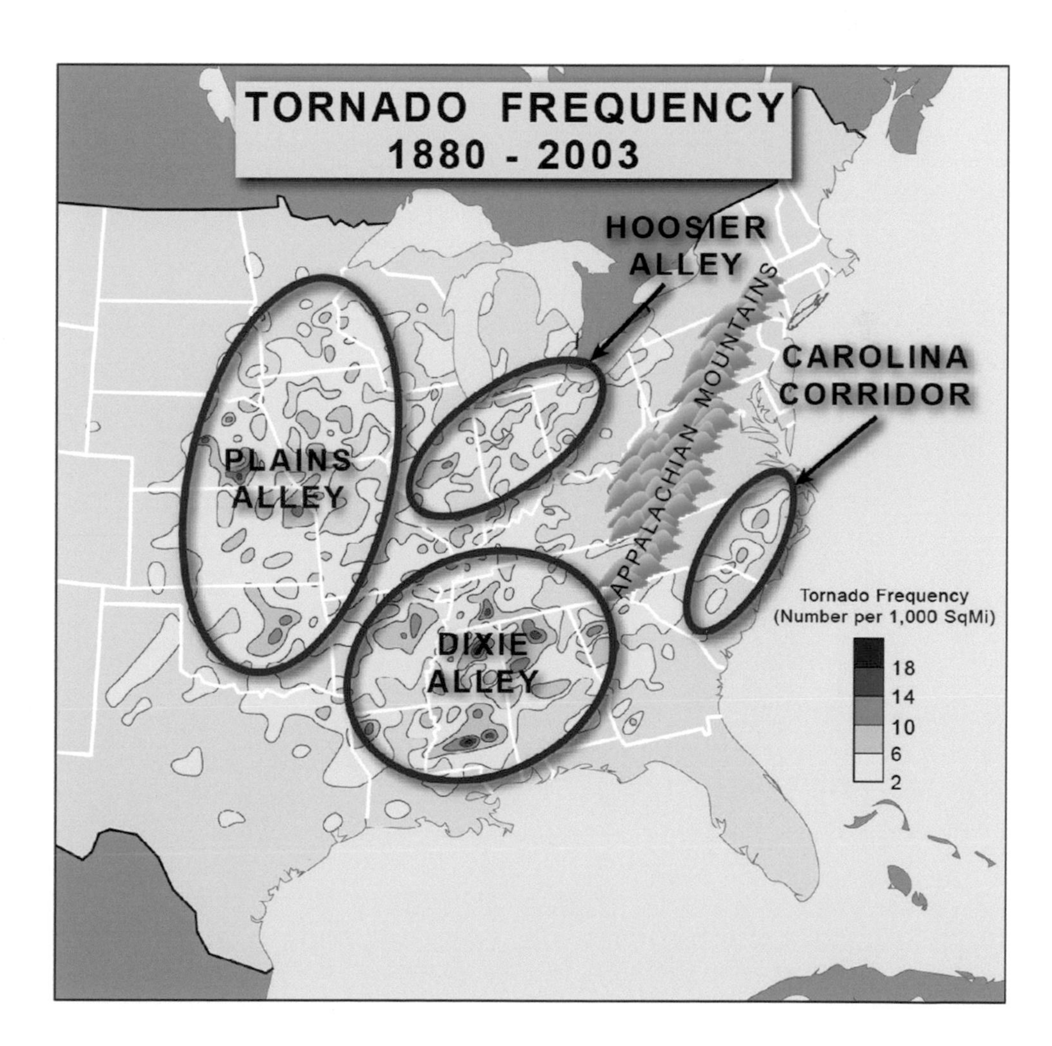

Figure 12.9 Regions of high tornado activity. The frequency of tornadoes per 2590 km^2 (1000 square miles) F is used to identify four key regions (alleys/corridors) of high tornado activity. The Plains Alley (often called Tornado Alley) stretches from Oklahoma to Minnesota. Hoosier Alley cuts across Illinois, Indiana, and Ohio. Dixie Alley is found in the deep south and is a "hot spot" for tornado activity early in the tornado season. Finally, a less known corridor of tornado activity is located in the eastern portions of the Carolinas. (Adapted from Broyles et al., 2004.)

constrained by the Rockies and Appalachians, there are in fact several "alleys" across the United States:

- The classic Great Plains Alley stretches from Oklahoma to Minnesota.
- A large Dixie Alley is centered on Mississippi, Arkansas, and Alabama (the Mid-South).
- A small region called Hoosier Alley straddles Illinois and Indiana.
- Finally, a narrow Carolina Corridor is located across the Piedmont and coastal plain of the Carolinas.

The peculiar vulnerability of the U.S. Mid-South (encompassing the lower Mississippi and upper Tennessee valleys) is receiving increased media exposure. Unlike the Great Plains, Dixie Alley is under the threat of tornadoes throughout the entire year, not just in the spring. Compared to the Plains, Dixie Alley experiences more strong and violent tornadoes and twice as many tornado outbreak days. Most importantly, Dixie Alley experiences more killer tornadoes (3:1). The specific aspects of Dixie Alley render the region particularly vulnerable. These aspects include (1) higher population density for Dixie Alley vs Plains Alley (90 persons/mile^2 vs 44 persons/mile^2); (2) more mobile homes vs permanent homes; (3) a greater tendency for nocturnal tornadoes (warnings are less effective when the population is asleep); and (4) tornadoes that are harder to observe (because of tree lines, hills, and roads that do not follow a farmland grid pattern). The last factor also happens because tornadoes are obscured by heavy rainfall (high precipitation supercells are more common in the Mid-South).

It is also noteworthy that several recent studies point to an eastward geographical shift in maximum tornado incidence, out of the Southern Plains and into the Mid-South region. Any link to climate change remains uncertain and warrants further investigation.

Tornadoes Through the Course of the Year

The geographic distribution of concentrated U.S. tornado activity is not fixed throughout the course of the year. As Figure 12.10 shows, heavy activity begins over the Mid-South in early April (late March over the Carolinas), shifting north and west into the Plains during May and June. During midsummer, the greatest activity extends across the far

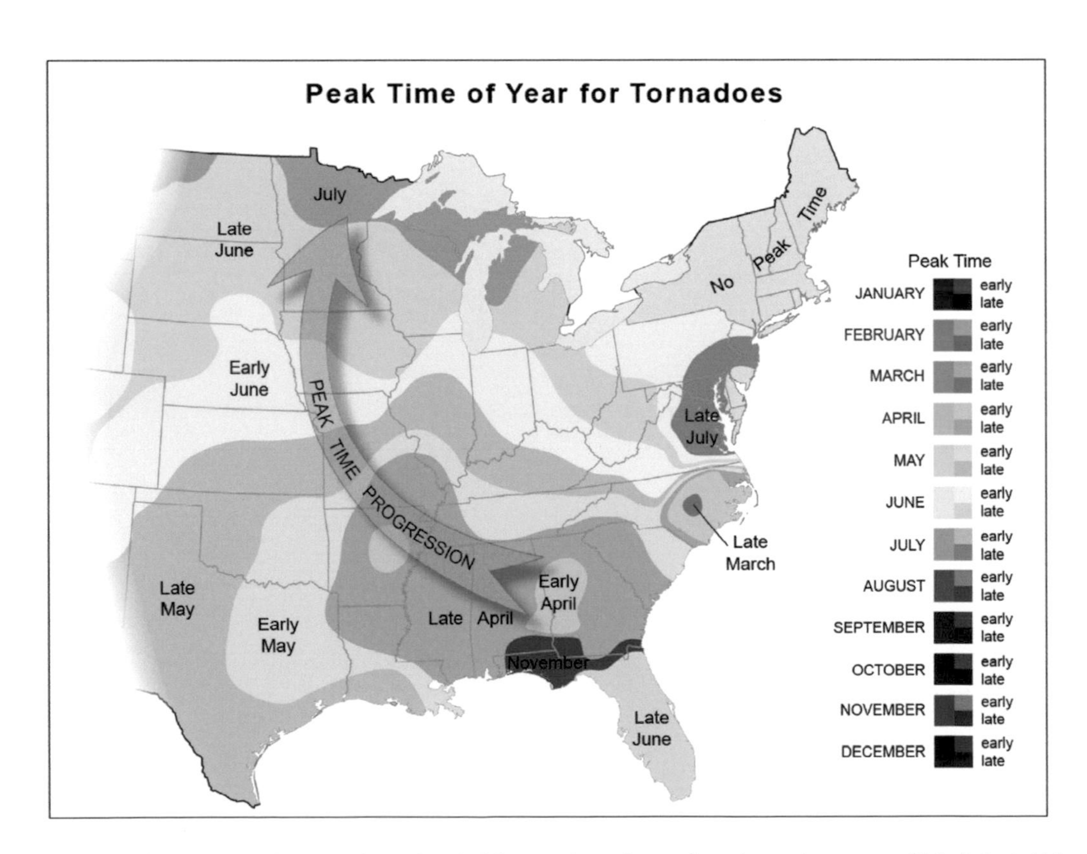

Figure 12.10 Tornado activity throughout the year. The peak period for tornadoes migrates from the southern states of Mississippi, Alabama, and Georgia to the upper Midwest states of Minnesota, Wisconsin, and Michigan. This transition begins in early April and progresses through May, June, and July. (Adapted from Brooks et al., 2003.)

northern Plains, Great Lakes, and Mid-Atlantic. In the fall, concentrated activity envelops the Gulf Coast.

Average tornado frequency in the United States peaks during the springtime between April and June (Figure 12.11), with a maximum in May. However, this distribution is purely statistical; in any year, a large tornado outbreak can develop outside the expected month of peak activity. Nor does the distribution in Figure 12.11 capture seasonal differences between the Great Plains and Dixie Alleys. Figure 12.12(a) reveals that Dixie Alley tornadoes peak earlier than those over the Plains. In fact, tornadoes in the Mid-South often occur during the winter months. Dixie Alley tornadoes increase abruptly in November.

Following is a season-by-season synopsis describing how the principal ingredients of severe thunderstorms and tornadoes – namely, wind shear and unstable air masses – interact, based on time of year.

Winter Months

- Weak surface heating > air masses are only weakly unstable.
- Wind shear is very strong (jet stream envelops southern United States).

Spring Months

- Strong surface heating and cold air aloft make air masses highly unstable.
- Strong shear remains over south-central United States, but the jet stream begins migrating northward

Summer Months

- Strong surface heating renders air masses highly unstable.
- Wind shear is very weak (jet stream shifts over northernmost United States and southern Canada and weakens considerably).

Fall Months

- Surface heating weakens; air masses are not as unstable.
- Strong shear begins developing as jet stream advances southward out of northern United States.

The greatest opportunity for tornado development occurs in regions where both strong instability and strong wind shear (generated by the jet stream) overlap. This rules out much of winter and summer. However, from March to May, the best "window of opportunity" develops east of the Rockies. In the fall, there is a weaker tornado season, mainly in October and November. A large fraction of these tornadoes are spawned by

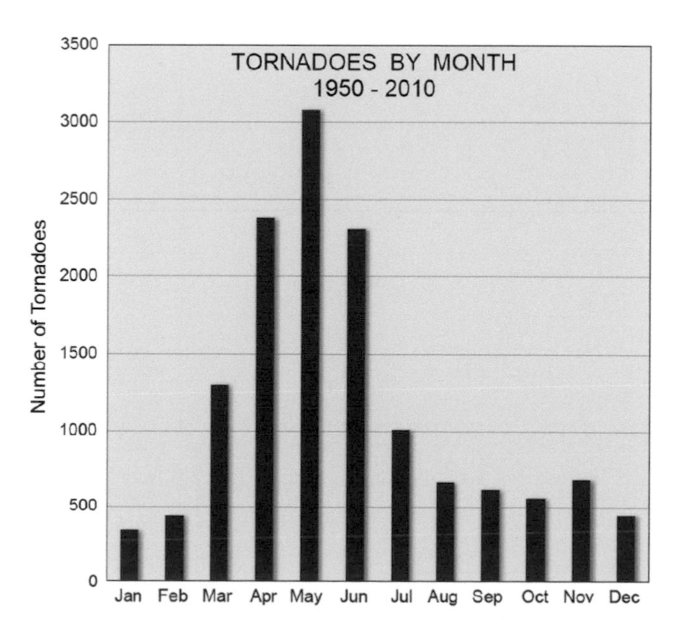

Figure 12.11 Peak tornado activity in the United States. The peak month for U.S. tornado activity is May. January has the fewest tornadoes.

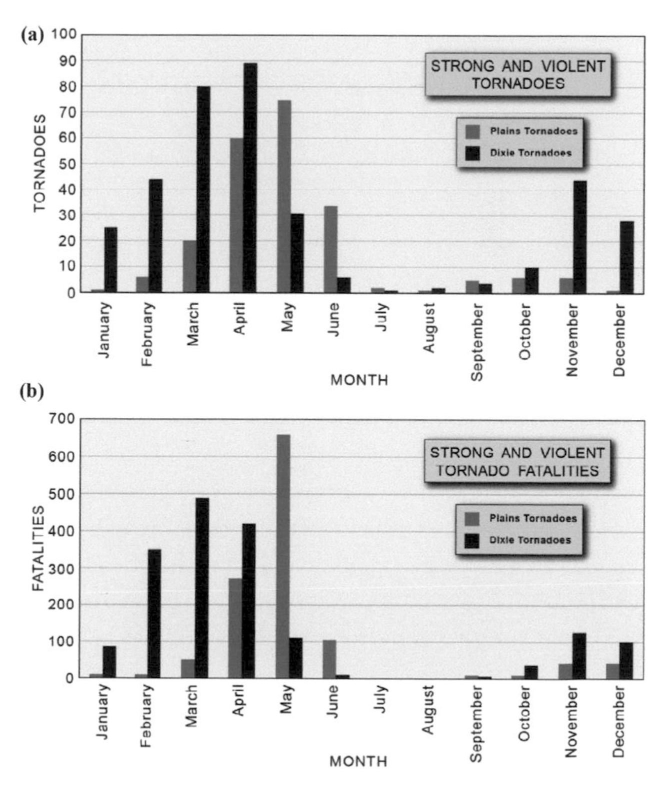

Figure 12.12 (a) Dixie tornadoes peak earlier in the spring than the Plains tornadoes. There is a clearly defined period of low activity between July and October. (b) Deaths peak in March for Dixie tornadoes and in May for Plains tornadoes. (Adapted from Gagan et al., 2010.)

tropical cyclones that strike the Gulf Coast, Southeast, and Mid-Atlantic regions.

Tornadoes Spawned by Tropical Cyclones

Tropical cyclones (TCs) weaken rapidly once their inner region makes landfall. Paradoxically, this is when they are most likely to spawn tornadoes. Figure 12.13 shows the distribution of all hurricane-generated tornadoes from 1950 to 2007. The number of TC tornadoes varies considerably from year to year, depending partly on the number of tropical cyclones making landfall. In some years the number of U.S. TC tornadoes exceeds 500. The highest frequencies are along the coast, dropping rapidly inland. The first 24–48 hours of landfall are critical for tornado formation. While wind speeds in the hurricane vortex remain intense aloft (it takes time for the storm to weaken), surface-level winds sharply decline from increased friction over land. The result is a shallow layer of strong vertical wind shear along the coast, from which TC tornadoes derive their spin (the spin is initially oriented in the horizontal, then becomes tipped into the vertical).

Tornadoes spawned by TCs account for 3–4% of all U.S. tornadoes. The fraction of tornadoes due to hurricanes is greatest in Virginia and South Carolina. The F-scale intensity distribution of these tornadoes is skewed toward the low end (49% are rated F0 vs 42% for all U.S. tornadoes; in terms of F2+ tornadoes, the numbers are 14 and 19%, respectively). F3 or stronger TC tornadoes are exceedingly rare. TC tornadoes spin up rapidly from small, shallow supercells (mini-supercells) embedded in TC rain bands. Doppler radar does not do a good job of detecting these tiny, transient circulations. Tornado lifetime is so short that by the time community-level warnings are put in place, the tornadoes have dissipated.

TC tornadoes pose a significant forecast challenge, and their hazard remains significant. Some tropical cyclones at landfall have generated tornado counts in excess of 100. The current record holder is Hurricane Ivan (September 2004), which generated 119 tornadoes over a 3-day period across the Southeast and Mid-Atlantic.

Tornadoes by Time of Day

Most tornadoes occur during the midafternoon-early evening (3–9 p.m.) because supercell thunderstorms arise

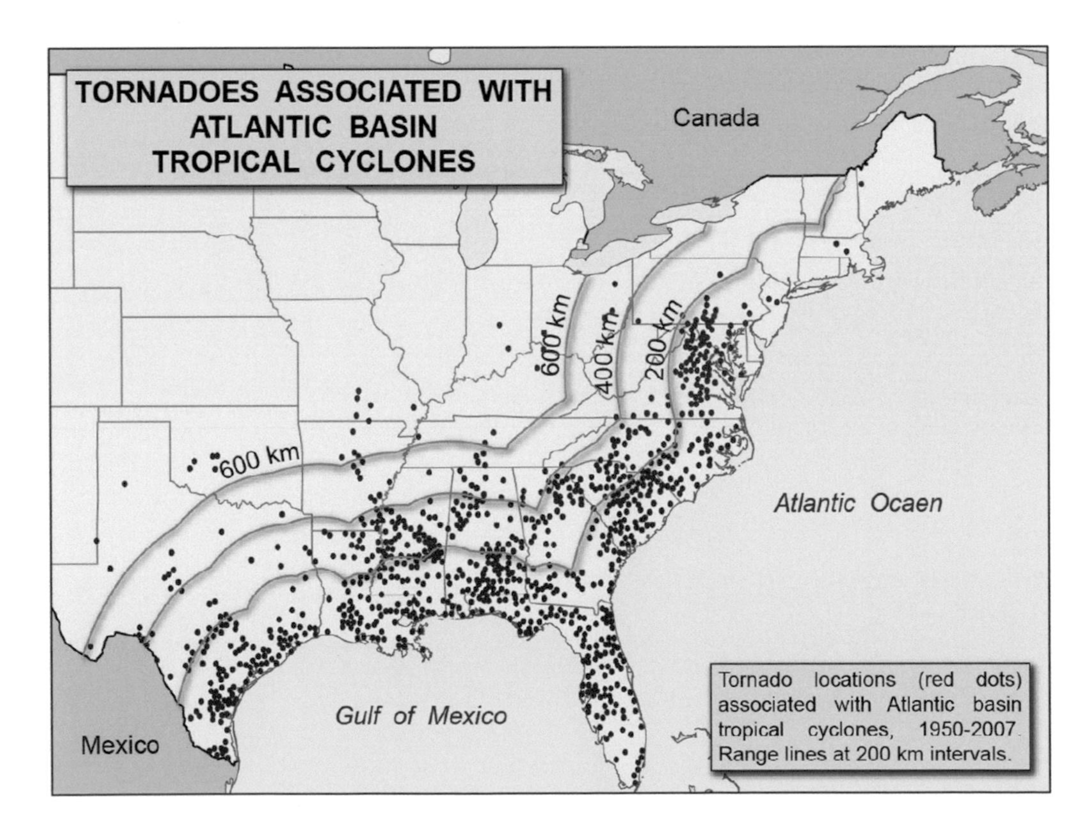

Figure 12.13 Distribution of hurricane-generated tornadoes, 1950–2007. The red dots represent tornadoes generated by tropical cyclones. Notice how quickly the number of tornadoes decreases with distance from the coast. This effect is a direct response to the rapid weakening of surface wind in the tropical cyclone when it encounters land. (Data courtesy of D. Cecil and L. Schmidt, ESSC, University of Alabama.)

from unstable air. Thunderstorms commonly initiate during the hottest afternoon hours (1–3 p.m.). After growing and strengthening, they generate severe weather between the hours of 4–6 p.m. If the atmosphere is exceptionally unstable, supercells and other severe storms may continue to develop after sundown, extending the tornado threat into the evening.

Figure 12.14 shows the 24-hour distribution of strong and violent (killer) tornadoes by time of day, for both the Great Plains and Dixie Alleys. Both regions show a prominent late afternoon peak, but Dixie Alley tornadoes continue initiating late into the night, even early morning. This distribution is very problematic from a community warning standpoint. While Doppler radar detection provides many minutes of lead time, field verification of an actual tornado on the ground (by trained spotters, local law enforcement, or citizens carrying cell phones) is crucial for the issuance of tornado warnings. Darkness makes visual tornado sightings impossible, and even with timely warnings, who is awake to receive them? NOAA Weather Radio, activated by a special warning alert tone, can notify a sleeping community oblivious to TV warnings and tornado sirens.

Long-Term Trends in Tornado Frequency, Property Damage, and Mortality

As population density increases and the amount of property rises exponentially, tornadoes have become more costly to society. The annual dollar damage varies markedly, depending on the number of significant tornado outbreaks and whether tornadoes move through densely populated regions. Average annual losses exceed $1 billion, but a given year can vary from less than $1 billion to $5 billion or more. Total economic losses are always greater than insured (property) losses because economic losses include all costs – the loss of jobs and community productivity, declining property values, medical costs, and so on. The year 2011 featured one of the most prolific tornado seasons ever, with total economic loss estimated at $26 billion.

Long-Term Trends in Tornado Counts

In the United States, average annual tornado occurrences now top 1200, but there is significant year-to-year variability. A low-count year might see 1000 tornadoes, while 2011 produced

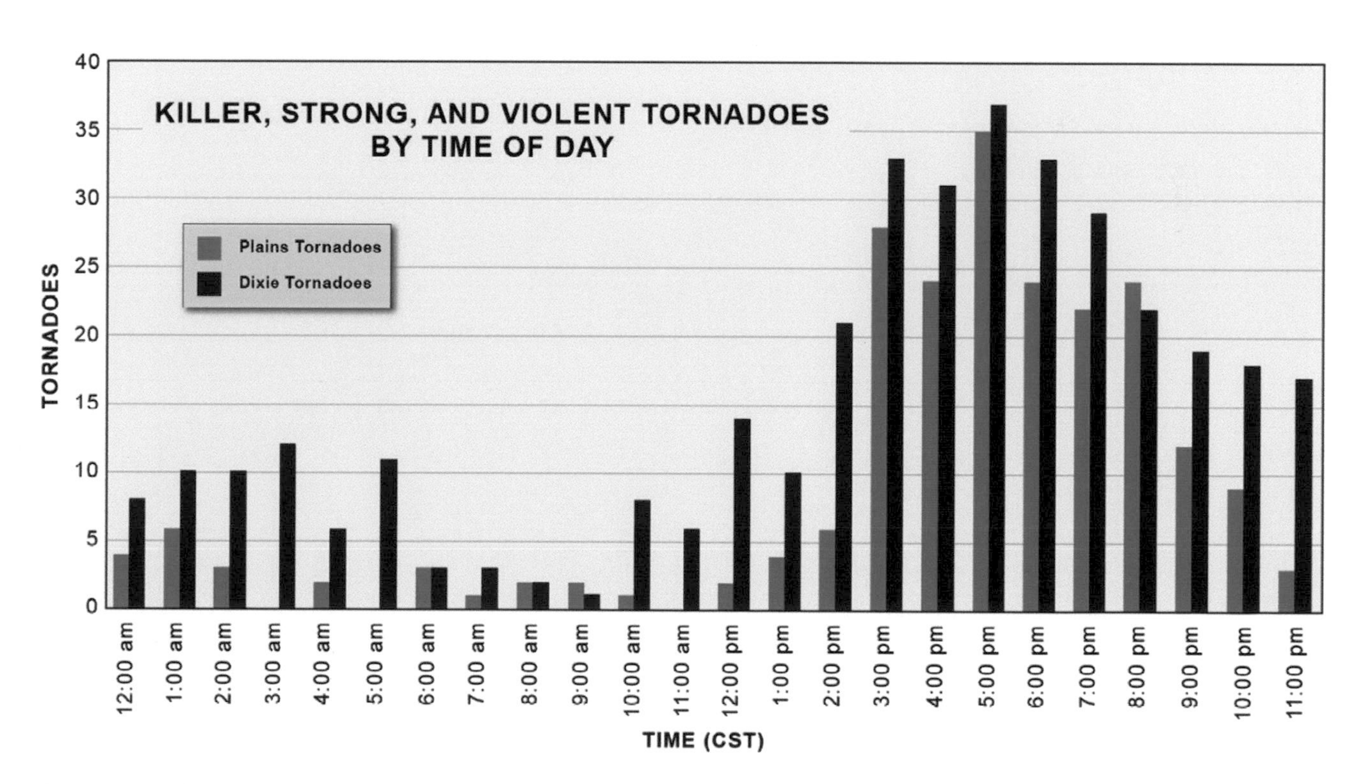

Figure 12.14 Plains and Dixie "killer" tornado frequency by the hour. While killer tornadoes in both regions tend to peak in late afternoon to early evening, Dixie tornadoes persist well into the night, increasing the likelihood that residents will not hear broadcast warnings because they are asleep, thereby elevating the possibility of fatalities. (Adapted from Gagan et al., 2010.)

1700 tornadoes. Annual tornado counts have changed over the past century. There is a substantial upward trend in reported tornadoes over the period, but does this trend reflect an actual increase in the tornado hazard? Likely not; it's instead due to increased detection and reporting of tornadoes, especially the weak EF0 and EF1 categories. Before 1940, fewer than 200 tornadoes were reported annually. As population rose, annual tornado numbers jumped to 800. More people were living in previously remote locations, lending more sets of eyeballs to notice and report tornadoes. In the 1970s and 1980s, the NWS launched efforts to record tornadoes systematically, in order to create an accurate national database. The number of annual tornado reports jumped again in the late 1980s/early 1990s when the Doppler radar network (NEXRAD) went online. Doppler technology greatly facilitates detection of rotation within supercell thunderstorms and in some cases can detect tornadoes. Nonetheless, tornado historian Tom Grazulis suggests that up to 50% of tornadoes still go undocumented. Recall that the vast majority of tornadoes are very small and short-lived (with path lengths of 1–3 km [0.6–1.8 miles], lasting just a few minutes). Many tornadoes come during darkness, are enshrouded in rain, have no visible funnel, are blocked from view by trees, or touch down in remote regions.

Long-term tornado deaths in the United States average about 60 per year, ranking third in weather-related fatalities (behind heat waves and flash floods). But long-term averages can be hugely misleading. For instance, 2011 saw 550 deaths and 5400 injuries resulting from tornadoes. Figure 12.15 shows the long-term trend in tornado mortality (smoothed data) from 1875 to 2010. Tornado fatalities are expressed as deaths per million persons to account for population growth. There is large year-to-year variation in tornado deaths. But in spite of the large U.S. population increase, the line has trended steadily downward since 1925. Deaths averaged 3 per million people in the 1920s, dropping to less than 1 per million in 2010. In the meantime, the U.S. population has more than quadrupled. A significant year was 1950, when the Weather Service began issuing public tornado warnings.

Conventional (non-Doppler) radar detection of tornadoes increased around 1960. Through the 1970s and 1980s, faster modes of communication disseminated warnings more rapidly. The NEXRAD Doppler radar network went online in the 1990s, and many TV stations purchased their own Doppler radar. Networks of storm spotters and amateur storm chasers have become commonplace. And improved weather-prediction

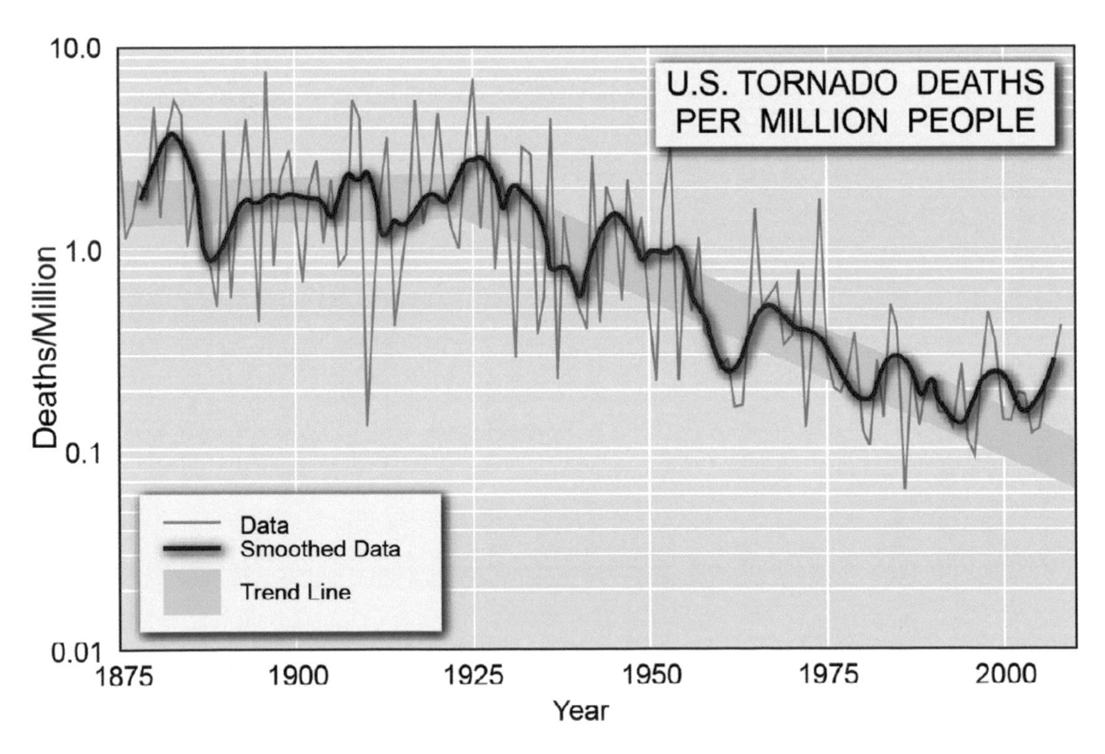

Figure 12.15 U.S. tornado fatalities per million people, 1875–2010. Despite significant population increases, we have seen an overall downward trend in fatalities per million persons. Much of this can be attributed to better detection and early warnings of potential tornado outbreaks or actual sightings of tornadoes. (Adapted from Brooks and Doswell, 2002.)

models have made it possible to accurately predict tornado outbreaks 3 days in advance.

Cyclic Variations in Tornado Frequency Due to El Niño and La Niña

Recall that El Niño refers to an alteration of ocean currents and wind patterns across the tropical Pacific Ocean, associated with the anomalous warming of eastern Pacific surface waters every 2 to 5 years. But in some years, El Niño's effects extend into the higher latitudes, including North America. La Niña refers to an abnormally cold ocean in the tropical eastern Pacific.

El Niño and La Niña shift the location and intensity of the subtropical jet stream (see Chapter 7, Figure 7.26). These changes are most pronounced during the winter season. Jet streams produce vertical wind shear, which tornado-producing supercells require. Additionally, the subtropical jet governs the frequency and track of extratropical cyclones. These cyclones harbor the warm sector of unstable air in which supercells thrive.

Several studies have examined the link between El Niño and subsequent springtime tornado outbreaks. Figure 12.16 illustrates the findings of one study. During La Niña events, the subtropical jet and tornado activity shift northward into the Mid-South and eastern Great Plains, relative to non-La Niña years. During El Niño events, southward displacement of the subtropical jet (and its intensification) enhances tornado activity along the Gulf Coast. However, other studies fail to demonstrate these types of changes. Clearly, more work is needed to clarify the El Niño–tornado relationship.

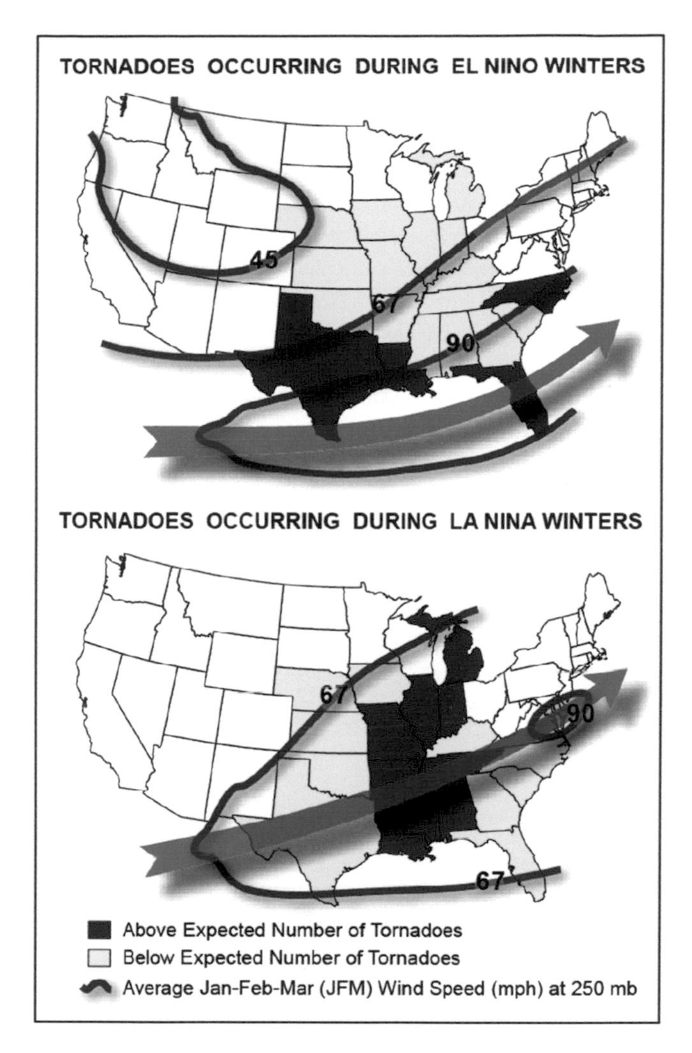

Figure 12.16 El Niño and La Niña effects on tornado distribution: results of one study. These maps reflect the results of a study to determine whether El Niño and La Niña conditions in the Pacific affect the distribution of tornadoes. The study suggests that during El Niño periods, the subtropical jet is farther south, which concentrates tornado occurrence along the Gulf Coast. When La Niña conditions prevail, the subtropical jet is farther north, thereby increasing the tornado count in the U.S. Central and Midwest regions. (Adapted from Cook and Schaefer, 2008.)

Tornado Outbreaks, Past and Present: When Twisters Make the Evening News

Isolated tornadoes are newsworthy events, particularly in communities that rarely experience them. But large, concentrated swarms of tornadoes occasionally ravage one or more states, sometimes for days. There is no universally accepted definition of a tornado outbreak. However, meteorologists agree that the term refers to multiple tornadoes, usually 6–10, that develop in the same region, over a 24-hour period, from the same large-scale weather system. In the most common situation, the warm sector of a springtime extratropical cyclone spawns numerous supercell thunderstorms, which in turn create a barrage of tornadoes. Intense quasi-linear convective systems (bow echoes) can also generate tornadoes.

When tropical cyclones make landfall, their convective rain bands frequently breed large numbers of tornadoes. Understanding the behavior of tornado outbreaks is important because historical records indicate 80% of all U.S. tornado fatalities occur in connection with tornado outbreaks.

A tornado outbreak sequence, sometimes called simply an outbreak, occurs when tornadoes occur over two or more consecutive days, from the same parent weather system. Outbreak sequences can impact large regions of the United

States. The term "super outbreak" is reserved for the most deadly and prolific outbreaks or outbreak sequences.

Big tornado outbreaks do not occur every year, and fewer than 4% of days per year experience an outbreak. However, outbreaks account for 80% of all tornado fatalities. Recent studies demonstrate that tornadoes have been clustering into outbreaks more in recent decades than in the past. The most intense tornadoes (EF4 and EF5) rarely occur outside a tornado outbreak. Whether a particular outbreak generates a substantial number of fatalities depends on (1) the number of strong-violent tornadoes, (2) whether these tornadoes strike densely populated regions, and (3) whether people are in especially vulnerable situations (mobile vs permanent homes, daylight vs overnight hours).

In the next sections, we focus on two time periods when intense tornadoes struck the U.S. – April 1974 and April 2011. The event in 1974 was a single, 2-day super outbreak. During 2011, successive outbreaks raked the United States over a 2-month period. These events featured a large proportion of violent, killer tornadoes, and underscore the vulnerability of Dixie Alley across the Mid-South.

April 3–4, 1974, Super Outbreak: The Greatest Tornado Event in U.S. History

The greatest single tornado outbreak in terms of strong-violent tornadoes, fatalities, and injuries (since 1875) occurred on April 3–4, 1974. A total of 148 tornadoes developed, 64% of which were rated significant (F2+) and 20% of which were rated F4–F5. No other outbreak boasts such large fractions of high-intensity tornadoes. There were 335 fatalities and more than 6000 injuries. Ten states were declared federal disaster areas. Forecasters in Indiana could not keep track of the tornadoes that developed all at once, so they placed the entire state under a single tornado warning (the first and only time this has been done).

Figure 12.17 plots these tornadoes, including path length and F-scale intensity. The outbreak straddled the Hoosier and Dixie Alleys and featured a large number of long-track tornadoes. Supercells that spawned these tornadoes developed within the warm sector of an extratropical cyclone over Kansas, which tracked northeastward toward the Great Lakes as the outbreak progressed.

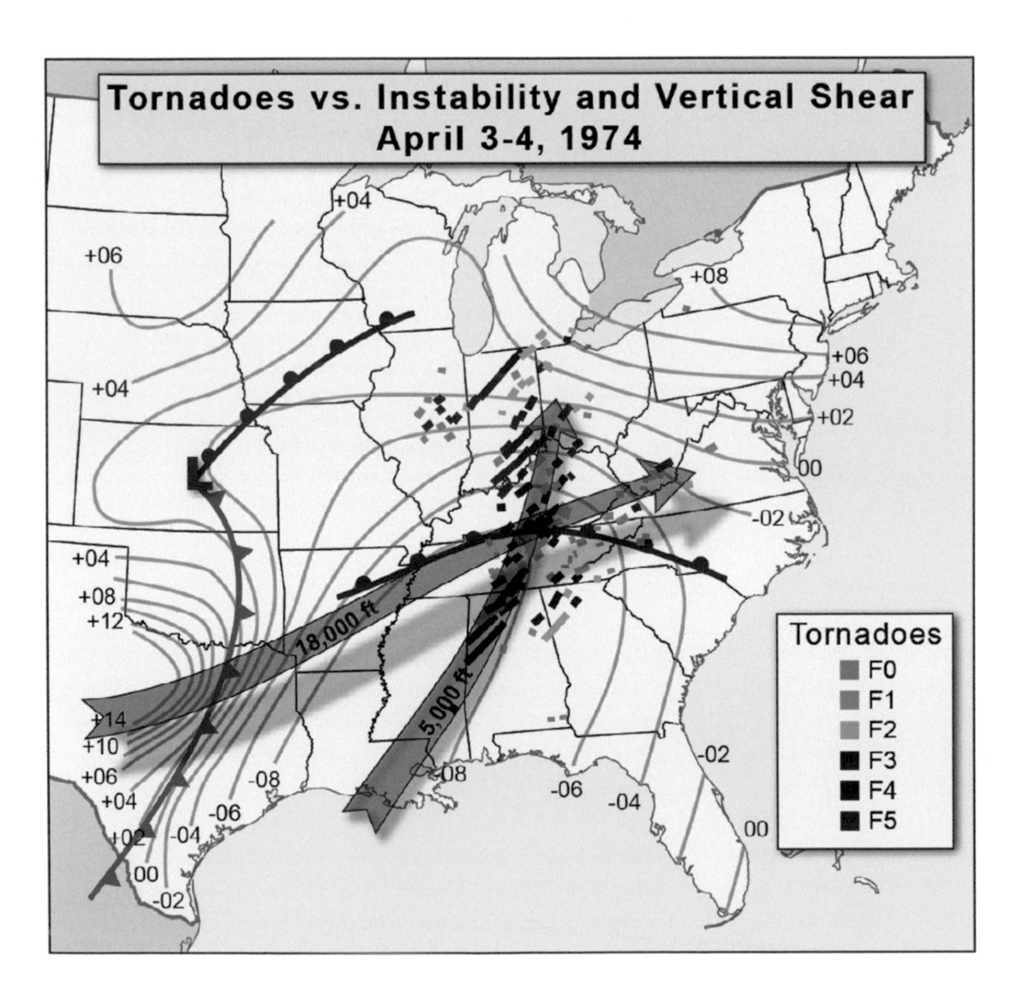

Figure 12.17 April 3–4, 1974 super outbreak. This super outbreak produced a record number of strong-violent tornadoes in the Midwest. The perfect set of ingredients (warm moist air and strong wind shear) came together to create 148 tornadoes, of which about 30 were rated as F4 or F5. (Adapted from Hoxit and Chappell, 1975 and Corfidi et al., 1975.)

Figure 12.17 also shows how a highly unstable air mass and strong vertical wind shear came together in the outbreak region. The Lifted Index measures the buoyancy of air in storm updrafts. The more negative the value, the stronger the updrafts. Values of −8 are extreme and reflect very warm, moist air streaming northward into the cyclone's warm sector. The winds at different altitudes (shown by purple arrows) contained strong speed and directional shear (clockwise turning). The shear created mesocyclones in the supercell updrafts. Most tornadoes formed in the zone of overlap between large instability and strong shear. This combination of synoptic-scale meteorological factors is typical of large, intense tornado outbreaks.

2011: An Exceptional U.S. Tornado Year and a Multibillion-Dollar Tornado

After 1971, the death toll from a single tornado had not exceeded 50. But in spring of 2011, three deadly tornadoes broke this longstanding record. These included an EF5 tornado that struck northern Alabama (April 27, 72 fatalities), an EF4 in Tuscaloosa, Alabama (April 27, 64 deaths), and an EF5 in Joplin, Missouri (May 22, 159 deaths). The violent Joplin tornado, with a price tag of $2.9 billion, became the costliest on record since 1950. The late April tornadoes were part of an extremely violent tornado outbreak sequence during April 25–28, 2011. During these four days, intense tornadic activity spread across the U.S. Southeast and Mid-Atlantic. Especially large and violent tornadoes raked Dixie Alley; 74% of the outbreak deaths occurred in Alabama alone. Figure 12.18 maps many of these tornadoes on the most intense day of this sequence, April 27.

The outbreak occurred in the warm sector of an extratropical cyclone. A total of 358 tornadoes developed over 3 days, a new U.S. record for tornadoes spawned in a continuous outbreak. In this outbreak, fewer than 5% of the tornadoes were rated EF4–EF5 – in contrast to 20% on April 3–4, 1974. However, the death toll in both outbreaks was comparable (325 tornado fatalities for April 3–4, 1974, vs 324 total for April 25–28, 2011).

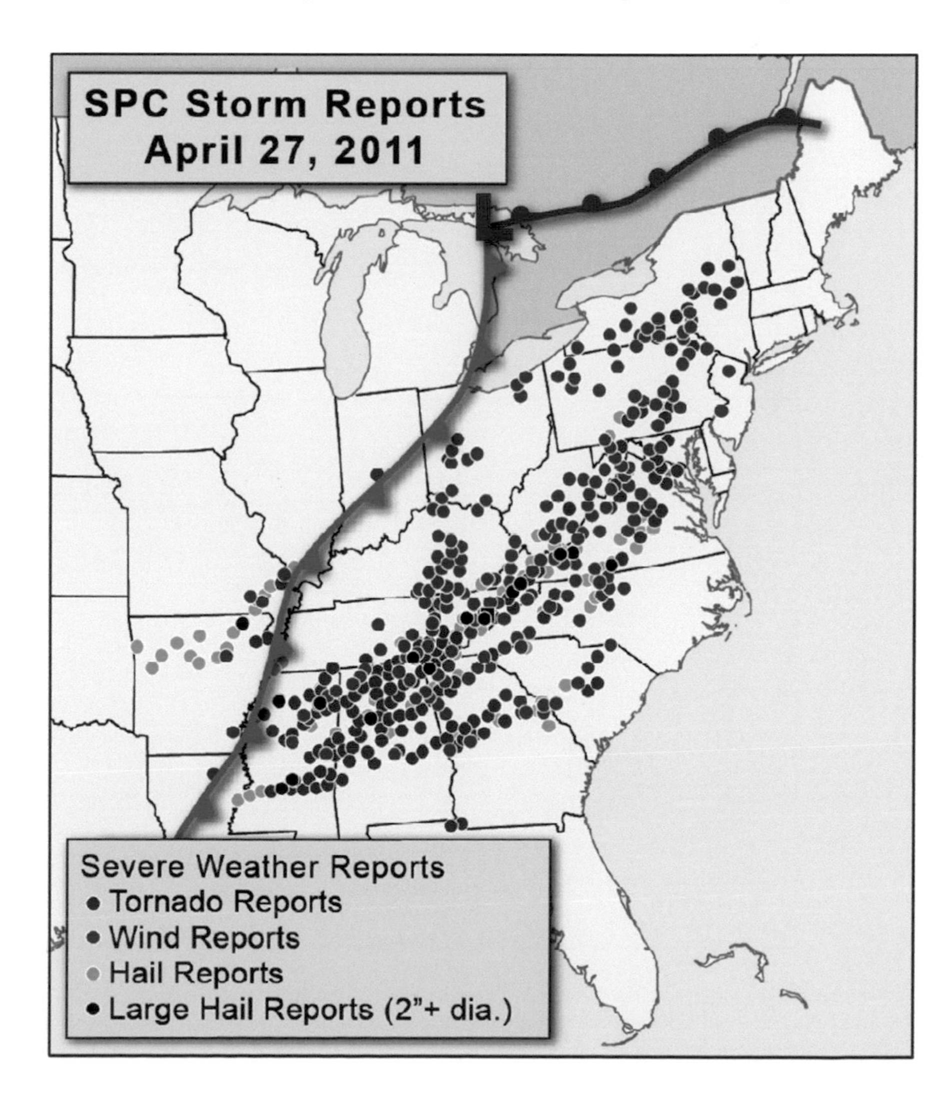

Figure 12.18 Contemporary tornado super outbreak, April 27, 2011. The April 25–28, 2011, tornado outbreak ranks as one of the worst outbreaks in U.S. history. Tornadoes prevailed throughout the eastern third of the country, but Mississippi and Alabama were hit especially hard. Over 350 tornadoes and many fatalities were reported during this outbreak. This figure maps many of the tornadoes on the most intense day of the outbreak, April 27. (Adapted from NOAA.)

Both the 1974 and 2011 outbreaks featured multiple, long-track supercells. Many of the tracks exceeded 100 miles. Doppler weather radar easily detects large, intense supercells, allowing for advance warning (10 or more minutes). Figure 12.19 shows an example of a single cell from April 11, 2011. The image is a composite of many radar scans as the storm tracked across four states. This supercell was a prolific tornado producer. As we've seen, long-lived supercells can produce tornado families. Tornado families occur when several tornadoes are produced in a cyclic (repeating) manner, usually by separate mesocyclones in the same supercell.

Extreme tornado outbreaks can severely stress detection and warning systems. When multiple NWS forecast offices and media outlets issue a barrage of warnings, emergency planners and the public can feel overwhelmed and even confused. The 3-day outbreak sequence on April 14–16, 2011, underscores the unrelenting nature of an intense, springtime extratropical cyclone. The daily weather maps in Figure 12.20 show the cyclone's warm sector tracking across the United States. A swarm of severe weather erupts on each day, hammering a different region.

Tornado deaths in 2011 stand at 552, the most since 1936 (555 deaths). Once again, the year's activity underscores the vulnerability of Dixie Alley: 242 of the deaths occurred in Alabama, 159 in Missouri, 33 in Tennessee, and 32 in Mississippi. Has 2011's high death toll erased decades of progress made in warning and protecting the public? Or was the year just an unusual one? In their book *Deadly Season*, tornado researchers Kevin Simmons and Daniel Sutter examined both the hazard and risk. They concluded that 2011 was an unusually hazardous year in terms of shear intensity and number of tornadoes. Throughout 2011, the deadly outbreaks across Dixie Alley did conform to the usual pattern of lethal tornadoes that characterizes Mid-South tornado outbreaks. This pattern includes (1) tornadoes after dark, (2) tornadoes during the fall and winter months, and (3) a high percentage of people living in mobile homes. However, the April 11, 2011 outbreak was more of a Great Plains–style outbreak – it came in April, mainly during the afternoon and evening, with more fatalities in permanent homes. The tornado warnings issued by the NWS were timely, but one factor may have reduced their effectiveness. An overnight and morning outbreak of severe weather across Alabama, Mississippi, and Tennessee created widespread power outages.

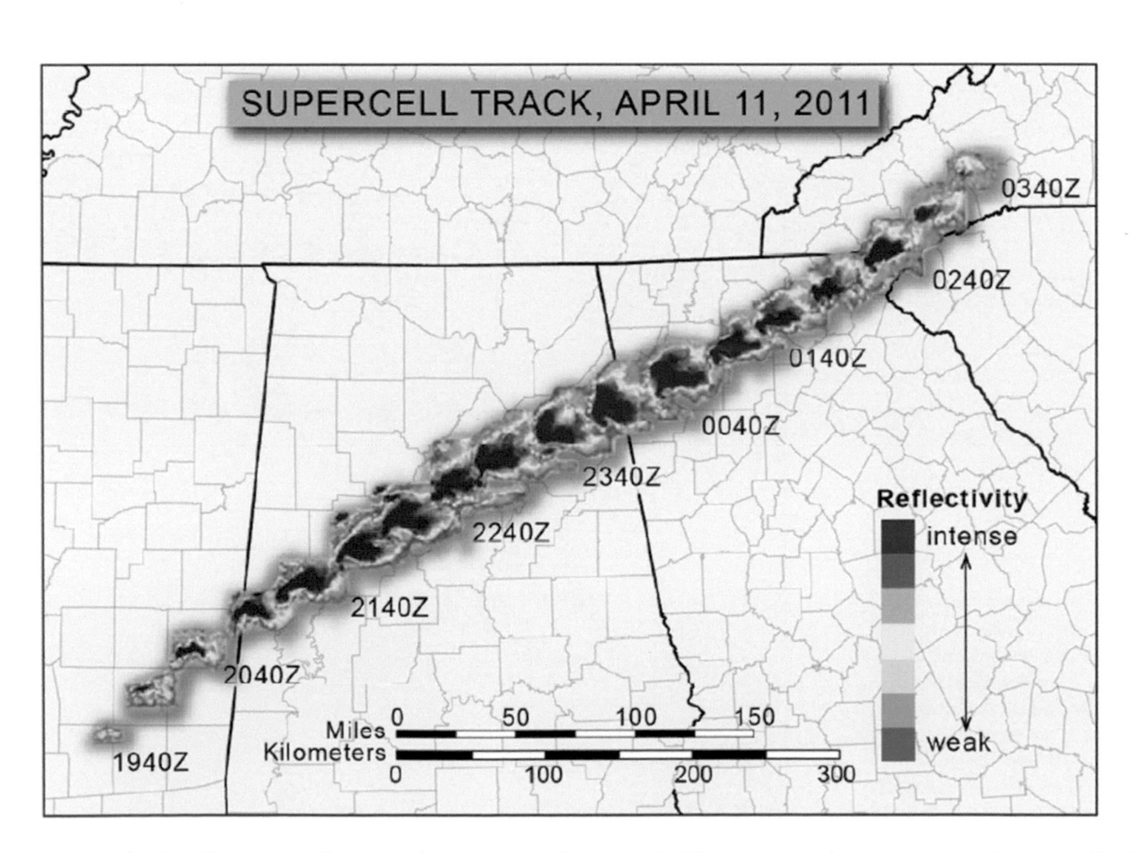

Figure 12.19 Long track of a solitary supercell: a supercell as it traverses four states. Its life cycle is clearly evident as it begins as a small cell, grows into a powerful supercell as it approaches the Alabama–Georgia border (2240Z–2340Z) and then slowly dissipates by 0340Z. (Adapted from National Center for Atmospheric Research [NCAR].)

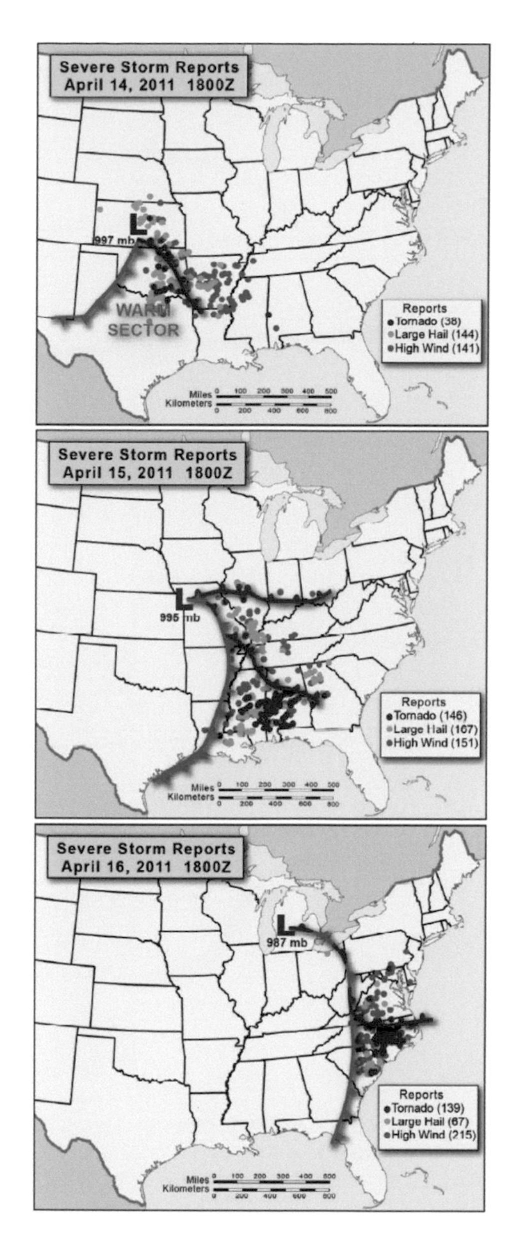

Figure 12.20 Daily Weather Maps, April 14–16, 2011. The outbreak that occurred during April 14–16, 2011, generated so many watches and warnings that it became difficult for people to keep track of all the events. Information overload can cause people to dismiss life-saving information. (Adapted from NOAA.)

These outages may have contributed to deaths from the second, more intense wave of tornadoes that afternoon and evening – because warnings never reached the public.

Figure 12.21 shows annual U.S. tornado fatalities from 1880 to 2000. Fatalities have averaged less than 100 per year since about 1965. Improved technologies, enabling better detection and warning, have certainly helped keep numbers low despite

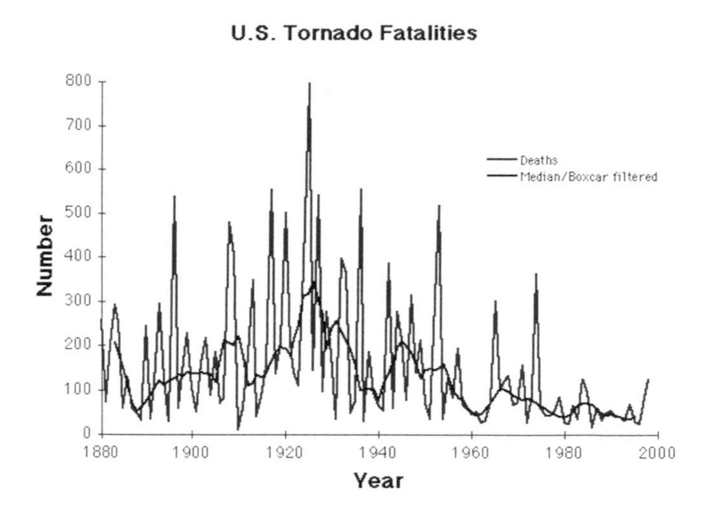

Figure 12.21 U.S. tornado fatalities per year, 1880–2000. The average death toll has declined in the past half century (smoothed black curve). (National Severe Storms Laboratory.)

growing populations that place more people at risk. But there are still big years that remind us of our occasional vulnerability. It's clear that 2011 – with 552 fatalities – was indeed an outlier year, a very sobering number that was last seen in the late 1950s. The excessive death toll stemmed both from an unusually large number of tornado outbreaks and high-end tornadoes that hit very populated regions. Is this the future, in which strong and violent tornadoes will occasionally strike towns and cities that have increasingly large suburban sprawl and thus "footprint"? Population targets are getting larger. Such unfortunate hits on big population centers can generate hundreds of fatalities and several billion dollars in damage.

Tornado Detection, Watches, and Warning

Undoubtedly, the general trend of decreasing tornado fatalities is in large part due to vastly improved tornado detection and warning systems. In the sections that follow, we take a closer look at how this technology works.

Predicting the Next Tornado Outbreak

Successful forecasts of tornadoes got their start in the 1950s when meteorologists discovered key relationships between unstable air masses, low-level moisture, frontal boundaries, and wind shear. Based on surface and weather balloon observations, they constructed composite weather analysis charts, allowing forecasters to pinpoint susceptible regions a day or two in advance. Today's satellites rapidly scan supercells at 1-minute intervals. Automated networks of wind profilers monitor changing wind shear profiles. All of these tools feed gigabytes of data into numerical prediction models, which can reliably predict regions that will experience severe thunderstorms up to 3 days in advance.

NOAA's Storm Prediction Center (SPC) is located in Oklahoma City, Oklahoma. SPC meteorologists issue daily outlooks for severe local storms, along with severe thunderstorm and tornado watches. Figure 12.22, which we call the tornado forecast funnel, outlines the multiday, three-step process for defining severe weather threat areas. The concept of the "funnel" relates to the shrinking spatial dimension, by which the threat region becomes better localized over time.

Day 3 Prior to Outbreak

- SPC issues one or more Convective Outlooks, defining risk levels (mild = green, moderate = yellow, high = magenta).

Day 2 Prior to Outbreak

- SPC updates outlooks issued on Day 3, reflecting updated numerical weather prediction guidance and atmospheric trends.

Day 1 Prior to Outbreak

- Risk areas are finalized, narrowing down specific geographic regions and times.
- High Risk category may be issued in the final hours before an outbreak.
- Particularly Dangerous Situation (PDS) bulletins may be issued.
- SPC issues Tornado Watches in the final hours before an outbreak – defining individual counties at risk.

During the Outbreak – Tornadoes Suspected or Confirmed (Doppler radar and storm spotters)

- Local NWS offices issue Tornado Warnings for portions of counties and towns, valid for up to 30 minutes.
- For violent tornadoes with the potential for great loss of life and property, a Tornado Emergency may be declared.

Let's elaborate more on this process. In a given year, very few days experience High Risk conditions, but these foretell a potentially deadly and highly destructive outbreak. When issuing a High Risk, SPC works with local NWS forecast offices and media outlets to issue media statements called Particularly Dangerous Situations (PDS). PDS bulletins contain strong language that raises public awareness to the highest possible alert level. Tornado Watches cover about 20,000 square miles and are typically valid for 6–7 hours. Watches are designed to prepare the public for the possibility of severe weather. They provide a fairly diffuse alert (typically 99% of the watch area does not experience a tornado), but they do raise the awareness for possible later tornado warnings. Schools may take the opportunity to practice tornado safety drills. Vendors hosting large events (concerts, ball games) review contingency plans. Aviation officials prepare to issue alternate flight tracks that

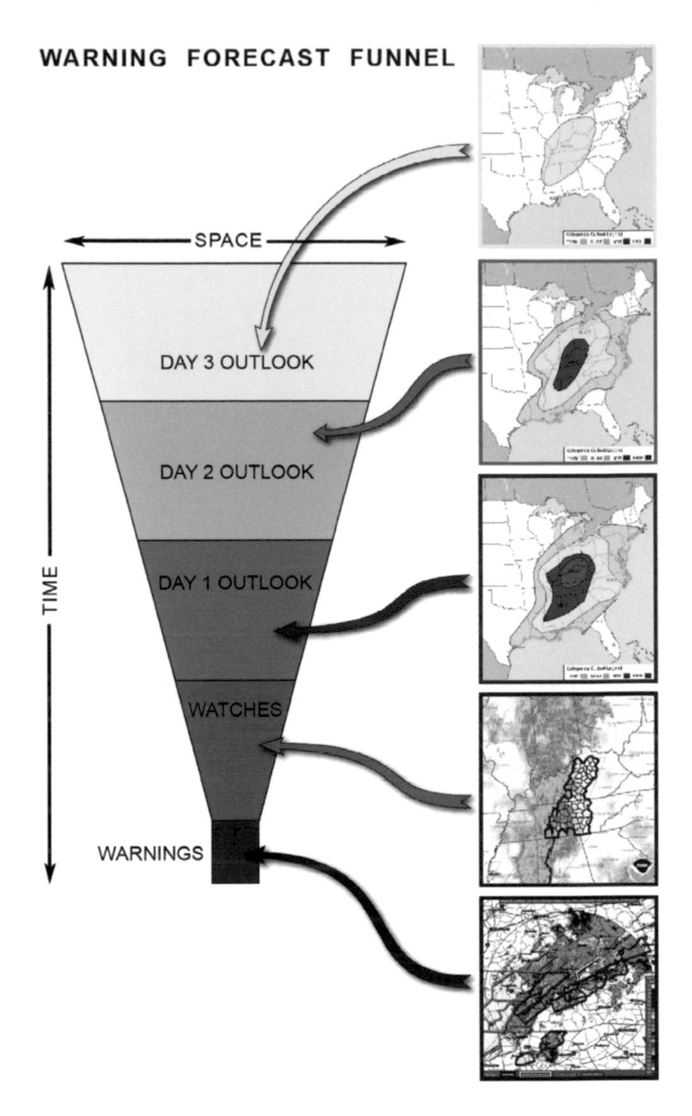

Figure 12.22 Tornado forecast funnel. In the early stages of forecasting, predictions are made several days out (Day 3). As the storm comes closer, the forecasts become increasingly more detailed (Days 2 and 1). On the day of the storm, watches are issued for regions that may experience severe weather. As the National Weather Service actually detects events happening, warnings are issued for areas that are under immediate threat. As the storm gets closer and closer, the geographic area of predictions, watches, and warnings tends to become focused on progressively smaller regions.

route air traffic away from severe thunderstorms. Networks of storm spotters are alerted. As the day's severe weather threat evolves, watches may be added, existing ones expanded, or older ones canceled in response to the formation and movement of convective storms.

Many sectors of society, including industry, farming, and transportation, benefit from the advance notification. Figure 12.23 shows the average annual number of tornado watches issued by the SPC by county across the United States. The magenta-colored regions receive the highest annual number of watches. Both the Great Plains and Dixie Alleys stand out. Certain counties in southern Mississippi and Alabama receive tornado watches 17–18 times per year. Few watches are issued along the spine of the Appalachians, the northern Great Lakes, and New England. West of the Rockies, tornado watches are almost never issued. The majority of tornadoes, about two-thirds, occur within a tornado or severe thunderstorm watch region. While severe thunderstorm watches alert people to the possibility of specific nontornado hazards (wind gusts > 50 kts [58 MPH] and/or hail > 2.5 cm [1.0 inch] diameter), there is always the possibility that a weak tornado or two will arise from a severe thunderstorm.

Weather radars and spotters feed information directly to local NWS Weather Service Forecast Offices (WSFOs). When Tornado Warnings are necessary, WSFOs also activate the Emergency Broadcast System (EBS). You are probably familiar with the EBS, which is heralded by alert tones that interrupt radio and TV broadcasts. Warnings (1) are county-specific, (2) identify cities and towns in the path of a tornado, (3) require immediate action by the public, and (4) are typically valid for less than 30 minutes. Warnings cover relatively small areas (counties or portions of counties) and are polygon shaped. They are issued when a tornado is possible based on Doppler radar velocity or confirmed (either visually or based on a special combination of Doppler signatures, as discussed later in the chapter). Their shape and orientation arise from extrapolation of the observed or radar-detected tornado. They provide the final and definitive word in the forecast funnel. In the very small warning area, an actual tornado strike is imminent. During the warning stage, WSFOs switch from the process of forecasting to one of *nowcasting* – that is, using Doppler radars to track ongoing tornadic supercells and computer programs that extrapolate storm location 30–60 minutes into the future. The fine granularity of these predictions is such that tornado warnings are now issued for individual towns and cities along

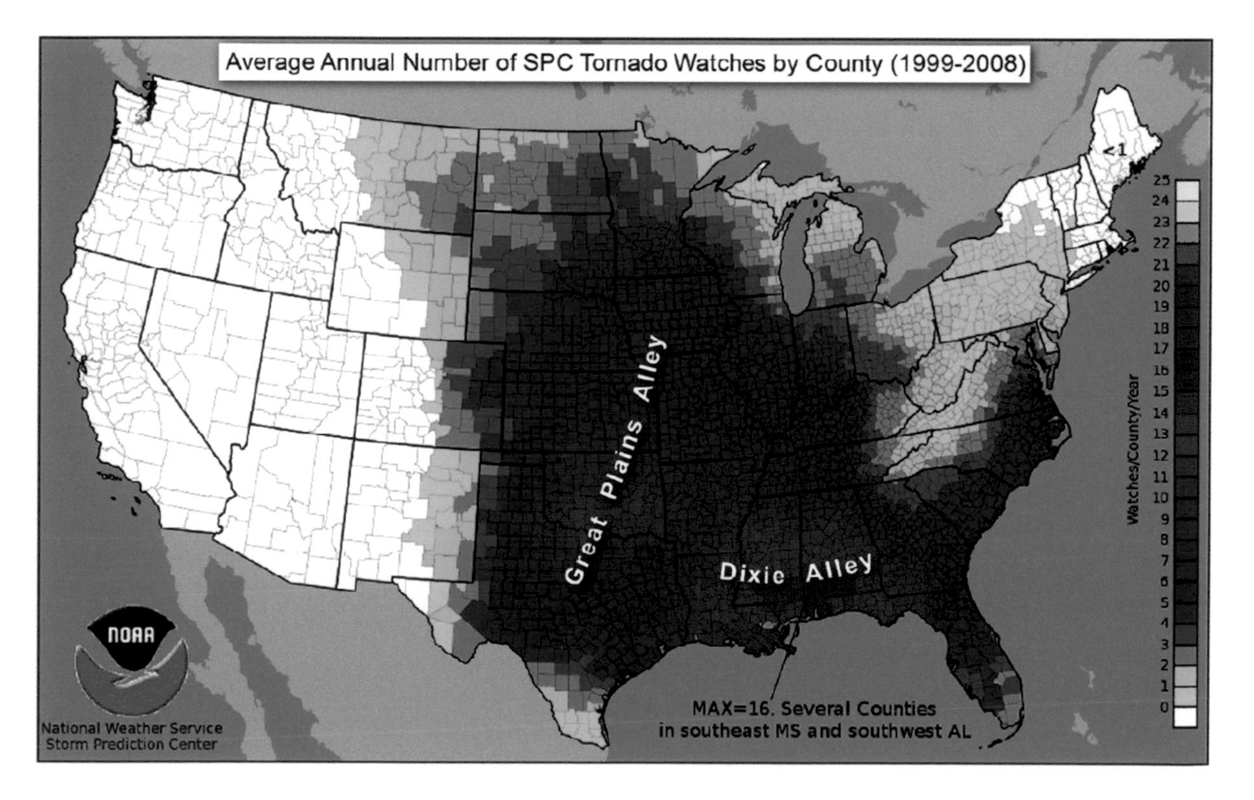

Figure 12.23 Annual tornado watches per year. Not unexpected is the concentration of tornado watches in the Great Plains and Dixie Alleys. Many of the counties in these regions receive a dozen or more tornado watches per year. Notice the absence of watches west of the Rocky Mountains. The Appalachian Mountains and New England also receive very few tornado watches. (NOAA.)

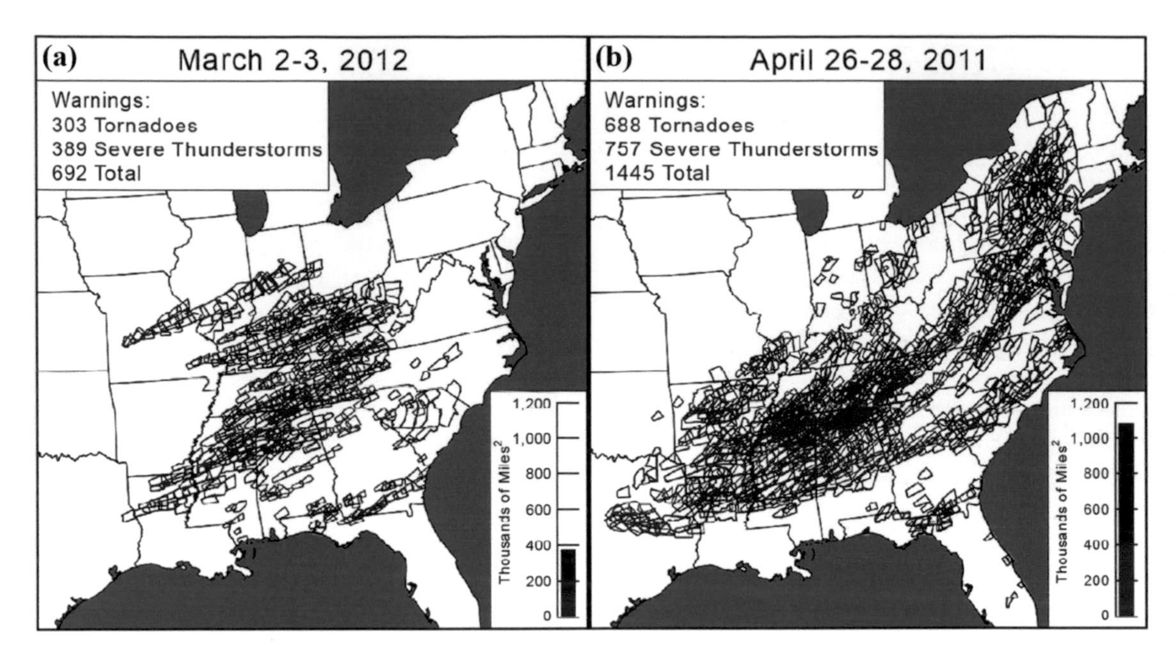

Figure 12.24 Tornado warnings and watches for two tornado events. (a) During the March 2–3, 2012, outbreak, nearly 700 tornado and severe thunderstorm watches were issued. (b) During the April 26–28, 2011, outbreak, a record 1445 watches were announced. (NOAA.)

the storm's path. When a confirmed large and violent tornado threatens one or more dense population centers – thus capable of producing catastrophic damage – WSFOs now issue a rare Tornado Emergency, which speaks to the high lethality of the tornado.

During a particularly intense tornado outbreak, dozens of WFOs issue hundreds of severe thunderstorm and tornado warnings. Figure 12.24 shows a graphic example. As we've seen, the April 26–28, 2011, outbreak was the most prolific tornado event in U.S. history. Panel (b) shows that nearly 700 individual tornado warnings were issued; clearly our warning network has become very sophisticated and responsive. The outbreak severely distressed the warning system, but it did not break down. Panel (a) shows the warning distribution for another deadly tornado outbreak in March 2012, for which 303 tornado warnings were issued across the Mid-South.

Radar Detection of Tornadoes

How exactly does weather radar "see" a tornado? Weather radars indirectly detect a tornado's most probable location. Only rarely does Doppler radar actually measure tornadic winds.

Before the days of widespread Doppler radar (pre-1990), tornado detection relied on recognition of certain radar reflectivity patterns. A supercell thunderstorm contains a helical updraft, the mesocyclone. The "meso" is a 6–10 km (4–6 mile) diameter vertical vortex extending from near the surface to the upper levels of the cloud. Most mesocyclones spin counterclockwise and are located in the supercell's southwest quadrant. Heavy rain and hail are drawn into the meso's circulation. The precipitation coils up into a spiral-shaped hook echo (Figure 12.25).

The hook echo is not a tornado. Rather, it implies that the supercell has a mesocyclone from which a tornado may develop. The hook is a key indicator of a supercell that has tornadic potential. Another important type of radar signature is a debris ball. Strong tornadoes are most likely to develop at the very end of the hook, which may enlarge into an apostrophe shape and enclose a core of very high radar reflectivity (the magenta region in Figure 12.25). Pieces of wind-borne debris (such as roofing, paneling, tree limbs, and even motor vehicles) scatter large amounts of radar energy. But debris balls are not very common, and their absence does not rule out the possibility of a strong or violent tornado.

Not all supercells produce hook echoes. Supercells in Dixie Alley are often the high-precipitation (HP) variety. HP supercells may instead have a prominent indentation or inflow notch along the southeastern edge. These storms contain mesocyclones, which are shrouded in heavy rain and hail. Also, supercells in hurricane rain bands can be particularly difficult to identify.

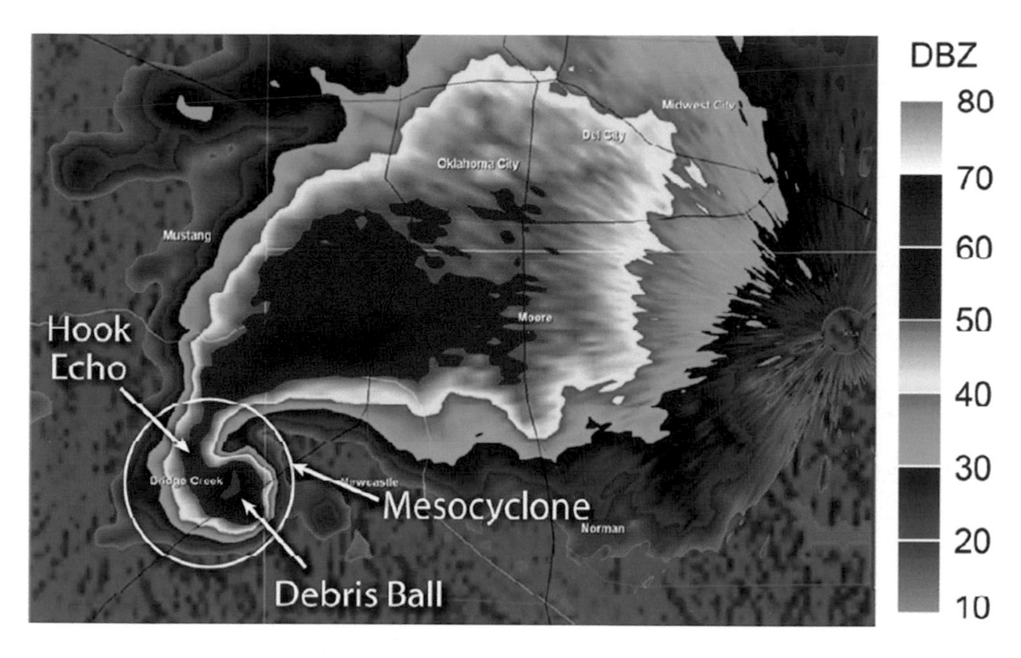

Figure 12.25 Hook echo. Before Doppler radar, meteorologists depended on the shape of rainfall patterns to identify mesocyclones that often harbored tornadoes. Hook echoes were often key to issuing a tornado warning. (NWS.)

These so-called mini supercells contain small mesocyclones, often lacking a hook echo. Mini-supercells are shallow, small, and shorter-lived than classic supercells. Meteorologists look for cores of strong, persistent reflectivity and Doppler detection of rotation (as described in the next section).

The U.S. Doppler radar network, NEXRAD, consists of 166 radar units. Doppler radar detects the frequency shift of microwave energy pulses. A shift in Doppler frequency reveals air motion within storm clouds. Doppler can only measure air velocity along the line of sight – that is, in the direction of the beam (the radial velocity) as the beam sweeps around in a 360° circle. Red colors are assigned to receding air, green to approaching air. The beam's circular sweep is stepped through a series of elevations, working from bottom to top; it takes about 8–10 minutes to complete a volume scan of air surrounding the radar antenna.

Figure 12.26 shows how the Doppler technique identifies mesocyclones within a supercell. The figure shows a typical supercell with a hook echo, but the technique can also identify rotation within bow echoes. Panel (a) shows the conventional radar depiction (precipitation reflectivity). We would expect to find a mesocyclone inside the hook echo, and indeed the Doppler wind analysis in Panel (b) confirms our expectation. The wind field contains a velocity couplet – a region of strong outbound winds (red) adjacent to strong inbound winds (green). Since the radar is located to the south and east of this storm, as we see in Panel (c), we infer from the radial velocity pattern that there must be a counterclockwise rotating vortex – a mesocyclone (Panel d).

Radar algorithms (automated computer programs) automatically identify velocity couplets and alert the radar operator to the presence of a mesocyclone. Threshold values must be met in terms of wind intensity, vertical depth, and persistence. When a meso is detected, the NWS warning-coordination meteorologist decides whether to issue a tornado warning. Not every mesocyclone produces a tornado; in fact, fewer than 30% do. For this reason, storm spotter reports are critical, confirming a funnel cloud or tornado on the ground, beneath the Doppler-detected meso. But Doppler radar does provide valuable lead time, particularly for strong and violent tornadoes that develop downward from an intense mesocyclone aloft. This lead time is currently about 8–9 minutes.

Doppler radar saves lives. It is one of the reasons why tornado fatalities have declined since the late 1980s. However, the Doppler is a remote sensing device, and it is an imperfect tool. Figure 12.27 illustrates some of its shortcomings. As panel (a) shows, the first major limitation is Earth's curvature. The radar beam travels on a straight path away from the antenna. When a supercell is close to the radar (within 40 miles), the beam passes through the lower part of the cloud, where tornado-bearing mesocyclones are strongest. When a supercell lies far from the radar (beyond 100 miles) the radar beam overshoots much of

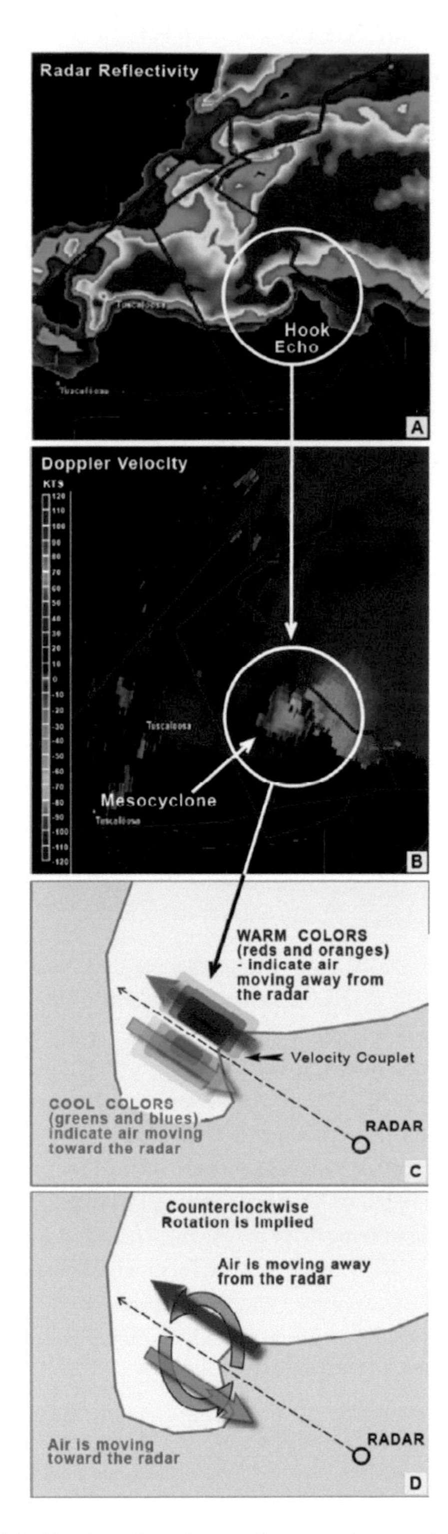

Figure 12.26 How Doppler radar reveals mesocyclones and direction of rotation. (a) Conventional radar image. The hook echo is evident, but Doppler radar must be used to determine the degree and direction of rotation, as shown in (b). The red tones indicate winds that are blowing away from the radar position. Green indicates wind that is moving toward the radar location. (c) The velocity couplet shown in panel (b). Panel (d) shows how the couplet's color scheme indicates the direction of rotation.

the storm, missing the mesocyclone. Also, if a supercell is very close to or above the radar antenna, storm features go undetected (the radar beam cannot point vertically) or become lost in ground clutter. In this case the NWS must rely on adjacent, more distant radars for guidance.

Panel (b) shows how the radar beam widens as it moves away from the antenna. At large distances from the radar, the beam's resolving power is lost. Close by, the narrow beam can adequately sample both halves of the meso's velocity couplet. In a distant supercell, the beam cannot distinguish inbound from outbound components, and the meso may go undetected.

Sometimes a large and intense mesocyclone will be close enough for the radar to detect the actual tornado circulation. In this case, an algorithm called the tornado vortex signature (TVS) is triggered. A TVS is a very concentrated, small-scale, and intense couplet of opposing radial velocities. It is either embedded within the parent mesocyclone or develops as the mesocyclone contracts into a tornado.

Doppler radars operated by the NWS are now equipped with polarimetric capability – meaning that radar pulses are transmitted with both horizontally and vertically polarized microwaves. Polarimetry allows a radar to better discern the shapes of precipitating rain and ice targets and also, in some cases, large pieces of debris lofted into the air. A small, circular region of debris captured on the polarimetric display is called a Tornado Debris Signature (TDS), which often coincides with a debris ball at the end of a supercell's hook echo. A debris ball is simply an enhanced region where large debris aloft scatters significant energy back toward the radar, regardless of its polarimetric properties. When both a hook echo and a TDS coincide with a Doppler velocity couplet, the likelihood that an actual tornado is on the ground increases to nearly 100%. An example of this "trifecta" of tornado detection is shown in three of the four panels in Figure 12.28.

Very recent research has shown promise in using Doppler radar observations to estimate tornado wind speed. In this technique, the strength of the velocity couplet is corelated against the height of the TDS (the maximum observed altitude of lofted debris). As both of these quantities increase, the probability of a more intense tornado on the ground increases. (NWS).

Measures of Tornado Warning Success

With the exception of 2011, tornado fatalities have greatly declined in the past 50 years, despite an exponentially growing population. Much of the success results from (1) increased public awareness of tornado hazards, (2) construction of home

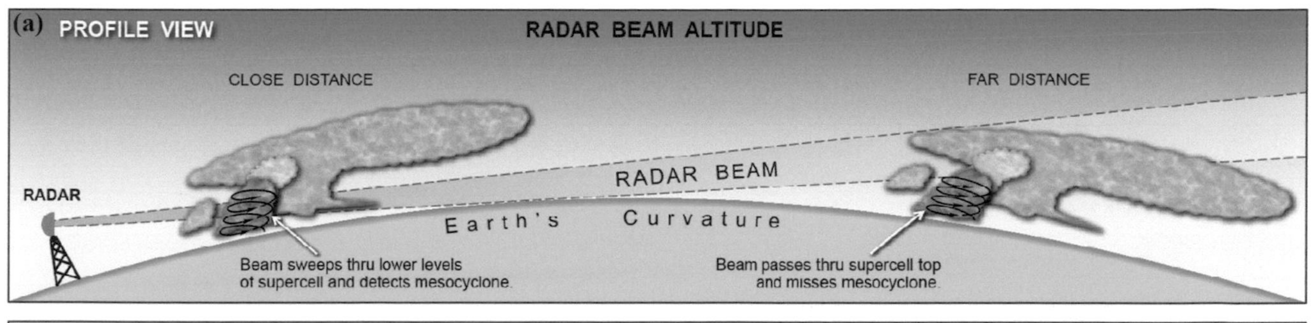

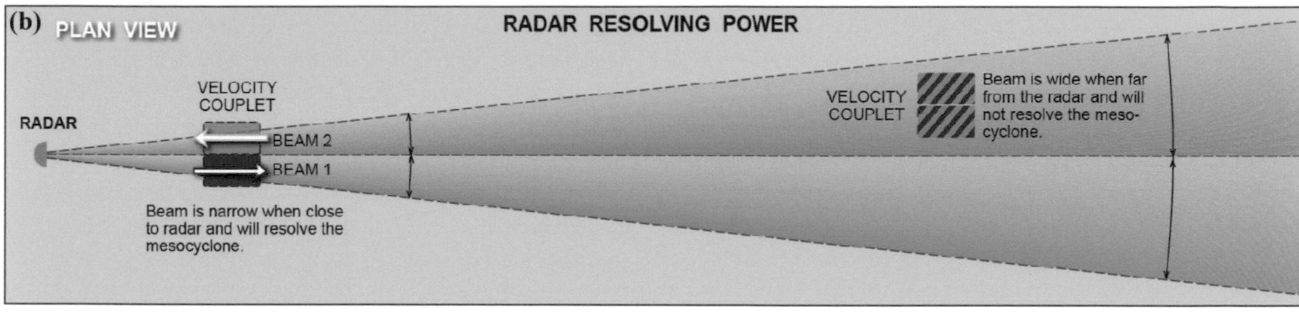

Figure 12.27 Limitations of Doppler radar. Doppler radar has limitations with regard to determining the presence of a mesocyclone. (a) The farther away a storm is, the less of the storm's lower portion is detected due to the Earth's curvature. (b) Doppler radar can determine velocity couplet rotation only if the couplet is close to the radar, and the beam width is small enough to detect both parts of the couplet individually. When the storm is farther from the radar, both parts of the couplet are detected simultaneously, preventing the radar from determining rotation.

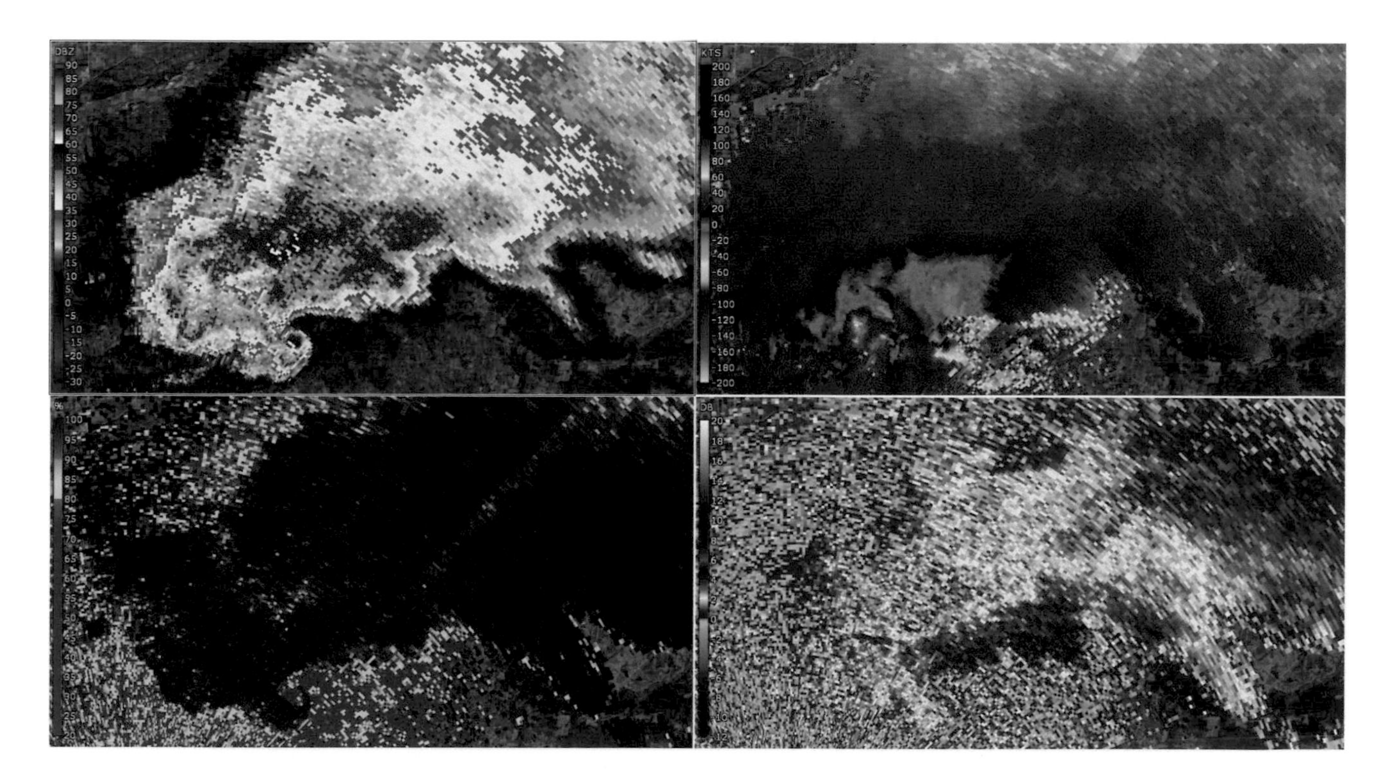

Figure 12.28 Three indicators of a tornado on the ground, using Doppler radar, for a supercell thunderstorm in Louisiana. Top left: Radar reflectivity showing hook echo with embedded debris ball (dark red circle) in the hook's tip. Top right: Doppler radar signature of the tornado; small magenta adjacent to green region indicates the velocity couplet aligned with the hook echo. Bottom left: Isolated dark blue, circular region within the hook echo indicates lofted debris in this polarimetric radar display. The three indicators (debris ball, velocity signature, and polarimetric debris signature) all perfectly coincide. Bottom right panel shows a characteristic called Doppler Spectrum Width, a measure of the degree of turbulent air motion.

and community storm shelters, (3) warning systems such as tornado sirens, (4) NOAA Weather Radio, (5) proliferation of Doppler radars among the broadcast media, (6) installation of the NWS NEXRAD Doppler network, and (7) a better scientific understanding of how tornadoes form. The aforementioned researchers Simmons and Sutter estimate that the NEXRAD Doppler network has reduced U.S. tornado fatalities by 34% since its inception.

Several statistical measures have tracked tornado forecasting success over decades (Figure 12.29). Probability of detection (POD) indicates the percentage of all tornadoes correctly warned for (i.e., three out of four tornadoes warned gives a POD = 75%). The warning lead time describes the elapsed time between Doppler detection of rotation aloft and tornado touchdown. As of 2006, lead time was around 13 minutes while probability of detection had risen to about 78%.

The past 30 years have witnessed a slight decline in the false alarm ratio (FAR; purple curve in Figure 12.29). The FAR is the percentage of unverified tornado warnings issued. In 2006, the nationally averaged FAR for tornadoes stood at approximately 75%, which is still quite high. Recall that many tornado warnings are based solely on Doppler detection of rotation in a thunderstorm, but few rotating thunderstorms create tornadoes. Although tornado warnings save lives, there is real concern that needless warnings desensitize the public; this is sometimes referred to as "warning fatigue." Since 2006, emphasis has been placed on reducing the FAR. As of 2015 (not shown), FAR had been reduced to 70%. However, the improvement in this critical metric may not be without consequences. During the period from 2006 to 2015, lead time had decreased from a high of 15 minutes in 2011 to 8 minutes, and POD had fallen to 58%. Perhaps our attempts to reduce false alarms has created some inadvertent, missed tornadoes. Additionally, the number of tornadoes arising from bow echoes and other types of squall lines has been on the rise. These tornadoes tend to be small, weak, and short-lived, spinning up in less than 5 minutes, thus causing overall tornado lead time metrics to suffer.

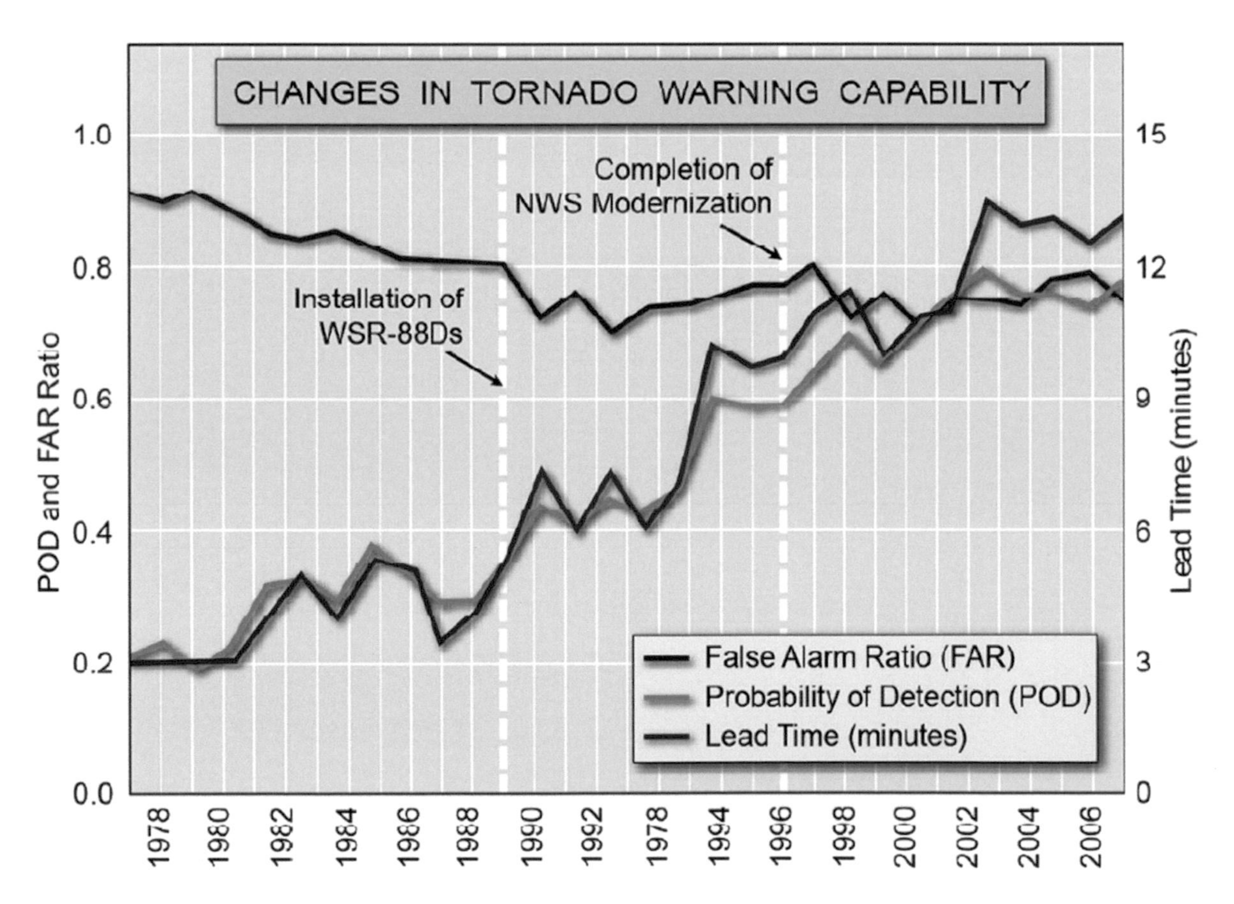

Figure 12.29 Measures of tornado forecasting success. The graph shows a modest downward trend for false alarms. The probability of detection has improved since the 1980s, and lead times for tornado warnings have steadily increased to a high point of 15 minutes in 2011 (not shown). (Adapted from NOAA, 2018.)

Tornadoes, Society, and the Media: Skywarn, Crowd Sourcing, and Pocket Doppler

The issuance of public tornado warnings is a complex process involving many agencies and a number of steps that must be executed quickly and unambiguously. Warnings are provided not just by the National Weather Service but also by the broadcast media and private weather companies. Local TV stations in tornado-prone regions maintain small Doppler radars, sometimes called pocket Dopplers. Indeed, a network that owns its own Doppler, which can provide high resolution, street-level detection and predictive tornado tracking, has a competitive advantage with its viewership. TV stations also install "towercams" and fly helicopters as a tornado approaches, providing live video coverage. Media outlets sponsor tornado spotter teams that race to the outskirts of town as a tornadic supercell approaches, scooping each other for an exclusive live feed. Local TV meteorologists have become legendary in tornado-prone regions, including Gary England (KWTV, Oklahoma City, Oklahoma) and James Spann (WBMA-LD, Birmingham, Alabama).

Another level of the tornado safety net is composed of dedicated legions of citizen tornado spotters (part of the NWS Skywarn network). Ham radio enthusiasts rapidly establish communication about a tornado's approach and assist in disaster recovery. Legions of citizens – armed with digital camcorders, cell phones, and tablets – monitor the skies and take storm footage, which they immediately upload to Facebook and YouTube. This can assist with both warning and post-storm verification that an actual tornado occurred. Many communities implement tornado sirens as a warning of last resort. Police and EMS units drive the streets, sounding sirens and announcing a tornado's approach. In locations such as the Great Plains that experience a very regular, seasonal tornado spike, community awareness is heightened, safety drills are practiced, and people automatically take shelter.

When does the tornado warning process break down? While the vast majority of tornado warnings are issued with several minutes' lead time, information is sometimes erroneously transliterated, mishandled, or misrouted as it changes hands between the issuing agency (NWS) and the broadcast media. There have been isolated instances where official warnings were not disseminated to the right community, warnings were dispatched too late, or the community received conflicting information. Sirens may malfunction or be turned off prematurely. Siren use policy varies widely among communities; in some local areas, they are used to warn for all severe thunderstorm hazards. The frequent use of sirens in unverified situations can desensitize the public and hinder future compliance; such overuse leads to warning fatigue.

While many timely tornado warnings are issued, a warning is effective only if it is properly perceived, understood, and acted upon. Failure to heed warnings is tied to the psychology of risk perception. In tornado-prone regions, long periods of time between tornado touchdowns can lead to apathy and complacency (tornado amnesia). In locations such as the Great Plains that experience a very regular, seasonal tornado spike, community awareness is heightened, safety drills are practiced, and people automatically take shelter. Inaction can stem from indecision. Sometimes people are forced to choose between two equally poor sheltering choices (i.e., stay in the mobile home vs run through the rain and lightning to a distant community shelter). Other times people simply believe "It won't happen to me" or allow bravado to get in the way of rational decision making. Indeed, the National Weather Service in recent years has been working with teams of social scientists to better understand how risk perception varies among individuals and how to better craft more compelling warning language that is disseminated to all decision makers in society.

Amazing Storms: Moore, Oklahoma – Tornado City, USA

It's often heard that "lightning never strikes twice" but can the same be said about tornadoes? There is a city in the heartland of Oklahoma, Moore, which has experienced the wrath of *23 documented tornadoes* since 1893. Some folks refer to Moore as "the tornado alley within tornado alley." The deadliest of these was the May 3, 1999, Moore-Bridge Creek tornado, rated a devastating F5, killing 36 and injuring an astounding 583, and the first billion-dollar tornado in the U.S. This tornado also has the distinction of exhibiting the highest measured winds in any single tornado, 263 kts (302 MPH), using portable Doppler radar. However, the term "measured" does not mean "direct" as in sampling by an anemometer; rather, the remote sensing techniques of Doppler radar were used. Half of the 23 tornadoes striking

Moore have occurred in May, with most of the others in the shoulder months of April and June.

Figure 12.30 is a remarkable composite of three weather radar snapshots, showing the approach of menacing supercell storms – their deadly hook echoes about to strike Moore – in May of 1999, 2003, and 2013. In each case, the hook echo contained a violent tornado (rated EF4 or EF5). This sequence in and of itself is a remarkable spate of bad luck spanning just 14 years.

The May 2013 EF5 tornado killed 24 and injured 212. Entire suburbs of well constructed wood frame homes were wiped clean of foundation slabs. No building code can offer protection from winds of an EF5 tornado, which is sometimes nicknamed "the grinder."

The Moore, Oklahoma, region is not the only spot in the United States to experience a very high frequency of tornado threats. Statistics compiled using the Iowa Environmental Mesonet show that the number one tornado-warning-issuing National Weather Service office in the United States resides in Dixie Alley, namely, the Jackson, Mississippi office. This office tallied an impressive 1031 tornado warnings in its multi-county coverage area between 2010 and 2020, compared to 741 in central Oklahoma.

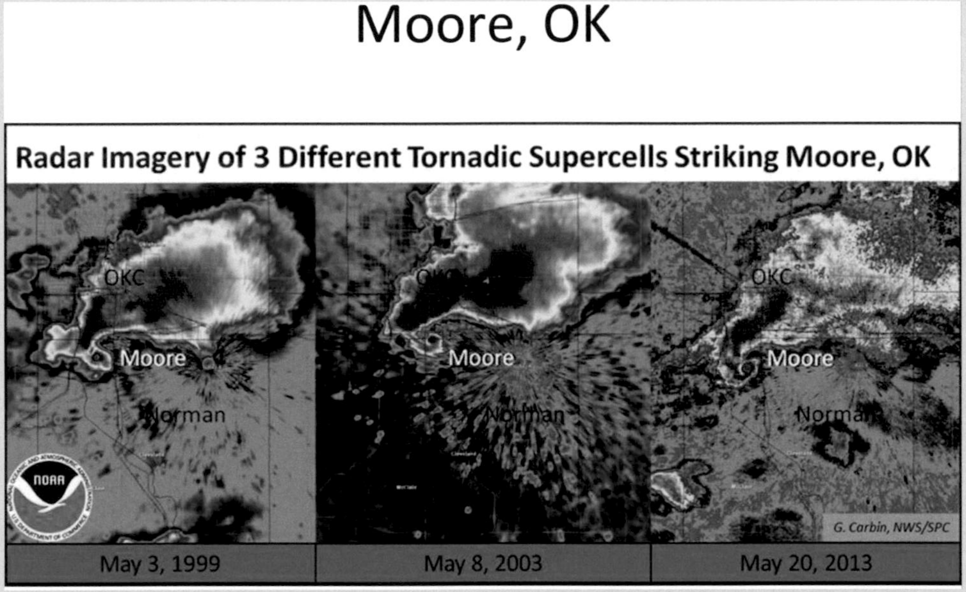

Figure 12.30 Hook echo bearing supercells detected by weather radar near Moore, Oklahoma, on three occasions from 1999 through 2003. In all three cases, note the debris balls (magenta colors) at the end of the hook appendage. This is evidence of large debris being lofted by significant tornadoes on the ground. (NWS.)

Digging Deeper: Novel Tornado Detection Technologies

Because of 2011's enormous tornado death toll, we still need to improve our understanding of tornado hazards and identify better ways to reduce societal risk. Our scientific understanding of tornado formation remains incomplete, particularly in our ability to discriminate between tornadic and nontornadic supercells. A promising new initiative is called Warn-on-Forecast. Currently, we warn for tornadoes only when detected. But numerical weather prediction models have become more sophisticated with finer spatial resolution and can now target localized regions down to individual counties. Soon it may be possible to predict the location of an incipient tornado 60+ minutes into the future, either before or just after the parent supercell has actually formed. This concept is presently being tested. Real-world radar observations of thunderstorms help provide the initial or "first guess" input into these models; every time a new radar scan becomes available, a new simulation can be run. The rapid blending of radar data and numerical modeling requires very fast computers.

Figure 12.31 illustrates the Warn-on-Forecast concept. The probable location of a future tornado (blue shades) will be used to issue warnings with long lead time. Such lead time will help evacuate large gatherings to safer locations, including sports arenas, airport terminals, hospitals, and nursing homes. Utility companies and first responders can be alerted, which will improve responsiveness.

A promising new radar technology is called Phased Array Radar (PAR). Envisioned as an eventual replacement for the NEXRAD Doppler network, a phased array is a flat panel antenna in which the radar beam is steered electronically, as opposed to a conventional radar antenna in which the dish is pivoted and swiveled mechanically. The electronically steered PAR beam can perform a complete scan of a supercell thunderstorm in as little as 1 minute, that is, in far less time than the 8–10 minutes required by NEXRAD. Since tornado formation can be a very rapid process, the PAR scanning strategy should significantly enhance warning lead time. Another goal with PAR is to decrease the spacing between radars, which will help surmount the problem of Earth curvature, so that tornadic detection can be enhanced in the lower portions of supercell clouds.

Several studies in the past decade have demonstrated how lightning networks can be used to predict the development of severe weather, such as hail, damaging straight-line wind and tornadoes. The key is monitoring total lightning, not just cloud-to-ground discharges. Total lightning count (intracloud plus cloud-to-ground) is often observed to rapidly increase, or "jump," several minutes before the onset of severe weather production in a thunderstorm.

Our conceptual model of thunderstorm cell growth (Chapter 9) explains why the lightning jump–severe weather connection makes sense. The rise in total lightning corresponds to the early phase of a thunderstorm's life cycle, when the cloud updraft is intensifying. Large hail requires tens of minutes to grow, requiring a powerful updraft to levitate stones; then several minutes transpire before the core of hail descends to the surface. The cloud downdraft develops several minutes after the updraft matures because this current is generated by the downward drag of air imparted by heavy rain and hail. Once downdrafts arrive at the surface, they create a damaging outflow of winds (Chapter 13) and are also a key component of tornadogenesis (Chapter 11).

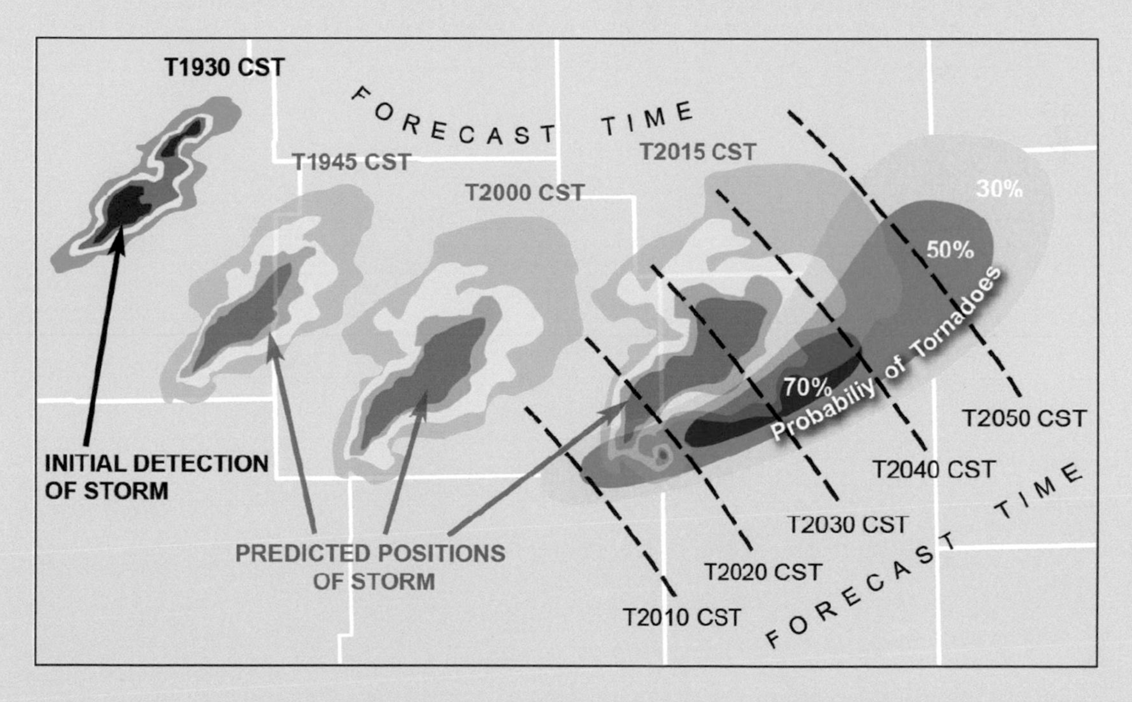

Figure 12.31 Warn-on-Forecast system, extending the lead time in tornado warning. A storm is initially detected on radar. A computer ingests the radar data, which, when combined with a host of other meteorological data, predicts where and how the storm will develop. A cone of probability for tornado development is then generated. (Adapted from Stensrud et al., 2009.)

Amazing Storms: March 2019 Dixie Alley Tornado Outbreak

As discussed in this chapter, the tornado-prone region known as Dixie Alley includes many of the states along the U.S. Gulf Coast. Alabama is often ground zero for late winter/early spring tornado outbreaks, and on March 13, 2019, the deadliest U.S. tornado outbreak since 2013 unfolded across the region. Figure 12.32 portrays the extent of the outbreak. The region shaded in light pink denotes the zone of tornado watches issued by the Storm Prediction Center, hours in advance of the outbreak. Red polygons show the location of tornado warnings issued by various National Weather Service (NWS) forecast offices. One particularly violent, long track tornado cut across far east-central Alabama into Mississippi. The tornado was rated EF4 with a 43 km (27-mile) path length, at times a mile wide, and it struck Beauregard, Alabama, during the mid-afternoon, killing 23 people and injuring 90. Such was the intensity of this tornado and parent supercell, that Doppler radar detected a plume of structural debris lofted over 3 miles in altitude inside the storm cloud. The tornado was tracked constantly using Doppler radar, and the NWS was able to provide up to 22 minutes lead time – an exceptionally long time (the national average is 9 minutes) – and issued a rare tornado emergency (described earlier in this chapter), given the storm's persistent and violent track. With such outstanding forecasts and warnings, why did so many people die during this outbreak?

The answers became apparent in the immediate days following these tornadoes. When geographical information system (GIS) tools were used to create an overlay of the tornado track and mobile home locations, it was found that the tornado cut through a very high density of manufactured homes. Even well engineered, foundation-anchored homes have trouble withstanding the 148 kts (170+ MPH)

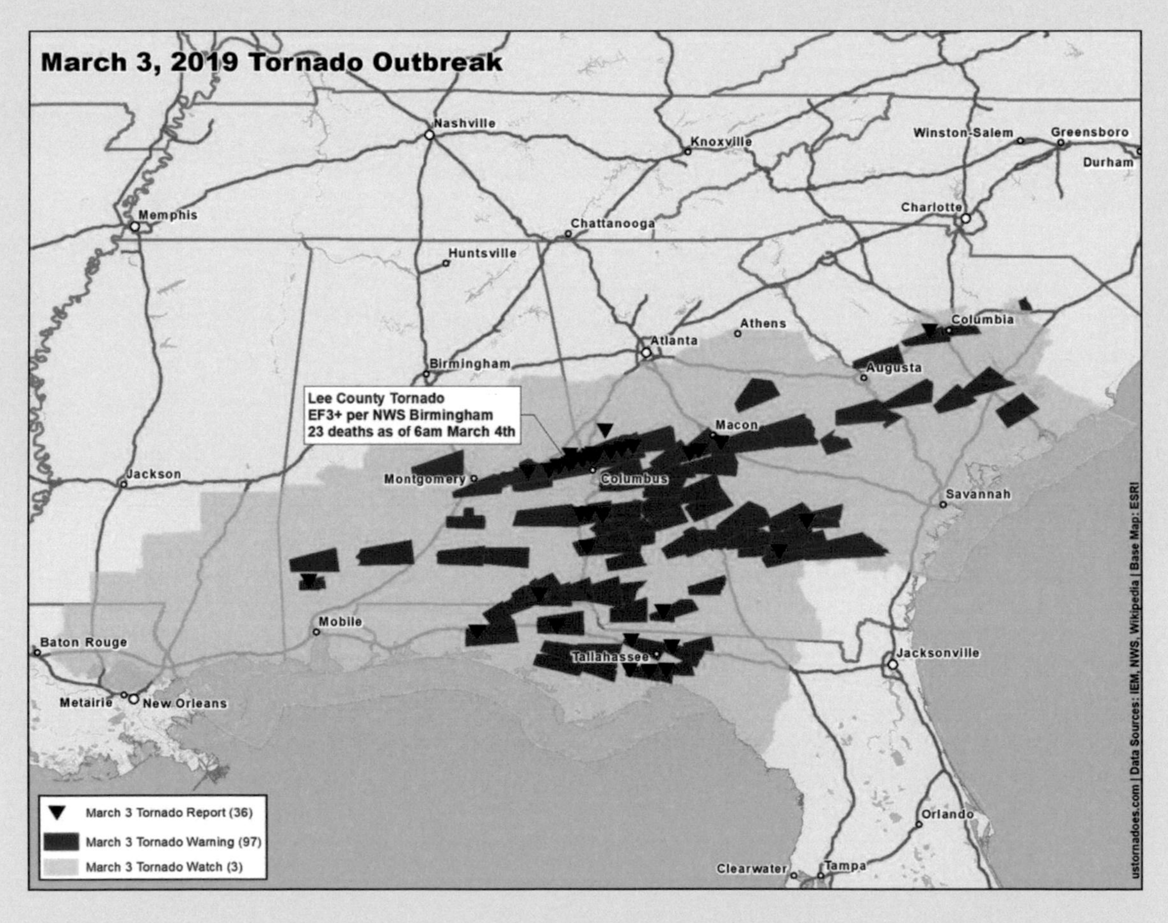

Figure 12.32 Depiction of tornado watches and warnings, along with an overlay of tornado reports, on March 13, 2019. (usatornadoes.com.)

winds of EF4, much less manufactured homes – which are very light structures anchored in a couple locations with metal tie-down straps (if at all). Additional analyses revealed that the region of the Beauregard tornado featured very few public tornado shelters, compared to other areas of the state. Simply put, despite the excellent warnings, there were simply no sturdy places available for people to seek shelter.

It's a serious misconception to think that tornadoes somehow "seek out" mobile home parks. One thing is for certain, however: As population density increases, and more large subdivisions of dwellings occupy the landscape, the size of the vulnerable "human target" is increasing. This makes future encounters between tornadoes and the built environment all the more likely . . . with concomitant escalation of insured losses and lives lost.

Summary

LO1 Discuss how tornadoes are rated and ranked according to the Enhanced Fujita wind damage scale and how meteorologists distinguish between tornado and downburst damage.

1. The Fujita tornado intensity scale assigns a range of wind speeds to six categories of tornado. Wind speeds are estimates based on severity and type of damage. The original scale was refined in 2007 and is called the Enhanced Fujita scale.
2. In the wake of a severe thunderstorm, the type of wind damage (tornadic vs nontornadic) must be assessed. Tornadoes produce a narrow, linear swath of convergent debris. Downbursts produce a localized, fan-shaped outburst of damage.
3. Strong (EF2 and EF3) and violent (EF4 and EF5) tornadoes are characterized by longer and wider paths than the weakest tornadoes (F0 and F1). Paths may be several tens of kilometers in length and 0.5–2 km in total width. The most intense damage, which creates cycloidal swaths along the track, is limited to just a few meters and is created by suction vortices (top winds ~ 328 kts [320 MPH]).
4. More than half of tornadoes that occur are rated F0 or F1; F4 and F5 tornadoes are exceedingly rare. However, deaths caused by F5 tornadoes outnumber F0 deaths by a factor of 20,000.

LO2 Discuss how annual tornado incidence varies geographically across the United States and also by time of day, recognizing that there is more than one U.S. "tornado alley."

1. In the United States, tornadoes are concentrated in several geographical regions, not just a single "Tornado Alley." The Great Plains, Mid-South (Dixie Alley), and lower Great Lakes states (Hoosier Alley) all experience concentrated tornado outbreaks.
2. Of all the tornado-spawning regions, Dixie Alley tornadoes are the most violent and kill the most people on an annual basis. Unlike the Great Plains, which experiences peak tornado activity in April–May and during the late afternoon–early evening, Dixie Alley tornadoes occur year-round, and many strike after midnight.
3. Tornadoes spawned by decaying tropical cyclones over land occur in the late summer and fall months. They account for about 4% of all U.S. tornadoes but can number 100 or more in the 1–3 day period post-landfall. These tornadoes are typically weak (F0 and F1), very short-lived, and difficult to detect by Doppler radar.

LO3 Discuss how the reporting of tornadoes, tornado fatalities, and damage have changed over the past 100 years, including influences by U.S. population growth and short-term climate fluctuations, as well as short-term climate fluctuations including El Niño and La Niña.

1. The number of detected and reported tornadoes has steadily increased since 1940, thanks to (1) increasing population, (2) dedicated scientific efforts to track and rank them, and (3) improved detection technology such as Doppler radar. The upward trend reflects only weak tornadoes; the annual frequency of tornadoes rated EF2 and higher has remained nearly constant over many decades.

2 An average of 1200 tornadoes per year occur in the United States, but this number varies between < 1000 and up to 1700. Dollar damage and fatalities can vary markedly from year to year, so long-term averages are less meaningful. For instance, whereas the average tornado death count is about 60 per year, there were over 550 fatalities in the year 2011 alone.

LO4 Distinguish between a tornado outbreak, outbreak sequence, super outbreak, and tornado family.

1 Tornado outbreaks refer to multiple (> 10) tornadoes in a 24-hour period, over the same geographical region, and from the same parent weather system (typically an extratropical cyclone). An outbreak sequence refers to 2 or more successive days of tornadic activity from the same parent system.

2 The April 3–4, 1974, super outbreak stands as the greatest in U.S. history, in terms of number of strong-violent tornadoes, fatalities, and injuries.

3 The spring 2011 season has challenged this record year in many ways. For instance, in 2011 there were three tornadoes (part of very large outbreaks), each of which produced greater than 60 fatalities; high fatality counts in excess of 50 per tornado have not occurred since 1971. Most of 2011's 550+ fatalities occurred in Dixie Alley, but it is unlikely that this region has become more vulnerable in terms of societal factors. The fatalities were driven by an unusually large number of strong and violent tornadoes that hit major population centers.

LO5 Illustrate how tornadoes are predicted, beginning 3 days before a major outbreak, down to the time of actual tornado touchdown, and discuss how the effectiveness of the tornado warning system is evaluated.

1 The NOAA Storm Prediction Center (SPC) issues daily Convective Outlooks, which broadly define the region that may experience a tornado outbreak, up to 3 days in advance. Within 6–12 hours of an outbreak, one or more tornado watches are issued, which narrow down the time and space window in terms of specific counties. Once probable tornadoes are detected using Doppler radar or visually confirmed by spotters in the field, tornado warnings are issued by NWS Weather Service Forecast Offices.

2 Doppler radar detects the speed and direction of airflow in thunderstorms and is used to identify regions of concentrated rotation. These regions are most often mesocyclones contained in supercells, which may or may not produce tornadoes (fewer than 30% do). The lead time between Doppler detection and tornado touchdown averages 13 minutes.

3 Doppler false alarm rate for tornado detection is on the order of 75%. Other limitations of Doppler include Earth curvature, the relatively long time interval between Doppler scans, and beam spreading, which degrades resolution of storm features at increasing distance.

LO6 Describe the ways that the broadcast media and ordinary citizens are using technology to improve tornado detection.

1 Aside from the National Weather Service, TV and radio stations have enlisted their own meteorologists and even small Doppler radars to issue tornado warnings.

2 Skywarn is a cooperative, citizen observer network, operated by the National Weather Service, providing trained tornado spotters to help verify Doppler radar indications of potential tornadoes – and to fill in gaps between sparse Doppler observations.

3 Citizens armed with portable electronic devices (cell phones, tablets) have enabled the "crowd sourcing" of tornadoes via Twitter and Facebook – helping to identify tornado threats and to provide valuable verification of tornado events to the National Weather Service.

References

Brooks, H., C.A. Doswell and M.P. Kay, 2003. Climatological estimates of local daily tornado probability for the United States. *Weather and Forcasting*, 18:626–640.

Brooks, H.E. and C.A. Doswell, 2002. Deaths in the 3 May 1999 Oklahoma City tornado from a historical perspective. *Weather and Forecasting*, 17:354–361.

Broyles, C. and C. Crosbie, 2004. *Evidence of smaller tornado alleys across the United States based on a long track F3 to F5 tornado climatology study from 1880 to 2003*. NOAA Storm Prediction Center, Norman, OK.

Cook, A.R. and J.T. Schaefer, 2008. The relation of El Nino-Southern Oscillation (ENSO) to winter tornado outbreaks. *Monthly Weather Review*, 136:3121–3137.

Corfidi, S.F., S.J. Weiss, J.S. Kain, S.J. Corfidi, R.M. Rabin and J.J. Levit, 1975. Revising the 3–4 April 1974 super outbreak of tornadoes. *Weather and Forecasting*, 25:465–510.

Fujita, T.T., 1970. The Lubbock tornadoes: A study of suction spots. *Weatherwise*, 23:160–197. Washington, DC: Heldref Publications.

Fujita, T.T., 1978. *Manual of downburst identification for Project Nimrod*. Satellite and Mesometeorology Research Project Paper No. 156, University of Chicago, Chicago.

Gagan, J., P.A. Gerard and J. Gordon, 2010. Historical and statistical comparison of "Tornado Alley" to "Dixie Alley". *National Weather Digest*, 34:145–155.

Hoxit, L.R. and C.F. Chappell, 1975. *Tornado outbreak of April 3–4, 1974; Synoptic analysis*. NOAA Technical Report ERL 338-APCL 37, Boulder, CO.

National Oceanographic and Atmospheric Administration (NOAA), 2018. *Tornado warnings (nation)*. https://verification.nws.noaa.gov/services/gpra/NWS_GPRA_Metrics.pdf.

Stensrud, D.J., et al., 2009. Convective-scale warn-on-forecast system. *Bulletin of American Meteorological Society*, 90:1487–1499.

CHAPTER 13

Violent Thunderstorm Downdrafts: Downbursts and Derechos

Learning Objectives

1. Describe a thunderstorm's straight-line winds, distinguishing between the downdraft, outflow, and gust front.
2. Explain how downbursts form, the damage they cause, and how they are detected.
3. What two types of severe thunderstorms commonly produce widespread and destructive downbursts? What type of marginal severe storm produces isolated downbursts?
4. What is a derecho, and how is this different from a downburst? Where do these events occur and at what times of the year? How are bow echo thunderstorm systems uniquely tied to derechos?
5. State several reasons why derechos are just as hazardous as violent tornadoes and many landfalling hurricanes. What type of societal problems do they create? How do they adversely impact large ecosystems such as forests?

Introduction

On June 29, 2012, a devastating summertime windstorm tracked across 10 states, from Iowa to the Atlantic Coast, laying down an 18-hour track of devastation. The large storm produced extreme wind gusts, ranging from 70 to 100 mph (61–87 kts). It knocked down millions of trees and left nearly 8 million people sweltering in an early summer heat wave. The National Weather Service received 1100 official reports of wind damage. The death toll from falling trees, windblown debris, and lightning was 22 – a remarkable number for a nontornadic line of severe thunderstorms.

This type of severe summer windstorm is called a derecho. Meteorologists define a derecho as a long-lived, convective windstorm that produces severe wind damage across an extensive region. By definition, a derecho's wind gusts must exceed 50 kts (58 MPH) along a path greater than 400 km (249 miles). The damage intensity is similar to that of a weak-moderate tornado but distributed over a broad corridor hundreds of miles in length. What makes a derecho so intense is its propensity for generating localized blasts of intense downdraft wind, called downbursts, in large number.

Figure 13.1 illustrates the June 30 derecho from space. The line of severe storms crossed the Ohio Valley during the day, weakened slightly while crossing the Appalachians, then struck Washington, D.C. and Baltimore late in the evening, with renewed vigor. As the derecho approached, observers described nearly continuous lightning, followed by a roaring "wall of wind" that caught many by surprise. Whole trees bent over. Horizontal sheets of rain mixed with limbs and debris flew through the air. Large portions of the electrical grid were demolished, with arcing wires entangled in felled limbs. Within just 30 minutes, the storm was over. Many businesses were shut down for days. Transportations networks were crippled, and cell towers became jammed with calls. Even vital 911 call centers were knocked out.

Figure 13.2 shows the amazing number of severe thunderstorm warnings issued by the National Weather Service – over 300 in all – as meteorologists struggled to stay ahead of the storm. As a result of the widespread media coverage, the curious term "derecho" finally became a household word across a large segment of the eastern United States.

In this chapter, we examine the geographical and seasonal distribution of derechos, as well as the mechanisms by which they generate destructive winds. We then discuss the geographic and seasonal distribution of derechos, as well as the societal impacts of derechos. A single derecho can be as devastating and costly as a tornado outbreak or landfalling hurricane. Our goal in this chapter is to demystify these damaging wind phenomena, which have only recently (in the past two decades) been appreciated as a unique class of weather hazard, very deserving of a chapter in this severe weather text.

DOI: 10.4324/9781003344988-16

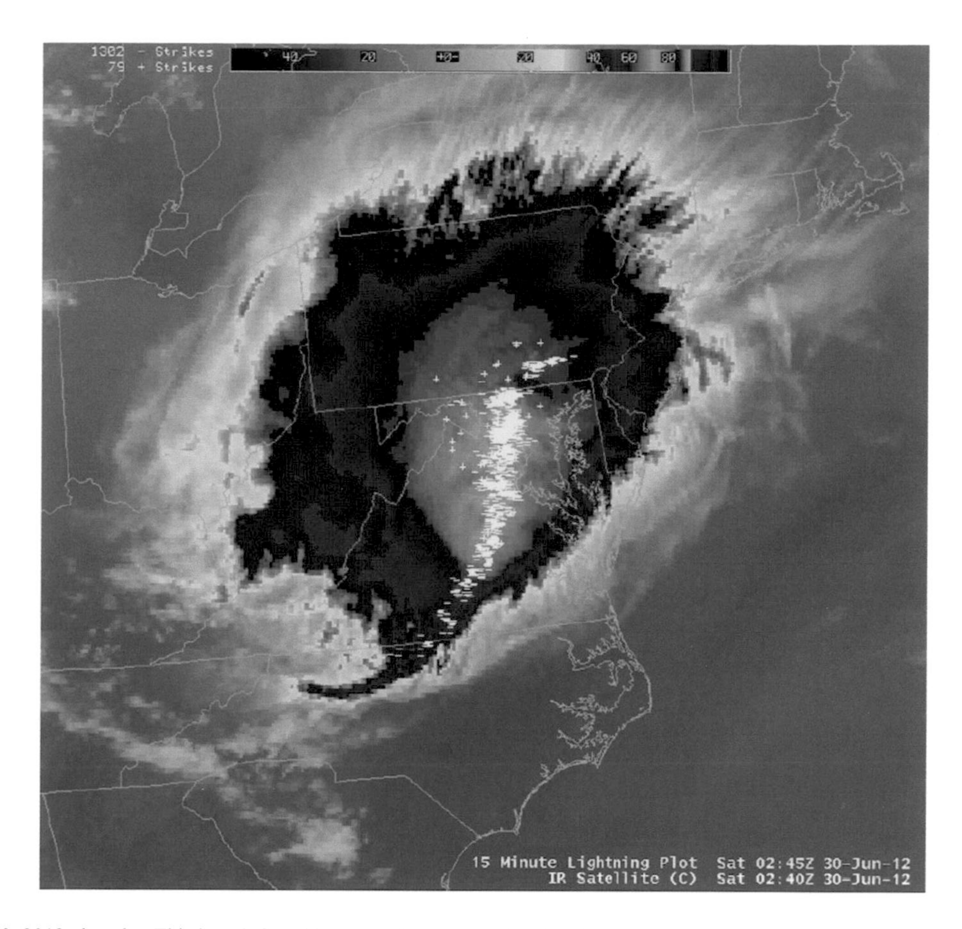

Figure 13.1 June 30, 2012, derecho. This is an infrared image of the derecho that struck the Washington-Baltimore region on June 30, 2012. (NOAA.)

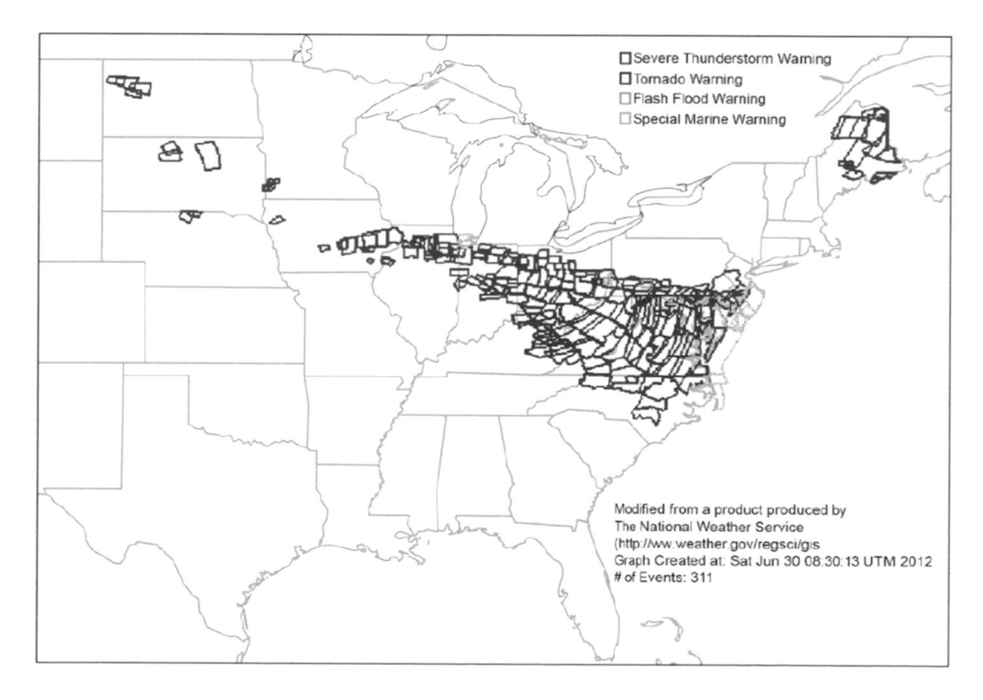

Figure 13.2 NWS severe thunderstorm warnings, June 30, 2012. Note the extraordinary number of severe thunderstorm warnings (dark blue boxes) issued by the National Weather Service during the derecho of June 30, 2012. (NWS.)

First Blast of Wind: The Thunderstorm Gust Front

In Chapters 11 and 12, we focused exclusively on tornadoes, which are highly localized, intense vortices spawned by severe thunderstorms. Thunderstorms generate three categories of severe winds (Figure 13.3), not all of which are tornadic. The first type is the thunderstorm's straight-line wind – also called the storm outflow (or cold pool) – a dense, chilled air current that fans out from the storm cell. This chilly blast of wind is perhaps the most familiar type. Wind speed rises dramatically, just minutes before the heavy rain arrives. Gusts of wind bend over trees, blow sheets of rain, and roar through treetops. The outflow is fed by downdrafts of rain-cooled air that strike the surface.

The leading edge of the outflow is termed the gust front (Figure 13.4). It is sometimes marked by a low-hanging arc of cloud, called the shelf cloud, which may appear dark gray or white (Figure 13.5). The shelf cloud forms when humid air below the cloud base mixes with rain-chilled air from the downdraft and then becomes lifted along the gust front. Moisture rapidly condenses into a menacing cloud below the main cloud base. Heavy rain, sometimes mixed with hail, forms a grayish-white curtain immediately behind the shelf cloud. The storm's outflow layer is about 1 km (0.6 mile) deep. During the storm's mature stage, the gust front remains close to the rain shaft, along the edge of the storm cloud. But as the storm dissipates, downdrafts dominate the overall storm circulation, propelling the outflow many kilometers ahead of the cell. During this stage, the storm may create a surge of strong, gusty winds but very little rain.

As Figure 13.4 shows, the gust front's head generates extreme turbulence. Its arrival is heralded by an abrupt drop in air temperature, caused by the evaporation of rain. The cold, dense air increases the air pressure by several millibars. But strong winds are the main feature. Sustained winds often approach 25–35 kts (29–40 MPH), remaining elevated for 15–30 minutes. Turbulence generates frequent gusts, sometimes exceeding 43 kts (49 MPH). In severe storm cells, gusts top the 52–61 kts (60–70 MPH) mark. Gusts may intensify with the arrival of the heavy rain and pockets of embedded hail. The buffeting wind (gust loading) creates structural vibration and damage, snapping tree limbs, bending signposts, and stripping off roof shingles. The heavy mass of rainwater and hail is slung sideways by turbulent air, adding to the wind's impact.

A region of high pressure beneath the storm cell generates outflow wind behind the gust front. Downdrafts create the

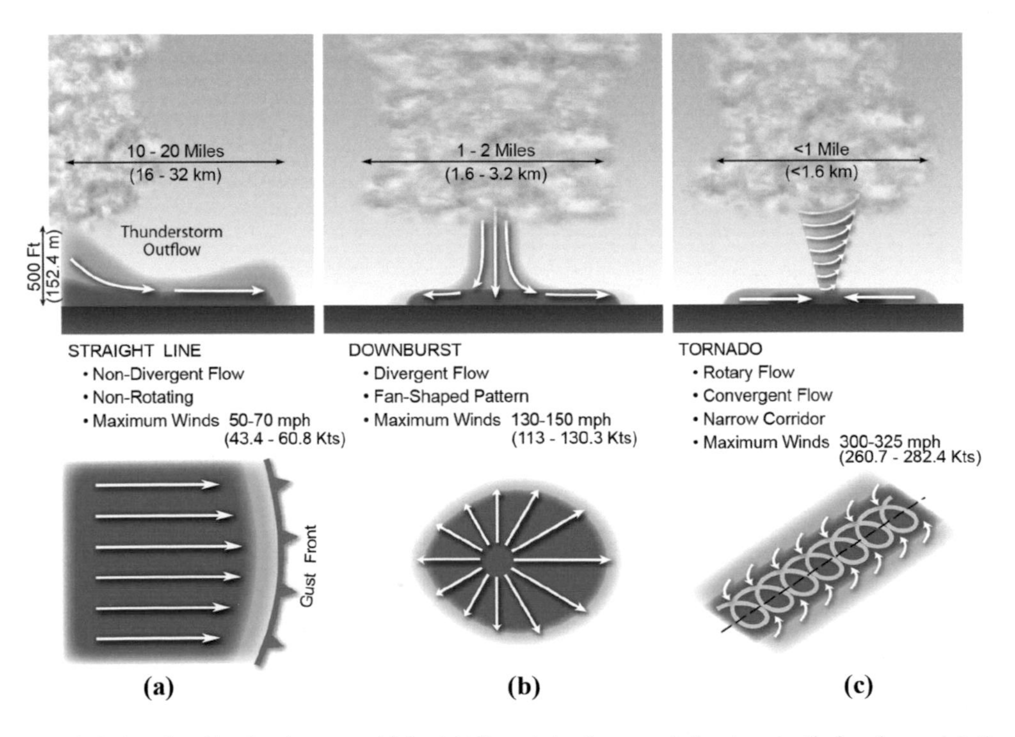

Figure 13.3 Three types of wind produced by thunderstorms. (a) Straight-line winds often precede the storm just before the precipitation arrives. (b) The downburst forms from a rapidly descending rain-chilled column of air within the thunderstorm. (c) Tornadoes, which are high-speed wind vortices, are caused by the complex interaction of updrafts, downdrafts, and shearing winds.

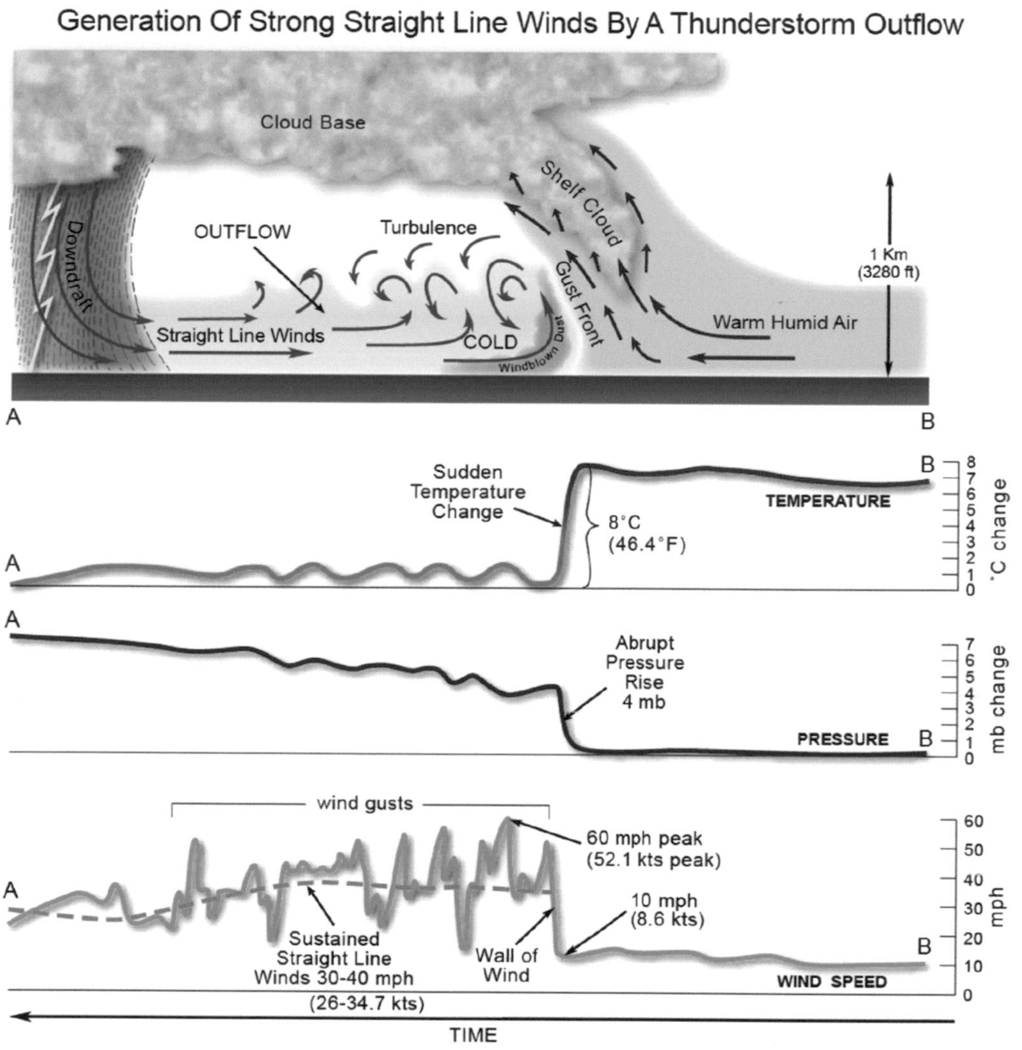

Generation Of Strong Straight Line Winds By A Thunderstorm Outflow

Cloud Base
Shelf Cloud
Downdraft
OUTFLOW
Turbulence
Gust Front
1 Km (3280 ft)
Straight Line Winds
COLD
Windblown Dust
Warm Humid Air
A
B

Sudden Temperature Change
8°C (46.4°F)
TEMPERATURE
°C change

Abrupt Pressure Rise 4 mb
PRESSURE
mb change

wind gusts
60 mph peak (52.1 kts peak)
10 mph (8.6 kts)
Sustained Straight Line Winds 30-40 mph (26-34.7 kts)
Wall of Wind
WIND SPEED
mph
TIME

Figure 13.4 Gust front. Downdrafts resulting from the cooling effect of rain cause straight-line winds to develop in advance of the rain curtain. The straight-line winds encounter inflowing warm humid air. The gust front is found at this boundary. Often a shelf cloud and windblown dust and light debris locate the position of the gust front. Notice how temperature, pressure, and wind speed change abruptly along the boundary of the gust front.

Figure 13.5 Shelf cloud. The shelf cloud is the low-hanging, band-like cloud structure that often marks the position of the gust front. (U.S. Government.)

pressure rise. The rain-cooled air is dense, raising the pressure. Some of the pressure rise is also due to a "dynamic" effect of air impacting the ground (Figure 13.6). A strong pressure gradient develops between the center of high pressure and the gust front, accelerating air outward as a broad, fan-shaped surge of straight-line wind. Over the Great Plains, turbulent airflow along the gust front lofts soil and dust hundreds of feet as billowing clouds. Over the Desert Southwest, fine sand particles billow upward along the gust front into a boiling, tan-colored cloud called a haboob (Figure 13.7).

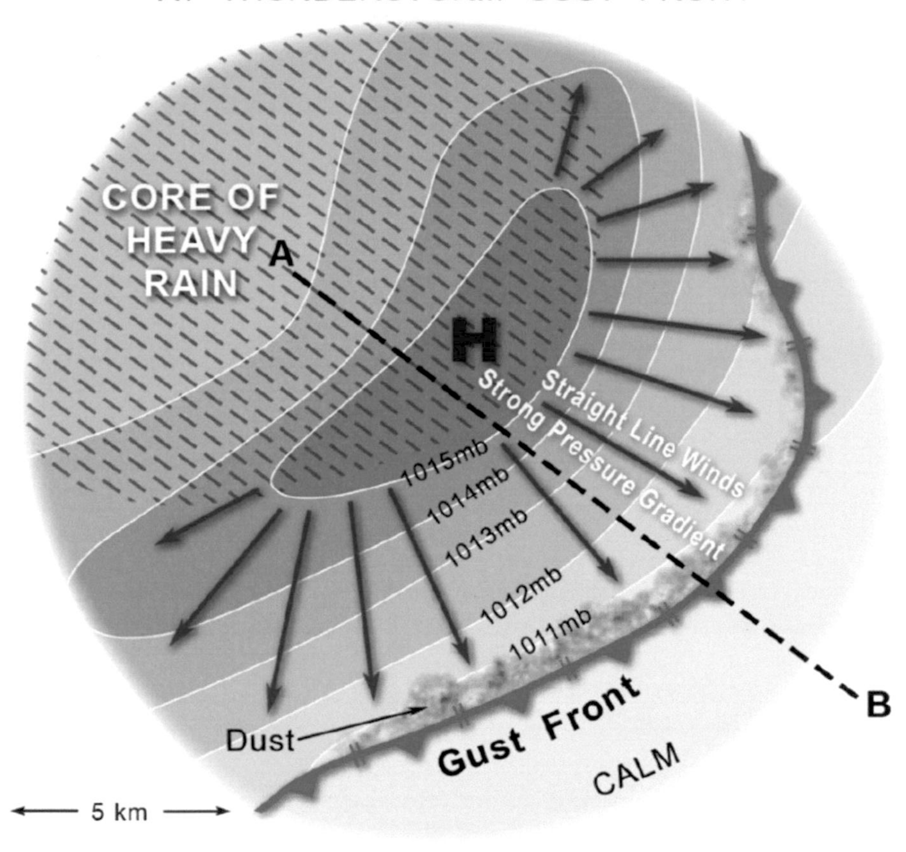

Figure 13.6 Downdrafts. The result of rain-cooled air, downdrafts elevate the pressure at the surface. As the air column cascades to the surface, it deflects outward in the form of a straight-line wind whose leading edge marks the gust front.

Figure 13.7 Haboobs. This roiling brown dust is a haboob, a result of the powerful winds along a gust front. Haboobs most often form in arid regions where large amounts of fine loose soil particles are exposed at the surface. (U.S. Government.)

Deadly Bubble of Destruction: The Downburst

Intense straight-line wind from a thunderstorm outflow creates a swath of damage along the storm's path, which may extend tens of kilometers. But the most extreme wind damage is almost always localized in the downbursts, which are pockets or "bubbles" just a few kilometers in diameter.

Basic Properties of Downbursts

A downburst is a rapidly descending bubble of rain-chilled air (Figure 13.3[b]). When it strikes the surface, the bubble fans out rapidly as a violent outburst of wind, with winds as high as 113–130 kts (130–150 MPH). This type of star-burst pattern is called a divergent outflow, and it contrasts with the parallel winds in a straight-line outflow. The pattern of downburst damage is quite distinct from that of a tornado. While the ground-level damage may appear the same, an aerial view reveals a fan-shaped, divergent damage pattern, particularly when large trees are involved (Figure 13.8). In contrast, tornadoes are known to create linear tracks or swaths, with damage converging inward toward a central axis. Cycloidal (curved) etchings may also be apparent along the tornado's damage track. Large trees are commonly uprooted, at times demolishing roofs and walls of adjacent homes. As trees mature in residential developments, they become increasingly more vulnerable to uprooting and snapping in the strong winds that accompany severe thunderstorms. Large roof sections may peel away, particularly when poorly attached to walls. Roof overhangs or eaves permit the wind to get "under" the roof and exert lift, contributing to or hastening the roof failure.

Recall that the most extreme manifestation of a tornado's wind is due to suction vortices rooted in the parent tornado vortex. The tornado, in turn, is contained within a larger, parent cyclone (the mesocyclone) of the supercell thunderstorm. There is a hierarchy of vortex scale, and the spinning wind increases with each stepwise reduction in vortex size.

Figure 13.9 shows that thunderstorm downdrafts have a similar scaling. The parent line of severe thunderstorms, which might assume the form of a bow-shaped squall line, extends 100 km (62 MPH) or more. A nearly continuous wall of straight-line wind, with an intense gust front, lies just ahead of the bow echo's leading edge. Winds may gust to 43–52 kts (49–60 MPH) behind the front. Embedded within the outflow, clusters of downbursts – a downburst family – develop. Downbursts are driven by small, intense downdrafts feeding the storm outflow. They are often 5–10 km (3.1–6.2 miles) in diameter, with winds in the 70–105 kts (81–121 MPH) range. Embedded within downbursts are even smaller pockets of wind called microbursts, which span just 1–2 km (0.6–1.2 miles) and are capable of generating winds up to 130 kts (150 MPH).

Figure 13.10 is a detailed damage survey of a downburst family, produced as a summertime bow echo raced across

Figure 13.8 Destruction from straight-line winds. Straight-line winds produce a divergent, fan-shaped damage pattern in forest areas, particularly when large trees are involved. (NWS.)

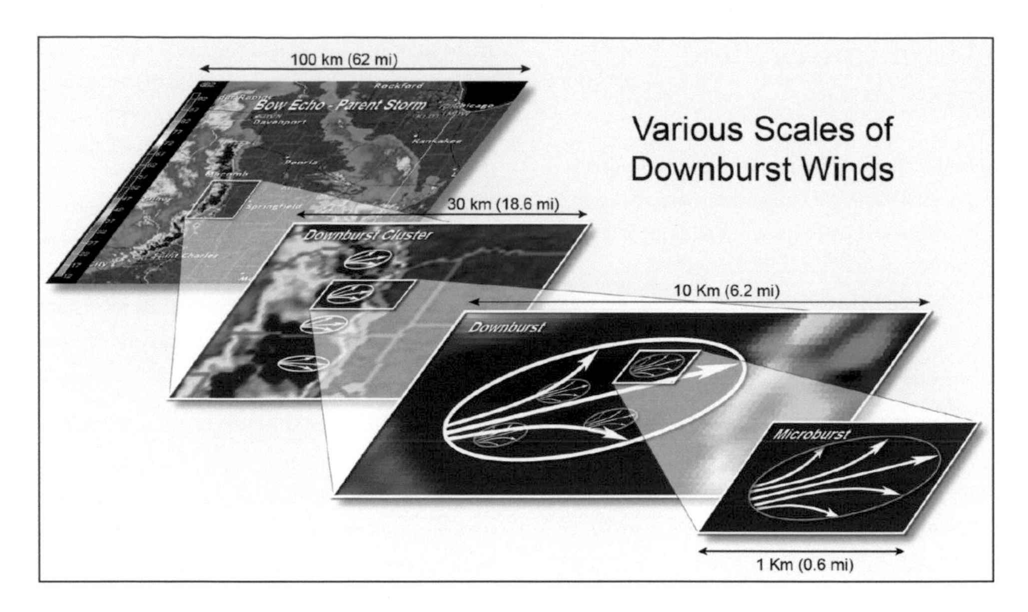

Figure 13.9 Scaling of thunderstorm downdrafts. As thunderstorm downdrafts progress from large bow echoes to downburst families, to individual downbursts and microbursts, the wind speed tends to accelerate. (Adapted from NOAA Storm Prediction Center.)

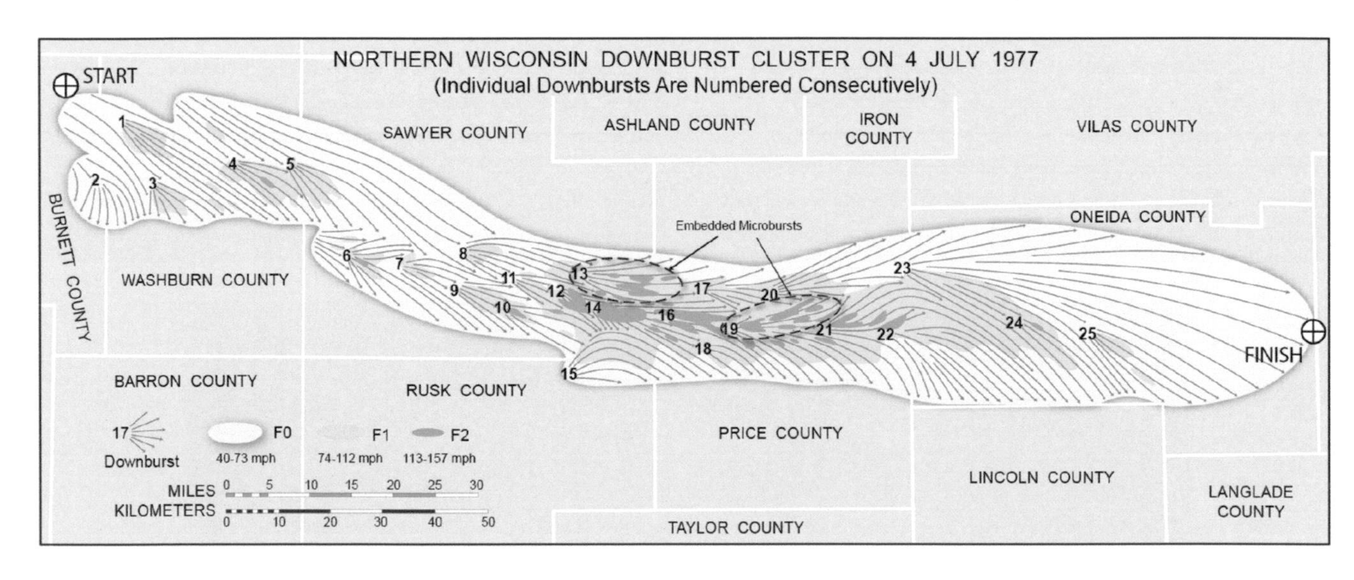

Figure 13.10 Damage survey from a downburst family. This map shows a detailed damage survey of a severe thunderstorm possessing a well-developed bow echo and numerous downbursts that traveled some 250 km across northern Wisconsin on July 4, 1977. (Adapted from Fujita, 1978.)

northern Wisconsin in 1977. The swath of wind damage was 250 km (155 miles) long by 30 km (19 miles) wide. Strong straight-line winds were continuous along this path, rated at F0 intensity at 35–63 kts (40–73 MPH). (See Chapter 12 for a discussion of the F-scale measurement of wind intensity.) Embedded within are families of downbursts, starting with the cluster of individual downbursts labeled 1–5. The next family begins with downburst 6. This family generated 18 downbursts. The downbursts had winds of F1 intensity (64–97 kts [74–112 MPH]). During the most intense phase, the fan-shaped envelopes of adjacent downbursts overlapped. The smallest pockets of wind damage, individual microbursts, generated F2 category winds (98–134 kts [113–154 MPH]).

Formation of Downbursts

How do downbursts develop? The process starts when an elevated bubble of negatively buoyant air forms near or below the base of the thunderstorm cloud (Figure 13.11). The evaporation of rain and/or melting of hail chills the air, making it dense. The dense air

DOWNBURST EVOLUTION

Cloud Base
Virga
Descending Evaporatively Chilled Air Pocket
Zone of Most Intense Surface Winds
Zone of Most Intense Surface Winds
Horizontal Vortex (Rotor)
Time 1 FORMATION
Time 2 IMPACT
Time 3 SPREAD and DISSIPATION

Figure 13.11 Development of a downburst. As rain and/or hail falls through a thunderstorm, it cools the air through evaporation, triggering a cold dense column of air that descends toward the ground. The downward drag on air exerted by heavy precipitation causes the air column to gather strength as it descends. As the air column strikes the surface, it spreads outward, forming a downburst.

has negative buoyancy, which means that the air pocket descends toward the surface, accelerating as it does so. Upon reaching the surface, the pocket generates a violent outburst of air. At the moment of impact, the most intense surface winds lie close to the point of impact. Airflow diverges away from the impact point. A zone of strong wind expands radially outward in a doughnut-shaped ring. Fast- moving air just above the ground curls up into a horizontal, ring-shaped vortex called a rotor.

Figure 13.12 illustrates further details of downburst structure. Downbursts are often asymmetric and fan-shaped. Because they are embedded in the surging outflow, they become elongated in the direction of the storm's motion. Sometimes it is possible to spot downbursts visually, particularly over the Great Plains. There, much of the rain may have evaporated before reaching the surface, but the wind whips up towering clouds of dust and soil. In other locations, such as the Southeast United States, downbursts are embedded in dense curtains of rain and are difficult to identify. If a downburst impacts the surface at the edge of a rain shaft, rainwater gets pulled into the rotor, at which time a downburst's identity is easier to discern.

How does the air inside a downburst become so intensely chilled? Downbursts fall into two categories, depending on how they form and how the air becomes so chilled. Figure 13.13 compares these two kinds of downbursts. A dry downburst (Figure 13.3[a]) is characteristic of the High Plains region of Colorado and Wyoming. The storm's cloud base is elevated because the air's dew point (moisture content) is low. Precipitation forms initially as tiny snowflakes, then sublimates (changes directly from ice to vapor) and evaporates during fallout into dry air. These processes chill the air very intensely. A descending bubble of air is formed, accelerating rapidly toward the surface.

TRANSLATING DOWNBURST

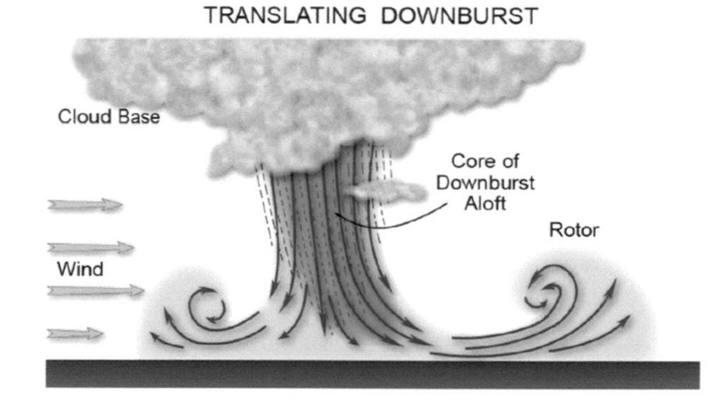

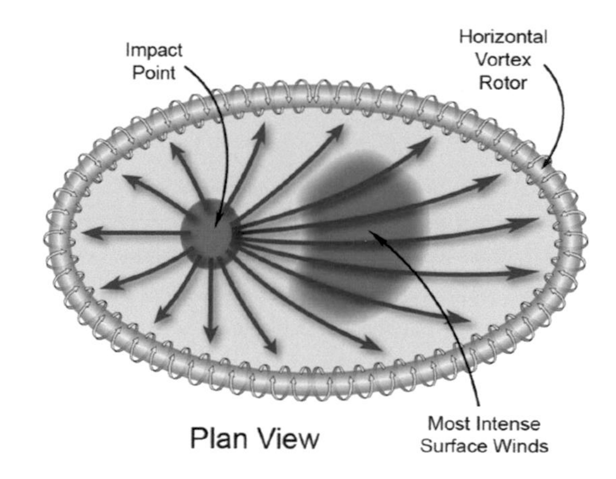

Figure 13.12 A more detailed view of the structure of downbursts. Notice the asymmetrical shape resulting from the storm movement.

In thunderstorms that form east of the Mississippi, where humid air prevails, wet downbursts (Figure 13.13[b]) are more common. The cloud base is close to the surface in high-dew-point (very moist) air. A core of very heavy rainfall and embedded hail descends through the cloud. This heavy mass of water exerts a strong drag

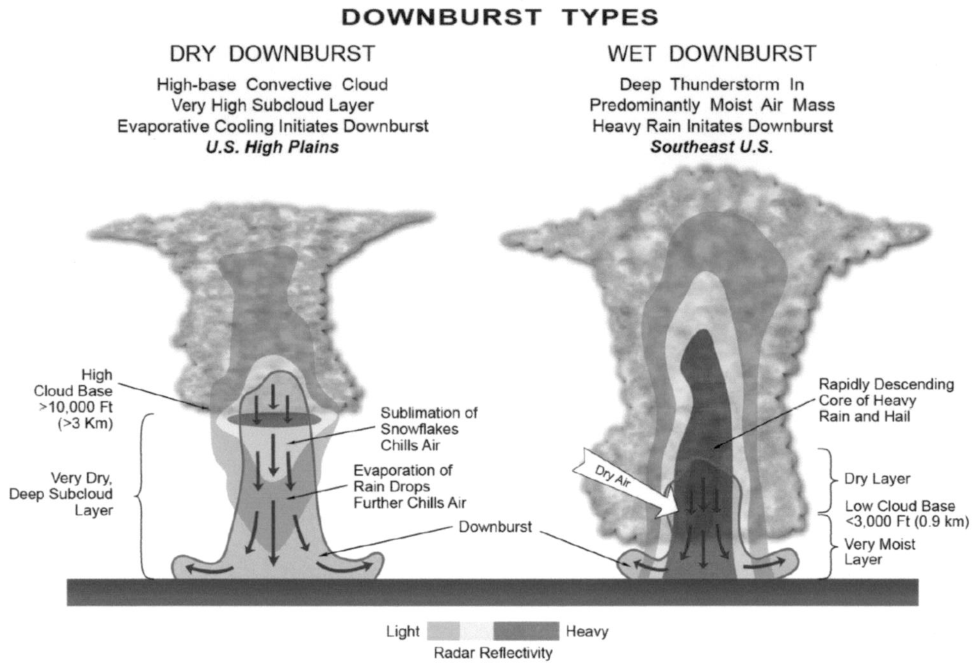

Figure 13.13 Two types of downbursts. Left: Note the high cloud base associated with the dry downburst and the lower levels of radar reflectivity. Under these conditions, evaporative cooling sets the downburst in motion. Right: In contrast, the wet downburst generally has a much taller cloud column and stronger radar reflectivity. Here, heavy rain is the primary initiator of the downburst.

on the surrounding air, pulling it downward. The cloud's downdraft pulls drier air from the cloud environment into its core, which chills the air through evaporation. When hail descends below the cloud base, it partially melts, contributing to further cooling. So a combination of water loading and evaporative cooling produces a rapidly descending bubble of air. Some radar observations reveal that the top of the storm cloud collapses in the wake of the sinking water mass, signaling that downbursts may be imminent.

A thunderstorm's updraft intensity is measured by its available buoyant energy (Chapter 9). Downdraft intensity can be similarly assessed through analysis of negative buoyant energy (Figure 13.14). Atmospheric soundings collected by radiosondes or those generated by forecast models can be used to assess the potential for downburst formation. On a day ripe for thunderstorms, air mass properties that may lead to downbursts include (1) a very dry air layer beneath the cloud base (facilitating evaporation) and (2) a strong decrease in temperature with height. If air above the surface is very warm, the difference in temperature between the chilled, descending bubble and its surroundings is increased – meaning it has large negative buoyancy (tendency to sink strongly) near the surface.

Detection of Downbursts

If we can't visually spot downbursts, how are they detected? The short answer is that a great many downbursts, especially microbursts, are not detected at all. Their presence is discovered only during post-storm damage assessment. However, Doppler radar can identify the wind signature associated with large and strong downbursts, when they occur close to the radar. Because the wind shears generated by downbursts pose aviation hazards, a network of Terminal Doppler Weather Radars (TDWR) has been installed at major U.S. airfields. These Dopplers use an extra-narrow beam and rapid scan strategy to maximize downburst detection within the immediate vicinity of air terminals. TDWRs are supplemented by surface networks of wind shear detectors and by trained observers in the control tower. Figure 13.15 shows an example of a downburst signature as seen by Doppler radar.

Large tornadoes have a very characteristic Doppler wind signature, consisting of a patch of inbound (green-shaded) and outbound (red-shaded) wind that lies on adjacent radar beams. The pattern shows the location and intensity of a small, rotating vortex. But the airflow in a downburst is highly divergent rather than spinning. The patch of inbound air (green) and outbound air (red) lie along the same radar beam and indicates airstreams spreading apart. In a downburst, the point of impact lies in the transition zone between inbound and receding winds (circled area in Figure 13.15).

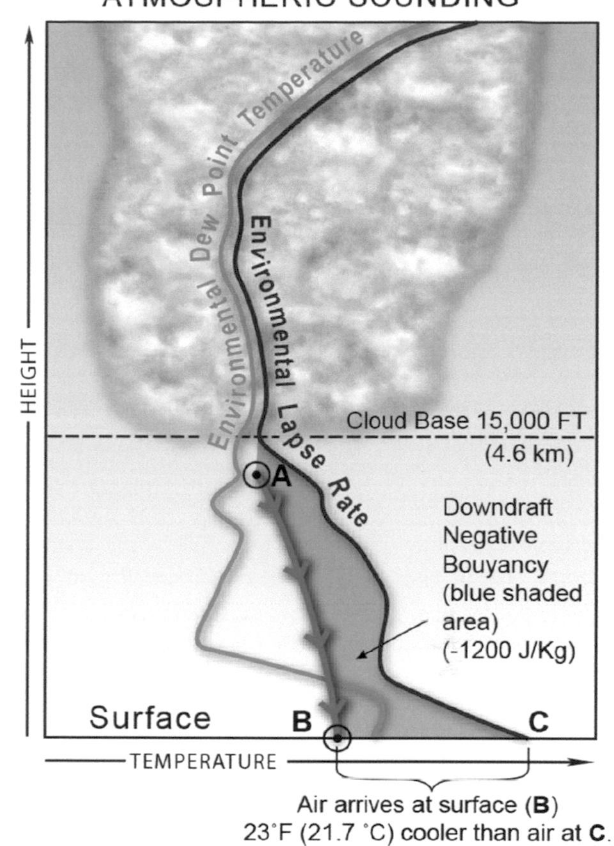

Figure 13.14 Negative buoyancy. The cloud air parcel is colder than the environment because of evaporation (Point A), and it begins sinking. Arriving at the surface (Point B), the air parcel is much colder than the surrounding air. Total area between the path of the sinking air parcel and the environmental temperature profile (shaded blue region) is proportional to negative buoyant energy. The larger the negative buoyancy, the greater the air's tendency to rapidly sink.

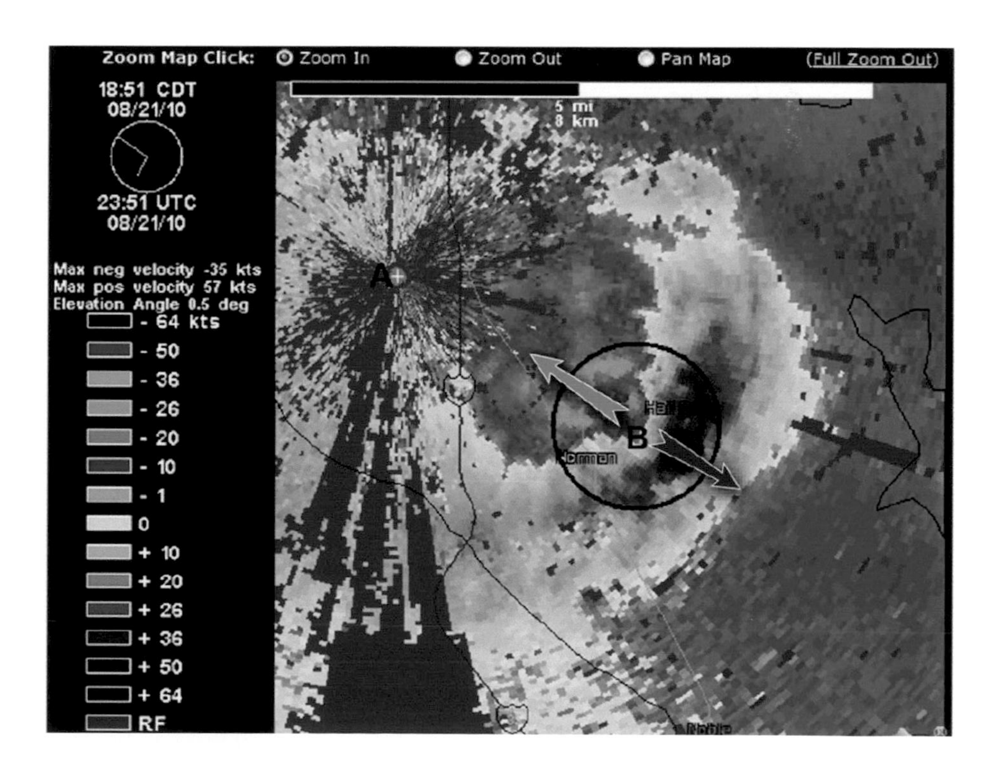

Figure 13.15 Doppler radar image displaying a downburst signature. The area circled shows wind (green) moving toward the radar system (Point A) and strong flow moving away from the system (red). The epicenter of the downburst is at the boundary of these color returns (Point B). (NWS.)

Types of Severe Thunderstorms Producing Downbursts

Downbursts are generated by a variety of severe thunderstorm types. Bow echoes are the most profligate downburst producers, but supercells – particularly the high precipitation (HP) variety – generate both downburst damage and severe straight-line wind damage. Even isolated ordinary storm cells may enter a brief severe phase, producing a pulse of downburst winds and hail – the class of so-called pulse-type severe thunderstorms. Figure 13.16 summarizes the storm types giving rise to downbursts.

Bow Echoes

Let's start with the bow echo. As described in Chapter 10, a bow echo storm system is quasi-linear and long-lived, with a characteristic protrusion or bowed-out leading edge. Families of downbursts cause the outflow winds to dramatically strengthen and surge ahead of the main portion of the line, which distorts the line into a bowed shape. When meteorologists identify bow-shaped echoes on radar, they often issue Severe Thunderstorm Warnings in advance of the storm, anticipating wind damage. These advance warnings are often required because bow echoes can accelerate to 52–61 kts (60–70 MPH) during their downburst-producing phase. The northern end of a bow echo occasionally wraps up into a distinct spiral, or comma head, because of a large cyclonic (counterclockwise-spinning) vortex. Small, short-lived tornadoes frequently occur in conjunction with downbursts, particularly along and north of the bow echo's apex. The tornadoes are very difficult to detect on Doppler radar, and their damage may be impossible to discern from the downburst damage.

Supercells

Supercells occasionally generate one or more downbursts along the rear flank of the rotating updraft (mesocyclone). They are inserted in a strong surge of outflow generated by the rear flank downdraft (RFD; described in Chapter 11). The RFD is spiral downdraft that wraps around the mesocyclone. If the downdraft becomes large and intense, it may dominate the supercell's overall circulation. Then the storm transitions from an updraft-dominated, rotating supercell to a downdraft-dominated bow echo. This transition happens most often in the HP-type supercells. Because of their heavy precipitation, HP supercells tend to develop widespread downdrafts and strong outflows early in their evolution.

Isolated Downbursts

Finally, isolated downbursts can develop even on days when organized severe thunderstorms are not explicitly forecast. When surface temperatures are very high and available buoyant energy favors rapid storm growth, a large volume of rainwater rapidly forms. When the updraft can no longer suspend the water mass, the

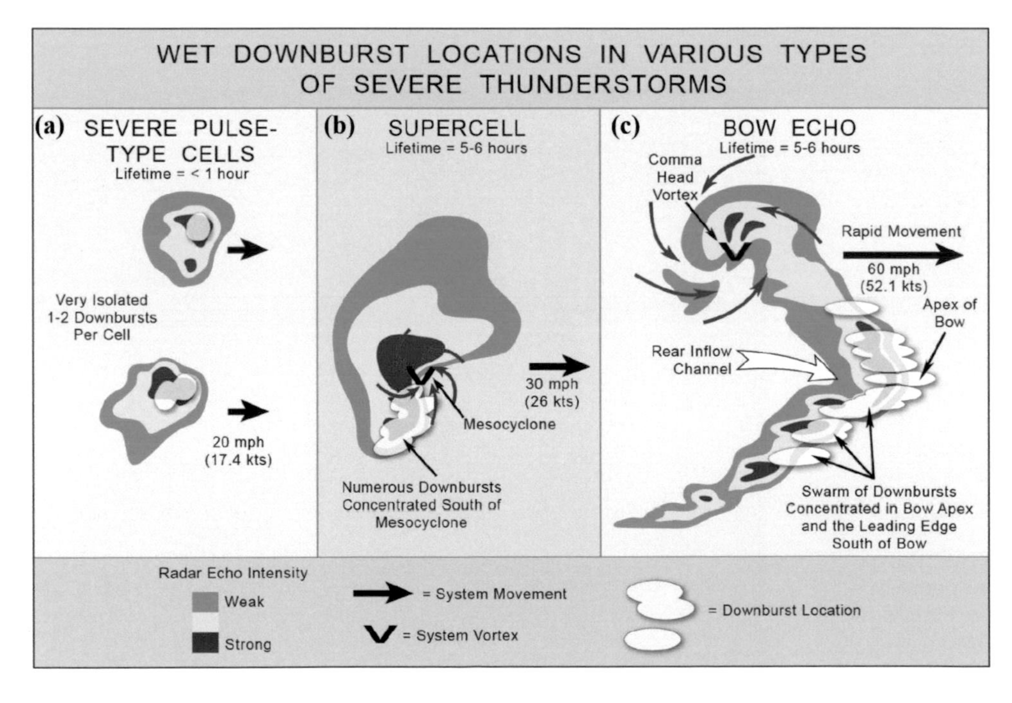

Figure 13.16 Storm types that give rise to downbursts. Downbursts may accompany (a) pulse-type severe thunderstorms, (b) supercells, and (c) powerful bow echo thunderstorms. While downbursts may be found almost anywhere within these storms, the diagrams indicate the most likely location of the downbursts for typical storms.

storm cloud collapses on itself, producing one or more downbursts. By the time the radar detects cell collapse and a warning is issued, the downburst is already in progress . . . and the storm dissipates. This pulse of damaging wind is very short-lived and localized.

Derechos In-Depth

In the long-lived, convective windstorm called a derecho, intense downburst families and severe straight-line winds are produced by one or more bow echoes. The bow echoes are exceptionally long-lived, sometimes lasting 12–18 hours.

Coining the Term "Derecho"

People who live on the Plains and in the Mid-South have always feared tornadoes, but derechos are equally formidable. They strike suddenly and generate devastation over widespread areas. The term "derecho" was coined by a chemist named Gustavus Hinrichs in 1888 to describe a severe Iowa windstorm. "Derecho" in Spanish means "straight" or "direct," distinguishing the phenomenon from the "twisted" winds of a tornado. The term fell out of vogue for nearly 100 years, until meteorologists Robert Johns and William Hirt reintroduced it in 1987. Since then, the term has slowly filtered into the public discourse on severe weather – but it still has not achieved widespread recognition as a distinct mode of severe weather.

Geographical and Seasonal Distribution of Derechos

Let's examine the annual geographic distribution of derechos across the United States (Figure 13.17), based on a 15-year climatology. You will note two corridors of high incidence: one across the Upper Midwest and Ohio Valley, the second across

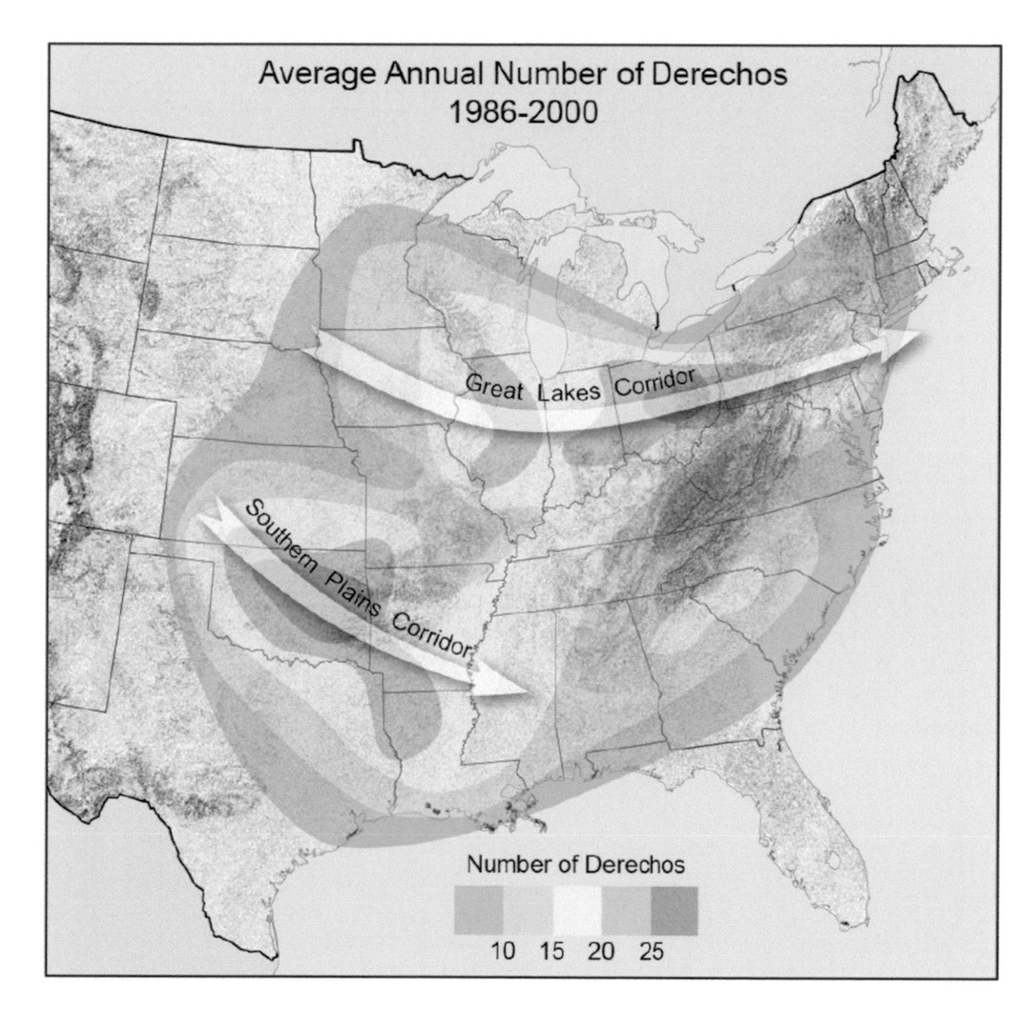

Figure 13.17 Geographical distribution of derechos. An analysis of 15 years of derecho records reveals two distinct corridors that derechos traverse across the United States. The Great Lakes Corridor drapes across the upper Midwest with the highest concentration of derechos in Ohio. The Southern Plains Corridor swings over Oklahoma and continues across the South. (Adapted from Bentley and Sparks, 2003.)

the southern Great Plains and Mid-South. There is a pronounced minimum over the Appalachians. Within the continental United States, derechos are most frequent over Ohio and Oklahoma.

Figure 13.18 divides the annual derecho distribution into warm (May–August) and cool (September–April) seasons. The geographical focus shifts markedly over the course of the year. The northern corridor experiences greatest summertime activity, while the southern corridor includes both frequent wintertime activity (over the Mid-South) and summertime action over the southern Great Plains.

Figure 13.19 shows another way to visualize the changing spatial distribution over the seasons. The figure shows

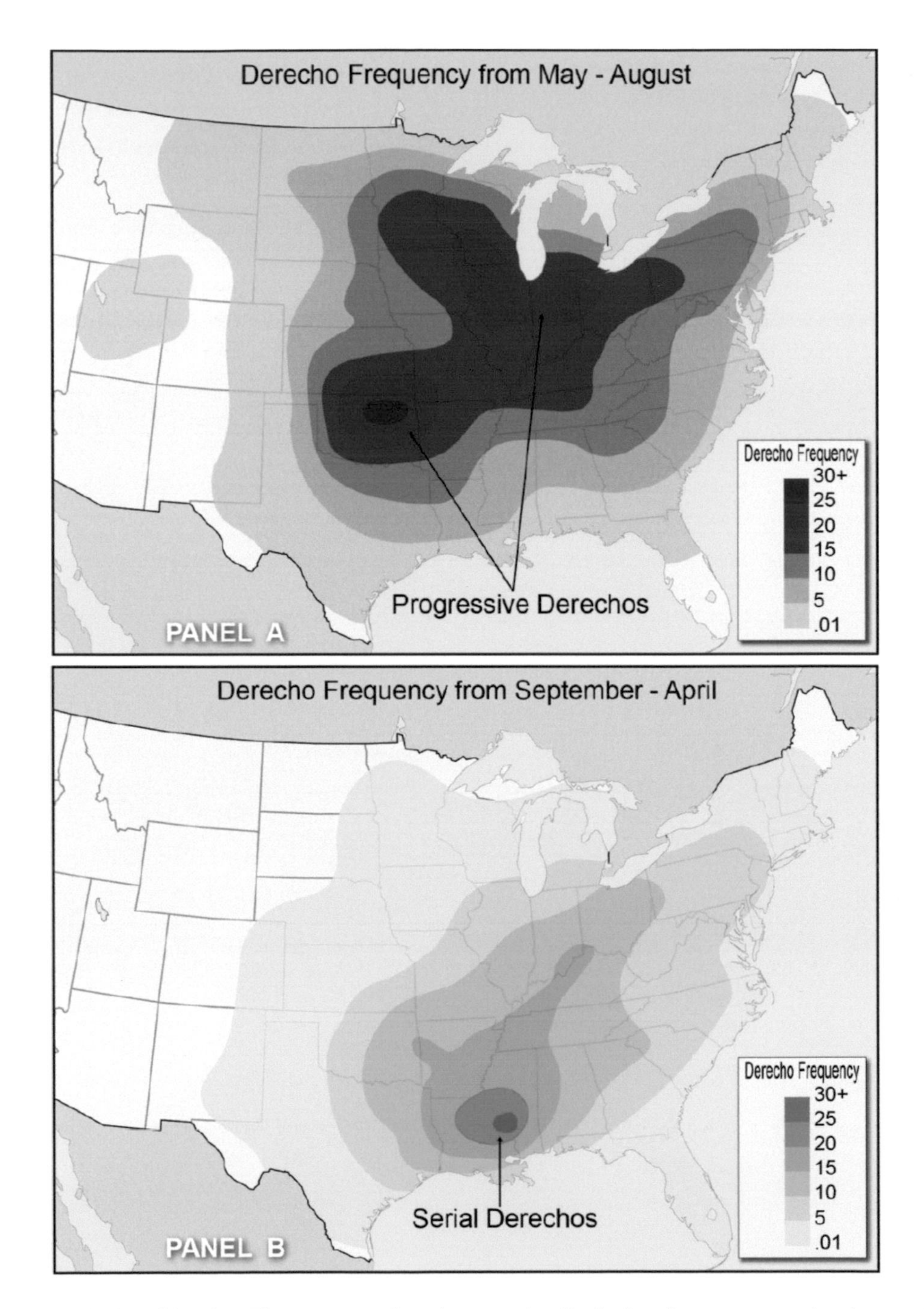

Figure 13.18 Seasonal distribution of derechos. These two maps show the contrasting distribution of warm-season and cool-season derechos. As heating intensifies with the progression of the northern hemisphere summer, derecho activity increases, and concentrations shift northward and westward. (Adapted from Ashley and Mote, 2005.)

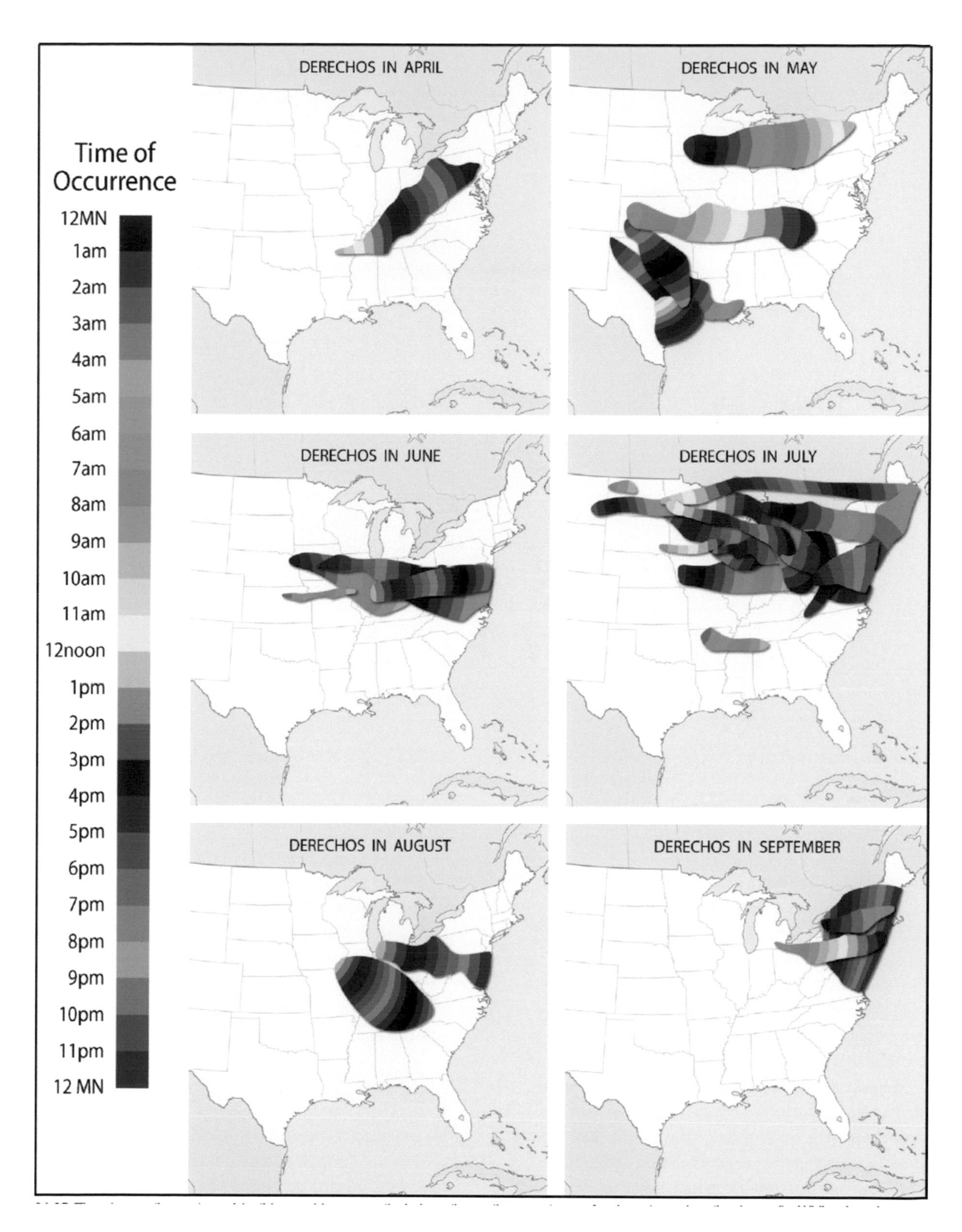

Figure 13.19 Active periods for derechos. The 6 months portrayed here are particularly active periods for derechos. Each swath of colors represents a derecho. Notice that the derechos exhibit great variation in their spatial coverage. Some tracks are exceedingly long, while others are comparatively short. Some tracks are quite wide, while others are very narrow. Additionally, a temporal component is revealed through the spectrum of colors (examine the legend).

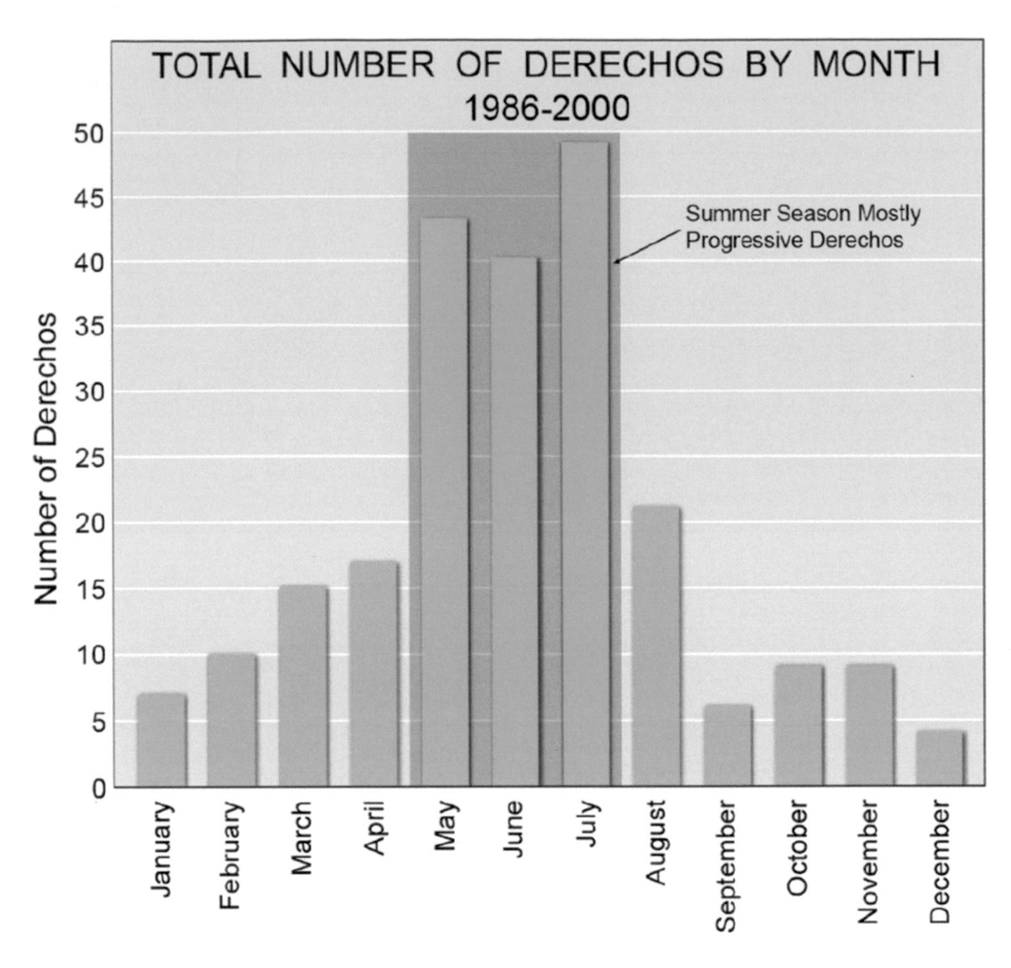

Figure 13.20 Derechos, 1986–2000. This graph plots derechos for a 15-year period from 1986 to 2000 by month. May, June, and July are the most active months. Most of the derechos that occur during the summer season are single, fast-moving arcs of thunderstorms called "progressive derechos." (Adapted from Bentley and Sparks, 2003.)

individual derecho tracks within larger climatological corridors. Six months are selected that highlight the seasonal concentration and progression. July stands out as an exceptionally active month and is clearly dominated by derechos across the northern tier. While there is considerable variation in track width and length, most derecho tracks are oriented west to east and some are exceptionally long, spanning half the continental United States. The figure's color scheme provides additional information on the timing of individual derechos. The local hour of occurrence steps chromatically from bright colors (daytime) to dark tones (nighttime). A great many derechos initiate during early afternoon and continue through the evening and overnight hours. Some derechos form late at night, persisting until the early morning. Derechos can remain very intense after sundown, continuing to produce damage overnight. These derechos often call for warnings because nocturnal storms have greater potential to harm a sleeping and thus largely unaware population.

Figure 13.20 presents the grand tally of derechos by month through the course of the 15-year climatological period. Once again, an early summer maximum (May–July) is quite apparent, with the largest numbers occurring during July. These 3 months account for 70% of all derechos. Compare this number with the tornado climatology, which shows its greatest incidence during late winter and spring.

Derecho Types

Meteorologists recognize two general categories of derecho, depending on the type of large-scale weather system involved. Cool-season months are dominated by serial derechos, so named because a sequence of small bow echoes develops repeatedly in the warm sector of an intense extratropical cyclone (Figure 13.21[a]). The cyclone and its warm sector move toward the east-northeast, sweeping an intense squall line ahead of the cold front. The short, embedded bow echoes tend to move from southwest to northeast along the squall line, propelled by strong winds in the lower and middle troposphere. Strong wind shear is more important than instability in organizing these bow echoes, as are the dynamics of the larger cyclone.

In contrast, the progressive derecho (Figure 13.21[b]) is the most common summertime derecho. It requires a very unstable atmosphere and exceptionally large values of lower atmospheric humidity. Progressive derechos feature one large, dominant bow echo that continually reforms while traveling along a stationary front. Conspicuously absent is an extratropical cyclone. Wind shear helps organize and intensify thunderstorms, but the shear is weaker than in winter. However, when winds aloft blow parallel to the frontal boundary, the bow echo is steering along the front, feeding off very hot and humid air pooled along it. In fact, summertime derechos are most likely during a heat wave. Progressive derechos translate at very high speed, from 43–61 kts (49–70 MPH). They present a significant warning challenge: National Weather Service offices ahead of the storm must efficiently coordinate a large number of warnings. These bow echoes travel quickly because they "ride the wind" along the front and because new cells continuously regenerate ahead of the storm's advancing gust front.

Figure 13.22 reveals the complex atmospheric setting of progressive derechos during a summertime heat wave. The primary weather system is an intense ridge of high pressure

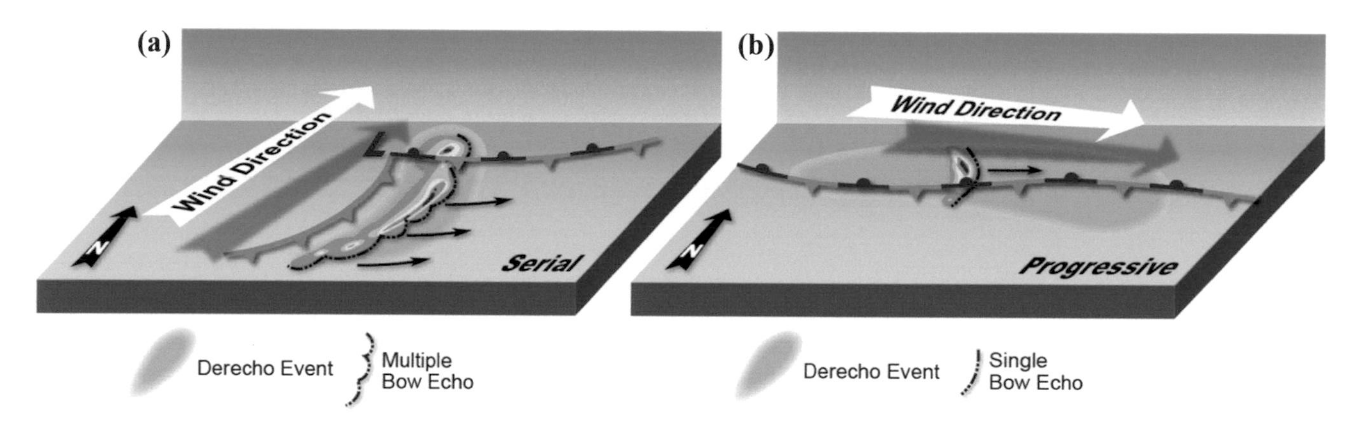

Figure 13.21 Serial and progressive derechos. This diagram shows the major differences between (a) a serial derecho and (b) a progressive derecho. Note the direction of winds aloft in relation to the storm complex's movement. The serial derecho forms in the warm sector of a large cyclone, while the progressive derecho derives its energy from a stationary front. (Adapted from NOAA Storm Prediction Center.)

MAJOR AIR MASS AND AIRFLOW FEATURES ASSOCIATED WITH SUMMERTIME PROGRESSIVE DERECHOS

WIND SHEAR
JET STREAM
TROUGH
TROUGH
ELEVATED HOT, DRY LAYER
PROGRESSIVE DERECHOS
COOL
COOL
HIGH HUMIDITY
SURFACE HEAT DOME
H
H
5,000 to 20,000 FT
(1.5 to 6.1 km)
ATMOSPHERIC SOUNDING LOCATION

Figure 13.22 Structure of a progressive derecho during a summertime heat wave. An intense region of high pressure covers much of the eastern half of the United States. An oppressive heat dome with high humidity settles beneath a stationary front. A layer of elevated hot, dry air streams in from the southwest, creating an inversion layer, or "lid," on top of the hot humid air situated along the frontal boundary, bottling it up like a pressure cooker. The trapped hot, humid air builds up until strong vertical updrafts punch through the inversion layer, releasing a large reservoir of buoyant energy that erupts into powerful, long-lived thunderstorms.

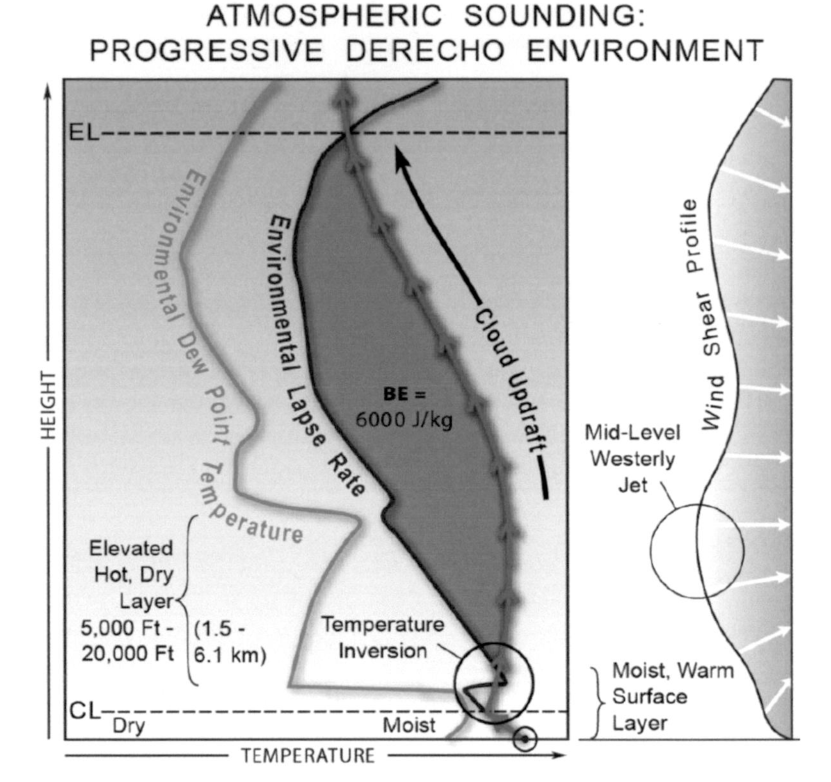

Figure 13.23 The summertime atmospheric "pressure cooker" condition leading to derecho formation. The filled-in red area on the graph represents the pent-up buoyant energy that is suddenly released when the updrafts punch through the hot, dry capping inversion (circled region).

sitting over the eastern two-thirds of the United States. The axis of the jet stream is pushed far to the north, streaming from west to east across southern Canada. An oppressive heat dome with daily high temperatures in the upper 90s or even 100+° F (38+° C) combines with high humidity pooled along the stationary front. The stationary front divides the heat dome to the south from the cooler Canadian air. Finally, an elevated plume of hot, dry air flows northeastward off the Mexican Plateau (above 1515 m or ~5,000 feet), then along the stationary front.

This elevated hot and dry layer plays a critical role in derecho formation because it helps increase the potential for explosive release of instability. This layer forms a temperature or capping inversion (rapid increase in temperature with altitude). Figure 13.23 shows a representative atmospheric sounding (vertical temperature and dew point profile) just south of the stationary front. The capping inversion bottles up the energetic air beneath it, allowing low-level heat and humidity to become excessive. During the early afternoon, cloud updrafts are not energetic enough to break through the cap. By late afternoon or evening, solar heating increases updraft buoyancy, and the updrafts can break through the cap. The updrafts suddenly tap an enormous reservoir of available buoyant energy residing above the cap, and a line of severe thunderstorms rapidly erupts along the stationary front.

The capping inversion can also be weakened by upper air disturbances. In Figure 13.23, notice that one or more shortwave troughs (upper air disturbances) pass through the jet stream. These ripples induce upward motion ahead of them, which lifts and cools the capping inversion. Individual bow echoes are thus organized by traveling shortwaves, two of which are shown in the figure. These shortwaves explain why progressive derechos tend to erupt in families, with two or three convective windstorms sweeping across the same geographical region over a few days. Some years experience more summertime derecho activity than others, but for reasons that depend on the strength and position of the heat dome and jet stream.

Families of derechos are the most extreme manifestation of summertime's so-called Ring of Fire. The Ring sets up along the northern periphery of a large heat dome across the eastern United States. Hot, humid air on the periphery of the dome erupts into thunderstorms. The anticyclone and its attendant heat wave are a regular feature of the U.S. summertime weather pattern. The Ring of Fire thus establishes a geographically

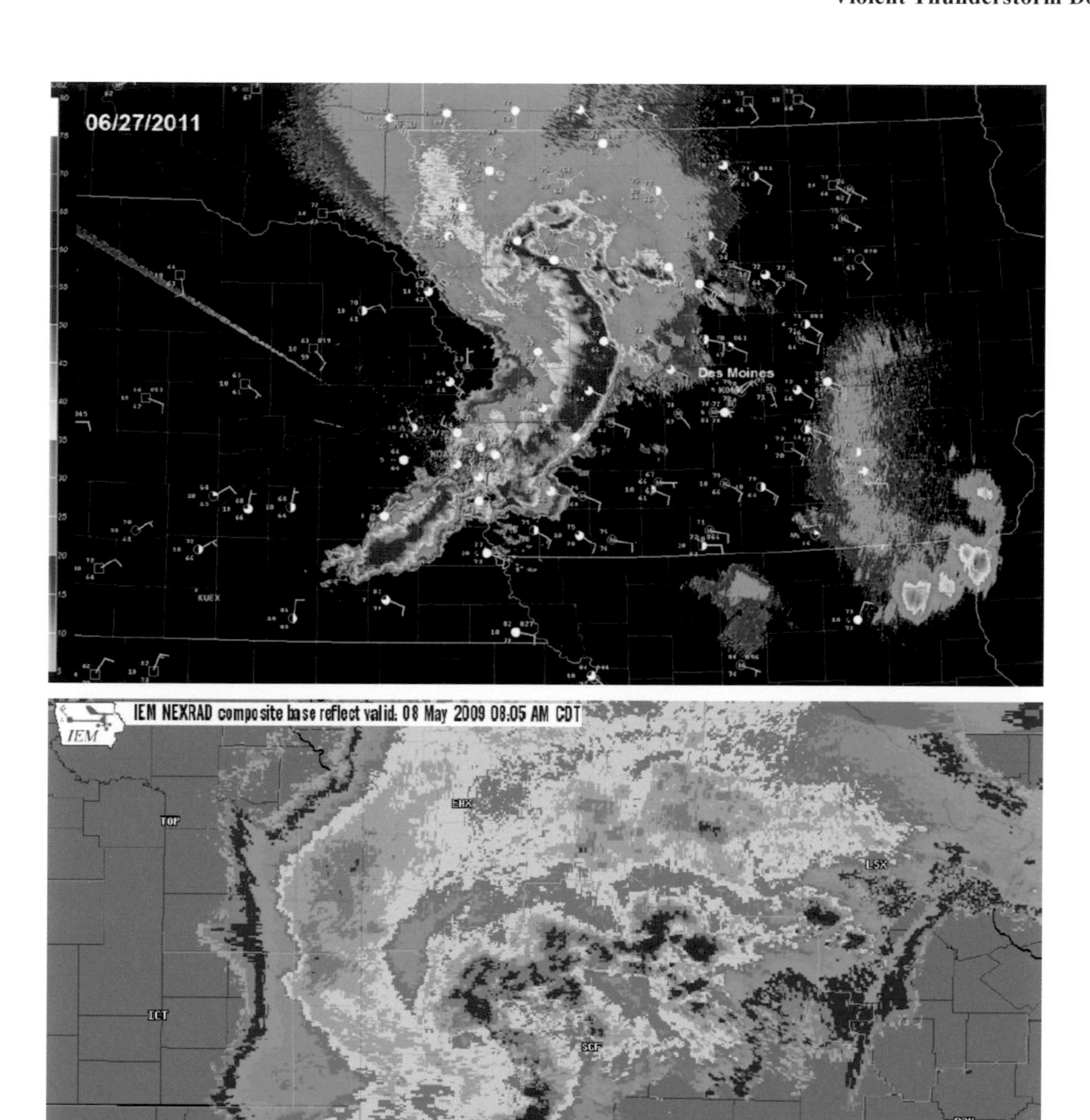

Figure 13.24 Bow echoes. These two radar images show clearly defined bow echo returns. The first image was obtained on June 6, 2011, as a severe storm was positioned over western Iowa. (NWS.) The second image reveals a strongly defined bow echo over southwest Missouri. (Iowa State University.)

favored corridor for repeated derecho activity across the Upper Midwest and Ohio Valley.

Structure and Evolution of Derechos

Bow echoes are fast-moving, arc-shaped lines of severe thunderstorms that create a long swath of destructive wind. The bows range in length from several tens to several hundreds of kilometers. Figure 13.24 shows examples of what bow echoes look like on weather radar. Figure 13.25 is a montage of radar snapshots, spaced at 1-hour intervals, along the derecho of May 8, 2009. This particular storm, called the Super Derecho, created severe straight-line and downburst winds, as well as 39 tornadoes, over a 16-hour period. The damage swath extended from Kansas to the Atlantic Ocean, with $500 million in damages and 4 fatalities. Figure 13.26 shows an interesting aspect of this bow, namely the

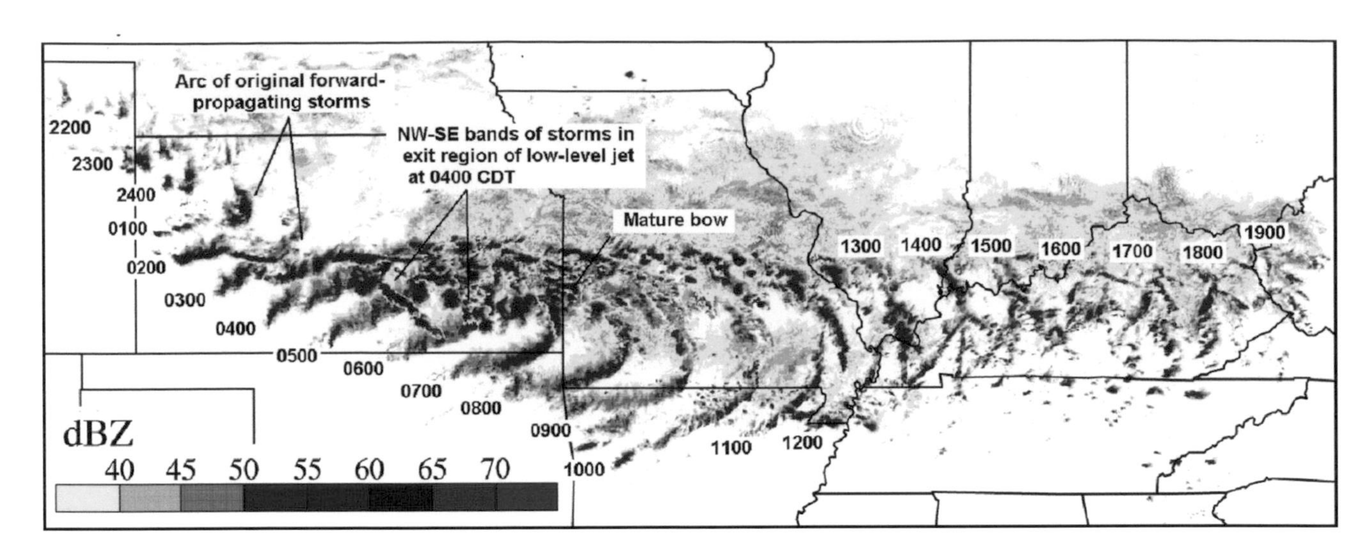

Figure 13.25 Track of a derecho. This figure uses a composite of radar images taken at 1-hour intervals to represent the track of a derecho that raked across the United States on May 8, 2009. (NOAA Storm Prediction Center.)

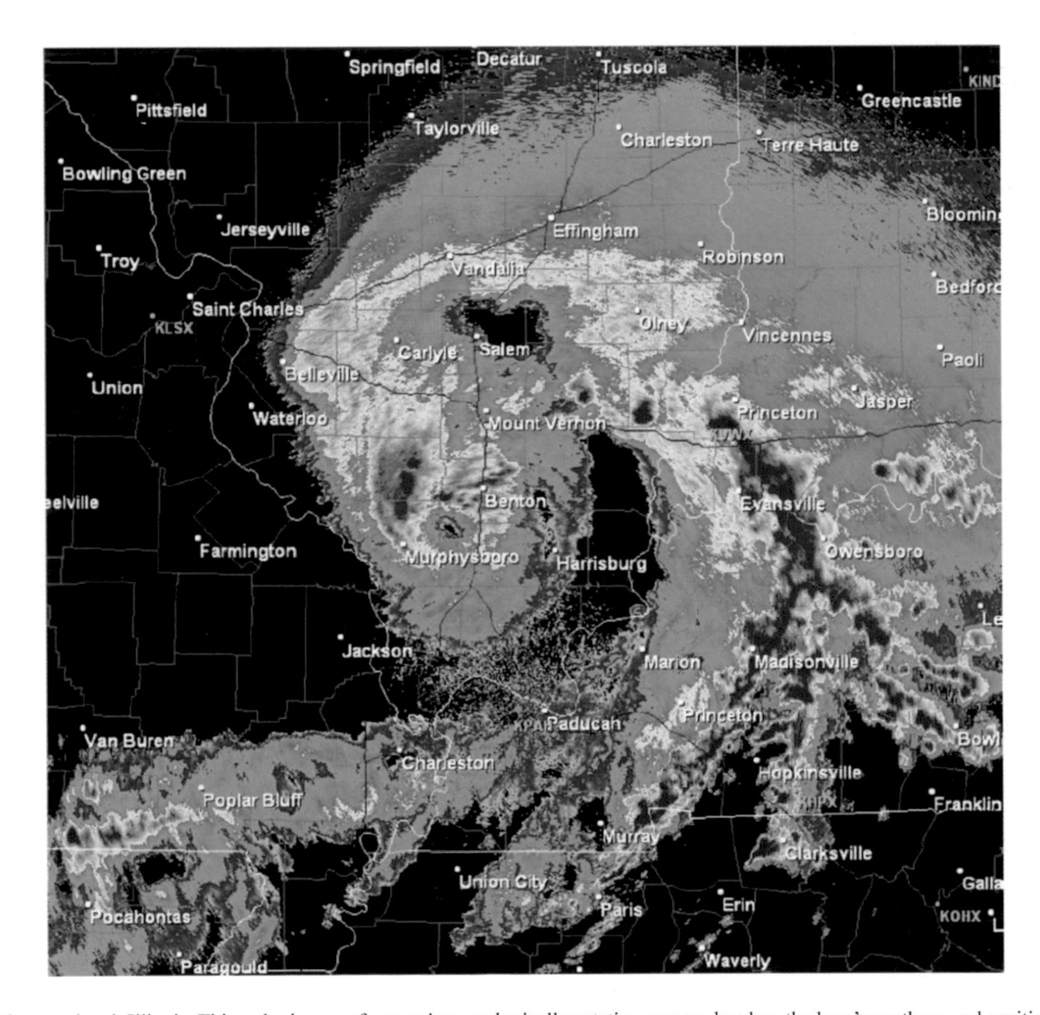

Figure 13.26 Comma head, Illinois. This radar image of a massive, cyclonically rotating comma head on the bow's northern end positioned over Illinois displays a vortex with a clear eye just southwest of Benton, Illinois. (NWS.)

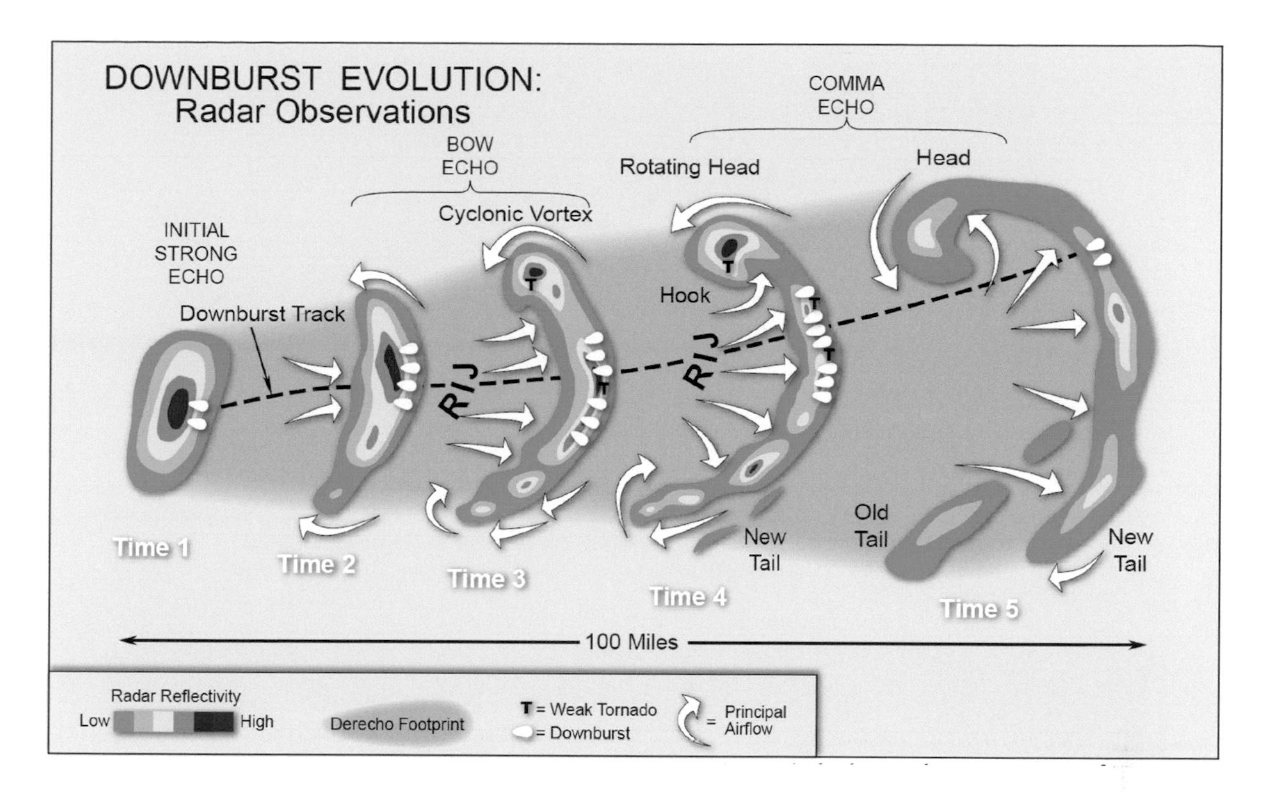

Figure 13.27 Development of a bow echo thunderstorm. At Time 1, the storm begins with an unusually dense and deep radar echo. At Time 2, the storm begins to bow out due to a strong surge of air entering the back of the storm. The air rapidly descends and blasts out the front of the storm line, creating a bow-shaped arc of thunderstorms (Time 3) accompanied by downbursts. At Time 4, rotation on the north end of the line intensifies into a large comma-shaped head. At Time 5, the storm is in a dissipating stage but may spin off new tail. (Adapted from Fujita, 1978.)

formation of a massive, cyclonically rotating comma head on the bow's northern end over Illinois. As it built down to the surface, the vortex created its own extensive swath of wind damage.

Bow echoes typically follow the structural evolution outlined schematically in Figure 13.27. The storm begins as an unusually intense and deep radar echo (Time 1), perhaps containing hail. Over the next 1–2 hours, the storm acquires a linear form (Time 2). A fast current of air entering the back of the line a few kilometers above the ground, called the rear inflow jet, plunges to the surface (Time 3). The result is a blast of intense straight-line wind, with embedded downbursts, that pushes out ahead of the main cloud line, bowing out the leading edge. This is the mature, damage-producing phase of the bow echo, and it accelerates forward. Additionally, a prominent comma-shaped head often develops on the north end, reflecting intensification of a cyclonic vortex at mid-levels, roughly 50–100 km (31–62 miles) in diameter. There is usually an anticyclonic (clockwise)-turning vortex on the south end of the line, although it is weak and not readily observed on radar. Together, the northern and southern vortices form a vertical vortex pair, termed bookend vortices, that are a defining characteristic of the bow echo's structure. Isolated tornadoes, generally weak and short-lived, sometimes develop along and north of the bow apex.

At Time 4, the bow echo is an intense, fast-moving, wind-producing machine. By Time 5, the northern line-end vortex has expanded into a large comma-like cloud structure, while the rest of the arc-shaped system fans out and starts to weaken. The time frame for complete evolution spans 3 or more hours, and the mature, damage-producing storm may persist in a quasi-steady state for 6 or more hours. The longest-lived bow echoes last 18–24 hours.

Bow echoes are three-dimensional cloud systems, and Figure 13.28 shows how they become dominated by an intensifying and deepening cold pool. The cold pool's strong surging straight-line winds helps propel the leading edge of the bow echo forward at high speed, regenerating the storm by lifting humid, unstable air along the gust front. Also depicted is the location of the rear inflow jet. The jet is generated internally when the storm enters its mature phase. If the jet contains dry environmental air, evaporation erodes precipitation along the back edge of the storm, forming the rear inflow notch (RIN) seen in radar imagery. Its presence, along with the bowed leading edge, implies that the storm is likely producing damaging surface winds (Figure 13.29).

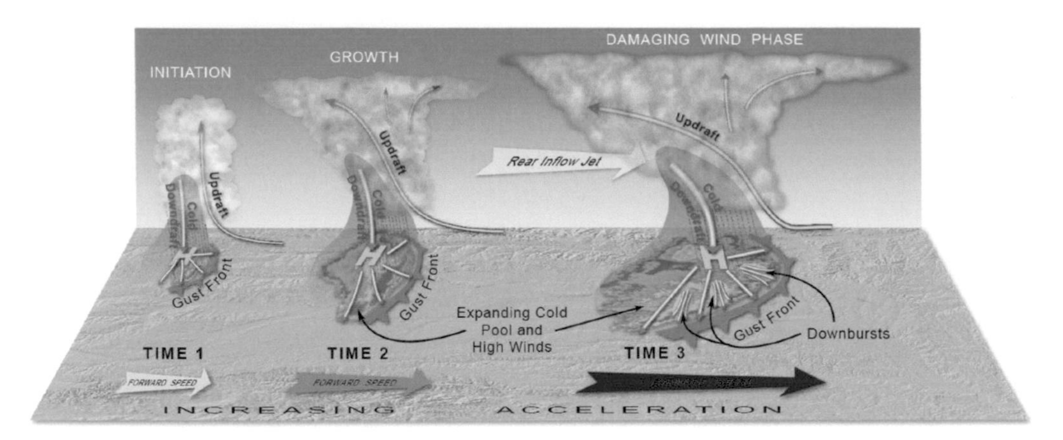

Figure 13.28 Intensifying bow echo system. Bow echo systems intensify as cold air increases the downdraft. The downdraft thrusts the gust front outward, forcing warm moist air to enter the storm system as an updraft along the leading edge of the storm. The new "fuel" then strengthens the storm. A strong rear inflow jet reinforces the cold downdraft. All of this positive feedback causes the storm to accelerate forward.

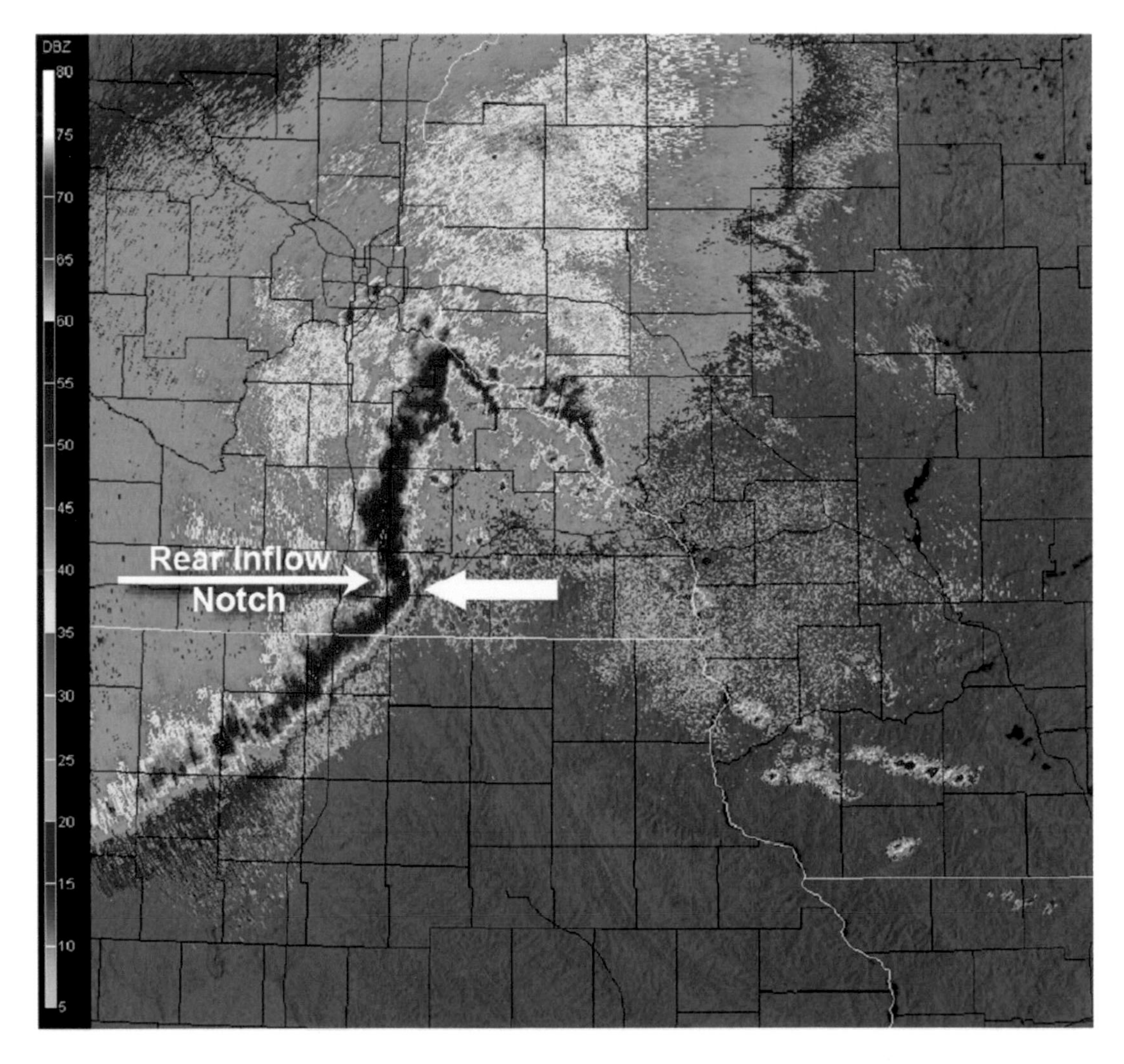

Figure 13.29 Damaging winds from the rear inflow notch (RIN). The white arrow in this image marks the position where the bow echo is being pushed outward by the rear inflow jet that produces the rear inflow notch (RIN). This area produces the highest likelihood of damaging winds. (NWS.)

Formation of Bow Echoes

How do bow echoes form? The details of supercell formation were presented in Chapter 10, and the formation of bow echoes uses some of the same mechanisms. These include (1) formation of horizontal vortex tubes in strongly sheared wind flow and (2) tilting of these tubes into a vertical position. Strong wind shear in the lower-middle troposphere (where wind speed increases with altitude) is a prerequisite for the formation of both bow echoes and supercells. In the case of supercells, however, the low-level winds must also veer with altitude. In bow echoes, the wind shear remains "unidirectional" with winds oriented from the same direction at all altitudes.

Bow echoes, particularly the progressive type, also require very unstable and humid air, which leads to powerful cloud updrafts, which in turn rapidly condense large amounts of water. The descent of the cloud's water mass leads to strong downdrafts, which cool further through evaporation.

Figure 13.30 illustrates the role of wind shear in the formation of a bow echo (the storm cloud has been omitted for clarity). Along the gust front, there is an opposing airflow between the storm's downdraft air and air surging inward toward the updraft. This opposition creates a shear in which air overturns, creating horizontal vortex tubes. Like rollers swept up an inclined plane, the loops are drawn upward into the storm and tilted vertically by the updraft. The vortex loop splits into two separate vortices becoming the pair of bookend vortices. The cyclonically rotating vortex is located on the north side of the storm, with anticyclonic spin on the south side.

The bookend vortices become incorporated into the ends of the storm's cloud line. As Figure 13.31 shows, they expand in width to 75–100 km (47–62 miles) and become several kilometers deep. Between the two vortices, air is drawn inward through the back edge of the storm cloud. This is the rear inflow jet (RIJ) described earlier. Up to 50–60% of the rear inflow's strength is thought to derive from the spinning action of these bookend vortices. In some storms, the rear inflow suddenly plunges to the surface along the storm's leading edge. The horizontal, forward-directed momentum of this jet is brought to the surface. A blast of air fans out, reinforcing the outward surge of wind created by downdrafts. The combined processes give rise to the damaging "wall of wind" that bows the leading edge outward.

Figure 13.32 depicts an example of an intense gust front along the bow echo responsible for the June 30, 2012, derecho. The left panel shows radar reflectivity (precipitation intensity), while the right panel illustrates Doppler wind velocity at the same

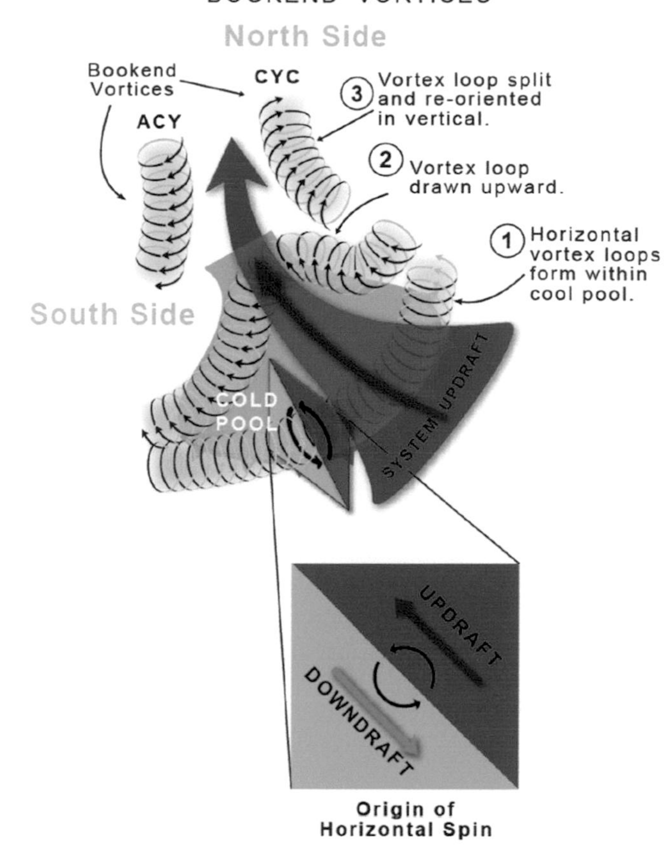

Figure 13.30 Role of wind shear in bow echo formation. Bookend vortices result from the shearing action between the cold outflow at the surface and the warm inflow along the storm's leading edge. This shearing produces a rotating cylinder of air along the leading edge. As the updraft strengthens, it drags the center section of the cylinder upward, rotating the two ends of the cylinder into a vertical orientation. These rotating vortices act to accelerate the rear inflow jet to speeds that can result in great destruction.

time (note the location of the radar, shown by the symbol; red = outbound wind, green = inbound wind). Widespread 43–52 kts (49–60 MPH) winds literally appear as a sharp "wall of wind," with embedded downbursts in the 61–70 kts (70–81 MPH) range (magenta shades, contained inside the yellow ellipse).

With time, the cyclonic vortex on the storm's north side grows in dimension and intensity because it spins in the same sense as the Earth's rotation (both counterclockwise). Earth's rotation becomes a significant player in shaping the storm dynamics as the size of the bow echo increases and the time frame lengthens.

The latest radar studies also suggest that much smaller, vertical vortices, with horizontal dimensions on the order of 5–10 km (3.1–6.2 miles), develop along the leading edge of the bow echo. The position of these vortices is revealed by S-shaped, wavelike

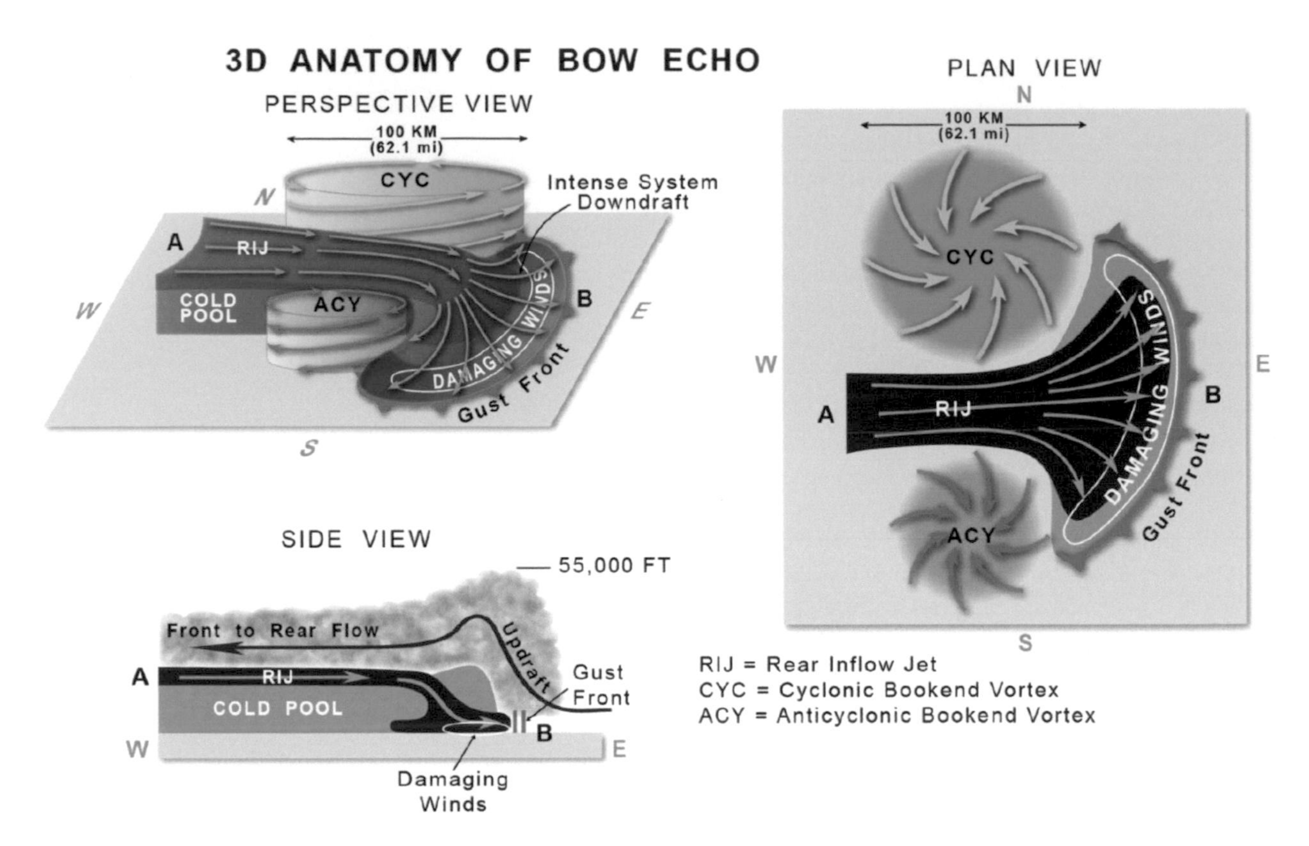

Figure 13.31 Rear inflow jet (RIJ). Perspective view: The rear inflow jet (RIJ) enters the storm from the rear and above the cold pool of air. It accelerates due to the rotational direction of the bookend vortices of the cyclonic vortex and the anticyclonic vortex. Side view: The storm's vertical structure; the plan view depicts the storm's spatial arrangement.

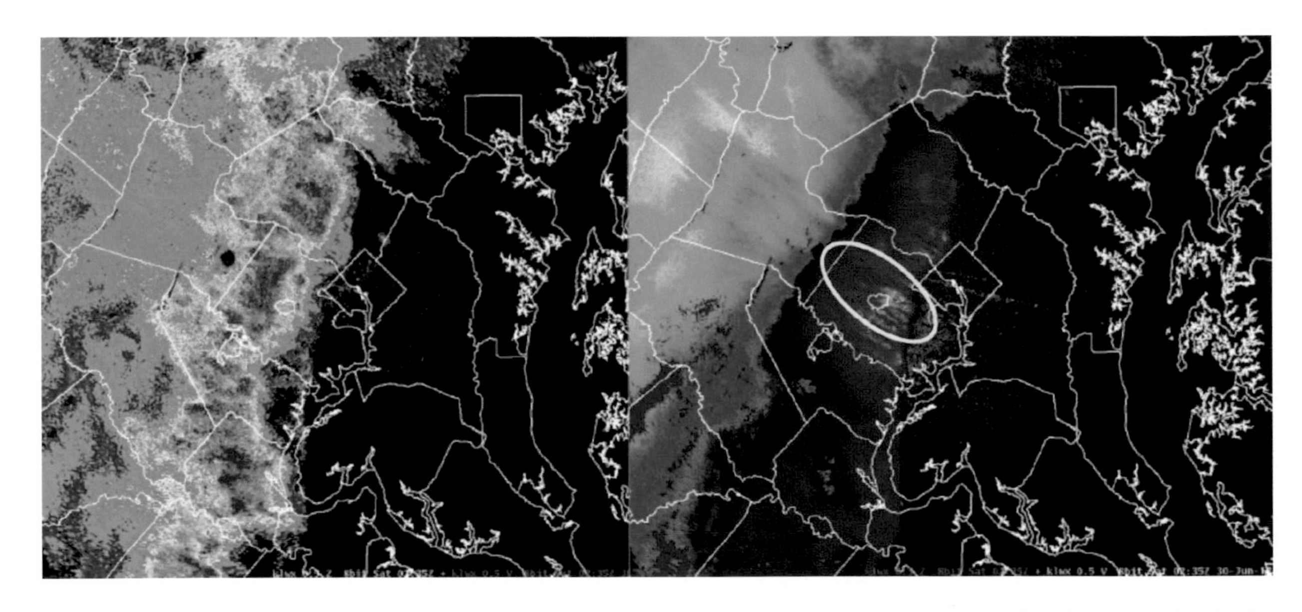

Figure 13.32 Radar images of the derecho that struck the Washington, D.C., area, June 30, 2012. The image on the left is conventional radar designed to show rainfall intensity. The image on the right is Doppler radar used to determine wind direction. Green signifies that the wind is blowing in the direction of the radar system (located on the left side of the imagery). Red indicates that the wind is blowing away from the radar system. The right edge of the red area marks the gust front; the demarcation between the red and green areas is due to the location of the radar. (NWS.)

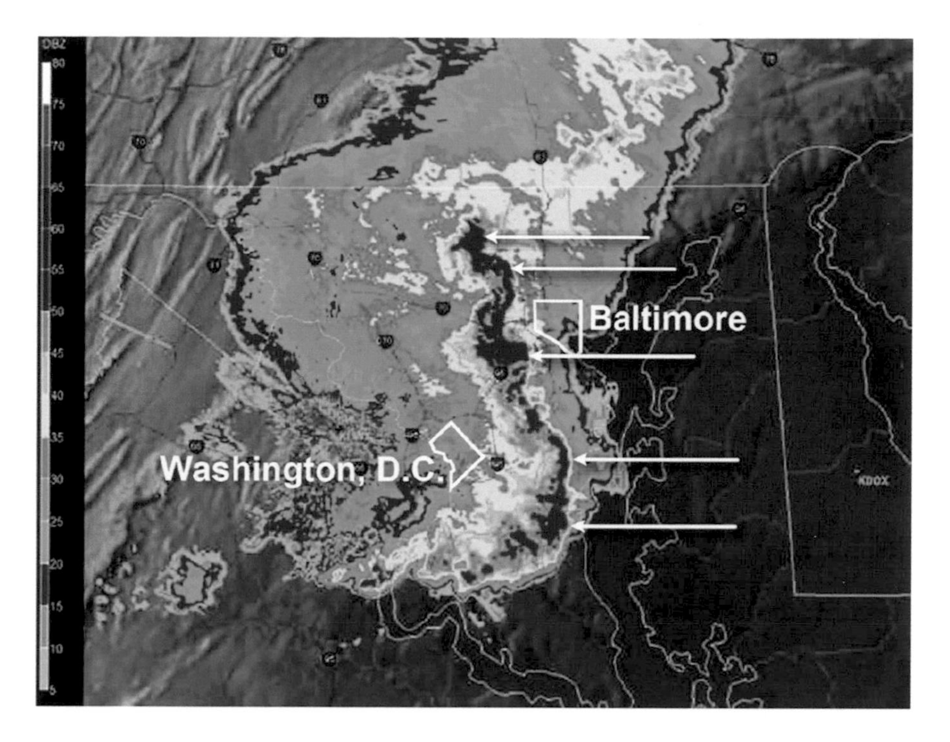

Figure 13.33 Smaller vortices along the bow echo line. Smaller but powerful vortices can develop along the bow echo line. The vortices are marked by a scalloped edge (arrows), and their formation is often associated with strong, damaging winds. (NWS.)

indentations along the storm's leading edge (Figure 13.33), sometimes called a lobe and cleft structure. Strong inflow into the vortices causes the clefts or indentations along the line's leading edge. The vortices can generate small, transient tornadoes that are difficult to detect in Doppler radar and sometimes so short-lived that they have dissipated by the time a warning gets issued.

Societal Impacts of Downbursts and Derechos

Historically, violent tornadoes and tornado outbreaks have commanded much of the media coverage of severe local storms. Many people are less familiar with downbursts and derechos, perhaps because tornadoes are much more visible than downbursts (downbursts east of the Mississippi are concealed by heavy rain). Additionally, while tornadoes have been described in newspapers and literature for almost 150 years (the oldest and most popular photograph dates to 1884), downbursts have only recently been scientifically documented with radar studies (in the 1980s).

Downbursts are capable of producing damage comparable to weak- and moderate-intensity tornadoes. The most violent tornadoes, those rated EF5, grow to 1–2 km (0.6–1.2 miles) width and travel 50 km (31 miles) or more. However, these events are highly isolated, impacting less than 1% of the land area of a typical county. In contrast, derechos sweep across much larger areas, covering 100,000 km^2 (62,000 mi^2) or more. Simply put, they cause severe, widespread wind damage, and they are capable of felling millions of trees in densely forested areas. Whole metropolitan areas lie completely in their path. The costs and impacts of an intense and long-track derecho are massive.

Indeed, a derecho event labeled as the "most damaging thunderstorm in U.S. history" occurred on August 10, 2020, across Iowa and Illinois, creating crop and structural damage estimated to be close to $8 billion. The derecho contained a winding swath exceeding 104 kts (120 MPH), numerous tornadoes, and winds that slammed the ground so hard their presence was detected by earthquake seismographs!

Derechos are most common during summer heat waves, causing power outages that affect millions. Derechos knock out portions of the power grid at a time when it is already stressed, when it is most needed to cool homes and buildings.

Downbursts pose a distinct threat to commercial aviation. Several major air disasters, involving hundreds of fatalities, have been attributed to encounters between aircraft and downbursts.

Many are familiar with the term "wind shear" in the context of aviation safety. Much progress has been made in the installation of wind-shear detection equipment at airports, including ground detection networks and terminal Doppler radar. Airline pilots are now given simulator training that teaches them how to avoid thunderstorm wind shear. In-flight wind shear detection equipment is now standard issue on all major U.S. air carriers. In response to the massive effort to avoid downburst encounters, the number of wind-shear-related aviation incidents has declined markedly.

Deaths on the ground from downbursts and derechos are due to crush-type injuries, from trees falling on cars, toppled walls, or breached mobile homes. High-profile vehicles such as tractor trailers are easily overturned by intense thunderstorm winds. Injuries from flying glass are common. Highways and railroads become impassible from the tangle of fallen trees. Those caught in wilderness areas when a derecho strikes are particularly prone to injury and death, particularly those on small boats in open water. The high incidence of derechos across the Great Lakes region during July is unfortunate because this area draws large numbers of people to the outdoors. Many are incommunicado (a goal of those seeking a true wilderness experience) and therefore do not receive any warning of an approaching windstorm.

A study by Professors Walker Ashley and Thomas Mote in the Bulletin of the American Meteorological Society underscores just how significant the U.S. derecho hazard is. Derecho fatalities per event can top 20, and injuries often number in the 100s. Annual derecho fatalities exceed those for 88% of all tornadoes reported annually and are comparable to deaths from a significant hurricane at landfall. The most common location for derecho fatalities is a road vehicle (30%), followed by a boat (19%). Damage from an individual derecho has reached $1.3 billion – a larger amount of damage than that caused by many landfalling hurricanes and major U.S. tornadoes – and several derecho events have created more than $100 million in damage. Ashley and Mote suggest that official reports of downburst and derecho wind damage may be underreported by a factor of 10.

Amazing Storms: The Boundary Waters Derecho – an Extensive Forest Blowdown

On July 4–5, 1999, an exceptionally long-track derecho cut across the northern Great Lakes region. Figure 13.34 (top map) shows the swath of the windstorm in solid magenta. The derecho initiated as a bow echo at 9 a.m. on July 4 over eastern North Dakota. Between noon and 1 p.m., the bow echo created a devastating, widespread tree blowdown across the Boundary Waters Canoe Area (BWCA) in the Superior National Forest. After destroying the forest, the bow raced eastward, arcing across the northern border of the Great Lakes and southeastern Canada. At 2 a.m. the next morning, the bow echo complex crossed the northern Adirondacks into upstate New York, then dissipated over central Maine by sunrise, July 5.

The sequence of radar snapshots portrayed in the figure suggests that the single bow echo remained exceptionally well organized and intense during its 22-hour traverse, spanning 2100 km (1305 miles). This is a classic example of a summertime derecho traveling along the convective Ring of Fire, as described earlier in this chapter.

The forest devastation left behind by this derecho was extensive, as the bottom panel of Figure 13.34 shows. Extensive tracts of forest suffered near-complete destruction, as trees were snapped and uprooted by downburst winds, with gusts exceeding 78 kts (90 MPH). Some 665,000 acres of BWCA's 1 million total suffered damage. An estimated 25 million trees were blown down in this single storm. Damage totaled $100 million, 4 lives were lost, and 70 people were injured. The storm's approach and passage were unusually swift, with only 30 minutes elapsing between arrival of the gust front and the return of clearing skies.

Survivors injured by debris and falling trees suffered from broken limbs, as well as spine and head trauma. Rescuers spent one week searching for and extricating stranded campers and canoeists. Because normal avenues of ground movement (trails, roads) were completely blocked by fallen trees, several rescues were made by helicopter. Communicating the severe weather threat to those in the wilderness was especially problematic. While the National Weather Service office in Duluth, Minnesota, issued several warnings with significant lead times, there was no mechanism to convey these warnings to campers, hikers, and those engaged in recreational

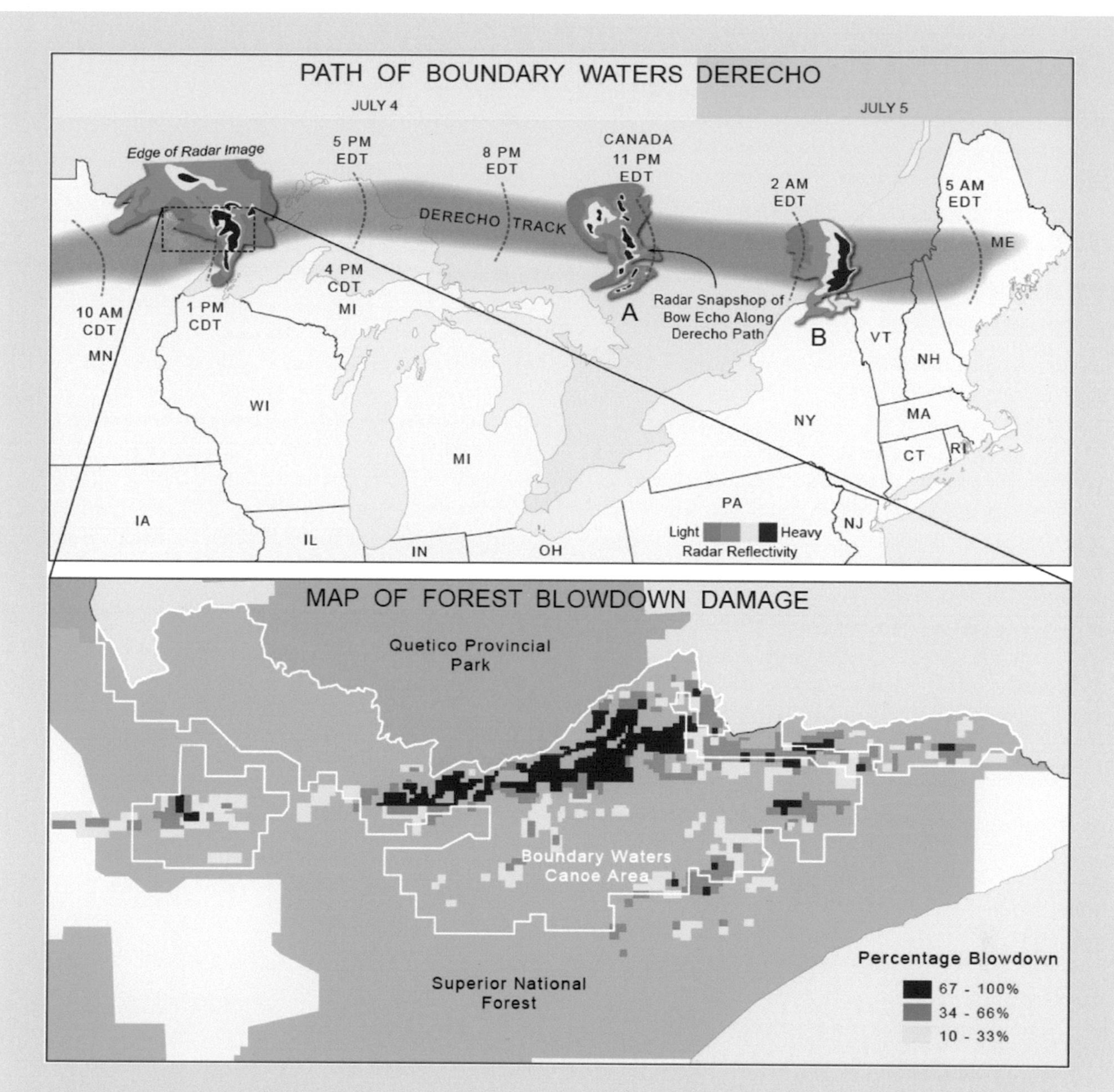

Figure 13.34 Track and timing of the Boundary Waters Derecho. The top map shows the track and timing of the Boundary Waters Derecho as it crossed northern Minnesota, then north of the Great Lakes and eventually into Maine. The bottom map illustrates the extensive forest blowdown that occurred in the Boundary Waters Canoe Area in northern Minnesota. (Adapted from NOAA Storm Prediction Center.)

sports on the water. People were simply out of reach of radio transmitters.

The immediate and delayed impacts on the forest ecosystem were extreme. In addition to high tree mortality, biodiversity was lost, and the dynamics of normal forest succession were vastly accelerated. Outside of wilderness areas such as BWCA, timber losses from derechos can amount to hundreds of millions of dollars. In the wake of the BWCA derecho, a debate ensued on whether to log the region or allow natural succession to run its course. The impetus to log was driven by the threat of extensive wildfire, given the tremendous amount of felled biomass (the dead fuel load was estimated at 60 tons per acre).

Amazing Storms: Pan Am's Fatal Encounter With a Microburst

One of the most dreaded weather-related mishaps in commercial aviation is a microburst encounter. A number of fatal crashes involving downbursts and microbursts have happened at airfields across the United States. Downbursts and associated wind shear (rapid changes in wind speed and/or direction over a short distance) pose the greatest hazard when aircraft are at a low altitude and therefore closest to the surface – that is, during the takeoff and landing phases of a flight.

In this Case Study, we examine the events leading to the crash of Pan Am Flight 759, a Boeing 727, during its takeoff from New Orleans in 1982. The graphic shown in Figure 13.35 is drawn to scale and shows the sequence of events leading to the aircraft's demise. The flight was bound from Miami to San Diego with a stopover in New Orleans. During takeoff from New Orleans, there were thunderstorms at the eastern end of the airport. The aircraft entered the microburst at the beginning of its takeoff roll. The aircraft became airborne but then struggled to maintain lift, reaching a peak altitude of only 45 m (148 feet). It then rapidly plunged to the ground, crashing into homes beyond the edge of the runway. Six houses were destroyed. Because the aircraft had been refueled before takeoff, there was a tremendous explosion and ground fire fed by the fuel. A total of 153 people perished, including 8 on the ground.

Unbeknownst to the aircrew, the takeoff was initiated into the leading edge of the microburst's rotor, which must have been invisible (Point 1 in the diagram). The takeoff roll was unusually short, given the strong headwind, which increases lift over the wings (Point 2). However, the aircraft quickly crossed into the core of the descending microburst, where airflow was directed into the ground (Point 3). At this point, all attempts to remain airborne became a losing prospect, as the strong headwind reversed to a tailwind. A tailwind robs the aircraft of lift. In spite of maximum climb power, the aircraft failed to recover and plunged into the surface (Point 4). The sudden change from a headwind of 15 kts (17 MPH) to a tailwind of 37 kts (43 MPH) changed the relative wind by nearly 52 kts (60 MPH) over a very short time interval. This sudden shift is what constitutes low-level wind shear. The pilots, apart from trying to recognize the predicament they faced, had only seconds to react in an attempt to save the flight.

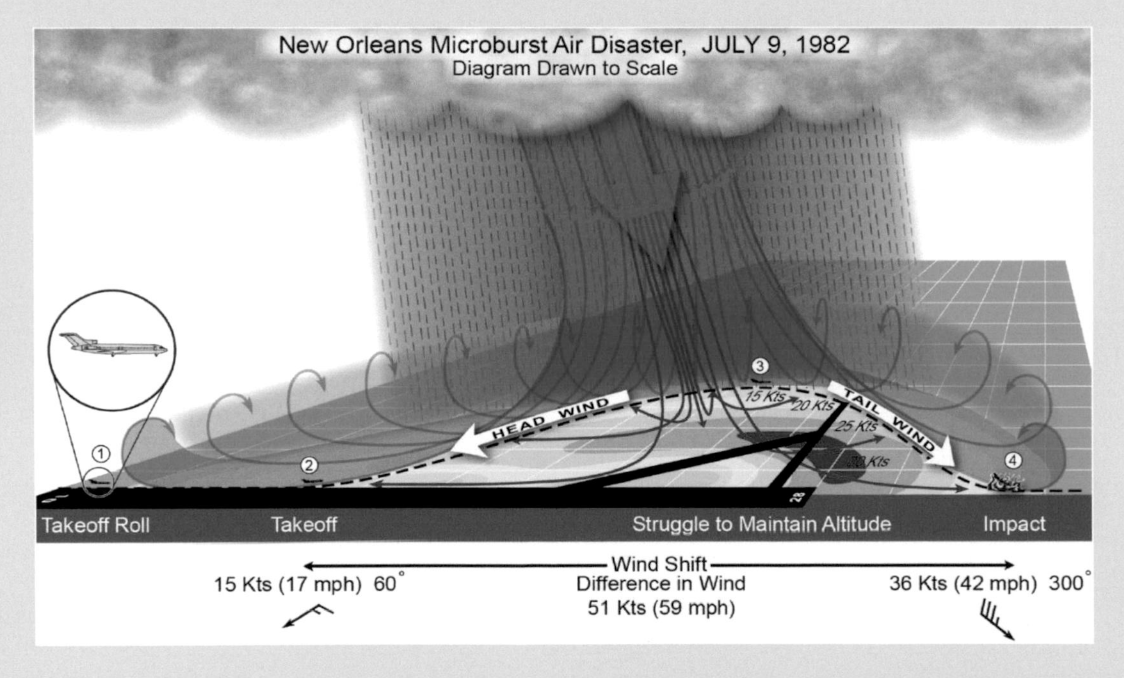

Figure 13.35 Pan Am microburst disaster. On July 9, 1982, Pan Am Flight 759 going from Miami to San Diego was taking off after refueling in New Orleans when it encountered a fatal microburst. This diagram illustrates how the thunderstorm and associated microburst caused the plane to crash. Notice how relative wind speed and direction forced the aircraft into a condition of sudden loss of lift and ultimately a disastrous crash. (Adapted from Job, 1996.)

Before the explicit hazard of downbursts was recognized, a pilot's instincts typically ran counter to what is required to survive a wind-shear encounter. Namely, when the sudden headwind phase is encountered, lift increases. The pilot lowers the nose and perhaps reduces engine thrust to decrease the climb angle. However, doing so puts the aircraft behind the power curve for the tailwind phase of the microburst. As tailwind increases, lift decreases and the aircraft begins to sink.

The pilot may raise the plane's nose in an attempt to climb, but with airspeed decreasing, the aircraft comes dangerously close to its stall margin. The throttles are slammed full forward, to gain airspeed and lift, but gas turbine engines require several seconds to "spool up" to maximum thrust. By then it is too late; the aircraft contacts the ground.

During the 1982 Pan Am incident, technology to detect wind shear had limited capability, and the phenomenon of downbursts was only then being discovered and integrated into the body of knowledge available to the aviation community. Today, with Terminal Doppler Weather Radar, advanced ground detectors, a high level of wind-shear awareness among pilots and controllers, mandatory simulator training on downburst recovery, and wind-shear detection equipment on every aircraft, encounters with downbursts are much less likely.

Summary

LO1 Describe a thunderstorm's straight-line winds, distinguishing between the downdraft, outflow and gust front.

1. A thunderstorm's straight-line wind or outflow is generated by the storm downdrafts, forming an outward-moving mass of gusty, rain-chilled air.
2. The leading edge of the thunderstorm outflow is called the gust front. It is characterized by an abrupt pressure rise, arrival of evaporatively chilled air, and strong wind gusts.

LO2 Explain how downbursts form, the damage they cause, and how they are detected.

1. The downburst is a highly divergent, localized outburst of strong wind at the surface, produced by the most intense type of thunderstorm downdraft.
2. Downbursts are capable of producing wind gusts to 130 kts (150 MPH) over a several square kilometer region, and the surface damage is often comparable to that of a tornado – including uprooted trees and significant structural damage.
3. Downbursts often produce an asymmetric, fan-shaped region of damage, and high winds spin up into a horizontal "rotor" along the edge of the expanding air.

LO3 What two types of severe thunderstorms commonly produce widespread and destructive downbursts? What type of marginal severe storm produces isolated downbursts?

1. Isolated pulse-type severe thunderstorms can produce downbursts for a brief (10–15 min) period during their collapse phase.
2. Supercells frequently generate multiple downbursts, particularly when the storm transitions into a bow-shaped echo.
3. The most prolific downburst-breeding thunderstorm is the bow echo, a large, quasi-linear squall line that concentrates families or clusters of downbursts along its arc-shaped leading edge.
4. Bow echoes travel at speeds of up to 50–60 kts (58–69 MPH) because they are propelled in part by the surging outflow of downburst wind, and new storm cells continuously regenerate ahead of the system's gust front.

LO4 What is a derecho, and how is this different from a downburst? Where do these events occur and at what times of the year? How are bow echo thunderstorm systems uniquely tied to derechos?

1. The derecho, or "line storm," is a long-lived convective windstorm, consisting of one or more bow echoes, with gusts exceeding 50 kts (58 MPH) over a path of at least 400 km (249 miles).
2. Derechos occur year-round in the United States, but there are two pronounced seasonal corridors of high concentration: a summertime, west-east

corridor extending from the upper Midwest across the Great Lakes and Mid-Atlantic, and a southern Plains track with activity during all months of the year.

3 Derechos can be categorized as either a cool-season, serial variety or a summertime progressive type. Serial derechos consist of multiple bow echoes that develop in the warm sector of a strong extratropical cyclone. Progressive derechos form along a summertime stationary front and feed on an exceptionally hot, humid, and unstable air mass (often associated with a summer heat wave).

4 Nearly 70% of all derechos occur during the warm-season months of May through July and are mostly progressive-type.

5 Bow echoes develop a pair of counter-rotating, line-end vortices that channel and accelerate a strong jet of inflowing air into the back of the storm. The descent of this jet to the surface helps create a powerful outflow of damaging wind containing multiple downbursts.

6 The surge of surface wind, or "wall of wind," in a bow echo can be identified on Doppler radar. Small-scale, wavelike undulations along the leading edge of the storm's gust front may harbor intense, nontornadic vortices that generate some of the most damaging winds in a derecho.

LO5 State several reasons why derechos are just as hazardous as violent tornadoes and many landfalling hurricanes. What type of societal problems do they create? How do they adversely impact large ecosystems such as forests?

1 The extreme devastation created by derechos has only recently been quantified and is comparable to the damage and fatalities caused by the most violent tornadoes and even landfalling hurricanes.

2 Derechos exact a heavy toll on large forests, causing widespread blowdown of mature trees, loss of species diversity, and future threat of wildfire (from the accumulation of massive amounts of dead timber "fuel"). Furthermore, derechos seriously injure and kill those caught unaware in forest wilderness areas and persons engaged in recreational activities on open water.

3 Downbursts were only discovered in the early 1980s, largely due to a series of fatal commercial airliner disasters. Today, all licensed airline pilots undergo rigorous simulator training in downburst avoidance and recovery, and sophisticated wind-shear detection equipment is standard on all airliners. Accordingly, the rate of aviation incidents involving downbursts has dropped dramatically since the early 1980s.

References

Ashley, W.S. and T.L. Mote, 2005. Derecho hazards in the United States. *Bulletin of the American Meteorological Society*, 86:1577–1592.

Bentley, M.L. and J.A. Sparks, 2003. A 15 yr climatology of derecho-producing mesoscale convective systems over the central and eastern United States. *Climate Research*, 24:129–139.

Fujita, T.T., 1978. *Manual of downburst identification for Project Nimrod*. Satellite and Mesometeorology Research Project Paper No. 156, University of Chicago.

Job, M.A., 1996. *Air Disaster, Vol 2*. Sydney, AU: Motorbooks International.

CHAPTER 14
Science of Flash Floods

Learning Objectives

1. Describe the geographical distribution of flash floods across the United States.
2. Distinguish between flash floods, aerial (river) floods, and coastal inundation.
3. Explain the meteorological factors that enable water vapor to build to excessive levels, priming the atmosphere for a flash flood.
4. Describe how multiple types of weather systems generate excessive rain – including complexes of thunderstorms and large-scale vortices such as tropical cyclones. What processes do many of these systems share?
5. Discuss how steep terrain contributes to heavy rain generation, then concentrates runoff water into a flash flood.
6. How do flash floods rank in terms of U.S. weather fatalities? Why is prediction of flash flooding so difficult, compared to other types of severe storms? What are some corollary (secondary) hazards of flash flooding?
7. Discuss some of the evidence suggesting flash flood hazards are increasing as a result of anthropogenic (human) factors, and describe the hypothesized reasons.

Introduction

On the evening of August 19, 1969, the deadliest U.S. flash flood of the 20th century hit Nelson County, Virginia. Rain fell so hard it was difficult to breathe, and up to 68 cm (27 inches) of water accumulated in just 8 hours across a rural county along central Virginia's Blue Ridge Mountains. The cause of the flood was the remnant of Hurricane Camille. Camille made landfall over Mississippi and Louisiana on August 17 as a fierce Category 5 hurricane. Camille rapidly weakened after landfall. On the 18th, the post-tropical vortex turned east across the lower Ohio Valley. Small pockets of 2.5–7.5 cm (1–3 inches) of rain fell along Camille's inland track from northern Mississippi to West Virginia. On August 19, the remnant low approached the Appalachians. The Weather Bureau's forecast for central Virginia called for "showers overnight, clearing in the morning."

Then, after sunset, a biblical rainstorm exploded over the Blue Ridge Mountains, centered on Nelson County. Sheets of blinding rain fell continuously through the entire night. Mountain slopes liquefied into fast-flowing sheets of mud. Soil, boulders, trees, thickets, and scrub flowed down the mountain. Houses and bridges were swept away. Roads and bridges disappeared. Unimaginable volumes of water, soil, and forest consolidated into roaring waves that permanently scoured away parts of Nelson County. Nearly 160 lives perished in the flood of rainwater, mud, rock, forest, and structures.

How could such a catastrophic flood unfold so swiftly? In this chapter, we discuss the science of flash floods, and the many ways that these deadly situations arise. Unlike most severe storms – such as tornado outbreaks or intense winter storms, which have preferred geographical regions – flash floods are a severe storm hazard across the country. No single region across the entire United States has escaped their wrath, even the desert. Of all the lethal severe local storms in the United States – including lighting, tornadoes, and hurricanes – flash floods rank number one, based on 30 years of data.

First, we describe the geographical incidence of flash floods across the United States and define various types of floods. Then we study the atmospheric ingredients that conspire to generate unusually heavy rain, exploring several storm patterns that commonly lead to flash flooding. Two case studies highlight how heavy rain-producing weather systems interact with terrain and urban landscapes to create catastrophe. The chapter concludes with challenges associated with flash flood forecasting and an important question: Will global warming lead to more flooding?

DOI: 10.4324/9781003344988-17

Geographical Distribution and Nature of Flash Floods

Flash floods are a widespread and common meteorological hazard. The basic mechanism leading to the production of heavy rain – the vigorous uplift of very humid air within small but intense weather systems or along mountain slopes – is common across the entire continental United States. Even arid regions such as the Desert Southwest (Arizona, New Mexico, southern California) are prone to flash floods, albeit just a few months a year. Flash floods have occurred in every state, and in some states several dozen events have occurred in the past several decades.

Figure 14.1 presents a partial census of where flash floods have occurred across the United States since 1850. While the figure does not include all of the floods that occurred over the period, the sample of 66 historically significant events illustrates the widespread nature of flash floods. Proportional circles illustrate number of deaths from each event on one side, with property loss (millions of dollars, unadjusted for inflation) on the other.

Note the relative proximity of flash floods to ocean basins or along tracts where oceanic air surges inland. The Mississippi Valley tract, in particular, reflects northward penetration of very humid air from the Gulf of Mexico. The large concentration of flash floods over the Mid-Atlantic arises from the Atlantic's moisture influx and the Appalachians, which lift humid air streams. Many of the eastern U.S. floods are in fact generated by tropical cyclone remnants. Along the Front Range of the Rockies, we see a few events tied to high terrain. This far inland, the lack of moisture limits frequent or widespread flash flooding. Along the U.S. West Coast, Pacific air is forced to ascend steep mountain slopes along the Sierra Nevada and Cascades.

Notice also the difference in flash flood impacts, with high-cost damage events clustering west of the Appalachians and high mortality along the East Coast. The latter outcome may reflect much higher population density in the east compared to the southern Plains.

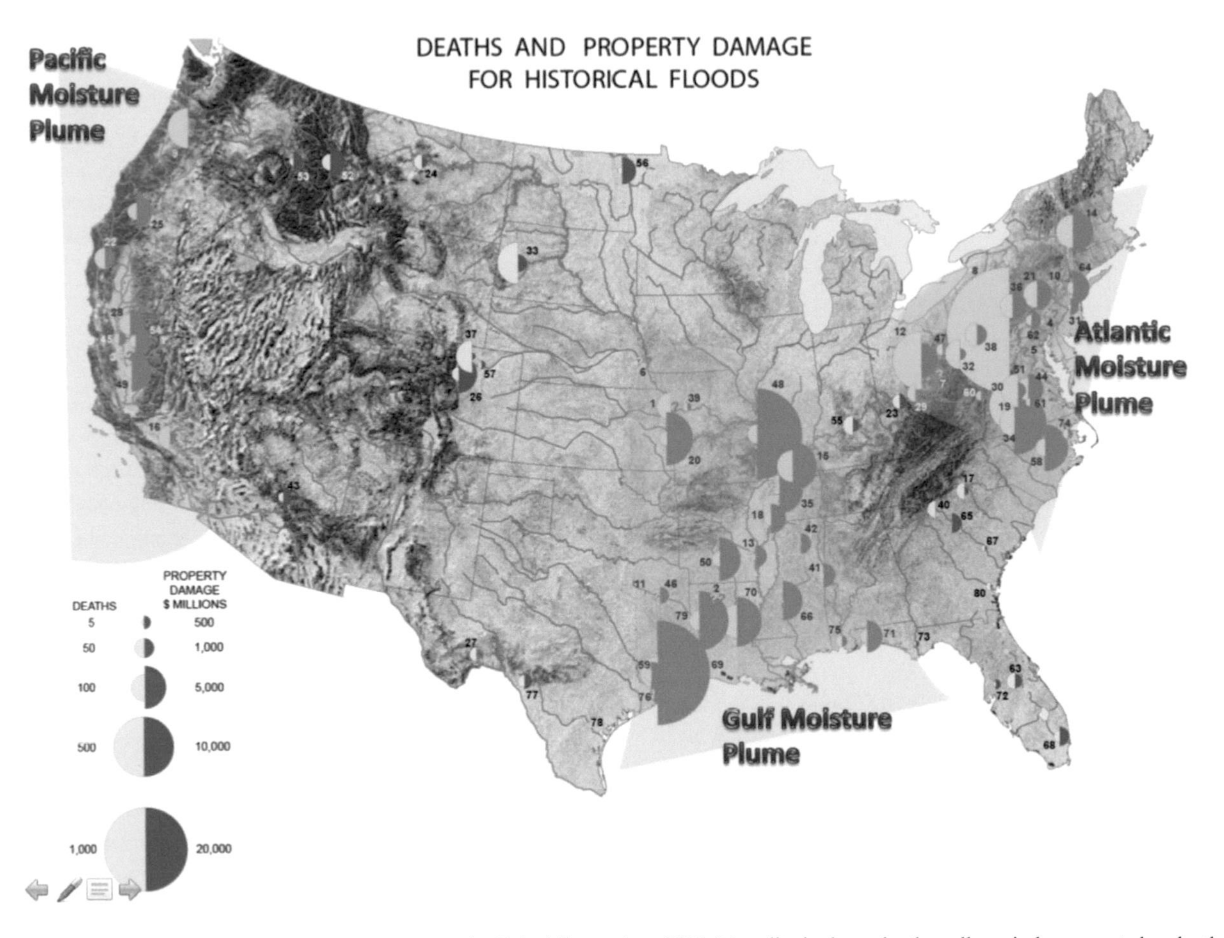

Figure 14.1 Sampling of 66 historical flood events across the United States since 1850. Mortality is shown by the yellow circle, property loss by the orange circle. Flash floods tend to cluster in regions exposed to influx of oceanic moisture and lifting along mountain barriers such as the Appalachians in the East and Sierra Nevada in the West.

Flood Types

Flash floods are one of several types of landscape response to excessive rainfall. There are several types of flood. Table 14.1 summarizes the characteristics and impacts of the three main categories: flash floods, aerial (river) floods, and coastal inundation. Let's briefly examine each type.

Flash floods are small-scale events of rapid onset, due to excessive rain. The timescale is brief, lasting minutes to hours. Flash floods impact regions of strong topographic relief and also small drainage systems (streams, culverts, and creeks). They are also common to urban regions featuring extensive networks of paved surfaces and stormwater drainage. The most common rain delivery system features one or more convective storm cells, including thunderstorms, although lightning and thunder may not occur. Rain rates in these cells typically approach 5–7.5 cm (2–3 inches) and, in extreme cases, 10–12.5 cm (4–5 inches) per hour.

Flash floods have high fatality rates. When you think flash flood, think "cloudburst" over a small area, so abrupt and intense that people are caught off guard. Corollary hazards include debris flows, mudslides, and sudden collapses of roadways and dwellings. Flash flood warnings are hampered by very short lead times. Some convective storms are poorly detected by weather radar, or there is uncertainty about how much rain has fallen. In fact, many flash floods are underway before the danger is recognized and warnings are broadcast.

Unlike flash floods, aerial or river floods involve days of successive, heavy rainstorms that track repeatedly across the same geographical region. Much larger drainage regions (catchments) are involved, often those that feed mainstem rivers and major tributaries. Rain rates are not necessarily extreme, but the prolonged nature of rainfall slowly accumulates a massive volume of water. Slow-moving extratropical cyclones, landfalling tropical cyclones, and even waves of summertime thunderstorms can lead to aerial flooding. Other factors may have preconditioned a large region for flooding, such as deep snowpack (which rapidly melts when it absorbs rain) or saturated soils from a previous heavy rain.

Aerial floods typically have longer lead time, sometimes several days. It takes time for the slowly moving flood crest or flood wave to travel the length of a mainstem river. The wave amplitude steadily increases as it moves out of the headwaters region, collecting more and more water feeding in from tributaries. Water levels take many hours to rise, then slowly recede as the crest passes through. During this time, low-lying floodplains become extensively flooded. Agricultural and property losses are particularly high. There is often time to evacuate and bolster defenses. However, sandbags, flood walls, berms, dams, and levees frequently fail to contain the rising water. The arrival of a flood crest is usually an anticipated event, with hydrological forecasts specifying time and amplitude of the flood stage at specific locations along a waterway.

Coastal floods, also called saltwater inundation, arise from storm surge, which is a wind-driven phenomenon. Storm surge is commonly generated by landfalling hurricanes, Pacific coastal cyclones, and East Coast Nor'easters during winter. It's not uncommon for coastal and even inland locations to receive 15–30 cm (6–12 inches) of rain from a tropical vortex. Much higher numbers, including 75–100 cm (30–40 inches), have been reported.

It's important to realize that some types of weather systems create both aerial floods and flash floods concurrently. Some of this rain arrives in the form of heavy tropical showers, at rates of

Table 14.1 Flood Types and Their Associated Properties

Property	Flash Flood	Aerial Flood	Coastal Flood
Type	Freshwater	Freshwater	Saltwater
Duration	Hours	Days	Hours to days
Area (S) impacted	Small streams; mountain catchments; urban areas; highways; underpasses	Mainstem rivers; tributaries	Coastal zones; barrier islands; bays; estuaries
Weather systems	Heavy convective showers; thunderstorms	Prolonged moderate to heavy rain from cyclone; snowmelt	Nor'easters; tropical cyclone
Primary losses	Fatalities; motor vehicle losses	Extensive property losses; agriculture losses	Extensive property loss
Associated hazards	Debris flows; mudslides; debris jams	Dam failures; levee breaches; contaminated drinking water; waterborne illnesses	Beach erosion; dune breaches; barrier islands channelized or shifted
Warning lead time	Minutes to hours	Days	Days

Table 14.2 Rain System Characteristics

Type of Rain	Characteristic Weather System	Rain Intensity	Rain Coverage	Notes
Convective	Thunderstorm cells characterized by small, intense updrafts	2.5–5 cm/h (1–2 inches/hour); up to 7.5–10 cm/h (3–4 inches//hour) in a cloudburst	Highly localized; features large gradients in rain accumulation	Often a "hit-or-miss" type of rain; duration is just tens of minutes.
Stratiform	Large-scale weather systems and fronts with sustained, gentle lifting of air	A few tenths of an inch per hour	Widespread, uniform rain depth from horizontally extensive layer clouds	Tending to be a "steady, soaking rain" lasting hours

Note: "/h" = per hour.

2.5–5 cm (1–2 inches) per hour, creating local flooding issues. Meanwhile, steadier, moderate-intensity rain (called stratiform rain; see Table 14.2) falls across a widespread region, setting up an aerial flood. Distinguishing between flash and river flooding may be of little practical significance when a region is under a hurricane warning.

Figure 14.2 summarizes the combination of factors that lead to of flash floods. Heavy rain is but one condition for a flash flood. The hydrological response to this rain depends on the features of the landscape. Waterlogged soils from previous heavy rain or springtime snowmelt precondition a region for a flood. Extensive urban landscapes prevent heavy rain from infiltrating. Steep, funnel-shaped mountain catchments rapidly convey rainwater runoff into small streams. In all of these cases, small stream channels or drainage culverts "flash over" or rapidly swell, spilling out of their banks. A highly hazardous situation can evolve in just tens of minutes.

The greatest susceptibility to flash floods occurs where multiple meteorological and hydrological aspects coincide. Many regions of the United States feature steep terrain, and our nation is becoming heavily urbanized (Figure 14.3). It's little wonder, then, that so much of the country is vulnerable to the ravages of flash floods.

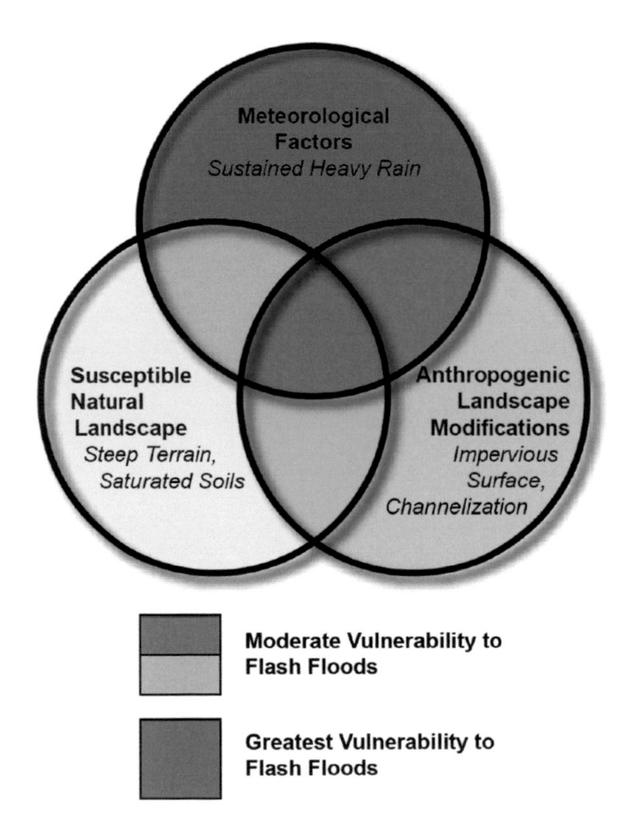

Figure 14.2 Floods are multifactorial hazards. Flash floods require a combination of two or more factors, not all of which involve excessive rain.

Gathering Atmospheric Moisture: "Priming the Keg" for a Flash Flood

If you think of a flash flood as an explosively developing situation – say, in response to a cloudburst – then the analogy of "priming the keg" with gunpowder is appropriate, where powder in this case equates to excessive atmospheric moisture. Water vapor is an invisible trace gas that normally accounts for a tiny fraction of the air's volume, typically much less than 1% on a global, whole-atmosphere basis. However, unlike many other uniformly distributed gasses, such as oxygen and nitrogen, water vapor tends to concentrate in some regions and not in others. Air above deserts contains a trace of vapor, but over a tropical ocean, vapor concentration increases a hundredfold. Furthermore, water vapor's concentration diminishes with increasing altitude in the atmosphere because vapor enters the atmosphere from evaporation (and transpiration – defined as water vapor emitted by vegetation). In other words, water vapor tends to concentrate most closely above its surface of origin.

Another important fact is that the air's humidity (simply, water vapor content; more involved definitions of humidity appear in Chapter 3), which is strongly controlled by air temperature. Figure 14.4(a) and (b) illustrates important processes related to gathering and accumulating atmospheric moisture. In Figure 14.4(a), two

Figure 14.3 Devastating flash flood during July 2017, in upstate New York.

air columns are shown. On the left is a wintertime air column at 4° C (39° F) close to saturation (dew point = 3° C [37° F]). On the right is a summertime air column with a temperature of 32° C (90° F) and dew point of 24° C (75° F). The summer air is the type of sticky air mass that many would label as "oppressive." But the summer air mass contains five times the vapor of the winter air mass (mixing ratio of 20 g/kg vs 4 g/kg), even though the temperature has only doubled! Warm air, with its greater energy content, can evaporate much more water, but the rise in vapor content due to the increase in temperature is not linear (one-for-one). For this reason, the majority of flash floods are generated during the warm season, when the highest humidity values prevail.

Figure 14.4(a) illustrates a convenient way to quantify all the water vapor in an air column. The parameter is termed total precipitable water (TPW), which is defined as the equivalent depth of rainwater after condensing *all* available water vapor in a "representative" 1 meter^2 air column, extending from surface to tropopause (12–14 km [7–9 miles] altitude). This calculation is routinely made during weather balloon launches, which occur twice daily at about 100 locations across the country. In the figure, the column on the left is warm and relatively dry, representing a desert location. It has a great capacity to hold a lot of vapor, but because of the dry land surface, its TPW yield is only18 cm (0.7 inch) of rain. In contrast, the air column on the right is also warm, but it sits along the Gulf Coast, where very moist, southerly winds import vast amounts of vapor off the ocean. Densely vegetated land surface also adds considerable moisture. The TPW value here is 5.5 cm (2.2 inches) – a large number for any U.S. location in the summer. In assessing flash flood potential across the United States, it's apparent that the Gulf Coast location is much more susceptible in summer.

If crops are abundant in a broad region, dense leaves transpire additional vapor into the lower air layer. (Some of the most humid air in the United States during summer occurs over the upper Midwest and Ohio Valley, also known as the Corn Belt, where dew points can exceed an astounding 27° C [80° F].)

Another process that boosts humidity is the horizontal movement of moist air from one location to another (Figure 14.4[b]). During spring and summer, it's common for a moist, southerly flow to sweep inland off the Gulf of Mexico. The air mass is already warm, which gives it large capacity to "hold" a lot of vapor, as we have just seen. This vapor transport from an oceanic source adds additional vapor to inland air. Convective clouds do not simply "rain out" or deplete the supply of TPW. As Figure 14.4[b] shows, vapor is drawn down locally, but there is continual resupply from airstreams converging over a very large area. As moist airstreams with high dew point flow in from the ocean, moisture is being gathered. This moisture tends to accumulate or "pool" along the warm side of frontal boundaries. Much of the vapor that feeds a convective cell only 10×10 km (6.2×6.2 miles) across may in fact have been sourced over a 10,000 km^2 (3861 sq mi) region! So

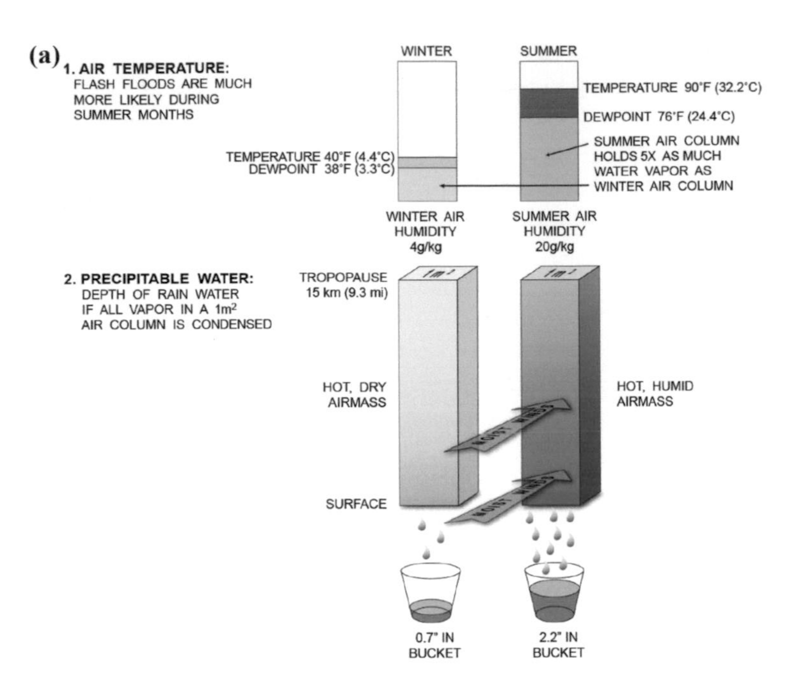

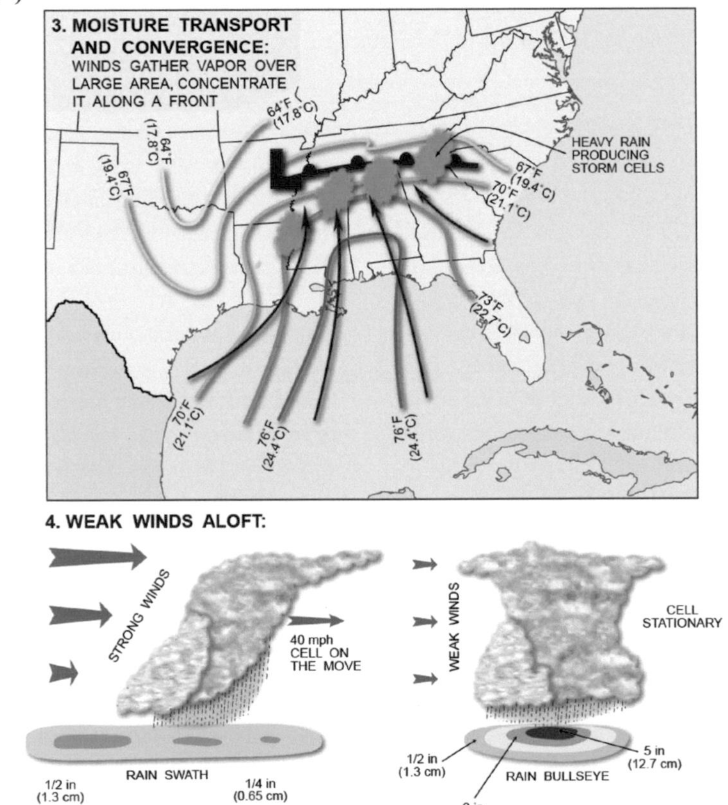

Figure 14.4 Accumulating atmospheric moisture. Many atmospheric processes work together to boost the humidity content of air, a necessary precondition for a flash flood. (a) Importance of air temperature and total precipitable water. (b) Large-scale moisture convergence along a frontal boundary, and the effect of weak winds on storm cell movement and thus rain accumulation.

while TPW is a local measure of available moisture, it does not tell the whole story; the measure does not account for continued moisture transport and resupply.

For this reason, localized convective storms can produce torrential rain for many hours. Rain accumulation may greatly exceed the region's local TPW content. For instance, hurricane remnants have led to 75–100 cm (30–40 inches) of rain accumulation in situations where the TPW never exceeded 3 inches. TPW is an important measure of flash flood potential, but it is only one component that determines a weather system's overall rainfall efficiency (the percentage of water vapor entering a storm system that is converted to rainfall).

Figure 14.4[b] concludes with an important characteristic of many flash floods, unrelated to the air's humidity: weak winds aloft. For a given volume of rain created by a convective cloud, the accumulation at any one location depends critically on the speed at which the storm cloud moves. On the left is a convective cloud on the move, translating at a brisk clip (35 kts [40 MPH]). While it's raining hard, perhaps 5 cm (2 inches) per hour, any one spot is subject to heavy rain for only 15 minutes, yielding 1.3 cm (0.5 inch) of rain for all locations along the cloud's swath. On the right side is the same cell but embedded in very weak wind flow. The cloud remains nearly stationary; it is not being propelled forward. As a consequence, all of the cloud's rainwater falls in just one location. For a storm that persists 2.5 hours, this equates to a bull's-eye of a 12.5 cm (5 inch) rainfall, producing a flash flood. Weak steering winds (upper-level flow in which convective cells are embedded) are very common in summer across the United States because the upper-level jet stream is displaced far to the north, across Canada. Weak steering, combined with warm, high humidity air, conspire to generate the worst flash flooding during the summer months.

Figure 14.5 puts many of the aforementioned elements together, showing the rain accumulation map for a potent tropical weather system along the U.S. Gulf Coast region – with over 1 m (3 feet) of rain accumulation!

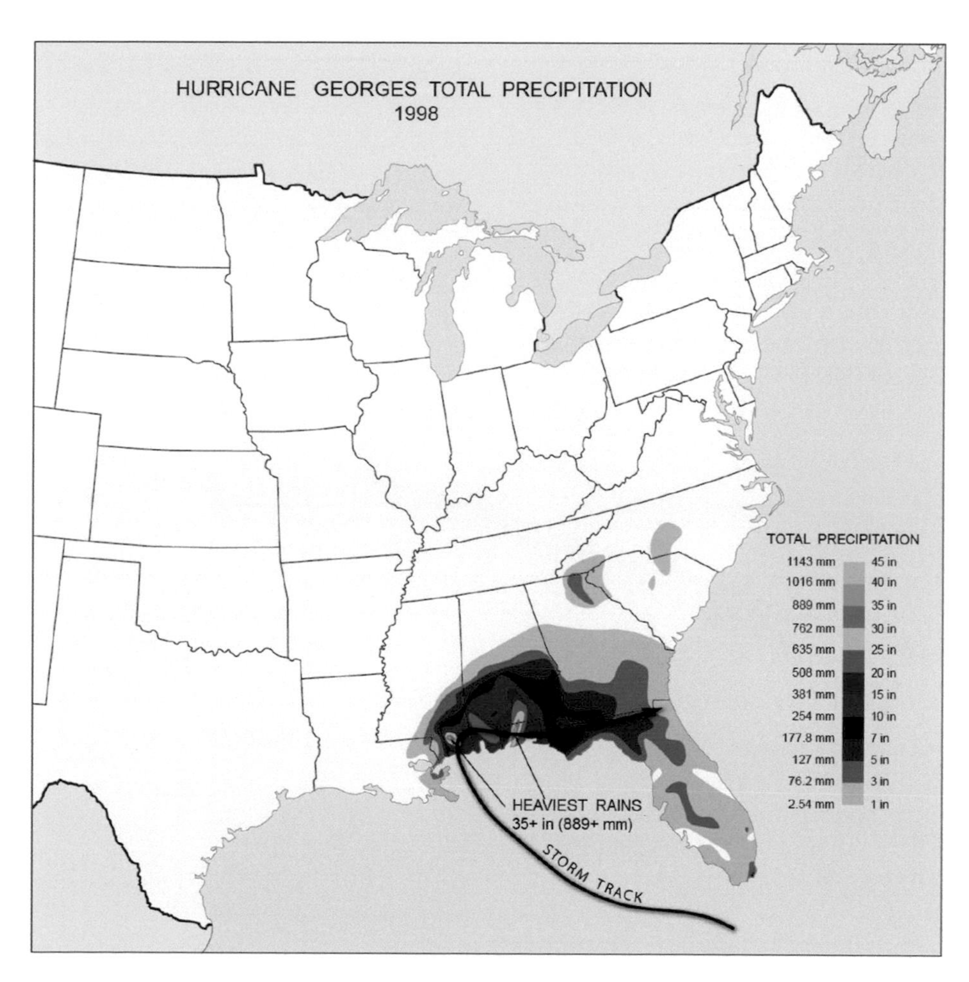

Figure 14.5 Heavy rain generated by Hurricane Georges, 1998. Tropical cyclones at landfall and their remnants are prolific rain generators. Hurricane Georges produced pockets of 35 inches of rainfall along the Florida, Alabama, and Mississippi coast. TPW here was very high, but other factors were involved too. (Adapted from NOAA Weather Prediction Center.)

Common Types of Storm Systems That Produce Flash Floods

Heavy rainfall leading to flash flooding is generated by a broad spectrum, or range, of weather system types – ranging from small but highly organized and efficient clusters of thunderstorms, up through large-scale vortices such as extratropical and tropical cyclones. In this section, we discuss the variety of heavy-rain-generating scenarios, while pointing out atmospheric processes common to most of the rain systems.

Mesoscale Convective Systems

As we learned in Chapter 9, isolated convective cells are the building blocks of larger mesoscale convective systems (MCS), such as multicell clusters and squall lines. Depending on wind shear, frontal boundaries, and other atmospheric characteristics, these larger aggregates can produce torrential rain and flash floods. In this section, we examine a few categories of flash-flood-generating MCS.

Ironically, brisk flow in the atmosphere – as opposed to the weak winds just described– can also lead to flash floods, provided the flow generates an "assembly line" of convective cells that pass over the same location. Figure 14.6 illustrates cell training, which is the cyclic regeneration of convective cells in the same spot for many hours.

In the diagram, the upper-level flow (solid purple arrow) is vigorous and flows parallel to a frontal boundary (a stationary front, shown by alternating blue and red segments). Moisture is transported toward the frontal boundary by a separate southwesterly current in low levels. Water vapor pools along the current's southern margin. New storm cells initiate at a specific location along the front, as indicated at the trigger location, where unstable, moist air is lifted. Each new cell is whisked off along the front, embedded in the upper-level flow. The diagram shows a train of five cells. As each cell passes over location A, it dumps an inch or more of rain. Rain accumulates over a few hours, leading to a flash flood. One could refer to this type of storm as the "rain train."

The small diagram portrays this cyclic process in terms of vectors. The blue arrow shows the motion of individual cells that track through the larger storm complex, with each cell carried by the wind. The purple arrow shows the direction of storm regeneration, indicating that new cell formation is occurring in the opposite direction of cell movement. This process is called back-building. When back-building is equal to and opposite cell movement, the storm complex as a whole

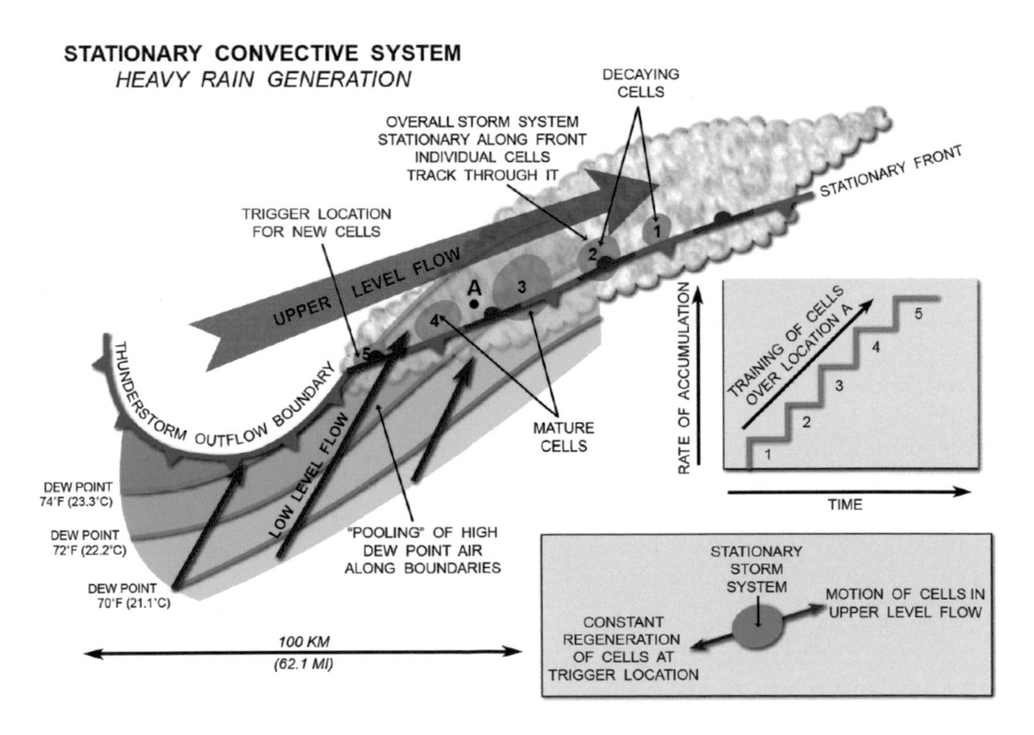

Figure 14.6 Cell training (a.k.a. the "rain train"). The figure shows how a back-building convective system generates heavy rain over a fixed location.

remains stationary. However, the storm's rainfall efficiency does have limits, and one important factor is the speed of the upper-level winds. If these winds are vigorous enough, some of the condensed water in the form of cloud droplets is swept downstream, away from the core of the storm, before the droplets have had time to grow into raindrops. Like many types of severe weather, an optimal balance of factors is needed for a cloudburst.

An extreme form of MCS is the mesoscale convective complex (MCC) of the Great Plains. In Figure 14.7, we see an MCC viewed from space. Cloud top temperatures are indicated by different colors, with dark red indicating the coldest and tallest cloud tops. The MCC is a massive complex of convective cells. According to the official definition, the cloud dimensions of MCCs must exceed 50,000 km^2 (19,305 sq mi) for a period of 6 or more hours. MCCs take time to congeal, initiating during the afternoon, when lines and clusters of thunderstorms begin to merge. The progenitor storms often develop along the Rocky Mountain Front Range and then drift eastward. After sunset the atmosphere cools. The aggregate of merged cells, now over eastern Colorado and western Kansas, develops an extensive, high-level cloud layer and its own internal circulation. The circulation features a large vortex that draws in copious moisture from the Gulf of Mexico.

Heavy rainfall in an MCC is widespread and peaks after midnight. Much of the rain is stratiform in nature, with embedded pockets of heavy convective rain. Figure 14.8 traces the evolution of an MCC through the course of a day. The red curves shows the MCC's overall rain volume, which peaks in the early evening. The green curve shows the MCC's rain-producing area, which expands until after midnight. As the rain area expands, overall rain intensity diminishes. But a large volume of rain continues to be produced, into the early morning hours, the effect of expanded area compensating for reduced rain rate (blue bars).

The heavy rain "footprint" left behind by an MCC is often elliptical if the complex is largely stationary or a broad swath if progressive. Large quantities of rainwater often fall across several states, with isolated pockets of up to 20–25+ cm (8–10+ inches) in some cases. Such heavy rains, while capable

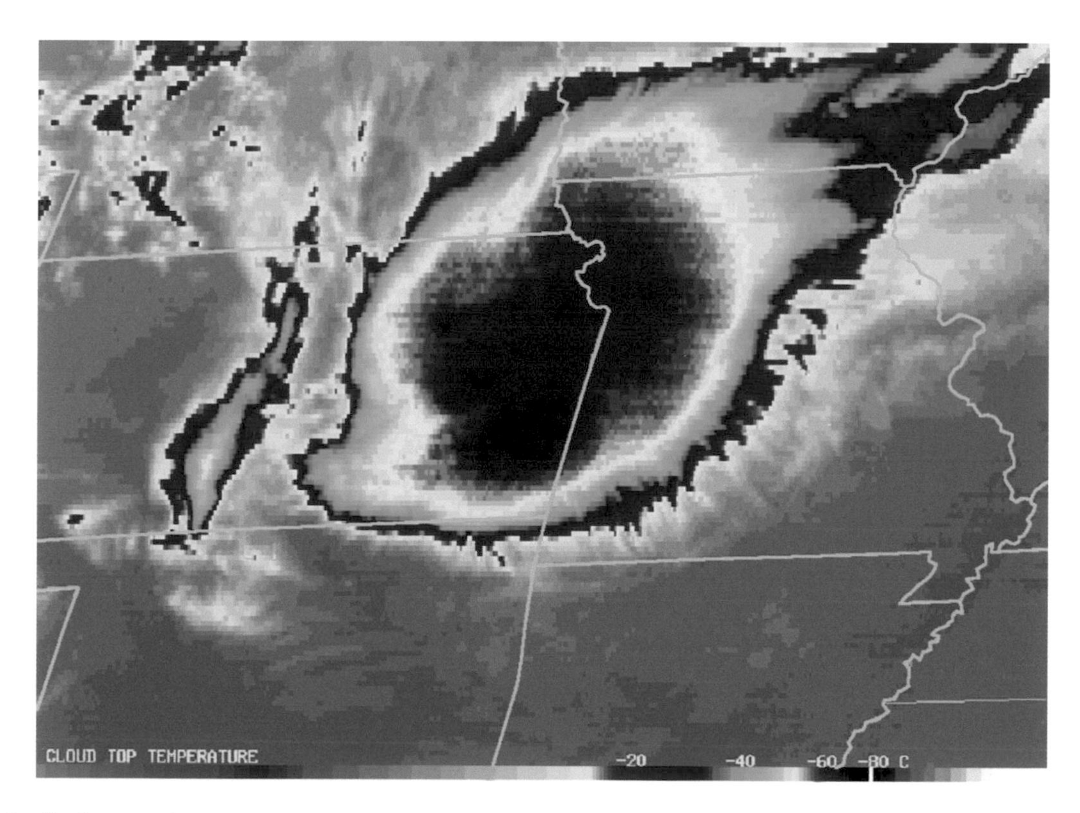

Figure 14.7 Satellite imagery of a mesoscale convective complex (MCC). An MCC is a highly efficient rain-producing cloud system common to the overnight hours over the Great Plains during summer. (NOAA.)

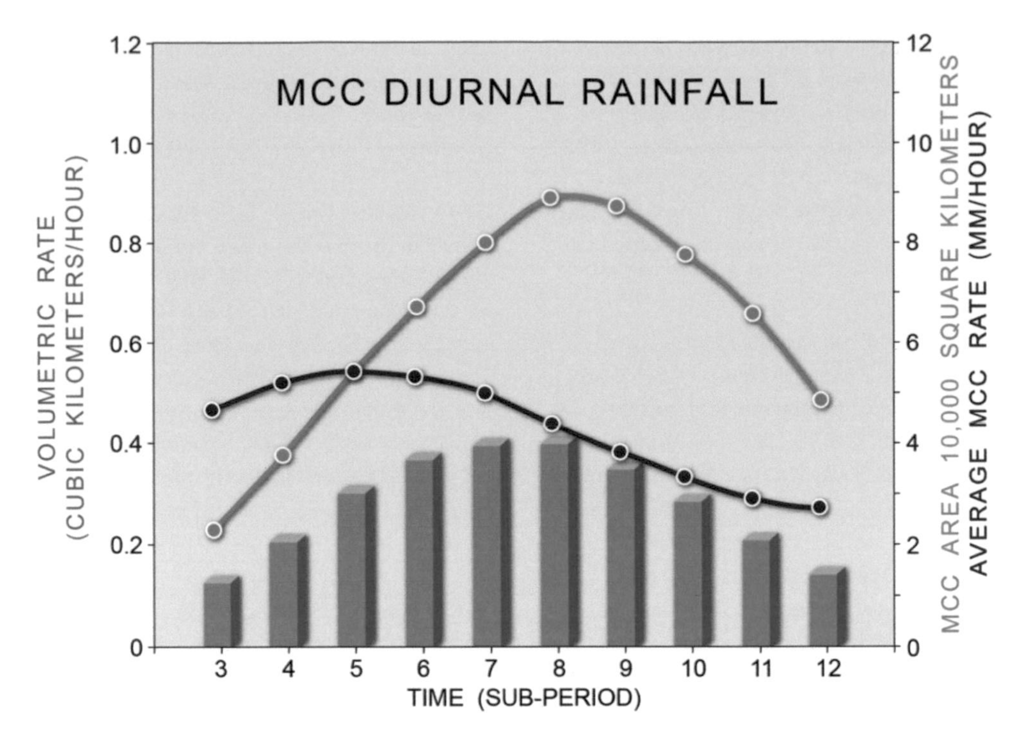

Figure 14.8 Evolution of an MCC: Relationship between MCC rain intensity and cloud area as a function of time. Rain intensity tapers after midnight (red), but the decrease in intensity is offset by an expansion of raining cloud area (green), such that the storm system's water output remains very large into the early morning (blue bars). (Adapted from McAnelly and Cotton, 1989.)

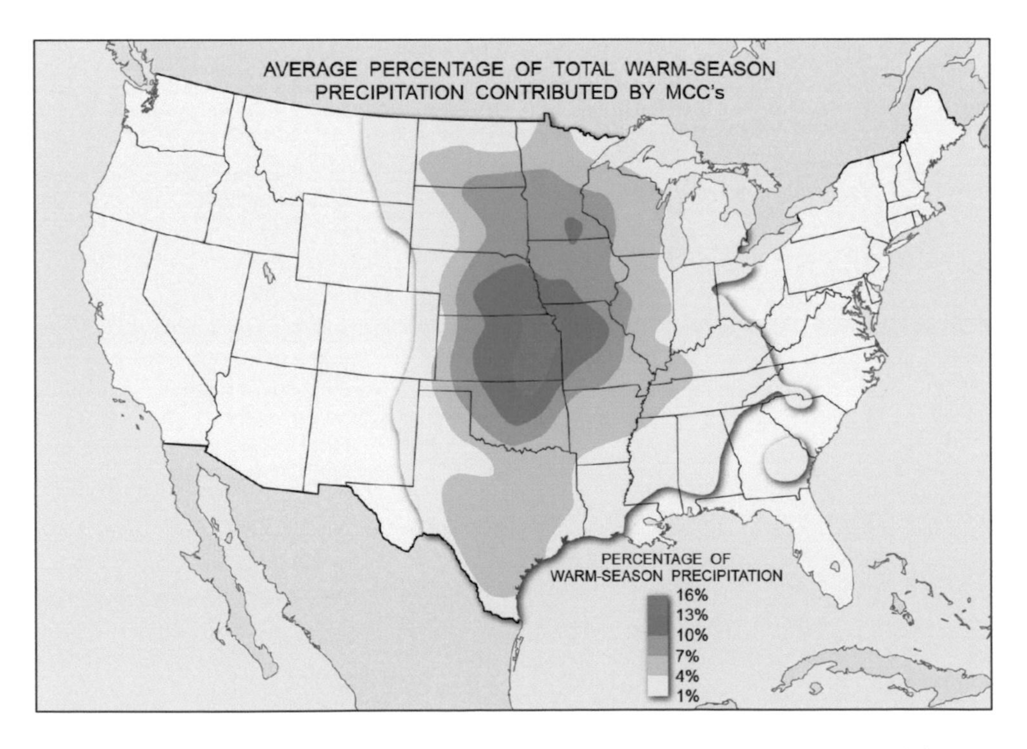

Figure 14.9 Total summertime rainfall from MCCs, U.S. Great Plains. MCCs contribute 10–15% of total summertime rainfall across a large segment of the Great Plains. (Adapted from Ashley et al., 2003.)

of producing flash floods, also benefit the region's enormous agricultural needs. MCCs deliver a significant fraction of summer rainfall to the Plains (Figure 14.9).

Why are MCCs such efficient rain producers? Figure 14.10 shows the internal structure of a typical MCC, identifying the major air circulation zones. The resemblance to a hurricane is a bit uncanny, but we stress that there is no such thing as an "inland hurricane"! A large, mid-level vortex organizes winds into an inflowing spiral. Bands of deep thunderstorm clouds (only some are shown here, for clarity) create corridors of heavy rainfall. Humid air streams into the MCC at the low and mid-levels, especially along a low-level jet (purple arrow), importing Gulf moisture. Downdrafts spreading from convective cells merge and form a massive pool of cool downdraft air (see Chapter 13) at the surface. This "cold dome" acts as a barrier, forcing inflowing, moist air to ascend. Because of the vortical nature of the primary circulation, a massive, circular cloud shield emerges. While the exact mechanism has not been identified, it is thought that the cloud canopy helps the storm system intensify after midnight; tall cloud tops cool very efficiently, like a giant radiator. The atmosphere destabilizes, sustaining convective updrafts through the night.

The mid-level vortex in an MCC (called a mesovortex) is so energetically efficient that it often outlasts the rest of the cloud system. During the morning, when convective clouds weaken and dissipate, remnant cloud layers reveal the vortex, which shows up as the "ghost of the MCC" in satellite imagery (Figure 14.11). Such vortices are quite persistent. As they drift eastward, they may trigger a new round of convective storms in the afternoon, often over the Great Lakes or Tennessee Valley.

Supercells and derechos are also capable of creating localized flash floods. Supercells, with their extremely powerful updrafts, are often heavy rain producers, especially high precipitation (HP) type of supercells (see Chapter 10). Derechos, while often very fast-moving, also include intense updrafts and tend to form in environments characterized by very high humidity. When flash floods occur, they develop in a region of cell training along the southern end of the derecho complex, called the tail of the system. Figure 14.12 illustrates the tendency of multiple cells to spread into a line along the tail. An example of a flash-flood-generating derecho is shown in Figure 13.26; there you will note an extended band of very heavy rain cells (red color) oriented from west to east along the southern end or tail of the storm.

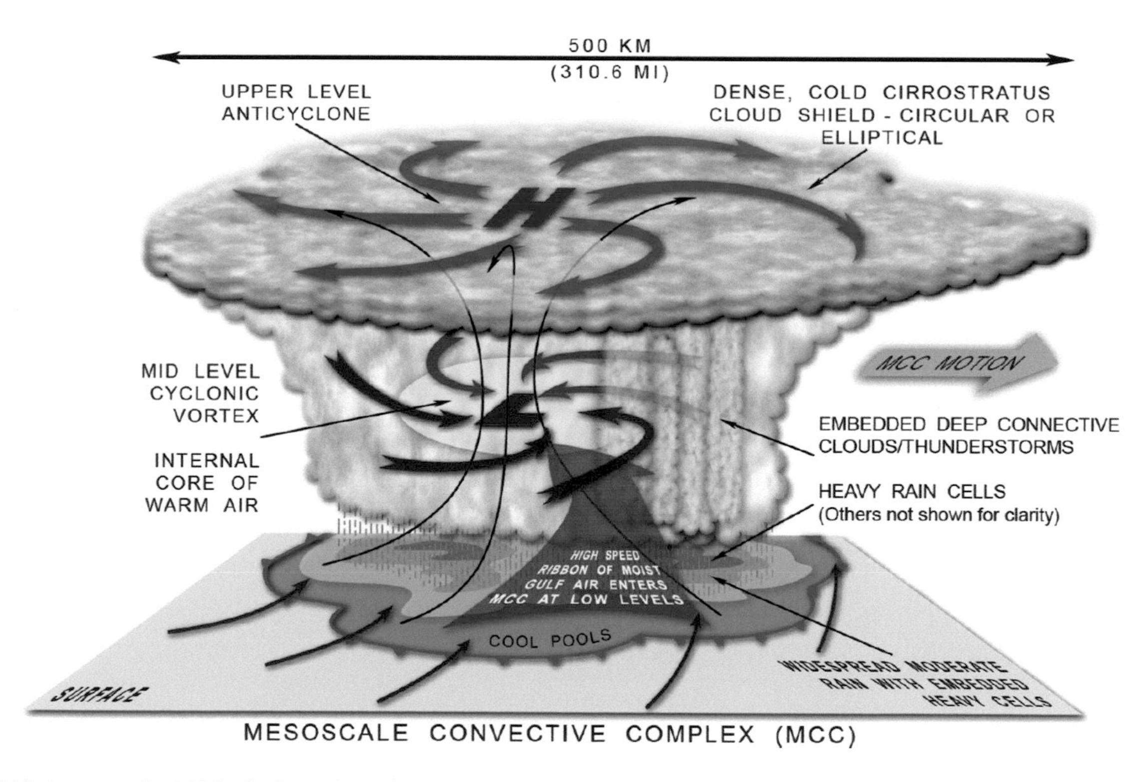

Figure 14.10 Anatomy of an MCC. The figure shows the vortical circulation in mid-levels, a low-level jet feeding oceanic moisture into the system, and columns of intense thunderstorm cells.

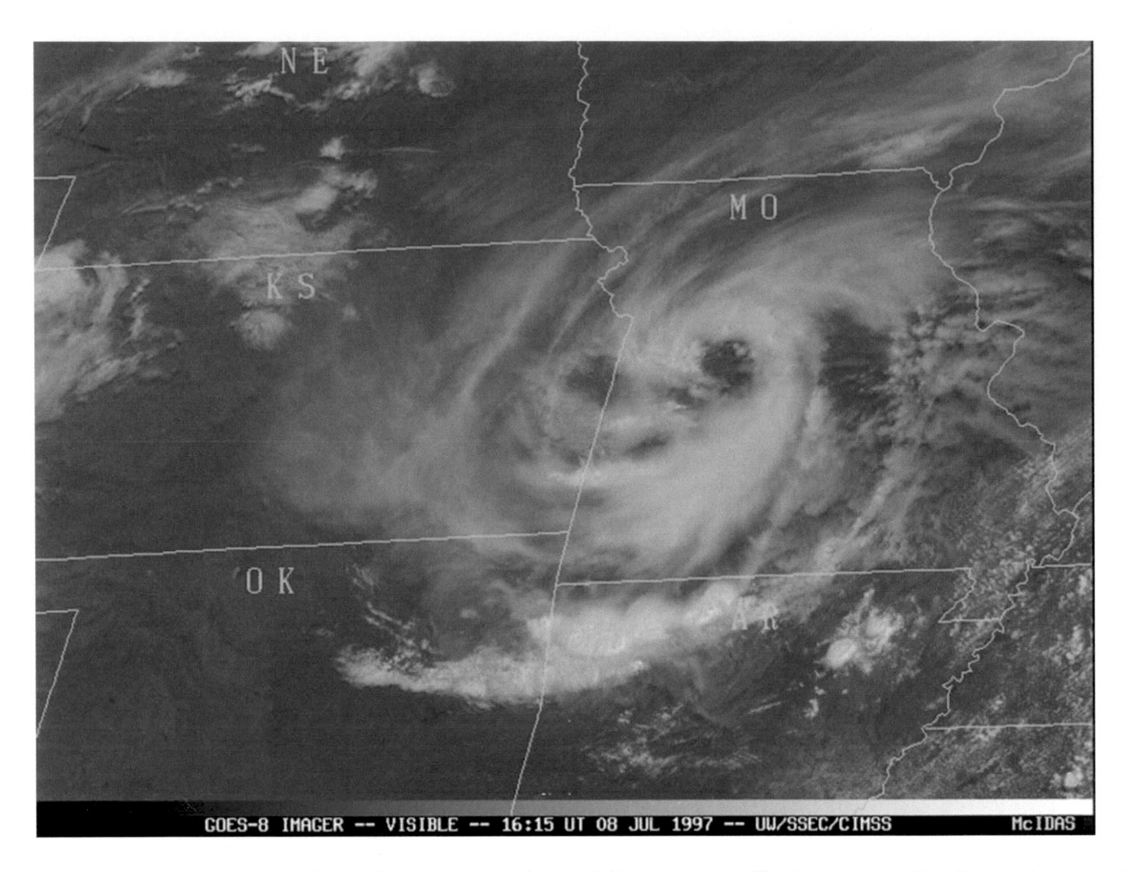

Figure 14.11 Ghost of a (former) MCC. The image shows a remnant cloud whirl in morning satellite imagery, revealing the massive vortex that helped organize the previous night's heavy rainfall. (NOAA.)

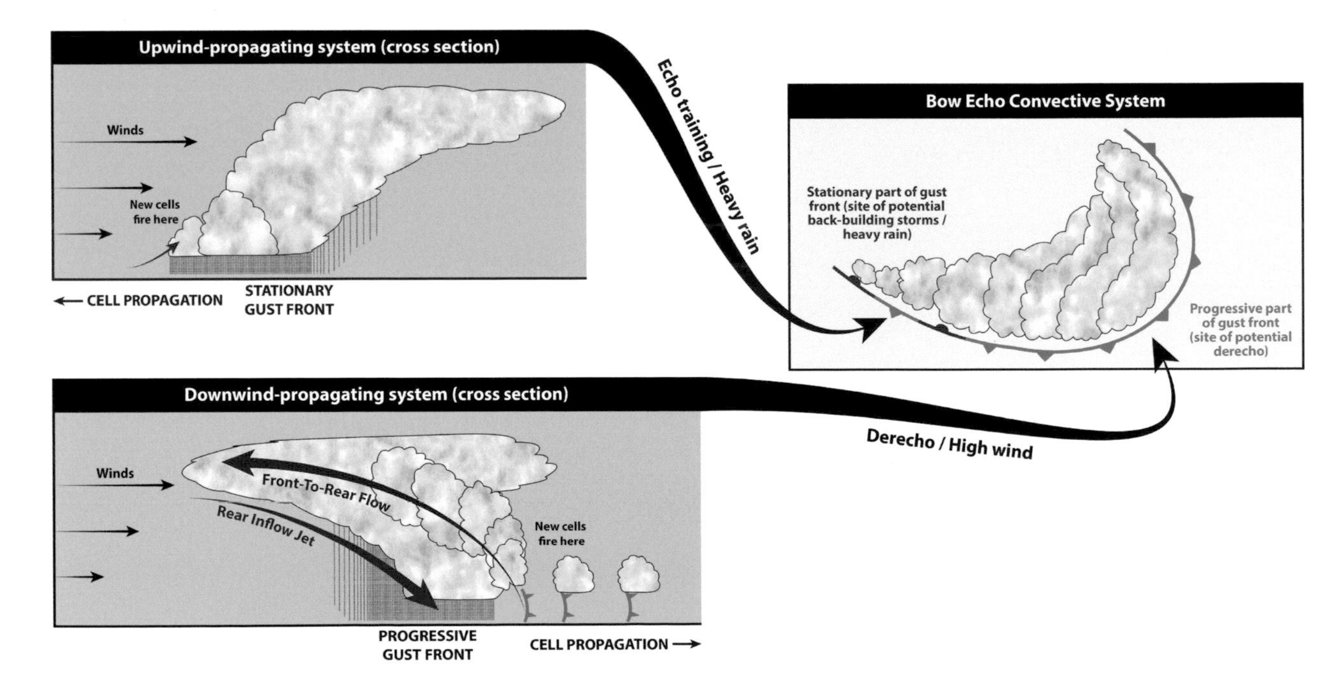

Figure 14.12 A zone of flash flooding can develop along the southern flank of a derecho cloud complex. A stream of convective cells redevelops along a stationary or stalled portion of the storm's gust front, passing repeatedly over the same locations. This contrasts with the forward, or progressive, part of the gust front, which surges rapidly ahead due to strong downburst winds. (Adapted from NOAA Storm Prediction Center.)

Synoptic Scale Storm Systems

Extratropical cyclones are the most common weather-generating systems in the mid-latitudes. During the cool-season months (October–April), these storms can produce locally heavy rain. Flash floods are less likely during winter because cold air holds much less moisture. Additionally, convective instability is weak during the cool season, so rain is generated not by thunderstorms but rather by gentler stratiform clouds. However, isolated convection can sometimes still develop in these large winter storms, particularly along the Gulf Coast.

Figure 14.13 shows how strong winds and a moist plume of air entering a cyclone's warm sector create a flash flood. The parent low is shown by the red "L" along the cold front and warm front. High dew point air enters the system's warm sector from the south, and airstreams converge as they flow north. The converging flow must ascend along a narrow conduit called the warm conveyor belt (purple arrow). Small amounts of instability generate convective cells (green centers, numbered) within the ascending conveyor. Once again, we see the making of a rain train – the repeated passage of cells over the same location – but on a larger scale than an individual, back-building thunderstorm complex. The key to the train is slow movement of the large-scale weather pattern. As long as the cyclone and its cold front advance slowly, the conveyor belt remains nearly stationary. The small graph (inset) shows cumulative rainfall at location A. When cell after cell passes repeatedly over A, the rain adds up, creating 7.5–12.5 cm (3–5 inches) after just a few hours.

During winter, powerful Nor'easter cyclones frequently develop along the Eastern Seaboard, tracking from the Carolina Outer Banks to New England. A unique combination of geographical factors can make these storms particularly heavy rain producers, as Figure 14.14 shows. When the air mass is too warm for snow, heavy rain is generated along the coastline and inland locations. These storms process a tremendous amount of water vapor because they overlie the warm Atlantic Ocean. Evaporation from the Gulf Stream ocean current is particularly intense because of the current's very warm temperature, even in the middle of winter. This evaporation creates a flow of high dew point air entering the storm.

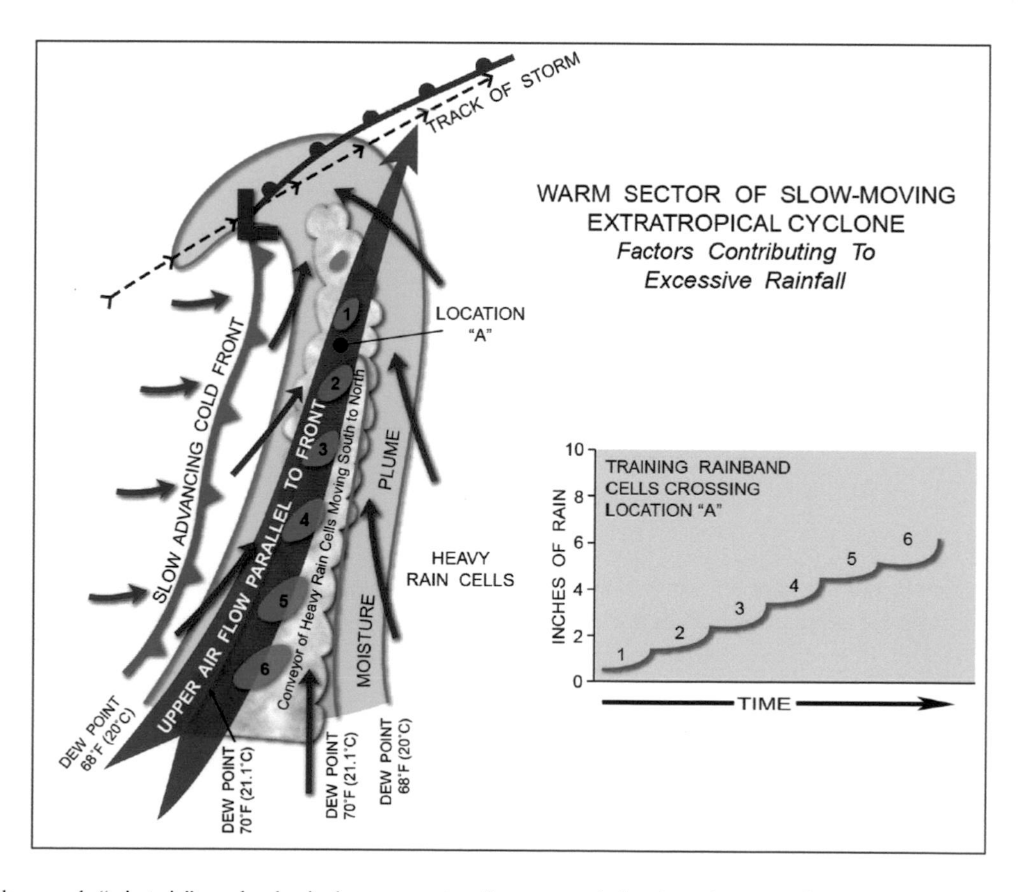

Figure 14.13 A large-scale "rain train" can develop in the warm sector of an extratropical cyclone. A stream of convective cells sweeps rapidly over the same locations, from the south, while the larger storm system moves very slowly toward the east.

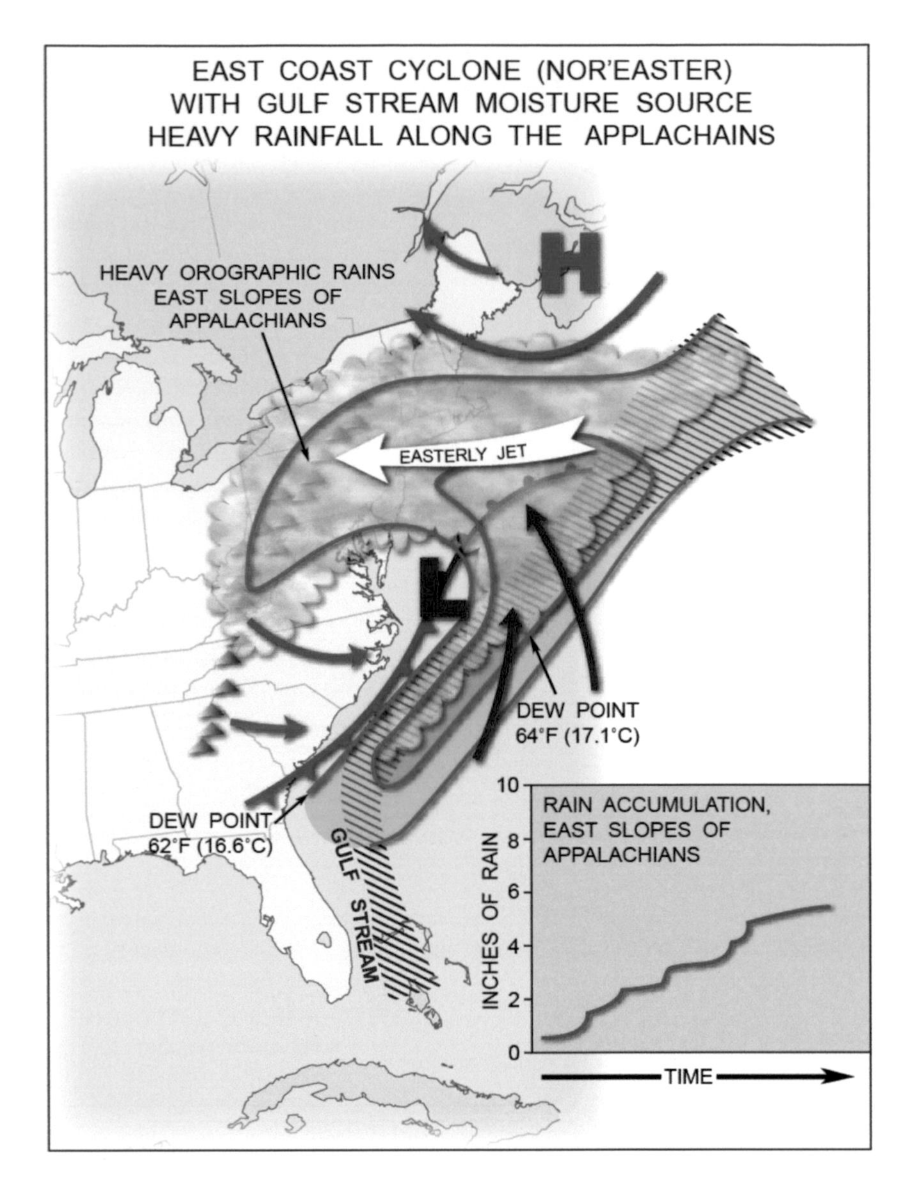

Figure 14.14 How Nor'easters create heavy rain in winter. Several geographical elements, including the Gulf Stream ocean current and Appalachian Mountains, create a heavy-rain-generating scenario unique to the wintertime Mid-Atlantic and New England regions.

The moisture-laden air is drawn inland by an easterly, low-level jet circulating around the northern edge of the low pressure center (white arrow). Other times, the jet impinges on the storm from the south, as a tropical plume sourced to the Caribbean. Bands of heavy rain develop just to the north and west of the low's center, where low-level air strongly converges and deforms (see Chapter 6 for more information about the formation of these precipitation bands). Even the Appalachian Mountains get into the act; as moist air impinges on the mountain barrier and ascends vigorously, leading to orographic rain enhancement on the eastern slopes. Orographic rain occurs when low-level, moist air is forced to ascend steep terrain; the abrupt uplift of air causes the moisture to condense into rainfall.

If the winter ground is frozen, rainwater infiltrates slowly (or not at all), leading to ponding and runoff. With snowpack on the ground, snow quickly melts and adds to the total runoff. Snowbanks along roadsides (created by plowing the roads) exacerbate the ponding and damming of water on roadways. These factors all lower the threshold for minor flash flooding.

A final type of synoptic weather system is the Pacific coastal low (Figure 14.15). These wintertime storms sweep in from the Gulf of Alaska, often at peak intensity. Their deep vertical circulations draw in copious amounts of Pacific moisture, which is forced to ascend the tall rampart of the Sierra Nevada and Cascades. Once again, orographic enhancement is at work, generating torrential rainfall along the lower mountain slopes (and often several feet of heavy, wet snow above the freezing line). A combination of slow storm movement, high-moisture content air, and steep terrain unleashes flash flooding. Corollary hazards, including landslides and debris flows, frequently develop in conjunction with rapidly rising water.

One unique aspect of Pacific systems is the tendency for water vapor to concentrate into massive, narrow ribbons termed atmospheric rivers. These moist channels originate in the tropical Pacific, far to the west, often in the vicinity of Hawaii – hence the term Pineapple Express. Figure 14.16 shows a plume of high atmospheric moisture streaming into high terrain over the U.S. West Coast. A single plume can transport more water (in the form of vapor) per unit time than the Amazon River, the largest on Earth. When these channels feed into wintertime extratropical cyclones and frontal systems along the U.S. West Coast, torrential rains and flooding often ensue. Imposing mountain barriers such as the Sierra Nevada and Cascades help lift the nearly saturated air, triggering extremely heavy rainfall along the western (leeward)) slopes. At times, lesser rains created by these rivers can prove beneficial to portions of the West Coast, contributing 30–50% of the annual water supply (rainfall and snowpack). To better assess the beneficial vs deleterious aspects of atmospheric rivers, a five category scale of their volumetric water vapor transport and effects has been developed, ranging from Category 1 (weak river, primarily beneficial rains) to Category 5 (exceptionally strong river, primarily hazardous).

Atmospheric rivers are not limited to the Pacific. It is now thought that some East Coast cyclones (Nor'easters) and oceanic storms impacting Europe occasionally develop river-like features, with tropical moisture originating from the Caribbean.

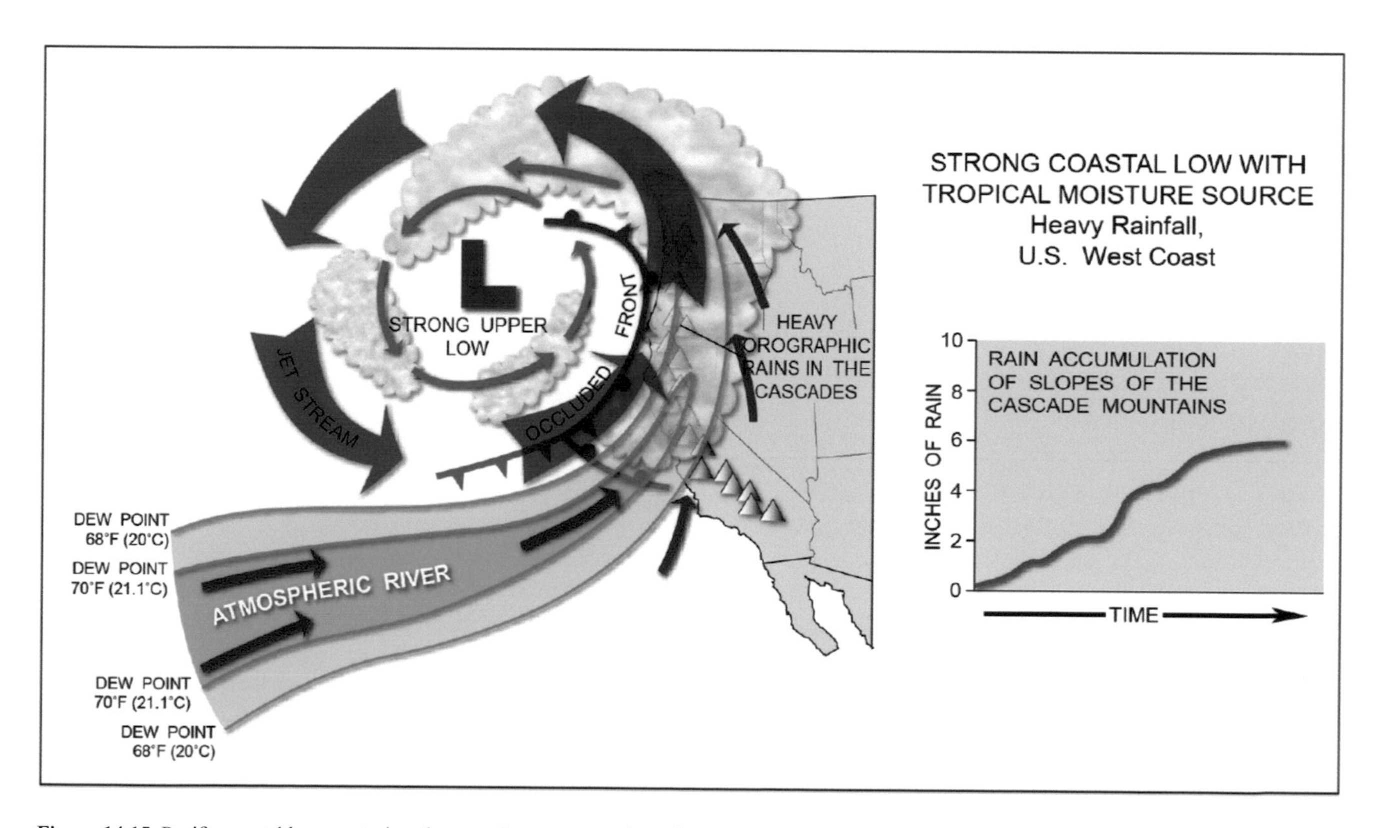

Figure 14.15 Pacific coastal lows: notorious heavy-rain generators along the coastline. In conjunction with landslides and debris flows, heavy rains create frequent wintertime hazards.

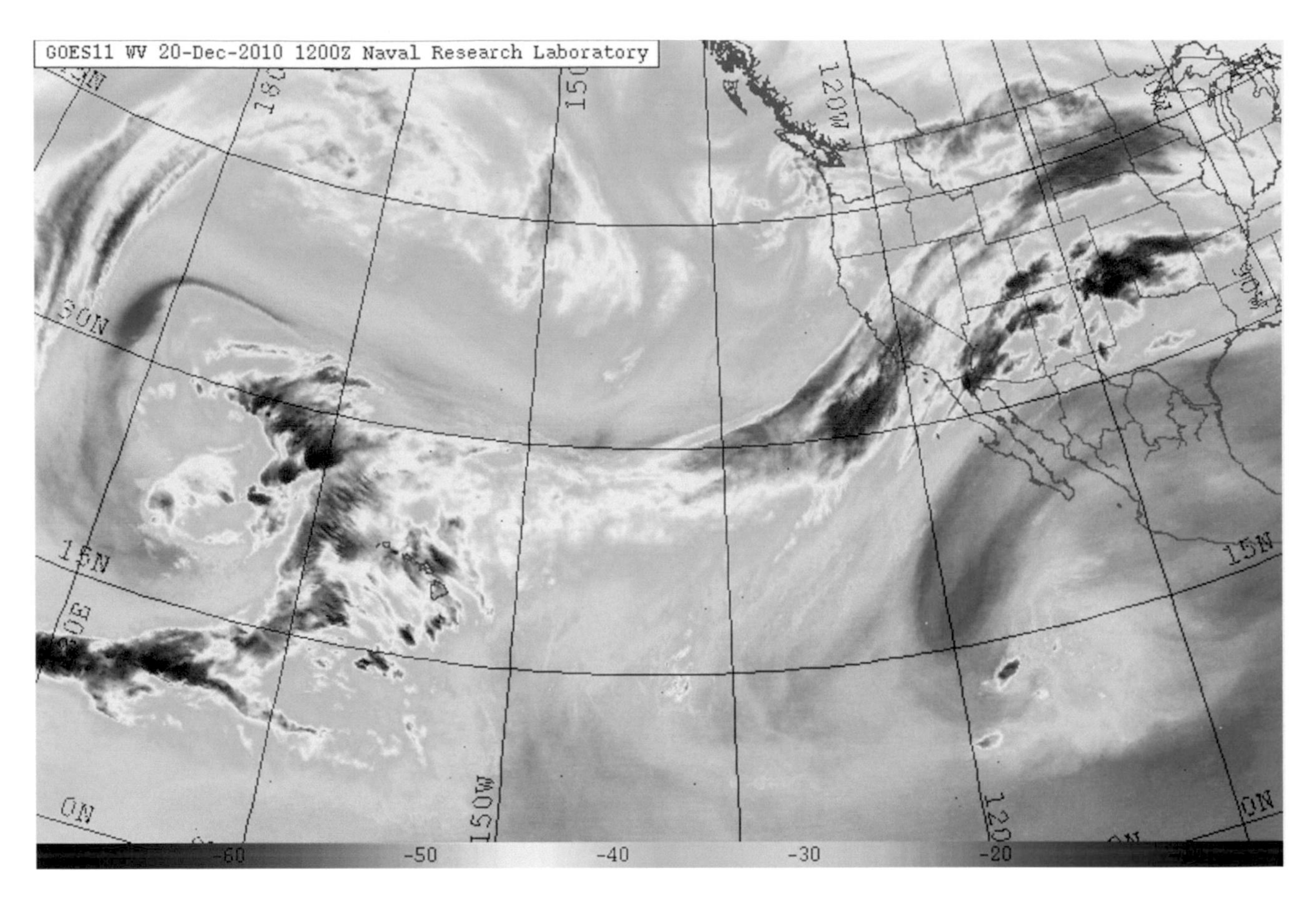

Figure 14.16 Atmospheric river. This satellite image reveals a veritable pipeline of high-humidity air aimed at the U.S. West Coast from the vicinity of Hawaii. The extreme moisture in this tropical plume contributed to significant flooding and heavy snowfall in December 2010 across southern and central California. (US Naval Research Laboratory.)

Summertime Mountain Flash Floods: Examples From the Rockies and Appalachians

Infrequently, atmospheric processes conspire during the summer months to create localized flash floods along the slopes of mountain ranges. The mountains need not be excessively tall, but moist, unstable airflow up steep terrain can unleash devastating convective floods.

In this section, we highlight two cases: the Madison County, Virginia, flood on June 27, 1995, and the Fort Collins, Colorado, flood on July 28, 1997. Both flood events, separated by 1800 miles (2897 km), were triggered primarily by orographic uplift, under strikingly similar meteorological patterns. Both featured small areas of extreme rain, just 100 km^2 (39 sq mi). However, the amount of rainfall produced and the impacts varied considerably. In the case of Fort Collins, a relatively flat urban landscape was destroyed by 28 cm (11 inches) of rain. Over Madison County, 75 cm (30 inches) of rain on hilly terrain produced landslides, mudflows, and debris flows over a sparsely populated region.

Fort Collins, Colorado (1997)

On the night of July 28, 1997, convective clouds blossomed over the foothills just to the northwest of Fort Collins. The storms did not produce hail or frequent lightning, like so many midsummer thunderstorms along the Front Range. But the rains kept coming and coming, and over 25 cm (10 inches) fell over the city's Quail Hollow region in just 6 hours. Stormwater facilities were overwhelmed, and irrigation ditches topped over. The Colorado State University (CSU) campus was inundated, and 5 died in mobile home parks along Spring Creek. Mobile homes were swept away by surging, muddy floodwater. Water rescues, nearly 450 in all, were carried out through the night, with rescuers plucking people from rooftops and trees. Over 60 people were seriously injured, and over 2200 homes and businesses were damaged or destroyed. The widespread flooding caused $500 million in damage across the city.

Figure 14.17 shows a map of the rain accumulation. Rain cells initiated along the highest portions of the ridgeline west of the city. However, heavy rain did not begin to fall until the cells moved a few kilometers east of the terrain because it takes a finite amount of time (10–15 minutes) for cloud updrafts to generate a core of heavy rain. The bull's-eye for rain accumulation was located just east of Horsetooth Reservoir, in the Spring Creek drainage basin. The figure typifies the way that rain creates a flash flood: highly localized, with pronounced gradients of rainfall.

A succession of five convective cells, at intervals of about an hour, erupted along the steep terrain bordering Fort Collins, parading across the city from the southwest. As each cell moved off the mountains, it dumped its core of heavy rain over the relatively flat, densely populated landscape. Paved roads, parking lots, and culverts quickly became overwhelmed by storm runoff. Rainwater rapidly ponded, then began to carve its own course through streets, yards, and business lots. In some locations, swiftly flowing water, carrying tangles of debris,

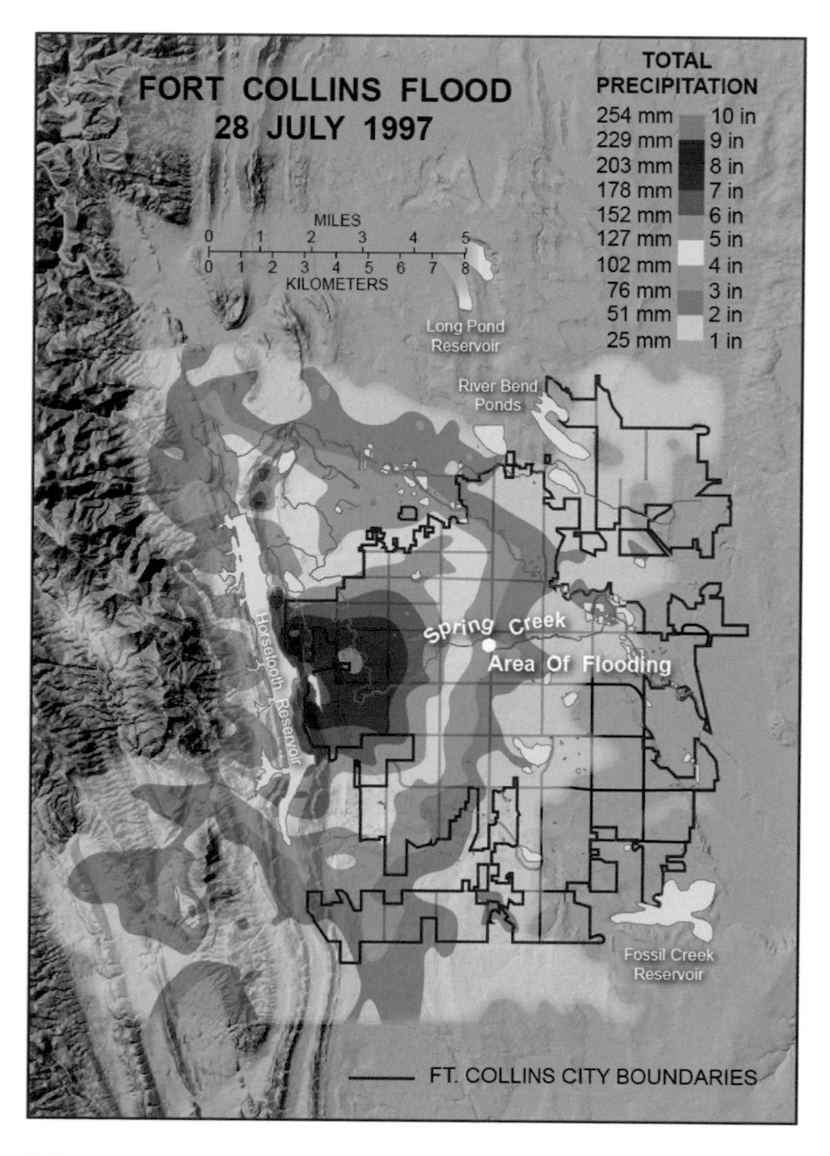

Figure 14.17 Rainfall in Fort Collins, Colorado, July 28, 1997. An amazingly intense, concentrated zone of heavy rain accumulated in just 6 hours across the relatively flat, urbanized landscape of Fort Collins. (Adapted from Petersen et al., 1999.)

converged from multiple directions. This is the training cell or back-building mechanism described earlier.

What factors conspired to generate this intense rainstorm? The key elements are shown in Figure 14.18. During the summer months, eastern Colorado receives an influx of oceanic moisture from the Gulf of California, as well as the Gulf of Mexico, via persistent southwest winds. The wind pattern is part of a regional circulation across the Desert Southwest called the North American Monsoon (NAM), which develops every summer. Monsoon moisture arrived at the mid-levels (purple arrows), while low-level moisture streamed in from the east (black arrows). A cold front was positioned to the southeast of Fort Collins, and high dew point air was pooled along this boundary. Air in the easterly current was nearly saturated, having ascended up the High Plains, a process that gradually cools the air.

An unstable, late afternoon air mass promoted the growth of convective clouds. These cells were triggered where the low-level easterly winds impinged along the Rocky Mountain Front Range. A shortwave trough in the upper level airflow generated additional ascent in the vicinity of Fort Collins. The convergence of moist airstreams at all levels ensured that these cells had abundant moisture to process. A small, back-building cluster of thunderstorms erupted in one spot, parked over Fort Collins. Cells were continuously regenerated along the ridgetop, moved off to the northeast, and then were immediately replaced by new cells. In this sense, the back-building storm became terrain locked for many hours.

Two years earlier, a similar terrain-locked storm complex had developed nearly 1800 miles away, in a completely different mountain setting – the lush, vegetated Appalachians of Virginia. This storm created a fierce flash flood, which we discuss next.

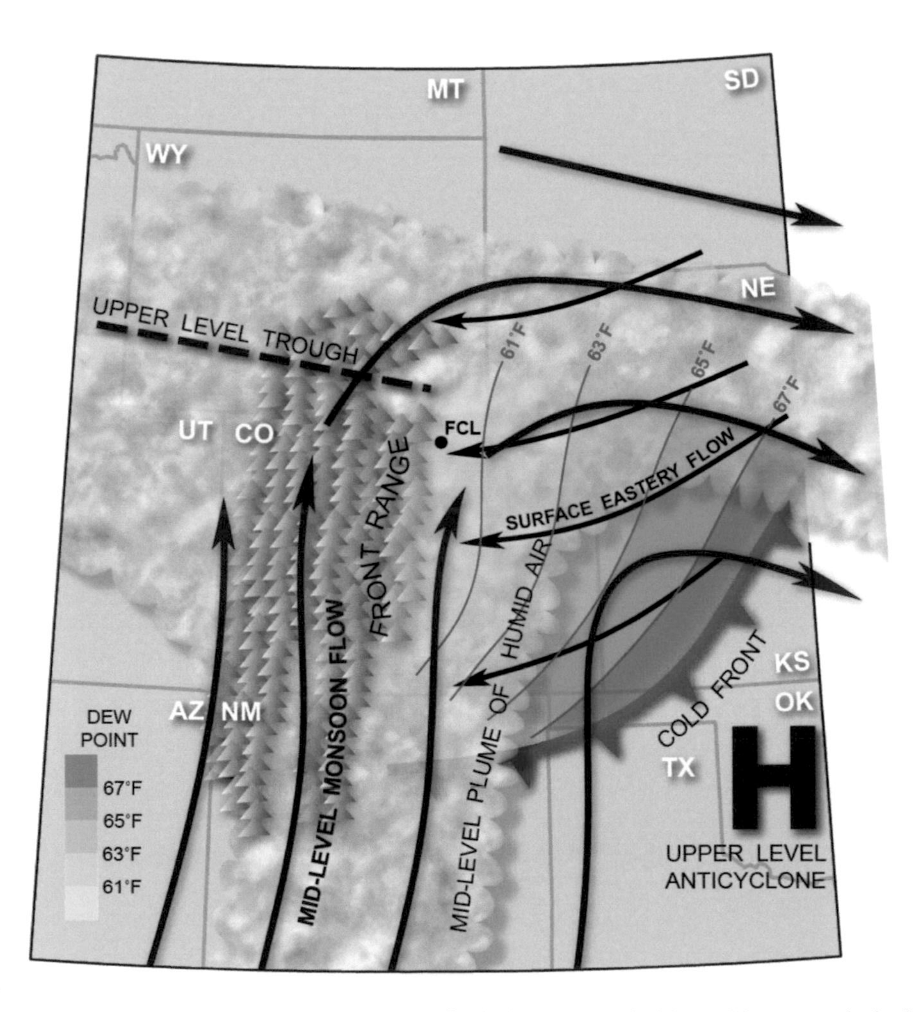

Figure 14.18 Convergence of atmospheric elements leading to flash flooding in the western United States. Many atmospheric elements converged over the mountains bordering Fort Collins on July 28, 1997, including a monsoon airstream, a humid easterly flow, a cold front, and an upper-level trough.

Madison County, Virginia (1995)

The Blue Ridge forms a 1067–1219 m (3500–4000 foot) rampart along the eastern margin of Virginia's Appalachian Mountains. On June 27, 1995, a cold front slipped into central Virginia from the north and draped across the mountains in central Virginia (Figure 14.19). Meanwhile, a shortwave trough in the upper-level flow approached the region from the southwest. Moist, low-level air pooled along the front and a strong low-level jet (green arrow) parallel to the front drew in moisture from the Atlantic. These factors coalesced along the eastern slopes of the Blue Ridge, over Madison County – a largely rural setting with a scattering of small towns, farmland, and orchards.

Convective cells began blossoming in moderately unstable air, over the ridgeline of the Blue Ridge. As in Fort Collins, cells drifted off the mountains to the northeast, rapidly matured, then dumped heavy rain over the same spot in Madison County. Here is yet another example of a back-building, terrain-locked rainstorm, anchored to one location for nearly 6 hours. The bull's-eye of rain was incredible: nearly 75 cm (30 inches), centered on the small town of Graves Mill, about 10 km (6 miles) east of the Blue Ridge's scenic Skyline Drive (Figure 14.20).

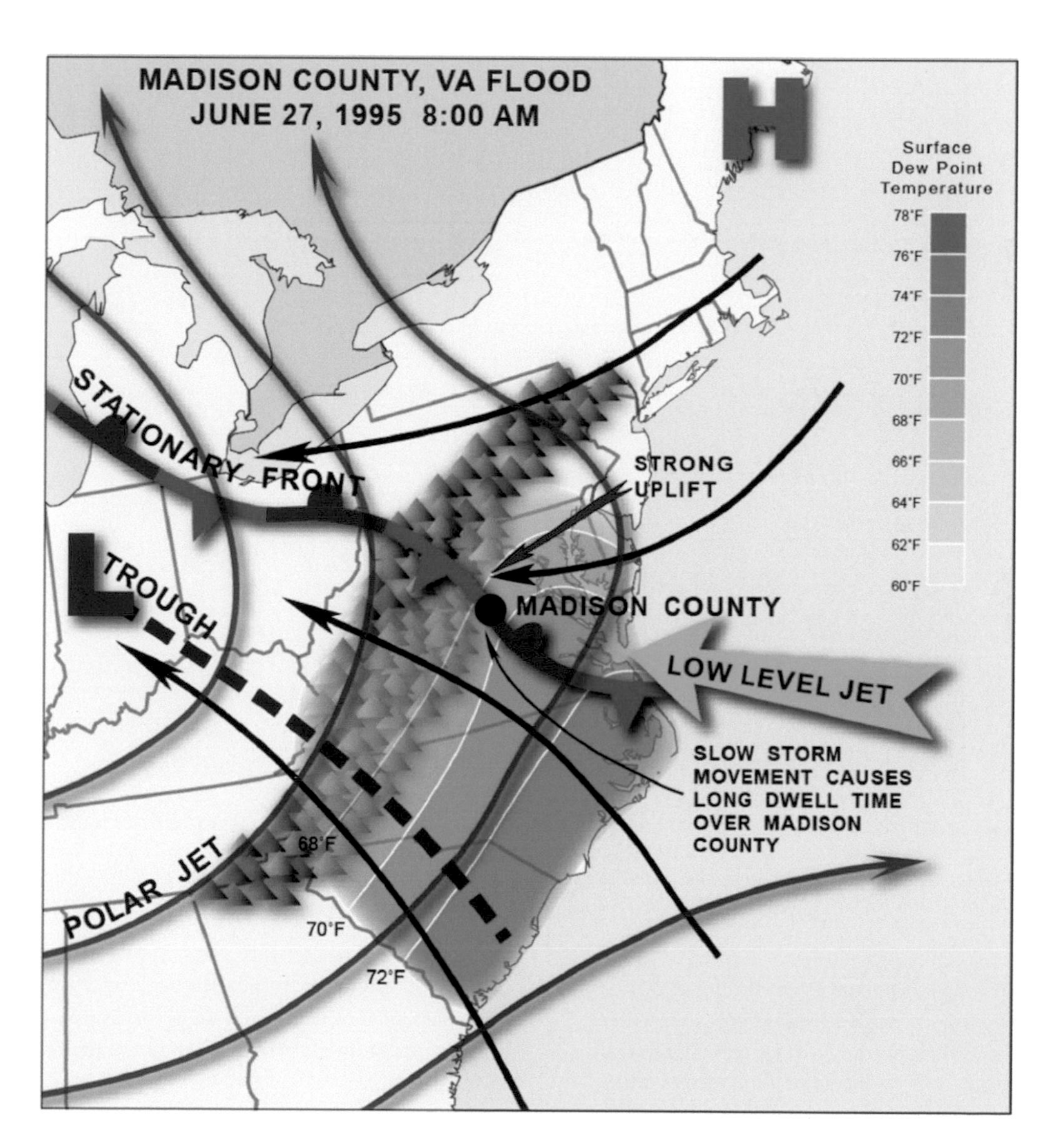

Figure 14.19 Madison County (Virginia) flash flood of 1995. As in the Fort Collins flash flood, the Madison County rainstorm developed from an interaction among steep terrain, copious moisture, a stalled front, and an approaching upper-air disturbance.

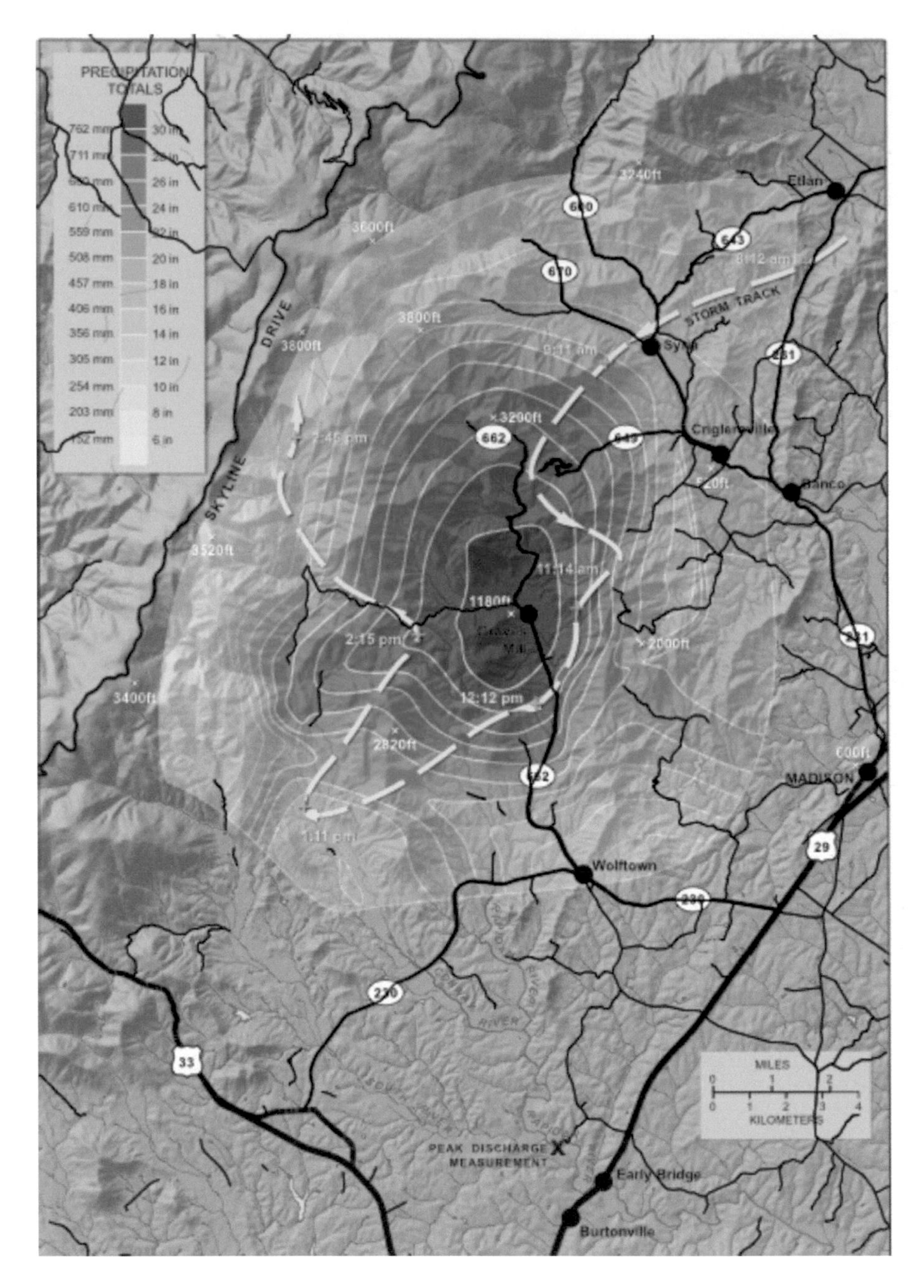

Figure 14.20 Rain accumulation and storm drift, Madison County, 1995. Spectacular rain accumulation over a small area led to extreme flash flooding, mudslides, debris flows, and record river discharge. Peak discharge on the Rapidan River is shown by the black "X" south of the heavy rain region. The dashed yellow line shows the slow drift of the convective storm complex over several hours. (Adapted from USGS.)

The flood's aftermath was a grim scene of devastation. Massive scars down to bedrock revealed locations where slopes liquefied and flowed into valleys, carrying water, mud, boulders, trees, and remnants of farm homes. Mud and debris moved rapidly downslope, fanning into valleys, destroying everything along the path. There were 3 fatalities, 20 injuries, and 800 evacuations. Over 400 roads were closed, 80 bridges were washed out, and 2000 homes were damaged or destroyed. Total losses exceeded $200 million, including severe damage to the region's agricultural infrastructure.

The USGS mapped the debris flows after the flood. Figure 14.21 depicts the results of this survey. The flows are fairly short,

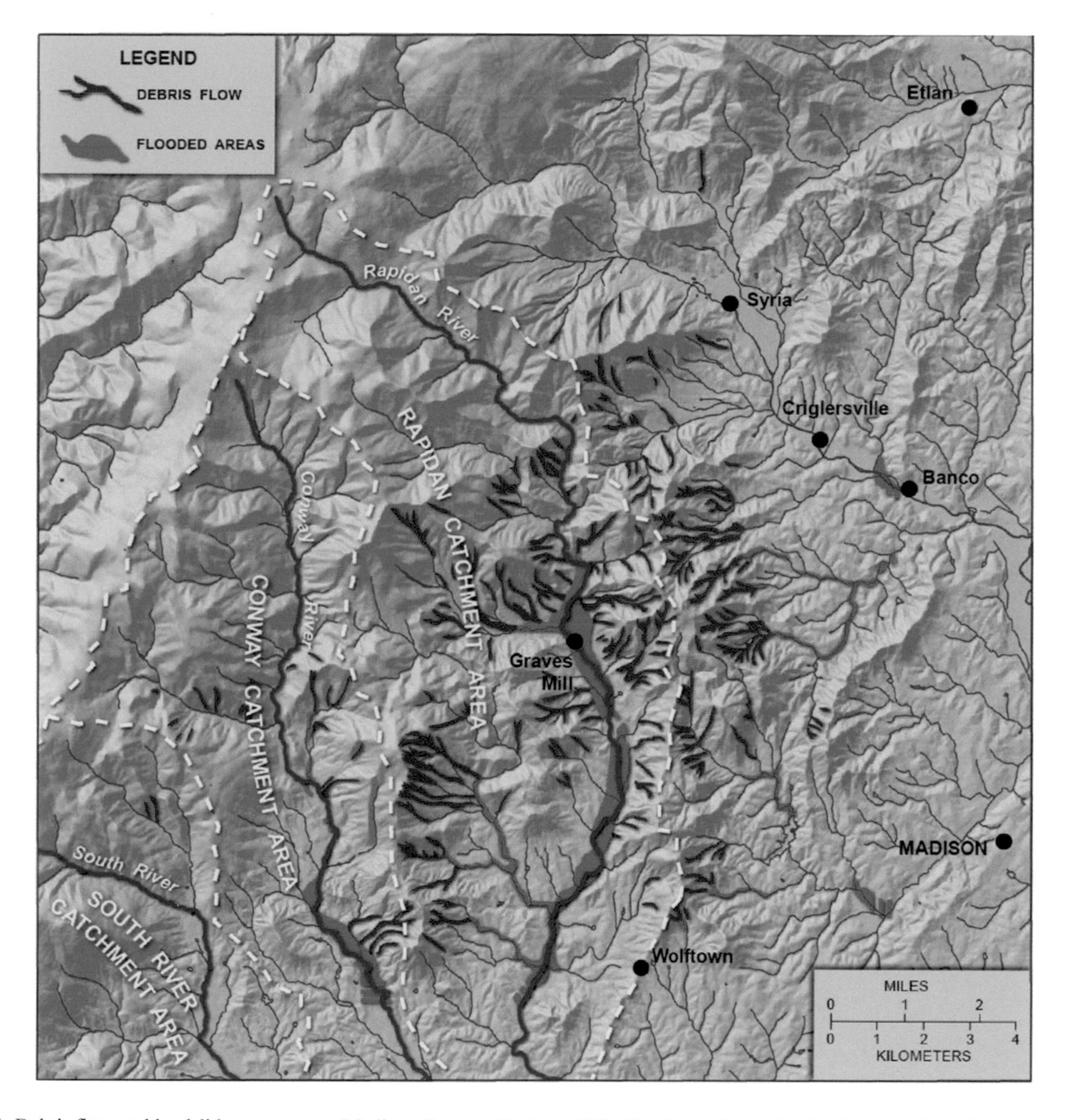

Figure 14.21 Debris flow and landslide survey map, Madison County, Virginia, 1995. The figure shows the distribution of flood water (orange) and debris flows (red) in relation to terrain and Madison County's stream network. (Adapted from USGS.)

converging down slopes that feed the watershed of the Rapidan River (outlined by yellow dashed lines). The region of greatest flooding is shown in orange, centered on Graves Mill, along the course of the Rapidan River, which drains to the south. The vertical relief here is modest, on the order of 457–610 (1500–2000 feet). The flood's epicenter is offset from the crest of the Blue Ridge, 8 km (5 miles) to the west. The region is densely forested with a deep layer of organic material and soil. This thick soil profile was quickly saturated in the heavy rain, then liquefied. Entire slabs of forest sloughed off bedrock, flowed downhill, converged along hollows, and spilled into valley floors.

The Rapidan River rose nearly 9.1 m (30 feet) in 1 hour. A peak river discharge of 3000 m^3/s was recorded at Ruckersville, Virginia, about 15 km (9 miles) downstream from the rainfall epicenter. To put this in perspective, the peak discharge from this tiny stream was half the low-stage discharge of the mighty Mississippi River! In fact, the Madison County flood established a new record for flood discharge for any U.S. stream located east of the Mississippi.

Figure 14.22 illustrates the time sequence of heavy rain and river discharge. You can compare river catchments in steep terrain to giant funnels: Rain enters the funnel from above (green curve),

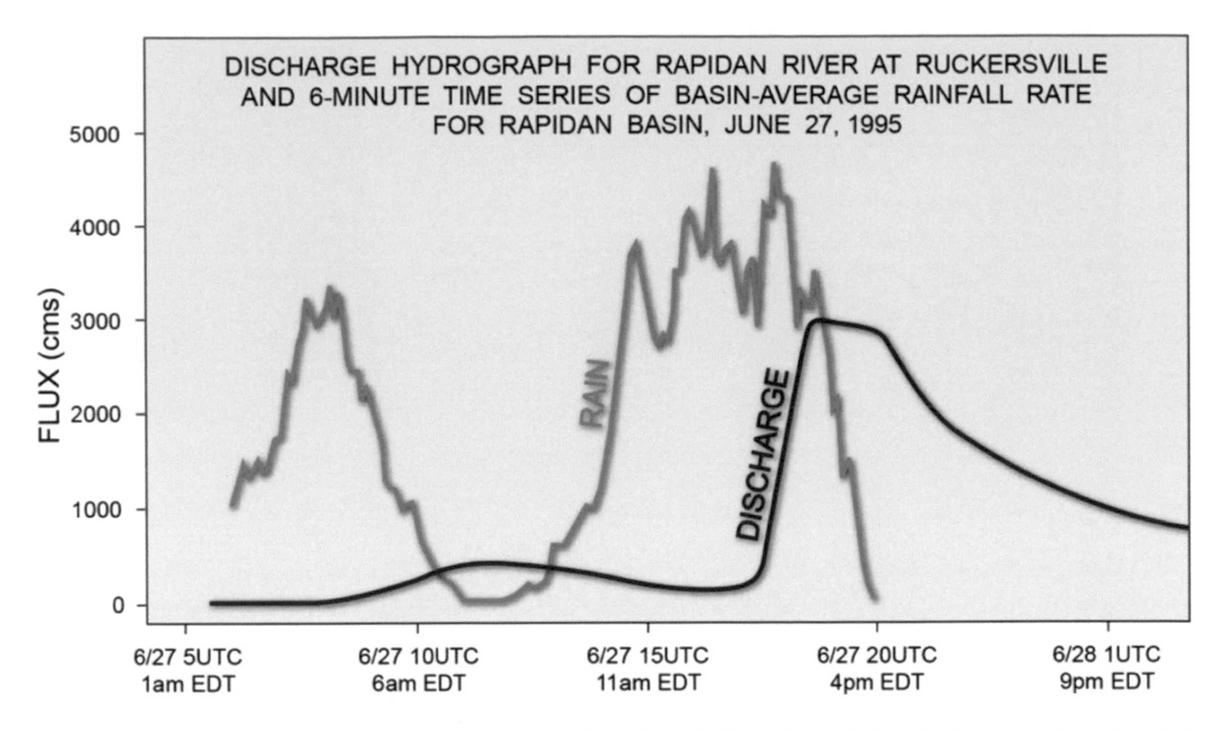

Figure 14.22 Stages of a catastrophic flash flood, Madison County, Virginia, 1995. The figure shows rain intensity (green curve) and river discharge (red curve). The earlier rain event likely primed the watershed for the flood event later that day. (Adapted from Smith et al., 1996.)

and an intense stream of water must flow out the nozzle (red curve). The stream hydrograph is a record of discharge, which rose abruptly halfway through the rainstorm. The time evolution of rain volume showed distinct fluctuations, timed to the training of individual storm cells.

Note the earlier occurrence of a separate rainstorm, during the overnight hours. This system produced significant rainfall over the Rapidan River watershed, not enough to trigger a flood, but it did prime the landscape for a flash flood later in the day. This antecedent rainfall often lowers the threshold for flash flooding. The earlier rain episode saturated the soils and filled the streams to capacity, eliminating all reserve capacity in the catchment basin (like soaking a sponge just to the point of dripping). The second rainstorm later that day quickly pushed the hydrological system beyond its flood threshold.

The previous two examples – Fort Collins and Madison County – share many meteorological and hydrological attributes, suggesting a common flash-flood-generating mechanism that can occur anywhere there is steep terrain and abundant moisture during the summer. In fact, the Camille flood in Nelson County, Virginia – described in the chapter introduction – also developed according to this model, in the manner of a back-building, terrain-locked rainstorm.

Flash Flood Damages, Detection, and Trends

In this section we examine our efforts to get ahead of the flash flood hazard, through improved detection and monitoring of incipient flash flooding, as well as better prediction.

Flash Flooding Casualties – The Problem Continues

According to a 20-year NWS study ending in 2006, an average of 95–100 people in the United States die every year from floods and flash floods. This number has remained constant, in spite of decreased mortality from other types of severe storms, such as tornadoes. Many U.S. floods are caused by tropical cyclone remnants – which include tropical storms and hurricanes – and these floods can unfold hundreds of miles inland. Freshwater flooding is the number one U.S. hurricane killer, accounting for more than half of all fatalities, and 60% of these deaths are inland.

Dam failure is another cause of death tied to flash floods. The United States has nearly 76,000 dams, 80% of which are earthen. These dams can fail when subjected to intense and/or prolonged rains. In fact, one of the deadliest flash flood scenarios (237 fatalities) was caused by the failure of an earthen dam in Rapid City, South Dakota, in 1972. We classify dam failure as a corollary hazard – a direct consequence of the flash flood but not the actual flood hazard itself.

Half of all flash flood fatalities are auto-related. People often think it's safe to drive through ponded water, or what appears to be only shallow water flowing across a road or bridge. Automobiles, particularly large SUVs, give people a false sense of protection, even invincibility. Simply turning around and trying a different route may seem inconvenient. But many people cannot accurately judge the depth of water, and they underestimate its force. Water weighs 1000 kg per cubic meter (63 pounds per cubic foot), and flowing water, even at a "leisurely" 4.3–8.7 kts (5–10 MPH), has tremendous momentum. Dense debris transported in the water adds to its momentum.

When a vehicle stalls in floodwater, the momentum of the flowing water is transferred to the car, about 227 kg (500 pounds) of lateral force for each foot of water rise. Buoyancy makes the car weigh 682 kg (1500 pounds) less for each foot of water elevation. In just 0.7 m (2 feet) of water, the combined lateral force and buoyancy will sweep the car away. People become trapped, unable to open doors against water pressure, and power window equipment will short out and fail. The car may tumble on its side or roof. In very cold water, people quickly develop hypothermia. If death does not come from outright drowning, the victim may be crushed. The way to protect yourself in a flood situation is to follow the simple but effective advice: "Turn Around, Don't Drown."

The Challenge of Flash Flood Prediction and Warning

It is particularly difficult to forecast flash floods and make people understand the dangers. From a prediction standpoint, the challenge is twofold: Forecasters must predict not only the occurrence of the event (heavy rain) in a given location, which is inherently difficult, but also the magnitude of heavy precipitation. The quantitative precipitation forecast (QPF) is widely recognized as the least reliable of all forecasts. Yet rainfall amount transforms an otherwise benign, commonplace weather event into a deadly flash flood.

Flash flood prediction is also challenging because it involves the interaction of meteorology with a region's hydrological characteristics. As you have learned in this chapter, many additional factors can precondition or render a region susceptible to a flash flood, including antecedent rainfall, the nature of the drainage basin, topography, and urban land cover. Meteorologists must work closely with hydrologists to make an accurate flash flood prediction.

Flash floods are inherently small-scale hazards, generated by convective clouds. However, the triggering and focusing mechanisms for storm initiation are often not identified in advance because our observation network doesn't have the required spatial resolution. Forecasting models run at a 4 km (2.5 mile) grid resolution, but even these simulations fail to accurately pinpoint the location and timing of small storm complexes, such as back-building cloud systems. Furthermore, because a diverse set of meteorological settings lead to flash floods, it is difficult for meteorologists to recognize all the possible meteorological patterns. Many times, the large-scale flow appears relatively benign, such as a ridge in the jet stream pattern, which we often associate with fair weather. Summertime ridging is deceptive because the suppressed storm pattern allows large amounts of humidity and instability to accumulate, much like charging an electrical capacitor. Then along comes a very subtle trigger (a weak shortwave trough and/or a weak frontal boundary) that creates a slow-moving or training storm complex, converting all the stored water vapor into heavy rain.

During the flood event itself, convective cells often appear benign on radar, and the flash flood threat is not immediately recognized. These cells often have shallow cloud tops and may not even generate lightning and thunder. A striking example is from the flash flood over Madison County, Virginia (on June 27, 1995) as a nearby thunderstorm generated more recognizable severe weather. The more vigorous storm a few tens of miles away on the Virginia Piedmont had all the attributes of a classic, severe thunderstorm, including cloud tops to 16 km (10 miles), compared to only 12 km (7 miles) for the Madison County storm; heavy precipitation extending through a much deeper region; and frequent lightning. In the Fort Collins storm, radar-observed cloud tops remained below 12–13 km (7–8 miles), nor did the system generate significant lightning or damaging wind gusts. Thus, flash-flood-generating complexes need special scrutiny, at times more benign in appearance, but their persistence over the same location can have devastating consequences.

Instead of focusing on snapshots of cell intensity and height, meteorologists must look for dynamic patterns such as cell training, overall storm organization, and storm intensity trends. They must also be cognizant of hydrological factors and the nature of underlying terrain. Radar estimates of rainfall must be frequently monitored and verified against ground truth (data and analysis to validate a remote sensing device, such as weather radar). Weather radar remains an imperfect tool for estimating the amount of rain; for many reasons, radars often underestimate the true rainfall by up to 50%. Of great value are members of law enforcement, Skywarn spotters, or ordinary citizens who convey actual gauge measurements of rain and report on rapidly rising streams and flooded roads. For a meteorologist sitting in front of a radar screen a hundred miles away, this information helps to create a more complete picture of the hazard.

The foregoing discussion identifies shortcomings in our ability to accurately forecast flash floods. But when a flash flood warning is issued, how is it perceived? How quickly do people take action? To be effective, warnings must be taken seriously by the public. The problem is that rainfall is an ordinary event. The vast majority of time, rainfall is benign. It can be very difficult to rouse public concern when rainfall becomes life-threatening, so much so that warning for flash-flood hazards has been called "the formidable challenge." Other severe storm hazards, such as tornadoes, damaging wind, and large hail, quickly raise public concern.

Severe thunderstorms often produce multiple, simultaneous weather hazards. It is easy to get distracted during a situation that includes a strong tornado, baseball-sized hail . . . and a flash flood warning. The media are invariably drawn to the "high drama" of the tornado, and everyone becomes intensely focused on this hazard. For example, the Mayfest Storm, which took place in Fort Worth, Texas, in May 1995, created catastrophic, windblown hail damage estimated at $1 billion. However, that same storm – a severe supercell – also generated flash floods, killing 13 people – a lesser-known fact compared to the drama of the destructive hailstorm.

IFLOWS: Example of a Network Used to Identify Flash Flood Threats

There are now technologies that integrate the meteorological and hydrological facets of flash flood prediction. One example is the Integrated Flood Observing and Warning System (IFLOWS). IFLOWS is a federally funded, computer-linked network of 1500 rain and stream stage sensors. It was created in the wake of a disastrous flood event in 1977 over Kentucky, Virginia, and West Virginia. Today the network encompasses a 12-state region along the Appalachian Mountains.

The goal of IFLOWS is to provide significantly improved lead time in identifying local flash flood threats to vulnerable populations and infrastructure such as key bridges, roadways, utilities, and communications systems. Fully automated stream and rain gauges transmit data over VHF radio frequency to computers that analyze and map data in real time. Every 15 minutes, an updated analysis of rainfall and stream response is made available to NWS offices and emergency management officials across the IFLOWS region. Warnings are automatically triggered when thresholds are exceeded. The NWS can also issue guidance on the amount of rain needed to trigger a flash flood at 1, 6, 12, and 24 hour advance intervals.

Flash Floods and Global Warming

There are important questions to ask: With the steady increase in global temperature over the past 30+ years, are flash floods increasing? Will they become more common and destructive in the future? These are difficult questions to answer because flash floods are a uniquely multifactorial weather hazard. With warmer surface air temperatures, particularly over the ocean, larger amounts of liquid water evaporate into the air. Weather balloon data have documented a small global rise in the average water vapor content of the atmosphere. In this sense, we can state that global warming may be "accelerating" the hydrological cycle by promoting increased evaporation. But does this acceleration translate into excessive rainfall?

It is difficult to ascribe individual weather events such as a single, catastrophic flash flood to climate change, as singular events are categorized as "weather." However, several studies present a consensus view that, statistically over the longer term and on a regional basis, extreme rain events are increasing, including over North America, in the range of 10–40% depending on region, over the past several decades. It is more difficult to tie these increases exclusively to global warming, when other factors may be contributing – including changes in jet stream pattern and intensity (which influences storminess and extreme weather events), changes in El Niño activity impacting the United States, and changes in regional evaporation stemming from land use change, including agriculture.

In summary: There is evidence that rainstorms have become more extreme, concurrent with the increased humidity content of the air. But in assessing future flash flood risk, bear in mind that land use changes are also a factor. Over the past 30 years, we have greatly expanded the reach of urban centers, creating vast amounts of impermeable surface, including highways, streets, parking lots, concrete channels, and rooftops. These changes directly impact the likelihood of flash flooding by causing excessive ponding of water and increase in runoff. So it is difficult to unequivocally state that global warming will be directly responsible for future increases in flash floods – when at least one other important factor, namely urbanization – will be increasing at the same time.

Digging Deeper: West Virginia Coal Mining and Flood Vulnerability

West Virginia possesses a rugged beauty – an endless vista of rolling mountains cloaked in hardwood forests. But the state's close proximity to the Atlantic Ocean creates a unique vulnerability to flooding from rainstorms. Atlantic moisture is frequently lifted along heating mountain slopes during the summer months, leading to thunderstorms and torrential downpours. West Virginia is ranked the third most flood-prone state. Flood disasters in West Virginia have increased during the past 10 years, as has spending related to floods. An ongoing debate revolves around whether human alteration of the landscape, through widespread mountaintop mining, has heightened the state's vulnerability to flash floods.

Flood vulnerability has two components, one natural and one anthropogenic. The natural component is the mountain-triggered cloudbursts during the summer months. The anthropogenic component involves human decisions and behaviors. The widespread deforestation of hill slopes and removal of mountaintops to access veins of coal fall into the realm of human decision making. There is evidence that these land use changes increase the frequency and intensity of flash floods, as well as the incidence of mudslides and debris flows during heavy rain events. In addition, the expansion of mining towns into flood-prone locations, along narrow stream valleys, has increased the exposure of people and property to the flood hazard.

A study by Dr. Laura Merner of the counties of southern West Virginia explores the complex equation of flood vulnerability. Many townships in southern West Virginia are located in winding, narrow stream valleys, at the foot of steep mountains. During summertime thunderstorms, flash floods are common, as streams become rapidly overwhelmed by heavy rain flowing into these narrow gorges. Residents of West Virginia know too well the inherent danger of these floods. For more than 200 years, many have lived in constant fear of what the water brings.

Coal mining is the region's economic lifeblood, and it has been common practice to gain access to coal seams by stripping away vast tracts of forest, soil, and bedrock along ridge lines. When hundreds of acres at each site are stripped to bedrock, heavy rain can no longer infiltrate (Figure 14.23). Instead, it runs off, downslope, exacerbating the magnitude of flash flooding. Additionally, soil and vegetation loosened along the mine's margins can easily become mobilized by the floodwater, creating mudflows and debris flows that choke hollows and stream gorges.

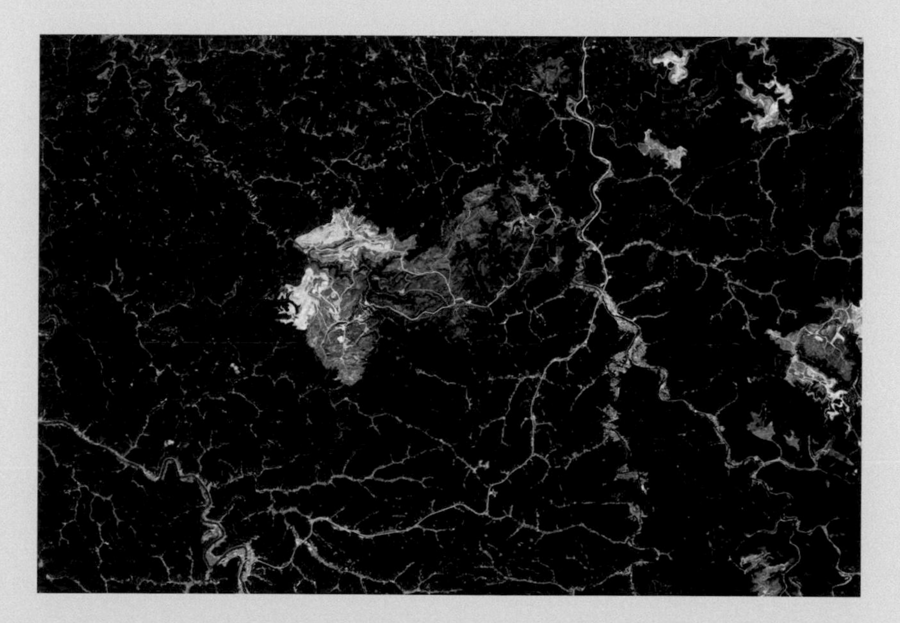

Figure 14.23 Mountaintop removal, Boone County, West Virginia, satellite view in 2015. Mountaintop removal is a key part of coal mining across southern West Virginia. This practice drastically alters the nature of the landscape, likely enhancing rainwater runoff and promoting the mass movement of soil and rock downslope during heavy rain. (NASA Earth Observatory.)

Many of the citizens of southern West Virginia are engaged in lawsuits against mining companies, which they blame for heightening the flood potential. As reported by one Mingo County, West Virginia, resident, there is a shared sentiment among citizens that the practice of mountaintop removal has changed the character of the floods:

> My family has been here for generations and I have lived here my entire life – about here – there have always been flood water, but it used to just wash up and go away, no real damage. When I was a kid the water came in clear and left with no worries. Now things are different. The waters are muddy and come down with logs and rocks. The floods rip up the road, destroy homes and yards – destroy the water lines. Things have changed with the mines.
>
> *(Excerpt from Merner, 2014)*

In summary: While the mining industry is clearly not causing the rainfall, it does likely play a role in altering the surface and flow of waters that result in increased flood impacts.

Summary

LO1 Describe the geographical distribution of flash floods across the United States.

1. Flash floods are the most ubiquitous type of severe local storm, impacting every state in the continental United States, in a variety of landscapes and meteorological settings.
2. Flash floods are a multifactorial hazard, meaning the meteorological factors producing heavy rain conspire with attributes of the landscape (saturated soils, impermeable ground, steep terrain) to create an overwhelmingly rapid hydrological response.

LO2 Distinguish between flash floods, aerial (river) floods, and coastal inundation.

1. Flash floods are small-scale events, of rapid onset, in which rainfall greatly exceeds the ability of water to infiltrate the surface – instead, running off and collecting in low spots and streams.
2. Flash floods are usually generated by convective rains common during summer months: specifically, locally heavy rain showers (often thunderstorms) characterized by short duration and large gradient in rain accumulation (think "cloudburst"). This is in contrast to stratiform rain, which is a widespread, moderate rain lasting several hours from a large-scale weather system (i.e., a soaking rain).
3. Compared to river (aerial) floods, flash floods have high fatality rates, with over half of the deaths occurring in automobiles. Of all severe local storm hazards (including tornadoes, hail, wind, and lightning), flash floods kill the most people annually in the United States, averaging nearly 100 deaths per year.

LO3 Explain the meteorological factors that enable water vapor to build to excessive levels, priming the atmosphere for a flash flood.

1. Water vapor in the atmosphere typically accumulates to excessive levels prior to a flash flood, including moisture locally contained in the air column and humid air transported from the ocean. The holding capacity of warm summer air for water vapor is nearly five times that of cold, winter air. Convective clouds feed on the local moisture and also draw from moisture sequestered over a large area.
2. Weak winds in the atmosphere (most common during summer) create flash flood situations because convective clouds can remain nearly stationary for several hours, allowing the rain to accumulate in one location.

LO4 Describe how multiple types of weather systems generate excessive rain – including complexes of thunderstorms and large-scale vortices such as tropical cyclones. What processes do many of these systems share?

1. The back-building type of convective storm remains stationary because new storm cells continuously redevelop over the same location, "training" one after another as they move away and dissipate.
2. Mesoscale convective complexes (MCCs) are enormous, nocturnal rainstorms common to the

Great Plains during summer – accounting for 10–15% of the region's rainfall and thus critical for growth of crops.

3 Synoptic-scale storm systems that occasionally generate heavy rain and flash flooding include (1) extratropical cyclones with a vigorous warm conveyor belt, (2) East Coast Nor'easters, and (3) Pacific coastal lows. The first type generates flash flooding through the training of rain cells; the latter two types create flash floods via orographic rain enhancement.

LO5 Discuss how steep terrain contributes to heavy rain generation, then concentrates runoff water into a flash flood.

1 A special type of back-building convective storm, called a terrain-locked storm, develops during the summer months, leading to flash floods along the Appalachians and Rockies.

2 Corollary (secondary) hazards of flash floods include mudslides and debris flows in regions of steep, forested terrain and also the rupture of earthen dams.

LO6 How do flash floods rank in terms of U.S. weather fatalities? Why is the prediction of flash flooding so difficult, compared to other types of severe storms? What are some corollary (secondary) hazards of flash flooding?

1 Flash floods remain a daunting severe weather hazard. Compared with other types of severe local storm (e.g., tornadoes), the annual death toll from flash floods has not decreased in the past several decades.

2 Flash flood prediction is particularly challenging because one must forecast the occurrence of this severe storm and its magnitude, and quantitative rain forecasts are considered the least reliable of all meteorological forecasts.

3 At times, it is very difficult to rouse public concern about the flash flood hazard, because rainfall per se is quite an ordinary event and rarely life-threatening.

4 The Integrated Flood Observing and Warning System (IFLOWS) is a 12-state network of automatic rain and stream gauges, designed to monitor flash flood onset across the Appalachian Mountain chain.

LO7 Discuss some of the evidence suggesting that flash flood hazards are increasing as a result of anthropogenic (human) factors, and describe the hypothesized reasons.

1 While global warming has increased the water vapor content of the Earth's lower atmosphere, whether this has accelerated flash flooding remains difficult to prove. Several studies have demonstrated an increase in the frequency of heavy to extreme rainstorms across the United States. However, drastic land use changes (including increased urbanization) arguably have also increased the nation's flash flood vulnerability.

References

Ashley, W.A., T.L. Mote, P.G. Dixon, S.L. Trotter, E.J. Powell, J.D. Durkee and A.J. Grundstein, 2003. Distribution of mesoscale convective complex rainfall in the United States. *Monthly Weather Review*, 131:3003–3017.

McAnelly, R.L. and W.R. Cotton, 1989. The precipitation life cycle of mesoscale convective complexes over the central United States. *Monthly Weather Review*, 117:784–808.

Merner, L., 2014. *Power and Knowledge*, PhD thesis, University of Maryland Baltimore County.

Petersen, W.A., et al., 1999. Mesoscale and radar observations of the Fort Collins flash flood of 28 July 1997. *Bulletin of the American Meteorological Society*, 80:191–216.

Smith, J.A., M.A. Baech, M. Steiner and A.J. Miller, 1996. Catastrophic rainfall from an upslope thunderstorm in the central Appalachians: The Rapidan storm of June 27, 1995. *Water Resources Research*, 32:2099–3113.

Afterword

Severe storms and high impact meteorological events exact a terrible toll in the United States (and rest of the globe) in terms of property loss, human fatalities, and injuries. At the time of this writing, on October 10, 2023, the United States has experienced an exceptional year in terms of the number of billion-dollar weather/climate disasters. We've had 24 events to date, far exceeding the 1980–2022 annual average of eight. These data, compiled by the National Oceanographic and Atmospheric Administration (NOAA), reveal that the United States has experienced $372 billion+ events since 1980, totaling $2.6 *trillion* in losses!

The book you have just read hopefully has provided you with a thorough scientific background on the many ways that atmospheric phenomena create societal havoc. This compendium, however, is a mere launching point for more intensive study of the various hazards. The knowledge constitutes just part of a newly recognized field of specialized study referred to as "disasterology." Hurricanes remain at the top of all U.S. natural hazards in terms of dollar damage and lives lost; future annual losses, unfortunately, are expected to grow as the density of vulnerable infrastructure and population increases along coastal zones. From August 29–31, Category 3 Hurricane Idalia razed a rapidly growing region of west-central Florida, with preliminary loss estimates of $2–5 billion. Months earlier, much of central California found itself deeply under water from a devastating, long-lived series of atmospheric-river-fueled Pacific cyclones. These "big storms" (which were detailed in Chapters 5–8) – whether originating in the tropics or extratropical latitudes – must never be underestimated in terms of the total disaster burden across the United States.

Scientific consensus has emerged that global warming is contributing to some facets of how intense and damaging these large vortices are becoming. Small increases in hurricane wind speed multiply damage in a nonlinear way; NOAA calculations demonstrate that increasing a minimal Cat 1 storm's winds from 65 to 70 kts (75 MPH to 80 MPH) results in a damage multiplier of 1.6 and that going from 65 to 74 kts (75 to 85 MPH) increases that multiplier to 2.9. More details of these computations can be found at the NOAA website.[1] And with the increase in ocean temperature has come an increase in atmospheric water vapor (again bearing a nonlinear signature), such that the potential for flooding rains in these events has certainly increased.

Small-scale storms, a.k.a. severe local storms (detailed in Chapters 9–14 and the Appendix on heat waves), also continue to inflict harm in novel ways. In the United States, heat waves have been occurring in unconventional locations, such as the deadly heat episode that struck the Pacific Northwest and southwestern Canada for a couple of weeks in June–July 2021. The event generated a maximum surface temperature of 50 C (121° F) and an estimated 1400 fatalities. A fierce downslope, Chinook windstorm during December 2021 produced wind gusts to 85 kts (100 MPH) over Boulder, Colorado. The storm created the most destructive wildfire in Colorado history. The fire raced through grasslands and wiped out large tracts of closely spaced homes in multiple subdivisions. And in late September 2023, the vulnerability of our heavily urbanized regions to flooding rains once again became apparent, with a flash flood of unprecedented disaster unfolding across five boroughs of New York City. Such floods are a conspiracy of global-warming-fueled increases in atmospheric water vapor and rampant land use change that converts the once natural, water-absorbing surface into an impervious barrier that can no longer absorb water.

And so we expect great tragedies such as these to continue to unfold in some surprising ways, across the globe. We cannot control the natural forces, but knowledge in these circumstances is power. It is our hope that you – the student – will use your newfound, scientific knowledge about these events to help mitigate their societal impacts. So please do go out and spread the word!

Note

1 www.noaa.gov/jetstream/tc-potential#:~:text=A%2010%20mph%20(16%20km,)%20hurricane%20to%2021%2Dtimes

Appendix: Heat Waves

As discussed in Chapter 1, episodes of extreme heat during the summer account for the greatest fraction of weather-related deaths, on average, in the United States. Not every year experiences a highly fatal heat wave, but in some years, death tolls have been in the hundreds, with thousands of hospitalizations. There is no exact definition of the term "heat wave," and different criteria are used in different regions by the National Weather Service, in terms of temperature, duration, and attendant humidity levels. Outside of the Desert Southwest, a heat wave usually refers to a combination of very high temperature *and* humidity exceeding 2 days duration. High humidity exacerbates the effects of the heat, such that sweat droplets on skin cannot readily evaporate, reducing the body's ability to stay cool.

The combined effect of high temperature and humidity is expressed as the Heat Index, or Humiture. A diagram showing calculated heat index values, from NOAA, is presented in Figure A.1. As excessive heat index values are approached, the NWS may issue a Heat Advisory during the afternoon hours.

NOAA national weather service: heat index

Relative humidity	Temperature															
	80 °F (27 °C)	82 °F (28 °C)	84 °F (29 °C)	86 °F (30 °C)	88 °F (31 °C)	90 °F (32 °C)	92 °F (33 °C)	94 °F (34 °C)	96 °F (36 °C)	98 °F (37 °C)	100 °F (38 °C)	102 °F (39 °C)	104 °F (40 °C)	106 °F (41 °C)	108 °F (42 °C)	110 °F (43 °C)
40%	80 °F (27 °C)	81 °F (27 °C)	83 °F (28 °C)	85 °F (29 °C)	88 °F (31 °C)	91 °F (33 °C)	94 °F (34 °C)	97 °F (36 °C)	101 °F (38 °C)	105 °F (41 °C)	109 °F (43 °C)	114 °F (46 °C)	119 °F (48 °C)	124 °F (51 °C)	130 °F (54 °C)	136 °F (58 °C)
45%	80 °F (27 °C)	82 °F (28 °C)	84 °F (29 °C)	87 °F (31 °C)	89 °F (32 °C)	93 °F (34 °C)	96 °F (36 °C)	100 °F (38 °C)	104 °F (40 °C)	109 °F (43 °C)	114 °F (46 °C)	119 °F (48 °C)	124 °F (51 °C)	130 °F (54 °C)	137 °F (58 °C)	
50%	81 °F (27 °C)	83 °F (28 °C)	85 °F (29 °C)	88 °F (31 °C)	91 °F (33 °C)	95 °F (35 °C)	99 °F (37 °C)	103 °F (39 °C)	108 °F (42 °C)	113 °F (45 °C)	118 °F (48 °C)	124 °F (51 °C)	131 °F (55 °C)	137 °F (58 °C)		
55%	81 °F (27 °C)	84 °F (29 °C)	86 °F (30 °C)	89 °F (32 °C)	93 °F (34 °C)	97 °F (36 °C)	101 °F (38 °C)	106 °F (41 °C)	112 °F (44 °C)	117 °F (47 °C)	124 °F (51 °C)	130 °F (54 °C)	137 °F (58 °C)			
60%	82 °F (28 °C)	84 °F (29 °C)	88 °F (31 °C)	91 °F (33 °C)	95 °F (35 °C)	100 °F (38 °C)	105 °F (41 °C)	110 °F (43 °C)	116 °F (47 °C)	123 °F (51 °C)	129 °F (54 °C)	137 °F (58 °C)				
65%	82 °F (28 °C)	85 °F (29 °C)	89 °F (32 °C)	93 °F (34 °C)	98 °F (37 °C)	103 °F (39 °C)	108 °F (42 °C)	114 °F (46 °C)	121 °F (49 °C)	128 °F (53 °C)	136 °F (58 °C)					
70%	83 °F (28 °C)	86 °F (30 °C)	90 °F (32 °C)	95 °F (35 °C)	100 °F (38 °C)	105 °F (41 °C)	112 °F (44 °C)	119 °F (48 °C)	126 °F (52 °C)	134 °F (57 °C)						
75%	84 °F (29 °C)	88 °F (31 °C)	92 °F (33 °C)	97 °F (36 °C)	103 °F (39 °C)	109 °F (43 °C)	116 °F (47 °C)	124 °F (51 °C)	132 °F (56 °C)							
80%	84 °F (29 °C)	89 °F (32 °C)	94 °F (34 °C)	100 °F (38 °C)	106 °F (41 °C)	113 °F (45 °C)	121 °F (49 °C)	129 °F (54 °C)								
85%	85 °F (29 °C)	90 °F (32 °C)	96 °F (36 °C)	102 °F (39 °C)	110 °F (43 °C)	117 °F (47 °C)	126 °F (52 °C)	135 °F (57 °C)								
90%	86 °F (30 °C)	91 °F (33 °C)	98 °F (37 °C)	105 °F (41 °C)	113 °F (45 °C)	122 °F (50 °C)	131 °F (55 °C)									
95%	86 °F (30 °C)	93 °F (34 °C)	100 °F (38 °C)	108 °F (42 °C)	117 °F (47 °C)	127 °F (53 °C)										
100%	87 °F (31 °C)	95 °F (35 °C)	103 °F (39 °C)	112 °F (44 °C)	121 °F (49 °C)	132 °F (56 °C)										

Key to colors: Caution | Extreme caution | Danger | Extreme danger

Figure A.1 Chart of temperature vs humidity levels, which creates "feels like" temperatures called humitures. (NOAA.)

Local criteria vary, but when heat index values are forecast to exceed 46 C (105° F) for longer than 3 hours or to exceed 46° C (115° F) for any duration of time, a Heat Warning is issued. Some heat waves are extended in duration, lasting 3–5 days. In these situations, with many nights of elevated temperatures, there is little to no relief from the daily heat stress, with a cumulative stress buildup on the body. Those without access to air conditioning can suffer serious – even fatal – health consequences.

The mechanism behind heat waves reflects broad ridging in the upper-level jet stream, such that the core of the jet curves to the north, allowing maritime tropical air to invade a large region of the United States. Downwind of the ridge axis, a strong and deep anticyclone becomes established. The anticyclone is sometimes a westward extension of the semipermanent Bermuda High over the western Atlantic Ocean. Deep, southerly flow causes very warm and humid air to surge northward over the eastern and central United States At the same time, air from high altitude sinks broadly downwind of the jet stream's ridge axis. This causes air in the middle and lower troposphere to warm adiabatically and dry out. A warm, dry air layer that is elevated above the surface is called a temperature inversion. Inversions usually lie a few thousand feet above the ground.

The inversion is also a stable air layer, so it "caps" the lower atmosphere, preventing convective currents from moving hot, humid air upward and away from the surface. The subsiding air in the anticyclone also prevents widespread cloud formation (recall that air must rise in order to form clouds), allowing the full power of the Sun to warm the surface. The massive volume of anticyclonically flowing hot and humid air close to the ground is often termed a heat dome. These processes are illustrated in Figure A.2.

Heat wave days are often poor air quality days. As ground-level air becomes stagnant beneath the temperature inversion, pollutants generated locally, or blown in from distant regions, become trapped. Long hours of intense sunlight (under clear skies) promote photochemical reactions that generate smog. A whitish, opaque haze is often established as water vapor condenses onto tiny droplets around pollutant aerosols.

Derechos, the severe windstorms most common in summer (Chapter 13), are at times linked to heat waves. Progressive derechos often form along the northern rim of the heat dome, moving from west to east or northwest to southeast, across the upper Mississippi Valley, Ohio Valley, Mid-Atlantic, and New England. The heat dome's inversion layer builds high

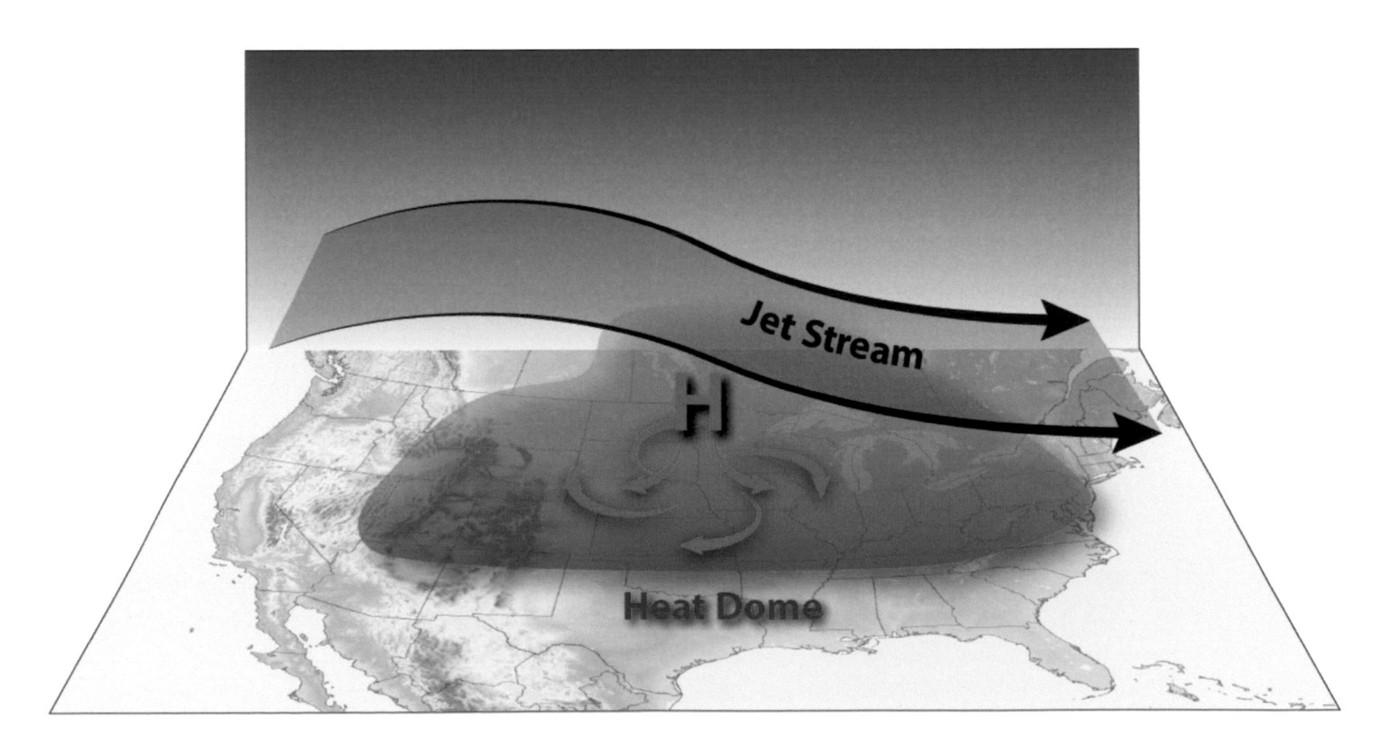

Figure A.2 Formation of heat dome beneath a sprawling anticyclone across the United States during the summer months. A low-level inversion layer traps heat, humidity, and pollutants close to the surface.

levels of heat and humidity near the surface. If the inversion becomes eroded by an approaching wave in the jet stream, the atmosphere rapidly destabilizes and becomes primed for severe thunderstorms. The inversion due to subsiding air is often augmented by a hot, dry, elevated air layer blowing off the high terrain of the Desert Southwest, arcing around the western and northern limbs of the heat dome. As discussed in Chapter 13, several long-track, progressive-type derechos in the United States have been connected directly to heat waves – including those in 1983, 1995, and 2012. The devastation created by a derecho – trees and utility lines down over a widespread area – often means extended power outages. This only magnifies the discomfort and suffering of an ongoing heat wave, as millions out of power no longer have air conditioning to stay cool!

In the past several decades, heat waves have become more frequent and intense over North America. This is likely a manifestation of global warming, and an example of how extreme weather events are actually linked to climate change. Why is this the case? Figure A.3 shows that the connection between climate change and heat waves is statistical. The distribution of anomalously warm and cold days is shown by the curve on the leftmost panel, constructed from the years 1951–1980. Each subsequent panel shows temperature anomalies over the following decades. For each later 10-year period, there have been more anomalously warm than cold days. In other words, global warming has been steadily shifting the temperature distribution toward warmer values; there are now many more anomalously warm days (the days of heat waves) in the 2001–2011 period, compared to earlier periods.

An infamous example of a severe U.S. heat wave was the 1995 Chicago Heat Wave during July 12–16. This was an extended, 5-day period of extreme heat and humidity, impacting much of the upper Midwest. Many locations topped the 38° C (100° F) mark, with extremely high dew point values, and nighttime temperatures not dipping below the low to mid-20s° C (70s–low 80s° F) For instance, Appleton, Wisconsin, reached 41° C (106° F) in the day on July 14, accompanied by a peak dew point of 32° C (90° F) (dew point is discussed in Chapter 3). (Any reading in excess of 27° C [80° F] is uncommon and extremely uncomfortable; values exceeding 27° C [80° F] are most likely to occur in extensively irrigated croplands during the summer months). The combination of searing heat and an almost unheard of humidity value pushed the heat index to 64° C (148° F)! The death toll was 729, with many of the fatalities occurring within and around the Chicago urban center. Urban heat island effects can accentuate heat waves at night, causing temperatures to remain many degrees warmer than the surrounding, vegetated countryside. Additionally, the dense congregation of poor and elderly – who are more likely to lack access to air conditioning – represent an especially vulnerable population in urban areas. Additionally, the elderly suffer disproportionately under conditions of heat stress, due to preexisting diseases and poor health, a diminished physiological coping response to heat, and isolated living conditions. In the poorer sections of cities, windows are often kept closed at night, out of fear of crime.

Heat illness occurs along a spectrum of physiological conditions, ranging from muscle cramps to heat exhaustion and heat stroke. Muscle cramps are common when exercising vigorously in the heat, as salts are depleted along with water in body sweat. As a person sweats more vigorously, dehydration can occur, which leads to the symptoms of heat exhaustion: lightheadedness or fainting, nausea, weakness, and skin that is cool, clammy, and pale. The person should immediately stop exertion, get into a cool place, rest, and slowly rehydrate with a diluted sports drink or plain water. More severe cases may require intravenous rehydration in a hospital. Athletes exercising in the heat and

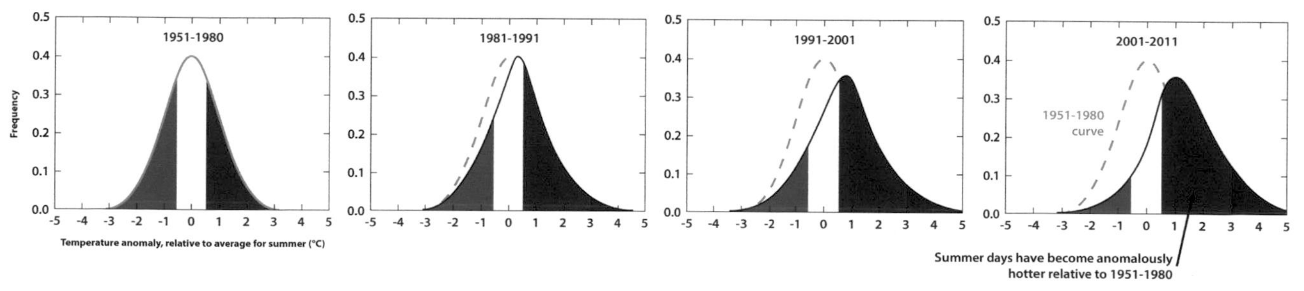

Figure A.3 Shift in frequency of exceptionally hot, summer days due to the effects of ongoing global warming. (NASA.)

firefighters in heavy turnout gear are uniquely vulnerable to heat exhaustion.

Under the most extreme heat exposure, the body stops sweating; the skin is hot, red, and dry; the person may have an altered mental state, become unconscious, and begin seizing. These are all signs of heat stroke, a medical emergency. As the core temperature rises beyond 40–41° C (105–106° F), death becomes significantly more likely. The person must be rapidly cooled (but not to the point of shivering) and rushed to the hospital.

Index

Note: Page numbers in *italics* indicate a figure and page numbers in **bold** indicate a table on the corresponding page.